AF213366

Semiconductor Optics and Transport Phenomena

Springer-Verlag Berlin Heidelberg GmbH

Physics and Astronomy ONLINE LIBRARY

http://www.springer.de/phys/

Advanced Texts in Physics

This program of advanced texts covers a broad spectrum of topics which are of current and emerging interest in physics. Each book provides a comprehensive and yet accessible introduction to a field at the forefront of modern research. As such, these texts are intended for senior undergraduate and graduate students at the MS and PhD level; however, research scientists seeking an introduction to particular areas of physics will also benefit from the titles in this collection.

Wilfried Schäfer   Martin Wegener

Semiconductor Optics and Transport Phenomena

With 151 Figures, 10 Tables,
114 Problems and Hints for Solutions

Springer

Professor Dr. Wilfried Schäfer
Forschungszentrum Jülich GmbH
John von Neumann - Institut für Computing
52425 Jülich
Germany
E-mail: wi.schaefer@fz-juelich.de

Professor Dr. Martin Wegener
Universität Karlsruhe (TH)
Institut für Angewandte Physik
Kaiserstrasse 12
76128 Karlsruhe
Germany
E-mail:
martin.wegener@physik.uni-karlsruhe.de

ISBN 978-3-642-08271-9 ISBN 978-3-662-04663-0 (eBook)
DOI 10.1007/978-3-662-04663-0

Library of Congress Cataloging-in-Publication Data.
Schäfer, Wilfried, 1951-
Semiconductor optics and transport phenomena /
Wilfried Schäfer and Martin Wegener.
p.cm. – (Advanced texts in physics, ISSN 1439-2674)
Includes bibliographical references and index.

1. Semiconductors–Optical properties. 2. Transport theory. I. Wegener, Martin, 1961- II. Title. III. Series.
QC611.6.O6 S43 2002 537.6'226–dc21 2001049393

This work is subject to copyright. All rights are reserved, whether the whole or part of the material is concerned, specifically the rights of translation, reprinting, reuse of illustrations, recitation, broadcasting, reproduction on microfilm or in any other way, and storage in data banks. Duplication of this publication or parts thereof is permitted only under the provisions of the German Copyright Law of September 9, 1965, in its current version, and permission for use must always be obtained from Springer-Verlag. Violations are liable for prosecution under the German Copyright Law.

http://www.springer.de

© Springer-Verlag Berlin Heidelberg 2002
Originally published by Springer-Verlag Berlin Heidelberg New York in 2002.
Softcover reprint of the hardcover 1st edition 2002
The use of general descriptive names, registered names, trademarks, etc. in this publication does not imply, even in the absence of a specific statement, that such names are exempt from the relevant protective laws and regulations and therefore free for general use.

Typesetting: Data prepared by the authors using a Springer TEX macropackage
Cover design: *design & production* GmbH, Heidelberg

Printed on acid-free paper SPIN 10493132 56/3141/di 5 4 3 2 1 0

Preface

Whenever a physicist visits the physics faculty in Dortmund, he/she is bound to hear the success story of the so-called integrated course, a four-semester introduction to physics. These lectures are given by two professors simultaneously, one experimentalist and one theorist. After having asked the common question, "How many professors have killed each other?", the visitor usually realizes that this is an excellent way of presenting a coherent introduction to both experimental and theoretical physics. We decided to try this concept in an advanced course on semiconductor physics. At that point the typical student has already had an introductory course in solid-state physics and solid-state theory. The aim of the lectures was to repeat some of the most important, well-known classics of semiconductor optics and transport and eventually guide the students to topics of current interest in research. When preparing the lectures, we did not find a textbook addressing all these aspects: experiment *and* theory in semiconductor optics *and* transport – which made us write this book. This book presents the phenomenology and a simple, intuitive understanding of many effects and, in addition, attempts to explain the underlying physics on a consistent theoretical footing. Calculations are presented such that a student should be able to follow them with a pencil and a piece of paper. It is our hope that this synthesis of experiment and theory will help to prepare young scientists to contribute something new at the current frontiers of semiconductor physics. After all, the optical and transport properties of semiconductors are among the most important aspects of this particular class of solids for the purposes of their applications – and there are many such applications in everyday life.

This book is organized as follows. A brief introduction to the material systems, we are concerned with in this book is given in Chap. 1. A reader familiar with the general aspects of semiconductor physics and with a basic knowledge of solid-state theory may easily skip Chaps. 2 and 3 without problems. These chapters are merely intended as a reminder of our basic knowledge of electrodynamics, quantum mechanics, statistical physics, and solid-state physics. In addition, they define our nomenclature. We then start discussing the optical properties of semiconductors in Chap. 4, where the important Coulomb correlations are neglected. Nevertheless, this allows us to understand basic experimental techniques, semiconductor photodetectors, and lasers. In Chaps. 5 and 6, Coulomb correlations are treated to first order, such that two-particle correlations are taken into account. This treatment constitutes the semiconductor Bloch equations and modifies the physics of Chap. 4. The dependence on static external fields is discussed in Chaps. 7 and 8, which enables us to understand most types of electro-optic modulators.

A method to include higher order correlations, in particular the formation of four-particle bound states, is introduced in Chap. 9. In order to take care also of other important many-particle interaction processes, such as scattering or screening, we head towards the quantum kinetic equations. The necessary mathematical background is derived in Chap. 10 and extended in Chap. 11 to include scattering from phonons. On a first reading, Chap. 10 (which contains some fairly spicy mathematics) might be skipped. A survey of basic experimental facts and of applications of the theory is presented in Chap. 12. The physics of semiconductor lasers is a nice example where semiconductor optics and transport merge. The physics of these technologically important devices is discussed in Chap. 13.

Classical transport (Chap. 14) is nothing but the strong-scattering limit of the Boltzmann equation. This limit is contrasted with the weak-scattering limit of transport, which is realized in mesoscopic systems. An introduction to this issue is given in Chaps. 15 and 16. Current topics, such as for example, single-electron charging effects, the fractional quantum Hall effect, and magnetotransport through dot and antidot lattices, are dealt with.

We use SI units consistently throughout this book. Nevertheless, a few reminders of fundamental constants in SI units are included in Chaps. 2 and 3 for those using other systems.

Jülich,
Karlsruhe
December 2001

Wilfried Schäfer
Martin Wegener

Contents

1. Some Basic Facts on Semiconductors

Semiconductors have entered our everyday life to such a degree that the notion of a "silicon age" has been employed. Silicon is in fact the most important material as far as commercial applications of semiconductors are concerned. However, while silicon satisfies most of our current needs for electronics, it is only of limited use for optoelectronic applications. Semiconductor lasers, which are at the heart of compact disc players (present in most households), laser printers, and light modulators, the key to today's telecommunication systems, require a direct band gap. Hence, many other semiconductor materials are subjects of current interest. Moreover, today's scientists are no longer satisfied with the variety of bulk materials provided by nature, but have become artists who design semiconductor heterostructures and mesoscopic semiconductor devices corresponding to their needs and interests. This often results in surprising and quite remarkable material properties. Many of these structures, and their optical and transport properties will be discussed in this book.

Semiconductors at room temperature are neither good conductors nor good insulators. They are defined by the presence of an energy gap E_g (throughout the entire Brillouin zone) between the lowest fully occupied bands (at $T = 0$ K), the *valence bands*, and the higher-energy empty bands, the *conduction bands*. It is this gap which makes semiconductors so different from metals. Figure 1.1 shows the computed band structure of gallium arsenide (GaAs). If the extrema of the bands are located at the same point in reciprocal space, the material is referred to as a *direct-gap semiconductor*; if they occur at different points, the term *indirect-gap semiconductor* is used. Silicon is an example of an indirect-gap and GaAs an example of a direct-gap semiconductor. Ideal semiconductors at zero temperature are perfect insulators, materials with large band-gap energies are insulating even at room temperature. Hence, crystals with a room-temperature conductivity less than 10^{-10} $(\Omega\ \mathrm{cm})^{-1}$ are not considered to be semiconductors, but rather *insulators*.

Besides the well-known elementary group IV semiconductor materials Si and Ge (with covalent bonding), a rich variety of binary, ternary, and quaternary semiconductors can be realized. Combinations of, for example, Ga with As (from groups III and V, respectively), Cd with S (from groups II

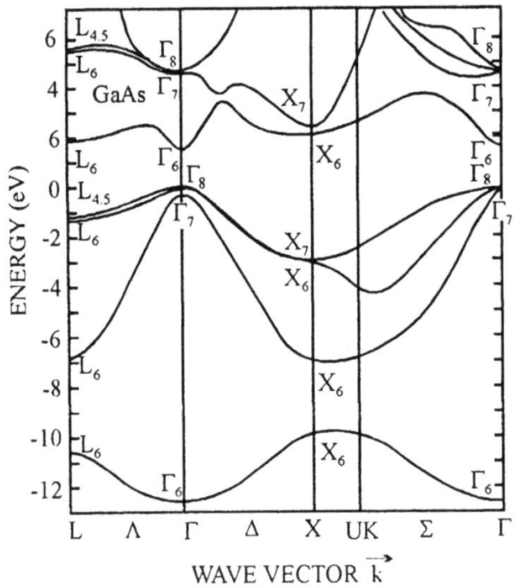

Fig. 1.1. Calculated band structure of the direct-gap semiconductor GaAs. Taken from [1]

and VI), or Cu with Cl (from groups I and VII) yield a rather similar electronic configuration. In the usual jargon, one speaks of III–V, II–VI, and I–VII semiconductors. Similarly, a combination of three elements such as Ga, Al, and As makes a ternary compound and a combination of four elements such as In, Ga, Al, and As constitutes a quaternary compound.

1.1 Semiconductor Heterostructures

One attractive feature of these ternaries and quaternaries is that the band-gap energy can be tailored by choice of the composition. The resulting range of accessible band-gap energies is shown in Fig. 1.2. It is usually a good first-order approximation to interpolate linearly between different materials, i.e. to write for the band gap of a material $A_{1-x}B_xC$ (with x ranging from 0 to 1) $E_g(A_{1-x}B_xC) = (1-x)\,E_g(AC) + xE_g(BC)$. If two semiconductor materials are to be grown on top of each other in the form of a single crystal, *a semiconductor heterostructure* (Fig. 1.3), the respective lattice constants have to be equal or closely similar in order for an unstrained single crystal to be obtained. Still, there is a lot of freedom of choice, as can be seen from Fig. 1.2. One of the most frequently investigated material systems is $GaAs/Al_{1-x}Ga_xAs$ because of the almost perfect lattice matching for any choice of x. The semiconductor lasers in most of today's CD players are made from this particular type of heterostructure. A transmission electron micrograph (TEM) (Fig. 1.4) reveals the position of individual atoms and demonstrates the amazing level of perfection to which epitaxial growth has advanced in this material system, and in several others also.

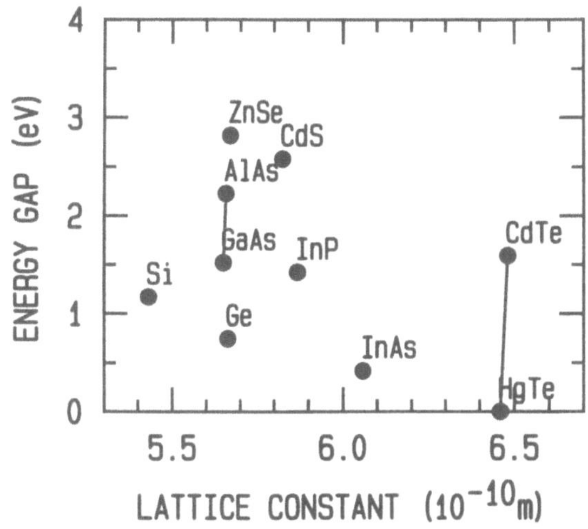

Fig. 1.2. Energy gap (lowest) versus lattice constant for a number of semiconductor materials at low temperature, $T \leq 4.4$ K. HgTe is a semimetal. Numbers taken from [2]

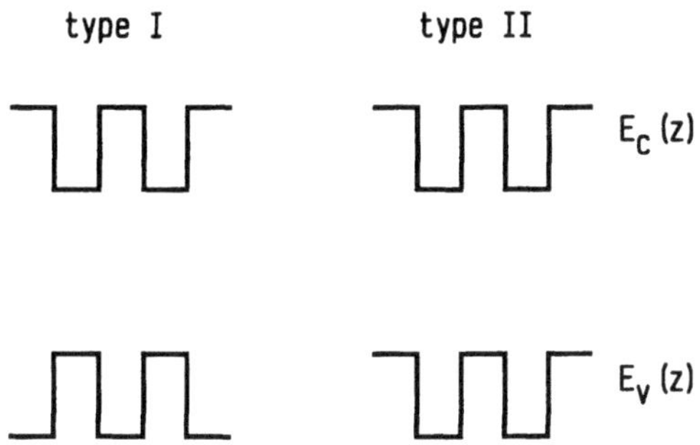

Fig. 1.3. Schematic illustration of the conduction band edge and the valence band edge as a function of the growth direction z in a type I and a type II heterostructure

Fig. 1.4. Chemical lattice image, a technique based on transmission electron microscopy, of a GaAs/AlGaAs heterostructure, revealing the positions of individual atoms of the various elements. Taken from [3]

A sequence of layers of different semiconductor materials results in a spatially varying band-gap energy, and hence can be described by a spatially varying potential for the electrons and holes within the structure. This approximation remains reasonable even down to layer thicknesses of the order of a few atomic monolayers. If the resulting minima, *quantum wells*, of this potential for electrons and holes are located within the same material we have a *type I* heterostructure; if they are in different spatial positions, we have a *type II* heterostructure.

The physics of semiconductor optics has benefited tremendously from the availability of such high-quality semiconductor heterostructures.

1.2 Doped and Modulation-Doped Semiconductors

In many situations, the poor conductivity of semiconductors is undesired and one would like to introduce additional electrons (or holes) into the conduction (or valence) band. In Si or Ge this is easily achieved by introducing a group V (or group III) element into the melt. Technically relevant dopants at low concentrations are well described by shallow, bound donor (or acceptor) energy levels below (or above) the corresponding band. At very low temperatures the electrons (or holes) are trapped in these levels, and are thermally activated only at elevated temperatures. This picture, however, breaks down at higher doping concentrations (e.g. $\approx 10^{16}$ cm^{-3} in Sb-doped Ge), and impurity bands are formed as the impurity wavefunctions exhibit increasing overlap. Owing to the resulting mobility, the conductivity is finite even at very low temperatures, a behavior that is well established as the Mott transition. Eventually, at yet higher doping concentrations, the impurity bands can hybridize with the conduction (or valence) bands.

An obvious disadvantage of doping is the inherent concentration of impurities, which serve as rather undesired scattering centers. This problem, which is unavoidable in bulk semiconductors, can be elegantly circumvented in semiconductor heterostructures. Here certain regions of the structure can be doped selectively in different ways, one example can be seen in Fig. 1.5.

The electrons in such structures originate from the doped, large-band-gap AlGaAs region and are located in the triangular-shaped energy minimum within the low-band-gap material (GaAs). It is clear that the steady-state solution (without additional bias) depicted is defined by a spatially and temporally constant electron chemical potential μ_e; details will be discussed in Chap. 15. Most importantly, electrons are now spatially separated from the impurities, hence reducing the scattering by orders of magnitude. Additionally, the electrons are confined to a region comparable to their de Broglie wavelength, resulting in a (quasi-) two-dimensional electron gas.

The physics of transport in semiconductors has greatly benefited from this concept of *modulation doping*.

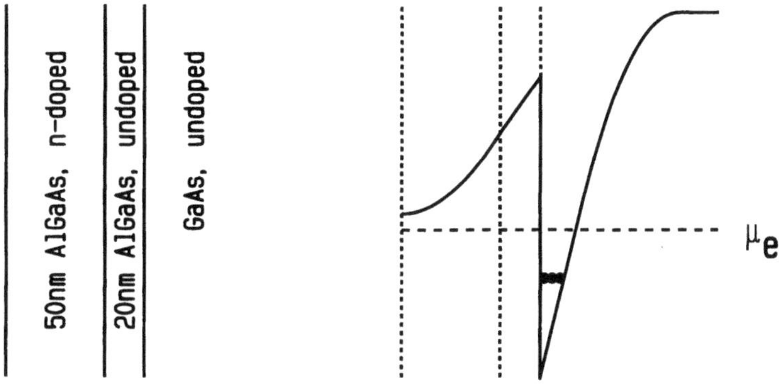

Fig. 1.5. Schematic illustration of a semiconductor heterojunction with a typical layer sequence (*left*) and the conduction band edge as a function of the growth direction (*right*). The carriers confined to the triangular-shaped potential minimum give rise to a highly mobile two-dimensional electron gas

2. Interaction of Matter and Electromagnetic Fields

Both semiconductor optics and transport are concerned with the intimate relationship between electromagnetic fields and matter. Charged particles and the resulting currents give rise to electromagnetic fields via the *Maxwell equations* [4]. In turn, these fields lead to an interaction between the particles. For a closed system that is isolated from the outside world, these interactions can be completely incorporated into the equations of motion of charged particles – in classical mechanics simply via Newton's law and the Lorentz force. On a quantum mechanical level, the coupling of matter and electromagnetic fields results again in an interacting many-particle system, which is described by the *interaction Hamiltonian of fields and matter.* In solid-state physics, the interaction is usually decomposed into a longitudinal, i.e. static, Coulomb interaction and a transverse, i.e. dynamic, part mediated by transverse photons. The range of these interactions extends from atomic scales up to the macroscopic dimensions of the system. Owing to the charge neutrality of the system as a whole, however, macroscopic fields usually do not exist in a closed system. Under the action of external sources, however, such macroscopic fields are generated, which carry a wealth of information about the interacting many-particle system.

External fields induce charge and current densities within the system which again give rise to electromagnetic fields and so forth. The external fields sum with the induced fields to give the total macroscopic fields, which are obtained from the *macroscopic Maxwell equations.* Thus the basic task is to find the relationship between external perturbations and the induced charge and current densities, a relation which may be either linear or nonlinear. In the linear range this allows one to reformulate the above problem as the task of evaluating the *dielectric tensor* of the system.[1] Equivalently, a number of related quantities, such as the conductivity tensor, the susceptibility and the macroscopic polarization can be computed. In the nonlinear case, one possibility is to insist on perturbational concepts and introduce nonlinear response functions. As in the linear case, these quantities depend on intrinsic properties of the unperturbed system only. Alternatively we can go beyond the perturbational approach and try to obtain the induced charge and cur-

[1] We restrict ourselves to nonmagnetic semiconductors.

rent densities to arbitrary order in the external fields. In this approach, the intrinsic properties become renormalized by the action of the external fields.

The bulk of this book is concerned with both experimental and theoretical concepts and techniques to determine the linear or nonlinear response of a semiconductor system, within or beyond the perturbational regime, under various circumstances. Calculations on a microscopic footing are rarely possible without approximations. Depending on the experiments to be addressed, different levels of sophistication are appropriate.

In the following, we shall at first recall the basic concepts, including the description of an interacting many-particle system within the framework of second quantization, the quantum mechanical description of the light field and the coupling between a particle system and external sources. In the second part of this chapter, we consider the various types of response functions which are suitable for determining the sources in the macroscopic Maxwell equations and the relationships between them. In the last part, the simplest relations between response functions and measurable quantities are discussed.

2.1 Microscopic Maxwell Equations

Let us start by considering the *microscopic Maxwell equations* for the internal charge density $\rho(\boldsymbol{r},t)$ and current density $\boldsymbol{j}(\boldsymbol{r},t)$ only. In SI[2] units, we have[3]

$$\begin{aligned}
\nabla \boldsymbol{E}(\boldsymbol{r},t) &= \frac{\rho(\boldsymbol{r},t)}{\varepsilon_0}\,,\\
\nabla \times \boldsymbol{E}(\boldsymbol{r},t) &= -\frac{\partial \boldsymbol{B}(\boldsymbol{r},t)}{\partial t}\,,\\
\nabla \boldsymbol{B}(\boldsymbol{r},t) &= 0\,,\\
\nabla \times \boldsymbol{B}(\boldsymbol{r},t) &= \mu_0\varepsilon_0\frac{\partial \boldsymbol{E}(\boldsymbol{r},t)}{\partial t} + \mu_0 \boldsymbol{j}(\boldsymbol{r},t)\,.
\end{aligned} \tag{2.1}$$

Maxwell's equations guarantee *charge conservation*, which leads owing to $\nabla(\nabla \times \boldsymbol{B}) = 0$, to the *continuity equation*

$$\frac{\partial \rho}{\partial t} + \nabla \boldsymbol{j} = 0\,. \tag{2.2}$$

Formal solutions of Maxwell's equations can be immediately found by introducing the vector potential $\boldsymbol{A}$ and the scalar potential ϕ, from which the electric field $\boldsymbol{E}$ and the magnetic field $\boldsymbol{B}$ are obtained according to

---

[2] Système Internationale d'Unités.

[3] $\varepsilon_0 = 8.8542 \times 10^{-12}$ A s/(V m) is the electric-field constant and $c_0 = 1/\sqrt{\varepsilon_0\mu_0} = 2.997925 \times 10^8$ m/s is the vacuum velocity of light, related via the magnetic field constant $\mu_0 = 4\pi \times 10^{-7}$ V s/(A m).

$$E = -\frac{\partial A}{\partial t} - \nabla\phi\,, \tag{2.3}$$

$$B = \nabla \times A\,. \tag{2.4}$$

Inserting (2.3) into (2.1), we find

$$\Delta\phi + \frac{\partial}{\partial t}\nabla A = -\frac{\rho}{\varepsilon_0}\,, \tag{2.5}$$

and, taking the curl of (2.4) and the time derivative of (2.3), we find, together with (2.1),

$$\nabla \times (\nabla \times A) + \frac{1}{c_0^2}\left(\frac{\partial^2 A}{\partial t^2} + \nabla\frac{\partial\phi}{\partial t}\right) = \mu_0 j\,. \tag{2.6}$$

Formally, only three of the four field components $(A_x, A_y, A_z, \phi/c_0)$ are linearly independent and as consequence of this ambiguity (2.3)–(2.6) are invariant under the gauge transformation

$$A \to A + \nabla\chi$$

$$\phi \to \phi - \frac{\partial\chi}{\partial t}\,, \tag{2.7}$$

i.e. the fields E and B are not affected by this transformation. Thus we have the freedom to choose an additional constraint on the field components. Several gauges are commonly used. In, for example, quantum electrodynamics it is most natural to work with the Lorentz-invariant Lorentz gauge

$$\nabla A = -\frac{1}{c_0^2}\frac{\partial\phi}{\partial t}\,, \tag{2.8}$$

in which both the vector potential and the scalar potential obey retarded wave equations. This gauge reflects the fact that at long distances the interaction between two charges is always retarded. In solids the retardation of the interaction is usually less important, and it is more convenient to choose a gauge in which the dominant contribution to the interaction is instantaneous. This can be realized in two ways. Either one chooses $\phi = 0$ or one chooses $\nabla A = 0$. In both cases one obtains an instantaneous relation between the longitudinal field and the charge density from (2.5). Whereas in the gauge $\phi = 0$ both longitudinal and transverse components of the field are contained in the vector potential, they are strictly separated in the second case. This favors the so-called *Coulomb*, or *transverse gauge*

$$\nabla A = 0\,. \tag{2.9}$$

As consequence, the time dependence in (2.5) vanishes and we obtain the Poisson equation

$$\Delta\phi = -\frac{\rho}{\varepsilon_0}\,, \tag{2.10}$$

with solution

$$\phi(\boldsymbol{r}) = \frac{1}{4\pi\varepsilon_0}\int \mathrm{d}^3r' \frac{\rho(\boldsymbol{r}')}{|\boldsymbol{r}-\boldsymbol{r}'|} = \int \mathrm{d}^3r' v(\boldsymbol{r}-\boldsymbol{r}')\rho(\boldsymbol{r}') \,, \tag{2.11}$$

where we have introduced the static Coulomb potential[4]

$$v(\boldsymbol{r}) = \frac{1}{4\pi\varepsilon_0|\boldsymbol{r}|} \,. \tag{2.12}$$

The evaluation of (2.6) is a little bit more complicated. We can use (2.12) together with the continuity equation (2.2) to eliminate the scalar potential. Furthermore, as the divergence of $\boldsymbol{A}$ vanishes and, hence $\nabla\times(\nabla\times\boldsymbol{A}) = -\Delta\boldsymbol{A}$, (2.6) takes the form of a wave equation,

$$\Delta\boldsymbol{A} - \frac{1}{c_0^2}\frac{\partial^2\boldsymbol{A}}{\partial t^2} = -\frac{\boldsymbol{j}_\mathrm{t}}{c_0^2\varepsilon_0} \,. \tag{2.13}$$

In the Coulomb gauge, longitudinal and transverse interactions are separated. The scalar potential determines the instantaneous longitudinal interaction between the charged particles, whereas the vector potential is transverse owing to the cancellation of the longitudinal contribution to (2.6). Correspondingly, the source term in (2.13) represents the transverse current, which is characterized by vanishing divergence and determined by the projection

$$\boldsymbol{j}_\mathrm{t} = \int \mathrm{d}^3r' \,\underline{\delta}^\mathrm{t}(\boldsymbol{r}-\boldsymbol{r}')\boldsymbol{j}(\boldsymbol{r}') \,, \tag{2.14}$$

where the components of the transverse delta function are defined according to

$$\delta^\mathrm{t}_{nm}(\boldsymbol{r}-\boldsymbol{r}') = \delta_{nm}\delta(\boldsymbol{r}-\boldsymbol{r}') + \frac{1}{4\pi}\frac{\partial^2}{\partial r_n \partial r_m}\frac{1}{|\boldsymbol{r}-\boldsymbol{r}'|} \,. \tag{2.15}$$

So far, we have reduced the solution of Maxwell's equations to the solution of the Poisson equation and a wave equation. The distribution of charged particles determines the scalar potential and thus the electrostatic interaction energy in the system, whereas the vector potential is responsible for transverse interaction effects. Both contributions enter into the quantum mechanical description of the particle system, as we shall see next.

2.2 The Many-Particle Hamiltonian

We consider a system of N electrons in a rigid periodic lattice coupled to an electromagnetic field. The interaction between matter and an electromagnetic field can be introduced by the requirement that Schrödinger's equation [5] is invariant under the gauge transformation (2.7) and a simultaneous gauge transformation of the wavefunction according to

[4] v is the Green's function of the Poisson equation, with dimensions $[v] = \mathrm{V/(A\,s)}$.

$$\psi(\boldsymbol{r}_1, ..., \boldsymbol{r}_N, t) \to \exp\left(\mathrm{i}\frac{e}{\hbar}\left[\chi(\boldsymbol{r}_1, t) + ... + \chi(\boldsymbol{r}_{N_\mathrm{e}}, t)\right]\right) \psi(\boldsymbol{r}_1, ..., \boldsymbol{r}_N, t) . \tag{2.16}$$

This invariance is guaranteed if we introduce the coupling of matter and the electromagnetic field via the replacements

$$\begin{aligned} \boldsymbol{p}_i &\to \boldsymbol{p}_i - ez_i \boldsymbol{A}(\boldsymbol{r}_i) , \\ \frac{\partial}{\partial t} &\to \frac{\partial}{\partial t} + \mathrm{i}\frac{e}{\hbar}\sum_i \phi(\boldsymbol{r}_i) , \end{aligned} \tag{2.17}$$

where the ith particle carries the charge[5] ez_i, and z_i is an integer. Thus the Hamiltonian is given by

$$H = \sum_i \frac{1}{2m_i}\left[\boldsymbol{p}_i - ez_i\boldsymbol{A}(\boldsymbol{r}_i)\right]^2 + \frac{\varepsilon_0}{2}\int \mathrm{d}^3\boldsymbol{r}\,\left(\boldsymbol{E}^2 + c_0^2\boldsymbol{B}^2\right) , \tag{2.18}$$

where $\boldsymbol{p}_i$ is the momentum operator of the ith particle,

$$\boldsymbol{p}_i = -\mathrm{i}\hbar\nabla_{\boldsymbol{r}_i} , \tag{2.19}$$

and the last term describes the energy of the electromagnetic field which still contains a contribution from longitudinal fields. These can easily be separated by using (2.3). As that contribution to the field energy which mixes the scalar and the vector potential vanishes in the transverse gauge, we have the following for the longitudinal component $\boldsymbol{E}_\mathrm{l}$ of the electric field:

$$\frac{\varepsilon_0}{2}\int \mathrm{d}^3\boldsymbol{r}\,\boldsymbol{E}_\mathrm{l}^2 = \frac{\varepsilon_0}{2}\int \mathrm{d}^3\boldsymbol{r}\,(\nabla\phi)^2 . \tag{2.20}$$

After partial integration and use of (2.10), with the charge density of the nuclei and electrons given by

$$\rho(\boldsymbol{r}) = \sum_i z_i e\delta(\boldsymbol{r} - \boldsymbol{r}_i) , \tag{2.21}$$

we obtain, using (2.11) and (2.12),

$$\frac{\varepsilon_0}{2}\int \mathrm{d}^3\boldsymbol{r}\,(\nabla\phi)^2 = \frac{1}{2}\sum_{ij} e^2 z_i z_j v(\boldsymbol{r}_i - \boldsymbol{r}_j) . \tag{2.22}$$

At the moment, we are considering a rigid lattice, and hence we can immediately separate the contribution of the nuclei from the total wavefunction. In the electronic Hamiltonian, we can summarize the nuclei–electron interaction by a potential $U(\boldsymbol{r})$, which has the periodicity of the lattice. In practice, $U(\boldsymbol{r})$ also includes the contribution of core electrons which are tightly bound to the nuclei.[6] We arrive at the following for the electronic part of the Hamiltonian:

[5] The fundamental charge is $e = 1.6021 \times 10^{-19}$ A s.

[6] For many applications, the resulting atomic core potentials can be represented by pseudopotentials (see Sect. 3.3).

$$H = \sum_i \left(\frac{1}{2m_i} [\boldsymbol{p}_i - ez_i \boldsymbol{A}(\boldsymbol{r}_i)]^2 + U(\boldsymbol{r}_i) \right) \\ + \frac{e^2}{2} \sum_{ij} v(\boldsymbol{r}_i - \boldsymbol{r}_j) + \frac{\varepsilon_0}{2} \int \mathrm{d}^3\boldsymbol{r} \, (\boldsymbol{E}_\mathrm{t}^2 + c_0^2 \boldsymbol{B}^2) \, . \tag{2.23}$$

In this form, (2.23) describes a system of interacting electrons in a rigid lattice coupled to a transverse electromagnetic field, which can be described either classically or quantum mechanically. Which of the two descriptions leads to physically reasonable results depends on the problem. In many cases the quantum mechanical nature of light becomes unimportant and we can assume that the total vector potential can be approximated by a classical field, which is induced by an externally controlled vector potential. In the case of vanishing external field, there is no pendant to a transverse, classical field in the system. There exist, however, fluctuations of the internal vector potential, which give rise to transverse interaction effects. Although quite small in general, the nonclassical properties of light are essential features of a number of linear and nonlinear optical phenomena. A rigorous treatment of the corresponding effects require a quantum mechanical description of the light field. Before we introduce the basic concept, i.e. the quantization of the electromagnetic field [6], let us consider first a mathematically more elegant formulation of the many-particle problem.

2.3 Second Quantization for Particles

Let us ignore the existence of transverse electromagnetic fields for the moment[7] and consider the Hamiltonian (2.23) for a system of N identical particles which obey Fermi–Dirac statistics [7]:

$$H = \sum_i h(\boldsymbol{r}_i) + \frac{e^2}{2} \sum_{i \neq j} v(\boldsymbol{r}_i - \boldsymbol{r}_j) = H_1 + H_2 \, , \tag{2.24}$$

where $h(\boldsymbol{r})$ is the one-particle Hamiltonian. As the particles are indistinguishable, even the simplest approximate solution of the corresponding Schrödinger equation has to obey the *Pauli exclusion principle* [8, 9, 10]. This principle states that no more than one electron is allowed to occupy a given state specified by a complete set of one-particle quantum numbers. If the many-particle wavefunction is completely antisymmetric under the interchange of any two particles, this requirement is fulfilled. The simplest wavefunction of this type is given by the *Slater determinant* [11] ,

[7] We shall consider the coupling to transverse fields by taking off-diagonal elements of the one-particle Hamiltonian into account.

$$\Phi^N_{E_1^0,...,E_N^0}(x_1,...,x_N) = \frac{1}{\sqrt{N!}}\sum_{\hat{\mathcal{P}}}(-1)^{\hat{\mathcal{P}}}\hat{\mathcal{P}}\,\varphi_{E_1^0}(x_1)...\varphi_{E_N^0}(x_N)$$

$$= \frac{1}{\sqrt{N!}}\begin{vmatrix} \varphi_{E_1^0}(x_1) & ... & \varphi_{E_1^0}(x_N) \\ \varphi_{E_2^0}(x_1) & ... & \varphi_{E_2^0}(x_N) \\ ... & ... & ... \\ \varphi_{E_N^0}(x_1) & ... & \varphi_{E_N^0}(x_N) \end{vmatrix} . \tag{2.25}$$

The one-particle eigenstates have the form

$$\varphi_{E_j^0}(x_{\rm i}) = \varphi_j(\boldsymbol{r}_i)\eta_{s_j}(s_{\rm i}) . \tag{2.26}$$

Here $x_{\rm i} = (\boldsymbol{r}_{\rm i}, s_{\rm i})$ represents the coordinate of the ith particle and consists of both the spatial coordinate and the spin variable, with spin states $\eta_{s_j}(s)$, which fulfill the orthogonality relation

$$\sum_s \eta_{s_j}(s)\eta_{s_k}(s) = \delta_{s_j s_k} . \tag{2.27}$$

The subscripts $E_j^0 \in \{E_1, E_2, ..., E_\infty\}$ denote a complete set of one-particle quantum numbers, including the spin quantum number s_j, where the orthonormal basis $\{\varphi_{E_1}(x)...\varphi_{E_\infty}(x)\}$ spans a one-particle *Hilbert space.* Thus $\{E_1^0, E_2^0, ..., E_N^0\}$ represents an arbitrary subset of N one-particle states, which is selected from the orthonormal basis. The antisymmetry of the Slater determinants implies that all quantum numbers must be different. Further, we have introduced the permutation operator $\hat{\mathcal{P}}$, which generates from the product $\varphi_{E_1^0}(x_1)..\varphi_{E_N^0}(x_N)$ of the N one-particle eigenstates $N!$ permutations. The Slater determinants constructed from these one-particle states again represent a complete set of antisymmetric N-particle wavefunctions, and obey the orthogonality relation [14]

$$\langle\Phi^N_{E_1^0,...,E_N^0}|\Phi^N_{E_1',...,E_N'}\rangle = \sum_{\hat{\mathcal{P}}}(-1)^{\hat{\mathcal{P}}}\hat{\mathcal{P}}\delta_{E_1^0,E_1'}...\delta_{E_N^0,E_N'}$$

$$= \begin{vmatrix} \delta_{E_1^0,E_1'} & ... & \delta_{E_1^0,E_N'} \\ \delta_{E_2^0,E_1'} & ... & \delta_{E_2^0,E_N'} \\ ... & ... & ... \\ \delta_{E_N^0,E_1'} & ... & \delta_{E_N^0,E_N'} \end{vmatrix} , \tag{2.28}$$

and the completeness relation

$$\frac{1}{N!}\sum_{E_1',...,E_N'} |\Phi^N_{E_1',...,E_N'}\rangle\langle\Phi^N_{E_1',...,E_N'}| = 1 . \tag{2.29}$$

Thus we can expand the solution of the Schrödinger equation according to

$$\Psi(x_1,...,x_N) = \sum_{E_1',...,E_N'} C(E_1',...,E_N',t)\,\Phi_{E_1',...,E_N'}(x_1,...,x_N) \tag{2.30}$$

In principle, the range of each summation index is unrestricted and runs over the set of quantum numbers $\{E_1, ..., E_\infty\}$. However, owing to the antisymmetry of the Slater determinants, any contributions that have the same states occupied are equal within a minus sign. This allows us to arrange all quantum numbers in an increasing sequence $\{E_1^0 < E_2^0 < ... < E_N^0\}$ and to restrict the sums in (2.30) from the very beginning to ordered subsets. Correspondingly, we admit coefficients with ordered arguments only. In order to perform the expansion with the Hamiltonian (2.24), we insert (2.30) into the Schrödinger equation and project onto an N-particle Slater determinant, with the result

$$\mathrm{i}\hbar\frac{\partial}{\partial t}C(E_1^0, ..., E_N^0, t) = \sum_{E_1', ..., E_N'} \langle \Phi^N_{E_1^0,...,E_N^0} | H | \Phi^N_{E_1',...,E_N'} \rangle C(E_1', ..., E_N', t) . \tag{2.31}$$

Let us consider first the matrix elements of H_1. We expand the N-particle Slater determinants with respect to the ith column, which yields a sum of $(N-1)$-particle determinants:

$$\Phi^N_{E_1^0,...,E_N^0}(x_1, ..., x_N) = \frac{1}{\sqrt{N}} \sum_{k=1}^{N} (-1)^{\mathrm{i}+k} \varphi_{E_k^0}(x_i)$$

$$\times \Phi^{(N-1)}_{E_1^0,...,E_{k-1}^0, E_{k+1}^0,...,E_N^0}(x_1, ..., x_{\mathrm{i}-1}, x_{\mathrm{i}+1}, ..., x_N) . \tag{2.32}$$

This allows us to write the matrix element of the one-particle Hamiltonian $h(\boldsymbol{r}_i)$ as

$$\sum_{s_1,...,s_N} \int \mathrm{d}^3\boldsymbol{r}_1 ... \mathrm{d}^3\boldsymbol{r}_N \, \Phi^N_{E_1^0,...,E_N^0}(x_1, ..., x_N) h(\boldsymbol{r}_i) \Phi^N_{E_1',...,E_N'}(x_1, ..., x_N)$$

$$= \frac{1}{N} \sum_{jk} (-1)^{j+k} \langle E_j^0 | h | E_k' \rangle \langle \Phi^{(N-1)}_{E_1^0,...,E_{j-1}^0, E_{j+1}^0,...,E_N^0} | \Phi^{(N-1)}_{E_1',...,E_{k-1}', E_{k+1}',...,E_N'} \rangle , \tag{2.33}$$

where the inner product is defined in (2.28) and the one-particle matrix element are given by

$$\langle E_j | h | E_k \rangle = \sum_{s_i} \int \mathrm{d}^3\boldsymbol{r}_i \, \varphi^*_{E_j}(x_i) h(\boldsymbol{r}_i) \varphi_{E_k}(x_i) . \tag{2.34}$$

Owing to the inner product in (2.33), there are various contributions to the expansion coefficients in (2.31). However, we have to select only those coefficients for which E_k occurs at its proper place in the ordered sequence. Further, all one-particle operators in H_1 give the same contribution. Thus the factor $1/N$ cancels and the resulting equation can be cast into the form

$$\mathrm{i}\hbar\frac{\partial}{\partial t}C(E_1^0, ..., E_N^0, t) = \sum_{j=1}^{N} \sum_{E_k'} \langle E_j^0 | h | E_k' \rangle C(E_1^0, ..., E_{j-1}^0, E_k', E_{j+1}^0, ..., E_N^0)$$

$$+ \sum_{j=1,k<j}^{N} \sum_{E'_k} \langle E_j^0|h|E'_k\rangle(-1)^{j+k}\mathcal{C}(E_1^0, ..., E'_k, ..., E_{j-1}^0, E_{j+1}^0, ..., E_N^0)$$
$$+ \sum_{j=1,k>j}^{N} \sum_{E'_k} \langle E_j^0|h|E'_k\rangle(-1)^{j+k}\mathcal{C}(E_1^0, ..., E_{j-1}^0, E_{j+1}^0, ..., E'_k, ..., E_N^0) \,, \tag{2.35}$$

where E'_k runs over those values of the infinite set of quantum numbers for which the resulting sequence is ordered. Even without the Coulomb interaction, it is obvious that the bookkeeping of occupied and unoccupied states and of the phase factors can become rather cumbersome. However, apart from these factors, the structure of (2.35) is rather simple. Owing to the action of an off-diagonal element of the one-particle Hamiltonian, a state with energy E_j^0 contributing on the left-hand side is removed from the coefficient and replaced by states with energy E'_k which have not contributed before. Now we are going to simplify the bookkeeping as well as the notation by means of the occupation number representation. This representation allows us to formalize the handling of many-particle wavefunctions with help of *creation and annihilation operators.* As we have defined the expansion coefficients on an ordered subset of quantum numbers, we can identify the sequence of quantum numbers $\{E_1^0, E_2^0, ..., E_N^0\}$ with a sequence of occupation numbers $\{n_1, n_2, ..., n_\infty\}$, which are defined on the complete sequence of quantum numbers $\{E_1, E_2, ..., E_\infty\}$. The occupation numbers $n_k \in \{0, 1\}$ are equal to one if the corresponding eigenstate with energy E_k occurs in a given subset of energies and are zero otherwise. Hence they obey the condition $\sum_k n_k = N$. This arrangement allows us to replace the expansion coefficients according to

$$\mathcal{C}(E_1^0 < E_2^0 < .... < E_N^0 t) = c(n_1, n_2, ..., n_\infty, t) \,. \tag{2.36}$$

Correspondingly, we can replace the Slater determinants by *occupation number states,*

$$|\Phi_{E_1^0,...,E_N^0}\rangle = |n_1, n_2, ..., n_\infty\rangle \,. \tag{2.37}$$

The introduction of these occupation number states generalizes the concept of the *N-particle Hilbert space* $\mathcal{H}_N$, as this basis spans a Hilbert space that contains states of varying numbers of particles [12, 13]. For different numbers of particles the states are orthogonal to each other. More formally, these states are a complete orthonormal set and span the direct sum $\mathcal{H}_0 \oplus \mathcal{H}_1 \oplus ... \oplus \mathcal{H}_n \oplus ...$ of n-particle Hilbert spaces, which defines the *Fock space* [14]. The replacement (2.37) allows us to rewrite the expansion (2.30) as

$$|\psi(t)\rangle = \sum_{n_1,...,n_\infty} c(n_1, ..., n_\infty, t)|n_1, ..., n_\infty\rangle \,. \tag{2.38}$$

The phase factors in (2.35) can be calculated by counting the number of occupied states between the state E_j^0, which is removed and the state E'_k, which becomes occupied. These ingredients allow us to transform (2.35) into the the form[8]

$$
\begin{aligned}
\mathrm{i}\hbar\frac{\partial}{\partial t}c(n_1,n_2,...,n_\infty,t) &= \sum_j \langle E_j|h|E_j\rangle\,\delta_{n_j,1}\,c(n_1,n_2,...,n_j,...,n_\infty,t)\\
&+\sum_{j,k<j}\langle E_j|h|E_k\rangle(-1)^{n_{k+1}+...+n_{j-1}}c(n_1,...,n_k+1,...,n_j-1,...,n_\infty,t)\\
&+\sum_{j,k>j}\langle E_j|h|E_k\rangle(-1)^{n_{j+1}+...+n_{k-1}}c(n_1,...,n_j-1,...,n_k+1,...,n_\infty,t)\,.
\end{aligned}
\tag{2.39}
$$

Thus, from (2.38) and (2.39), we obtain the representation

$$
\begin{aligned}
\mathrm{i}\hbar\frac{\partial}{\partial t}|\psi(t)\rangle = &\sum_{n_1,...,n_\infty}\Bigg(\sum_j\langle E_j|h|E_j\rangle c(n_1,...,n_\infty,t)|n_1,...,n_\infty\rangle\delta_{n_j,1}\\
&+\sum_{j,k<j}\langle E_j|h|E_k\rangle(-1)^{\xi_j-\xi_{k+1}}\delta_{n_k,1}\delta_{n_j,0}\\
&\times c(n_1,...,n_k,...,n_j,...,n_\infty)|n_1,...,n_k-1,...,n_j+1,...,n_\infty\rangle\\
&+\sum_{j,k>j}\langle E_j|h|E_k\rangle(-1)^{\xi_k-\xi_{j+1}}\delta_{n_k,1}\delta_{n_j,0}\\
&\times c(n_1,...,n_j,...,n_k,...,n_\infty,t)|n_1,...,n_j+1,...,n_k-1,...,n_\infty\rangle\Bigg)
\end{aligned}
\tag{2.40}
$$

where we have used the fact that, for example, $n_{k+1}+...+n_{j-1}=\xi_j-\xi_{k-1}$, where $\xi_j=n_1+...+n_{j-1}$. In order to obtain a closed equation, it is necessary to generate the same state vector on the right hand side as that which occurs on the left-hand side. This is accomplished by introducing creation and annihilation operators, which act on the state vector in occupation number space and change the occupation numbers from one to zero or vice versa. From (2.40), we obtain the result that such operators have to fulfill the conditions

$$c_j^\dagger|n_1,...,n_j,...,n_\infty\rangle=(-1)^{\xi_j}\delta_{n_j,0}|n_1,...,n_j+1,...,n_\infty\rangle\,, \tag{2.41}$$

$$c_k|n_1,...,n_k,...,n_\infty\rangle=(-1)^{\xi_k}\delta_{n_k,1}|n_1,...,n_k-1,...,n_\infty\rangle\,. \tag{2.42}$$

Repeating the action of the operators, we obtain for the diagonal elements the following:

$$c_j^\dagger c_j^\dagger|n_1,...,n_j,...,n_\infty\rangle=0\,, \tag{2.43}$$

$$c_jc_j|n_1,...,n_j,...,n_\infty\rangle=0\,. \tag{2.44}$$

[8] Note that the summation indices run over the infinite set of quantum numbers and not only over the occupied states as in (2.35).

These conditions guarantee that two particles never occupy the same state, and thus they reflect the Fermi statistics of electrons. Correspondingly, the occupation number operator for the jth state $c_j^\dagger c_j$ can only have one of the two eigenvalues zero and one:

$$c_j^\dagger c_j |n_1, ...n_j..., n_\infty\rangle = n_j |n_1, ...n_j..., n_\infty\rangle \ , \tag{2.45}$$

$$c_j c_j^\dagger |n_1, ...n_j..., n_\infty\rangle = (1 - n_j)|n_1, ...n_j..., n_\infty\rangle \ . \tag{2.46}$$

The relations (2.43)–(2.46) are obviously fulfilled if the operators obey the anticommutation rules [9]

$$[c_j, c_k^\dagger]_+ = c_j c_k^\dagger + c_k^\dagger c_j = \delta_{jk} \ , \tag{2.47}$$

$$[c_j^\dagger, c_k^\dagger]_+ = [c_j, c_k]_+ = 0 \ . \tag{2.48}$$

With these definitions of fermion operators, the conditions (2.41) and (2.42) are also fulfilled for $k \neq j$. This can immediately be shown by noting that an occupation number state can be generated by the action of creation operators on the vacuum state $|0\rangle$:

$$|n_1, ..., n_\infty\rangle = \prod_i^\infty (c_i^\dagger)^{n_i} |0\rangle \ . \tag{2.49}$$

Acting with c_j or $c_j^\dagger$ onto this state and using the anticommutation properties, we obtain (2.41) or (2.42), respectively.

With the fermion operators as ingredients, we can write (2.40) as

$$i\hbar \frac{\partial}{\partial t} |\psi(t)\rangle = \sum_{jk} \langle E_j|h|E_k\rangle c_j^\dagger c_k \, |\psi(t)\rangle, \tag{2.50}$$

which represents the Schrödinger equation for noninteracting particles in the so-called *second quantization*. To complete the representation of the Hamiltonian (2.24) in terms of creation and annihilation operators, we have to consider the two-particle interaction H_2. Without going through a lengthy derivation we can elegantly reduce this problem to the result obtained for the one-particle contribution. From the derivation of (2.50) it is obvious that any one-particle operator can be represented in the same manner. In particular, for the density operator

$$\hat{n}(\boldsymbol{r}) = \sum_i \delta(\boldsymbol{r} - \boldsymbol{r}_i) \ , \tag{2.51}$$

we obtain in second quantization

$$\hat{n}(\boldsymbol{r}) = \sum_{jk} \varphi_{E_j}^*(\boldsymbol{r}) \varphi_{E_k}(\boldsymbol{r}) c_j^\dagger c_k \ . \tag{2.52}$$

The two-particle contribution to (2.24) can be written as

$$H_2 = \frac{e^2}{2} \int \mathrm{d}^3\boldsymbol{r}\, \mathrm{d}^3\boldsymbol{r}'\, v(\boldsymbol{r}-\boldsymbol{r}')\Big(\hat{n}(\boldsymbol{r})\hat{n}(\boldsymbol{r}') - \delta(\boldsymbol{r}-\boldsymbol{r}')\hat{n}(\boldsymbol{r})\Big)\,, \tag{2.53}$$

where the second term on the right-hand side cancels the self-interaction contribution, i.e. the contribution with $i = j$, which occurs in the first term. Thus, with the help of (2.52), we almost obtain the final result for the many-particle Hamiltonian

$$H = \sum_{jk} \langle E_j|h|E_k\rangle c_j^\dagger c_k + \frac{e^2}{2} \sum_{jklm} \langle E_j E_k|v|E_l E_m\rangle c_j^\dagger c_k^\dagger c_m c_l\,, \tag{2.54}$$

where we have used $[c_k, c_{j'}^\dagger]_+ = \delta_{kj'}$, $[c_k, c_{k'}]_+ = 0$ and have applied the completeness relation for the one-particle states to cancel the self-interaction contribution. The two-particle matrix element is given by

$$\langle E_j E_k|v|E_l E_m\rangle = \sum_{ss'} \int \mathrm{d}^3\boldsymbol{r}\mathrm{d}^3\boldsymbol{r}'\, \varphi_{E_j}^*(x)\varphi_{E_k}^*(x')v(\boldsymbol{r}-\boldsymbol{r}')\varphi_{E_l}(x)\varphi_{E_m}(x'). \tag{2.55}$$

Remembering that the quantum numbers E_j include the spin, we obtain, using (2.26), the following more explicit forms for the one- and two-particle matrix elements:

$$\langle j|h|k\rangle = \int \mathrm{d}^3\boldsymbol{r}\, \varphi_j^*(\boldsymbol{r})h(\boldsymbol{r})\varphi_k(\boldsymbol{r})\delta_{s_j,s_k}\,, \tag{2.56}$$

$$\langle jk|v|lm\rangle = \delta_{s_j,s_l}\delta_{s_k,s_m} \int \mathrm{d}^3\boldsymbol{r}\, \mathrm{d}^3\boldsymbol{r}'\, \varphi_j^*(\boldsymbol{r})\varphi_k^*(\boldsymbol{r}')v(\boldsymbol{r}-\boldsymbol{r}')\varphi_l(\boldsymbol{r})\varphi_m(\boldsymbol{r}')\,, \tag{2.57}$$

where we have introduced an obvious short-hand notation for the matrix elements. This brings us to the final form of the material part of the Hamiltonian, described by

$$H = \sum_{jk} \langle j|h|k\rangle c_j^\dagger c_k + \frac{e^2}{2} \sum_{jklm} \langle jk|v|lm\rangle c_j^\dagger c_k^\dagger c_m c_l\,. \tag{2.58}$$

Sometimes it is more convenient to start with a real-space representation of this Hamiltonian. An equivalent formulation results if we introduce *field operators* $\psi^\dagger(x)$ and $\psi(x)$ according to the expansions

$$c_j^\dagger = \sum_s \int \mathrm{d}^3\boldsymbol{r}\, \varphi_j(\boldsymbol{r})\eta_{s_j}(s)\psi^\dagger(x) \tag{2.59}$$

and

$$c_j = \sum_s \int \mathrm{d}^3\boldsymbol{r}\, \varphi_j^*(\boldsymbol{r})\eta_{s_j}^*(s)\psi(x)\,. \tag{2.60}$$

Using (2.47) and (2.48), we find that the fermion field operators again obey anticommutation rules,

$$[\psi(x), \psi^\dagger(x')]_+ = \delta(\boldsymbol{r} - \boldsymbol{r}')\delta_{s,s'} \tag{2.61}$$

and

$$[\psi(x), \psi(x')]_+ = [\psi^\dagger(x), \psi^\dagger(x')]_+ = 0\,. \tag{2.62}$$

By means of the transformations (2.59) and (2.60), we obtain the real space representation of the many-particle Hamiltonian (2.58):

$$\begin{aligned} H = &\sum_s \int \mathrm{d}^3\boldsymbol{r}\, \psi^\dagger(x) h(\boldsymbol{r}) \psi(x) \\ &+ \frac{e^2}{2} \sum_{ss'} \int \mathrm{d}^3\boldsymbol{r}\, \mathrm{d}^3\boldsymbol{r}'\, \psi^\dagger(x)\psi^\dagger(x') v(\boldsymbol{r} - \boldsymbol{r}') \psi(x')\psi(x)\,. \end{aligned} \tag{2.63}$$

Mathematically, the Hamiltonian (2.58) represents a problem which is completely equivalent to the original formulation (2.24). Whenever we have found an approximate solution in either formulation, we can represent the result equivalently in the other formulation. Creation and annihilation operators represent a mathematically powerful tool, but lead to no new physical features. This is due to the fact that in a nonrelativistic theory the Hamiltonian commutes with the particle number operator

$$\hat{N} \hat{=} \sum_j c_j^\dagger c_j\,, \tag{2.64}$$

and thus the total particle number is conserved. In relativistic field theories this property is missing and creation and annihilation operators aquire an immediate physical meaning. In the context of solid state theory, the notion *second quantization* is somewhat misleading. The physics is completely identical to that of a many-particle problem quantized according to the usual rules of quantum mechanics.

In order to obtain a consistent quantum mechanical description of matter interacting with electromagnetic fields, we now turn to the quantization of the fields.

2.4 Quantization of Electromagnetic Fields

The transition from a classical to a quantum mechanical description is managed in two steps. First we find the pairs of canonical conjugate variables and then we postulate their metamorphosis into noncommuting operators. Usually, the canonical conjugate momentum is found from the Lagrangian of the system. Here we shall shortcut the derivation [15] and start with the Hamiltonian of the transverse electromagnetic radiation, which we obtain from (2.23) with the help of (2.3) and (2.4) as

$$H_{\text{rad}} = \frac{\varepsilon_0}{2} \int d^3 r \left\{ \left(\frac{\partial}{\partial t} A(r,t) \right)^2 + c_0^2 \left(\nabla \times A(r,t) \right)^2 \right\} . \tag{2.65}$$

If we take the vector potential as the canonical variable, the equations of motion for the free field, which follow from Maxwell's equations, have the following form:

$$\frac{\delta H}{\delta \boldsymbol{\Pi}(\boldsymbol{r},t)} = \frac{\partial}{\partial t}\boldsymbol{A}(\boldsymbol{r},t) , \tag{2.66}$$

$$\frac{\delta H}{\delta \boldsymbol{A}(\boldsymbol{r},t)} = -\frac{\partial}{\partial t}\boldsymbol{\Pi}(\boldsymbol{r},t) . \tag{2.67}$$

Thus the canonical conjugate momentum $\boldsymbol{\Pi}$ is given by

$$\boldsymbol{\Pi} = \varepsilon_0 \frac{\partial}{\partial t}\boldsymbol{A} , \tag{2.68}$$

where $\boldsymbol{A}(\boldsymbol{r},t)$ and $\boldsymbol{\Pi}(\boldsymbol{r},t)$ are canonical conjugate variables, i.e. their equations of motion are invariant with respect to infinitesimal canonical transformations. Analogously to the quantization rules of quantum mechanics, we postulate a nonvanishing commutator

$$[A_i(\boldsymbol{r},t), \Pi_j(\boldsymbol{r}',t)] = \mathrm{i}\hbar\delta^{\mathrm{t}}_{ij}(\boldsymbol{r}-\boldsymbol{r}') . \tag{2.69}$$

The appearance of the transverse δ function, which has already been defined in (2.15), is a consequence of the transverse gauge, where only two components of the vector potential are independent. In order to account for this restriction, we have to introduce a set of independent dynamical variables. Generally, we can write $\boldsymbol{A}(\boldsymbol{r},t)$ as a *Fourier sum*

$$\boldsymbol{A}(\boldsymbol{r},t) = \sum_{\boldsymbol{q}} \mathrm{e}^{\mathrm{i}\boldsymbol{q}\boldsymbol{r}} \boldsymbol{A}(\boldsymbol{q},t) \to \frac{V}{(2\pi)^3} \int d^3\boldsymbol{q}\, \mathrm{e}^{\mathrm{i}\boldsymbol{q}\boldsymbol{r}} \boldsymbol{A}(\boldsymbol{q},t) , \tag{2.70}$$

where $V = L_x L_y L_z$ is the volume of a box under consideration. We impose *periodic boundary conditions*, which restrict the allowed values of the components of the wave vector $\boldsymbol{q}$ to

$$q_i = \frac{2\pi}{L_i} n \tag{2.71}$$

where n is an integer. The last step in (2.70) is obtained from taking the continuum limit (see Sect. 3.6.), i.e. $L_i \to \infty$. The inverse Fourier transformation is given by

$$\boldsymbol{A}(\boldsymbol{q},t) = \frac{1}{V} \int d^3\boldsymbol{r}\, \mathrm{e}^{-\mathrm{i}\boldsymbol{q}\boldsymbol{r}} \boldsymbol{A}(\boldsymbol{r},t) . \tag{2.72}$$

We rewrite the expansion of $\boldsymbol{A}(\boldsymbol{r},t)$ into plane waves as

$$\boldsymbol{A}(\boldsymbol{r},t) = \sum_{\lambda \boldsymbol{q}} \boldsymbol{e}_{\lambda \boldsymbol{q}} \left[u_\lambda(\boldsymbol{q},t) + u^\dagger_\lambda(-\boldsymbol{q},t) \right] \mathrm{e}^{\mathrm{i}\boldsymbol{q}\boldsymbol{r}} . \tag{2.73}$$

The polarization vectors $\boldsymbol{e}_{\lambda\boldsymbol{q}}$ account for the two orthogonal transverse modes ($\lambda = 1, 2$). From the transversality condition $\boldsymbol{q}\boldsymbol{e}_{\lambda\boldsymbol{q}} = 0$, we have the relation

$$\sum_{\lambda}(\boldsymbol{e}_{\lambda\boldsymbol{q}})_i(\boldsymbol{e}_{\lambda\boldsymbol{q}})_j = \delta_{ij} - \frac{q_i q_j}{q^2} \,, \tag{2.74}$$

where the right-hand side is just the Fourier transform of the transverse δ function. The physical meaning of the new operators $u_\lambda(\boldsymbol{q})$ becomes more transparent if we consider their equation of motion. As the vector potential is a solution of the wave equation (2.13), we find that $u_\lambda(\boldsymbol{q}, t)$ is a solution of

$$q^2 u_\lambda(\boldsymbol{q}, t) + \frac{1}{c_0^2}\frac{\partial^2}{\partial t^2} u_\lambda(\boldsymbol{q}, t) = \frac{1}{V c_0^2 \varepsilon_0} \int \mathrm{d}^3\boldsymbol{r}\; \boldsymbol{e}_{\lambda\boldsymbol{q}} \boldsymbol{j}_t(\boldsymbol{r}, t) \mathrm{e}^{\mathrm{i}\boldsymbol{q}\boldsymbol{r}} \,. \tag{2.75}$$

Thus the $u_\lambda(\boldsymbol{q}, t)$ behave like classical driven harmonic oscillators, and we can easily eliminate the time dependence by the replacement

$$u_\lambda(\boldsymbol{q}, t) \to u_\lambda(\boldsymbol{q}) \mathrm{e}^{\mathrm{i}\omega_{\boldsymbol{q}} t} \,, \tag{2.76}$$

together with the dispersion relation

$$\omega_{\boldsymbol{q}} = |\boldsymbol{q}| c_0 \,. \tag{2.77}$$

Using this time dependence of the operators $u_\lambda(\boldsymbol{q}, t)$, we can immediately calculate the time derivative of the vector potential, and obtain

$$\frac{\partial}{\partial t}\boldsymbol{A}(\boldsymbol{r}, t) = -\frac{\mathrm{i}}{V}\sum_{\lambda\boldsymbol{q}} \boldsymbol{e}_{\lambda\boldsymbol{q}} \omega_{\boldsymbol{q}} \left(u_\lambda(\boldsymbol{q}, t) - u_\lambda^\dagger(-\boldsymbol{q}, t)\right) \mathrm{e}^{\mathrm{i}\boldsymbol{q}\boldsymbol{r}} \,. \tag{2.78}$$

Equations (2.73) and (2.78) can be inverted, with the result

$$u_\lambda(\boldsymbol{q}) = \frac{1}{2V}\int \mathrm{d}^3\boldsymbol{r}\, \mathrm{e}^{-\mathrm{i}\boldsymbol{q}\boldsymbol{r}} \left(\boldsymbol{e}_{\lambda\boldsymbol{q}}\boldsymbol{A}(\boldsymbol{r}, t) + \frac{\mathrm{i}}{\omega_{\boldsymbol{q}}}\boldsymbol{e}_{\lambda\boldsymbol{q}}\frac{\partial}{\partial t}\boldsymbol{A}(\boldsymbol{r}, t)\right) \,, \tag{2.79}$$

$$u_\lambda^\dagger(-\boldsymbol{q}) = \frac{1}{2V}\int \mathrm{d}^3\boldsymbol{r}\, \mathrm{e}^{-\mathrm{i}\boldsymbol{q}\boldsymbol{r}} \left(\boldsymbol{e}_{\lambda\boldsymbol{q}}\boldsymbol{A}(\boldsymbol{r}, t) - \frac{\mathrm{i}}{\omega_{\boldsymbol{q}}}\boldsymbol{e}_{\lambda\boldsymbol{q}}\frac{\partial}{\partial t}\boldsymbol{A}(\boldsymbol{r}, t)\right) \,. \tag{2.80}$$

In the next step, we calculate from (2.79) and (2.80) the commutator of $u_\lambda(\boldsymbol{q})$ and $u_\lambda^\dagger(\boldsymbol{q}')$ using the commutation relation (2.69), with the result

$$[u_\lambda(\boldsymbol{q}), u_\lambda^\dagger(\boldsymbol{q}')] = \frac{\hbar}{2V\varepsilon_0\omega_{\boldsymbol{q}}}\delta_{\lambda\lambda'}\delta_{\boldsymbol{q}\boldsymbol{q}'} \,. \tag{2.81}$$

A more familiar representation is obtained if we introduce the normalized operators:

$$a_{\boldsymbol{q}\lambda} = \sqrt{\frac{2\varepsilon_0\omega_{\boldsymbol{q}}V}{\hbar}}\, u_\lambda(\boldsymbol{q}) \,, \tag{2.82}$$

which obviously obey the usual boson commutation relations of creation and annihilation operators, well known from the treatment of harmonic oscillators:

$$[a_{q\lambda}, a^\dagger_{q'\lambda'}] = \delta_{\lambda\lambda'}\delta_{qq'} \ , \tag{2.83}$$

$$[a_{q\lambda}, a_{q'\lambda'}] = [a^\dagger_{q\lambda}, a^\dagger_{q'\lambda'}] = 0 \ . \tag{2.84}$$

In terms of the creation and annihilation operators, the expansion of the vector potential (2.73) takes the form

$$\boldsymbol{A}(\boldsymbol{r},t) = \frac{1}{\sqrt{V}} \sum_{\lambda q} \sqrt{\frac{\hbar}{2\varepsilon_0 \omega_q}} \boldsymbol{e}_{\lambda q} \mathrm{e}^{\mathrm{i}\boldsymbol{q}\boldsymbol{r}} \left(a_{q\lambda} \mathrm{e}^{-\mathrm{i}\omega_q t} + a^\dagger_{-q\lambda} \mathrm{e}^{\mathrm{i}\omega_q t} \right) . \tag{2.85}$$

Given these results, it is a straightforward exercise to calculate the Hamiltonian of the transverse electromagnetic field. Inserting (2.85) into (2.65) and using the dispersion relation (2.77), we finally obtain

$$H_{\mathrm{rad}} = \sum_{q\lambda} \hbar\omega_q \left(a^\dagger_{q\lambda} a_{q\lambda} + \frac{1}{2} \right) . \tag{2.86}$$

Thus the Hamiltonian of the light field is obtained as a sum of decoupled harmonic oscillators. Correspondingly, a general eigenstate of the field can be represented as a linear combination of occupation-number states $|n_{q\lambda}\rangle$:

$$|\Psi_{q\lambda}\rangle = \sum_{n_{q\lambda}} c(n_{q\lambda}) \, |n_{q\lambda}\rangle \ , \tag{2.87}$$

where $|c(n_{q\lambda})|^2$ denotes the probability of finding $n_{q\lambda}$ photons in the field. As in the case of the harmonic oscillator, we obtain from (2.83) and (2.84)

$$a^\dagger_{q\lambda} \, |n_{q\lambda}\rangle = \sqrt{n_{q\lambda}+1} \, |n_{q\lambda}+1\rangle \tag{2.88}$$

and

$$a_{q\lambda} \, |n_{q\lambda}\rangle = \sqrt{n_{q\lambda}} \, |n_{q\lambda}-1\rangle \ . \tag{2.89}$$

The relationship to oscillator states allows yet another representation by eigen functions, which are referred to as *Glauber* or *coherent states.* By means of these states, coherence properties of light, otherwise defined empirically by the contrast of interference patterns, can be given a precise mathematical meaning. Thus coherent states are important from a conceptual as well as from a practical point of view. We shall consider their basic features as an example of a representation of the light field by eigenfunctions.

2.4.1 Coherent States

Coherent states $|\alpha\rangle$ were considered for the first time by Schrödinger [16], and later by Glauber [17, 18, 19], and are defined as eigen functions of the annilation operator according to

$$a\,|\alpha\rangle = \alpha\,|\alpha\rangle. \tag{2.90}$$

For simplicity we restrict ourselves to a single photon mode and omit all subscripts from operators, eigenvalues, and eigenfunctions. The coherent states can be represented by an expansion in terms of the occupation-number states (2.88) according to

$$|\alpha\rangle = \sum_n \langle n|\alpha\rangle |n\rangle \; . \tag{2.91}$$

Making use of (2.90) and (2.89), we find the recursion formula

$$\langle n+1|\alpha\rangle = \frac{\alpha}{\sqrt{n+1}} \langle n|\alpha\rangle \; , \tag{2.92}$$

which results in

$$\langle n|\alpha\rangle = \frac{\alpha}{\sqrt{n!}} \; . \tag{2.93}$$

Performing the sum in (2.91), we find, together with the normalization condition

$$\langle\alpha|\alpha\rangle = |\langle 0|\alpha\rangle|^2 \mathrm{e}^{|\alpha|^2} = 1 \; , \tag{2.94}$$

the coherent states in the form

$$|\alpha\rangle = \mathrm{e}^{-1/2|\alpha|^2} \sum_n \frac{\alpha^n}{\sqrt{n!}} |n\rangle \; . \tag{2.95}$$

An equivalent representation can be found by means of the unitary displacement operator

$$\hat{U}(\alpha) = \hat{U}^\dagger(-\alpha) = \exp(\alpha\, a^\dagger - \alpha^* a) \; , \tag{2.96}$$

which allows us to generate the coherent states according to

$$|\alpha\rangle = \hat{U}(\alpha)|0\rangle \; . \tag{2.97}$$

As coherent states are eigenfunctions of a non-Hermitian operator, they are obviously nonorthogonal, but they form a complete set. Thus we can use them to expand an arbitrary state of the photon field.

Let us consider the expectation value of the operator of the electric field for an arbitrary single mode.[9] From (2.85) we obtain, using $\langle\alpha|a|\alpha\rangle = \alpha$ and $\langle\alpha|a^\dagger|\alpha\rangle = \alpha^*$,

$$\langle\alpha_{\lambda\boldsymbol{q}}|\boldsymbol{E}(\boldsymbol{r},t)|\alpha_{\lambda\boldsymbol{q}}\rangle = \mathrm{i}\sqrt{\frac{\hbar\omega_{\boldsymbol{q}}}{2\varepsilon_0 V}}\boldsymbol{e}_{\lambda\boldsymbol{q}}\, \mathrm{e}^{\mathrm{i}\boldsymbol{q}\boldsymbol{r}} \left(\alpha_{\boldsymbol{q}\lambda}\mathrm{e}^{-\mathrm{i}\omega_{\boldsymbol{q}}t} - \alpha^*_{-\boldsymbol{q}\lambda}\mathrm{e}^{\mathrm{i}\omega_{\boldsymbol{q}}t}\right) . \tag{2.98}$$

This is a plane wave with complex amplitude $\alpha_{\boldsymbol{q}\lambda}$, which determines the phase and the absolute value of the field strength. As the expectation value of the photon number operator in one mode is just given by

$$N_{\boldsymbol{q}\lambda} = a^\dagger_{\boldsymbol{q}\lambda} a_{\boldsymbol{q}\lambda} = |\alpha_{\boldsymbol{q}\lambda}|^2 \; , \tag{2.99}$$

[9] If more than a single mode is considered, we can define the global coherent state as the direct product of the coherent states of each single mode.

the amplitude is determined by the square root of the mean photon number. This quantity determines the standard deviation of the photon number, too:

$$\langle \alpha_{\lambda \boldsymbol{q}} | \Delta N^2_{\boldsymbol{q}\lambda} | \alpha_{\lambda \boldsymbol{q}} \rangle = \langle \alpha_{\lambda \boldsymbol{q}} | N^2_{\boldsymbol{q}\lambda} | \alpha_{\lambda \boldsymbol{q}} \rangle - \langle \alpha_{\lambda \boldsymbol{q}} | N_{\boldsymbol{q}\lambda} | \alpha_{\lambda \boldsymbol{q}} \rangle^2 = |\alpha_{\boldsymbol{q}\lambda}|^2 \,. \quad (2.100)$$

The standard deviation of the field strength results is obtained as

$$\left(\langle \alpha_{\lambda \boldsymbol{q}} | \Delta E^2_{\boldsymbol{q}\lambda} | \alpha_{\lambda \boldsymbol{q}} \rangle \right)^{1/2} = \frac{\hbar \omega_{\boldsymbol{q}}}{\varepsilon_0 V} \,. \quad (2.101)$$

If we measure this quantity in units of the spatially averaged square of the field strength (2.98), we see that these relative field fluctuations decrease with increasing values of $|\alpha|^2$. Thus, with increasing photon number or, equivalently, with increasing standard deviation of the photon number, these relative fluctuations of the field strength become small. Indeed, in this limit, we have a strong resemblance between the expectation value of the field for a coherent state and a classical field. This illustrates the fact that the classical limit requires large photon numbers, which corresponds to high intensities. From the representation (2.95), it is obvious that the photon number in a coherent state is not well defined, but we find the probability of measuring the photon number n to be

$$|\langle n | \alpha \rangle|^2 = \frac{|\alpha|^{2n}}{n!} \mathrm{e}^{-|\alpha|^2} \,. \quad (2.102)$$

This is a *Poisson distribution*, with the mean value and the deviation both determined by $|\alpha|^2$. Finally, a further remarkable feature is the uncertainty relation. Calculation of the standard deviations of the conjugate field variables $\boldsymbol{A}(\boldsymbol{r},t)$ and $\boldsymbol{\Pi}(\boldsymbol{r},t)$ for coherent states shows that the minimum value is realized in the uncertainty relation. This is again the maximum amount of classical behaviour which can be expected for a quantum mechanical system.

Thus far, we have arrived at a quantum mechanical description for particles and the electromagnetic field in terms of fermion and boson operators, respectively. The corresponding treatment of the interaction of fields and matter is, however, still missing and will be given next.

2.5 The Interaction Hamiltonian of Fields and Particles

In Sect. 2.2, we have introduced the coupling between electromagnetic fields and matter. As can be seen from (2.23), the interaction with the transverse fields has the structure of a one-particle Hamiltonian and can be treated in the same way as the one-particle contribution in (2.24). Thus we can immediately apply the result (2.50) of Sect. 2.3 and write the interaction Hamiltonian as[10]

$$H_{\mathrm{e,rad}} = \sum_{jk} \left(\frac{e}{m} \langle j | \boldsymbol{p}\boldsymbol{A} | k \rangle + \frac{e^2}{2m} \langle j | \boldsymbol{A}^2 | k \rangle \right) c_j^\dagger c_k \,. \quad (2.103)$$

[10] Note that for electrons, we have $z_i = -1$ in (2.23) and m denotes the free-electron mass.

Inserting (2.85) into (2.103), we obtain

$$
\begin{aligned}
H_{\mathrm{e,rad}} &= \sum_{jk\lambda\boldsymbol{q}} \frac{e}{m}\sqrt{\frac{\hbar}{2\varepsilon_0\omega_{\boldsymbol{q}}V}}\langle j|\mathrm{e}^{\mathrm{i}\boldsymbol{q}\boldsymbol{r}}\boldsymbol{p}\boldsymbol{e}_{\lambda\boldsymbol{q}}|k\rangle\{a_{\boldsymbol{q}\lambda}+a^{\dagger}_{-\boldsymbol{q}\lambda}\}c^{\dagger}_{j}c_{k} \\
&+ \sum_{jk\lambda\lambda'\boldsymbol{q}\boldsymbol{q}'} \frac{e^2\hbar}{4m\varepsilon_0 V\sqrt{\omega_{\boldsymbol{q}}\omega_{\boldsymbol{q}'}}}\langle j|\mathrm{e}^{\mathrm{i}(\boldsymbol{q}+\boldsymbol{q}')\boldsymbol{r}}|k\rangle\boldsymbol{e}_{\lambda\boldsymbol{q}}\boldsymbol{e}_{\lambda'\boldsymbol{q}'} \\
&\times \left(a_{\boldsymbol{q}\lambda}+a^{\dagger}_{-\boldsymbol{q}\lambda}\right)\left(a_{\boldsymbol{q}'\lambda'}+a^{\dagger}_{-\boldsymbol{q}'\lambda'}\right)c^{\dagger}_{j}c_{k},
\end{aligned}
\tag{2.104}
$$

where we have incorporated the explicit time dependence in (2.85) into the photon operators. The complicated form of the interaction Hamiltonian can be simplified immediately if we assume that the spatial extent L of the system is small compared with the optical wavelength. This allows for an expansion with respect to the wave vector of the photon, provided the condition $|\boldsymbol{q}|L \ll 1$ is fulfilled. This condition is hardly met in extended systems. But we can make use of the translational invariance of crystalline solid. As will be seen in more detail later, L characterizes the scale on which the electronic properties in a solid vary. The latter is of the order of a few lattice constants and therefore the wavelength is several orders of magnitude larger than this scale, thus allowing for a multipole expansion. Even the lowest order, i.e. *the dipole approximation*, in which the optical field is considered to be constant over one lattice cell, is sufficient in most cases. This, however, does not imply that the spatial dependence of the electric field vanishes completely. Calculating the matrix elements in (2.104) with extended states, shows that the variation is, rather, described on a macroscopic scale, which is well separated from the microscopic scale. We shall see in Sect. 3.6 how such coarse-grained macroscopic scales can be introduced formally. At the moment we do not worry about the complications, owing to extended states, and approximate the exponential plane-wave factor by unity.[11]

Using this approximation, no spatial dependence of the field is left, and the interaction Hamiltonian reduces to

$$
\begin{aligned}
H_{\mathrm{e,rad}} &= -\sum_{jk\lambda\boldsymbol{q}} \frac{e}{m}\sqrt{\frac{\hbar}{2\varepsilon_0\omega_{\boldsymbol{q}}V}}\langle j|\boldsymbol{p}\boldsymbol{e}_{\lambda\boldsymbol{q}}|k\rangle\left(a_{\boldsymbol{q}\lambda}+a^{\dagger}_{-\boldsymbol{q}\lambda}\right)c^{\dagger}_{j}c_{k} \\
&+ \sum_{\lambda\boldsymbol{q}} \frac{\hbar\omega_{\mathrm{p}}^2}{4\omega_{\boldsymbol{q}}}\left(a_{\boldsymbol{q}\lambda}+a^{\dagger}_{-\boldsymbol{q}\lambda}\right)\left(a_{-\boldsymbol{q}\lambda}+a^{\dagger}_{\boldsymbol{q}\lambda}\right).
\end{aligned}
\tag{2.105}
$$

[11] Whether we deal with extended sytems or not is of minor importance for the following analysis. However, it should be noted that transition matrix elements of the dipole operator with extended states are not well defined for intraband transitions (compare Sect. 3.6). Quantities of the form $\sum_j |\boldsymbol{d}_{jk}|^2(E_j - E_k)$, however, which occur in the calculation of physical response functions, have a well-defined meaning (compare Sect. 2.6 and also Problems 2.4 and 5.5).

The second term, the so-called diamagnetic contribution, has been obtained by taking into account the fact, that the particle number is conserved and hence the particle-number operator can be replaced by the particle number. The plasma frequency ω_{p} is defined via

$$\omega_{\mathrm{p}}^2 = \frac{n_e e^2}{\varepsilon_0 m} , \tag{2.106}$$

where $n_{\mathrm{e}} = N/V$ is the total electron density in the system. In the dipole approximation there is obviously no contribution of the $\boldsymbol{A}^2$ term in (2.105) that would lead to optical transitions between different electronic bands. The remaining intraband contribution of the $\boldsymbol{A}^2$ term seems to be a tedious constant which can be neglected. Although this assertion holds in most applications, it is instructive to see how the $\boldsymbol{A}^2$ term can be eliminated exactly by introducing an alternative representation for the transition matrix element in (2.105). The corresponding transition from momentum to dipole transition matrix elements is well known in atomic physics, but is more involved for a many-particle system. Therefore we shall perform this transformation for the many-particle case within the framework of second quantization. We start our reasoning with a heuristic consideration.

The momentum transition matrix element can be related to the dipole transition matrix element according to the relation

$$[\boldsymbol{r}, h(\boldsymbol{r})] = \frac{\mathrm{i}\hbar \boldsymbol{p}}{m} , \tag{2.107}$$

and we can write the $\boldsymbol{pA}$ coupling in (2.103) as

$$\frac{e}{m} \boldsymbol{p}_{jk} \boldsymbol{A} = -\frac{\mathrm{i}}{\hbar} (E_j - E_k) \boldsymbol{d}_{jk} \boldsymbol{A} , \tag{2.108}$$

where we have chosen $h(\boldsymbol{r})$ such that the one-particle states in (2.104) are eigenstates, and have introduced $\boldsymbol{p}_{jk}$ and $\boldsymbol{d}_{jk}$, where $\boldsymbol{d} = -e\boldsymbol{r}$, as abbreviations for the transition matrix elements. Instead of the vector potential, we can now introduce the electric field $\boldsymbol{E} = -\dot{\boldsymbol{A}}$ on the right-hand side, and obtain

$$\frac{e}{m} \boldsymbol{p}_{jk} \boldsymbol{A} = -\boldsymbol{d}_{jk} \boldsymbol{E} - \frac{\mathrm{i}}{\hbar} \left(\frac{\hbar}{\mathrm{i}} \frac{\partial}{\partial t} + E_j - E_k \right) \boldsymbol{d}_{jk} \boldsymbol{A} . \tag{2.109}$$

If we consider resonant contributions only and approximate the time dependence of the vector potential by a plane wave with center frequency ω_0, the difference between the $\boldsymbol{pA}$ and $\boldsymbol{dE}$ couplings is proportional to $(\hbar\omega_0 + E_j - E_k)/\hbar\omega_0$. This factor can be neglected, provided we have resonant excitation conditions and the range of transition energies $(E_j - E_k)$ is much smaller than the central frequency of the field. In nonlinear optics there are, however, situations (e.g. any kind of multi-photon process) where the central frequency of the field is far from any resonances of the system. In such cases the correction in (2.109) can become extremely large and the question arises whether it is physically meaningful or not. Indeed, the answer is negative, as the corrections in (2.109), together with the $\boldsymbol{A}^2$ term, can be eliminated exactly by means of a unitary transformation of the many-particle wavefunction. The contribution of the time derivative in (2.109) suggests that this phase transformation should be given by

$$|\Psi(t)\rangle = \mathrm{e}^{\mathrm{i}\xi(t)}|\tilde{\Psi}(t)\rangle \tag{2.110}$$

where

$$\xi(t) = \frac{1}{\hbar}\sum_{jk}\langle j|\boldsymbol{dA}(\boldsymbol{r},t)|k\rangle c_j^{\dagger}c_k \,. \tag{2.111}$$

We shall now apply this transformation directly to the many-particle Schrödinger equation

$$\mathrm{i}\hbar\frac{\partial}{\partial t}|\Psi(t)\rangle = H|\Psi(t)\rangle \,, \tag{2.112}$$

where H is given by (2.58) with the interaction contribution (2.103). Using (2.110) in (2.112), we obtain

$$\begin{aligned}\mathrm{i}\hbar\frac{\partial}{\partial t}|\tilde{\Psi}(t)\rangle &= \left(\mathrm{e}^{-\mathrm{i}\xi(t)}H\mathrm{e}^{\mathrm{i}\xi(t)} + \hbar\frac{\partial}{\partial t}\xi(t)\right)|\tilde{\Psi}(t)\rangle \\ &= \left(H_{\mathrm{t}} - \sum_{jk}\langle j|\boldsymbol{dE}(\boldsymbol{r},t)|k\rangle c_j^{\dagger}c_k\right)|\tilde{\Psi}(t)\rangle = \tilde{H}\;|\tilde{\Psi}(t)\rangle \,. \end{aligned}\tag{2.113}$$

In order to calculate the transformed Hamiltonian we expand H_{t} with respect to ξ, and obtain

$$H_{\mathrm{t}} = H - \mathrm{i}[\xi,H] - \frac{1}{2}[\xi,[\xi,H]] + \ldots \,. \tag{2.114}$$

This expansion is finite if one of the multiple commutators vanishes. At this point, the dipole approximation comes into play. Neglecting the variation of the optical field within the range of a lattice cell, the series is truncated after the second order contribution. A straightforward calculation, using the completeness relation, shows that the commutator can be reduced to

$$\begin{aligned}[\xi,H] &= \frac{1}{\hbar}\sum_{jk}\langle j|\Big[\frac{\boldsymbol{p}^2}{2m} + \frac{e}{m}\boldsymbol{pA},\boldsymbol{dA}\Big]|k\rangle c_j^{\dagger}c_k \\ &+ \frac{e^2}{\hbar}\sum_{ijkln}\Big(\langle kl|v|in\rangle\langle i|\boldsymbol{dA}|j\rangle - \langle k|\boldsymbol{dA}|i\rangle\langle il|v|jn\rangle\Big)c_k^{\dagger}c_l^{\dagger}c_n c_j. \end{aligned}\tag{2.115}$$

If we use the completeness relation again in the Coulomb term in (2.115), this contribution vanishes, as it should. The one-particle contribution yields, using the commutator (2.107),

$$[\xi,H] = -\mathrm{i}\sum_{jk}\langle j|\frac{e}{m}\boldsymbol{pA} + \frac{e^2}{m}|\boldsymbol{A}|^2|k\rangle c_j^{\dagger}c_k \,. \tag{2.116}$$

Correspondingly, by means of

$$[\boldsymbol{dA}, \boldsymbol{pA}] = \mathrm{i}e\hbar|\boldsymbol{A}|^2 \,, \tag{2.117}$$

we obtain the double commutator as

$$[\xi, [\xi, H]] = -\frac{e^2}{m} \sum_{jk} \langle j||\boldsymbol{A}|^2|k\rangle c_j^\dagger c_k \,. \tag{2.118}$$

Summarizing these results, we obtain the following for $\tilde{H}$ in (2.113):

$$\tilde{H} = \sum_{jk} \langle j|h|k\rangle c_j^\dagger c_k + \frac{e^2}{2} \sum_{jklm} \langle jk|v|lm\rangle c_j^\dagger c_k^\dagger c_m c_l - \sum_{jk} \langle j|\boldsymbol{dE}|k\rangle \, c_j^\dagger c_k \,, \tag{2.119}$$

where the matrix element $\langle j|\boldsymbol{dE}|k\rangle$ has to be evaluated within the dipole approximation (compare Sect. 3.6). Beyond the dipole approximation, we obtain also higher-order electric multipole transitions. Moreover, the gauge transformation generates magnetic multipoles. However, if such contributions become important, it is more convinient to use the $\boldsymbol{pA}$ coupling.

Up to this point, we have restricted our treatment to the description of intrinsic microscopic properties of the system. Electromagnetic fields, as well as charge densities and currents have been considered as internal quantities. To obtain a coupling of the system to the outside world, we have to introduce an external charge densitiy ρ_{ext} and current densitiy $\boldsymbol{j}_{\mathrm{ext}}$, which give rise to additional, explicitly time-dependent contributions to the Hamiltonian,

$$H_{\mathrm{ext}}(t) = \int \mathrm{d}^3\boldsymbol{r} \Big(\hat{\rho}_{\mathrm{ext}}(\boldsymbol{r})\phi(\boldsymbol{r},t) - \hat{\boldsymbol{j}}_{\mathrm{ext}}(\boldsymbol{r})\boldsymbol{A}(\boldsymbol{r},t) \Big) \,, \tag{2.120}$$

which are invariant under the gauge transformation considered above. The external potential ϕ_{ext} is related to ρ_{ext} by the Poisson equation. Correspondingly, we could make use of the fact that the total scalar potential and the density are related by the Poisson equation. This allows us to replace the coupling of the external potential to the charge density by a coupling of the external charge density to the total potential. The same is valid for the coupling of the total current to an external vector potential, which is equivalent to the coupling of the total vector potential $\boldsymbol{A}$ to an external current density. Thus the coupling to external sources in (2.120) is equivalent to

$$H_{\mathrm{ext}}(t) = \int \mathrm{d}^3\boldsymbol{r} \Big(\hat{\rho}(\boldsymbol{r})\phi_{\mathrm{ext}}(\boldsymbol{r},t) - \hat{\boldsymbol{j}}(\boldsymbol{r})\boldsymbol{A}_{\mathrm{ext}}(\boldsymbol{r},t) \Big) \,, \tag{2.121}$$

where the charge density operator is given by

$$\hat{\rho} = -e\hat{n} \,, \tag{2.122}$$

using the one-particle density operator introduced in (2.52). The current operator is defined as

$$\begin{aligned}\hat{\boldsymbol{j}}(\boldsymbol{r}) &= -\frac{e}{2m}\sum_{jk}\Big[\varphi_j^*(\boldsymbol{r})\Big(\hat{\boldsymbol{p}}+e\boldsymbol{A}(\boldsymbol{r},t)\Big)\varphi_k(\boldsymbol{r}) \\ &\quad -\varphi_k(\boldsymbol{r})\Big(\hat{\boldsymbol{p}}-e\boldsymbol{A}(\boldsymbol{r},t)\Big)\varphi_j^*(\boldsymbol{r})\Big]c_j^\dagger c_k \\ &= \hat{\boldsymbol{j}}_0(\boldsymbol{r}) - \frac{e^2}{m}\hat{n}(\boldsymbol{r})\boldsymbol{A}(\boldsymbol{r},t), \end{aligned} \tag{2.123}$$

where $\hat{\boldsymbol{j}}_0(\boldsymbol{r})$ denotes the contribution for vanishing vector potential. In the classical limit, we replace the operators of the transverse electromagnetic field by their expectation values. These expectation values, however, vanish for vanishing external sources. Thus, without external fields, only field correlation functions contribute to the light–matter interaction. If external fields are present, the electric field in (2.119) is just the total field, i.e. the sum of the quantum mechanical average of the field and the external field. Thus even in the classical limit, a consistent treatment of Maxwell's equations and the Hamiltonian is required. This intimate coupling can be avoided if the change of the external field due to the internal field is negligible. In that case we can identify the total electromagnetic field in (2.119) with the external field.

2.6 Macroscopic Maxwell Equations and Response Functions

Any external field gives rise to induced charge and current densities which in turn produce macroscopic fields. This response of the system to external perturbations contains detailed information about physical properties of the system and is directly related to measurable quantities. Electromagnetic fields which result from the response of the system contain contributions which are detectable outside or at the boundary of a sample. Therefore the response of the system is generally macroscopic. In order to separate macroscopic and microscopic properties of a solid, we have to average over distances which are large in comparison with one lattice constant, but are small compared with the typical wavelengths of the fields involved. As a result of such an averaging procedure, only the long-range contributions to the fields survive. Formally this is most easily achieved by considering the Maxwell equations in Fourier space, where the contributions with small wave vector just correspond to long-range contributions in real space. Noting that the internal charge and current densities do not contribute in the macroscopic limit, owing to the charge neutrality of the system, Maxwell's equations can be written as

$$
\begin{aligned}
\boldsymbol{q}\,\boldsymbol{E}(\boldsymbol{q},\omega) &= -\frac{\mathrm{i}}{\varepsilon_0}[\rho_{\mathrm{ext}}(\boldsymbol{q},\omega)+\rho_{\mathrm{ind}}(\boldsymbol{q},\omega)]\,,\\
\boldsymbol{q}\times\boldsymbol{E}(\boldsymbol{q},\omega) &= \omega\boldsymbol{B}(\boldsymbol{q},\omega)\,,\\
\boldsymbol{q}\,\boldsymbol{B}(\boldsymbol{q},\omega) &= 0\,,\\
\boldsymbol{q}\times\boldsymbol{B}(\boldsymbol{q},\omega) &= -\frac{\omega}{c_0^2}\boldsymbol{E}(\boldsymbol{q},\omega)-\mathrm{i}\mu_0[\boldsymbol{j}_{\mathrm{ext}}(\boldsymbol{q},\omega)+\boldsymbol{j}_{\mathrm{ind}}(\boldsymbol{q},\omega)]\,.
\end{aligned}
\tag{2.124}
$$

In the following, we shall consider nonmagnetic materials only, i.e. external fields do not lead to an internal magnetization.[12]

There are two possible ways to proceed from this point. We start with the simplest one, which uses the basic definitions of charge and current densities (2.122) and (2.123).

2.6.1 Direct Calculation of Induced Charges and Currents

The expectation value of the density operator (2.122) yields, after spatial Fourier transformation (see (2.72)),

$$
\rho(\boldsymbol{q},t) = -\frac{e}{V}\sum_{ij}\langle i|\mathrm{e}^{-\mathrm{i}\boldsymbol{q}\boldsymbol{r}}|j\rangle\langle c_i^\dagger c_j\rangle\,. \tag{2.125}
$$

Correspondingly, the current density is found from (2.123) as

$$
\boldsymbol{j}(\boldsymbol{q},t) = -\frac{e}{2mV}\sum_{ij}\Big(\langle i|\mathrm{e}^{-\mathrm{i}\boldsymbol{q}\boldsymbol{r}}\boldsymbol{p}+\boldsymbol{p}\mathrm{e}^{-\mathrm{i}\boldsymbol{q}\boldsymbol{r}}|j\rangle + 2e\langle i|\mathrm{e}^{-\mathrm{i}\boldsymbol{q}\boldsymbol{r}}\boldsymbol{A}|j\rangle\Big)\langle c_i^\dagger c_j\rangle\,. \tag{2.126}
$$

We are interested in the induced part of the charge and current densities in the long-wavelength limit. The 1 in the expansion of the exponential factor for ρ yields the total electronic charge density, which is compensated by the charge density of the nuclei. Therefore the leading contribution to the induced charge density is proportional to $\boldsymbol{q}$. In contrast, the leading contribution to the current density results from the $\boldsymbol{q}=0$ contribution. Thus we have, in the long-wavelength limit,

$$
\rho_{\mathrm{ind}}(\boldsymbol{q},t) = \frac{\mathrm{i}e}{V}\sum_{ij}\boldsymbol{q}\langle i|\boldsymbol{r}|j\rangle\langle c_i^\dagger c_j\rangle = -\frac{\mathrm{i}}{V}\sum_{ij}\boldsymbol{q}\boldsymbol{d}_{ij}\langle c_i^\dagger c_j\rangle \tag{2.127}
$$

and

$$
\boldsymbol{j}_{\mathrm{ind}}(\boldsymbol{q},t) = -\frac{e}{mV}\sum_{ij}\langle i|\boldsymbol{p}|j\rangle\langle c_i^\dagger c_j\rangle + \frac{e^2}{mV}\sum_i\boldsymbol{A}\langle c_i^\dagger c_i\rangle\,. \tag{2.128}
$$

The induced charge and current densities are proportional to off-diagonal and diagonal elements of $\langle c_i^\dagger c_j\rangle$, referred to as the *one-particle density matrix.*

---

[12] In other textbooks, the $\boldsymbol{H}$ field is introduced. Nonmagnetic materials lead to the relation $\boldsymbol{B}=\mu_0\boldsymbol{H}$.

Equation (2.127) suggests that we should introduce the *polarization* according to

$$\rho_{\mathrm{ind}}(\boldsymbol{q},t) = -\mathrm{i}\boldsymbol{q}\boldsymbol{P}(\boldsymbol{q},t) \,, \tag{2.129}$$

where

$$\boldsymbol{P}(\boldsymbol{q},t) = \frac{1}{V}\sum_{ij} \boldsymbol{d}_{ij}\langle c_i^\dagger c_j\rangle \equiv \frac{1}{V}\sum_{ij} \boldsymbol{d}_{ij}\, n_{ij}(t) \,. \tag{2.130}$$

Within the dipole approximation, the $\boldsymbol{q}$ dependence of $\boldsymbol{P}$ enters only via a parametrical $\boldsymbol{q}$ dependence of the density matrix, which is induced by the optical field.

Alternatively, we can relate the current to the polarization by using the continuity equation (2.2) in the form

$$\mathrm{i}\frac{\partial}{\partial t}\rho_{\mathrm{ind}}(\boldsymbol{q},t) = \boldsymbol{q}\boldsymbol{j}_{\mathrm{ind}}(\boldsymbol{q},t) \,. \tag{2.131}$$

Inserting (2.129) we obtain

$$\boldsymbol{j}_{\mathrm{ind}}(\boldsymbol{q},t) = \frac{\partial}{\partial t}\boldsymbol{P}(\boldsymbol{q},t) = -\frac{1}{V}\sum_{ij}\boldsymbol{d}_{ij}\frac{\partial}{\partial t}\langle c_i^\dagger(t)c_j(t)\rangle. \tag{2.132}$$

As we have introduced the continuity equation by means of a consistency requirement on the Maxwell equations, we must, finally, verify that (2.132) is indeed equivalent to (2.128). We can use the Heisenberg equation of motion[13] to write (2.131) in the form

$$\langle[\hat{\rho}(\boldsymbol{q},t),H]\rangle = \hbar\boldsymbol{q}\boldsymbol{j}(\boldsymbol{q},t) \,, \tag{2.133}$$

which can be regarded as a microscopic version of the continuity equation. The commutator of the density operator with the Hamiltonian (2.58), together with the light–matter interaction described by (2.103), leads to

$$\begin{aligned}\langle[\hat{\rho}(\boldsymbol{q},t),H]\rangle &= \frac{e}{V}\sum_{ij}(E_i-E_j)\langle i|\mathrm{e}^{-\mathrm{i}\boldsymbol{q}\boldsymbol{r}}|j\rangle\langle c_i^\dagger(t)c_j(t)\rangle \\ &-\frac{e^2}{mV}\sum_{ijj'}\langle i|\Big(\boldsymbol{A}\boldsymbol{p}|j'\rangle\langle j'|\mathrm{e}^{-\mathrm{i}\boldsymbol{q}\boldsymbol{r}}-\mathrm{e}^{-\mathrm{i}\boldsymbol{q}\boldsymbol{r}}|j'\rangle\langle j'|\boldsymbol{A}\boldsymbol{p}\Big)|j\rangle\langle c_i^\dagger(t)c_j(t)\rangle \,.\end{aligned} \tag{2.134}$$

For simplicity, we have already neglected the Coulomb interaction, which cancels similarly to the way it does in (2.115) in the final equation. The first term on the right-hand side of (2.134) can be transformed by noting that

$$\begin{aligned}(E_i-E_j)\langle i|\mathrm{e}^{-\mathrm{i}\boldsymbol{q}\boldsymbol{r}}|j\rangle &= -\langle i|[\mathrm{e}^{-\mathrm{i}\boldsymbol{q}\boldsymbol{r}},h]|j\rangle \\ &= -\frac{\hbar}{2m}\langle i|\boldsymbol{q}\boldsymbol{p}\mathrm{e}^{-\mathrm{i}\boldsymbol{q}\boldsymbol{r}}+\mathrm{e}^{-\mathrm{i}\boldsymbol{q}\boldsymbol{r}}\boldsymbol{q}\boldsymbol{p}|j\rangle \,,\end{aligned} \tag{2.135}$$

[13] The equation-of-motion method will be introduced later. At the moment we use the necessary relations without proof and postpone details to Chap. 4.

which is the generalization of the relation (2.107) for a finite $\boldsymbol{q}$ vector and represents the continuity equation of elementary quantum mechanics. The second contribution on the right-hand side of (2.134) can be simplified by means of the completeness relation. Thus we find

$$\langle[\hat{\rho}(\boldsymbol{q},t),H]\rangle = -\frac{e\hbar}{2mV}\sum_{ij}\langle i|\boldsymbol{qp}\mathrm{e}^{-\mathrm{i}\boldsymbol{qr}} + \mathrm{e}^{-\mathrm{i}\boldsymbol{qr}}\boldsymbol{qp}|j\rangle\langle c_i^\dagger(t)c_j(t)\rangle$$
$$-\frac{e^2}{mV}\sum_{ij}\langle i|\boldsymbol{Ap}\mathrm{e}^{-\mathrm{i}\boldsymbol{qr}} - \mathrm{e}^{-\mathrm{i}\boldsymbol{qr}}\boldsymbol{Ap}|j\rangle\langle c_i^\dagger(t)c_j(t)\rangle\,. \qquad (2.136)$$

Comparison with (2.126) completes the desired proof of the continuity equation (2.133), which we could have used in this way to introduce the current density.

By means of (2.127) and (2.132), we have related both the induced charge density and the induced current density to dipole transition matrix elements and off-diagonal elements of the one-particle density matrix. This is nothing but the transition from the $\boldsymbol{pA}$ coupling to the $\boldsymbol{dE}$ coupling, which we introduced in the last section for the interaction Hamiltonian, now performed for the coupling of light and matter in Maxwell's equations.

As already indicated in (2.134), we can write down equations of motion for the density matrices and try to solve them. This procedure has several advantages in nonlinear optics if external fields have to be taken into account to arbitrary order and the time evolution of these fields comes into play. If, however, we want to study intrinsic optical properties to low orders of the external fields, it is more convenient to get rid of the functional dependence on the fields. In this case, it is advantageous to introduce response functions, which determine a perturbational expansion of the induced charge and current densities with respect to the external field. With increasing order in the field, one obtains higher-order correlation functions of the particle system in this way. In the linear case, the response function is a two-particle correlation function.

2.6.2 Phenomenological Theory of Linear Response

On a phenomenological level, it is convenient to define the response with respect to the total field rather than to the external field. Thus we relate the induced current density to the total field by the relation

$$\boldsymbol{j}_{\mathrm{ind}}(\boldsymbol{q},\omega) = \underline{\sigma}(\boldsymbol{q},\omega)\boldsymbol{E}(\boldsymbol{q},\omega)\,, \qquad (2.137)$$

where the *conductivity tensor* $\underline{\sigma}$ can be written as sum of a longitudinal contribution

$$\sigma_{ij}^{\mathrm{l}}(\boldsymbol{q},\omega) = \sum_k \frac{q_i q_k}{q^2}\sigma_{kj}(\boldsymbol{q},\omega) \qquad (2.138)$$

and a transverse contribution

$$\sigma^{\mathrm{t}}_{ij}(\boldsymbol{q},\omega) = \sum_k \left(\delta_{ik} - \frac{q_i q_k}{q^2}\right)\sigma_{kj}(\boldsymbol{q},\omega) \equiv \sum_k \delta^{\mathrm{t}}_{ik}(\boldsymbol{q})\sigma_{kj}(\boldsymbol{q},\omega)\,. \tag{2.139}$$

Inserting (2.137) into (2.124) we can relate the external and the total field[14] according to

$$\boldsymbol{E}_{\mathrm{ext}}(\boldsymbol{q},\omega) = \left(\underline{1} + \frac{\mathrm{i}}{\varepsilon_0\omega}\underline{\sigma}(\boldsymbol{q},\omega)\right)\boldsymbol{E}(\boldsymbol{q},\omega) =: \underline{\varepsilon}(\boldsymbol{q},\omega)\boldsymbol{E}(\boldsymbol{q},\omega)\,, \tag{2.140}$$

where we have also defined the *dielectric tensor* $\underline{\varepsilon}$ of the system; this can again be decomposed into a longitudinal and a transverse contribution in a manner similar to (2.138) and (2.139). $\underline{1}$ is the unit matrix. Using the continuity equation for the longitudinal current in the form

$$\boldsymbol{q}\boldsymbol{j}(\boldsymbol{q},\omega) = \omega\rho(\boldsymbol{q},\omega)\,, \tag{2.141}$$

we can introduce the external field into the macroscopic Maxwell equations and obtain, instead of (2.124), the more familiar formulation

$$\begin{aligned}
\boldsymbol{q}\,\boldsymbol{E}_{\mathrm{ext}}(\boldsymbol{q},\omega) &= -\frac{\mathrm{i}}{\varepsilon_0}\rho_{\mathrm{ext}}(\boldsymbol{q},\omega)\,,\\
\boldsymbol{q}\times\boldsymbol{E}(\boldsymbol{q},\omega) &= \omega\boldsymbol{B}(\boldsymbol{q},\omega)\,,\\
\boldsymbol{q}\,\boldsymbol{B}(\boldsymbol{q},\omega) &= 0\,,\\
\boldsymbol{q}\times\boldsymbol{B}(\boldsymbol{q},\omega) &= -\frac{\omega}{c_0^2}\boldsymbol{E}_{\mathrm{ext}}(\boldsymbol{q},\omega) - \mathrm{i}\,\mu_0\boldsymbol{j}_{\mathrm{ext}}(\boldsymbol{q},\omega)\,.
\end{aligned} \tag{2.142}$$

In optics, external sources or fields are externally controlled quantities. Thus a knowledge of the longitudinal dielectric function is sufficient to determine the total longitudinal fields. For the transverse fields, we obtain from (2.142) and (2.140) the wave equation

$$\left(q^2 - \frac{\omega^2}{c_0^2}\underline{\varepsilon}^{\mathrm{t}}(\boldsymbol{q},\omega)\right)\boldsymbol{E}^{\mathrm{t}}(\boldsymbol{q},\omega) = \mathrm{i}\mu_0\omega\boldsymbol{j}^{\mathrm{t}}_{\mathrm{ext}}(\boldsymbol{q},\omega) \tag{2.143}$$

or, equivalently, a wave equation for the vector potential

$$\left(q^2 - \frac{\omega^2}{c_0^2}\underline{\varepsilon}^{\mathrm{t}}(\boldsymbol{q},\omega)\right)\boldsymbol{A}(\boldsymbol{q},\omega) = \mu_0\boldsymbol{j}^{\mathrm{t}}_{\mathrm{ext}}(\boldsymbol{q},\omega)\,. \tag{2.144}$$

The task is now to calculate the induced charge and current densities which describe the response of the system to external perturbations, from a microscopic theory. Although in this section we are mainly interested in the macroscopic limits of response functions, the basic definitions given in the following are completely general and are also the starting point for calculating the microscopic response of the system. The practical determination of

[14] In other textbooks, the $\boldsymbol{D}$ field is introduced, which is simply connected to $\boldsymbol{E}_{\mathrm{ext}}(\boldsymbol{q},\omega)$ via $\boldsymbol{D}(\boldsymbol{q},\omega) = \varepsilon_0\,\boldsymbol{E}_{\mathrm{ext}}(\boldsymbol{q},\omega)$.

response functions under various circumstances is one of the principal objectives of this book and will be dealt with in various chapters. In the following we first will recall some basic results of time-dependent perturbation theory.

2.6.3 Time-Dependent Perturbation Theory

The time evolution of a system under the action of a Hamiltonian $H+H_{\mathrm{ext}}(t)$ is determined by the time evolution operator, which causes a state vector at a time t_0, sufficiently long before the perturbation is switched on, to evolve to a time t, according to

$$|\Psi(t)\rangle = U(t,t_0)|\Psi(t_0)\rangle \, , \tag{2.145}$$

where U obeys the equation

$$\mathrm{i}\hbar\frac{\partial}{\partial t}U(t,t_0) = [H+H_{\mathrm{ext}}(t)]U(t,t_0) \, . \tag{2.146}$$

The time evolution under the action of H can be eliminated by introducing

$$S(t,t_0) = \mathrm{e}^{-\mathrm{i}H(t-t_0)/\hbar}U(t,t_0) \, , \tag{2.147}$$

which is a solution of

$$\mathrm{i}\hbar\frac{\partial}{\partial t}S(t,t_0) = \bar{H}_{\mathrm{ext}}(t)S(t,t_0) \, , \tag{2.148}$$

where the bar denotes the time evolution under the action of H, i.e.

$$\bar{H}_{\mathrm{ext}} = \mathrm{e}^{\mathrm{i}H(t-t_0)/\hbar}H_{\mathrm{ext}}(t)\mathrm{e}^{-\mathrm{i}H(t-t_0)/\hbar} \, . \tag{2.149}$$

Integration of (2.148) yields, with the condition $S(t_0,t_0)=1$,

$$S(t,t_0) = 1 - \frac{\mathrm{i}}{\hbar}\int_{t_0}^{t}\mathrm{d}t'\,\bar{H}_{\mathrm{ext}}(t')S(t',t_0) \, . \tag{2.150}$$

A formal solution of (2.150) can be obtained by iteration. The general treatment will be discussed in Chap. 10. At the moment we restrict ourselves to the linear case, which yields

$$S(t,t_0) = 1 - \frac{\mathrm{i}}{\hbar}\int_{t_0}^{t}\mathrm{d}t'\,\bar{H}_{\mathrm{ext}}(t') \, . \tag{2.151}$$

The expectation value of any operator $\mathcal{O}$ can be expressed according to

$$\langle\mathcal{O}(t)\rangle = \langle\Psi(t_0)|S^{\dagger}(t,t_0)\bar{\mathcal{O}}S(t,t_0)|\Psi(t_0)\rangle. \tag{2.152}$$

Thus the linear induced change $\langle\mathcal{O}\rangle_{\mathrm{ind}} = \langle\mathcal{O}(t)\rangle - \langle\Psi(t_0)|\bar{\mathcal{O}}|\Psi(t_0)\rangle$ due to the action of H_{ext} is obtained from

$$\langle\mathcal{O}\rangle_{\mathrm{ind}} = \frac{\mathrm{i}}{\hbar}\int_{t_0}^{t}\mathrm{d}t'\,\langle[\bar{H}_{\mathrm{ext}}(t'),\bar{\mathcal{O}}(t)]\rangle \, . \tag{2.153}$$

A formula of this type which describes the linear response by means of a correlation function is usually referred to as a *Kubo formula* [20, 21]. Now we can apply this general result of linear response theory to the specific case of the density and current response functions which occur in the macroscopic Maxwell equations. Using the external perturbations introduced in (2.121), we obtain the following for the induced density:

$$\rho_{\text{ind}}(\boldsymbol{r},t) = -\frac{\mathrm{i}}{\hbar}\int_{t_0}^{t} \mathrm{d}t'\,\mathrm{d}^3\boldsymbol{r}'\,\langle[\hat{\rho}(\boldsymbol{r},t),\hat{\rho}(\boldsymbol{r}',t')]\rangle\phi_{\text{ext}}(\boldsymbol{r}',t'). \tag{2.154}$$

This result shows that the induced density is determined by just a density–density correlation function. The induced current density is obtained according to (2.123), as a sum of two contributions. Using (2.153), we obtain now a response function which is determined by a current–current correlation function [21] and the $n\boldsymbol{A}$ contribution, sometimes referred to as the Drude term:[15]

$$\begin{aligned}\boldsymbol{j}_{\text{ind}}(\boldsymbol{r},t) = {} & \frac{\mathrm{i}}{\hbar}\int_{t_0}^{t} \mathrm{d}t'\,\mathrm{d}^3\boldsymbol{r}'\,\langle[\hat{\boldsymbol{j}}_0(\boldsymbol{r},t),\boldsymbol{A}_{\text{ext}}(\boldsymbol{r}',t')\hat{\boldsymbol{j}}_0(\boldsymbol{r}',t')]\rangle \\ & - \frac{e^2}{m}n(\boldsymbol{r})\boldsymbol{A}(\boldsymbol{r},t)\,.\end{aligned} \tag{2.155}$$

An immediate comparison of these results with the phenomenological relation (2.137) is only possible if the total field is taken as the external field. The basic definitions of the microscopic response functions are necessarily related to the external perturbations. Let us consider the relation between the external and total fields. In the case of a homogeneous system, this is rather trivial because we can use (2.140). In inhomogeneous systems such as solids, the problem is more complicated. As we shall see in the following, the complication arises from the fact that a macroscopic perturbation induces not only macroscopic but also microscopic fields, which in turn influence the macroscopic response. This will be discussed for the longitudinal case first.

2.6.4 Longitudinal Response Functions

Within the restriction to the linear approximation, we can use (2.154) to express the correlation function by means of variational derivatives. We write (2.154) in the form

$$\rho_{\text{ind}}(\boldsymbol{x}) = \frac{\varepsilon_0}{V}\int \mathrm{d}^4\boldsymbol{x}'\,\chi(\boldsymbol{x},\boldsymbol{x}')\phi_{\text{ext}}(\boldsymbol{x}')\,, \tag{2.156}$$

where we have used the notation $\boldsymbol{x} := \boldsymbol{r}, t$. Apart from prefactors, the density–density correlation function is the same as the dielectric susceptibility $\chi(\boldsymbol{x},\boldsymbol{x}')$, which is defined as the variational derivative of the density with respect to the external potential:

[15] Note that in the transformation from the current–current correlation function in the long-wavelength limit to the dipole–dipole correlation function, the Drude term is cancelled exactly (compare Problem 5.5).

$$\chi(\boldsymbol{x},\boldsymbol{x}') = -\frac{\mathrm{i}V}{\hbar\varepsilon_0}\theta(t-t')[\hat{\rho}(\boldsymbol{x}),\hat{\rho}(\boldsymbol{x}')] \,. \tag{2.157}$$

The induced charge density results, via the Poisson equation, in an effective potential

$$\phi_{\mathrm{eff}}(\boldsymbol{x}) = \phi_{\mathrm{ext}}(\boldsymbol{x}) + \int \mathrm{d}^4\boldsymbol{x}'\, v(\boldsymbol{x}-\boldsymbol{x}')\rho_{\mathrm{ind}}(\boldsymbol{x}') \,, \tag{2.158}$$

where $v(\boldsymbol{x}-\boldsymbol{x}') = v(\boldsymbol{r}-\boldsymbol{r}')\delta(t-t')$. This effective potential itself depends on the induced charge density and thus is related to the external potential by an integral equation, which is obtained from (2.156) and (2.158):

$$\phi_{\mathrm{eff}}(\boldsymbol{x}) = \phi_{\mathrm{ext}}(\boldsymbol{x}) + \frac{\varepsilon_0}{V}\int \mathrm{d}^4\boldsymbol{x}'\,\mathrm{d}^4\boldsymbol{x}''\, v(\boldsymbol{x}-\boldsymbol{x}')\chi(\boldsymbol{x}',\boldsymbol{x}'')\phi_{\mathrm{ext}}(\boldsymbol{x}'') \,. \tag{2.159}$$

This equation suggests that one can shortcut the mutual influence of the induced charge densities and the corresponding potentials resulting from the Coulomb interaction by defining a quantity which describes the variation of the charge density with respect to the effective potential, through the relation

$$\rho_{\mathrm{ind}}(\boldsymbol{x}) = \frac{\varepsilon_0}{V}\int \mathrm{d}^4\boldsymbol{x}' P(\boldsymbol{x},\boldsymbol{x}')\phi_{\mathrm{eff}}(\boldsymbol{x}') \,. \tag{2.160}$$

The quantity P is frequently called the *polarization propagator.* Inserting (2.159) into (2.160) and calculating the variational derivative with respect to $\phi_{\mathrm{ext}}(x')$, we obtain the following for the susceptibility:

$$\boldsymbol{\chi(x,x') = P(x,x') + \frac{\varepsilon_0}{V}\int \mathrm{d}^4 x_1\,\mathrm{d}^4 x_2\, P(x,x_1)v(x_1-x_2)\chi(x_2,x')} \,. \tag{2.161}$$

An iterative solution of (2.161) illustrates (see Fig. 2.1) the process by which an induced charge density gives rise to a change in the potential via the Coulomb interaction, which in turn yields an induced charge density and so forth. The result of this infinite series of charge redistribution processes is the screening of the external perturbation. In compact form, this screening is described by the inverse microscopic dielectric function

$$\begin{aligned}\varepsilon^{-1}(\boldsymbol{x},\boldsymbol{x}') &= \frac{\delta\phi_{\mathrm{eff}}(\boldsymbol{x})}{\delta\phi_{\mathrm{ext}}(\boldsymbol{x}')} \\ &= \delta(\boldsymbol{x}-\boldsymbol{x}') + \frac{\varepsilon_0}{V}\int \mathrm{d}^4\boldsymbol{x}''\, v(\boldsymbol{x}-\boldsymbol{x}'')\chi(\boldsymbol{x}'',\boldsymbol{x}') \,,\end{aligned} \tag{2.162}$$

which allows us to relate the susceptibility and the polarization propagator by

$$\chi(\boldsymbol{x},\boldsymbol{x}') = \int \mathrm{d}^4\boldsymbol{x}'' P(\boldsymbol{x},\boldsymbol{x}'')\,\varepsilon^{-1}(\boldsymbol{x}'',\boldsymbol{x}') \,. \tag{2.163}$$

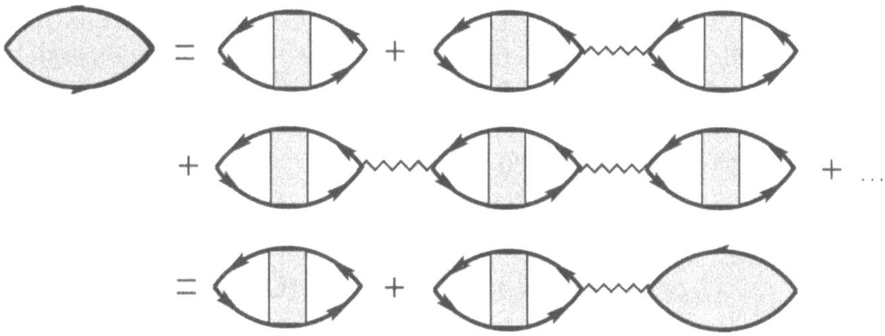

Fig. 2.1. The iteration scheme of charge redistribution can be illustrated diagramatically (and quite intuitively) by the bubbles, as depicted. The susceptibility (*shaded bubbles*) is an infinite sum over polarizations P (*open bubbles*) multiplied by the Coulomb interaction v (*wavy lines*). These graphs also have a precise mathematical meaning, as will be described in Chap. 10

The physical meaning of this inverse microscopic dielectric function is at least twofold. On the microscopic level, it describes the screening of the interaction between charge carriers in the system. In the long-wavelength limit, it determines the macroscopic dielectric function, which gives rise to screening of external perturbations. The microscopic aspect will be studied extensively throughout this book, but at the moment we are interested in the macroscopic limit which enters the Maxwell equations.

In homogeneous systems we can use translational invariance, i.e. the susceptibility and the polarization propagator depend on difference coordinates only. The convolution integrals in real space which occur for example in (2.161) can be reduced to products in Fourier space. This yields simple algebraic equations and the macroscopic limit becomes trivial. In solids we have translational invariance on a lattice only, i.e. $P(\boldsymbol{r},\boldsymbol{r}') = P(\boldsymbol{r}+\boldsymbol{L},\boldsymbol{r}'+\boldsymbol{L})$ and $\chi(\boldsymbol{r},\boldsymbol{r}') = \chi(\boldsymbol{r}+\boldsymbol{L},\boldsymbol{r}'+\boldsymbol{L})$, where $\boldsymbol{L}$ is an arbitrary lattice vector. Thus the Fourier transform of (2.161) with respect to $t-t'$, $\boldsymbol{r}$, and $\boldsymbol{r}'$ can be cast into the form

$$\chi(\boldsymbol{G}+\boldsymbol{q},\boldsymbol{G}'+\boldsymbol{q},\omega) = P(\boldsymbol{G}+\boldsymbol{q},\boldsymbol{G}'+\boldsymbol{q},\omega)$$
$$+\varepsilon_0 V \sum_{\boldsymbol{G}''} P(\boldsymbol{G}+\boldsymbol{q},\boldsymbol{G}''+\boldsymbol{q},\omega)\, v(\boldsymbol{G}''+\boldsymbol{q})\ ,\chi(\boldsymbol{G}''+\boldsymbol{q},\boldsymbol{G}'+\boldsymbol{q},\omega)\ ,(2.164)$$

where $\boldsymbol{G}$ are the vectors of the reciprocal lattice, $\boldsymbol{q}$ is restricted to the first Brillouin zone, and

$$v(\boldsymbol{q}) = \frac{1}{\varepsilon_0 V \boldsymbol{q}^2} \tag{2.165}$$

is the Fourier transform of the Coulomb potential. The macroscopic limit of the susceptibility is obtained from the contributions for small wave vectors i.e. the components with $G = G' = 0$. As the Coulomb interaction itself is long-ranged, the $1/q^2$ singularity in (2.164) has to be treated separately. For this purpose, we split the sum over reciprocal-lattice vectors into the singular and the nonsingular contribution and obtain the following for the macroscopic susceptibility $\chi(\boldsymbol{q}, \boldsymbol{q}, \omega) \equiv \chi(\boldsymbol{q}, \omega)$:

$$\chi(\boldsymbol{q},\omega) = P(\boldsymbol{q},\omega) + \varepsilon_0 V P(\boldsymbol{q},\omega)\, v(\boldsymbol{q})\, \chi(\boldsymbol{q},\omega)$$

$$+ \varepsilon_0 V \sum_{\boldsymbol{G}\neq 0} P(\boldsymbol{q}, \boldsymbol{G}+\boldsymbol{q},\omega)\, v(\boldsymbol{G}+\boldsymbol{q})\, \chi(\boldsymbol{G}+\boldsymbol{q},\boldsymbol{q},\omega)\,. \tag{2.166}$$

Defining

$$\tilde{\chi}(\boldsymbol{q},\omega) = \frac{\chi(\boldsymbol{q},\omega)}{1+\varepsilon_0 V v(\boldsymbol{q})\chi(\boldsymbol{q},\omega)} = \chi(\boldsymbol{q},\omega)/\varepsilon^{-1}(\boldsymbol{q},\omega)\,, \tag{2.167}$$

where $\varepsilon^{-1}(\boldsymbol{q}\,\omega)$ denotes the long-wavelength limit of the inverse dielectric function, we obtain from (2.166)

$$\tilde{\chi}(\boldsymbol{q},\omega) = P(\boldsymbol{q},\omega) + \varepsilon_0 V \sum_{\boldsymbol{G}\neq 0} P(\boldsymbol{q},\boldsymbol{G}+\boldsymbol{q},\omega) v(\boldsymbol{G}+\boldsymbol{q})\tilde{\chi}(\boldsymbol{G}+\boldsymbol{q},\boldsymbol{q},\omega)\,, \tag{2.168}$$

where $\tilde{\chi}(\boldsymbol{G}+\boldsymbol{q},\boldsymbol{q},\omega)$ is a solution of

$$\tilde{\chi}(\boldsymbol{G}+\boldsymbol{q},\boldsymbol{q},\omega) = P(\boldsymbol{G}+\boldsymbol{q},\boldsymbol{q},\omega)$$

$$+\varepsilon_0 V \sum_{\boldsymbol{G}'\neq 0} P(\boldsymbol{G}+\boldsymbol{q},\boldsymbol{G}'+\boldsymbol{q},\omega)\, v(\boldsymbol{G}'+\boldsymbol{q})\, \tilde{\chi}(\boldsymbol{G}'+\boldsymbol{q},\boldsymbol{q},\omega)\,. \tag{2.169}$$

The $\boldsymbol{G} \neq 0$ contributions in (2.168) and (2.169) i.e. the short-range parts of the Coulomb interaction, are the so-called local-field corrections which are responsible for the complication that the macroscopic limit of the inverse dielectric function does not coincide with the inverse of the macroscopic limit of the dielectric function. In terms of the susceptibility $\tilde{\chi}(\boldsymbol{q},\omega)$, which incorporates the local-field corrections, the definition of the macroscopic dielectric function is straightforward. From (2.167), we obtain as a result

$$\chi(\boldsymbol{q},\omega) = \frac{\tilde{\chi}(\boldsymbol{q},\omega)}{1-\varepsilon_0 V v(\boldsymbol{q})\,\tilde{\chi}(\boldsymbol{q},\omega)}\,, \tag{2.170}$$

or the equivalent relation

$$\varepsilon^{-1}(\boldsymbol{q},\omega) = \frac{1}{\varepsilon(\boldsymbol{q},\omega)}\,, \tag{2.171}$$

where the macroscopic dielectric function is given by

$$\varepsilon(\boldsymbol{q},\omega) = 1 - \frac{1}{\boldsymbol{q}^2}\,\tilde{\chi}(\boldsymbol{q},\omega)\,. \tag{2.172}$$

A schematic illustration of the real part of the macroscopic dielectric function from the infrared to the ultraviolet range is given in Fig. 2.2. As there is a gap between the occupied and unoccupied states in semiconductors, in contrast to metals, no redistribution of the charge density is possible in the limit of vanishing energy and momentum transfer. Thus one expects, for general reasons – details will be given in Chap. 5 – that an expansion for small wave vectors starts at the second order:

$$\tilde{\chi}(\boldsymbol{q},\omega) = \boldsymbol{q}\,\underline{\tilde{\chi}}(\boldsymbol{q},\omega)\,\boldsymbol{q}\,. \tag{2.173}$$

The $1/\boldsymbol{q}^2$ singularity of the Coulomb potential in (2.172) cancels out, and the dielectric function of a semiconductor is regular in the zero frequency and long-wavelength limit.

Now the macroscopically induced charge density can immediately be expressed in terms of the effective potential:

$$\begin{aligned}\rho_{\text{ind}}(\boldsymbol{q},\omega) &= \frac{\varepsilon_0\tilde{\chi}(\boldsymbol{q},\omega)}{1-\varepsilon_0 V v(\boldsymbol{q})\tilde{\chi}(\boldsymbol{q},\omega)}\,\phi_{\text{ext}}(\boldsymbol{q},\omega)\\ &= \varepsilon_0\tilde{\chi}(\boldsymbol{q},\omega)\phi_{\text{eff}}(\boldsymbol{q},\omega)\,.\end{aligned} \tag{2.174}$$

Using

$$\boldsymbol{E}(\boldsymbol{q},\omega) = -\mathrm{i}\boldsymbol{q}\,\phi_{\text{eff}}(\boldsymbol{q},\omega)\,, \tag{2.175}$$

and the first equation of (2.124), we finally obtain

$$\boldsymbol{q}\left[\underline{1} - \underline{\tilde{\chi}}(\boldsymbol{q},\omega)\right]\boldsymbol{E}(\boldsymbol{q},\omega) = -\frac{\mathrm{i}}{\varepsilon_0}\rho_{\text{ext}}(\boldsymbol{q},\omega)\,, \tag{2.176}$$

which is nothing but the first Maxwell equation (2.142). Comparison with (2.140) yields

$$\underline{\sigma}(\boldsymbol{q},\omega) = -\mathrm{i}\varepsilon_0\,\omega\,\underline{\tilde{\chi}}(\boldsymbol{q},\omega)\,. \tag{2.177}$$

The components of the conductivity tensor are thus related to the macroscopic limit of the longitudinal response function, which obviously contains all

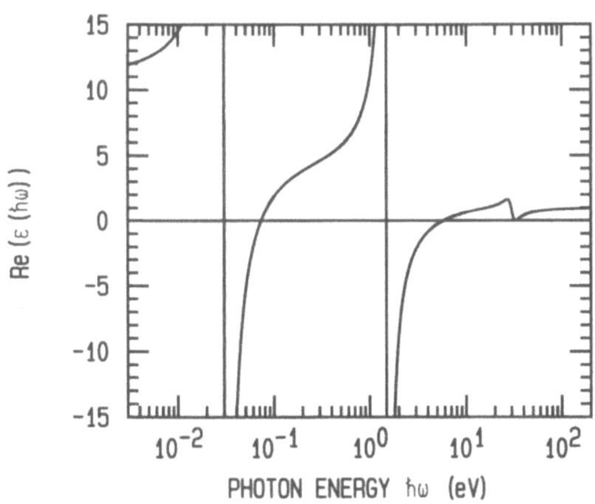

Fig. 2.2. Schematic illustration of the real part of the dielectric function from the infrared to the ultraviolet range. The structure on the low-energy side corresponds to resonances, owing to lattice vibrations. The structure on the high-energy side corresponds to electronic transitions

information required to calculate the transverse response, too. As we have already shown that the dipole coupling $\boldsymbol{dE}$, which governs the longitudinal response in the long-wavelength limit, is unitarily equivalent to the $\boldsymbol{pA}$ coupling, we could omit an explicit discussion of the transverse case. It is, however, instructive to see the way, in which local-field corrections enter the transverse response. Moreover, in the following we shall introduce some basic properties of the transverse modes, which result from their dynamical degree of freedom that is missing in the longitudinal case.

2.6.5 Transverse Response Functions

Considering transverse fields, our task is to derive (2.137) again, now, however, as a relation between the transverse current and the total transverse field, from the microscopic definition of the transverse response function. We start with the wave equation for the vector potential,

$$\left(\boldsymbol{q}^2 - \frac{\omega^2}{c_0^2}\right)\boldsymbol{A}(\boldsymbol{q},\omega) = \mu_0 \left(\boldsymbol{j}^t_{\mathrm{ext}}(\boldsymbol{q},\omega) + \boldsymbol{j}^t_{\mathrm{ind}}(\boldsymbol{q},\omega)\right) , \tag{2.178}$$

which is obtained from (2.124), and proceed similarly to the longitudinal case. First we introduce the transverse susceptibility as the variational derivative of the current density with respect to the external vector potential by the definition

$$\boldsymbol{j}_{\mathrm{ind}}(\boldsymbol{x}) = \frac{1}{\mu_0 V}\int \mathrm{d}^4x'\, \underline{\chi}^{\mathrm{t}}(\boldsymbol{x},\boldsymbol{x}')\boldsymbol{A}_{\mathrm{ext}}(\boldsymbol{x}') . \tag{2.179}$$

Making use of (2.155) and applying the chain rule for variational derivatives, we can relate the transverse susceptibility to the response function with respect to the total vector potential:

$$\begin{aligned}&\underline{\chi}^{\mathrm{t}}(\boldsymbol{x},\boldsymbol{x}')\\ &= \int \mathrm{d}^4x'' \left(\underline{P}^{\mathrm{t}}(\boldsymbol{x},\boldsymbol{x}'')\frac{\delta \boldsymbol{A}(\boldsymbol{x}'')}{\delta \boldsymbol{A}_{\mathrm{ext}}(\boldsymbol{x}')} + \boldsymbol{P}^{\mathrm{tl}}(\boldsymbol{x},\boldsymbol{x}'')\frac{\delta \phi_{\mathrm{eff}}(\boldsymbol{x}'')}{\delta \boldsymbol{A}_{\mathrm{ext}}(\boldsymbol{x}')}\right) ,\end{aligned} \tag{2.180}$$

where we have introduced the *transverse polarization propagator*

$$\underline{P}^{\mathrm{t}}(\boldsymbol{x},\boldsymbol{x}') = -\frac{\mu_0 e^2 V}{m} n(\boldsymbol{r})\delta(\boldsymbol{x}-\boldsymbol{x}') + \mu_0 V \frac{\delta \boldsymbol{j}_0(\boldsymbol{x})}{\delta \boldsymbol{A}(\boldsymbol{x}')} \tag{2.181}$$

and a mixed term

$$\boldsymbol{P}^{\mathrm{tl}}(\boldsymbol{x},\boldsymbol{x}') = \mu_0 V \frac{\delta \boldsymbol{j}_0(\boldsymbol{x})}{\delta \phi_{\mathrm{eff}}(\boldsymbol{x}')} , \tag{2.182}$$

which couples longitudinal and transverse excitations and contributes only in inhomogeneous systems, in which translational invariance is absent. In the case of the longitudinal response discussed earlier, this coupling would have resulted in a transverse correction to the local-field corrections and was therefore neglected. The longitudinal–transverse coupling in (2.180) can be further transformed by means of the Poisson equation, yielding

$$\frac{\delta\phi_{\text{eff}}(\boldsymbol{x})}{\delta\boldsymbol{A}_{\text{ext}}(\boldsymbol{x}')} = \frac{1}{\mu_0 V}\int \mathrm{d}^4\boldsymbol{x}_1\,\mathrm{d}^4\boldsymbol{x}_2\,v(\boldsymbol{x}-\boldsymbol{x}_1)\,\chi^{\text{lt}}(\boldsymbol{x}_1,\boldsymbol{x}_2)\,\frac{\delta\boldsymbol{A}(\boldsymbol{x}_2)}{\delta\boldsymbol{A}_{\text{ext}}(\boldsymbol{x}')}\,, \tag{2.183}$$

where

$$\chi^{\text{lt}}(\boldsymbol{x},\boldsymbol{x}') = \boldsymbol{P}^{\text{lt}}(\boldsymbol{x},\boldsymbol{x}') + \frac{\varepsilon_0}{V}\int \mathrm{d}^4\boldsymbol{x}_1\,\mathrm{d}^4\boldsymbol{x}_2\,P(\boldsymbol{x},\boldsymbol{x}_1)v(\boldsymbol{x}_1-\boldsymbol{x}_2)\chi^{\text{lt}}(\boldsymbol{x}_2,\boldsymbol{x}') \tag{2.184}$$

and

$$\boldsymbol{P}^{\text{lt}}(\boldsymbol{x},\boldsymbol{x}') = \mu_0 V\frac{\delta\rho(\boldsymbol{x})}{\delta\boldsymbol{A}(\boldsymbol{x}')}\bigg|_{\phi_{\text{eff}}=\text{const}}\,. \tag{2.185}$$

Here we have used the fact that the charge density is a functional of both $\boldsymbol{A}$ and $\phi_{\text{eff}}[\boldsymbol{A}]$. In order to calculate the variational derivatives with respect to the external vector potential in (2.180) and (2.182), we solve the wave equation (2.178). This yields, after Fourier transformation, the solution

$$\boldsymbol{A}(\boldsymbol{x}) = \int \mathrm{d}^4\boldsymbol{x}'\,\underline{\mathcal{D}}_0^{\text{r}}(\boldsymbol{x},\boldsymbol{x}')\Big(\boldsymbol{j}_{\text{ext}}(\boldsymbol{x}') + \boldsymbol{j}_{\text{ind}}(\boldsymbol{x}')\Big)\,, \tag{2.186}$$

where we have introduced the retarded free-photon Green's function

$$\underline{\mathcal{D}}_0^{\text{r}}(\boldsymbol{x},\boldsymbol{x}') = \frac{\mu_0}{V}\sum_{\boldsymbol{q}}\int\frac{\mathrm{d}\omega}{2\pi}\mathrm{e}^{\mathrm{i}[\boldsymbol{q}(\boldsymbol{r}-\boldsymbol{r}')-\omega(t-t')]}\underline{\delta}^{\text{t}}(\boldsymbol{q})\mathcal{D}_0^{\text{r}}(\boldsymbol{q},\omega)\,. \tag{2.187}$$

We define for later reference

$$\mathcal{D}_0^{\text{r}}(\boldsymbol{q},\omega) = \frac{c_0^2}{\omega_{\boldsymbol{q}}^2 - (\omega+\mathrm{i}\delta)^2}\,, \tag{2.188}$$

where $\omega_{\boldsymbol{q}} = c_0|\boldsymbol{q}|$. The photons, which mediate the transverse interaction, constitute a separate degree of freedom, in contrast to the longitudinal static Coulomb interaction. Thus the interaction is dynamic, i.e. frequency-dependent. Now we can use the fact that, according to (2.186), the external current and vector potential are related by

$$\boldsymbol{A}_{\text{ext}}(\boldsymbol{x}) = \int \mathrm{d}^4\boldsymbol{x}'\,\underline{\mathcal{D}}_0^{\text{r}}(\boldsymbol{x},\boldsymbol{x}')\boldsymbol{j}_{\text{ext}}(\boldsymbol{x}')\,, \tag{2.189}$$

and obtain the following for the variational derivative:

$$\frac{\delta\boldsymbol{A}(\boldsymbol{x})}{\delta\boldsymbol{A}_{\text{ext}}(\boldsymbol{x}')} = \delta(\boldsymbol{x}-\boldsymbol{x}') + \frac{1}{V}\int \mathrm{d}^4\boldsymbol{x}''\,\underline{\mathcal{D}}_0^{\text{r}}(\boldsymbol{x},\boldsymbol{x}'')\underline{\chi}^{\text{t}}(\boldsymbol{x}''\boldsymbol{x}')\,, \tag{2.190}$$

which is the transverse analogue of (2.162). Combining (2.180) with (2.183) and (2.190), we obtain as a result

$$\underline{\chi}^{\text{t}}(\boldsymbol{x},\boldsymbol{x}') = \underline{\tilde{\chi}}^{\text{t}}(\boldsymbol{x},\boldsymbol{x}') + \frac{1}{\mu_0 V}\int \mathrm{d}^4\boldsymbol{x}_1\,\mathrm{d}^4\boldsymbol{x}_2\,\underline{\tilde{\chi}}^{\text{t}}(\boldsymbol{x},\boldsymbol{x}_1)\underline{\mathcal{D}}_0^{\text{r}}(\boldsymbol{x}_1,\boldsymbol{x}_2)\underline{\chi}^{\text{t}}(\boldsymbol{x}_2,\boldsymbol{x}')$$

$$= \underline{\tilde{\chi}}^{\mathrm{t}}(\boldsymbol{x},\boldsymbol{x}') + \frac{1}{\mu_0 V}\int \mathrm{d}^4\boldsymbol{x}_1\,\mathrm{d}^4\boldsymbol{x}_2\,\underline{\tilde{\chi}}^{\mathrm{t}}(\boldsymbol{x},\boldsymbol{x}_1)\underline{\mathcal{D}}^{\mathrm{r}}(\boldsymbol{x}_1,\boldsymbol{x}_2)\underline{\tilde{\chi}}^{\mathrm{t}}(\boldsymbol{x}_2,\boldsymbol{x}')\,, \quad (2.191)$$

where

$$\underline{\tilde{\chi}}^{\mathrm{t}}(\boldsymbol{x},\boldsymbol{x}') = \underline{P}^{\mathrm{t}}(\boldsymbol{x},\boldsymbol{x}') + \frac{1}{\mu_0 V}\int \mathrm{d}^4\boldsymbol{x}_1\mathrm{d}^4\boldsymbol{x}_2\boldsymbol{P}^{\mathrm{tl}}(\boldsymbol{x},\boldsymbol{x}_1)v(\boldsymbol{x}_1-\boldsymbol{x}_2)\chi^{\mathrm{lt}}(\boldsymbol{x}_2,\boldsymbol{x}')\,. \quad (2.192)$$

In the second line of (2.191), we have rearranged the integral equation and introduced, instead of the renormalized tranverse susceptibility, the renormalized transverse photon Green's function, which obeys the equation

$$\underline{\mathcal{D}}^{\mathrm{r}}(\boldsymbol{x},\boldsymbol{x}') = \underline{\mathcal{D}}_0^{\mathrm{r}}(\boldsymbol{x},\boldsymbol{x}') + \frac{1}{\mu_0 V}\int \mathrm{d}^4\boldsymbol{x}_1\mathrm{d}^4\boldsymbol{x}_2\underline{\mathcal{D}}_0^{\mathrm{r}}(\boldsymbol{x},\boldsymbol{x}_1)\underline{\tilde{\chi}}^{\mathrm{t}}(\boldsymbol{x}_1,\boldsymbol{x}_2)\underline{\mathcal{D}}^{\mathrm{r}}(\boldsymbol{x}_2,\boldsymbol{x}')\,. \quad (2.193)$$

For the purpose of a quantum mechanical description of the electromagnetic field, we have to replace the classical vector potential by the expectation value of (2.85). All other basic definitions and steps of the derivation remain unchanged. In terms of *field operators*, the retarded photon Green's function is defined by

$$\underline{\mathcal{D}}^{\mathrm{r}}(x,x') = -\frac{\mathrm{i}}{\hbar}\Theta(t-t')\langle \boldsymbol{A}(x)\boldsymbol{A}(x') - \boldsymbol{A}(x')\boldsymbol{A}(x)\rangle\,. \quad (2.194)$$

A more detailed discussion of this and related Green's functions will be given in Chaps. 12 and 13, together with some applications.

The macroscopic limit can be taken as in the longitudinal case, and we shall only discuss the results without giving all equations in detail. The mixed contributions $\boldsymbol{P}^{tl}$ and χ^{lt} both vanish in the long-wavelength limit and for $\boldsymbol{G} = \boldsymbol{G}' = 0$. There are, however, contributions for $\boldsymbol{G} \neq 0$. These lead to purely transverse interaction corrections, which can be neglected. The short-range contributions from the Coulomb potential lead to just the same local-field corrections to the transverse susceptibility as those already described for the longitudinal case (compare (2.168) and (2.169)).

Now we can proceed further with the derivation of the transverse macroscopic dielectric function. As the $\boldsymbol{G} \neq 0$ contributions are already incorporated into the definition of $\underline{\tilde{\chi}}^{\mathrm{t}}$, we can immediately take the macroscopic limit, and obtain from (2.191)

$$\underline{\chi}^{\mathrm{t}}(\boldsymbol{q},\omega) = \left(\underline{1} - \underline{\tilde{\chi}}^{\mathrm{t}}(\boldsymbol{q},\omega)\mathcal{D}_0^{\mathrm{r}}(\boldsymbol{q},\omega)\right)^{-1}\underline{\tilde{\chi}}^{\mathrm{t}}(\boldsymbol{q},\omega)\,. \quad (2.195)$$

In the the long-wavelength limit, we obtain the following from (2.179) and (2.189):

$$\boldsymbol{j}_{\mathrm{ind}}^{\mathrm{t}}(\boldsymbol{q},\omega) = \underline{\chi}^{\mathrm{t}}(\boldsymbol{q},\omega)\mathcal{D}_0^{\mathrm{r}}(\boldsymbol{q},\omega)\boldsymbol{j}_{\mathrm{ext}}^{\mathrm{t}}(\boldsymbol{q},\omega)\,. \quad (2.196)$$

Inserting (2.196) together with (2.195) into the wave equation (2.178), we finally obtain the result

$$\left(q^2 - \frac{\omega^2}{c_0^2}\left(1 - \frac{c_0^2}{\omega^2}\underline{\tilde{\chi}}^{\mathrm{t}}(q\omega)\right)\right) \boldsymbol{A}(\boldsymbol{q},\omega) = \mu_0 \boldsymbol{j}^{\mathrm{t}}_{\mathrm{ext}}(\boldsymbol{q},\omega)\,, \tag{2.197}$$

which is obviously equivalent to the wave equation (2.144). Comparison with (2.140) now yields the relation between the conductivity tensor and the transverse response,

$$\underline{\sigma}(\boldsymbol{q},\omega) = -\frac{\mathrm{i}}{\mu_0\omega}\underline{\tilde{\chi}}^{\mathrm{t}}(\boldsymbol{q},\omega)\,. \tag{2.198}$$

As can be seen from (2.181), the conductivity contains in addition to the correlation function, a contribution which stems from the $n\boldsymbol{A}$ term of the induced current. If no intraband transitions are present, this term cancels exactly. Otherwise, together with the intraband contributions to the correlation function, it leads to the Drude–Lorentz model of the dielectric function [22]. From (2.198) and (2.177), we obtain also a relation between the susceptibility tensors obtained from the transverse and longitudinal response functions:

$$\underline{\tilde{\chi}}^{t}(\boldsymbol{q},\omega) = \frac{\omega^2}{c_0^2}\underline{\tilde{\chi}}(\boldsymbol{q},\omega)\,. \tag{2.199}$$

As the longitudinal response involves dipole transition matrix elements, in accordance with the $\boldsymbol{dE}$ coupling, and the transverse response involves momentum matrix elements, the identity (2.199) is far from trivial, as the relation can only be enforced by using the continuity equation in the form of (2.134). The identity (2.199) is thus an important test of the internal consistency of any microscopic theory of response functions. There is yet another important feature of the wave equation (2.197). Equation (2.186) shows that the solution for the vector potential can be expressed as the product of the free-photon Green's function and the total current. Equation (2.186), together with (2.196), allows us to write the same result in a different manner, i.e. as a product of a photon Green's function renormalized by the induced current density and the external current. Thus the effects of the light–matter interaction can be incorporated either into the currents, i.e. the matter, or into the photons, i.e. the light field. The two possibilities are completely equivalent.

2.7 Measurable Quantities in Optics

The theory of response functions discussed so far is best suited as a starting point for the calculation of linear optical and transport properties from a microscopic theory. Beyond the linear regime, the dielectric tensor itself can be a complicated functional of the electromagnetic field. In this case the wave equation (2.197) becomes nonlinear and thus constitutive of nonlinear optics. A rich variety of phenomena concerning the propagation of electromagnetic fields in matter is described by this equation. Depending on the objective of interest, various simplifications are possible. In order to get rid of the external sources, we can replace their action by a boundary condition at the surface of the sample. Thus they do not occur explicitly in the equation to be solved. A further simplification is obtained in a cubic lattice, where the off-diagonal elements vanish. For symmetry reasons, all diagonal elements are equal, and in the long-wavelength limit, no dependence on the wave vector remains. In this case the longitudinal and transverse response functions are strictly equivalent. Thus we can introduce a susceptibility $\chi(\omega)$, which may be either identified with the diagonal elements of the susceptibility tensor in (2.173) according to

$$\chi(\omega) = \tilde{\chi}_{xx}(\omega) = \tilde{\chi}_{yy}(\omega) = \tilde{\chi}_{zz}(\omega) \tag{2.200}$$

or calculated in a similar way from the transverse response tensor $\underline{\tilde{\chi}}^{\mathrm{t}}(\omega)\omega^2/c_0^2$. Using the definition (2.129) together with (2.174) and (2.175), we can express the product of the susceptibility and the field by the polarization

$$\boldsymbol{P}(\boldsymbol{q},\omega) = \varepsilon_0 \chi(\omega) \boldsymbol{E}(\boldsymbol{q},\omega) \tag{2.201}$$

and write the wave equation as

$$\left(\boldsymbol{q}^2 - \frac{\omega^2}{c_0^2}\right) \boldsymbol{E}(\boldsymbol{q},\omega) = \frac{\omega^2}{c_0^2} \chi(\omega) \boldsymbol{E}(\boldsymbol{q},\omega) = \frac{\omega^2}{\varepsilon_0 c_0^2} \boldsymbol{P}(\boldsymbol{q},\omega) . \tag{2.202}$$

Obviously this result is equivalent to the simple form

$$\left(\boldsymbol{q}^2 - \frac{\omega^2}{c_0^2} \varepsilon(\omega)\right) \boldsymbol{E}(\boldsymbol{q},\omega) = 0 . \tag{2.203}$$

The representation (2.202) is best suited for dealing with the nonlinear case. The right-hand side is just the linear contribution in the $\boldsymbol{q} \to 0$ limit of the polarization. The same form is valid in the nonlinear case, where the polarization may be an arbitrary functional of the field. We can either try to calculate this dependence nonperturbatively or perform the expansion of (2.150) to higher-order and introduce nonlinear susceptibilities. Both approaches will be applied in later chapters.

In many cases it is sufficient to consider only a few transition energies of the system in the vicinity of a frequency ω_0. This allows us to split the linear dielectric function into a resonant part $\chi^{(1)}(\omega)$, describing just the contribution of those few transition energies, and a part describing all other off-resonant contributions. These latter are lumped together into the background dielectric constant ε_{b}, which, even in the nonlinear case, should be independent of the electromagnetic field. Thus we can write

$$\varepsilon(\omega) = \varepsilon_{\mathrm{b}} + \chi^{(1)}(\omega) = \varepsilon_{\mathrm{b}} \left[1 + \chi^{(1)}(\omega)/\varepsilon_{\mathrm{b}}\right] . \tag{2.204}$$

The background dielectric constant renormalizes the velocity of light c_0 to the velocity of light in the system i.e. $c = c_0/\sqrt{\varepsilon_b} = c_0/n_b$, where the background refractive index is n_b. Correspondingly, we have to divide $\chi^{(1)}$ by ε_b. As the susceptibility is proportional to e^2 (compare (2.157)), this division simply means that the bare charge is screened by the background dielectric function.[16]

The renormalization of the light velocity allows us to define the dielectric function according to

$$\varepsilon(\omega) =: 1 + \chi^{(1)}(\omega)/\varepsilon_b\,, \tag{2.205}$$

and to write (2.203) as

$$\left(\boldsymbol{q}^2 - \frac{\omega^2}{c^2}\varepsilon(\omega)\right)\boldsymbol{E}(\boldsymbol{q},\omega) = 0\,. \tag{2.206}$$

Thus we have found a reduced wave equation in which *the velocity of light and bare charges are renormalized by* ε_b, and only a few resonant contributions of the response function contribute. Nontrivial solutions of (2.206) clearly require

$$c^2\boldsymbol{q}^2 = \omega^2\varepsilon(\omega) = \omega^2 n^2(\omega)\,. \tag{2.207}$$

As the dielectric function $\varepsilon = \varepsilon' + \mathrm{i}\varepsilon''$ is generally complex, this also holds for the wave vector and the refractive index $n = n' + in''$. The solution of (2.207) is a wave vector $\boldsymbol{q}(\omega)$, and hence (2.206) is solved by

$$\boldsymbol{E}(\boldsymbol{q},\omega) = \boldsymbol{E}_\pm(\boldsymbol{q},\omega)\,\delta(\boldsymbol{q} \pm \boldsymbol{q}(\omega))\,, \tag{2.208}$$

where $\boldsymbol{E}_\pm(\boldsymbol{q},\omega)$ has to be determined from the initial and boundary conditions. Depending on these conditions, a wealth of solutions in real space and in the time-domain are possible, which can be represented as a Fourier integral of (2.208). After interaction with the sample, the emitted fields carry detailed information about the dielectric function of the system and are the central quantity to be measured. If we consider a monochromatic field with frequency ω_0, the corresponding solutions are plane waves, which are given by

$$\begin{aligned} \boldsymbol{E}(\boldsymbol{r},t) &= \boldsymbol{E}_0 \cos(\boldsymbol{qr} - \omega_0 t) = \frac{1}{2}\left(\boldsymbol{E}_0 \mathrm{e}^{\mathrm{i}\boldsymbol{qr}}\mathrm{e}^{-\mathrm{i}\omega_0 t} + \boldsymbol{E}_0\mathrm{e}^{-\mathrm{i}\boldsymbol{qr}}\mathrm{e}^{\mathrm{i}\omega_0 t}\right) \\ &=: \frac{1}{2}\left(\tilde{\boldsymbol{E}}(\boldsymbol{r})\mathrm{e}^{-\mathrm{i}\omega_0 t} + \tilde{\boldsymbol{E}}^*(\boldsymbol{r})\mathrm{e}^{\mathrm{i}\omega_0 t}\right). \end{aligned} \tag{2.209}$$

We shall assume this type of solution in the following and relate the dielectric function to a number of quantities commonly used in optics.

[16] Indeed, this screening is obtained also from a formal decomposition of the susceptibility into resonant and nonresonant transitions. A detailed treatment of the decomposition of the various contributions to the dielectric function is given in Sect. 10.7.1.

2.7.1 Linear Optical Susceptibility and Macroscopic Polarization

In many cases it is more convinient to consider the wave equation in real space and in the time domain. Fourier transforming (2.206) back into real space leads to

$$\Delta \boldsymbol{E}(\boldsymbol{r},\omega)+\frac{\omega^2}{c^2}\boldsymbol{E}(\boldsymbol{r},\omega)=-\omega^2\mu_0\,\varepsilon_0\,\chi^{(1)}(\omega)\boldsymbol{E}(\boldsymbol{r},\omega)\,. \tag{2.210}$$

If we transform back into the time domain also the right-hand side becomes in general a convolution of the susceptibility and the field. We shall consider first the simplest case, in which the time dependence of the field is dominated by rapidly oscillating phase factors as in (2.209). The time dependence of the amplitudes is assumed to be negligible. In that case we may insert the Fourier transform of (2.209),

$$\boldsymbol{E}(\boldsymbol{r},\omega)=\pi\left(\tilde{\boldsymbol{E}}(\boldsymbol{r})\,\delta(\omega-\omega_0)+\tilde{\boldsymbol{E}}^*(\boldsymbol{r})\,\delta(\omega+\omega_0)\right), \tag{2.211}$$

into the right-hand side of (2.210). Fourier transformation back into the time domain delivers

$$\begin{aligned}&\Delta \boldsymbol{E}(\boldsymbol{r},t)-\frac{1}{c^2}\frac{\partial^2}{\partial t^2}\boldsymbol{E}(\boldsymbol{r},t)\\ &=\frac{\mu_0}{2}\frac{\partial^2}{\partial t^2}\left(\varepsilon_0\chi^{(1)}(\omega_0)\tilde{\boldsymbol{E}}(\boldsymbol{r})\,\mathrm{e}^{-\mathrm{i}\omega_0 t}+\varepsilon_0\chi^{(1)}(-\omega_0)\tilde{\boldsymbol{E}}^*(\boldsymbol{r})\,\mathrm{e}^{\mathrm{i}\omega_0 t}\right).\end{aligned} \tag{2.212}$$

As $\chi^{(1)}(-\omega)=\left[\chi^{(1)}(\omega)\right]^*$, which holds because $\chi^{(1)}(t-t')$ is real, the right-hand side is real, too. According to (2.202), the contribution to the macroscopic polarization (2.201) which corresponds to our selected transition energies can also be represented as

$$\boldsymbol{P}(\boldsymbol{r},t)=\frac{\varepsilon_0}{2}\chi^{(1)}(\omega_0)\tilde{\boldsymbol{E}}(\boldsymbol{r})\mathrm{e}^{-\mathrm{i}\omega_0 t}\ +\ \mathrm{c.c.}\,, \tag{2.213}$$

which yields instead of (2.212)

$$\Delta \boldsymbol{E}(\boldsymbol{r},t)-\frac{1}{c^2}\frac{\partial^2}{\partial t^2}\boldsymbol{E}(\boldsymbol{r},t)=\mu_0\frac{\partial^2}{\partial t^2}\boldsymbol{P}(\boldsymbol{r},t)\,. \tag{2.214}$$

This is formally the same as the Fourier transform of (2.202), but with renormalized charges and a renormalized velocity of light. Despite our assumption of a linear susceptibility, it is more general and, in particular, valid in the nonlinear case.

Let us consider solutions of (2.214) which are once more of the type (2.209), but with a weak dependence of the amplitude $\boldsymbol{E}_0$ on the space and time variables. We split the polarization into the sum of a linear and a nonlinear contribution, $\boldsymbol{P}^{\mathrm{nl}}$. It can be shown that the linear contribution leads to an additional renormalization of the light velocity. To this end, we approximate the convolution of the field and the linear susceptibility as follows:

$$\int \mathrm{d}t' \int \mathrm{d}\omega\, \chi^{(1)}(\omega) \mathrm{e}^{-\mathrm{i}\omega(t-t')} \Big(\boldsymbol{E}_0(\boldsymbol{r}, t', \omega_0) \mathrm{e}^{\mathrm{i}(\boldsymbol{q}\boldsymbol{r}-\omega_0 t')} + \text{c.c.} \Big)$$
$$= \mathrm{e}^{\mathrm{i}(\boldsymbol{q}\boldsymbol{r}-\omega_0 t)} \Big(\chi^{(1)} \boldsymbol{E}_0(\boldsymbol{r}, t, \omega_0) + \mathrm{i} \frac{\partial}{\partial \omega_0} \chi(\omega_0) \frac{\partial}{\partial t} \boldsymbol{E}_0(\boldsymbol{r}, t, \omega_0) \Big) + \text{c.c.}, \quad (2.215)$$

where we have used a first-order Taylor expansion

$$\boldsymbol{E}_0(\boldsymbol{r}, t', \omega_0) = \boldsymbol{E}_0(\boldsymbol{r}, t, \omega_0) + (t' - t) \frac{\partial}{\partial t} \boldsymbol{E}_0(\boldsymbol{r}, t, \omega_0) \,. \quad (2.216)$$

All derivatives of the amplitude of higher than first order are neglected. We apply the same approximation with respect to all spatial and temporal derivatives in (2.214). Using (2.215), we find that the amplitude obeys the equation

$$\left(\boldsymbol{q} \nabla + \frac{|\boldsymbol{q}|}{v_\mathrm{g}} \frac{\partial}{\partial t} \right) \boldsymbol{E}_0(\boldsymbol{r}, t, \omega_0) = -\frac{\mathrm{i}\mu_0}{2} \mathrm{e}^{-\mathrm{i}(\boldsymbol{q}\boldsymbol{r}-\omega_0 t)} \frac{\partial^2}{\partial t^2} \boldsymbol{P}^{\mathrm{nl}}(\boldsymbol{r}, t) \,, \quad (2.217)$$

where we have introduced the *group velocity* of the wave, given by

$$\frac{1}{v_\mathrm{g}} = \frac{\sqrt{\varepsilon(\omega_0)}}{c} + \frac{\omega_0}{2c\sqrt{\varepsilon(\omega_0)}} \frac{\partial}{\partial \omega_0} \varepsilon(\omega_0) \,. \quad (2.218)$$

Equation (2.217) represents the wave equation in the *slowly-varying-amplitude approximation*, which is a common starting point for studying pulse propagation and a rich variety of related phenomena. This approximation is reasonable as long as the central frequency of the pulse is large in comparison with its spectral width or its inverse duration. As is obvious from (2.218), however, not only the parameters of the light pulse, but also the frequency dependence of the susceptibility determine the quality of this approximate treatment. Propagation effects become important with increasing thickness of the sample. If the propagation length is sufficiently short, it follows from a formal integration of (2.217) that the fields which are emitted from the sample are proportional to P^{nl}. Such conditions allow us to neglect propagation effects and to identify the outgoing fields with the polarization determined by the incoming fields.

2.7.2 Absorption Coefficient

Equation (2.207) constitutes the dispersion relation of light in the medium. Generally, the wave vector is not simply proportional to the frequency, because of the frequency dependence of $n(\omega)$. The wave vector in the medium is given by

$$q = \frac{\omega}{c} n(\omega) = \frac{\omega}{c_0} n_\mathrm{b}\, n(\omega) \,. \quad (2.219)$$

Consequently, the electric field propagates, say in the z direction, according to

$$\begin{aligned} \boldsymbol{E}(z,t) &= \boldsymbol{E}(0,t)\,\mathrm{e}^{\mathrm{i}qz} = \boldsymbol{E}(0,t)\,\exp\left(\mathrm{i}\frac{\omega}{c_0}n_{\mathrm{b}}n(\omega)z\right) \\ &= \boldsymbol{E}(0,t)\,\exp\left(\mathrm{i}\frac{\omega}{c_0}n_{\mathrm{b}}n'(\omega)z\right)\exp\left(-\frac{\omega}{c_0}n_{\mathrm{b}}n''(\omega)z\right). \end{aligned} \tag{2.220}$$

The last exponential describes damping due to the imaginary part of the refractive index, $\mathrm{Im}[n(\omega)] = n''$. The absolute value of the Poynting vector $S(z,t) = |\boldsymbol{S}(z,t)| = |\boldsymbol{E}(z,t) \times \boldsymbol{H}(z,t)| = \sqrt{\varepsilon_0/\mu_0}|\boldsymbol{E}(z,t)|^2$, averaged over the rapidly oscillating light frequency, is called the *intensity* of the wave. The intensity decays according to

$$\langle S(z)\rangle = \langle S(0)\rangle\,\exp\left(-2\frac{\omega}{c_0}n_{\mathrm{b}}n''(\omega)z\right) = \langle S(0)\rangle\,\mathrm{e}^{-\alpha(\omega)z}\,, \tag{2.221}$$

which is nothing but *Beer's law*, with the *absorption coefficient* given by

$$\alpha(\omega) = 2\frac{\omega}{c_0}n_{\mathrm{b}}n''(\omega)\,. \tag{2.222}$$

This can be evaluated further as follows:

$$\begin{aligned} \alpha(\omega) &= 2\frac{\omega}{c_0}n_{\mathrm{b}}\,\mathrm{Im}\left(\sqrt{1+\chi^{(1)}(\omega)/\varepsilon_{\mathrm{b}}}\right) \\ &\approx 2\frac{\omega}{c_0}n_{\mathrm{b}}\,\mathrm{Im}\left(1+\frac{\chi^{(1)}(\omega)}{2\varepsilon_{\mathrm{b}}}\right) \\ &= \frac{\omega}{c_0\sqrt{\varepsilon_{\mathrm{b}}}}\,\mathrm{Im}\left(\chi^{(1)}(\omega)\right) \end{aligned} \tag{2.223}$$

in the limit $\chi^{(1)}(\omega)/\varepsilon_{\mathrm{b}} \ll 1$, by Taylor expansion of the square root. This approximation is often used in the literature. Care has to be taken, however, because $\chi^{(1)}(\omega)$ can become quite large in the spectral vicinity of narrow resonances, in which case the general form (2.222) has to be employed.

2.8 Problems

2.1. Consider a system of N interacting electrons in a homogeneous background of positive charges which guarantee charge neutrality. Show that in the thermodynamic limit ($V \to \infty$, $N \to \infty$, N/V = const) there is no contribution of the $\boldsymbol{q} \equiv 0$ component of the Coulomb potential to the Hamiltonian. Hint: You should start the proof with a Yukawa potential $v(\boldsymbol{q}) = 1/[\varepsilon_0 V(\boldsymbol{q}^2 + \lambda^2)]$ and take the limit $\lambda \to 0$ after the thermodynamic limit.

2.2. Show that the displacement operator (2.96) effects the transformation $\hat{U}^\dagger(\alpha)\,a\,\hat{U}(\alpha) = a + \alpha$. Use this property to show that a coherent state may be considered to be the ground state of a shifted harmonic oscillator.

2.3. Derive a formal expression for the second order density response function by means of (2.150).

2.4. Prove the f-sum rule

$$\int_0^\infty \mathrm{d}\omega\, \omega\, \mathrm{Im}[\chi(\boldsymbol{q},\omega)] = \frac{N}{V}\frac{\boldsymbol{q}^2}{2m} \tag{2.224}$$

for the longitudinal response function. Hint: Evaluate the density–density correlation function by means of exact eigenstates of the many-particle Hamiltonian and calculate the double commutator $[[\hat{n}(\boldsymbol{q}), H], \hat{n}(-\boldsymbol{q})]$, where $\hat{n}(\boldsymbol{q})$ is the Fourier transform of the particle density operator (2.51).

2.5. Calculate the linear susceptibility $\chi(\boldsymbol{q},\omega)$ for a noninteracting, homogeneous electron gas by means of (2.156).

2.6. Use the continuity equation (2.133) to relate the longitudinal contribution to the current–current correlation function to the density–density correlation function. Show that the Drude term in (2.155) is canceled exactly.

2.7. Start with (2.194) and make use of the representation (2.85) of the vector potential, to derive a representation of the photon Green's function in terms of the photon annilation and creation operators $a_{\boldsymbol{q}\lambda}$ and $a^\dagger_{\boldsymbol{q}\lambda}$.

2.8. Show that the refractive index and the dielectric function are related by

$$n = \frac{1}{\sqrt{2}}\left\{[|\varepsilon| + \mathrm{Re}(\varepsilon)]^{1/2} + \mathrm{i}\,[|\varepsilon| - \mathrm{Re}(\varepsilon)]^{1/2}\right\}. \tag{2.225}$$

3. One-Particle Properties

Naively, one might expect that at least approximate solutions of the many-particle problem in the form of the eigenvalues and eigenfunctions of the Hamiltonian, as introduced in the last chapter, would be an unambigous condition for proceeding further. Indeed, there are many experimental data, ranging from ground-state properties to response functions, which can be directly related to this information. But, apart from the problem that there is no hope of solving the many-particle problem exactly, there would also be no chance of dealing with the tremendous wealth of mostly superflous information. Fortunately, this is not the only way to proceed. Concerning *macroscopic properties* as measured experimentally, most of the microscopic information is redundant.

A semiconductor crystal consists of many electrons in outer shells and of electrons tightly bound to the nuclei. In a variety of experimental techniques, one detects particles or their emission spectra outside the system, after the particles have passed through it. Clearly the outgoing particles are not bare, but carry a wealth of information resulting from their interaction with the system. In this way, the process of measurement projects out of the many-particle problem the properties of a few effective particles only. As consequence of this reduction, the corresponding states are in general described by a *non-Hermitian* effective Hamiltonian. Compared with the states of free particles, they are renormalized, i.e. the positions of the spectral peaks become *shifted* and their linewidth *broadened*. The latter effect is equivalent to a finite lifetime of the effective particles. The introduction of particles *renormalized* owing to the interaction with their surroundings and which are subject to a weak residual interaction only is the basic idea behind the concept of *quasiparticles* [23, 24]. In principle, this concept can be applied at the level of an arbitrary number of effective particles.[1] In practice, consideration of the one- and two-particle levels is feasible and sufficient in most cases in semiconductor physics.

---

[1] A general and mathematically rigorous introduction to the quasiparticle concept will be given in Chap. 10.

3.1 Hartree–Fock Theory for Zero Temperature

In the following we shall introduce the quasiparticle concept at the most elementary level for the one-particle case, in which we try to incorporate at least certain contributions of the mutual interaction into new effective particles. The simplest version follows intuitively from the *Poisson equation.* The charge density of electrons gives rise to an electrostatic potential as already described in Chap. 2. Describing the charge density in terms of occupied one-particle states by

$$\rho(\boldsymbol{r}) = -e\sum_i |\varphi_i(\boldsymbol{r})|^2 , \tag{3.1}$$

one obtains the Hartree potential

$$\phi_{\text{eff}}(\boldsymbol{r}) = e^2\sum_i \int \mathrm{d}^3\boldsymbol{r}' v(\boldsymbol{r}-\boldsymbol{r}')|\varphi_i(\boldsymbol{r}')|^2 , \tag{3.2}$$

which acts as an electrostatic potential in addition to the potential of the nuclei. Therefore the potential of the nuclei is screened by the electronic charge density. In this description, every electron is influenced only via a mean potential generated by all the others. No information about the *Pauli exclusion principle*, however, can be included in this potential, and we have to perform a more systematic derivation of an effective one-particle approximation.

The *Ritz variational theorem* offers a way to derive an equation for the one-particle states that enter into the matrix elements in the many-particle Hamiltonian. If we calculate the expectation value of the energy according to

$$E = \frac{\langle\Psi|\hat{H}|\Psi\rangle}{\langle\Psi|\Psi\rangle} , \tag{3.3}$$

then any state $|\Psi\rangle$ that makes the functional $E[\Psi]$ stationary – provided $|\Psi\rangle$ is allowed to vary over the complete Hilbert space – is an eigenstate of H with eigenvalue E. In particular, for any state $|\Psi\rangle$, we have the inequality $E[\Psi] \geq E[\Psi_0]$, where $|\Psi_0\rangle$ is the ground state. Thus the Ritz variational principle offers a systematic way to obtain upper bounds for the ground-state energy. Application of this variational principle can be performed at various degrees of sophistication. The wavefunction might be a Slater determinant, a superposition of Slater determinants or some other construction which accounts for correlation effects other than the pure quantum statistical ones. Of course, the choice of such a wavefunction remains an art. For the moment we restrict ourselves to the simplest case, i.e. a single Slater determinant, and consider the Hamiltonian (2.58) but with a diagonal one-particle contribution only, i.e.

$$H = \sum_j \langle j|h|j\rangle c_j^\dagger c_j + \frac{e^2}{2}\sum_{jklm}\langle jk|v|lm\rangle c_j^\dagger c_k^\dagger c_m c_l . \tag{3.4}$$

Minimizing the total energy with respect to the one-particle states will yield the effective one-particle equation. We perform the calculation of the total energy with a single Slater determinant (2.37), denoted by $|\Phi_0\rangle$, in which the lowest N states are occupied and all others are empty. With this approximation, the matrix elements in (3.4) can easily be evaluated:

$$\langle\Phi_0|c_j^\dagger c_j|\Phi_0\rangle = n_j \,, \tag{3.5}$$

where the occupation number $n_j = 1$ for occupied states and zero otherwise. The two-particle matrix elements reduce to

$$\begin{aligned}\langle\Phi_0|c_j^\dagger c_{j'}^\dagger c_{k'} c_k|\Phi_0\rangle &= \delta_{jk}\delta_{j'k'}\langle\Phi_0|c_j^\dagger c_k|\Phi_0\rangle\langle\Phi_0|c_{j'}^\dagger c_{k'}|\Phi_0\rangle \\ &\quad - \delta_{jk'}\delta_{j'k}\langle\Phi_0|c_j^\dagger c_{k'}|\Phi_0\rangle\langle\Phi_0|c_{j'}^\dagger c_k|\Phi_0\rangle \\ &= (\delta_{jk}\delta_{j'k'} - \delta_{jk'}\delta_{j'k})n_k n_{k'} \,. \end{aligned} \tag{3.6}$$

Equation (3.6) illustrates the property that the expectation value of any operator acting on a Slater determinant can be expressed in terms of one-particle expectation values alone. This results from the simple structure of Slater determinants (2.37) and is the essence of the *Hartree–Fock approximation* [25, 26, 27]. As an alternative we can consider expectation values for arbitrary wavefunctions and apply a factorization scheme according to (3.6). This procedure is also equivalent to the Hartree–Fock approximation. Thus we obtain for the total ground-state energy of the N-particle system

$$\begin{aligned} E_0^{(N)} &= \langle\Phi_0|H|\Phi_0\rangle \\ &= \sum_j \langle j|h|j\rangle n_j + \frac{e^2}{2}\sum_{jj'}\Big(\langle jj'|v|jj'\rangle - \langle jj'|v|j'j\rangle\Big)n_j n_{j'} \,. \end{aligned} \tag{3.7}$$

Considering the constraint that the one-particle functions have to be normalized, we have to minimize the functional

$$\bar{E} = E_0^{(N)} + \sum_j E_j(1 - \langle j|j\rangle) \,, \tag{3.8}$$

where E_j are Lagrange multipliers. The fact that we have taken only this constraint into account anticipates the result that the resulting effective one-particle Hamiltonian will be Hermitian. Thus orthogonality of the eigenfunctions is guaranteed without introducing an explicit constraint. Because the one-particle functions are complex, we can independently vary the real and imaginary parts or, equivalently, the variations of φ^* and φ are independent. Taking the variational derivative of (3.8) with respect to $\varphi_n^*(\boldsymbol{q})$, for example, where $\boldsymbol{q} = \{\boldsymbol{r}, s\}$, we obtain

$$\begin{aligned} h(\boldsymbol{r})\varphi_n(\boldsymbol{q}) + e^2\sum_j \int \mathrm{d}^3\boldsymbol{q}\Big(&v(\boldsymbol{r}-\boldsymbol{r}')\varphi_j^*(\boldsymbol{q}')\varphi_j(\boldsymbol{q}')\varphi_n(\boldsymbol{q}) \\ &-v(\boldsymbol{r}-\boldsymbol{r}')\varphi_j^*(\boldsymbol{q}')\varphi_n(\boldsymbol{q}')\varphi_j(\boldsymbol{q})\Big) = E_n\varphi_n(\boldsymbol{q}) \,. \end{aligned} \tag{3.9}$$

These are the *Hartree–Fock equations*, which contain in addition to the electrostatic Hartree potential, an exchange contribution. Performing the summation over spin indices, one sees immediately that the exchange potential acts only on particles with the same spin, as is expected from the Pauli exclusion principle. In contrast, the spin summation in the Hartree contribution yields a factor of 2. The resulting effective one-particle Hamiltonian is obviously Hermitian; thus the one-particle states are mutually orthogonal and the Lagrange multipliers which enter the equation as eigenvalues are real. The physical meaning of these multipliers becomes clear if we consider the total energy of an $N-1$ particle system, i.e. where, for example, the nth one-particle state is removed from the system. Assuming that the remaining particles are unchanged – this assumption becomes exact in the limit $N \to \infty$ – we obtain

$$E_n^{(N-1)} = \sum_{j \neq n} \langle j|h|j\rangle n_j + \frac{e^2}{2} \sum_{j,j' \neq n} \Big(\langle jj'|v|jj'\rangle - \langle jj'|v|j'j\rangle \Big) n_j n_{j'} \,. \quad (3.10)$$

Using (3.9), we have

$$E_n = \langle n|h|n\rangle + e^2 \sum_j \Big(\langle nj|v|nj\rangle - \langle nj|v|jn\rangle \Big) n_j \,. \quad (3.11)$$

Comparison of (3.10) and (3.11) with (3.7) yields, for $n \leq N$, the result

$$E_n = E_0^{(N)} - E_n^{(N-1)} \,. \quad (3.12)$$

This is *Koopmans's theorem* [28], which shows that an eigenvalue of the Hartree–Fock equation determines directly the energy we need to remove a particle from an orbital φ to outside the system. If one assumes that there is no reorganization of orbitals associated with the removal of an electron, this is just the ionization energy. Similarly, we find, for $n \geq N+1$,

$$E_n = E_n^{(N+1)} - E_0^{(N)} \,. \quad (3.13)$$

A remarkable further feature of these energies results from the Hartree–Fock equations (3.9), which allows us to express the total energy as

$$E_0^{(N)} = \sum_j E_j n_j - \frac{e^2}{2} \sum_{jj'} \Big(\langle jj'|v|jj'\rangle - \langle jj'|v|j'j\rangle \Big) n_j n_{j'} \,. \quad (3.14)$$

The eigenenergy of each occupied state includes the interaction with all other occupied electron states. Thus the interaction contributions are counted twice in the sum over E_j. The interaction contribution in (3.14) compensates this double counting.

Up to now, we have discussed ground-state properties only. The question arises, of how to calculate *excited states* within the Hartee–Fock approach. It could be suggested that one might describe an excited state, in which an occupied state n is excited to a state n', by a Slater–determinant again:

$$|\Phi_{nn'}\rangle = c^{\dagger}_{n'} c_n |\Phi_0\rangle \, . \tag{3.15}$$

Performing the same calculations as before, we find the following for the excitation energy:

$$\begin{aligned} \delta E_{nn'} &= \langle \Phi_{nn'} | H | \Phi_{nn'} \rangle - E_0^{(N)} \\ &= E_{n'} - E_n - e^2 \Big(\langle n'n | v | n'n \rangle - \langle n'n | v | nn' \rangle \Big) n_n (1 - n_{n'}) \, . \end{aligned} \tag{3.16}$$

This result illustrates the fact that optical transition energies, which are just determined by $\delta E_{nn'}$, are not simply given by a difference of quasiparticle energies, but that there is still an interaction between these one-particles states. The excited electron and the removed electron, i.e. the hole state, represent a two-particle excitation which is coupled by the Coulomb interaction. In many systems this residual interaction becomes negligible. In semiconductors, however, it remains long-ranged and may even lead to bound states between electrons and holes, which are called *excitons.* We shall consider these two-particle correlations in detail in Chap. 6. It should be mentioned that excitation energies calculated within this approximation scheme are not rigorous upper bounds on the exact energies. This property is valid only for the ground-state energy.

A further generalization of the Hartree–Fock equation is obtained if we consider finite temperatures. The occurrence of the occupation numbers in, for example (3.7) suggests that a modification of the theory for finite temperatures could simply replace the occupation numbers by the one-particle distribution function for a finite temperature. This is indeed the case. We shall formally derive this result and, in the process, briefly recall some basic results and methods of quantum statistics [14, 29].

3.2 Hartree–Fock Theory for Finite Temperature

When our knowledge of the state of a system is incomplete, we describe it by means of the statistical operator

$$\hat{\varrho} = \frac{\mathrm{e}^{-\beta(\hat{H} - \mu \hat{N})}}{\mathrm{Tr}\left\{ \mathrm{e}^{-\beta(\hat{H} - \mu \hat{N})} \right\}} \, , \tag{3.17}$$

where $\beta \equiv 1/k_{\mathrm{B}}T$ and the trace is performed in Fock space, i.e.

$$\mathrm{Tr}\{...\} = \sum_{n_1 \ldots n_\infty} \langle n_1, ..., n_\infty | ... | n_1, ..., n_\infty \rangle \, . \tag{3.18}$$

This form of the statistical operator can be derived from the principle of maximum entropy, which states that the entropy

$$S = -k_{\mathrm{B}} \mathrm{Tr} \left\{ \hat{\varrho} \ln \hat{\varrho} \right\} \tag{3.19}$$

should reach a maximum for the true statistical operator, provided that the average energy and particle number are determined by

$$E = \mathrm{Tr}\left\{\hat{\varrho}\hat{H}\right\}, \tag{3.20}$$

$$N = \mathrm{Tr}\left\{\hat{\varrho}\hat{N}\right\}. \tag{3.21}$$

This is obviously true if we minimize the thermodynamic potential [30]

$$\Omega = \mathrm{Tr}\left\{\hat{\varrho}\left(\frac{1}{\beta}\ln\hat{\varrho} + \hat{H} - \mu\hat{N}\right)\right\}, \tag{3.22}$$

where the constraints (3.19) and (3.20) are taken into account by means of Lagrange multipliers, i.e. β and the chemical potential μ. Their meaning becomes immediately obvious when we rewrite (3.22) in the familiar form

$$\Omega = E - \mu N - TS\,. \tag{3.23}$$

One can easily check that (3.17) minimizes the thermodynamic potential, which is obtained as

$$\Omega = -\frac{1}{\beta}\ln Z\,, \tag{3.24}$$

where the grand partition function is

$$Z = \mathrm{Tr}\left\{\mathrm{e}^{-\beta(\hat{H}-\mu\hat{N})}\right\}. \tag{3.25}$$

The great advantage of introducing the statistical operator by means of a variational principle is that it allows one to construct upper bounds for the thermodynamic potential. In the following we apply this method to derive the Hartree–Fock equation for finite temperature. In order to obtain again an effective one-particle equation, an appropriate trial form for the statistical operator is

$$\hat{\varrho} = \frac{\mathrm{e}^{-\beta H_{\mathrm{tr}}}}{\mathrm{Tr}\left\{\mathrm{e}^{-\beta H_{\mathrm{tr}}}\right\}}, \tag{3.26}$$

where the trial Hamiltonian is

$$H_{\mathrm{tr}} = \sum_j (E_j - \mu)\,\hat{n}_j\,; \tag{3.27}$$

the unknown one-particle energies E_j are variational parameters. Using

$$\ln\hat{\varrho} = -\beta H_{\mathrm{tr}} - \ln\left(\mathrm{Tr}\left\{\mathrm{e}^{-\beta H_{\mathrm{tr}}}\right\}\right), \tag{3.28}$$

we have, according to (3.22), to minimize

$$\Omega = -\frac{1}{\beta}\ln\left(\mathrm{Tr}\left\{\mathrm{e}^{-\beta H_{\mathrm{tr}}}\right\}\right) + \frac{\mathrm{Tr}\left\{\mathrm{e}^{-\beta H_{\mathrm{tr}}}(H - H_{\mathrm{tr}})\right\}}{\mathrm{Tr}\left\{\mathrm{e}^{-\beta H_{\mathrm{tr}}}\right\}}\,. \tag{3.29}$$

Using the Fock space representation of the trace, the traces of one-particle operators can easily be evaluated. First we consider the trace of the trial statistical operator

$$\begin{aligned}\mathrm{Tr}\left\{\mathrm{e}^{-\beta H_{\mathrm{tr}}}\right\} &= \sum_{n_1\ldots n_\infty} \langle n_1,\ldots,n_\infty|\mathrm{e}^{-\beta\sum_j (E_j-\mu)\hat{n}_j}|n_1,\ldots,n_\infty\rangle \\ &= \prod_{j=1}^{\infty}\sum_{n_j=0,1} \mathrm{e}^{-\beta(E_j-\mu)n_j} = \prod_{j=1}^{\infty}(1+\mathrm{e}^{-\beta(E_j-\mu)})\,. \end{aligned} \tag{3.30}$$

Correspondingly, we obtain, for the trace containing the trial Hamiltonian,

$$\begin{aligned}&\mathrm{Tr}\left\{\mathrm{e}^{-\beta H_{\mathrm{tr}}}H_{\mathrm{tr}}\right\} \\ &= \sum_{n_1,\ldots,n_\infty} \langle n_1,\ldots,n_\infty|\sum_i (E_i-\mu)\hat{n}_i\mathrm{e}^{-\beta\sum_j (E_j-\mu)\hat{n}_j}|n_1,\ldots,n_\infty\rangle \\ &= \sum_{i,n_i}(E_i-\mu)n_i\mathrm{e}^{-\beta(E_i-\mu)n_i}\prod_{j\neq i}^{\infty}\sum_{n_j=0,1}\mathrm{e}^{-\beta(E_j-\mu)n_j} \\ &= \sum_i \frac{(E_i-\mu)\mathrm{e}^{-\beta(E_i-\mu)}}{1+\mathrm{e}^{-\beta(E_i-\mu)}}\prod_{j=1}^{\infty}(1+\mathrm{e}^{-\beta(E_j-\mu)})\,. \end{aligned} \tag{3.31}$$

Using (3.30), we obtain

$$\frac{\mathrm{Tr}\left\{\mathrm{e}^{-\beta H_{\mathrm{tr}}}H_{\mathrm{tr}}\right\}}{\mathrm{Tr}\left\{\mathrm{e}^{-\beta H_{\mathrm{tr}}}\right\}} = \sum_i E_i f_i\,. \tag{3.32}$$

Thus the statistical average in Fock space is reduced to a trace of one-particle energies, weighted with the one-particle distribution function, i.e. the *Fermi distribution function*

$$f_i = \frac{1}{1+\mathrm{e}^{\beta(E_i-\mu)}}\,. \tag{3.33}$$

This also follows directly from the general definition of the one-particle distribution function

$$f_i = \mathrm{Tr}\left\{\hat{\varrho}\,\hat{n}_i\right\}\,, \tag{3.34}$$

if the statistical operator is approximated by (3.26). Now we proceed to evaluate (3.29) and calculate the contribution of the many-particle Hamiltonian. As before, the trace in Fock space of the one-particle contribution yields a trace in one-particle Hilbert space, i.e.

$$\frac{\mathrm{Tr}\left\{\sum_i\langle i|\hat{h}|i\rangle\mathrm{e}^{-\beta H_{\mathrm{tr}}}\right\}}{\mathrm{Tr}\left\{\mathrm{e}^{-\beta H_{\mathrm{tr}}}\right\}} = \sum_i\langle i|\hat{h}|i\rangle f_i = \sum_i h_i f_i\,. \tag{3.35}$$

The trace over the interaction contribution is obtained by applying the Hartree–Fock factorization (3.6), as follows:

$$
\begin{aligned}
&\mathrm{Tr}\left\{\sum_{ij} v_{ij}\hat{n}_i\hat{n}_j \mathrm{e}^{-\beta H_{\mathrm{tr}}}\right\} \\
&= \sum_{n_1,\ldots,n_\infty} \langle n_1,\ldots,n_\infty| \sum_{ij} v_{ij}\hat{n}_i\hat{n}_j \mathrm{e}^{-\beta\sum_k (E_k-\mu)\hat{n}_k} |n_1\ldots n_\infty\rangle \\
&= \sum_{ij} v_{ij} \prod_{k\neq i,j} (1+\mathrm{e}^{-\beta(E_k-\mu)}) = \sum_{ij} v_{ij} f_i f_j \mathrm{Tr}\left\{\mathrm{e}^{-\beta H_{\mathrm{tr}}}\right\} ,
\end{aligned} \tag{3.36}
$$

where we have used the abbreviation

$$
v_{ij} = \langle ij|v|ij\rangle - \langle ij|v|ji\rangle . \tag{3.37}
$$

Thus the statistical average of the Hamiltonian takes the form

$$
\frac{\mathrm{Tr}\left\{\mathrm{e}^{-\beta h_{\mathrm{tr}}} H\right\}}{\mathrm{Tr}\left\{\mathrm{e}^{-\beta h_{\mathrm{tr}}}\right\}} = \sum_i h_i f_i + \frac{e^2}{2}\sum_{ij} v_{ij}\, f_i f_j . \tag{3.38}
$$

Lumping these results together, we see that the *thermodynamic potential* is given by

$$
\begin{aligned}
\Omega = &-\frac{1}{\beta}\sum_i \ln\left(1+\mathrm{e}^{-\beta(E_i-\mu)}\right) \\
&+\sum_i (h_i - E_i) f_i + \frac{e^2}{2}\sum_{ij} (\langle ij|v|ij\rangle - \langle ij|v|ji\rangle) f_i f_j .
\end{aligned} \tag{3.39}
$$

Evaluating the derivative with respect to the variational parameter E_i, we obtain, as a condition for the stationarity of the thermodynamic potential,

$$
h_i + e^2 \sum_j (\langle ij|v|ij\rangle - \langle ij|v|ji\rangle) f_j = E_i , \tag{3.40}
$$

which shows that we have solved the problem if the one-particle energies are determined from a Hartree–Fock equation similar to (3.9) in which, however, the occupation numbers have been replaced by Fermi functions. The resulting equation constitutes a rather complicated problem of self-consistency, which we shall now rewrite in a more compact form:

$$
[h(\boldsymbol{r}) + \phi_{\mathrm{eff}}(\boldsymbol{r})]\varphi_i(\boldsymbol{r}) + \int \mathrm{d}^3\boldsymbol{r}'\, \Sigma(\boldsymbol{r},\boldsymbol{r}')\, \varphi_i(\boldsymbol{r}') = E_i \varphi_i(\boldsymbol{r}) , \tag{3.41}
$$

where ϕ_{eff} is the Hartree potential and we have introduced the so-called *exchange self-energy*

$$
\Sigma(\boldsymbol{r},\boldsymbol{r}') = e^2 \sum_j v(\boldsymbol{r}-\boldsymbol{r}')\varphi_j^*(\boldsymbol{r})\varphi_j(\boldsymbol{r}') f_j . \tag{3.42}
$$

As will be shown later, the exchange self-energy is just a very special approximation of the exact one-particle self-energy which can be introduced by means of Green's function techniques. In contrast to the Hartree–Fock Hamiltonian, the exact one-particle Hamiltonian is no longer Hermitian. *Thus quasiparticle states become damped, i.e. their lifetime is finite.*

Even at the Hartree–Fock level, the solution of the one-particle problem constitutes a very involved problem. This is especially true for finite temperatures, as the Hartee–Fock energies E_i both determine the one-particle distribution function and are determined by the f_i according to (3.33) Obviously, such a problem can be attacked only by iterative procedures. We shall not go into a discussion of the technical problems of these procedures but shall discuss some essential physical implications of the results derived so far. Comparing the Hartree–Fock results for zero and finite temperature, we see that for finite temperature the system becomes thermally excited, i.e. a proportion of the electrons occupies the conduction band, thus generating missing electron states in the valence band. This is illustrated in Fig. 3.1 in the vicinity of the band gap of a direct-gap semiconductor. It is instructive to quantitatively calculate the proportion of electrons which is thermally excited. From the expression determining the total particle number

$$N = \sum_i f_i \, , \tag{3.43}$$

we obtain the following in a low temperature expansion:

$$N = \sum_{i,E_i<\mu} 1 - \mathrm{e}^{-\beta(\mu - E_i)} + \sum_{i,E_i>\mu} \mathrm{e}^{-\beta(E_i-\mu)} \, . \tag{3.44}$$

As the sum over occupied states is just N, we have the condition

$$\sum_{i,E_i<\mu} \mathrm{e}^{-\beta(\mu - E_i)} = \sum_{i,E_i>\mu} \mathrm{e}^{-\beta(E_i-\mu)} \, , \tag{3.45}$$

which states nothing but particle-number conservation. The particles removed from low-lying states, which are occupied at zero temperature, now occupy higher states, which are unoccupied at zero temperature. As will be discussed in detail in Sect. 3.5, the chemical potential of an undoped semiconductor at zero temperature is in the middle of the energy gap. Small corrections for finite temperature depend on the curvature of the bands. The change of the occupation functions for finite temperature corresponds to a change of the one-particle potential and thus leads to new one-particle energies. This is also indicated in Fig. 3.1, where the difference between the dotted and the full lines corresponds to the difference between the Hartree–Fock self-energies for zero and finite temperature. In reality, this temperature effect due to the change of the one-particle occupation is usually dominated by the action of lattice vibrations, i.e. *phonons*, which will be introduced in Chap. 11. Nevertheless, our analysis has shown that a change in the occupation of the one-particle states gives rise to a change of the corresponding energies. This is an important result, as a change of the one-particle occupation can be accomplished not only by increasing the temperature, but also by an *optical*

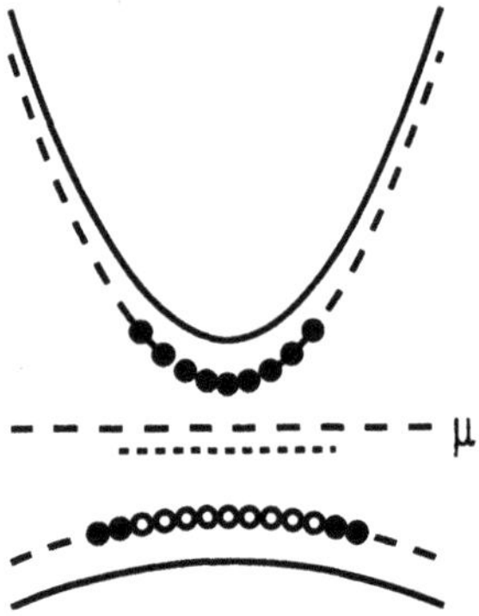

Fig. 3.1. Schematic illustration showing that the electronic eigenenergies change owing to thermal occupation of high-energy states (*dashed lines*). The *solid lines* are the energies at zero temperature

excitation of the system. In this case, however, the one-particle distributions are nonequilibrium ones and are determined by the action of the external excitation. Before we start to attack such problems, some basic features of the one-particle states in a solid will be introduced in the following sections.

3.3 Band Structure and Ground-State Properties

Using the Hartree–Fock approach, the original many-body problem of interacting electrons moving in the periodic potential of the nuclei has been approximated by effective new particles without mutual interaction moving in an energy-dependent nonlocal potential. For the purpose of quantitative calculations of one-particle energies in a solid, even this rather complicated method has turned out to be insufficient.[2] As will be shown later in detail, an improved version requires the screening of the exchange interaction by the inverse dielectric function of the system. This leads to extremely entangled equations, and most practical bandstructure calculations are based on a different approximation scheme, i.e. the density functional theory.

3.3.1 The Local-Density Approximation

The origin of the density functional theory goes back to the early work of Thomas and Fermi [31, 32], who showed that the total energy of an atom can be expressed in terms of the electron density alone. A theorem of Hohenberg and Kohn [33] proves that, under certain conditions [34], the use of the electron density as the basic variable in calculations of the total energy is indeed exact. As in the case of the Hartree–Fock approximation, one can apply a variational principle, which states that for any density the energy functional is an upper limit of the true ground-state energy. Now one can proceed similarly to the way we proceeded in the case of the Hartree–Fock

[2] This will be discussed in more detail in Sect. 10.6.

equations. The density is represented in a basis of one-particle states. Variation of the energy functional with respect to these states yields a one-particle equation. Its solution allows one to calculate the ground-state density, and thus the energy functional and the resulting ground-state properties. Owing to the detour via the one-particle basis, the kinetic energy is treated exactly, but the explicit form of the exchange correlation energy functional is not known. The strategy for overcoming this difficulty is as follows [35]. In a homogeneous system the density is constant, so we can derive the contribution of exchange and correlation to the ground-state energy at a certain level, e.g. the Hartree–Fock approximation, where by definition correlation effects are neglected. If, in an inhomogeneous system, the spatial variation of the density is sufficiently weak, one can take the energy functional found for the homogeneous system and apply it locally for the inhomogeneous system. The exchange–correlation energy functional constructed in this way is just the leading contribution to an infinite expansion in terms of gradients of the density. Neglecting all gradient corrections is the essence of the local-density approximation, which is usually applied in band structure calculations, but it leads to serious difficulties. Let us consider the chemical potential of an N-particle sytem, which is given by

$$\mu^{(N)} = E_0^{(N)} - E_0^{(N-1)} \,. \tag{3.46}$$

Correspondingly, we have for an $(N+1)$-particle system

$$\mu^{(N+1)} = E_0^{(N+1)} - E_0^{(N)} \,. \tag{3.47}$$

The transition from N to the $N+1$ particles is continuous in a homogeneous system. When passing to an inhomogeneous system, however, symmetry breaking of the translational invariance generates energy gaps in the one-particle energies. In a semiconductor $\mu^{(N)}$ is just the top of the valence band, whereas $\mu^{(N+1)}$ is the bottom of the conduction band; these are separated by the gap energy. Correspondingly, there exists a discontinuity of the exchange–correlation potential [36, 37]. Without additional corrections, the gap energies usually turn out to be much too small.

To illustrate the local-density approximation, let us consider the case of the Hartree–Fock equation for a homogeneous system. From (3.7), we obtain the following for the exchange energy in a homogeneous electron gas:

$$\begin{aligned} E_{\mathrm{exc}} &= -\frac{e^2}{2V^2} \sum_{ss'\boldsymbol{k}\boldsymbol{k}'} \Theta(k_{\mathrm{F}} - |\boldsymbol{k}|)\Theta(k_{\mathrm{F}} - |\boldsymbol{k}'|) \\ &\quad \times \delta_{s,s'} \int \mathrm{d}^3\boldsymbol{r}\,\mathrm{d}^3\boldsymbol{r}'\, v(\boldsymbol{r}-\boldsymbol{r}')\mathrm{e}^{\mathrm{i}(\boldsymbol{k}-\boldsymbol{k}')(\boldsymbol{r}-\boldsymbol{r}')} \,, \end{aligned} \tag{3.48}$$

where k_{F} is the Fermi wavenumber, which can be introduced by noting that the total particle number is given by

$$N = 2\sum_{\boldsymbol{k}} \theta(k_{\mathrm{F}} - |\boldsymbol{k}|) = \frac{k_{\mathrm{F}}^3}{3\pi^2} V \,. \tag{3.49}$$

This can easily be calculated in the thermodynamic limit, in which we can replace the summations according to[3]

$$\frac{1}{V}\sum_{\boldsymbol{k}} \rightarrow \int \frac{\mathrm{d}^3\boldsymbol{k}}{(2\pi)^3} \,. \tag{3.50}$$

Thus, the wavenumber of the highest occupied state is given by

$$k_\mathrm{F} = \left(3\pi^2 n\right)^{1/3} \,, \tag{3.51}$$

where the density $n = N/V$. Performing the spatial integrations, the Coulomb interaction given by (2.12) is replaced by its Fourier transform (2.165), and we obtain

$$\begin{aligned} E_\mathrm{exc} &= -e^2 \sum_{\boldsymbol{k}\boldsymbol{k}'} \theta(k_\mathrm{F} - |\boldsymbol{k}|)\theta(k_\mathrm{F} - |\boldsymbol{k}'|)v(\boldsymbol{k} - \boldsymbol{k}') \\ &=: \sum_{\boldsymbol{k}} \theta(k_\mathrm{F} - |\boldsymbol{k}|)\varepsilon_\mathrm{exc}(\boldsymbol{k}) \,, \end{aligned} \tag{3.52}$$

where ε_exc is just the exchange contribution to the one-particle energies. Integration yields

$$\varepsilon_\mathrm{exc}(\boldsymbol{k}) = -\frac{e^2}{8\varepsilon_0\pi^2} k_\mathrm{F} \left[1 + \left(\frac{k_\mathrm{F}}{2|\boldsymbol{k}|} - \frac{|\boldsymbol{k}|}{2k_\mathrm{F}}\right)\right] \ln\left|\frac{k_\mathrm{F} + |\boldsymbol{k}|}{k_\mathrm{F} - |\boldsymbol{k}|}\right| \,. \tag{3.53}$$

Performing also the second $\boldsymbol{k}$ integration, we find for the exchange energy

$$\frac{E_\mathrm{exc}}{V} = -\frac{e^2}{16\varepsilon_0\pi^4} k_\mathrm{F}^4 \,. \tag{3.54}$$

By means of (3.51) we obtain

$$\frac{E_\mathrm{exc}}{V} = -\frac{3e^2}{16\pi\varepsilon_0}\left(\frac{3}{\pi}\right)^{1/3} n^{4/3} \,. \tag{3.55}$$

Thus we have obtained an explicit expression for the exchange energy. Now the local-density approximation asserts that the exchange potential in the one-particle equation for an inhomogeneous system is given by

$$V_\mathrm{exc}(\boldsymbol{r}) = \frac{1}{V}\frac{\partial E_\mathrm{exc}(n)}{\partial n}\bigg|_{n=n(\boldsymbol{r})} = -\frac{e^2}{4\pi\varepsilon_0}\left(\frac{3}{\pi} n(\boldsymbol{r})\right)^{1/3} \,, \tag{3.56}$$

which is nothing, but the variation of the exchange energy with the homogeneous density taken at a certain density of the inhomogeneous system.[4]

[3] This transition is discussed in more detail in Sect. 3.7.

[4] If we apply the homogeneous-electron-gas approximation in the Hartree–Fock equations instead for the ground-state functional, we obtain in (3.56) a prefactor 2/3. This simplification of the Hartree–Fock theory, invented by Slater [38], is different from the basic idea of the local-density approximation.

Although there are a number of much more refined versions of the local-density approximation, which allow one to account for correlation effects in the homogeneous electron gas [39, 40, 41, 42], the basic procedure is always the same. First, one has to solve a one-particle equation of the type

$$\left[-\frac{\hbar}{2m} + V_{\text{loc}}\Big(\boldsymbol{r}, [n(\boldsymbol{r})]\Big) \right] \varphi_\alpha(\boldsymbol{r}) = E_\alpha \, \varphi_\alpha(\boldsymbol{r}) \,, \tag{3.57}$$

where the local one-particle potential is given by the sum of the potentials of the nuclei or ion cores, the Hartree potential and an exchange–correlation contribution, according to

$$V_{\text{loc}}\Big(\boldsymbol{r}, [n(\boldsymbol{r})]\Big) = U(\boldsymbol{r}) + \phi_{\text{eff}}(\boldsymbol{r}) + V_{\text{exc}}(\boldsymbol{r}) \,, \tag{3.58}$$

where the density is given by

$$n(\boldsymbol{r}) = \sum_\alpha f_\alpha \, \varphi^*_\alpha(\boldsymbol{r}) \varphi_\alpha(\boldsymbol{r}) \,. \tag{3.59}$$

For the purpose of practical calculations the potentials of the ion cores have to be replaced by pseudopotentials, which were introduced originally to study atomic systems [43]. Making use of the orthogonality of valence states and core states, the core potential can be replaced by a much weaker pseudopotential [44, 45], which yield the same eigenenergies. Pseudopotentials constructed this way, however, are non-unique, and additonal specifications are necessary. The most important one, is the correct normalization of the charge density outside the core region. Such norm-conserving pseudopotential can be constructed from calculations for atomic systems with a high degree of accuracy [46].

When calculating the ground-state energy and related properties by means of the one-particle energies, we have, as in the Hartree–Fock theory, to take care of the double counting of the interaction contributions. Thus the electronic contribution to the ground-state energy is determined by

$$\begin{aligned} E_{\text{el}}[n] = \sum_\alpha f_\alpha E_\alpha \; - \frac{e^2}{2} \int \mathrm{d}^3\boldsymbol{r}\, \mathrm{d}^3\boldsymbol{r}' \, v(\boldsymbol{r} - \boldsymbol{r}') \, n(\boldsymbol{r}) \, n(\boldsymbol{r}') \\ - \int \mathrm{d}^3\boldsymbol{r} \, V_{\text{exc}}[n(\boldsymbol{r})] \, n(\boldsymbol{r}) + \frac{1}{V} \int \mathrm{d}^3\boldsymbol{r} \, E_{\text{exc}}[n(\boldsymbol{r})] \,. \end{aligned} \tag{3.60}$$

In the simplest case, $V_{\text{exc}}[n(\boldsymbol{r})]$ and $E_{\text{exc}}[n(\boldsymbol{r})]$ are given by (3.56) and (3.55), respectively. Improved energy functionals, including correlational contributions, can be constructed similarly to the exchange contributions from more refined theories of a homogeneous electron gas at various levels of sophistication. In particular, quantum mechanical Monte–Carlo simulations of a homogeneous electron gas [39] allow one the construction of highly accurate exchange–correlation functionals [40]. The total ground-state energy is given by the sum of (3.60) and the contribution owing to the ion-ion Coulomb interaction.[5] There is a wealth of literature, which deals with the numerical

[5] If pseudopotentials are applied an additional correction is necessary [47].

evaluation of the local-density theory for numerous materials (see for example [48] and references therein). In particular, norm-conserving pseudopotentials [46] are best suited to obtain accurate results, and allow one straightforward numerical calculations in a plane-wave basis [47].

As the density gradients in real solids are large, it is rather surprising that the local-density approximation yields excellent results for the ground-state properties. A typical example of a valence charge density is shown in Fig. 3.2 for GaAs [49]. Ground-state energies, calculated for a few selected semiconductors with norm-conserving pseudopotentials and the exchange–correlation functional proposed in [40], are shown in Fig. 3.3 in dependence on the lattice constant. Comparison of the equilibrium lattice constant, the bulk modulus, and the minimal total energy with experimental data[6] as listed in Table 3.1, illustrates the striking accuracy of the local-density approximation. The deviations are less than 2% with respect to the lattice constant, and up to 3% with respect to the bulk modulus and the crystal energy.

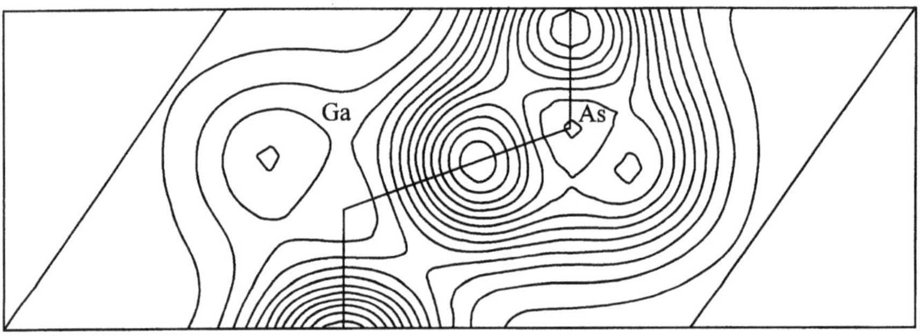

Fig. 3.2. Valence charge density distribution of GaAs. Taken from [49]

Given the auxiliary nature of the one-particle states, however, there is no simple physical meaning of these states. Nevertheless, with some refinements the one-particle energies obtained in the local-density approximation can serve as a reasonable starting point to explain the optical properties of solids. In particular, they serve as a starting point for more rigorous quasi-particle bandstructure calculations [54, 55, 56, 57, 58] (compare Sect. 10.6). Now we focus our interest on the consequences of the lattice-symmetry. The following considerations are valid for both local and nonlocal potentials. For simplicity we restrict ourselves to the local case.

[6] Note that the experimental values for the minimal energy of the crystal must be calculated from the sum of atomic valence ionization energies [51], the cohesive energy [52] and the zero-point vibration energy [53] (see for example [50]).

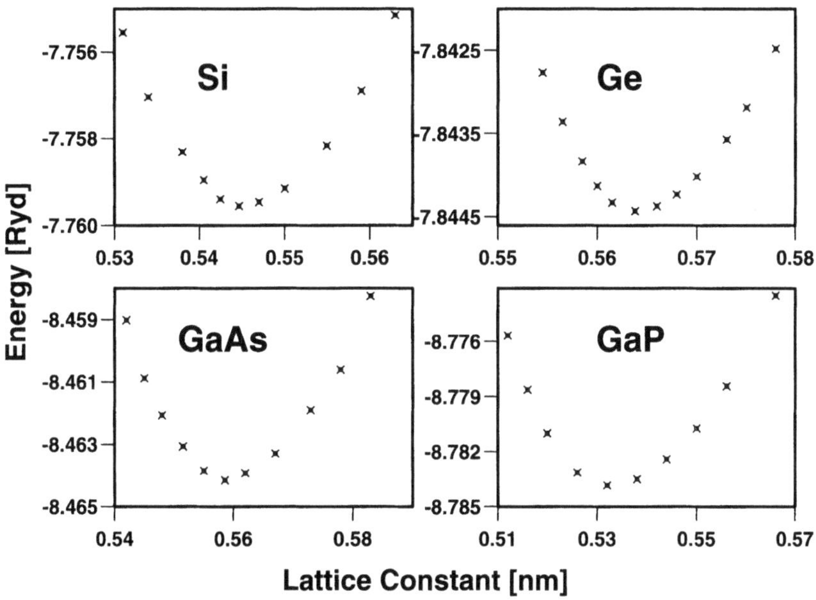

Fig. 3.3. Calculated ground-state energies (per atom) of Si, Ge, GaAs and GaP in dependence on the lattice constant

Table 3.1. Calculated equilibrium lattice constant (a), bulk modulus (B) and crystal energy (E_{min}) of a few selected semiconductors in comparison with experimental data (in parentheses). The experimental data for a,B and E_{min} are taken from [49], [51], and [51, 52, 53], respectively

	Si	Ge	GaAs	GaP
a (nm)	5.447	5.638	5.586	5.320
	(5.431)	(5.657)	(5.654)	(5.451)
B (kbar)	963	754	761	895
	(988)	(734)	(748)	(887)
E_{min} (Ryd)	−7.759	−7.844	−8.464	−8.784
	(−7.919)	(−7.910)	(−8.527)	(8.836)

3.3.2 Lattice Periodicity

Owing to the periodicity of the lattice, the effective potential has translational invariance, and hence the one-particle Hamiltonian $h(\boldsymbol{r})$ also remains invariant when translated by a lattice translation $\boldsymbol{L}$:

$$T_{\boldsymbol{L}} h(\boldsymbol{r}) := h(\boldsymbol{r} + \boldsymbol{L}) = h(\boldsymbol{r}) \, . \tag{3.61}$$

Consequently, the Hamiltonian commutes with the translation operator:

$$[T_{\boldsymbol{L}}, h] = T_{\boldsymbol{L}} h - h T_{\boldsymbol{L}} = 0 \, . \tag{3.62}$$

Hence, they have a common system of eigenfunctions. While it is usually difficult to solve the eigenvalue problem even for a Hamiltonian with an effective potential, it is straightforward to determine the general structure of the eigenvalues $\xi(\boldsymbol{L})$ and the eigenfunctions $\varphi(\boldsymbol{r})$ of the translation operator, such that

$$T_{\boldsymbol{L}}\varphi(\boldsymbol{r}) = \xi(\boldsymbol{L})\,\varphi(\boldsymbol{r})\,. \tag{3.63}$$

We apply the translation operator for arbitrary lattice translations $\boldsymbol{L}_1$ and $\boldsymbol{L}_2$:

$$\begin{aligned} T_{\boldsymbol{L}_1+\boldsymbol{L}_2}\varphi(\boldsymbol{r}) &= \varphi(\boldsymbol{r}+\boldsymbol{L}_1+\boldsymbol{L}_2) = \xi(\boldsymbol{L}_1+\boldsymbol{L}_2)\,\varphi(\boldsymbol{r}) \\ &= T_{\boldsymbol{L}_1}\varphi(\boldsymbol{r}+\boldsymbol{L}_2) = \xi(\boldsymbol{L}_1)\,\varphi(\boldsymbol{r}+\boldsymbol{L}_2) \\ &= \xi(\boldsymbol{L}_1)T_{\boldsymbol{L}_2}\varphi(\boldsymbol{r}) = \xi(\boldsymbol{L}_1)\,\xi(\boldsymbol{L}_2)\,\varphi(\boldsymbol{r})\,. \end{aligned} \tag{3.64}$$

Thus we conclude that

$$\xi(\boldsymbol{L}_1+\boldsymbol{L}_2) = \xi(\boldsymbol{L}_1)\xi(\boldsymbol{L}_2) \quad \forall\ \boldsymbol{L}_1,\boldsymbol{L}_2\,. \tag{3.65}$$

This identity has to be fulfilled for any combination of lattice translations $\boldsymbol{L}_1$ and $\boldsymbol{L}_2$, which implies the following general form for the eigenvalue:

$$\xi(\boldsymbol{L}) = \mathrm{e}^{\mathrm{i}\boldsymbol{k}\boldsymbol{L}} \tag{3.66}$$

where at this point, the vector $\boldsymbol{k}$ is arbitrary. This results in the following property of the wavefunctions:

$$\varphi_n(\boldsymbol{k},\boldsymbol{r}+\boldsymbol{L}) = \mathrm{e}^{\mathrm{i}\boldsymbol{k}\boldsymbol{L}}\varphi_n(\boldsymbol{k},\boldsymbol{r})\,, \tag{3.67}$$

which is referred to as the *Bloch theorem* [59]. It is satisfied by the *Bloch functions*

$$\varphi_n(\boldsymbol{k},\boldsymbol{r}) = \frac{1}{\sqrt{V}}u_n(\boldsymbol{k},\boldsymbol{r})\mathrm{e}^{\mathrm{i}\boldsymbol{k}\boldsymbol{r}}\,, \tag{3.68}$$

which form a complete set of orthonormal functions, where n labels different bands. We may use a discrete normalization according to

$$\begin{aligned} \frac{1}{V}\int \mathrm{d}^3\boldsymbol{r}\,\varphi_n^*(\boldsymbol{k},\boldsymbol{r})\varphi_{n'}(\boldsymbol{q},\boldsymbol{r}) &= \frac{1}{V}\int \mathrm{d}^3\boldsymbol{r}\,\mathrm{e}^{-\mathrm{i}(\boldsymbol{k}-\boldsymbol{q})\boldsymbol{r}}u_n^*(\boldsymbol{k},\boldsymbol{r})u_{n'}(\boldsymbol{q},\boldsymbol{r}) \\ = \frac{1}{\mathcal{N}\Omega}\sum_{\boldsymbol{L}}\int_{\text{cell}} \mathrm{d}^3\boldsymbol{r}\,\mathrm{e}^{-\mathrm{i}(\boldsymbol{k}-\boldsymbol{q})(\boldsymbol{r}-\boldsymbol{L})}&u_n^*(\boldsymbol{k},\boldsymbol{r})u_{n'}(\boldsymbol{q},\boldsymbol{r}) = \delta_{n,n'}\delta_{\boldsymbol{k},\boldsymbol{q}}\,. \end{aligned} \tag{3.69}$$

Here $\mathcal{N}$ is the number of unit cells. In the last step we have made use of the invariance of the functions $u_n(\boldsymbol{k},\boldsymbol{r}) = u_n(\boldsymbol{k},\boldsymbol{r}+\boldsymbol{L})$ with respect to lattice translations, which implies the orthonormality relation

$$\frac{1}{\Omega}\int_{\text{cell}} \mathrm{d}^3\boldsymbol{r}\,u_n^*(\boldsymbol{k},\boldsymbol{r})u_{n'}(\boldsymbol{k},\boldsymbol{r}) = \delta_{n,n'}\,. \tag{3.70}$$

The integration is performed over a single unit cell, whose volume is Ω. For an empty lattice, clearly $u_n(\boldsymbol{k},\boldsymbol{r}) = 1$, and we obtain plane waves $\varphi_n(\boldsymbol{k},\boldsymbol{r}) = \mathrm{e}^{\mathrm{i}\boldsymbol{k}\boldsymbol{r}}$, allowing us to identify $\boldsymbol{k}$ with the electron wave vector in this case.

There is a large variety of different methods to treat the one-particle problem in a lattice periodic potential, which, owing to the dependence of the Hartree and exchange potentials on the occupied states, requires a self-consistent solution. In principle, one can distinguish between two opposite approaches. The lattice periodicity of the function $u_n(\boldsymbol{k}, \boldsymbol{r})$ allows us to expand them into a plane-wave basis, which explicitly takes the extended nature of Bloch functions into account. The opposite method starts with localized, atomic-like wavefunctions as the basis set. We shall illustrate the latter, rather intuitive approach in some detail.

Let us imagine that we are able to move the individual atoms. If we move them far apart from each other, the wavefunction of the system can certainly be well approximated by the atomic wavefunctions. If we keep the lattice periodicity even for this large separation, the total wavefunction of the system will still have to obey the Bloch theorem. Consequently, one expects a relation between Bloch states and localized atomic-like wavefunctions. Indeed, an exact expansion of this kind exists,

$$\varphi_n(\boldsymbol{k}, \boldsymbol{r}) = \frac{1}{\sqrt{\mathcal{N}}} \sum_{\boldsymbol{L}} \mathrm{e}^{\mathrm{i}\boldsymbol{k}\boldsymbol{L}} w_n(\boldsymbol{r} - \boldsymbol{L}) \;, \tag{3.71}$$

which satisfies the Bloch theorem. This expansion defines the *Wannier functions* [60, 61] $w_n(\boldsymbol{r}-\boldsymbol{L})$, which can immediately be calculated from the inversion of (3.71). This yields

$$\boldsymbol{w_n(r - L) = \frac{1}{\sqrt{\mathcal{N}}} \sum_{k} \mathrm{e}^{-\mathrm{i}kL} \varphi_n(k, r)} \,, \tag{3.72}$$

where we have used

$$\sum_{\boldsymbol{k}} \mathrm{e}^{\mathrm{i}\boldsymbol{k}\boldsymbol{L}} = \mathcal{N} \delta_{\boldsymbol{L},0} \;. \tag{3.73}$$

The orthogonality relation for Wannier functions follows from the orthogonality of Bloch functions in the form

$$\int \mathrm{d}^3\boldsymbol{r}\, w_n^*(\boldsymbol{r} - \boldsymbol{L}) w_{n'}(\boldsymbol{r} - \boldsymbol{L}') = \delta_{nn'} \delta_{\boldsymbol{L}\boldsymbol{L}'} \;. \tag{3.74}$$

Although, at least in the core region Wannier functions and atomic functions are closely related, the orthogonality relation constitutes an essential difference. Whereas Wannier functions on different lattice sites are mutually orthogonal, atomic wavefunctions have a finite overlap, which may be interpreted as the origin of chemical bonding. Apart from this difference, the relations between Wannier functions and atomic functions depend on the symmetry properties of the bands under consideration. Only if a single, isolated band corresponds to one atomic energy level, is there a one to one correspondence between the two functions. This case is realized for atomic orbitals with s symmetry only. In most relevant cases there is a correspondence between Wannier functions and a linear combination of atomic orbitals. Inserting the expansion (3.71) into the one-particle Schrödinger equation

$$h(\boldsymbol{r})\varphi_n(\boldsymbol{k},\boldsymbol{r}) = E_n(\boldsymbol{k})\varphi_n(\boldsymbol{k},\boldsymbol{r}) \ , \tag{3.75}$$

we obtain, after multiplication by $w_n^*(\boldsymbol{r})$ and spatial integration,

$$\sum_{\boldsymbol{L}} \mathrm{e}^{\mathrm{i}\boldsymbol{k}\boldsymbol{L}} \int \mathrm{d}^3\boldsymbol{r}\, w_n^*(\boldsymbol{r})h(\boldsymbol{r})w_n(\boldsymbol{r}-\boldsymbol{L}) = E_n(\boldsymbol{k}) \ , \tag{3.76}$$

where we have used the orthogonality relation (3.74). This expansion of the Schrödinger equation is still exact and can serve as the starting point for realistic band structure calculations.

Example (Tight-Binding Model). Let us consider a simple cubic lattice. We restrict ourselves to the case of one isolated s band. In this case (3.76) yields

$$\begin{aligned} &E_n(\boldsymbol{k}) \\ &= E_n^0 + \sum_{l=1}^{\infty} E_n^l \left(\mathrm{e}^{\mathrm{i}k_x la} + \mathrm{e}^{-\mathrm{i}k_x la} + \mathrm{e}^{\mathrm{i}k_y la} + \mathrm{e}^{-\mathrm{i}k_y la} + \mathrm{e}^{\mathrm{i}k_z la} + \mathrm{e}^{-\mathrm{i}k_z la}\right) , \end{aligned} \tag{3.77}$$

where a is the lattice constant. The first contribution on the right-hand side is the on-site matrix element, where both Wannier functions are localized at the same lattice site. In the simplest case it corresponds to an atomic energy, which is renormalized owing to the difference between the atomic and the crystal potential. The other contributions E_n^l are the transfer matrix elements between lth nearest neighbors as shown in Fig. 3.4. As $h(\boldsymbol{r})$ has the symmetry of the lattice and $w_n(\boldsymbol{r})$ has s symmetry, these transfer matrix elements depend only on the distance between the sites and thus are equal for all nearest neighbors, and correspondingly for all second nearest neighbors, and so forth. Asymptotically, Wannier functions are exponentially localized and one expects that the sum in (3.77) converges rapidly with respect to l. If

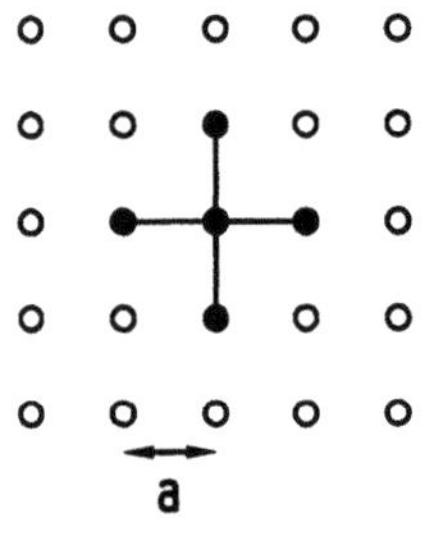

Fig. 3.4. Nearest neighbors of the simple cubic lattice

we restrict ourselves to nearest neighbors only, the energy dispersion is given by

$$E_n(\boldsymbol{k}) = E_n^0 + 2E_n^1\Big[\cos(k_x a) + \cos(k_y a) + \cos(k_z a)\Big] , \tag{3.78}$$

which is visualized in Fig. 3.5 within the *first Brillouin zone* of the simple cubic lattice, i.e. for $-\pi/a < k_{x,y,z} < \pi/a$ in the case of two bands, the upper and lower of which have $E_\mathrm{c}^1 > 0$ and $E_\mathrm{v}^1 < 0$, respectively. A more realistic band structure has already been shown in Chap. 1.

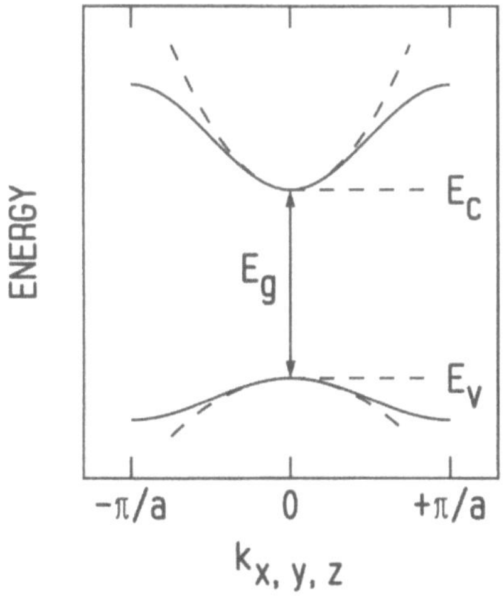

Fig. 3.5. Energy band structure of a simple cubic lattice. The *dashed parabolas* illustrate the effective-mass approximation

3.4 The Effective-Mass Approximation

In most applications in semiconductor optics and transport, the physically interesting quantities are generated by carriers with energies close to those of symmetry points in the Brillouin zone. Therefore we can restrict ourselves to the vicinity of band extrema. In particular, in direct gap semiconductors, the band structure is often only required near the Γ point, i.e. $\boldsymbol{k} = 0$. For small $\boldsymbol{k}$ we can expand a band energy $E_n(\boldsymbol{k})$ into a Taylor series (see Fig. 3.5) and introduce the *tensor of the effective mass*

$$(m_n^{-1})_{ij} = \frac{1}{\hbar^2}\frac{\partial^2}{\partial k_i \partial k_j} E_n(\boldsymbol{k}) . \tag{3.79}$$

If this tensor is diagonal and all diagonal elements are equal (compare, for example, (3.78)), we end up with the very intuitive picture of electrons with effective masses moving through the semiconductor. The valence and conduction band energies are thus given by

$$E_c(\boldsymbol{k}) = E_c + \frac{\hbar^2 \boldsymbol{k}^2}{2m_c}, \tag{3.80}$$

$$E_v(\boldsymbol{k}) = E_v + \frac{\hbar^2 \boldsymbol{k}^2}{2m_v}, \tag{3.81}$$

where we have introduced the minimum and maximum of the conduction and valence band $E_c = E_c(0)$ and$E_v = E_v(0)$, respectively. The *band gap energy* is $E_g = E_c - E_v$. Some examples are given in Table 3.2.

The effective mass tensor defined in (3.79) can be obtained either from experimental data or from band structure calculations. The simplest way

Table 3.2. Band gap energies of several semiconductor materials. Taken from [2]

Material	Type of band gap	E_g/eV ($T \leq 4.2$ K)	E_g/eV ($T = 300$ K)
Si	Indirect	1.17	1.110
Ge	Indirect	0.744	0.664
GaAs	Direct	1.52	1.43
AlAs	Indirect	2.228	2.14
ZnSe	Direct	2.82	2.7
CdS	Direct	2.582	2.5
CuCl	Direct	3.395	–

to calculate effective masses is based on the fact that the wavefunctions $e^{i(\boldsymbol{k}-\boldsymbol{k}_0)\boldsymbol{r}}\varphi_n(\boldsymbol{k}_0, \boldsymbol{r})$ form a complete orthonormal set. If we know all wavefunctions and energies at a certain point $\boldsymbol{k}_0$, we can expand the one-particle Schrödinger equation according to

$$\varphi_n(\boldsymbol{k}, \boldsymbol{r}) = \sum_{n'} c_{nn'}(\boldsymbol{k}) e^{i(\boldsymbol{k}-\boldsymbol{k}_0)\boldsymbol{r}} \varphi_{n'}(\boldsymbol{k}_0, \boldsymbol{r}) . \tag{3.82}$$

If the self-consistency of the one-particle potential is taken seriously, a knowledge of the Bloch functions and energies at the Γ point requires a band structure calculation throughout the whole Brillouin zone and the expansion (3.82) seems to be rather superfluous. But even if the expansion is not actually known, it can be used to analyze properties of the band structure near band extrema in great detail. Inserting (3.82) into the Schrödinger equation (3.75), we obtain the eigenvalue equation

$$\sum_{n'} \left\{ \left(\frac{\hbar^2}{2m}(\boldsymbol{k}^2 - \boldsymbol{k}_0^2) + E_m(\boldsymbol{k}_0) - E_n(\boldsymbol{k}) \right) \delta_{mn'} \right.$$
$$\left. + \frac{\hbar}{m}(\boldsymbol{k} - \boldsymbol{k}_0)\boldsymbol{p}_{mn'}(\boldsymbol{k}_0) \right\} c_{n'n}(\boldsymbol{k}) = 0 , \tag{3.83}$$

where

$$\boldsymbol{p}_{nm}(\boldsymbol{k}) = \frac{1}{\Omega} \int_{\text{cell}} d^3\boldsymbol{r}\, u_m^*(\boldsymbol{k}, \boldsymbol{r}) \boldsymbol{p}\, u_n(\boldsymbol{k}, \boldsymbol{r}) \tag{3.84}$$

is a matrix element. The scalar products $(\boldsymbol{k}-\boldsymbol{k}_0)\boldsymbol{p}_{nm}(\boldsymbol{k}_0)$ thus determine the Bloch energies and Bloch functions throughout the entire Brillouin zone, provided they are known at a certain $\boldsymbol{k}=\boldsymbol{k}_0$. Standard second order perturbation theory leads to

$$
\begin{aligned}
E_n(\boldsymbol{k}) = {} & E_n(\boldsymbol{k}_0) + \frac{\hbar^2(\boldsymbol{k}^2-\boldsymbol{k}_0^2)}{2m} + \frac{\hbar}{m}(\boldsymbol{k}-\boldsymbol{k}_0)\boldsymbol{p}_{nn}(\boldsymbol{k}_0) \\
& + \frac{\hbar^2}{m^2}\sum_{n'\neq n}\frac{(\boldsymbol{k}-\boldsymbol{k}_0)\boldsymbol{p}_{nn'}(\boldsymbol{k}_0)\,(\boldsymbol{k}-\boldsymbol{k}_0)\boldsymbol{p}_{n'n}(\boldsymbol{k}_0)}{E_n(\boldsymbol{k}_0)-E_{n'}(\boldsymbol{k}_0)} . \qquad (3.85)
\end{aligned}
$$

Correspondingly, lattice-periodic functions are obtained from

$$
\begin{aligned}
u_n(\boldsymbol{k},\boldsymbol{r}) = {} & u_n(\boldsymbol{k}_0,\boldsymbol{r}) \\
& + \frac{\hbar}{m}\sum_{n'\neq n}\frac{(\boldsymbol{k}-\boldsymbol{k}_0)\boldsymbol{p}_{n'n}(\boldsymbol{k}_0)\,u_{n'}(\boldsymbol{k}_0,\boldsymbol{r})}{E_n(\boldsymbol{k}_0)-E_{n'}(\boldsymbol{k}_0)} . \qquad (3.86)
\end{aligned}
$$

Although introduced perturbationally, (3.85) is exact up to order $\boldsymbol{k}^2$ and yields rigorous results for the first and second derivatives of the Bloch energy with respect to the wave vector. Taking the first derivative, we obtain

$$
\boldsymbol{p}_{nn}(\boldsymbol{k}_0) = \frac{m}{\hbar^2}\nabla_{\boldsymbol{k}}E_n(\boldsymbol{k})|_{\boldsymbol{k}_0} . \qquad (3.87)
$$

At $\boldsymbol{k}_0 = 0$, the gradient vanishes owing to symmetry. At a general point, however, (3.87) is an essential result, as it relates the gradients of the Bloch energies to the quantum mechanical expectation value of the momentum operator, which can be regarded as the product of group velocity and a mass. In spite of its interaction with the lattice potential, *a Bloch electron moves without energy dissipation*. Obviously, this is a result of the lattice periodicity of the potential, which is incorporated into the properties of Bloch electrons. Consequently, one expects that any perturbation of the perfect symmetry by lattice imperfections, lattice vibrations, or impurities, for example, lead to dissipative processes. This is indeed the case and this point will be discussed in more detail in Chap. 11. Computing the second derivative according to (3.79) at $\boldsymbol{k}_0 = 0$, we obtain the effective-mass tensor

$$
(m_n^{-1})_{ij} = \frac{1}{m}\delta_{ij} + \frac{1}{m^2}\sum_{n'\neq n}\frac{p^i_{nn'}p^j_{n'n} + p^j_{nn'}p^i_{n'n}}{E_n(\boldsymbol{k}_0)-E_{n'}(\boldsymbol{k}_0)} . \qquad (3.88)
$$

This constitutes not only a prescription for calculating the effective masses but also a sum rule, and is therefore often referred to as the effective-mass sum rule. For localized electrons, i.e. if the bands become completely flat, the effective masses approach infinity and (3.88) reduces to the usual f-sum rule of atomic physics. In the following we consider only the diagonal elements for simplicity. Some qualitative conclusions can already be drawn from (3.88). The contributions of lower and higher bands tend to decrease and increase, respectively, the effective mass. For semiconductors or insulators one expects a large effective mass for a large energy gap and a small effective mass for a small energy gap. The simplest case is that of a two-band model, which yields the following for the effective masses of the valence and conduction bands:

$$m_{\mathrm{v}} = -m\left(\frac{2\hbar^2|p_{\mathrm{vc}}|^2}{mE_{\mathrm{g}}} - 1\right)^{-1} , \tag{3.89}$$

$$m_{\mathrm{c}} = m\left(\frac{2\hbar^2|p_{\mathrm{vc}}|^2}{mE_{\mathrm{g}}} + 1\right)^{-1} . \tag{3.90}$$

As in typical semiconductors the kinetic energy corresponding to the momentum transition matrix element is more than an order of magnitude larger than the gap energy, this simple result explains immediately the fact that the effective mass of the valence band is negative. One should, however, be cautious with a quantitative application of (3.89) and (3.90), as the two equations imply the relation

$$\frac{1}{m_{\mathrm{v}}} + \frac{1}{m_{\mathrm{c}}} = \frac{2}{m} , \tag{3.91}$$

which is usually *not* fulfilled. Thus we should regard (3.89) and (3.90) merely as rules of thumb.

A missing electron with wave vector $\boldsymbol{k}$ in the valence band is usually described as a *hole* with wave vector $-\boldsymbol{k}$. For convenience, the mass of the hole is defined as the negative valence band mass, such that

$$\begin{aligned} m_{\mathrm{h}} &:= -m_{\mathrm{v}} > 0 , \\ m_{\mathrm{e}} &:= m_{\mathrm{c}} . \end{aligned} \tag{3.92}$$

The mass of the hole is then again a positive number. Finally, we define the energy of a hole as the missing energy in the valence band; hence

$$E_{\mathrm{h}}(\boldsymbol{k}) := -E_{\mathrm{v}}(-\boldsymbol{k}) = -E_{\mathrm{v}} + \frac{\hbar^2\boldsymbol{k}^2}{2m_{\mathrm{h}}} , \tag{3.93}$$

$$E_{\mathrm{e}}(\boldsymbol{k}) := E_{\mathrm{c}}(\boldsymbol{k}) = E_{\mathrm{c}} + \frac{\hbar^2\boldsymbol{k}^2}{2m_{\mathrm{e}}} . \tag{3.94}$$

Whenever we switch to the electron and hole picture, we shall also replace the conduction band index "c" on masses and energy dispersions by "e", as already done in (3.92) and (3.93). A selection of typical effective masses for a few semiconductors is given in Table 3.3.

Our previous treatment of effective masses within the framework of $\boldsymbol{kp}$ perturbation theory implies that the bands at $\boldsymbol{k} = 0$ are nondegenerate. Unfortunately, in most cases this is not true. Apart from the spin degeneracy, valence bands are degenerate as a consequence of the symmetry properties of the Bloch functions. A simple example occurs in semiconductors with the zinc blende structure, where the uppermost valence band states are mainly built from atomic $\mathrm{p_x}$,$\mathrm{p_y}$ and $\mathrm{p_z}$ orbitals. Owing to the cubic symmetry, all

Table 3.3. tive masses (in units of the free-electron mass $m = 9.1091 \times 10^{-31}$ kg) of several semiconductor materials at $T \leq 4.2$ K. Taken from [2]

Material	m_c/m	m_v/m
Si	0.1905 ($\perp$), 0.9163 ($\parallel$)	0.537 (hh), 0.153 (lh)
Ge	0.08152 ($\perp$), 1.588 ($\parallel$)	0.28 (hh), 0.0438 (lh)
GaAs	0.0665	0.475 (hh), 0.087 (lh)
AlAs	0.124	0.5 (hh), 0.26 (lh)
ZnSe	0.13 – –0.16	0.78 (hh, $\langle 100 \rangle$), 0.145 (lh)
CdS	0.2	0.7 ($\perp$), 5 ($\parallel$)
CuCl	≈ 0.4 (Γ_{6c})	≈ 13 (Γ_{7v})

the corresponding spatial directions are equivalent and therefore one expects sixfold degenerate bands at the Γ point. This degenerate case will be treated in the next section.

3.5 kp Perturbation Theory for Degenerate Bands

There are two possible ways to treat degenerate bands within the effective-mass approximation. We can assume that we know the solution of the one-particle problem and thus the effective masses. In this case the one-particle Hamiltonian is diagonal. Alternatively, we can take the coupling between degenerate or neighboring bands explicitly into account and work in a basis in which the one-particle Hamiltonian is off-diagonal [62]. Concerning one-particle energies, both possibilities are strictly equivalent and are related by a unitary transformation resulting from the solution of (3.83). With respect to the Bloch functions, the effective-mass approximation asserts, in lowest order, that we set

$$\varphi_n(\boldsymbol{k}, \boldsymbol{r}) = \mathrm{e}^{\mathrm{i}\boldsymbol{kr}} u_n(0, \boldsymbol{r}) \tag{3.95}$$

and assume $u_n(0, \boldsymbol{r}) = 1$, i.e., in a plane-wave expansion, only the long-range $\boldsymbol{G} = 0$ contribution is taken into account. As can be seen by a comparison of the expansion (3.82) with the approximation (3.95), the latter is no longer valid in the case of band degeneracy, where we have to deal with a linear combination of the degenerate states. Thus, whenever wavefunctions come into play as, for example, in the calculation of transition matrix elements or matrix elements of the Coulomb interaction, degenerate bands require consideration of band-mixing effects.

To proceed, we consider the case of a set of degenerate valence bands, denoted by indices $v \in \mathcal{V}$ and treat their contribution in (3.83) separately. In many interesting cases, the only momentum matrix elements that are nonvanishing are those between the degenerate subset and states of different energies. In this case we can solve (3.82) in two steps. For the states $n \ni \mathcal{V}$, we have

$$\left(\frac{\hbar^2\boldsymbol{k}^2}{2m} + E_n(0) - E_m(\boldsymbol{k})\right) c_{nm}(\boldsymbol{k}) + \frac{\hbar}{m}\sum_{\mathrm{v}\in\mathcal{V}} \boldsymbol{k}\boldsymbol{p}_{n\mathrm{v}} c_{\mathrm{v}m}(\boldsymbol{k}) = 0\,. \qquad (3.96)$$

The equation for the degenerate states reads

$$\left(\frac{\hbar^2\boldsymbol{k}^2}{2m} + E_{\mathrm{v}}(0) - E_m(\boldsymbol{k})\right) c_{\mathrm{v}m}(\boldsymbol{k}) + \frac{\hbar}{m}\sum_{n\ni\mathcal{V}} \boldsymbol{k}\boldsymbol{p}_{\mathrm{v}n} c_{nm}(\boldsymbol{k}) = 0\,. \qquad (3.97)$$

Inserting the solution of (3.96) into (3.97) we arrive at the eigenvalue problem

$$\sum_{v'\in\mathcal{V}} h_{\mathrm{vv}'} c_{\mathrm{v}'m}(\boldsymbol{k}) = E_m(\boldsymbol{k}) c_{\mathrm{v}m}(\boldsymbol{k})\,, \qquad (3.98)$$

where we have defined

$$h_{\mathrm{vv}'} = \left(E_{\mathrm{v}}(0) + \frac{\hbar^2\boldsymbol{k}^2}{2m}\right)\delta_{\mathrm{vv}'} + \frac{\hbar^2}{m^2}\sum_{n\ni\mathcal{V}} \frac{\boldsymbol{k}\boldsymbol{p}_{\mathrm{v}n}\,\boldsymbol{k}\boldsymbol{p}_{n\mathrm{v}'}}{E_{\mathrm{v}}(0) - E_n(0)} \qquad (3.99)$$

and have already replaced $E_m(\boldsymbol{k})$ by $E_{\mathrm{v}}(0)$ in the denominator of (3.99), which is correct to the order of $\boldsymbol{k}^2$. The remaining task is to reduce the sums over matrix elements and energy denominators by means of symmetry arguments to a few independent material parameters. We shall work through this method for the example of semiconductors with the zinc blende structure. First we need the transformation properties of the Bloch functions for the degenerate valence band. In the language of group theory, these states transform according to the irreducible representation of the cubic point group, which in our case corresponds to eigenstates of the operator of the total angular momentum $\mathcal{J}$, with eigenvalue $J = 3/2$. The $J = 1/2$ states may be neglected, as they are energetically separated owing to spin–orbit interaction. Thus we have to deal with four states only, which transform according to

$$\left|j_z = \frac{3}{2}\right\rangle = \frac{1}{\sqrt{2}}|x + \mathrm{i}y\rangle s_+\,, \qquad (3.100)$$

$$\left|j_z = \frac{1}{2}\right\rangle = \frac{1}{\sqrt{6}}(|x + \mathrm{i}y\rangle s_- - 2|z\rangle s_+)\,, \qquad (3.101)$$

$$\left|j_z = -\frac{1}{2}\right\rangle = \frac{1}{\sqrt{6}}(|x - \mathrm{i}y\rangle s_+ + 2|z\rangle s_-)\,, \qquad (3.102)$$

$$\left|j_z = -\frac{3}{2}\right\rangle = \frac{1}{\sqrt{2}}|x - \mathrm{i}y\rangle s_-\,, \qquad (3.103)$$

where s_+ and s_- denote the spin states with spin up and spin down, respectively. The states $|x\rangle$, $|y\rangle$ and $|z\rangle$ are, for example, realized by states which transform in the same way as the $\mathrm{p_x}$, $\mathrm{p_y}$, and $\mathrm{p_z}$ functions. Apart from these transformation properties, no further information is needed to find out which matrix elements in (3.99) are different. As the sum over intermediate states is invariant under arbitrary symmetry transformations, it is sufficient to consider matrix elements of the form

$$\langle j_z|m|j'_z\rangle = \alpha \sum_{ij} k_i k_j \langle j_z|p_i p_j|j'_z\rangle \,, \tag{3.104}$$

in (3.99), where α is a suitably chosen constant. In the following, only three different types of matrix elements occur:

$$m_1 = \alpha\langle x|p_x^2|x\rangle \,, \tag{3.105}$$

$$m_2 = \alpha\langle x|p_y^2|x\rangle = \langle x|p_z^2|x\rangle \,, \tag{3.106}$$

and

$$m_3 = \alpha\langle x|p_x p_y|y\rangle \,. \tag{3.107}$$

All other matrix elements can be reduced to one of these three types by using the cubic symmetry. Using, for example, (3.100) in (3.104), we find

$$\begin{aligned}\left\langle\frac{3}{2}\middle|m\middle|\frac{3}{2}\right\rangle &= \frac{\alpha}{2}\sum_{ij} k_i k_j \langle x - \mathrm{i}y|p_i p_j|x + \mathrm{i}y\rangle \\ &= \frac{1}{2}\sum_{i} k_i^2 \langle x - \mathrm{i}y|p_i^2|x + \mathrm{i}y\rangle \\ &= \frac{1}{2}\left[m_1(k_x^2 + k_y^2) + m_2(k_x^2 + k_y^2 + 2k_z^2)\right] \,.\end{aligned} \tag{3.108}$$

Correspondingly, for the other matrix elements we find

$$\left\langle\frac{3}{2}\middle|m\middle|\frac{1}{2}\right\rangle = \frac{2}{\sqrt{3}} m_3 (k_x - \mathrm{i}k_y) k_z \,, \tag{3.109}$$

$$\left\langle\frac{3}{2}\middle|m\middle|-\frac{1}{2}\right\rangle = \frac{1}{\sqrt{12}}\left[(m_1 - m_2)(k_x^2 - k_y^2) - 4\mathrm{i}m_3 k_x k_y\right] \,, \tag{3.110}$$

$$\left\langle\frac{1}{2}\middle|m\middle|\frac{1}{2}\right\rangle = \frac{1}{6}\left[(m_1(k_x^2 + k_y^2 + 4k_z^2) + m_2(5(k_x^2 + k_y^2) + 2k_z^2)\right] \tag{3.111}$$

and

$$\left\langle\frac{3}{2}\middle|m\middle|-\frac{3}{2}\right\rangle = \left\langle\frac{1}{2}\middle|m\middle|-\frac{1}{2}\right\rangle = 0 \,. \tag{3.112}$$

No other different matrix elements occur in the evaluation of (3.99). The structure of the resulting Hamiltonian matrix can be further simplified by introducing the *Luttinger parameters* [63, 64]

$$\gamma_1 = 1 - \frac{1}{3}(m_1 + 2m_2) \,, \tag{3.113}$$

$$\gamma_2 = -\frac{1}{6}(m_1 - m_2) \,, \tag{3.114}$$

$$\gamma_3 = \frac{1}{3} m_3 \,. \tag{3.115}$$

Thus the valence band matrix (3.99) can be written in the form

$$\underline{h} = E_{\mathrm{v}}(0)\,\underline{1} + \begin{pmatrix} a_+ & b^* & c^* & 0 \\ b & a_- & 0 & c^* \\ c & 0 & a_- & -b^* \\ 0 & c & -b & a_+ \end{pmatrix} , \tag{3.116}$$

where we have ordered the elements according to the sequence of states $|3/2\rangle$, $|1/2\rangle$, $|-1/2\rangle$ and $|-3/2\rangle$. The matrix elements are defined by

$$a_\pm = -\frac{\hbar^2}{2m}[(\gamma_1 \pm \gamma_2)\boldsymbol{k}^2 \mp 3\gamma_2 k_z^2] , \tag{3.117}$$

$$b = \frac{\hbar^2}{2m} 2\sqrt{3}\gamma_3(k_x - \mathrm{i}k_y)k_z , \tag{3.118}$$

$$c = -\frac{\hbar^2}{2m}\sqrt{3}[\gamma_2(k_x^2 - k_y^2) - 2\mathrm{i}\gamma_3 k_x k_y]\frac{\hbar^2}{2m} . \tag{3.119}$$

Diagonalizing the Hamiltonian matrix (3.116), we find that the twofold degenerate eigenvalues[7] are determined by

$$[a_+ - E_{\mathrm{v}}(\boldsymbol{k})][a_- - E_{\mathrm{v}}(\boldsymbol{k})] - |b|^2 - |c|^2 = 0 , \tag{3.120}$$

which yields the following for the valence band energies:

$$E_{\mathrm{v}}(\boldsymbol{k}) = E_{\mathrm{v}}(0) - \frac{\hbar^2}{2m}\left(\gamma_1 \boldsymbol{k}^2 \pm \sqrt{4\gamma_2^2\boldsymbol{k}^4 + 12(\gamma_3^2 - \gamma_2^2)(k_x^2k_y^2 + k_x^2k_z^2 + k_y^2k_z^2)}\right) . \tag{3.121}$$

With $E_{\mathrm{h}}(\boldsymbol{k}) = -E_{\mathrm{v}}(-\boldsymbol{k})$ (3.121) defines also the hole band energies.

As illustrated in Fig. 3.5, the valence bands depend in general on the direction of the wave vector. These so-called *warping effects* are depicted in Fig. 3.6. They vanish in the limit $\gamma_2 = \gamma_3$, which is actually not too bad an approximation in the case of most III–V semiconductors. Replacing γ_2 and γ_3 by $\gamma = (\gamma_2 + \gamma_3)/2$, we finally obtain

$$E_{\mathrm{v}}(\boldsymbol{k}) = E_{\mathrm{v}}(0) - \frac{\hbar^2}{2m}(\gamma_1 \pm 2\gamma_2)\boldsymbol{k}^2 . \tag{3.122}$$

This result defines a heavy-hole band and a light-hole band with energy dispersions

$$E_{\mathrm{hh}}(\boldsymbol{k}) = -E_{\mathrm{v}} - \frac{\hbar^2\boldsymbol{k}^2}{2m_{\mathrm{hh}}} , \tag{3.123}$$

$$E_{\mathrm{lh}}(\boldsymbol{k}) = -E_{\mathrm{v}} - \frac{\hbar^2\boldsymbol{k}^2}{2m_{\mathrm{lh}}} . \tag{3.124}$$

[7] This degeneracy, which is also present if spin–orbit interaction is taken into account, is a general consequence of time-reversal symmetry and is usually referred to as the Kramers degeneracy.

For many optical and transport phenomena, this is a sufficient description of the uppermost valence bands. In particular, the transformation properties of the corresponding wavefunctions are treated exactly. This allows a detailed analysis of the selection rules for transition matrix elements. In many cases further simplification is possible. Optical transitions from two different valence bands which are degenerate at $\boldsymbol{k} = 0$ can be simulated by a transition from one valence band with an appropriately chosen mass. We shall come back to this point in Chap. 4.

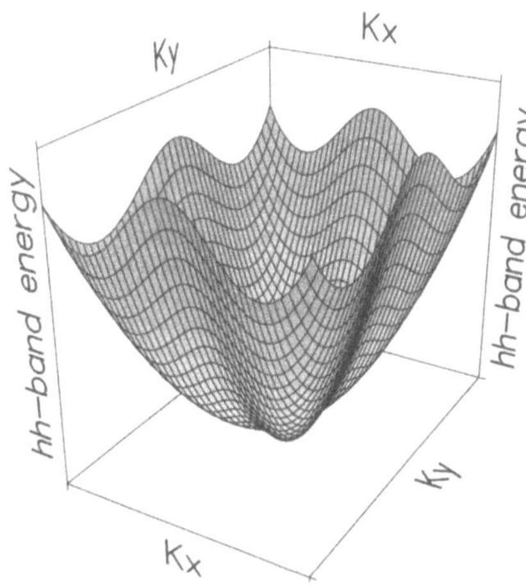

Fig. 3.6. Warping of the heavy-hole band of GaAs in the k_x-k_y plane. The wave vectors cover the range $-\pi/(50\,a) < k_x < \pi/(50\,a)$, $-\pi/(50\,a) < k_y < \pi/(50\,a)$ in the Brillouin zone

3.6 Transition Matrix Elements

The effective-mass approximation discussed in the preceding sections can be applied if we consider length scales which are large compared with the atomic spacing. A property which cannot be described in this limit is the optical transition matrix element, for which we require not the asymptotic but the complete information about the Bloch functions. Moreover, if we have to deal with extended states, the dipole approximation introduced in Sect. 2.5 is not trivial and deserves some comments.

Let us consider first the matrix element in (2.119) which has to be evaluated within the dipole approximation. Using the Wannier representation (3.71), we obtain for the matrix element

$$\begin{aligned}&\langle n', \boldsymbol{k}' | \boldsymbol{dE} | n, \boldsymbol{k} \rangle \\ &= -\frac{e}{\mathcal{N}} \sum_{\boldsymbol{L},\boldsymbol{L}'} \mathrm{e}^{-\mathrm{i}(\boldsymbol{k}'\boldsymbol{L}' - \boldsymbol{k}\boldsymbol{L})} \int \mathrm{d}^3\boldsymbol{r}\, w^*_{n'}(\boldsymbol{r} - \boldsymbol{L}')\, \boldsymbol{r}\boldsymbol{E}(\boldsymbol{r})\, w_n(\boldsymbol{r} - \boldsymbol{L})\end{aligned}$$

$$= -\frac{e}{\mathcal{N}} \sum_{\boldsymbol{L},\boldsymbol{L}'} \mathrm{e}^{\mathrm{i}[\boldsymbol{kL}-(\boldsymbol{k}'-\boldsymbol{k})\boldsymbol{L}']} \int \mathrm{d}^3\boldsymbol{r}\, w_{n'}^*(\boldsymbol{r})\, (\boldsymbol{r}+\boldsymbol{L}')\boldsymbol{E}(\boldsymbol{L}')\, w_n(\boldsymbol{r}-\boldsymbol{L})\,, \tag{3.125}$$

where in the last step, the transformations $\boldsymbol{r} \to \boldsymbol{r}+\boldsymbol{L}'$ and $\boldsymbol{L} \to \boldsymbol{L}+\boldsymbol{L}'$ have been used. Further, we have assumed that the Wannier functions are localized within the range of a few lattice constants; hence, we are allowed to apply the dipole approximation (see Sect. 2.5) which implies $\boldsymbol{E}(\boldsymbol{r}+\boldsymbol{L}') \simeq \boldsymbol{E}(\boldsymbol{L}')$. Owing to the orthogonality of the Wannier functions, the contribution increasing with $\boldsymbol{L}'$ comes into play only for intraband transitions and is proportional to the derivative of the Fourier transform of the field. Considering interband transitions only, we obtain

$$\begin{aligned}\langle n',\boldsymbol{k}'|\boldsymbol{dE}|n,\boldsymbol{k}\rangle &= -e\sum_{\boldsymbol{L}} \mathrm{e}^{\mathrm{i}\boldsymbol{kL}} \int \mathrm{d}^3\boldsymbol{r}\, w_{n'}^*(\boldsymbol{r})\, \boldsymbol{r}\boldsymbol{E}(\boldsymbol{k}'-\boldsymbol{k})\, w_n(\boldsymbol{r}-\boldsymbol{L}) \\ &\equiv \boldsymbol{d}_{n'n}(\boldsymbol{k})\boldsymbol{E}(\boldsymbol{k}'-\boldsymbol{k})\,. \end{aligned} \tag{3.126}$$

Thus, within the dipole approximation the interband transition matrix element becomes diagonal with respect to the wave vector and off-diagonal contributions occur only in the transverse field.

For the purpose of other applications, it is instructive to consider transition matrix elements without the detour via Wannier states. Let us consider a plane-wave transition matrix element between Bloch states in the limit of small wave vector:

$$\lim_{\boldsymbol{q}\to 0} \langle n,\boldsymbol{k}|\, \mathrm{e}^{-\mathrm{i}\boldsymbol{qr}}\, |n',\boldsymbol{k}+\boldsymbol{q}\rangle = \lim_{\boldsymbol{q}\to 0} \frac{1}{V} \int \mathrm{d}^3\boldsymbol{r}\, u_n^*(\boldsymbol{k},\boldsymbol{r}) u_{n'}(\boldsymbol{k}+\boldsymbol{q},\boldsymbol{r})\,. \tag{3.127}$$

In the limit of vanishing wave vector, $u_{n'}(\boldsymbol{k}+\boldsymbol{q},\boldsymbol{r})$ is obtained exactly by means of $\boldsymbol{kp}$ perturbation theory. Making use of (3.86), we obtain

$$\begin{aligned}&\lim_{\boldsymbol{q}\to 0} \langle n,\boldsymbol{k}|\, \mathrm{e}^{-\mathrm{i}\boldsymbol{qr}}\, |n',\boldsymbol{k}+\boldsymbol{q}\rangle \\ &= \delta_{n,n'} + \frac{\hbar}{m} \frac{\boldsymbol{q}\,\boldsymbol{p}_{n'n}(\boldsymbol{k})}{E_{n'}(\boldsymbol{k})-E_n(\boldsymbol{k})} (1-\delta_{n,n'})\,. \end{aligned} \tag{3.128}$$

Expressing the interband momentum matrix element by the dipole matrix element according to (2.108), we have the alternative representation

$$\lim_{\boldsymbol{q}\to 0} \langle n,\boldsymbol{k}|\, \mathrm{e}^{-\mathrm{i}\boldsymbol{qr}}\, |n',\boldsymbol{k}+\boldsymbol{q}\rangle = \delta_{n,n'} - \mathrm{i}\boldsymbol{q}\,\boldsymbol{r}_{n'n}(\boldsymbol{k})\,(1-\delta_{n,n'}) \tag{3.129}$$

The integration required to obtain the momentum matrix elements as well as for the dipole matrix elements, is taken again over a single unit cell. Considering again interband transitions $n \neq n'$ only, (3.128) shows that the plane-wave transition matrix elements to lowest order, are linear in the wave vector. This implies, as already mentioned in Sect. 2.6, that an expansion of the density–density correlation starts at the second order and the Coulomb singularity in the dielectric function is canceled (see Sect. 5.2).

The relation (3.129) can be simplified to lowest order in the gap energy, by replacing $\boldsymbol{p}_{n'n}(\boldsymbol{k})$ by $\boldsymbol{p}_{n'n}(0)$. This yields the following for the interband dipole matrix element:

$$\boldsymbol{d}_{n'n}(\boldsymbol{k}) = -\frac{e\hbar^2}{m}\frac{\boldsymbol{p}_{n'n}(0)}{E_{n'}(\boldsymbol{k}) - E_n(\boldsymbol{k})} = \boldsymbol{d}_{n'n}(0)\,\frac{E_{n'}(0) - E_n(0)}{E_{n'}(\boldsymbol{k}) - E_n(\boldsymbol{k})} \,. \tag{3.130}$$

Thus, for parabolic bands, within the effective-mass approximation, we have

$$\boldsymbol{d}_{n'n}(\boldsymbol{k}) = \boldsymbol{d}_{n'n}(0)\,\frac{2m_{\mathrm{r}}\,E_{\mathrm{g}}}{2m_{\mathrm{r}}\,E_{\mathrm{g}} + \hbar^2\boldsymbol{k}^2}\,, \tag{3.131}$$

where the reduced mass m_r is defined by $m_r^{-1} = m_{n'}^{-1} - m_n^{-1}$. If high-lying continuum states are excited, the dependence of the dipole matrix elements on the wave vector introduces a cutoff. This cutoff, however, becomes effective only for large $\boldsymbol{k}$ values in the Brillouin zone, for which the kinetic energy is of the same order of magnitude as the gap energy. In the vicinity of the band gap, the $\boldsymbol{k}$ dependence becomes negligible.

We shall also consider other types of transition matrix elements, for which the wavefunctions can be factorized into a lattice-periodic contribution and a function varying slowly on the scale of the lattice constant. This structure occurs, for example, in quantum well systems and under the action of static magnetic or electric fields and allows us again, to make use of the separability of length scales. Let us consider a wavefunction of the type

$$\varphi_{mn}(\boldsymbol{r}, \boldsymbol{k}) = \frac{1}{\sqrt{A}}\,\mathrm{e}^{\mathrm{i}\boldsymbol{k}\boldsymbol{\rho}} g_{mn}(z) u_n(0, \boldsymbol{r})\,, \tag{3.132}$$

where $\boldsymbol{k}$ and $\boldsymbol{\rho}$ denote vectors perpendicular to the z direction, and the envelope function $g_{nm}(z)$ takes any additional structure in the one-particle Hamiltonian with quantum number m in the nth band into account. The dipole transition matrix element takes the form

$$\begin{aligned}\langle m'n', \boldsymbol{k}|\boldsymbol{d}|mn, \boldsymbol{k}\rangle &= -\frac{\mathrm{e}}{A}\int \mathrm{d}^3\boldsymbol{r}\, u_{n'}^*(0,\boldsymbol{r}) g_{m'n'}^*(z)\boldsymbol{r} u_n(0,\boldsymbol{r}) g_{mn}(z) \\ &= -\frac{\mathrm{e}}{A}\int_{\text{cell}} \mathrm{d}^3\boldsymbol{r} \sum_{\boldsymbol{L}} u_{n'}^*(0,\boldsymbol{r})(\boldsymbol{r} + \boldsymbol{L}) u_n(0,\boldsymbol{r}) g_{m'n'}^*(z + L_z) g_{mn}(z + L_z)\,.\end{aligned} \tag{3.133}$$

According to our assumption that $g_{mn}(z)$ is a slowly varying envelope function, z may be neglected in comparison with L_z and the lattice vector can be treated as a continous variable. Taking only interband transitions into account, we obtain the following for the matrix element:

$$\langle m'n', \boldsymbol{k}|\boldsymbol{d}|mn, \boldsymbol{k}\rangle = d_{n'n}(0)\delta_{m'm}\,. \tag{3.134}$$

The wavefunction (3.132) is appropriate for describing one-particle states in quantum wells. In other problems, the structure may be more complicated. The basic procedure for calculating transition matrix elements within the *envelope function approximation* is always the same, however.

3.7 Density of States

In a number of problems one has to evaluate the sum over all states (with the electron (or hole) wave vector $\boldsymbol{k}$ and the spin s as quantum numbers) of some function $f[E(\boldsymbol{k})]$. Such a sum can be rearranged according to

$$\begin{aligned}\sum_{\boldsymbol{k},\, s} f[E_\mathrm{e}(\boldsymbol{k})] &= \sum_{\boldsymbol{k},\, s} f[E_\mathrm{e}(\boldsymbol{k})]\left(\int_{-\infty}^{\infty} \mathrm{d}E\, \delta[E - E_\mathrm{e}(\boldsymbol{k})]\right) \\ &= \int_{-\infty}^{\infty} \mathrm{d}E \sum_{\boldsymbol{k},\, s} f[E_\mathrm{e}(\boldsymbol{k})]\delta[E - E_\mathrm{e}(\boldsymbol{k})] \\ &= \int_{-\infty}^{\infty} \mathrm{d}E\, f(E)\, D_\mathrm{e}(E)\,, \end{aligned} \tag{3.135}$$

where we have introduced the electron *density of states*

$$D_\mathrm{e}(E) = \sum_{\boldsymbol{k},\, s} \delta[E - E_\mathrm{e}(\boldsymbol{k})]\,. \tag{3.136}$$

A similar expression is obtained for the holes. For a finite d-dimensional crystal of spatial extent L in all directions, we use periodic boundary conditions, i.e. $\varphi_n(\boldsymbol{k}, x) = \varphi_n(\boldsymbol{k}, x + L) = u_n(\boldsymbol{k}, x + L)\mathrm{e}^{\mathrm{i}k_x(x+L)} = \varphi_n(\boldsymbol{k}, x)\mathrm{e}^{\mathrm{i}k_x L}$; hence $k_x L = 2\pi N$, where N is an integer. The spacing between adjacent states in $\boldsymbol{k}$ space is $2\pi/L$ and using the same argument for the other directions, the volume of one state in a d dimensional $\boldsymbol{k}$-space becomes $(2\pi)^d/L^d$. Consequently, the sum (3.136) for densely spaced $\boldsymbol{k}$ can be rewritten as a more convenient integral,

$$\sum_{\boldsymbol{k},\, s} \delta[E - E_\mathrm{e}(\boldsymbol{k})] \longrightarrow \left(\frac{L}{2\pi}\right)^d \sum_s \int \delta[E - E_\mathrm{e}(\boldsymbol{k})]\mathrm{d}^d k\,. \tag{3.137}$$

The sum $\sum_s$ reduces to a mere factor of two. The integral in (3.137) is easily worked out within the effective-mass approximation (3.93), employing the usual mathematics of δ functions and we obtain the following:

- *Density of states in three dimensions*:

$$D_\mathrm{e}(E) = \frac{L^3}{2\pi^2}\left(\frac{2m_\mathrm{e}}{\hbar^2}\right)^{3/2} \Theta[E - E_\mathrm{e}(0)]\sqrt{E - E_\mathrm{e}(0)}\,. \tag{3.138}$$

- *Density of states in two dimensions*:

$$D_\mathrm{e}(E) = \frac{L^2}{2\pi}\frac{2m_\mathrm{e}}{\hbar^2}\,\Theta[E - E_\mathrm{e}(0)]\,. \tag{3.139}$$

• *Density of states in one dimension:*

$$D_e(E) = \frac{L\sqrt{2m_e}}{\pi\hbar}\,\Theta[E - E_e(0)]\,\frac{1}{\sqrt{E - E_e(0)}}\,. \tag{3.140}$$

3.8 Position of the Chemical Potential

It is now straightforward to compute the chemical potential of the electrons (or holes) in a three-dimensional, undoped semiconductor. Starting from the electron density $n_e = N/L^3$ (3.43), we obtain

$$n_e = \frac{1}{L^3}\sum_{\boldsymbol{k},s} f_e(\boldsymbol{k},s) = \frac{1}{L^3}\int_{-\infty}^{\infty} \mathrm{d}E\, D_e(E) f_e(E)\,. \tag{3.141}$$

Employing a low-occupation expansion for the Fermi function (3.33), namely $f_e(E) \ll 1 \Rightarrow f_e(E) \approx \exp[-\beta(E-\mu)]$, usually referred to as the *Boltzmann approximation*, and inserting the electron density of states $D_e(E)$ given by (3.138) leads to

$$n_e = \frac{1}{2\pi^2}\left(\frac{2m_e}{\hbar^2\beta}\right)^{3/2} \exp\left[\beta(\mu - E_c)\right]\,, \tag{3.142}$$

where we have substituted $x = \beta\,(E - E_c)$. Using the corresponding expression for the density of holes n_h in the low-occupation limit

$$n_h = \frac{1}{2\pi^2}\left(\frac{2m_h}{\hbar^2\beta}\right)^{3/2} \exp\left[\beta(E_v - \mu)\right]\frac{\sqrt{\pi}}{2}\,, \tag{3.143}$$

and the fact that the total number of electrons in the crystal is conserved, i.e. $n_h = n_e$ (as already discussed in Sect. 3.2), yields the result

$$\mu = \frac{E_c + E_v}{2} + \frac{3}{4\beta}\ln\left(\frac{m_h}{m_e}\right)\,. \tag{3.144}$$

At zero temperature the chemical potential is exactly in the middle of the forbidden gap; it rises linearly with temperature because the hole mass is usually larger than the electron mass. As the mass ratio rarely exceeds 10, the prefactor of $k_B T$ is of the order of two, and hence the chemical potential rises by about 50 meV at room temperature, which has to be compared with the band gap energy of typically a few eV.

For undoped semiconductors at large carrier densities or in case of doped semiconductors, the Boltzmann approximation is not justified. Unfortunately, the integration (3.141) cannot be performed analytically in general. In the *degenerate limit*, the electron density is so large that the Fermi distribution function can be approximated by a step function, i.e. $f_e(E) \approx \Theta(\mu - E)$ (similarly to metals). In this limit the electron density (using (3.141)) becomes

$$n_e = \frac{1}{3\pi^2}\left(\frac{2m_e}{\hbar^2}\right)^{3/2} (\mu - E_c)^{3/2} . \tag{3.145}$$

A numerical solution is shown in Fig. 3.7 (solid line). It is apparent that the Boltzmann approximation is justified for $(\mu - E_c)/(k_B T) < -3$, and the degenerate limit for $(\mu - E_c)/(k_B T) > 3$ (see dashed lines). As a knowledge of the relation between electron density and chemical potential is important in many situations, a rich variety of approximation schemes exists which avoid numerical evaluation of (3.141).

Nature is more kind to us in the case of two-dimensional semiconductors, where n_e is easily obtained from integration of

$$n_e = \frac{1}{L^2}\int_{-\infty}^{+\infty} \mathrm{d}E\, D_e(E) f_e(E) . \tag{3.146}$$

Inserting the two-dimensional density of states (3.139) yields

$$n_e = \frac{m_e}{\pi\hbar^2\beta} \ln\left(\frac{\exp[\beta(E_c - \mu)] + 1}{\exp[\beta(E_c - \mu)]}\right) . \tag{3.147}$$

This result is depicted in Fig. 3.7 (dotted line) and is qualitatively very similar to the three-dimensional case.

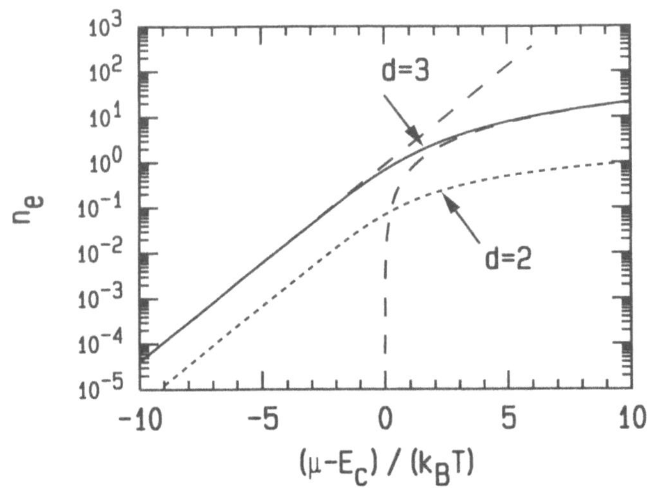

Fig. 3.7. Electron density n_e of a three dimensional semiconductor in units of $(2m_e/\hbar^2)^{3/2}(k_B T)^{3/2}/(2\pi^2)$ versus chemical potential. The *dashed lines* correspond to the Boltzmann and the degenerate approximation. The *dotted line* shows the electron density for a two-dimensional semiconductor in units of $m_e k_B T/(\pi\hbar^2)$, shifted down by one order of magnitude for clarity

3.9 Problems

3.1. Apply the Hartree–Fock theory for $T \neq 0$ to derive the ground-state energy of a homogeneous electron gas.

3.2. The gap energy of a semiconductor is given within the Hartree–Fock approximation at zero temperature. Now assume a finite temperature is switched on, which yields a finite occupation of conduction band states and thus a redistribution of the charge density. Use effective-mass theory to calculate the change of the gap energy.

3.3. Show that the breaking of translational invariance, i.e. the existence of a lattice-periodic potential, leads to energy gaps in the band structure. Hint: Make use of a plane-wave expansion and treat the lattice-periodic potential as a weak perturbation.

3.4. Show that any function which obeys Bloch's theorem has the property $\varphi_n(\boldsymbol{k}, \boldsymbol{r}) = \varphi_n(\boldsymbol{k} + \boldsymbol{G}, \boldsymbol{r})$, where $\boldsymbol{G}$ is an arbitrary vector of the reciprocal lattice.

3.5. We can use a superposition of atomic orbitals $\phi_i(\boldsymbol{r})$ to represent Bloch functions according to

$$\varphi_n(\boldsymbol{k}, \boldsymbol{r}) = \sum_{i,\boldsymbol{L}} c_{ni}(\boldsymbol{k}) \mathrm{e}^{\mathrm{i}\boldsymbol{k}\boldsymbol{L}} \phi_i(\boldsymbol{r} - \boldsymbol{L}) . \tag{3.148}$$

Derive a relation between Wannier functions and atomic orbitals. Why are atomic orbitals more localized than Wannier functions?

3.6. Derive the band energies for a complex of three atomic p orbitals in a simple cubic lattice. Construct again the Wannier functions.

3.7. Show that the momentum matrix element $\langle n', \boldsymbol{k}' | \boldsymbol{p} | n, \boldsymbol{k} \rangle$ is diagonal with respect to the wave vector for interband transitions. Hint: Make use of the Wannier representation.

3.8. Derive the density-of-states formulas (3.138)–(3.140) from (3.137).

3.9. Calculate the combined density-of-states for a transition between a valence band maximum and a conduction band minimum in three dimensions; the band extrema are separated by a wave vector $\boldsymbol{q}$ in the Brillouin zone and can be described within the effective mass approximation. Hint: Start with the generalization of (3.136), which is given by

$$D_{\mathrm{e}}(E) = \sum_{\boldsymbol{k},\, s} \delta[E - E_{\mathrm{c}}(\boldsymbol{k} + \boldsymbol{q}) + E_{\mathrm{v}}(\boldsymbol{k})] . \tag{3.149}$$

3.10. Repeat Problem 3.9 for a two-dimensional and a one-dimensional system.

4. Uncorrelated Optical Transitions

In a semiconductor, there is a wealth of possible transitions between electronic states mediated by a photon. Figure 4.1 summarizes a number of them. As typical photon wave vectors are negligible with respect to the extent of the Brillouin zone, the photon looks practically like a vertical line in the energy dispersion diagram.[1] *Direct interband transitions* (Fig. 4.1a,b), *direct inter-valence-band transitions* (Fig. 4.1d) and similarly *direct inter-conduction-band transitions*, and *photoemission* (Fig. 4.1f) directly connect an occupied initial state (filled circles) to an empty final state (open circles). *Intra-conduction-band* (Fig. 4.1c) and similarly *intra-valence-band transitions*, and *indirect interband transitions* (Fig. 4.1e) on the other hand, require a phonon to be emitted or absorbed in addition. Clearly, at low temperatures, only emission

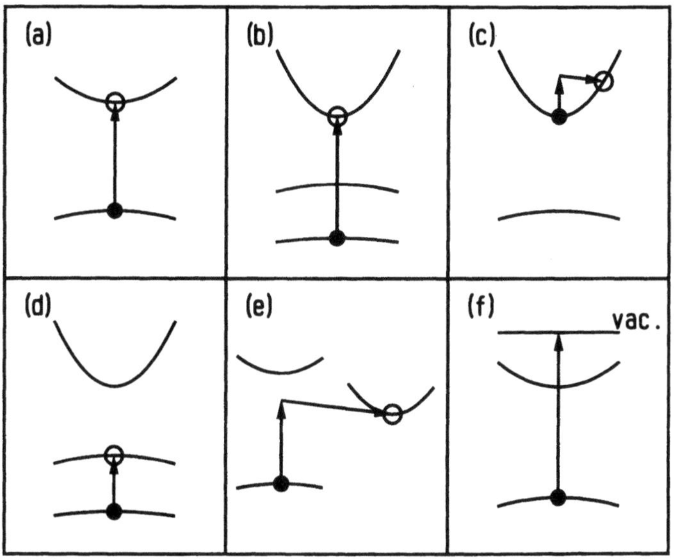

Fig. 4.1a–f. A selection of possible optical transitions in a semiconductor, shown schematically

[1] See Sect. 3.6 and Problem 3.7.

remains (only this channel has been drawn in Fig. 4.1). Transitions requiring both a photon and a phonon are generally much weaker than the others and are usually only of relevance when other transitions are absent in a particular energy regime.

4.1 The Optical Bloch Equations

Let us consider the simplest type of interaction of an electromagnetic field with a semiconductor – (allowed) direct interband transitions – sketched in Fig. 4.2. Neglecting the small (but finite) shift of electronic states, owing to the wave vector of the photon, within the dipole approximation[2], the transition matrix elements become diagonal. Hence, the dipole matrix elements (3.125) determine the *optical selection rules.* As, for example, a circularly polarized photon carries angular momentum $\hbar$, the angular momentum of the electronic states involved must change by that amount, and hence the parity changes. Otherwise, the transition is energetically possible, but dipole forbidden. In this case, higher order perturbation theory can still lead to finite transition strengths, as, for example, in *quadrupole transitions.*

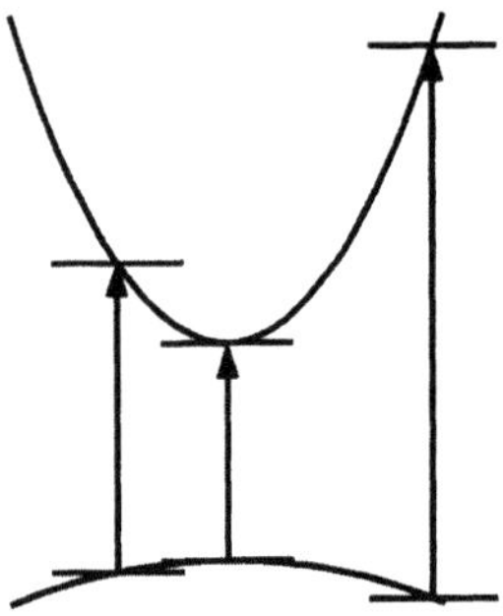

Fig. 4.2. Schematic illustration of direct optical transitions

A much more severe approximation, which we shall use throughout this chapter, is the following: We assume the knowledge of one-particle states, which are calculated within a suitable approximation (see Chapt. 3), but neglect the Coulomb interaction between excited one-particle states. More complete treatments of the problem will be given successively in the following chapters. Even though this way of proceeding is not free of redundancy, it allows us a lot of physical insight at this level without suffering from too much mathematical difficulty. Replacing in (2.119) h by an effective one-

[2] See Sect. 2.5 and Sect. 3.6

particle Hamiltonian $h_{\rm eff}$ and neglecting the remaining Coulomb interaction[3] we start from the Hamiltonian

$$H = \sum_{jk} \langle j|h_{\rm eff}|k\rangle c_j^\dagger c_k - \sum_{jk} \langle j|\boldsymbol{dE}|k\rangle\, c_j^\dagger c_k\,. \tag{4.1}$$

On this level the one-particle properties should be described correctly; hence, the Bloch states, discussed in the preceeding chapter are the most appropriate basis. Thus, we choose $h_{\rm eff}$ such that the Bloch states are eigenstates and express the quantum numbers by the wave vector $\boldsymbol{k}$ and the band indices c and v. Using the dipole approximation (3.126) for the matrix element and restricting ourselves to a two-band model, (4.1) leads to the form

$$\begin{aligned} H = &\sum_{\boldsymbol{k}} E_{\rm c}(\boldsymbol{k})\, c_{{\rm c}\boldsymbol{k}}^\dagger c_{{\rm c}\boldsymbol{k}} + \sum_{\boldsymbol{k}} E_{\rm v}(\boldsymbol{k})\, c_{{\rm v}\boldsymbol{k}}^\dagger c_{{\rm v}\boldsymbol{k}} \\ &- \sum_{\boldsymbol{k},\boldsymbol{q}} \left(\boldsymbol{d}_{\rm cv}(\boldsymbol{k})\boldsymbol{E}(\boldsymbol{q},t)\, c_{{\rm c}\boldsymbol{k}}^\dagger c_{{\rm v}\boldsymbol{k}+\boldsymbol{q}} + \boldsymbol{d}_{\rm cv}^*(\boldsymbol{k})\boldsymbol{E}(\boldsymbol{q},t)\, c_{{\rm v}\boldsymbol{k}}^\dagger c_{{\rm c}\boldsymbol{k}+\boldsymbol{q}} \right)\,, \end{aligned} \tag{4.2}$$

where the one-particle energies are described within the effective-mass approximation. In the simplest case, we neglect the dependence of the field on the wave vector from the very beginning and assume that $\boldsymbol{E}(\boldsymbol{q},t) \propto \delta_{\boldsymbol{q},0}$. In this case, the quantities of interest are the elements of the one-particle density matrix, introduced in Sect. 2.6, for which we shall use the following notation:

- The *distribution function* of electrons in the conduction band

$$f_{\rm c}(\boldsymbol{k}) = \langle c_{{\rm c}\boldsymbol{k}}^\dagger c_{{\rm c}\boldsymbol{k}}\rangle\,, \tag{4.3}$$

- the distribution function of electrons in the valence band

$$f_{\rm v}(\boldsymbol{k}) = \langle c_{{\rm v}\boldsymbol{k}}^\dagger c_{{\rm v}\boldsymbol{k}}\rangle\,, \tag{4.4}$$

- the *transition amplitude* from the conduction to the valence band

$$p_{\rm vc}(\boldsymbol{k}) = \langle c_{{\rm v}\boldsymbol{k}}^\dagger c_{{\rm c}\boldsymbol{k}}\rangle\,, \tag{4.5}$$

- the *transition amplitude* from the valence to the conduction band

$$p_{\rm cv}(\boldsymbol{k}) = \langle c_{{\rm c}\boldsymbol{k}}^\dagger c_{{\rm v}\boldsymbol{k}}\rangle\,. \tag{4.6}$$

If, we take the $\boldsymbol{q}$ dependence of the field in (4.2) serious, it is obvious that such a dependence is transferred to the density matrices. Formally, this means that we have to consider also off-diagonal elements with respect to the wave vector in addition to the quantities defined above. Thus, our starting point has to be more general. Intuitively, one expects that the field induces a center-of-mass motion of excited pairs of carriers. Therefore, we start with a real space representation and introduce center-of-mass variables $\boldsymbol{r} = (\boldsymbol{r}_{\rm c} + \boldsymbol{r}_{\rm c}')/2$ and

[3] A corresponding splitting of the Coulomb interaction can be obtained, by adding and substracting an appropriate chosen one-particle potential. A systematic treatment will be given in Sect. 5.1.

difference variables $\Delta\boldsymbol{r} = \boldsymbol{r}_\mathrm{c} - \boldsymbol{r}'_\mathrm{c}$, which allow us to define the generalization of (4.3) as the Fourier transform[4]

$$f_\mathrm{c}(\boldsymbol{k},\boldsymbol{q}) \equiv \langle c^\dagger_{\mathrm{c}\boldsymbol{k}+\boldsymbol{q}/2} c_{\mathrm{c}\boldsymbol{k}-\boldsymbol{q}/2}\rangle = \int \mathrm{d}^3\Delta\boldsymbol{r}\,\mathrm{d}^3\boldsymbol{r}\,\mathrm{e}^{\mathrm{i}\boldsymbol{k}\Delta\boldsymbol{r}}\mathrm{e}^{\mathrm{i}\boldsymbol{q}\boldsymbol{r}} f_\mathrm{c}(\Delta\boldsymbol{r},\boldsymbol{r}) , \quad (4.7)$$

where the wave vectors in the exponential are obtained from $(\boldsymbol{k}+\boldsymbol{q}/2)/2 + (\boldsymbol{k}-\boldsymbol{q}/2)/2 = \boldsymbol{k}$ and $(\boldsymbol{k}+\boldsymbol{q}/2) - (\boldsymbol{k}-\boldsymbol{q}/2) = \boldsymbol{q}$. For electron–hole pairs, the center-of-mass coordinate is given by $\boldsymbol{r} = \beta\boldsymbol{r}_\mathrm{c} - \alpha\boldsymbol{r}_\mathrm{v}$, with the mass ratios $\alpha = m_\mathrm{v}/(m_\mathrm{c} - m_\mathrm{v})$ and $\beta = m_\mathrm{c}/(m_\mathrm{c} - m_\mathrm{v})$ (compare Problem 4.1). Thus, the corresponding definition for p_vc takes the form

$$p_\mathrm{vc}(\boldsymbol{k},\boldsymbol{q}) \equiv \langle c^\dagger_{\mathrm{v}\boldsymbol{k}-\alpha\boldsymbol{q}} c_{\mathrm{c}\boldsymbol{k}-\beta\boldsymbol{q}}\rangle = \int \mathrm{d}^3\Delta\boldsymbol{r}\,\mathrm{d}^3\boldsymbol{r}\,\mathrm{e}^{\mathrm{i}\boldsymbol{k}\Delta\boldsymbol{r}}\mathrm{e}^{\mathrm{i}\boldsymbol{q}\boldsymbol{r}} p_\mathrm{vc}(\Delta\boldsymbol{r},\boldsymbol{r}) , \quad (4.8)$$

where we have used $\beta(\boldsymbol{k}-\alpha\boldsymbol{q}) - \alpha(\boldsymbol{k}-\beta\boldsymbol{q}) = \boldsymbol{k}$ and $(\boldsymbol{k}-\alpha\boldsymbol{q}) - (\boldsymbol{k}-\beta\boldsymbol{q}) = \boldsymbol{q}$. Corresponding results are valid for $p_\mathrm{cv}(\boldsymbol{k},\boldsymbol{q})$ and $f_\mathrm{v}(\boldsymbol{k},\boldsymbol{q})$. Knowledge of the transition amplitudes allows immediate computation of the *macroscopic polarization* $\boldsymbol{P}(\boldsymbol{r},t)$ (the Fourier transform of (2.130)) due to *direct interband transitions*,

$$\boldsymbol{P}(\boldsymbol{r},t) = \frac{1}{V}\sum_{\boldsymbol{k}} [\boldsymbol{d}^*_\mathrm{cv} p_\mathrm{vc}(\boldsymbol{k},\boldsymbol{r},t) + \boldsymbol{d}_\mathrm{cv} p_\mathrm{cv}(\boldsymbol{k},\boldsymbol{r},t)] , \quad (4.9)$$

where $p_\mathrm{vc}(\boldsymbol{k},\boldsymbol{r},t)$ and $p_\mathrm{cv}(\boldsymbol{k},\boldsymbol{r},t)$ are the Fourier transforms of $p_\mathrm{vc}(\boldsymbol{k},\boldsymbol{q},t)$ and $p_\mathrm{cv}(\boldsymbol{k},\boldsymbol{q},t)$ with respect to $\boldsymbol{q}$. Thus, the transition amplitudes contain the physics of both the linear and the nonlinear optical properties.

In order to follow the temporal evolution of the macroscopic polarization we employ the Heisenberg equation of motion, which, for an arbitrary operator $\mathcal{O}$, is given by

$$-\mathrm{i}\hbar\frac{\partial}{\partial t}\mathcal{O} = [H,\mathcal{O}] . \quad (4.10)$$

Obviously this requires evaluation of a number of commutators. We use the anticommutator algebra for fermions (see Sect. 2.3),

$$[c_{\mathrm{c}\boldsymbol{k}}, c^\dagger_{\mathrm{c}\boldsymbol{k}'}]_+ = \delta_{\boldsymbol{k}\boldsymbol{k}'} , \qquad [c_{\mathrm{v}\boldsymbol{k}}, c^\dagger_{\mathrm{v}\boldsymbol{k}'}]_+ = \delta_{\boldsymbol{k}\boldsymbol{k}'} ; \quad (4.11)$$

all other anticommutators are zero. Using (4.2), we obtain the following:

$$-\mathrm{i}\hbar\frac{\partial}{\partial t}c^\dagger_{\mathrm{c}\boldsymbol{k}} = E_\mathrm{c}(\boldsymbol{k})\,c^\dagger_{\mathrm{c}\boldsymbol{k}} - \sum_{\boldsymbol{q}} \boldsymbol{d}^*_\mathrm{cv}(\boldsymbol{k}-\boldsymbol{q})\boldsymbol{E}(\boldsymbol{q},t)\,c^\dagger_{\mathrm{v}\boldsymbol{k}-\boldsymbol{q}} , \quad (4.12)$$

$$-\mathrm{i}\hbar\frac{\partial}{\partial t}c_{\mathrm{c}\boldsymbol{k}} = -E_\mathrm{c}(\boldsymbol{k})\,c_{\mathrm{c}\boldsymbol{k}} + \sum_{\boldsymbol{q}} \boldsymbol{d}_\mathrm{cv}(\boldsymbol{k}+\boldsymbol{q})\boldsymbol{E}(\boldsymbol{q},t)\,c_{\mathrm{v}\boldsymbol{k}+\boldsymbol{q}} , \quad (4.13)$$

---

[4] Note that the use of continous space variables rather than discrete lattice vectors is implied by the effective-mass approximation (see Problem 5.3).

$$-\mathrm{i}\hbar\frac{\partial}{\partial t}c^{\dagger}_{\mathrm{v}\boldsymbol{k}} = E_{\mathrm{v}}(\boldsymbol{k})\,c^{\dagger}_{\mathrm{v}\boldsymbol{k}} - \sum_{\boldsymbol{q}} \boldsymbol{d}_{\mathrm{cv}}(\boldsymbol{k}-\boldsymbol{q})\boldsymbol{E}(\boldsymbol{q},t)\,c^{\dagger}_{\mathrm{c}\boldsymbol{k}-\boldsymbol{q}}\,, \tag{4.14}$$

$$-\mathrm{i}\hbar\frac{\partial}{\partial t}c_{\mathrm{v}\boldsymbol{k}} = -E_{\mathrm{v}}(\boldsymbol{k})\,c_{\mathrm{v}\boldsymbol{k}} + \sum_{\boldsymbol{q}} \boldsymbol{d}^{*}_{\mathrm{cv}}(\boldsymbol{k}+\boldsymbol{q})\boldsymbol{E}(\boldsymbol{q},t)\,c_{\mathrm{c}\boldsymbol{k}+\boldsymbol{q}}\,. \tag{4.15}$$

Now, we are ready to derive the equations of motion. For the transition amplitude (4.8), for example, we have to use the chain rule

$$\frac{\partial}{\partial t}\left\langle c^{\dagger}_{\mathrm{v}\boldsymbol{k}-\alpha\boldsymbol{q}}c_{\mathrm{c}\boldsymbol{k}-\beta\boldsymbol{q}}\right\rangle = \left\langle\left(\frac{\partial}{\partial t}c^{\dagger}_{\mathrm{v}\boldsymbol{k}-\alpha\boldsymbol{q}}\right)c_{\mathrm{c}\boldsymbol{k}-\beta\boldsymbol{q}}\right\rangle + \left\langle c^{\dagger}_{\mathrm{v}\boldsymbol{k}-\alpha\boldsymbol{q}}\frac{\partial}{\partial t}c_{\mathrm{c}\boldsymbol{k}-\beta\boldsymbol{q}}\right\rangle . \tag{4.16}$$

and obtain by means of (4.13), (4.14), (4.8), and (4.7) the following:

$$\begin{aligned} -\mathrm{i}\hbar\frac{\partial}{\partial t}p_{\mathrm{vc}}(\boldsymbol{k},\boldsymbol{q}) &= [E_{\mathrm{v}}(\boldsymbol{k}-\alpha\boldsymbol{q}) - E_{\mathrm{c}}(\boldsymbol{k}-\beta\boldsymbol{q})]\,p_{\mathrm{vc}}(\boldsymbol{k},\boldsymbol{q}) \\ &- \sum_{\boldsymbol{q}'} \boldsymbol{d}_{\mathrm{cv}}(\boldsymbol{k}-\alpha\boldsymbol{q}-\boldsymbol{q}')\boldsymbol{E}(\boldsymbol{q}')\,f_{\mathrm{c}}\left(\boldsymbol{k}-\frac{\alpha+\beta}{2}\boldsymbol{q}-\frac{\boldsymbol{q}'}{2},\boldsymbol{q}-\boldsymbol{q}'\right) \\ &+ \sum_{\boldsymbol{q}'} \boldsymbol{d}_{\mathrm{cv}}(\boldsymbol{k}-\beta\boldsymbol{q}+\boldsymbol{q}')\boldsymbol{E}(\boldsymbol{q}')\,f_{\mathrm{v}}\left(\boldsymbol{k}-\frac{\alpha+\beta}{2}\boldsymbol{q}-\frac{\boldsymbol{q}'}{2},\boldsymbol{q}-\boldsymbol{q}'\right) . \end{aligned} \tag{4.17}$$

In the last two contributions the relative motion and the center-of-mass motion are still coupled. The corresponding scales can, however, be separated, as the modulation of the wave vector, which describes the relative motion, by the photon wave vector is negligible. Likewise, we can neglect this modulation of the one-particle energies and of the dipole matrix elements. Taking finally the Fourier transform of (4.17) with respect to $\boldsymbol{q}$, we obtain for the transition amplitude

$$\begin{aligned} &\left(\hbar\frac{\partial}{\partial t} + \mathrm{i}\,[E_{\mathrm{c}}(\boldsymbol{k}) - E_{\mathrm{v}}(\boldsymbol{k})]\right)p_{\mathrm{vc}}(\boldsymbol{k},\boldsymbol{r}) \\ &\quad = \mathrm{i}\boldsymbol{d}_{\mathrm{cv}}(\boldsymbol{k})\boldsymbol{E}(\boldsymbol{r},t)\,(f_{\mathrm{v}}(\boldsymbol{k},\boldsymbol{r}) - f_{\mathrm{c}}(\boldsymbol{k},\boldsymbol{r}))\,. \end{aligned} \tag{4.18}$$

For the equation of motion of the one-particle distribution of the conduction band electrons we obtain in the same manner

$$\begin{aligned} \hbar\frac{\partial}{\partial t}f_{\mathrm{c}}(\boldsymbol{k},\boldsymbol{r}) &= \mathrm{i}\,[\boldsymbol{d}_{\mathrm{cv}}(\boldsymbol{k})\boldsymbol{E}(\boldsymbol{r},t)\,p^{*}_{\mathrm{vc}}(\boldsymbol{k},\boldsymbol{r}) - \boldsymbol{d}^{*}_{\mathrm{cv}}(\boldsymbol{k})\boldsymbol{E}(\boldsymbol{r},t)\,p_{\mathrm{vc}}(\boldsymbol{k},\boldsymbol{r})] \\ &= -2\,\mathrm{Im}\left[\boldsymbol{d}_{\mathrm{cv}}(\boldsymbol{k})\boldsymbol{E}(\boldsymbol{r},t)\,p^{*}_{\mathrm{vc}}(\boldsymbol{k},\boldsymbol{r})\right] . \end{aligned} \tag{4.19}$$

Here we have already employed $p_{\mathrm{vc}}(\boldsymbol{k},\boldsymbol{r}) = p^{*}_{\mathrm{cv}}(\boldsymbol{k},\boldsymbol{r})$. This relation and that of particle number conservation $f_{\mathrm{c}}(\boldsymbol{k},\boldsymbol{r}) + f_{\mathrm{v}}(\boldsymbol{k},\boldsymbol{r}) = 1$ follow directly from the equations of motion. Owing to our separation of macroscopic and microscopic length scales, which is consistent with the dipole approximation, the macroscopic variable $\boldsymbol{r}$ appears merely as a parameter. We shall see, however, that this dependence becomes crucial for the investigation of a variety of experimental configurations.

The set of equations (4.18) and (4.19) is known as the *optical Bloch equations.* In an appropriate notation, these equations turn out to be equivalent to Bloch's equations describing the dynamics of the magnetization vector in a time-dependent magnetic field [65]. It is often convenient to switch to the *electron and hole picture*, i.e. to replace $f_{\mathrm{v}}(\boldsymbol{k})$ by $1 - f_{\mathrm{h}}(-\boldsymbol{k}) = 1 - f_{\mathrm{h}}(\boldsymbol{k})$ and $E_{\mathrm{v}}(\boldsymbol{k})$ by $-E_{\mathrm{h}}(\boldsymbol{k})$[5] and to replace the index "c" by "e", leading to

$$\begin{aligned}\left(\hbar\frac{\partial}{\partial t} + \mathrm{i}\left[E_{\mathrm{e}}(\boldsymbol{k}) + E_{\mathrm{h}}(\boldsymbol{k})\right] + \hbar\gamma_2\right)p_{\mathrm{vc}}(\boldsymbol{k}) \\ = \mathrm{i}\,\boldsymbol{d}_{\mathrm{cv}}\boldsymbol{E}(\boldsymbol{r},t)(1 - f_{\mathrm{e}}(\boldsymbol{k}) - f_{\mathrm{h}}(\boldsymbol{k}))\end{aligned} \tag{4.20}$$

and

$$\begin{aligned}\hbar\frac{\partial}{\partial t}f_{\mathrm{e,h}}(\boldsymbol{k}) + \hbar\gamma_1^{\mathrm{e,h}}\left(f_{\mathrm{e,h}}(\boldsymbol{k}) - f_{\mathrm{e,h}}^{(0)}(\boldsymbol{k})\right) \\ = -2\,\mathrm{Im}\left[\boldsymbol{d}_{\mathrm{cv}}\boldsymbol{E}(\boldsymbol{r},t)p_{\mathrm{vc}}^{*}(\boldsymbol{k})\right].\end{aligned} \tag{4.21}$$

Finally, we have added the two *phenomenological damping rates* $\gamma_1^{\mathrm{e,h}}$ for the occupation numbers of electrons and holes, respectively, and γ_2 for the transition amplitude. For the sake of simplicity we shall assume that the relaxation rates of electrons and holes are equal, i.e. $\gamma_1^{\mathrm{e}} = \gamma_1^{\mathrm{h}} = \gamma_1$. These damping rates, with the corresponding time constants $T_1 = \gamma_1^{-1}$ and $T_2 = \gamma_2^{-1}$, have great physical importance. T_1 describes the relaxation of an arbitrary initial distribution function to a (quasi-)equilibrium Fermi function $f_{\mathrm{e}}^{(0)}$, and T_2 the damping of the oscillation of the transition amplitude $p_{\mathrm{vc}}(\boldsymbol{k})$ and hence of the macroscopic polarization $\boldsymbol{P}(\boldsymbol{r},t)$. We come back to the microscopic origin of these terms in Chaps. 10 and 11. The term $1 - f_{\mathrm{e}}(\boldsymbol{k}) - f_{\mathrm{h}}(\boldsymbol{k})$ on the right-hand side of (4.20) is often called the *Pauli blocking factor* and is an immediate consequence of the Pauli exlusion principle.

The optical Bloch equations have a rather simple mathematical form. They represent a set of coupled, first-order, *nonlinear* differential equations, with the external electric field $\boldsymbol{E}(\boldsymbol{r},t)$ giving rise to the inhomogeneities. Note, however, that the equations are uncoupled with respect to the wavenumber $\boldsymbol{k}$, which is an immediate consequence of the fact that we have neglected the Coulomb interaction.

---

[5] Here, we have employed the assumption of a symmetric distribution function and a symmetric energy dispersion; further, we shall omit the parametrical dependence on $\boldsymbol{r}$.

4.2 Linear Optical Properties

Let us discuss the optical Bloch equations in detail. In an intrinsic semiconductor at zero temperature, $f_{\rm e}^{(0)}(\boldsymbol{k}) = 0$, and for linear optics, i.e. arbitrarily weak incident light fields, we can neglect the distribution functions of electrons and holes, i.e. $f_{\rm e}(\boldsymbol{k}) \ll 1$ and $f_{\rm h}(\boldsymbol{k}) \ll 1$. To solve (4.18) or (4.20) we merely have to perform a Fourier transformation, which leads to

$$p_{\rm vc}(\boldsymbol{k},\omega) = \frac{\boldsymbol{d}_{\rm cv}\boldsymbol{E}(\boldsymbol{r},\omega)}{E_{\rm e}(\boldsymbol{k}) + E_{\rm h}(\boldsymbol{k}) - \hbar\omega - {\rm i}\hbar\gamma_2} \tag{4.22}$$

and correspondingly, $p_{\rm cv}(\boldsymbol{k},\omega) = [p_{\rm vc}(\boldsymbol{k},-\omega)]^*$ for the nonresonant contribution. For an incident monocromatic field (2.211) with frequency ω_0, the back transformation is trivial[6] and, according to (4.9), the polarization is given by

$$\begin{aligned}\boldsymbol{P}(\boldsymbol{r},t) = \frac{1}{2V}\Bigg(&\sum_{\boldsymbol{k}} \frac{\boldsymbol{d}_{\rm cv}^*\left(\boldsymbol{d}_{\rm cv}\tilde{\boldsymbol{E}}(\boldsymbol{r})\right)}{E_{\rm e}(\boldsymbol{k}) + E_{\rm h}(\boldsymbol{k}) - \hbar\omega_0 - {\rm i}\hbar\gamma_2} \\ &+ \sum_{\boldsymbol{k}} \frac{\boldsymbol{d}_{\rm cv}\left(\boldsymbol{d}_{\rm cv}^*\tilde{\boldsymbol{E}}^*(\boldsymbol{r})\right)}{E_{\rm e}(\boldsymbol{k}) + E_{\rm h}(\boldsymbol{k}) + \hbar\omega_0 + {\rm i}\hbar\gamma_2}\Bigg) {\rm e}^{-{\rm i}\omega_0 t} + {\rm c.c.}\end{aligned} \tag{4.23}$$

For cubic symmetry, all elements of the vector $\boldsymbol{d}_{\rm cv}$ are equal. Correspondingly, the linear susceptibility tensor that we have introduced in Sect. 2.6 becomes diagonal and reduces to a scalar quantity (compare (2.200)). Making use of (2.213) or, alternatively, of (2.201) in the frequency domain, we obtain for the *linear optical susceptibility*

$$\begin{aligned}\chi^{(1)}(\omega) = \frac{1}{\varepsilon_0 V}\sum_{\boldsymbol{k}}\Bigg(&\frac{|d_{\rm cv}|^2}{E_{\rm e}(\boldsymbol{k}) + E_{\rm h}(\boldsymbol{k}) - \hbar\omega - {\rm i}\hbar\gamma_2} \\ &+ \frac{|d_{\rm cv}|^2}{E_{\rm e}(\boldsymbol{k}) + E_{\rm h}(\boldsymbol{k}) + \hbar\omega + {\rm i}\hbar\gamma_2}\Bigg) .\end{aligned} \tag{4.24}$$

Under resonant conditions, i.e. $E_{\rm e}(\boldsymbol{k}) + E_{\rm h}(\boldsymbol{k}) - \hbar\omega \approx 0$, the second term in (4.24), which stems from the $\exp(+{\rm i}\omega t)$ contribution to $\boldsymbol{E}(\boldsymbol{r},t)$, is negligible with respect to the first term.

As discussed in Sect. 2.7, the real part of the linear optical susceptibility is directly related to the refractive index of the medium, and the imaginary part to the absorption coefficient. The real and imaginary parts can be separated by means of the relation

$$\frac{1}{x - {\rm i}\gamma_2} = \frac{x}{x^2+\gamma_2^2} + {\rm i}\frac{\gamma_2}{x^2+\gamma_2^2} . \tag{4.25}$$

In the limit of vanishing dephasing, this reduces to the Dirac identity

[6] As we are considering the linear case, it is obvious that we could assume an arbitrary time-dependence for the incident field.

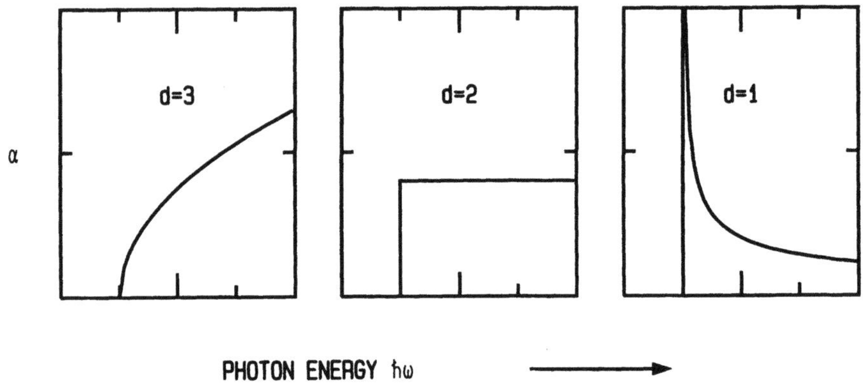

Fig. 4.3. Schematic absorption spectra of d=3, d=2 and d=1–dimensional semiconductors, direct allowed transitions, no Coulomb correlations

$$\lim_{\gamma_2 \to 0} \frac{1}{x - \mathrm{i}\gamma_2} = \mathcal{P}\frac{1}{x} + \mathrm{i}\pi\delta(x) , \tag{4.26}$$

which, as usual, is only meaningful under an integral; $\mathcal{P}$ is the principal value. With (4.24), the *absorption coefficient* becomes

$$\alpha(\omega) = \frac{\omega}{c_0\sqrt{\varepsilon_\mathrm{b}}}\frac{\pi}{\varepsilon_0}\frac{1}{V}\sum_{\boldsymbol{k}} |d_\mathrm{cv}|^2 \delta[E_\mathrm{e}(\boldsymbol{k}) + E_\mathrm{h}(\boldsymbol{k}) - \hbar\omega] . \tag{4.27}$$

This is a result which one might have guessed intuitively. The absorption coefficient is proportional to the density of states at $\hbar\omega$ with respect to the transition energy $E_\mathrm{e}(\boldsymbol{k}) + E_\mathrm{h}(\boldsymbol{k})$, referred to as the *combined density of states.* Within the effective-mass approximation, we obtain $E_\mathrm{e}(\boldsymbol{k}) + E_\mathrm{h}(\boldsymbol{k}) = (\hbar^2\boldsymbol{k}^2)/(2m_\mathrm{r}) + E_\mathrm{g}$, where the reduced effective mass is given by $m_\mathrm{r}^{-1} = m_\mathrm{e}^{-1} + m_\mathrm{h}^{-1}$. Depending on the dimensionality d of the system, we obtain the absorption coefficient via the usual density-of-states formulas (3.138)–(3.140):

$$\alpha(\omega) \propto \omega\,\Theta(\hbar\omega - E_\mathrm{g})\,(\hbar\omega - E_\mathrm{g})^{d/2-1} , \tag{4.28}$$

a result sketched in Fig. 4.3. The prefactor ω can often be considered as a constant, because usually one is only interested in a rather narrow spectral region around the band gap energy E_g. Transitions from two different valence bands that are degenerate at $\boldsymbol{k} = 0$ can be lumped together into a transition from one averaged valence band. In d dimensions the absorption coefficient scales with the reduced mass according to (see (3.138)–(3.140))

$$\alpha \propto (m_\mathrm{r})^{d/2}|d_\mathrm{cv}|^2 . \tag{4.29}$$

If both heavy-hole (hh) and light-hole (lh) transitions (see Sect. 3.5) are accounted for, one has the sum of two contributions

$$\alpha \propto \left((m_{\mathrm{r,\,hh}})^{d/2} |d_{\mathrm{cv,\,hh}}|^2 + (m_{\mathrm{r,\,lh}})^{d/2} |d_{\mathrm{cv,\,lh}}|^2 \right) , \tag{4.30}$$

where $m_{\mathrm{r,\,hh}}^{-1} = m_{\mathrm{e}}^{-1} + m_{\mathrm{hh}}^{-1}$ and $m_{\mathrm{r,\,lh}}^{-1} = m_{\mathrm{e}}^{-1} + m_{\mathrm{lh}}^{-1}$, and the dipole matrix elements for the two transitions are $d_{\mathrm{cv,\,hh}}$ and $d_{\mathrm{cv,\,lh}}$, respectively. This prefactor (4.30) can obviously be expressed by one single, averaged valence band mass. In GaAs, for example, $|d_{\mathrm{cv,\,hh}}|^2 = 3\,|d_{\mathrm{cv,\,lh}}|^2$.

Although an experimental determination of the absorption coefficient is conceptually simple, it is not always trivial in practice. Measuring the transmission coefficient $T(\omega) = \exp[-\alpha(\omega)L]$ of an antireflection-coated piece of semiconductor with thickness L is only easy if $\alpha(\omega)L$ is neither too large nor too small, i.e. $0.1 < \alpha(\omega)L < 10$. Otherwise, alternative techniques such as photocurrent or excitation spectroscopy have to be employed. Figure 4.4 shows a measured absorption spectrum of germanium which roughly resembles the expected square-root-like increase. Close to the band edge, however, a small peak is observed. We come back to the physics of systematic deviations of this kind in Chap. 6.

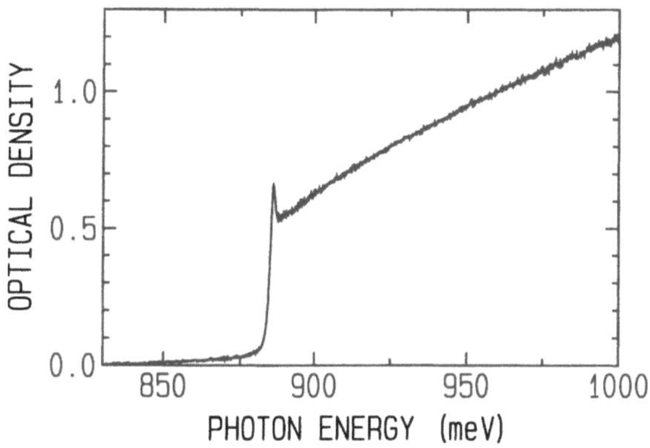

Fig. 4.4. Measured absorption spectrum at the direct gap of bulk germanium at a lattice temperature $T = 1.8\,\mathrm{K}$. The thickness of the platelet smaple, with an antireflection coating on both sides, was $L = 4.5\,\mu\mathrm{m}$. The optical density n_{OD} is related to the absorption coefficient α via $\mathrm{e}^{-\alpha L} = 10^{-n_{OD}}$. Data taken from [340]

In a semiconductor laser, where quasi-equilibrium Fermi functions $f_{\mathrm{e}}^{(0)}$ and $f_{\mathrm{h}}^{(0)}$ are generated via current injection, the optical spectra detected by a weak field $\boldsymbol{E}$ change. Repeating our calculation making the replacement $\boldsymbol{d}_{\mathrm{cv}} \to \boldsymbol{d}_{\mathrm{cv}}\,(1 - f_{\mathrm{e}}^{(0)} - f_{\mathrm{h}}^{(0)})$ (see (4.20)), immediately leads us to the absorption coefficient in d dimensions,

$$\begin{aligned}\alpha(\omega) &= -\,g(\omega) \\ &\propto \omega\, \Theta(\hbar\omega - E_{\mathrm{g}})\, (\hbar\omega - E_{\mathrm{g}})^{d/2-1} \left(1 - f_{\mathrm{e}}^{(0)}(\omega) - f_{\mathrm{h}}^{(0)}(\omega)\right) , \end{aligned} \tag{4.31}$$

where $f_{\mathrm{e,h}}^{(0)}(\omega) = f_{\mathrm{e,h}}^{(0)}[\boldsymbol{k}(\omega)]$ and $\boldsymbol{k}(\omega)$ is obtained from $E_{\mathrm{e}}(\boldsymbol{k})+E_{\mathrm{h}}(\boldsymbol{k})-\hbar\omega = 0$. For large carrier densities, the inversion factor $1 - f_{\mathrm{e}}^{(0)} - f_{\mathrm{h}}^{(0)}$ and, hence, the absorption coefficient become negative and we have amplification due to stimulated emission, i.e. gain. In this case, the *gain coefficient* $g(\omega) = -\alpha(\omega)$ is positive. The behavior of the gain coefficient is exemplified in Fig. 4.5 for $d = 2$.

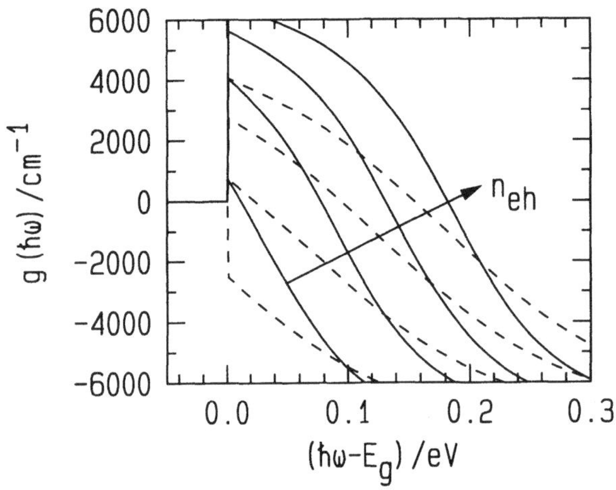

Fig. 4.5. Gain coefficient for a two-dimensional system, for $T = 300\,\mathrm{K}$ (*solid lines*) and $T = 600\,\mathrm{K}$ (*dashed lines*). The carrier density n_{eh} increases from $1 \times 10^{12}\,\mathrm{cm}^{-2}$ (*left*) to $4 \times 10^{12}\,\mathrm{cm}^{-2}$ (*right*) in steps of $10^{12}\,\mathrm{cm}^{-2}$ in each case; parameters GaAs have been used

4.3 Nonlinear Optical Properties

Deviations from the simple limit of linear optics give rise to a rich variety of physical phenomena, such as harmonic generation, photon echoes, and wave mixing, which will be discussed in the following. While it is possible to switch back and forth from the time domain to the frequency domain at any point in linear optics, care has to be taken in nonlinear optics since the coupled optical Bloch equations (4.18), (4.19) are a set of *nonlinear* differential equations.

4.3.1 Perturbation Analysis in the Frequency Domain

We start by discussing phenomena where the envelope of the electric field is constant in time,[7] i.e.

$$\boldsymbol{E}(\boldsymbol{r},t) = \frac{1}{2}\left(\tilde{\boldsymbol{E}}(\boldsymbol{r})\,\mathrm{e}^{-\mathrm{i}\omega t} + \tilde{\boldsymbol{E}}^*(\boldsymbol{r})\,\mathrm{e}^{\mathrm{i}\omega t}\right) , \tag{4.32}$$

as already employed in the case of linear optics. The frequency ω can be considered as a parameter in a spectroscopic experiment. In many situations it is sufficient to consider small deviations from linear optics, allowing a perturbation approach. We make the replacement

$$\boldsymbol{E}(\boldsymbol{r},t) \to \lambda \boldsymbol{E}(\boldsymbol{r},t) \tag{4.33}$$

and expand the transition amplitude

$$p_{\mathrm{vc}} = \lambda p_{\mathrm{vc}}^{(1)} + \lambda^2 p_{\mathrm{vc}}^{(2)} + \lambda^3 p_{\mathrm{vc}}^{(3)} + \ldots \tag{4.34}$$

and the occupation functions

$$f_{\mathrm{h}} = f_{\mathrm{e}} = \lambda f_{\mathrm{e}}^{(1)} + \lambda^2 f_{\mathrm{e}}^{(2)} + \lambda^3 f_{\mathrm{e}}^{(3)} + \ldots \tag{4.35}$$

into a perturbation series with an *ordering parameter* $\lambda \ll 1$. Here we have already chosen $f_{\mathrm{e}}^{(0)} = 0$. Introducing this ansatz into the optical Bloch equations (4.20), (4.21) and ordering all terms according to λ leads to the following:

- Order λ^1:

$$\left(\hbar\frac{\partial}{\partial t} + \mathrm{i}\Big(E_{\mathrm{e}}(\boldsymbol{k}) + E_{\mathrm{h}}(\boldsymbol{k})\Big) + \hbar\gamma_2\right) p_{\mathrm{vc}}^{(1)}(\boldsymbol{k}) = \mathrm{i}\boldsymbol{d}_{\mathrm{cv}}\boldsymbol{E}(\boldsymbol{r},t) , \tag{4.36}$$

$$\left(\hbar\frac{\partial}{\partial t} + \hbar\gamma_1\right) f_{\mathrm{e}}^{(1)}(\boldsymbol{k}) = 0 . \tag{4.37}$$

- Order λ^2:

$$\left(\hbar\frac{\partial}{\partial t} + \mathrm{i}\Big(E_{\mathrm{e}}(\boldsymbol{k}) + E_{\mathrm{h}}(\boldsymbol{k})\Big) + \hbar\gamma_2\right) p_{\mathrm{vc}}^{(2)}(\boldsymbol{k}) = 0 , \tag{4.38}$$

$$\left(\hbar\frac{\partial}{\partial t} + \hbar\gamma_1\right) f_{\mathrm{e}}^{(2)}(\boldsymbol{k}) = -2\,\mathrm{Im}\left(\boldsymbol{d}_{\mathrm{cv}}\boldsymbol{E}(\boldsymbol{r},t) p_{\mathrm{vc}}^{(1)\,*}(\boldsymbol{k})\right) . \tag{4.39}$$

- Order λ^3:

$$\left(\hbar\frac{\partial}{\partial t} + \mathrm{i}\Big(E_{\mathrm{e}}(\boldsymbol{k}) + E_{\mathrm{h}}(\boldsymbol{k})\Big) + \hbar\gamma_2\right) p_{\mathrm{vc}}^{(3)}(\boldsymbol{k}) = -\mathrm{i}2\boldsymbol{d}_{\mathrm{cv}}\boldsymbol{E}(\boldsymbol{r},t) f_{\mathrm{e}}^{(2)}(\boldsymbol{k}) , \tag{4.40}$$

$$\left(\hbar\frac{\partial}{\partial t} + \hbar\gamma_1\right) f_{\mathrm{e}}^{(3)}(\boldsymbol{k}) = 0 . \tag{4.41}$$

[7] We shall see later that the field need not be constant but merely has to vary slowly compared with the relaxation time γ_2^{-1} or the inverse detuning from resonance.

And so forth. It is obvious that all even orders of the transition amplitudes and all odd orders of the occupation functions vanish. Even though this set of equations looks somewhat lengthy, it has a very simple mathematical structure. In particular, the equations are not truly coupled, and hence they can be solved in an iterative scheme: $p_{\mathrm{vc}}^{(1)}(\boldsymbol{k})$ appears merely as an inhomogeneity in the equation for $f_{\mathrm{e}}^{(2)}(\boldsymbol{k})$ which drives $p_{\mathrm{vc}}^{(3)}(\boldsymbol{k})$ and so forth. We now introduce (4.32) into (4.37), (4.39) and (4.41). Obviously, the first order in $p_{\mathrm{vc}}(\boldsymbol{k})$ is identical to that in linear optics. The second order, however, is already quite lengthy:

$$\left(\hbar\frac{\partial}{\partial t}+\hbar\gamma_1\right)f_{\mathrm{e}}^{(2)}(\boldsymbol{k})=\mathrm{i}^2\frac{|\boldsymbol{d}_{\mathrm{cv}}|}{4}$$
$$\times\left\{\frac{|\tilde{\boldsymbol{E}}(\boldsymbol{r})|^2}{E_{\mathrm{e}}(\boldsymbol{k})+E_{\mathrm{h}}(\boldsymbol{k})-\hbar\omega+\mathrm{i}\hbar\gamma_2}+\frac{[\tilde{\boldsymbol{E}}^*(\boldsymbol{r})]^2\mathrm{e}^{2\mathrm{i}\omega t}}{E_{\mathrm{e}}(\boldsymbol{k})+E_{\mathrm{h}}(\boldsymbol{k})-\hbar\omega+\mathrm{i}\hbar\gamma_2}\right.$$
$$+\frac{[\tilde{\boldsymbol{E}}(\boldsymbol{r})]^2\mathrm{e}^{-2\mathrm{i}\omega t}}{E_{\mathrm{e}}(\boldsymbol{k})+E_{\mathrm{h}}(\boldsymbol{k})+\hbar\omega+\mathrm{i}\hbar\gamma_2}+\frac{|\tilde{\boldsymbol{E}}(\boldsymbol{r})|^2}{E_{\mathrm{e}}(\boldsymbol{k})+E_{\mathrm{h}}(\boldsymbol{k})+\hbar\omega+\mathrm{i}\hbar\gamma_2}$$
$$-\frac{|\tilde{\boldsymbol{E}}(\boldsymbol{r})|^2}{E_{\mathrm{e}}(\boldsymbol{k})+E_{\mathrm{h}}(\boldsymbol{k})-\hbar\omega-\mathrm{i}\hbar\gamma_2}+\frac{[\tilde{\boldsymbol{E}}^*(\boldsymbol{r})]^2\mathrm{e}^{2\mathrm{i}\omega t}}{E_{\mathrm{e}}(\boldsymbol{k})+E_{\mathrm{h}}(\boldsymbol{k})+\hbar\omega-\mathrm{i}\hbar\gamma_2}$$
$$\left.+\frac{[\tilde{\boldsymbol{E}}(\boldsymbol{r})]^2\mathrm{e}^{-2\mathrm{i}\omega t}}{E_{\mathrm{e}}(\boldsymbol{k})+E_{\mathrm{h}}(\boldsymbol{k})-\hbar\omega-\mathrm{i}\hbar\gamma_2}+\frac{|\tilde{\boldsymbol{E}}(\boldsymbol{r})|^2}{E_{\mathrm{e}}(\boldsymbol{k})+E_{\mathrm{h}}(\boldsymbol{k})+\hbar\omega-\mathrm{i}\hbar\gamma_2}\right\}. \quad (4.42)$$

Introducing this expression into the third-order equation (4.41) will lead to 16 terms in general. Let us consider rather an example and focus on the contributions to $p_{\mathrm{vc}}^{(3)}(\boldsymbol{k})$ which oscillate as $\exp(-3\mathrm{i}\omega t)$ and $\exp(3\mathrm{i}\omega t)$. Only the terms $\propto\exp(-2\mathrm{i}\omega t)$ and $\propto\exp(2\mathrm{i}\omega t)$ from $f_{\mathrm{e}}^{(2)}(\boldsymbol{k})$ can contribute. For this particular contribution, we find, using (4.9),

$$\boldsymbol{P}_{3\omega}(\boldsymbol{r},t)=\frac{1}{V}\sum_{\boldsymbol{k}}\frac{|d_{\mathrm{cv}}|^4}{4}$$
$$\times\left[\frac{[\tilde{\boldsymbol{E}}(\boldsymbol{r})]^3\mathrm{e}^{-3\mathrm{i}\omega t}}{[E_{\mathrm{e}}(\boldsymbol{k})+E_{\mathrm{h}}(\boldsymbol{k})-3\hbar\omega-\mathrm{i}\hbar\gamma_2]\,(2\hbar\omega+\mathrm{i}\hbar\gamma_1)}\right.$$
$$\times\left(\frac{1}{E_{\mathrm{e}}(\boldsymbol{k})+E_{\mathrm{h}}(\boldsymbol{k}+\hbar\omega)+\mathrm{i}\hbar\gamma_2}-\frac{1}{E_{\mathrm{e}}(\boldsymbol{k})+E_{\mathrm{h}}(\boldsymbol{k})-\hbar\omega-\mathrm{i}\hbar\gamma_2}\right)$$
$$-\frac{[\tilde{\boldsymbol{E}}^*(\boldsymbol{r})]^3\mathrm{e}^{3\mathrm{i}\omega t}}{[E_{\mathrm{e}}(\boldsymbol{k})+E_{\mathrm{h}}(\boldsymbol{k})+3\hbar\omega-\mathrm{i}\hbar\gamma_2]\,(2\hbar\omega-\mathrm{i}\hbar\gamma_1)}$$
$$\left.\times\left(\frac{1}{E_{\mathrm{e}}(\boldsymbol{k})+E_{\mathrm{h}}(\boldsymbol{k})-\hbar\omega+\mathrm{i}\hbar\gamma_2}-\frac{1}{E_{\mathrm{e}}(\boldsymbol{k})+E_{\mathrm{h}}(\boldsymbol{k})+\hbar\omega-\mathrm{i}\hbar\gamma_2}\right)\right]$$
$$+\mathrm{c.c.} \quad (4.43)$$

The contribution $\propto \exp(-3\mathrm{i}\omega t)$, the third harmonic, exhibits a resonant behavior for $E_\mathrm{e}(\boldsymbol{k}) + E_\mathrm{h}(\boldsymbol{k}) - 3\hbar\omega \approx 0$, while the contributions $\propto \exp(3\mathrm{i}\omega t)$ are very small under these conditions. Three photons of frequency ω generate one photon of frequency 3ω, resonant with interband transitions. Similarly fifth, seventh and higher harmonics can be generated for larger incident fields. As mentioned above, all even orders vanish.[8] In order to avoid having to deal with the usually rather lengthy expressions for the polarization, the *nonlinear optical susceptibilities* are introduced. We can write

$$\begin{aligned}
\boldsymbol{P}(\boldsymbol{r},t) &= \varepsilon_0\chi^{(1)}(\omega)\frac{\tilde{\boldsymbol{E}}(\boldsymbol{r})}{2}\,\mathrm{e}^{-\mathrm{i}\omega t} + \text{c.c.}\\
&+ \varepsilon_0\chi^{(3)}(3\omega;\omega,\omega,\omega)\left(\frac{\tilde{\boldsymbol{E}}(\boldsymbol{r})}{2}\right)^3 \mathrm{e}^{-3\mathrm{i}\omega t} + \text{c.c.}\\
&+ \varepsilon_0\chi^{(3)}(\omega;\omega,\omega,-\omega)\left(\frac{\tilde{\boldsymbol{E}}(\boldsymbol{r})|\tilde{\boldsymbol{E}}(\boldsymbol{r})|^2}{2^3}\right) \mathrm{e}^{-\mathrm{i}\omega t} + \text{c.c.}\\
&+ \varepsilon_0\chi^{(5)}(5\omega;\omega,\omega,\omega,\omega,\omega)\left(\frac{\tilde{\boldsymbol{E}}(\boldsymbol{r})}{2}\right)^5 \mathrm{e}^{-5\mathrm{i}\omega t} + \text{c.c.}\\
&+ \dots\,.
\end{aligned} \tag{4.44}$$

The first line is identical to linear optics, the second line abbreviates the 3ω contributions discussed in the above example, the third line contains the contributions of third order with frequency ω, and so forth. The notation $\chi^{(3)}(3\omega;\omega,\omega,\omega)$ anticipates the more general case of several incident fields with different frequencies ω_j in which case all combinations $\chi^{(3)}(\omega_1+\omega_2+\omega_3;\omega_1,\omega_2,\omega_3)$ have to be accounted for. Many aspects of nonlinear optics could be discussed at this point, as, for example, the interplay between the nonlinear polarization and the wave equation which leads to the *phase-matching* condition. We refer the reader to the extensive and excellent literature on this subject.

4.3.2 Introducing the Bloch Vector

Let us now rewrite the optical Bloch equations as an equation of motion for the *Bloch vector*, which is well known from the Bloch equations in nuclear magnetic resonance [66, 67], and discuss the case of negligible damping ($\gamma_1 = \gamma_2 = 0$).

The real and imaginary parts of $p_\mathrm{vc}(\boldsymbol{k})$, together with $f_\mathrm{e}(\boldsymbol{k}) = f_\mathrm{h}(\boldsymbol{k})$, form a set of three equations of motion for each individual $\boldsymbol{k}$ state. We have

[8] This can also be understood by considering that the optical Bloch equations describe a system with inversion symmetry.

already been faced twice, in (4.24) and in the previous section, with the fact that excitation with a field of the form

$$\boldsymbol{E}(\boldsymbol{r},t) = \frac{1}{2}\left(\tilde{\boldsymbol{E}}(\boldsymbol{r},t)\,\mathrm{e}^{-\mathrm{i}\omega t} + \tilde{\boldsymbol{E}}^*(\boldsymbol{r},t)\,\mathrm{e}^{\mathrm{i}\omega t}\right) \tag{4.45}$$

leads to resonant contributions (originating from the $\mathrm{e}^{-\mathrm{i}\omega t}$ term) and nonresonant contributions. $\tilde{\boldsymbol{E}}(\boldsymbol{r},t)$ is a function that varies slowly in time with respect to the light frequency ω. In order to get rid of the nonresonant terms right from the start we use the ansatz

$$p_{\mathrm{vc}}(\boldsymbol{k}) = \tilde{p}_{\mathrm{vc}}(\boldsymbol{k})\mathrm{e}^{-\mathrm{i}\omega t}, \tag{4.46}$$

account only for resonant terms in (4.20) and (4.21) (with $f_{\mathrm{e}}^{(0)} = 0$), and arrive at

$$\left(\hbar\frac{\partial}{\partial t} + \mathrm{i}\hbar\Delta + \hbar\gamma_2\right)\tilde{p}_{\mathrm{vc}}(\boldsymbol{k}) = \frac{\mathrm{i}}{2}\boldsymbol{d}_{\mathrm{cv}}\tilde{\boldsymbol{E}}(\boldsymbol{r},t)[1 - f_{\mathrm{e}}(\boldsymbol{k}) - f_{\mathrm{h}}(\boldsymbol{k})]\,, \tag{4.47}$$

$$\left(\hbar\frac{\partial}{\partial t} + \hbar\gamma_1\right)f_{\mathrm{e}}(\boldsymbol{k}) = -\operatorname{Im}\left(\boldsymbol{d}_{\mathrm{cv}}\tilde{\boldsymbol{E}}(\boldsymbol{r},t)\tilde{p}^*_{\mathrm{vc}}(\boldsymbol{k})\right)\,, \tag{4.48}$$

where the detuning Δ is given by

$$\hbar\Delta(\boldsymbol{k}) = E_{\mathrm{e}}(\boldsymbol{k}) + E_{\mathrm{h}}(\boldsymbol{k}) - \hbar\omega\,. \tag{4.49}$$

This procedure is called the *rotating-wave approximation* (RWA) and is well justified for typical resonant experimental conditions. For nonresonant phenomena such as two-photon absorption or third-harmonic generation, however, it is clearly not appropriate.

Using the definitions of the three components of the *Bloch vector* u, v, w

$$\begin{aligned} u(\boldsymbol{k}) &= 2\,\mathrm{Re}[\tilde{p}_{\mathrm{vc}}(\boldsymbol{k})]\,,\\ v(\boldsymbol{k}) &= 2\,\mathrm{Im}[\tilde{p}_{\mathrm{vc}}(\boldsymbol{k})]\,,\\ w(\boldsymbol{k}) &= -[1 - f_{\mathrm{e}}(\boldsymbol{k}) - f_{\mathrm{h}}(\boldsymbol{k})] \end{aligned} \tag{4.50}$$

and neglecting any damping ($\gamma_1 = \gamma_2 = 0$), we derive the equation of motion

$$\begin{aligned} \dot{u}(\boldsymbol{k}) &= \Delta(\boldsymbol{k})v\,,\\ \dot{v}(\boldsymbol{k}) &= -\Delta(\boldsymbol{k})u - \Omega_{\mathrm{R}}w\,,\\ \dot{w}(\boldsymbol{k}) &= \Omega_{\mathrm{R}}v\,, \end{aligned} \tag{4.51}$$

where the *Rabi frequency* Ω_{R} is equal to $\boldsymbol{d}_{\mathrm{cv}}\tilde{\boldsymbol{E}}(\boldsymbol{r},t)/\hbar$. The Bloch vector has an immediate physical meaning in nuclear magnetic resonance, where it represents the three spatial components of the magnetization vector [65]. The length $\sqrt{u^2+v^2+w^2}$ of the Bloch vector is conserved, as can be seen from considering the time derivative

$$\frac{\partial}{\partial t}\left(u^2+v^2+w^2\right)=2u(\Delta v)+2v(-\Delta u-\Omega_{\mathrm{R}}w)+2w(\Omega_{\mathrm{R}}v)=0\,. \quad (4.52)$$

For a constant Rabi frequency, it is easy to map the Bloch equations onto a set of harmonic oscillators. Taking the time derivative of the second Bloch equation and inserting the two other equations, we have

$$\ddot{v}(\boldsymbol{k})+\left([\Delta(\boldsymbol{k})]^2+\Omega_{\mathrm{R}}^2\right)v(\boldsymbol{k})=0\,. \quad (4.53)$$

The solution can be written by means of a rotation matrix

$$\hat{T}=\begin{pmatrix} \dfrac{\Omega_{\mathrm{R}}^2+\Delta^2\cos(\bar{\Omega}_{\mathrm{R}}t)}{\bar{\Omega}_{\mathrm{R}}^2} & \dfrac{\Delta}{\bar{\Omega}_{\mathrm{R}}}\sin(\bar{\Omega}_{\mathrm{R}}t) & \dfrac{\Delta\Omega_{\mathrm{R}}}{\bar{\Omega}_{\mathrm{R}}^2}[\cos(\bar{\Omega}_{\mathrm{R}}t)-1] \\ -\dfrac{\Delta}{\bar{\Omega}_{\mathrm{R}}}\sin(\bar{\Omega}_{\mathrm{R}}t) & \cos(\bar{\Omega}_{\mathrm{R}}t) & -\dfrac{\Omega_{\mathrm{R}}}{\bar{\Omega}_{\mathrm{R}}}\sin(\bar{\Omega}_{\mathrm{R}}t) \\ \dfrac{\Delta\Omega_{\mathrm{R}}}{\bar{\Omega}_{\mathrm{R}}^2}[\cos(\bar{\Omega}_{\mathrm{R}}t)-1] & \dfrac{\Omega_{\mathrm{R}}}{\bar{\Omega}_{\mathrm{R}}}\sin(\bar{\Omega}_{\mathrm{R}}t) & \dfrac{\Delta^2+\Omega_{\mathrm{R}}^2\cos(\bar{\Omega}_{\mathrm{R}}t)}{\bar{\Omega}_{\mathrm{R}}^2} \end{pmatrix}, \quad (4.54)$$

acting on an arbitrary initial Bloch vector (u_0, v_0, w_0) at any time t_0. $\bar{\Omega}_{\mathrm{R}}$ has the meaning of a generalized Rabi frequency and is defined as

$$\bar{\Omega}_{\mathrm{R}}=\sqrt{\Omega_{\mathrm{R}}^2+\Delta^2}\,. \quad (4.55)$$

Let us consider two limiting cases. If the Rabi frequency is small compared with the detuning ($\Omega_{\mathrm{R}} \ll \Delta$, and hence $\Omega_{\mathrm{R}} \approx 0$), we obtain

$$\begin{pmatrix} u \\ v \\ w \end{pmatrix}=\begin{pmatrix} \cos(\Delta t) & \sin(\Delta t) & 0 \\ -\sin(\Delta t) & \cos(\Delta t) & 0 \\ 0 & 0 & 1 \end{pmatrix}\begin{pmatrix} u_0 \\ v_0 \\ w_0 \end{pmatrix}, \quad (4.56)$$

which can be visualized as a rotation of the Bloch vector in the uv plane (Fig. 4.6). If the Rabi frequency is constant and large compared with the detuning ($\Omega_{\mathrm{R}} \gg \Delta$, hence $\Delta \approx 0$), we obtain

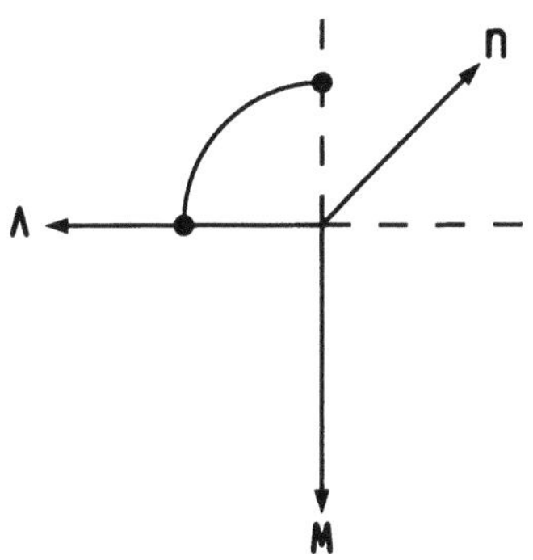

Fig. 4.6. Bloch vector rotating in the uv plane

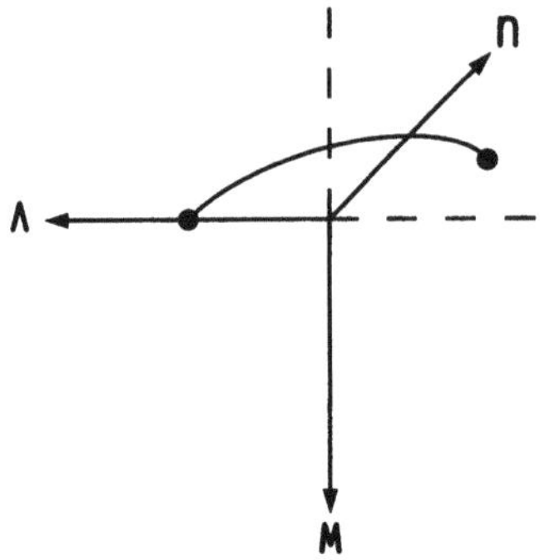

Fig. 4.7. Bloch vector rotating in the vw plane. Here a $\pi/2$ pulse has been applied to the ground state

$$\begin{pmatrix} u \\ v \\ w \end{pmatrix} = \begin{pmatrix} 1 & 0 & 0 \\ 0 & \cos(\Omega_R t) & -\sin(\Omega_R t) \\ 0 & \sin(\Omega_R t) & \cos(\Omega_R t) \end{pmatrix} \begin{pmatrix} u_0 \\ v_0 \\ w_0 \end{pmatrix} . \tag{4.57}$$

This solution describes a rotation of the Bloch vector in the vw plane, as illustrated in Fig. 4.7. After a period of time corresponding to $\Omega_R t = 2\pi$ the initial situation is restored. For a temporally varying Rabi frequencies and vanishing detuning, the solution has the same form as (4.56), but with the replacement

$$\Omega_R t \to \Theta_R(t) = \int_{-\infty}^{t} dt' \Omega_R(t') . \tag{4.58}$$

In the limit $t \to \infty$, Θ_R defines the pulse area, which is nothing but the angle of rotation of the Bloch vector in the vw plane. A pulse with area Θ_R is called a Θ_R pulse.

For general impulsive excitation conditions, the solutions discussed thus far are not valid. In particular, if an unbounded continuum of states is excited by an ultrashort pulse, the Rabi frequency varies rapidly in time and as the detuning is determined by the spectral width of the pulse, Ω_R/Δ is of the order of the pulse area. In such cases we have to rely on numerical solution. An example, which can be analyzed analytically with some simplifying assumptions, is the *photon echo*.

Example (Photon Echo). A system in its ground state ($f_e(\boldsymbol{k}) = f_h(\boldsymbol{k}) = 0 \Rightarrow w(0) = w_0 = -1$) is excited with a short $\pi/2$ pulse at $t = 0$; the pulse is defined by the Rabi frequency $\Omega_R = \pi/(2\tau)[\theta(t+\tau/2) - \theta(t-\tau/2)]$, where τ denotes the pulse duration. For the excitation, we assume that the detuning can be neglected in comparison with the Rabi frequency and that the Bloch vector rotates according to (4.57). Owing to this simplification, the information about the dependence of the occupation on the detuning is lost.[9] According to (4.56), states with $\Delta(\boldsymbol{k}) \gg \pi/(2\tau)$ (because $\Omega_R = \pi/(2\tau)$) are

[9] Strictly speaking, the condition $\Omega_R \gg \Delta(\boldsymbol{k})$ is not fulfilled, as the range of contributing states is of order $1/\tau$.

not excited. This corresponds to the general property, resulting from (4.51), that the occupation is determined by the spectral width of the pulse. Thus, we can repair our approximation by limiting the wave vector summation in the final result.

After the excitation with the $\pi/2$ pulse followed by a free evolution for a time delay T, the system is excited with a short π pulse and again evolves undisturbed for another period $t-T$. Multiplying the corresponding matrices is straightforward and leads to

$$\begin{aligned}\begin{pmatrix} u \\ v \\ w \end{pmatrix} &= \Theta(T)\Theta(t-T)\begin{pmatrix} \cos(\Delta\tau) & +\sin(\Delta\tau) & 0 \\ -\sin(\Delta\tau) & \cos(\Delta\tau) & 0 \\ 0 & 0 & 1 \end{pmatrix}\begin{pmatrix} 1 & 0 & 0 \\ 0 & -1 & 0 \\ 0 & 0 & -1 \end{pmatrix} \\ &\times \begin{pmatrix} \cos(\Delta T) & \sin(\Delta T) & 0 \\ -\sin(\Delta T) & \cos(\Delta T) & 0 \\ 0 & 0 & 1 \end{pmatrix}\begin{pmatrix} 1 & 0 & 0 \\ 0 & 0 & -1 \\ 0 & 1 & 0 \end{pmatrix}\begin{pmatrix} 0 \\ 0 \\ -1 \end{pmatrix} \\ &= \Theta(T)\Theta(t-T)\begin{pmatrix} -\sin[\Delta(t-2T)] \\ -\cos[\Delta(t-2T)] \\ 0 \end{pmatrix} ,\end{aligned} \tag{4.59}$$

where we have simplified the final result using the rules for trigonometrical functions. For the transition amplitude in the rotating-wave approximation, we find

$$\tilde{p}_{\mathrm{vc}}(\boldsymbol{k}) = \frac{1}{2}(u + \mathrm{i}v) = -\frac{\mathrm{i}}{2}\,\Theta(T)\Theta(t-T)\,\mathrm{e}^{-\mathrm{i}\Delta(\boldsymbol{k})(t-2T)} , \tag{4.60}$$

and hence, using (4.46), (4.49),

$$\begin{aligned} p_{\mathrm{vc}}(\boldsymbol{k}) &= \tilde{p}_{\mathrm{vc}}(\boldsymbol{k})\,\mathrm{e}^{-\mathrm{i}\omega t} \\ &= -\frac{\mathrm{i}}{2}\,\Theta(T)\Theta(t-T)\,\mathrm{e}^{-\mathrm{i}[E_{\mathrm{e}}(\boldsymbol{k})+E_{\mathrm{h}}(\boldsymbol{k})]\,(t-2T)/\hbar}\,\mathrm{e}^{-2\,\mathrm{i}\omega T} . \end{aligned} \tag{4.61}$$

Evaluating the macroscopic polarization (4.9), we map the $\boldsymbol{k}$ integration onto a density-of-states integration and obtain

$$\boldsymbol{P}(\boldsymbol{r},t) = \boldsymbol{d}_{\mathrm{cv}}^{*}\int_0^\infty D_{\mathrm{cv}}(E)p_{\mathrm{vc}}(E)\,\mathrm{d}E + \mathrm{c.c.} \tag{4.62}$$

The final integration with the joint density of states $D_{\mathrm{cv}}(E)$ is performed easily only for a two-dimensional system (similarly to (3.139)), where

$$D_{\mathrm{cv}}(E) = D_0\Theta(E - E_{\mathrm{g}}) . \tag{4.63}$$

Our application of (4.57) must now be repaired by introducing a *cutoff energy* E_{co} which corresponds to the spectral width of the pulse. Then we can rewrite the density of states as

$$D_{\mathrm{cv}}(E) = D_0\Theta(E - E_{\mathrm{g}})\,\Theta(E_{\mathrm{co}} - E) . \tag{4.64}$$

Evaluating the integral over the exponentials is simple, and we arrive at the final result (with $\boldsymbol{d}_{\mathrm{cv}}$ real):

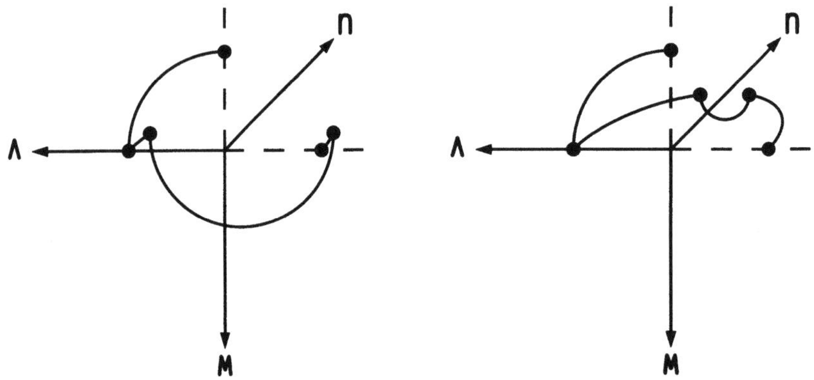

Fig. 4.8. Two individual fates of Bloch vectors with different detunings Δ. Starting from the ground state of the system (u, v, w)=(0, 0, -1), both eventually arrive at the same final point (0, -1, 0) (at $t = 2T$) after the $\pi/2$–π pulse sequence

$$\begin{aligned} \boldsymbol{P}(t) &= -2\frac{\boldsymbol{d}_{\mathrm{cv}}}{V}\hbar D_0\,\Theta(T)\Theta(t-T) \\ &\times \left(\sin[\Omega t - 2(\Omega-\omega)T]\,\frac{\sin[\delta\,\Omega(t-2T)]}{t-2T}\right), \end{aligned} \tag{4.65}$$

where we have introduced the abbreviations $\hbar\delta\Omega = (E_{\mathrm{co}} - E_{\mathrm{g}})/2$ and $\hbar\Omega = (E_{\mathrm{co}} + E_{\mathrm{g}})/2$. The polarization exhibits a peak at $t = 2T$, i.e. an *echo-like behavior* that becomes very sharp for large $\delta\Omega$ and approaches a δ function for $E_{\mathrm{co}} \gg E_{\mathrm{g}}$. The scenario is visualized in Figs. 4.8 and 4.9. This result is perfectly equivalent to the usual spin echo in nuclear magnetic resonance. The physics can be understood very intuitively in the Bloch picture. The $\pi/2$ pulse excites the Bloch vectors of all $\boldsymbol{k}$ states into the uv plane, where they rotate with their individual detuning frequencies $\Delta(\boldsymbol{k})$. After a brief period in time they run out of phase and the macroscopic polarization, which is the sum over all the Bloch vectors, becomes zero. This interference effect is called the *free polarization decay.* The π pulse after a period T rotates all individual Bloch vectors within the vw plane such that after another period T they all arrive at the same point. This is in fact only possible because the initial free polarization decay corresponds to a perfect *reversible decay* of the polarization. Different densities of states do not change the general picture, only the detailed shape of the echo, which is nothing but the Fourier transform of the density of states, is modified. Small deviations from the $\pi/2$–π pulse sequence are not critical. For large deviations, however, the strength of the echo gradually approaches zero. We leave it to the reader to repeat the above calculation for arbitrary pulse areas.

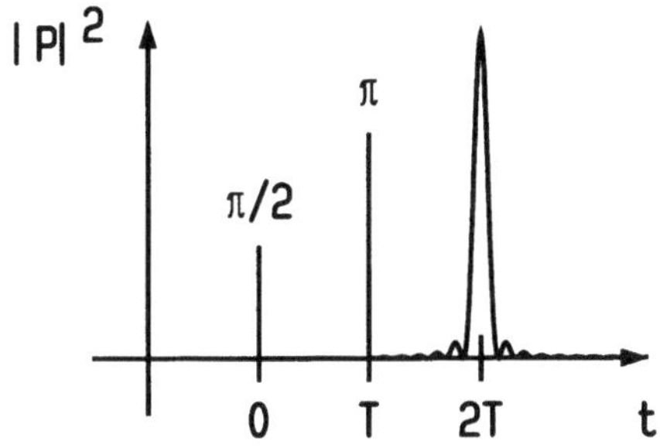

Fig. 4.9. Schematic illustrationof the photon echo obtained after excitation of the optical Bloch equations with a $\pi/2$–π pulse sequence using δ-shaped optical pulses. Only the slowly varying part is shown

In the language of atomic systems, the summation over all $\boldsymbol{k}$ states (4.9) is referred to as an *inhomogeneous broadening.* It should be clear, however, that the $\boldsymbol{k}$ summation is an intrinsic property of ideal semiconductors. Additional inhomogeneous broadening can occur because of local fluctuations of material parameters, for example, due to disorder, an effect which would be analogous to Doppler broadening of a line in atomic physics.

4.3.3 Perturbation Analysis in the Time Domain

In a number of experimental techniques, one uses intensities that are not even close to $\pi/2$ or π pulses (which correspond to a very large excitation intensity for a typical semiconductor) but, rather, one is interested in the first nonlinear corrections to linear optics.

Figure 4.10 shows two experimental techniques frequently employed to investigate relaxation processes, described by the two time constants T_1 and T_2 within our level of description. The idea of both techniques is very simple. In a *pump–probe* experiment the semiconductor is perturbed with a short, rather intense pump pulse which introduces carriers and hence changes the optical properties corresponding to the coefficient $1 - f_{\mathrm{e}}(\boldsymbol{k}) - f_{\mathrm{h}}(\boldsymbol{k})$ in the optical Bloch equations (4.20) and (4.21). The probe beam, which is detected by a slow detector and is much weaker than the pump pulse, probes the changes in the optical properties, which decay on a scale of T_1 as a function of time delay because the occupation functions approach quasi-equilibrium with a time constant T_1 (see (4.47)). Hence, such experiments deliver information on the *energy relaxation.*

Somewhat more subtly, a *four-wave-mixing* (FWM) experiment senses the *phase relaxation* (frequently referred to as dephasing), or, more precisely, the decay of the polarization. After impulsive excitation with the first pulse, the polarization in the medium oscillates coherently for a time comparable to T_2 (4.48) and is still able to interfere with the second, time-delayed pulse even for vanishing temporal overlap of the two pulses. In nonlinear optics, the resulting interference pattern, a grating, is able to diffract either of the beams similarly to diffraction by a physical grating in the usual linear optics.

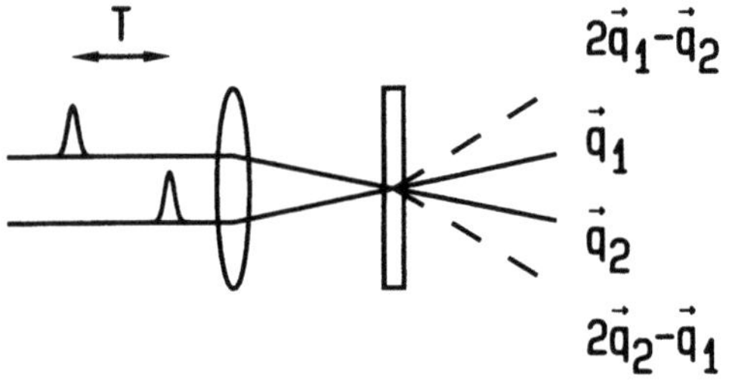

Fig. 4.10. Experimental arrangement for both four-wave-mixing and pump–probe experiments. Two pulses are delayed in time by T and focused onto a semiconductor structure. In the pump–probe geometry one detects one of the transmitted beams, and in the four-wave-mixing geometry one of the lowest-order diffracted beams (*dashed lines*)

The resulting diffracted signal hits the detector and is recorded as a function of time delay.

Small modifications of these two concepts form the basis of many experiments investigating electronic dynamics in semiconductors. The time delay is conveniently realized by physical movement of a mirror; for example, a time delay of 100 fs corresponds to a physical motion of about 33 μm, which is easily implemented mechanically.

Let us analyze these two experimental arrangements in more detail using the optical Bloch equations. First, we use the fact that all pulses involved are usually weak ($\Theta_{\mathrm{R}} \ll \pi$), which allows a perturbation approach in orders of the incident electric field. We can refer to our previous analysis, which has led us to (4.37)–(4.41). Now, however, we apply the RWA, which requires the replacement $\boldsymbol{E}(\boldsymbol{r},t) \to \tilde{\boldsymbol{E}}(\boldsymbol{r},t)\,\mathrm{e}^{-\mathrm{i}\omega t}/2$.

Coming back to the experimental arrangements, we are clearly interested in very short pulses. We employ the limit of δ-shaped pulses[10]

$$\tilde{\boldsymbol{E}}(\boldsymbol{r},t) = 2\Big(\tilde{\boldsymbol{E}}_1 \mathrm{e}^{\mathrm{i}\boldsymbol{q}_1\boldsymbol{r}}\,\delta(t) + \tilde{\boldsymbol{E}}_2 \mathrm{e}^{\mathrm{i}\boldsymbol{q}_2\boldsymbol{r}}\,\delta(t-T)\Big)\,, \tag{4.66}$$

where the index 1 denotes the pump pulse with (optical) wave vector $\boldsymbol{q}_1$, and 2 the probe pulse, with wave vector $\boldsymbol{q}_2$, arriving at the semiconductor after time delay T;[11] $\tilde{\boldsymbol{E}}_1$ and $\tilde{\boldsymbol{E}}_2$ are chosen to be real. The above scheme can now be integrated analytically. For the pump–probe arrangement, one is interested in contributions to $\tilde{p}_{\mathrm{vc}}^{(3)}$ with a phase factor $\mathrm{e}^{\mathrm{i}\boldsymbol{q}_2\boldsymbol{r}}$, i.e. in the direction of the probe beam; for the FWM arrangement, one is interested in contributions with phase factors $\mathrm{e}^{\mathrm{i}(2\boldsymbol{q}_2-\boldsymbol{q}_1)\boldsymbol{r}}$ or $\mathrm{e}^{\mathrm{i}(2\boldsymbol{q}_1-\boldsymbol{q}_2)\boldsymbol{r}}$, corresponding to the diffracted beams.

Working out these contributions is rather simple, but quite cumbersome. For the FWM experiment, we obtain the following for the transition amplitude in the RWA in the direction $2\boldsymbol{q}_2 - \boldsymbol{q}_1$:

[10] Note that in the nonperturbational range of intensities this limit is pathological. As can be shown from the integration of (4.51), the oscillatory behavior of the Bloch vector vanishes and inversion is only achieved asymptotically for arbitrarily high intensities.

[11] Note that because of the δ functions, $\tilde{\boldsymbol{E}}_1$ and $\tilde{\boldsymbol{E}}_2$ are in units of s V/m.

$$\tilde{p}_{\mathrm{vc}}^{(3)} = -\mathrm{i}d_{\mathrm{cv}}|d_{\mathrm{cv}}|^2\hbar^{-3}\tilde{\boldsymbol{E}}_2^2\tilde{\boldsymbol{E}}_1\mathrm{e}^{\mathrm{i}(2\boldsymbol{q}_2-\boldsymbol{q}_1)\boldsymbol{r}}\Theta(T)\Theta(t-T)\,\mathrm{e}^{-\gamma_2 t}\mathrm{e}^{-\mathrm{i}\Delta(\boldsymbol{k})(t-2T)}\,. \tag{4.67}$$

Hence the macroscopic polarization (4.9), to lowest order, in direction the $2\boldsymbol{q}_2 - \boldsymbol{q}_1$, is given by

$$\boldsymbol{P}(\boldsymbol{r},t) = \Bigg(-\mathrm{i}|d_{\mathrm{cv}}|^4\hbar^{-3}\tilde{\boldsymbol{E}}_2^2\tilde{\boldsymbol{E}}_1\mathrm{e}^{\mathrm{i}(2\boldsymbol{q}_2-\boldsymbol{q}_1)\boldsymbol{r}}\Theta(T)\Theta(t-T)\mathrm{e}^{-\gamma_2 t} \times \frac{1}{V}\sum_{\boldsymbol{k}} \mathrm{e}^{-\mathrm{i}[E_{\mathrm{e}}(\boldsymbol{k})+E_{\mathrm{h}}(\boldsymbol{k})]\,(t-2T)/\hbar}\mathrm{e}^{-2\,\mathrm{i}\omega T}\Bigg) + \mathrm{c.c.} \tag{4.68}$$

Again (see the previous section) the sum over $\boldsymbol{k}$ can be rewritten using the density of states. For the case of a two-dimensional system (see (4.63) and (4.64) and the corresponding definitions of Ω and $\delta\Omega$), we obtain

$$\boldsymbol{P}(\boldsymbol{r},t) = \frac{4}{V}D_0|d_{\mathrm{cv}}|^4\hbar^{-2}\tilde{\boldsymbol{E}}_2^2\tilde{\boldsymbol{E}}_1\Theta(T)\Theta(t-T) \times \Bigg(\sin[(2\boldsymbol{q}_2-\boldsymbol{q}_1)\boldsymbol{r} - \Omega(t-2T) - 2\omega T] \times \frac{\sin[\delta\Omega(t-2T)]}{t-2T}\Bigg)\mathrm{e}^{-\gamma_2 t}\,, \tag{4.69}$$

which resembles the photon echo discussed in Sect. 4.3.2. It should be noted, however, that an echo is obtained for arbitrarily weak incident pulses; this is quite distinct from the photon echo discussed in Sect. 4.3.2, where the two pulses were propagating in the same direction. In fact, detecting the field in only the direction $2\boldsymbol{q}_2 - \boldsymbol{q}_1$ serves as a filter that selectively detects the small amount of polarization that is recovered. The decay of the echo amplitude is given by T_2, hence allowing experimental investigation of phase relaxation. If only the energy of the echo $\propto \int \mathrm{d}t\,|\boldsymbol{P}|^2$ is detected, one observes a decay $\propto \{\exp[-(t=2T)/T_2]\}^2 = \exp(-4T/T_2)$. Figure 4.11 exhibits the result of an FWM experiment using pulses of 6 fs duration, exciting GaAs at about 0.6 eV above the direct gap [68]. The measured decay times have to be multiplied by a factor of four in order to obtain the dephasing time T_2. Depending on the carrier density, dephasing times in the range of some 10 fs are observed which are entirely governed by *carrier–carrier scattering*. Here our basic assumption of negligible influence of the Coulomb interaction fails. The microscopic physics of carrier–carrier scattering, which will be discussed in Chaps. 10 and 12, depends crucially on a variety of parameters and can be simulated only insufficiently by phenomenological dephasing rates.

For the pump–probe experiment, we obtain the following up to third order for weak probe pulses ($|\tilde{\boldsymbol{E}}_2| \ll |\tilde{\boldsymbol{E}}_1|$) in the direction of the probe pulse:

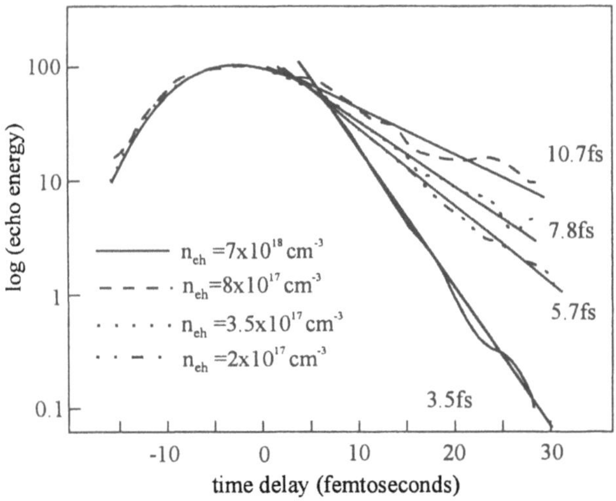

Fig. 4.11. Result of a transient four-wave-mixing experiment exciting the continuum of bulk GaAs at $T = 300\text{K}$ with optical pulses of 6 fs duration. Taken from [68]

$$\begin{aligned}\tilde{p}_{\text{vc}} &= \tilde{p}_{\text{vc}}^{(1)} + \tilde{p}_{\text{vc}}^{(3)} \\ &= \mathrm{i}\boldsymbol{d}_{\text{cv}}\hbar^{-1}\tilde{\boldsymbol{E}}_2\mathrm{e}^{\mathrm{i}\boldsymbol{q}_2\boldsymbol{r}}\Theta(t-T)\mathrm{e}^{-(\mathrm{i}\Delta+\gamma_2)(t-T)} \\ &\quad - 2\mathrm{i}d_{\text{cv}}|d_{\text{cv}}|^2\hbar^{-3}\tilde{\boldsymbol{E}}_2\tilde{\boldsymbol{E}}_1^2\mathrm{e}^{\mathrm{i}\boldsymbol{q}_2\boldsymbol{r}}\Theta(T)\Theta(t-T)\mathrm{e}^{-(\mathrm{i}\Delta+\gamma_2)(t-T)}\mathrm{e}^{-\gamma_1 T} \\ &\quad - \mathrm{i}d_{\text{cv}}|d_{\text{cv}}|^2\hbar^{-3}\tilde{\boldsymbol{E}}_2\tilde{\boldsymbol{E}}_1^2\mathrm{e}^{\mathrm{i}\boldsymbol{q}_2\boldsymbol{r}}\Theta(-T)\Theta(t)\mathrm{e}^{-(\mathrm{i}\Delta+\gamma_2)(t-T)} . \end{aligned} \tag{4.70}$$

This contains one term decaying with the energy relaxation time T_1, as expected intuitively, but also another one solely related to T_2. The latter contribution is similar to the one occuring in the discussion of the FWM experiment, except that here the zeroth-order diffracted beam, in the direction $(\boldsymbol{q}_2 - \boldsymbol{q}_1) + \boldsymbol{q}_1$, enters. In some of the early literature this contribution is referred to as the *coherent artefact*. The corresponding polarization (again using the cutoff energy and omitting the coherent contribution) becomes

$$\begin{aligned}\boldsymbol{P}(\boldsymbol{r},t) &= \frac{4}{V}D_0|d_{\text{cv}}|^2\tilde{\boldsymbol{E}}_2\Theta(t-T) \\ &\quad \times \left(\sin[\boldsymbol{q}_2\boldsymbol{r} - \Omega(t-T) - \omega T]\frac{\sin[\delta\Omega(t-T)]}{t-T}\right)\mathrm{e}^{-\gamma_2(t-T)} \\ &\quad \times \left(1 - 2|d_{\text{cv}}|^2\hbar^{-2}\tilde{\boldsymbol{E}}_1^2\Theta(T)\mathrm{e}^{-\gamma_1 T}\right) . \end{aligned} \tag{4.71}$$

In the experiment, one usually employs the so-called *differential transmission*, i.e. the difference between the values of the transmission with and without the pump, normalized by the transmission without the pump:

$$\frac{\Delta T}{T} = \frac{T(|\tilde{\boldsymbol{E}}_1| \neq 0) - T(|\tilde{\boldsymbol{E}}_1| = 0)}{T(|\tilde{\boldsymbol{E}}_1| = 0)} . \tag{4.72}$$

For small changes in the transmission ($2|d_{\rm cv}|^2\hbar^{-2}|\tilde{\boldsymbol{E}}_1|^2 \ll 1$) we obtain from the last line of (4.71) the following for the energy $\propto \int {\rm d}t\, |\boldsymbol{E}(t)|^2$ (see Sect. 4.4) seen by the detector:

$$\frac{\Delta T}{T} \propto \Theta(T)\, {\rm e}^{-\gamma_1 T} , \tag{4.73}$$

which is the result expected intuitively. An experimental example is given in Fig. 4.12 [69]. Here the initial decay of the differential transmisssion, with a 33fs time constant, is interpreted in terms of intervalley scattering of electrons from the Γ valley to the X and L side valleys. The longer 1.5 ps time constant is due to energy relaxation of the hot-carrier distributions via phonon emission.

This discussion has given an initial idea of how to interpret experiments on semiconductors using very short optical pulses. Close to the band edge, however, our analysis is not valid, because not only scattering processes, but also Coulomb correlations greatly modify the results, as will be discussed in Chap. 6.

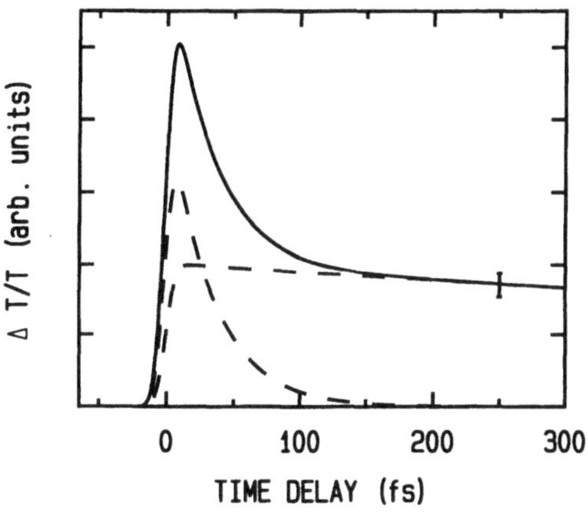

Fig. 4.12. Pump–probe experiment on bulk GaAs at room temperature using 6 fs optical pulses centered at 2.0 eV photon energy (about 570 meV above the direct gap). The excited carrier density is $3 \times 10^{18}\,{\rm cm}^{-3}$. The two underlying time constants have been decomposed and are shown as *dashed lines*. Drawn schematically after [69]

4.3.4 Alternative Approaches

The perturbation analysis discussed above is a helpful tool as long as the incident electric fields are weak. For stronger fields, however, the fifth order contribution eventually becomes comparable to the third order, and hence also the seventh order becomes comparable to the fifth and so forth. Consequently, the entire perturbation approach is essentially meaningless beyond the lowest, i.e. third-order, regime. For stronger fields, a *discrete spatial Fourier analysis* is a practical way to tackle the problem. The underlying idea is very simple. Considering again a two-beam geometry, the electric field can be written

$$\tilde{\boldsymbol{E}}(\boldsymbol{r},t) = \boldsymbol{E}^{10}(t)\exp(\mathrm{i}\boldsymbol{q}_1\boldsymbol{r}) + \boldsymbol{E}^{01}(t)\exp(\mathrm{i}\boldsymbol{q}_2\boldsymbol{r}) \ , \tag{4.74}$$

and the transition amplitudes and occupation numbers are expanded accordingly:

$$\tilde{p}_{\mathrm{vc}}(\boldsymbol{r},t) = \sum_{n_1,n_2;\, n_1+n_2=1} \tilde{p}_{\mathrm{vc}}^{n_1 n_2}(t)\exp[\mathrm{i}(n_1\boldsymbol{q}_1 + n_2\boldsymbol{q}_2)\boldsymbol{r}] \ , \tag{4.75}$$

$$f_{\mathrm{e}}(\boldsymbol{r},t) = \sum_{n_1,n_2;\, n_1+n_2=0} f_{\mathrm{e}}^{n_1 n_2}(t)\exp[\mathrm{i}(n_1\boldsymbol{q}_1 + n_2\boldsymbol{q}_2)\boldsymbol{r}] \ . \tag{4.76}$$

The spatial dependence is again a purely parametric one. The restrictions in the summations result from the rotating-wave approximation. It is straightforward to introduce these expansions into the optical Bloch equations and order them in terms of wave vectors, which leads to a coupled set of equations for the quantities $\tilde{p}_{\mathrm{vc}}^{n_1 n_2}(t)$ and $f_{\mathrm{e}}^{n_1 n_2}(t)$. This infinite set of equations has to be truncated at some $|n_1| + |n_2| \leq n_{\mathrm{max}}$. In many practical numerical calculations $n_{\mathrm{max}} = 3$ is sufficient. This already goes beyond the third order perturbation regime. In this case one has only the coefficients $\tilde{p}_{\mathrm{vc}}^{01}$, $\tilde{p}_{\mathrm{vc}}^{10}$, $\tilde{p}_{\mathrm{vc}}^{2\,-1}$, $\tilde{p}_{\mathrm{vc}}^{-1\,2}$, f_{e}^{00}, $f_{\mathrm{e}}^{-1\,1}$ and $f_{\mathrm{e}}^{1\,-1}$.

A third, alternative method to separate the nonlinear contributions propagating in different directions works as follows. We abbreviate the phase factors $\exp(\mathrm{i}\boldsymbol{q}_1\boldsymbol{r}) = \exp(\mathrm{i}\varphi_1)$ and $\exp(\mathrm{i}\boldsymbol{q}_2\boldsymbol{r}) = \exp(\mathrm{i}\varphi_2)$ and set $\varphi_1 = 0$ and $\varphi_2 = \varphi$. After the numerical solution of the Bloch equations in their original form ((4.47) and (4.48)), we obtain the polarization $\tilde{p}_{\mathrm{vc}}(\varphi)$, which depends on the artificial parameter φ. If, for example, we are interested in the four-wave-mixing signal propagating in the direction $2\boldsymbol{q}_2 - \boldsymbol{q}_1$, the dependence of the corresponding polarization is proportional to $\exp(\mathrm{i}2\varphi)$. Thus we can project this contribution out of $\tilde{p}_{\mathrm{vc}}(\varphi)$ by using

$$\tilde{p}_{\mathrm{vc},2\boldsymbol{q}_2-\boldsymbol{q}_1} = \frac{1}{2\pi}\int_0^{2\pi} \mathrm{d}\varphi\, \tilde{p}_{\mathrm{vc}}(\varphi)\exp(-\mathrm{i}2\varphi) \ . \tag{4.77}$$

Similarly, the transmitted signal in the direction $\boldsymbol{q}_1$ corresponds to a phase factor exp(0) and can be projected out of $\tilde{p}_{\mathrm{vc}}(\varphi)$ by using

$$\tilde{p}_{\mathrm{vc},\boldsymbol{q}_1} = \frac{1}{2\pi}\int_0^{2\pi} \mathrm{d}\varphi\, \tilde{p}_{\mathrm{vc}}(\varphi) \ , \tag{4.78}$$

and so forth. If these integrals can be evaluated by summation over only a few points, this procedure requires about the same memory on the computer as the discrete spatial Fourier analysis. The computer code, however, becomes quite transparent. This can be an advantage for equations which are more complicated than the optical Bloch equations.

4.4 Semiconductor Photodetectors

In a semiconductor photodetector incident photons, resonant with an interband transition, generate electron–hole pairs. These photogenerated carriers have to be removed from the active region by some means in order to obtain information about the temporal shape of the field $\boldsymbol{E}(\boldsymbol{r},t)$ incident on the detector. We consider an idealized detector, i.e. a very thin slice of semiconductor connected to two electrical contacts providing an electric field, as depicted in Fig. 4.13. We assume that all electrons and holes generated are rapidly removed from the active region into the contacts (corresponding to a time constant γ_1^{-1}) and thus contribute to the photocurrent in the z direction. Furthermore, we assume that all electrons propagate with $k_z > 0$, and all holes with $k_z < 0$. Under these assumptions, the photocurrent I_{pc} at time t becomes

$$I_{\mathrm{pc}}(t) = -e\gamma_1 \sum_{\boldsymbol{k}\,(k_z>0)} \left[f_{\mathrm{e}}(\boldsymbol{k},t) + f_{\mathrm{h}}(-\boldsymbol{k},t)\right] . \tag{4.79}$$

For a low incident flux,[12] $f_{\mathrm{e}}(\boldsymbol{k},t) = f_{\mathrm{h}}(-\boldsymbol{k},t) = f_{\mathrm{h}}(\boldsymbol{k},t) \approx f_{\mathrm{e}}^{(2)}(\boldsymbol{k},t)$ can be evaluated via a perturbation analysis of the optical Bloch equations. Using (4.37) and (4.39) within the RWA, this leads to

$$I_{\mathrm{pc}}(t) = -\frac{e\gamma_1}{\hbar^2} \sum_{\boldsymbol{k}\,(k_z>0)} \left(\int_{-\infty}^{\infty} \mathrm{d}t'\,\mathrm{d}t''\, \Theta(t-t')\,\mathrm{e}^{-\gamma_1(t-t')}\,\Theta(t'-t'') \right.$$
$$\left. \times\, \mathrm{Re}\left(\mathrm{e}^{[\mathrm{i}\Delta(\boldsymbol{k})-\gamma_2](t'-t'')} \boldsymbol{d}_{\mathrm{cv}}\tilde{\boldsymbol{E}}(\boldsymbol{r},t')\,\boldsymbol{d}_{\mathrm{cv}}^{*}\tilde{\boldsymbol{E}}^{*}(\boldsymbol{r},t'')\right) \right) . \tag{4.80}$$

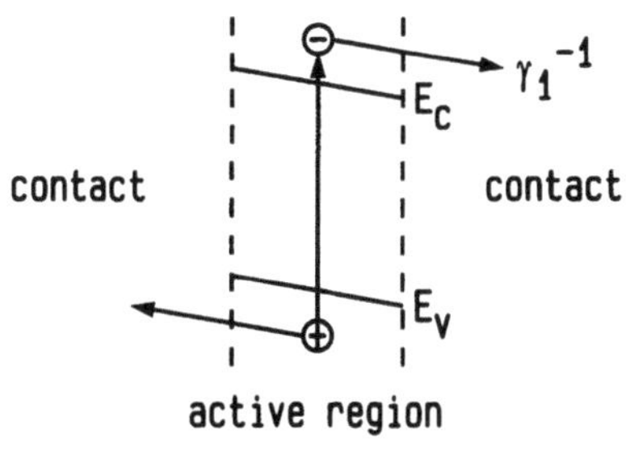

Fig. 4.13. Schematic illustration of an idealized semiconductor photodetector. All photogenerated electrons and holes are swept out into the contacts with a time constant γ_1^{-1}

[12] Only this limit is of practical interest. For larger incident fields one obtains undesired deviations from the behavior described above, i.e. saturation of the detector response.

In general, the photocurrent response is obviously given by a rather complicated double convolution integral of the quantity $\tilde{\boldsymbol{E}}(\boldsymbol{r},t')\,\tilde{\boldsymbol{E}}^*(\boldsymbol{r},t'')$.

4.4.1 The Field–Field Correlation Function and its Relation to Coherence

It is worth noting that our considerations up to now and in the following remain valid if the light field is treated quantum mechanically. According to Sect. 2.4, we have to replace the fields by the corresponding operators and introduce the quantum mechanical expectation value

$$\mathcal{D}(\boldsymbol{r}t',\boldsymbol{r}t'') = \langle \tilde{\boldsymbol{E}}^\dagger(\boldsymbol{r},t')\,\tilde{\boldsymbol{E}}(\boldsymbol{r},t'')\rangle\,, \tag{4.81}$$

which represents the first-order[13] field–field correlation function. Apart from prefactors, (4.81) is closely related to the retarded photon Green's function $\mathcal{D}^{\mathrm{r}}(\boldsymbol{r}t',\boldsymbol{r}t'')$ introduced in Sect. 2.6.5 (compare (2.194) and also Sect. 12.1). This correlation can also be understood intuitively in a semiclassical picture if we consider fluctuations of the (classical) light fields. In this case the (classical) ensemble average $\langle I_{\mathrm{pc}}\rangle$ with respect to the fluctuations is measured. This average can be moved under the integrals in (4.80) and the photocurrent is then related to the classical pendant of the field–field correlation function (4.81), because no other quantity in (4.80) exhibits fluctuations. The sum over $\boldsymbol{k}$ in (4.80) can again be worked out analytically for a two-dimensional system (similarly to the discussion of photon echoes; see (4.64)) which leads to

$$\lim_{E_{\mathrm{co}}\to\infty}\left(\sum_{\boldsymbol{k}\,(k_z>0)} \mathrm{e}^{(\mathrm{i}\Delta(\boldsymbol{k})-\gamma_2)(t'-t'')}\right) = \hbar^2\,|d_{\mathrm{cv}}|^{-2}\,I^0_{\mathrm{pc}}\,\delta(t'-t'')\,, \tag{4.82}$$

for a large cutoff energy in the density of states, i.e. $E_{\mathrm{co}}\to\infty$. The other convolution integral simplifies in the limit of fast carrier sweep-out, $\gamma_1\to\infty$

$$\lim_{\gamma_1\to\infty}\left(\gamma_1\,\Theta(t-t')\,\mathrm{e}^{-\gamma_1(t-t')}\right) = \delta(t-t')\,. \tag{4.83}$$

The photocurrent, in these limits, is simply proportional to the square of the electric field, and hence proportional to the intensity:

$$\langle I_{\mathrm{pc}}(t)\rangle = I^0_{\mathrm{pc}}\,\langle\tilde{\boldsymbol{E}}^\dagger(\boldsymbol{r},t)\tilde{\boldsymbol{E}}(\boldsymbol{r},t)\rangle\,. \tag{4.84}$$

If, on the other hand, the sweep-out process is slow compared with any variation in the electric field ($\gamma_1\to 0$), the photocurrent becomes proportional to the energy:

$$\langle I_{\mathrm{pc}}(t)\rangle = \gamma_1\,I^0_{\mathrm{pc}}\int_{-\infty}^{t}\mathrm{d}t'\,\langle\tilde{\boldsymbol{E}}^\dagger(\boldsymbol{r},t')\tilde{\boldsymbol{E}}(\boldsymbol{r},t')\rangle\,. \tag{4.85}$$

[13] Similarly, nth order correlation functions can be introduced.

Frequently, photodetectors are combined with spectrometers in order to obtain simultaneous spectral and temporal information (see Fig. 4.14). Any spectrometer can be represented by its filter function $F_{\omega_0}(\omega)$, which describes some more or less narrow spectral window around the spectrometer central frequency ω_0 that has been selected. The Fourier transform of the filter function is $F_{\omega_0}(t)$. Consequently, the electric field $\tilde{\boldsymbol{E}}(\boldsymbol{r},t)$ (with Fourier transform $\tilde{\boldsymbol{E}}(\boldsymbol{r},\omega)$) entering the spectrometer is altered by the spectrometer and becomes $\tilde{\boldsymbol{E}}_{\mathrm{det}}(\boldsymbol{r},\omega) = F_{\omega_0}(\omega)\tilde{\boldsymbol{E}}(\boldsymbol{r},\omega)$ after the spectrometer. This quantity is detected by the photodetector. Clearly, the product in frequency space becomes a convolution in the time domain,

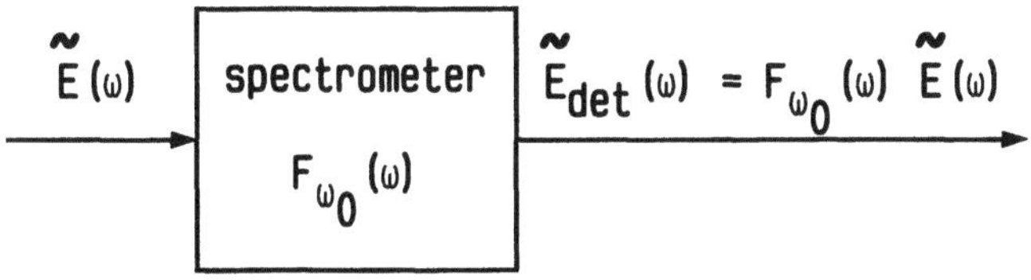

Fig. 4.14. A spectrometer alters the electric field incident on a detector behind the spectrometer

$$\tilde{\boldsymbol{E}}_{\mathrm{det}}(\boldsymbol{r},t) = \int_{-\infty}^{\infty} \mathrm{d}t'\, F_{\omega_0}(t-t')\, \tilde{\boldsymbol{E}}(\boldsymbol{r},t')\,, \tag{4.86}$$

and the detector photocurrent (4.84) is given by

$$\langle I_{\mathrm{pc}}(t,\omega_0)\rangle = I_{\mathrm{pc}}^0 \int_{-\infty}^{\infty} \mathrm{d}t'\,\mathrm{d}t''\, F_{\omega_0}(t-t')\, F_{\omega_0}^*(t-t'')\mathcal{D}(\boldsymbol{r},t',\boldsymbol{r},t'')\,, \tag{4.87}$$

which exhibits a general structure somewhat similar to (4.80). If $F_{\omega_0}(\omega)$ is given by a Lorentzian ($F_{\omega_0}(\omega) \propto 1/[(\omega-\omega_0)+\mathrm{i}\gamma]$), for example, its Fourier transform is a single-side exponential ($F_{\omega_0}(t) = \Theta(t)\,\mathrm{e}^{-\gamma t}\,\mathrm{e}^{\mathrm{i}\omega_0 t}$), which ensures causality. In the limit $\gamma \to 0$ (4.87) can be reduced to the *Wiener–Kintchine theorem* [70, 71],

$$\langle I_{\mathrm{pc}}(t,\omega)\rangle \propto \lim_{T\to\infty} \frac{1}{T} \int_0^T \mathrm{d}t\,\mathrm{d}t'\,\mathrm{e}^{-\mathrm{i}\omega(t-t')}\,\mathcal{D}(\boldsymbol{r},t,\boldsymbol{r},t')\,. \tag{4.88}$$

If $\mathcal{D}(\boldsymbol{r},t,\boldsymbol{r},t')$ depends on the time difference $\Delta t = t - t'$ only, the spectrum is nothing but the Fourier transform of the field–field correlation function.

Equations (4.80) and (4.87) are immediate consequences of the energy–time uncertainty principle. A narrow filter function $F_{\omega_0}(\omega)$ delivers good spectral resolution but bad temporal resolution via convolution of the electric field with the temporally broad Fourier transform $F_{\omega_0}(t-t')$ in (4.87), and

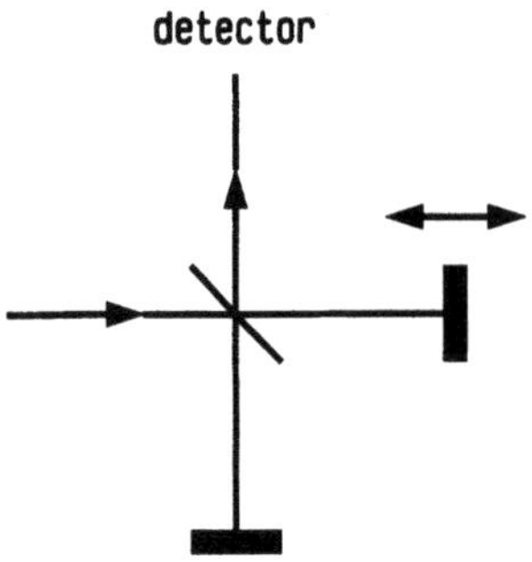

Fig. 4.15. Schematic illustration of a Michelson interferometer

vice versa. In any case, the detector measures a quantity directly related to the field–field correlation function.

The relation between the field–field correlation function and *coherence* can be made clear by considering the signal measured by a semiconductor photodetector at the output of a *Michelson interferometer* (Fig. 4.15). One arm of the interferometer can be translated by $\Delta\boldsymbol{r}$, which introduces the phase factor $\exp(\mathrm{i}\,\boldsymbol{q}\Delta\boldsymbol{r}) = \exp(\mathrm{i}\,\omega T)$; T is the corresponding time delay, related to $\Delta\boldsymbol{r}$ via the velocity of light. Thus the electric field incident on the detector positioned at the output of the interferometer is given by $\tilde{\boldsymbol{E}}_{\mathrm{det}}(\boldsymbol{r},t) = \tilde{\boldsymbol{E}}(\boldsymbol{r},t) + \tilde{\boldsymbol{E}}(\boldsymbol{r},t-T)\exp(\mathrm{i}\,\omega T)$, and the photocurrent of a fast detector (4.84) becomes

$$\begin{aligned}\langle I_{\mathrm{pc}}(t)\rangle &= I^0_{\mathrm{pc}}\Big(|\langle\tilde{\boldsymbol{E}}^\dagger(\boldsymbol{r},t)\tilde{\boldsymbol{E}}(\boldsymbol{r},t)\rangle| + |\langle\tilde{\boldsymbol{E}}^\dagger(\boldsymbol{r},t-T)\tilde{\boldsymbol{E}}(\boldsymbol{r},t-T)\rangle| \\ &\quad + 2\,\mathrm{Re}\Big(\langle\tilde{\boldsymbol{E}}^\dagger(\boldsymbol{r},t)\tilde{\boldsymbol{E}}(\boldsymbol{r},t-T)\rangle\mathrm{e}^{-\mathrm{i}\,\omega T}\Big)\Big\} \\ &= I^0_{\mathrm{pc}}\Big\{|\langle\tilde{\boldsymbol{E}}^\dagger(\boldsymbol{r},t)\tilde{\boldsymbol{E}}(\boldsymbol{r},t)\rangle| + |\langle\tilde{\boldsymbol{E}}^\dagger(\boldsymbol{r},t-T)\tilde{\boldsymbol{E}}(\boldsymbol{r},t-T)\rangle| \\ &\quad + 2\,\cos(\omega T-\varphi)\,|\langle\tilde{\boldsymbol{E}}^\dagger(\boldsymbol{r},t)\tilde{\boldsymbol{E}}(\boldsymbol{r},t-T)\rangle|\Big)\,. \end{aligned} \tag{4.89}$$

Here we have already separated the correlation function

$$\langle\tilde{\boldsymbol{E}}^\dagger(\boldsymbol{r},t)\tilde{\boldsymbol{E}}(\boldsymbol{r},t-T)\rangle = |\langle\tilde{\boldsymbol{E}}^\dagger(\boldsymbol{r},t)\tilde{\boldsymbol{E}}(\boldsymbol{r},t-T)\rangle|\,\exp(\mathrm{i}\varphi) \tag{4.90}$$

into its absolute value and a phase factor with phase $\varphi = \varphi(\boldsymbol{r},t,T)$. Obviously the last, oscillating term (containing the correlation function) in (4.89) describes the interference between the fields in the two arms. For vanishing field–field correlation functions, no interference is observed at all and the corresponding fields are completely incoherent.

In general, the field–field correlation function does not factorize, which means that it cannot be represented as a product,[14] i.e.

$$\langle\tilde{\boldsymbol{E}}^\dagger(\boldsymbol{r},t')\,\tilde{\boldsymbol{E}}(\boldsymbol{r},t'')\rangle \neq \mathcal{E}(\boldsymbol{r},t')\,\mathcal{E}^*(\boldsymbol{r},t'') \tag{4.91}$$

[14] In particular, $\langle\tilde{\boldsymbol{E}}^\dagger(\boldsymbol{r},t')\,\tilde{\boldsymbol{E}}(\boldsymbol{r},t'')\rangle \neq \langle\tilde{\boldsymbol{E}}^\dagger(\boldsymbol{r},t')\rangle\,\langle\tilde{\boldsymbol{E}}(\boldsymbol{r},t'')\rangle$ in general.

where $\mathcal{E}(\boldsymbol{r}, t)$ is some function not necessarily given by the electric field $\tilde{E}(\boldsymbol{r}, t)$. However, special electric fields exist for which the correlation function does factorize according to (4.91) as *coherent fields*. We have already introduced these in Sect. 2.4 as eigenstates of the annihilation operator of the light field. These coherent states or Glauber states do in fact correspond to factorizing field–field correlation functions and are therefore equivalent to maximum interference contrast in the Michelson interferometer. This merges the formal and experimental ways of expressing coherence. But, at the present level, neither of these concepts can answer the question of how coherent fields are built up in a system, nor of how coherence is lost owing to microscopic processes. We shall take up these questions again in Chaps. 10–13.

The concepts and methods developed in this chapter describe the physics of linear and nonlinear optical properties of semiconductors on the simplest footing. The *optical Bloch equations* neglect any Coulomb correlation completely and describe phase and energy relaxation on a phenomenological level. Although many basic ideas which we have introduced will survive, one must be aware of the fact that all results will be modified substantially owing to two-particle Coulomb correlations among electrons and holes (see Chap. 6) and also owing to higher-order correlations and a microscopic description of scattering processes (see Chaps. 9–12).

4.5 Problems

4.1. Transform the kinetic energy of an electron-hole pair $E_{\mathrm{c}}(\boldsymbol{k}) - E_{\mathrm{v}}(\boldsymbol{k}+\boldsymbol{q})$ to relative and center-of-mass momentum variables. Show that within the effective-mass approximation the contribution, which is linear in $\boldsymbol{q}$, vanishes.

4.2. Verify the *Kramers–Kronig relations* for the linear optical susceptibility

$$\mathrm{Re}\Big(\chi^{(1)}(\omega)\Big) = \mathcal{P}\int_{-\infty}^{\infty} \frac{1}{\pi} \frac{\mathrm{Im}\Big(\chi^{(1)}(\tilde{\omega})\Big)}{\tilde{\omega}-\omega}\, \mathrm{d}\tilde{\omega}\,,$$

and

$$\mathrm{Im}\Big(\chi^{(1)}(\omega)\Big) = -\mathcal{P}\int_{-\infty}^{\infty} \frac{1}{\pi} \frac{\mathrm{Re}\Big(\chi^{(1)}(\tilde{\omega})\Big)}{\tilde{\omega}-\omega}\, \mathrm{d}\tilde{\omega}.$$

Start with Cauchy's theorem

$$\chi^{(1)}(\omega) = \frac{1}{2\pi \mathrm{i}} \int_{\Gamma} \frac{\chi^{(1)}(\tilde{\omega})}{\tilde{\omega}-\omega}\, \mathrm{d}\tilde{\omega}\,.$$

for an arbitrary, closed, counterclockwise path Γ around ω. Choose an appropriate path and decompose the real and imaginary parts by means of the Dirac identity.

4.3. The linear susceptibility (4.24) represents the dipole approximation of the density–density correlation function. Make use of (2.107) to derive the corresponding current–current correlation function in the limit of vanishing wave vector.

4.4. Equation (2.212) can be simplified using the so-called *slowly-varying-amplitude approximation* (SVAA) with the ansatz

$$\boldsymbol{E}(\boldsymbol{r},t) = \bar{\boldsymbol{E}}(\boldsymbol{r},t)\exp[\mathrm{i}(\boldsymbol{q}_0\boldsymbol{r} - \omega_0 t)]$$

and

$$\boldsymbol{P}(\boldsymbol{r},t) = \bar{\boldsymbol{P}}(\boldsymbol{r},t)\exp[\mathrm{i}(\boldsymbol{q}_0\boldsymbol{r} - \omega_0 t)]$$

to obtain the form

$$\frac{\partial \bar{\boldsymbol{E}}(\boldsymbol{r},t)}{\partial z} = \frac{\mu_0\omega_0^2}{2q_0}\,\mathrm{i}\,\bar{\boldsymbol{P}}(\boldsymbol{r},t)\ .$$

Verify this. $\bar{\boldsymbol{E}}$ and $\bar{\boldsymbol{P}}$ vary slowly in space and time. Introduce the ansatz into the wave equation and neglect as many spatial and temporal derivatives of $\bar{\boldsymbol{E}}$ and $\bar{\boldsymbol{P}}$ as is meaningful to do.

4.5. Derive a relation between the density matrix elements $p_{\mathrm{vc}}(\boldsymbol{k})$ and $f_{\mathrm{e}}(\boldsymbol{k}) = f_{\mathrm{h}}(\boldsymbol{k})$ to lowest order in the transition amplitude for $\gamma_2 \neq 0$ and $\gamma_1 = 0$. Consider also the case $\gamma_2 \neq 0$, $\gamma_1 \neq 0$, and $f^{(0)}(\boldsymbol{k}) = 0$ to lowest order in the dephasing constants. Hint: Make use of the equations of motion.

4.6. In a stationary nondegenerate four-wave-mixing experiment (sometimes referred to as a hole-burning experiment), two beams (see Fig. 4.10), tuned above the band edge of a semiconductor, lead to a diffracted signal in the direction $2\boldsymbol{q}_2 - \boldsymbol{q}_1$. This signal is measured as a function of the detuning $\omega_2 - \omega_1$ between the two laser frequencies ω_1 and ω_2. Discuss qualitatively which time constants can be obtained from such an experiment. What is the frequency of the diffracted beam? Finally, analyze this arrangement within the perturbation approach up to third order.

4.7. Derive the generalization of the Bloch equations (4.51) for a system with an arbitrary number of valence and conduction bands. Hint: Derive first the generalization of (4.2) and of the Heisenberg equations of motion (4.12)–(4.15).

4.8. Specify the result of Problem 4.6 for a three-level system with two valence bands and one conduction band, and vice versa. Solve the equations for a four-wave-mixing configuration with δ pulses. Hint: Make use of the perturbation analysis in the time domain, which we have introduced in Sect. 4.3.3.

5. Correlated Transitions of Bloch Electrons

In solid-state physics, it is the Coulomb interaction which keeps the world together. In the last chapter the Coulomb interaction entered merely implicitly, owing to the occurrence of effective rather than free particle masses. However, not only are the correct one-particle energies determined by interaction effects, but also the optically excited electrons in the conduction band and the holes left in the valence band are influenced by their mutual interaction. To neglect these interactions is a serious omission, which will be remedied in this and the following chapter to the lowest level of approximation.

5.1 Equations of Motion in the Hartree–Fock Approximation

First we shall develop the Hartree–Fock theory in a general Bloch representation, which allows us to study the basic effects of interaction on optical transitions in the entire Brillouin zone and for various band combinations. Although exchange effects turn out to be overestimated in the Hartree–Fock theory, this theory provides a reasonable description of general trends in an energy range up to several eV above the band edge. After this excursion to correlated optical transitions of Bloch electrons, we shall concentrate again on optical properties in the vicinity of the absorption edge.

The aim is to obtain a generalization of the optical Bloch equations (4.20) and (4.21) which treats the Coulomb interaction consistently within the Hartree–Fock approximation. As the corresponding exchange self-energy (3.41) is linear in the Coulomb potential and thus instantaneous in time, we restrict ourselves to the static contribution of the interactions only. Higher-order contributions, which lead to noninstantaneous, that is, dynamic correlations, are missing in this approximation. Thus relaxation and dephasing processes, which result from such dynamic correlations, are treated as before via phenomenological rates γ_1 and γ_2.

The application of the effective-mass approximation, which will again be employed to obtain the final results, is not as trivial as suggested in the last chapter, because we have to distinguish between two different effects of the Coulomb interaction. One part of the interaction is independent of the light fields and determines the one-particle energies according to the Hartree–Fock equations, or a more refined one-particle theory, and thus the effective masses. In the last chapter, we have presupposed, without worrying about details, that this part can be separated, and neglected the second part of the interaction, which is induced by the light field. A separation of these two contributions cannot be performed systematically if the effective-mass approximation has already been used in the Hamiltonian. Thus we have to start from the very beginning, i.e. from the Hamiltonian (2.119), which we write in the form

$$H = \sum_{kl} \langle l|h|k\rangle \, c_l^\dagger c_k + \frac{1}{2} \sum_{klmn} \langle kl|v|mn\rangle \, c_k^\dagger c_l^\dagger c_n c_m, \tag{5.1}$$

where the one-particle contribution is defined as

$$h(\boldsymbol{r}) = \frac{\boldsymbol{p}^2}{2m_0} + U(\boldsymbol{r}) - \boldsymbol{dE} \, . \tag{5.2}$$

We proceed as as in the last chapter and consider first the Heisenberg equations of Fermi operators. Instead of (4.12)–(4.15), we have now the general results

$$-\mathrm{i}\hbar \frac{\partial}{\partial t} c_i^\dagger = \sum_l h_{li} \, c_l^\dagger + \frac{1}{2} \sum_{lmn} \Big(\langle|lm|v|in\rangle - \langle|lm|v|ni\rangle \Big) \, c_l^\dagger c_m^\dagger c_n \tag{5.3}$$

and

$$-\mathrm{i}\hbar \frac{\partial}{\partial t} c_j = -\sum_l h_{jl} \, c_l + \frac{1}{2} \sum_{lmn} \Big(\langle|jl|v|mn\rangle - \langle|lj|v|mn\rangle \Big) \, c_l^\dagger c_m c_n \, . \tag{5.4}$$

The diagonal and off-diagonal elements of the density matrix

$$n_{ij} = \langle c_i^\dagger c_j \rangle \tag{5.5}$$

thus obey the equation of motion

$$\begin{aligned} &\mathrm{i}\hbar \frac{\partial}{\partial t} n_{ij} + \sum_l (h_{li} \, n_{lj} - n_{il} \, h_{jl}) \\ &= \frac{1}{2} \sum_{lmn} \Big[\Big(\langle lm|v|ni\rangle - \langle lm|v|in\rangle \Big) \langle c_l^\dagger c_m^\dagger c_n c_j \rangle \\ &+ \Big(\langle lj|v|mn\rangle - \langle jl|v|mn\rangle \Big) \langle c_i^\dagger c_l^\dagger c_m c_n \rangle \Big] \\ &= \sum_{lmn} \Big(\langle lm|v|ni\rangle \langle c_l^\dagger c_m^\dagger c_n c_j \rangle - \langle jl|v|mn\rangle \langle c_i^\dagger c_l^\dagger c_m c_n \rangle \Big) , \end{aligned} \tag{5.6}$$

where we have used the symmetry property of the Coulomb matrix element $\langle lm|v|in\rangle = \langle ml|v|ni\rangle$. An alternative formulation of (5.6) is obtained from the real-space representation, which we have already introduced in Sect. 2.3.

For a number of applications, this is a more convenient starting point. We can either transform (5.6) by means of the fermion field operators in (2.59) and (2.60) or make use of the real-space representation of the Hamiltonian (2.63). As a result, we obtain

$$
\begin{aligned}
&i\hbar\frac{\partial}{\partial t}\langle\psi^{\dagger}_{s_1}(\boldsymbol{r}_1)\psi_{s_2}(\boldsymbol{r}_2)\rangle = [h(\boldsymbol{r}_2)-h(\boldsymbol{r}_1)]\,\langle\psi^{\dagger}_{s_1}(\boldsymbol{r}_1)\psi_{s_2}(\boldsymbol{r}_2)\rangle \\
&+\sum_{s'}\int \mathrm{d}^3r'\Big(v(\boldsymbol{r}_2-\boldsymbol{r}')\,\langle\psi^{\dagger}_{s_1}(\boldsymbol{r}_1)\psi^{\dagger}_{s'}(\boldsymbol{r}')\psi_{s'}(\boldsymbol{r}')\psi_{s_2}(\boldsymbol{r}_2)\rangle \\
&-v(\boldsymbol{r}_1-\boldsymbol{r}')\,\langle\psi^{\dagger}_{s_1}(\boldsymbol{r}_1)\psi^{\dagger}_{s'}(\boldsymbol{r}')\psi_{s'}(\boldsymbol{r}')\psi_{s_2}(\boldsymbol{r}_2)\rangle\Big)\,.
\end{aligned}
\tag{5.7}
$$

The real-space representation is an appropriate starting point to introduce a Wannier representation of the equations of motion, which allows us to analyze the various length scales, which are involved in the problem (see Problems 5.2 and 5.3). Even within in the dipole and the effective-mass approximation, the separation of the microscopic and macroscopic length scales is not as simple as the derivation in Sect. 4.1. Owing to the Coulomb interaction, different density matrices are spatially coupled. Spatial locality with respect to a macroscopic variable, can only be enforced by means of a gradient expansion. If the gradient corrections can be neglected, we are left again with a parametric spatial dependence of the density matrices, induced by the external field. This problem will be considered in detail in Chap. 10.

A challenging feature of (5.6) and (5.7) is the occurrence of an expectation value of four operators, often referred to as a *four-point function*. It is straightforward to write down the equations of motion for these four-point functions. If we do this, we find that they depend on six-point functions and so forth. In order to cut off the resulting infinite hierarchy of equations of motion, we have to develop a strategy for a systematic truncation scheme. There are several possible ways to proceed from here. The most systematic and general approach is the Green's function formalism, where, instead of the equal-time density matrices in (5.7) we deal with more general functions depending on two times. This allows us to close the set of equations already at the one-particle level by introducing an exact one-particle self-energy, which replaces the static Hartree–Fock approximation (3.41). Unfortunately but not surprisingly, there is no way to solve the equations that define the one-particle self-energy exactly. Nevertheless, the Green's function approach allows one to develop systematic approximations beyond the Hartree–Fock level. We shall use this approach extensively in Chap. 10 and the following chapters.

Another, completely different approach is appropriate in the coherent limit, in which the decay of the polarization can be neglected. As in a semiconductor at zero temperature any excited electron–hole pair results from the action of an external field, we can classify n-point functions with respect to the order of the external field. In a given order of the external field, the hierarchy of equations of motion is truncated correspondingly at a certain level. We shall introduce this approach in Chap. 9. In any case, the leading contributions to the interaction can be obtained from the Hartree–Fock approximation, which we apply in the following. We use the factorization (3.6) in (5.7) and introduce the one-particle *Hartree–Fock energies* given by (3.11) by adding and subtracting the corresponding interaction contributions. The result of this procedure is given by

$$\begin{aligned}&\left(\mathrm{i}\hbar\frac{\partial}{\partial t}+E_i-E_j\right)n_{ij}-\sum_l\left(\boldsymbol{d}_{li}\boldsymbol{E}\,n_{lj}-n_{il}\,\boldsymbol{d}_{jl}\boldsymbol{E}\right)\\&=e^2\sum_{lmn}\Big[\Big(\langle lm|v|ni\rangle-\langle ml|v|ni\rangle\Big)\Big(n_{ln}-\delta_{ln}\delta_{mi}\delta_{lv}\Big)\,n_{mj}\\&+\Big(\langle jl|v|mn\rangle-\langle jl|v|nm\rangle\Big)\Big(n_{ln}-\delta_{mj}\delta_{ln}\delta_{lv}\Big)\,n_{im}\Big]\,.\end{aligned} \tag{5.8}$$

To proceed further, we use a Bloch function representation of the one-particle states according to

$$\varphi_l(\boldsymbol{r})\equiv\varphi_l(\boldsymbol{k}_l\boldsymbol{r}) \tag{5.9}$$

and write the Coulomb matrix elements in the form

$$\langle l\boldsymbol{k}_l m\boldsymbol{k}_m|v|n\boldsymbol{k}_n i\boldsymbol{k}_i\rangle=\sum_{\boldsymbol{q}}v(\boldsymbol{q})\langle l\boldsymbol{k}_l|\mathrm{e}^{-\mathrm{i}\boldsymbol{qr}}|n\boldsymbol{k}_n\rangle\langle m\boldsymbol{k}_m|\mathrm{e}^{\mathrm{i}\boldsymbol{qr}}|i\boldsymbol{k}_i\rangle\,. \tag{5.10}$$

Expanding the Bloch functions by means of *Wannier functions* (3.71), we obtain the following for plane-wave matrix elements:

$$\begin{aligned}&\langle l\boldsymbol{k}_l|\mathrm{e}^{-\mathrm{i}\boldsymbol{qr}}|n\boldsymbol{k}_n\rangle\\&=\frac{1}{N}\sum_{\boldsymbol{L}\boldsymbol{L}'}\int\mathrm{d}^3\boldsymbol{r}\,w_l^\dagger(\boldsymbol{r}-\boldsymbol{L})\mathrm{e}^{-\mathrm{i}\boldsymbol{qr}}w_n(\boldsymbol{r})\mathrm{e}^{-\mathrm{i}\boldsymbol{k}_l\boldsymbol{L}}\mathrm{e}^{-\mathrm{i}(\boldsymbol{q}+\boldsymbol{k}_l-\boldsymbol{k}_n)\boldsymbol{L}'}\\&=\delta_{\boldsymbol{k}_l-\boldsymbol{k}_n+\boldsymbol{q},\boldsymbol{G}}\sum_{\boldsymbol{L}}\int\mathrm{d}^3\boldsymbol{r}\,w_l^\dagger(\boldsymbol{r}-\boldsymbol{L})\mathrm{e}^{-\mathrm{i}(\boldsymbol{qr}+\boldsymbol{k}_l\boldsymbol{L})}w_n(\boldsymbol{r})\,.\end{aligned} \tag{5.11}$$

This property of the matrix elements restricts the allowed values of wave vectors in (5.8). Within the dipole approximation, we restrict ourselves for the moment to the case of vanishing photon wave vector, in which the density matrix elements $n_{ij}(\boldsymbol{k},\boldsymbol{k}+\boldsymbol{q})$ become diagonal with respect to the wave vector and are replaced by $n_{ij}(\boldsymbol{k})$. For reasons, which will become obvious in the following, we keep, however, a residual dependence on the photon wave vector in the interaction matrix elements. Thus, in a *Bloch representation*, (5.8) takes the form

$$\begin{aligned}&\left(\mathrm{i}\hbar\frac{\partial}{\partial t}+E_i(\boldsymbol{k})-E_j(\boldsymbol{k})\right)n_{ij}(\boldsymbol{k})\\&-\sum_l\left(\boldsymbol{d}_{li}(\boldsymbol{k})\boldsymbol{E}\,n_{lj}(\boldsymbol{k})-n_{il}(\boldsymbol{k})\,\boldsymbol{d}_{jl}(\boldsymbol{k})\boldsymbol{E}\right)=e^2\lim_{\boldsymbol{q}\to0}\sum_{lmn\boldsymbol{k}'\boldsymbol{G}}v(\boldsymbol{G}+\boldsymbol{q})\\&\times\Big[\langle l,\boldsymbol{k}'|\mathrm{e}^{-\mathrm{i}(\boldsymbol{G}+\boldsymbol{q})\boldsymbol{r}}|n,\boldsymbol{k}'+\boldsymbol{q}\rangle\langle m,\boldsymbol{k}+\boldsymbol{q}|\mathrm{e}^{\mathrm{i}(\boldsymbol{G}+\boldsymbol{q})\boldsymbol{r}}|i,\boldsymbol{k}\rangle\end{aligned}$$

$$
\begin{aligned}
&\times \left(n_{ln}(\boldsymbol{k}') - \delta_{ln}\delta_{mi}\delta_{lv}\right) n_{mj}(\boldsymbol{k}) - \langle l, \boldsymbol{k}'|\mathrm{e}^{-\mathrm{i}(\boldsymbol{G}+\boldsymbol{q})\boldsymbol{r}}|n, \boldsymbol{k}'+\boldsymbol{q}\rangle \\
&\times \langle j, \boldsymbol{k}+\boldsymbol{q}|\mathrm{e}^{\mathrm{i}(\boldsymbol{G}+\boldsymbol{q})\boldsymbol{r}}|m, \boldsymbol{k}\rangle \left(n_{ln}(\boldsymbol{k}') - \delta_{ln}\delta_{mj}\delta_{lv}\right) n_{im}(\boldsymbol{k}) \Big] \\
&-e^2 \sum_{lmn\boldsymbol{k}'} v(\boldsymbol{k}-\boldsymbol{k}') \Big[\langle m, \boldsymbol{k}|\mathrm{e}^{\mathrm{i}(\boldsymbol{k}-\boldsymbol{k}')\boldsymbol{r}}|n, \boldsymbol{k}'\rangle \langle l, \boldsymbol{k}'|\mathrm{e}^{-\mathrm{i}(\boldsymbol{k}-\boldsymbol{k}')\boldsymbol{r}}|i, \boldsymbol{k}\rangle \\
&\times \left(n_{ln}(\boldsymbol{k}') - \delta_{ln}\delta_{mi}\delta_{lv}\right) n_{mj}(\boldsymbol{k}) - \langle j, \boldsymbol{k}|\mathrm{e}^{\mathrm{i}(\boldsymbol{k}-\boldsymbol{k}')\boldsymbol{r}}|n, \boldsymbol{k}'\rangle \\
&\times \langle l, \boldsymbol{k}'|\mathrm{e}^{-\mathrm{i}(\boldsymbol{k}-\boldsymbol{k}')\boldsymbol{r}}|m, \boldsymbol{k}\rangle \left(n_{ln}(\boldsymbol{k}') - \delta_{ln}\delta_{mj}\delta_{lv}\right) n_{im}(\boldsymbol{k}) \Big] . \qquad (5.12)
\end{aligned}
$$

The first two interaction terms on the right-hand side of (5.12) correspond to the Hartree contributions; the third and fourth term are the corresponding exchange contributions. The *Hartree terms* contain the $\boldsymbol{G}=0$ as well as the $\boldsymbol{G}\neq 0$ contributions. As will be seen in more detail for the linear case, to be discussed in the next section, the $\boldsymbol{G}\neq 0$ contributions give rise to the *local-field corrections* [72, 73] introduced in Sect. 2.6. The $\boldsymbol{G}=0$ term can be evaluated by expanding the plane waves up to first order in the photon wave vector and then taking the limit $\boldsymbol{q}\to 0$, which has already been taken for all other contributions. It is readily shown that only interband matrix elements contribute to the Hartree terms. This implies that the contributions $\sim 1/q^2$ vanish and that the singularity of the Coulomb potential is canceled. Hence the leading contributions are proportional to products of the interband dipole matrix elements

$$
\boldsymbol{d}_{ln}(\boldsymbol{k}) = -e\langle l\boldsymbol{k}|\boldsymbol{r}|n\boldsymbol{k}\rangle = -e\sum_{\boldsymbol{L}} \int \mathrm{d}^3\boldsymbol{r}\, w_l^\dagger(\boldsymbol{r}-\boldsymbol{L})\,\boldsymbol{r}\, w_n(\boldsymbol{r})\mathrm{e}^{-\mathrm{i}\boldsymbol{k}\boldsymbol{L}} , \qquad (5.13)
$$

which appear on the left-hand side of (5.12). In principle, the dipole matrix element also contains a divergent intraband part (l = n) (compare Sect. 3.6). Such contributions, however, cancel exactly, and do not contribute to (5.12). The $\boldsymbol{G}=0$ contributions calculated in this way describe the dipolar internal field, which has already occured in the calculation of the longitudinal response in Chap. 2. This relationship will become more evident in the following.

5.2 Linear Optical Properties: The Continuum of Interband Transitions

In our discussion of linear optical properties, we shall first exemplify the influence of the static Coulomb interaction on optical spectra over a large energy range, which requires the calculation of optical transitions throughout the whole Brillouin zone. Our starting point is (5.12), which can be considerably simplified in the linear case. We assume that for times long before the optical excitation, the system is in its ground state. This implies, as initial conditions, that all off-diagonal elements of the density matrix vanish, as well as the occupation of conduction states, and that all valence states are occupied. In the linear case these initial conditions imply that only off-diagonal elements coupling the valence and conduction bands contribute, whereas intraband contributions remain zero. Thus we obtain the following from (5.12) with $n_{\mathrm{vv}} \equiv f_{\mathrm{v}} = 1$ and $n_{\mathrm{cc}} \equiv f_{\mathrm{c}} = 0$ for the transition amplitude $p_{\mathrm{vc}}(\boldsymbol{k}, t)$:[1]

$$\begin{aligned}
&\left(\mathrm{i}\hbar\frac{\partial}{\partial t} + \mathrm{i}\hbar\gamma_2 + E_{\mathrm{v}}(\boldsymbol{k}) - E_{\mathrm{c}}(\boldsymbol{k})\right) p_{\mathrm{vc}}(\boldsymbol{k}, t) \\
&= -\boldsymbol{d}_{\mathrm{cv}}(\boldsymbol{k})\, \boldsymbol{E}(t) + e^2 \lim_{\boldsymbol{q}\to 0} \sum_{\mathrm{v'c'}\boldsymbol{k'}\boldsymbol{G}} v(\boldsymbol{G}+\boldsymbol{q}) \Big(\langle \mathrm{c}, \boldsymbol{k}+\boldsymbol{q} | \mathrm{e}^{\mathrm{i}(\boldsymbol{G}+\boldsymbol{q})\boldsymbol{r}} | \mathrm{v}, \boldsymbol{k} \rangle \\
&\times \langle \mathrm{v'}, \boldsymbol{k'} | \mathrm{e}^{-\mathrm{i}(\boldsymbol{G}+\boldsymbol{q})\boldsymbol{r}} | \mathrm{c'}, \boldsymbol{k'}+\boldsymbol{q} \rangle\, p_{\mathrm{c'v'}}(\boldsymbol{k'}, t) + \langle \mathrm{c}, \boldsymbol{k}+\boldsymbol{q} | \mathrm{e}^{\mathrm{i}(\boldsymbol{G}+\boldsymbol{q})\boldsymbol{r}} | \mathrm{v}, \boldsymbol{k} \rangle \\
&\times \langle \mathrm{c'}, \boldsymbol{k'} | \mathrm{e}^{-\mathrm{i}(\boldsymbol{G}+\boldsymbol{q})\boldsymbol{r}} | \mathrm{v'}, \boldsymbol{k'}+\boldsymbol{q} \rangle\, p_{\mathrm{v'c'}}(\boldsymbol{k'}, t) \Big) \\
&-e^2 \sum_{\mathrm{v'c'}\boldsymbol{k'}} v(\boldsymbol{k}-\boldsymbol{k'}) \Big(\langle \mathrm{c}, \boldsymbol{k} | \mathrm{e}^{\mathrm{i}(\boldsymbol{k}-\boldsymbol{k'})\boldsymbol{r}} | \mathrm{c'}, \boldsymbol{k'} \rangle \langle \mathrm{v'}, \boldsymbol{k'} | \mathrm{e}^{-\mathrm{i}(\boldsymbol{k}-\boldsymbol{k'})\boldsymbol{r}} | \mathrm{v}, \boldsymbol{k} \rangle p_{v'c'}(\boldsymbol{k'}, t) \\
&+ \langle \mathrm{c}, \boldsymbol{k} | \mathrm{e}^{\mathrm{i}(\boldsymbol{k}-\boldsymbol{k'})\boldsymbol{r}} | \mathrm{v'}, \boldsymbol{k'} \rangle \langle \mathrm{c'}, \boldsymbol{k'} | \mathrm{e}^{-\mathrm{i}(\boldsymbol{k}-\boldsymbol{k'})\boldsymbol{r}} | \mathrm{v}, \boldsymbol{k} \rangle p_{\mathrm{c'v'}}(\boldsymbol{k'}, t) \Big) \,. \qquad (5.14)
\end{aligned}$$

A corresponding equation is valid for $p_{\mathrm{cv}}(\boldsymbol{k}, t) = p^*_{\mathrm{vc}}(\boldsymbol{k}, t)$. This symmetry property follows immediately from (5.12). We can treat the time derivative in the linear case by means of Fourier transformation, which allows us to eliminate also the external field. Using (2.130), we obtain the following for the *linear polarization* in the frequency domain:

$$\boldsymbol{P}(\omega) = \frac{1}{V} \sum_{\mathrm{v}\mathrm{c}\boldsymbol{k}} \Big(\boldsymbol{d}_{\mathrm{vc}}(\boldsymbol{k})\, p_{\mathrm{cv}}(\boldsymbol{k}, \omega) + \boldsymbol{d}_{\mathrm{cv}}(\boldsymbol{k})\, p_{\mathrm{vc}}(\boldsymbol{k}, \omega) \Big) \,. \qquad (5.15)$$

This suggests the definition

$$p_{\mathrm{vc}}(\boldsymbol{k}, \omega) = \sum_{\mathrm{v'c'}\boldsymbol{k'}} \chi_{\mathrm{vcc'v'}}(\boldsymbol{k}, \boldsymbol{k'}, \omega)\, \boldsymbol{d}_{\mathrm{v'c'}}(\boldsymbol{k'}) \boldsymbol{E}(\omega) \,. \qquad (5.16)$$

From the Fourier transform of the Hermitian conjugate of (5.14) we see, correspondingly, that

$$p_{\mathrm{cv}}(\boldsymbol{k}, \omega) = \sum_{\mathrm{v'c'}\boldsymbol{k'}} \chi^*_{\mathrm{vcc'v'}}(\boldsymbol{k}, \boldsymbol{k'}, -\omega)\, \boldsymbol{d}_{\mathrm{c'v'}}(\boldsymbol{k'}) \boldsymbol{E}(\omega) \,. \qquad (5.17)$$

is valid. Considering the limit $\boldsymbol{q} \to 0$, we obtain the following from (5.14) with these definitions:

$$\begin{aligned}
&\chi_{\mathrm{vcc'v'}}(\boldsymbol{k}, \boldsymbol{k'}, \omega) = P^0_{\mathrm{vc}}(\boldsymbol{k}, \omega) \Big[\delta_{\mathrm{vv'}}\, \delta_{\mathrm{cc'}}\, \delta_{\boldsymbol{k}\boldsymbol{k'}} + \sum_{\mathrm{v''c''}\boldsymbol{k''}} v(\boldsymbol{q})\, \boldsymbol{q}\, \boldsymbol{d}_{\mathrm{cv}}(\boldsymbol{k}) \\
&\times \Big(\boldsymbol{q}\, \boldsymbol{d}_{\mathrm{c''v''}}(\boldsymbol{k''}) \chi_{\mathrm{v''c''c'v'}}(\boldsymbol{k''}, \boldsymbol{k'}, \omega) + \boldsymbol{q}\, \boldsymbol{d}_{\mathrm{v''c''}}(\boldsymbol{k''}) \chi^*_{\mathrm{v''c''c'v'}}(\boldsymbol{k''}, \boldsymbol{k'}, -\omega) \Big) \Big]
\end{aligned}$$

[1] Here and in the following we use the notation introduced in Sect. 4.1

$$
\begin{aligned}
&- e^2 P^0_{\mathrm{vc}}(\boldsymbol{k},\omega) \sum_{\mathrm{v''c''}\boldsymbol{k}''\boldsymbol{G}\neq 0} v(\boldsymbol{G})\Big(\langle \mathrm{c}\boldsymbol{k}|\mathrm{e}^{\mathrm{i}\boldsymbol{G}\boldsymbol{r}}|\mathrm{v}\boldsymbol{k}\rangle \langle \mathrm{c''}\boldsymbol{k}''|\mathrm{e}^{-\mathrm{i}\boldsymbol{G}\boldsymbol{r}}|\mathrm{v''}\boldsymbol{k}''\rangle \\
&\times \chi_{\mathrm{v''c''c'v'}}(\boldsymbol{k}'',\boldsymbol{k}',\omega) + \langle \mathrm{c}\boldsymbol{k}|\mathrm{e}^{\mathrm{i}\boldsymbol{G}\boldsymbol{r}}|\mathrm{v}\boldsymbol{k}\rangle \langle \mathrm{v''}\boldsymbol{k}''|\mathrm{e}^{-\mathrm{i}\boldsymbol{G}\boldsymbol{r}}|\mathrm{c''}\boldsymbol{k}''\rangle \\
&\times \chi^*_{\mathrm{v''c''c'v'}}(\boldsymbol{k}'',\boldsymbol{k}',-\omega)\Big) + e^2 P^0_{\mathrm{vc}}(\boldsymbol{k},\omega) \sum_{\mathrm{v''c''}\boldsymbol{k}''} v(\boldsymbol{k}-\boldsymbol{k}'') \\
&\times \Big(\langle \mathrm{c}\boldsymbol{k}|\mathrm{e}^{-\mathrm{i}(\boldsymbol{k}''-\boldsymbol{k})\boldsymbol{r}}|\mathrm{c''}\boldsymbol{k}''\rangle \langle \mathrm{v''}\boldsymbol{k}''|\mathrm{e}^{\mathrm{i}(\boldsymbol{k}''-\boldsymbol{k})\boldsymbol{r}}|\mathrm{v}\boldsymbol{k}\rangle\, \chi_{\mathrm{v''c''c'v'}}(\boldsymbol{k}'',\boldsymbol{k}',\omega) \\
&+ \langle \mathrm{c}\boldsymbol{k}|\mathrm{e}^{\mathrm{i}(\boldsymbol{k}-\boldsymbol{k}'')\boldsymbol{r}}|\mathrm{v''}\boldsymbol{k}''\rangle \langle \mathrm{c''}\boldsymbol{k}''|\mathrm{e}^{-\mathrm{i}(\boldsymbol{k}-\boldsymbol{k}'')\boldsymbol{r}}|\mathrm{v}\boldsymbol{k}\rangle\, \chi^*_{\mathrm{v''c''c'v'}}(\boldsymbol{k}'',\boldsymbol{k}',-\omega)\Big) . \qquad (5.18)
\end{aligned}
$$

The first quantity in the large parantheses on the right-hand side of (5.18) results from the long-range part of the Coulomb interaction, i.e. the $\boldsymbol{q} \to 0$ limit of the $\boldsymbol{G} = 0$ Hartree contribution in (5.14). We shall see that this contribution can be reduced to the long-wavelength limit of the *longitudinal inverse dielectric function* (compare Sect. 2.6). We postpone a more detailed discussion and the derivation of the dielectric function to the end of this section. Further, we have defined the *polarization function* of the noninteracting system as

$$
P^0_{\mathrm{vc}}(\boldsymbol{k},\omega) = \frac{1}{E_{\mathrm{c}}(\boldsymbol{k}) - E_{\mathrm{v}}(\boldsymbol{k}) - \hbar\omega - \mathrm{i}\hbar\gamma_2} . \qquad (5.19)
$$

Making use of (5.15)–(5.17) and the relationship between the macroscopic polarization and field

$$
\boldsymbol{P}(\omega) = \varepsilon_0\, \underline{\chi}(\omega)\boldsymbol{E}(\omega) , \qquad (5.20)
$$

we obtain from the solution of (5.18) the *linear susceptibility tensor*, given by

$$
\begin{aligned}
\underline{\chi}(\omega) = \frac{1}{\varepsilon_0 V} \sum_{\mathrm{vcv'c'}\boldsymbol{k}\boldsymbol{k}'} &\Big(\boldsymbol{d}_{\mathrm{cv}}(\boldsymbol{k})\, \chi_{\mathrm{vcc'v'}}(\boldsymbol{k},\boldsymbol{k}',\omega)\, \boldsymbol{d}_{\mathrm{v'c'}}(\boldsymbol{k}') \\
&+ \boldsymbol{d}_{\mathrm{vc}}(\boldsymbol{k})\, \chi^*_{\mathrm{vcc'v'}}(\boldsymbol{k},\boldsymbol{k}',-\omega)\, \boldsymbol{d}_{\mathrm{c'v'}}(\boldsymbol{k}')\Big) , \qquad (5.21)
\end{aligned}
$$

which contains all information about the longitudinal and the transverse response of the system.

5.2.1 The Bethe–Salpeter Equation

In order to make closer contact with the general theory of Chap. 2, we can decompose the contributions of the exchange and the Hartree interaction in (5.18). Owing to the coupling of resonant and off-resonant contributions, this must be done for the dipole and the $\boldsymbol{G} \neq 0$ contributions separately. We introduce an equation which determines the so-called *irreducible polarization propagator*,[2] and by definition includes all interaction effects except of the short-range Hartree interaction in (5.18):

$$\begin{aligned} P_{\mathrm{vcv'c'}}(\boldsymbol{k},\boldsymbol{k}',\omega) &= P^0_{\mathrm{vc}}(\boldsymbol{k},\omega)\,\delta_{\mathrm{vv'}}\,\delta_{\mathrm{cc'}}\,\delta_{\boldsymbol{k}\,\boldsymbol{k}'} \\ &+ \sum_{\mathrm{v''c''}\boldsymbol{k}''}\Big(P^0_{\mathrm{vc}}(\boldsymbol{k},\omega)V_{\mathrm{cc''v''v}}(\boldsymbol{k},\boldsymbol{k}'')P_{\mathrm{v''c''v'c'}}(\boldsymbol{k}'',\boldsymbol{k}',\omega) \\ &+ P^0_{\mathrm{vc}}(\boldsymbol{k},\omega)V_{\mathrm{cv''c''v}}(\boldsymbol{k},\boldsymbol{k}'')P^{*}_{\mathrm{v''c''}}(\boldsymbol{k}'',\boldsymbol{k}',-\omega)\Big)\,. \end{aligned} \tag{5.22}$$

Here we have defined the interaction matrix element

$$V_{\mathrm{cc'v'v}}(\boldsymbol{k},\boldsymbol{k}') = e^2\langle \mathrm{c}\boldsymbol{k}|\mathrm{e}^{-\mathrm{i}(\boldsymbol{k}'-\boldsymbol{k})\boldsymbol{r}}|\mathrm{c}'\boldsymbol{k}'\rangle v(\boldsymbol{k}-\boldsymbol{k}')\langle \mathrm{v}'\boldsymbol{k}'|\mathrm{e}^{\mathrm{i}(\boldsymbol{k}'-\boldsymbol{k})\boldsymbol{r}}|\mathrm{v}\boldsymbol{k}\rangle\,. \tag{5.23}$$

An equation of the type (5.22) is often referred to as the *Bethe–Salpeter equation* and describes two-particle correlations. Whereas this equation is usually derived by starting from two-particle functions, we have applied the one-particle picture only. This is possible owing to the explicit consideration of an external field which yields nonvanishing off-diagonal elements of the one-particle density matrix. As we have shown – at least at the Hartree–Fock level – the information resulting from these off-diagonal elements is completely equivalent to a description in terms of two-particle functions. Beyond the Hartree–Fock approximation, this statement is no longer valid in the nonlinear regime. If occupation of conduction band states comes into play, the Bethe–Salpeter equation derived from a two-particle function always involves additional interaction effects which are not present in the off-diagonal elements of the one-particle density matrix.

With the solution of (5.22) we can write the susceptibility tensor resulting from (5.18) and (5.21) in the form

$$\begin{aligned} \underline{\chi}(\omega) &= \underline{P}(\omega) + \frac{1}{q^2}[\underline{P}(\omega)\,\boldsymbol{q}]\,[\boldsymbol{q}\,\underline{\chi}(\omega)] \\ &\quad - \varepsilon_0 V e^2 \sum_{\boldsymbol{G}\neq 0} \boldsymbol{P}(0,\boldsymbol{G},\omega)v(\boldsymbol{G})\boldsymbol{\chi}(\boldsymbol{G},0,\omega)\,, \end{aligned} \tag{5.24}$$

where correspondingly, the short-range contribution to the susceptibility tensor is determined by

[2] A formal definition will be given in Chap. 10.

$$\chi(G,0,\omega) = P(G,0,\omega) + \frac{1}{q^2}[P(G,0,\omega)\,q]\,[q\,\underline{\chi}(\omega)] - \varepsilon_0 V e^2 \sum_{G'\neq 0} P(G,G',\omega)\,v(G')\,\chi(G',0,\omega)\,. \tag{5.25}$$

In (5.24) and (5.25) we have introduced the following definitions. The long-range polarization propagator is obtained from the solution of the Bethe–Salpeter equation as

$$\underline{P}(\omega) = \frac{1}{\varepsilon_0 V} \sum_{\mathrm{vcv'c'}\boldsymbol{kk'}} \Big(\boldsymbol{d}_{\mathrm{cv}}(\boldsymbol{k})\, P_{\mathrm{vcv'c'}}(\boldsymbol{k},\boldsymbol{k'},\omega)\, \boldsymbol{d}_{\mathrm{v'c'}}(\boldsymbol{k'}) + \boldsymbol{d}_{\mathrm{vc}}(\boldsymbol{k})\, P^*_{\mathrm{vcv'c'}}(\boldsymbol{k},\boldsymbol{k'},-\omega)\, \boldsymbol{d}_{\mathrm{c'v'}}(\boldsymbol{k'}) \Big)\,. \tag{5.26}$$

The short-range contribution is determined by

$$\boldsymbol{P}(0,\boldsymbol{G},\omega) = \frac{1}{\varepsilon_0 V} \sum_{\mathrm{vcv'c'}\boldsymbol{kk'}} \Big(\boldsymbol{d}_{\mathrm{vc}}(\boldsymbol{k}) P_{\mathrm{vcv'c'}}(\boldsymbol{k},\boldsymbol{k'},\omega) \langle \mathrm{c'}\boldsymbol{k'}|\mathrm{e}^{-\mathrm{i}\boldsymbol{Gr}}|\mathrm{v'}\boldsymbol{k'}\rangle + \boldsymbol{d}_{\mathrm{cv}}(\boldsymbol{k}) P^*_{\mathrm{vcv'c'}}(\boldsymbol{k},\boldsymbol{k'},-\omega) \langle \mathrm{v'}\boldsymbol{k'}|\mathrm{e}^{-\mathrm{i}\boldsymbol{Gr}}|\mathrm{c'}\boldsymbol{k'}\rangle \Big) \tag{5.27}$$

and corresponding equations are valid for $\boldsymbol{P}(\boldsymbol{G},0,\omega)$ and $P(\boldsymbol{G},\boldsymbol{G'},\omega)$. Thus the solution of the Bethe–Salpeter equation, together with (5.24) and (5.25), determines the longitudinal and transverse response functions. Approximations enter the Bethe-Salpeter equation, which we have introduced in its simplest version. An important improvement of the Hartree–Fock approximation is obtained from dynamical screening of the exchange interaction in (5.22).[3] At the moment we shall take this screening only roughly into account by making the replacement $v(\boldsymbol{k}) \to v(\boldsymbol{k})/\varepsilon_\infty$ in the Coulomb interaction, where ε_∞ is the static electronic dielectric function. Alternatively, we could make use of the background dielectric function ε_b, introduced in Chap. 2.

Although we are working on the Hartree–Fock level only, the practical evaluation of the theory discussed so far is rather involved. Not only is a knowledge of the band structure and Bloch functions throughout the Brillouin zone required, but also the solution of a complicated integral equation. Hence it is not surprising that most existing calculations of linear optical spectra are performed at the simplest level, in which interaction effects are completely neglected. The corresponding result for a two-band model has already been presented in the last chapter. Although there is the risk of some redundancy, we shall dicuss the dielectric function on a more general level in the following, which applies the general theory of Chap. 2 in the case of the Hartree–Fock approximation.

---

[3] This will be treated in detail in Chap. 10, where we shall see that the dynamical screened Coulomb interaction is determined by the inverse dielectric function $\varepsilon^{-1}(\boldsymbol{k},\omega)$

5.2.2 The Dielectric Function

In order to rediscover the various types of response functions introduced in Chap. 2, we could go through the derivation again, avoid the limit $\boldsymbol{q} \to 0$, and consider the $\boldsymbol{pA}$ or the $\rho\Phi$ coupling instead of the $\boldsymbol{dE}$-coupling. It is an easy task to show that all these types of response functions contain the same information and can immediately be derived for $\boldsymbol{q} \to 0$ from the solution of (5.18). Let us now consider the density–density correlation for finite wave vector, which enters the theory of screening, for example. Taking the dependence on the finite wave vector in (5.12) into account, we obtain in the linear case the following instead of (5.18):

$$\begin{aligned}
&\chi_{\mathrm{vcc'v'}}(\boldsymbol{k},\boldsymbol{k}+\boldsymbol{q},\boldsymbol{k}'+\boldsymbol{q},\boldsymbol{k}',\omega) = P^0_{\mathrm{vc}}(\boldsymbol{k},\boldsymbol{k}+\boldsymbol{q},\omega)\Bigg[\delta_{\mathrm{vv'}}\,\delta_{\mathrm{cc'}}\,\delta_{\boldsymbol{kk}'}\\
&-e^2\sum_{\mathrm{v''c''}\boldsymbol{k}''\boldsymbol{G}} v(\boldsymbol{G}+\boldsymbol{q})\,\langle \mathrm{c},\boldsymbol{k}+\boldsymbol{q}|\mathrm{e}^{\mathrm{i}(\boldsymbol{G}+\boldsymbol{q})\boldsymbol{r}}|\mathrm{v},\boldsymbol{k}\rangle\\
&\times\Bigg(\langle \mathrm{v''},\boldsymbol{k}''|\mathrm{e}^{-\mathrm{i}(\boldsymbol{G}+\boldsymbol{q})\boldsymbol{r}}|\mathrm{c''},\boldsymbol{k}''+\boldsymbol{q}\rangle\,\chi_{\mathrm{v''c''c'v'}}(\boldsymbol{k}'',\boldsymbol{k}''+\boldsymbol{q},\boldsymbol{k}'+\boldsymbol{q},\boldsymbol{k}',\omega)\\
&+\langle \mathrm{c''},\boldsymbol{k}''|\mathrm{e}^{-\mathrm{i}(\boldsymbol{G}+\boldsymbol{q})\boldsymbol{r}}|\mathrm{v''},\boldsymbol{k}''+\boldsymbol{q}\rangle\,\chi^*_{\mathrm{v''c''c'v'}}(\boldsymbol{k}''+\boldsymbol{q},\boldsymbol{k}'',\boldsymbol{k}',\boldsymbol{k}'+\boldsymbol{q},-\omega)\Bigg)\\
&+e^2\sum_{\mathrm{v''c''}\boldsymbol{k}''} v(\boldsymbol{k}-\boldsymbol{k}'')\langle \mathrm{c},\boldsymbol{k}+\boldsymbol{q}|\mathrm{e}^{-\mathrm{i}(\boldsymbol{k}''-\boldsymbol{k})\boldsymbol{r}}|\mathrm{c''},\boldsymbol{k}''+\boldsymbol{q}\rangle\\
&\times\langle \mathrm{v''},\boldsymbol{k}''|\mathrm{e}^{\mathrm{i}(\boldsymbol{k}''-\boldsymbol{k})\boldsymbol{r}}|\mathrm{v},\boldsymbol{k}\rangle\,\chi_{\mathrm{v''c''c'v'}}(\boldsymbol{k}'',\boldsymbol{k}''+\boldsymbol{q},\boldsymbol{k}'+\boldsymbol{q},\boldsymbol{k}',\omega)\\
&+e^2\sum_{\mathrm{v''c''}\boldsymbol{k}''} v(\boldsymbol{k}-\boldsymbol{k}'')\,\langle \mathrm{c},\boldsymbol{k}+\boldsymbol{q}|\mathrm{e}^{\mathrm{i}(\boldsymbol{k}-\boldsymbol{k}'')\boldsymbol{r}}|\mathrm{v''},\boldsymbol{k}''+\boldsymbol{q}\rangle\\
&\times\langle \mathrm{c''},\boldsymbol{k}''|\mathrm{e}^{-\mathrm{i}(\boldsymbol{k}-\boldsymbol{k}'')\boldsymbol{r}}|\mathrm{v},\boldsymbol{k}\rangle\,\chi^*_{\mathrm{v''c''c'v'}}(\boldsymbol{k}''+\boldsymbol{q},\boldsymbol{k}'',\boldsymbol{k}',\boldsymbol{k}'+\boldsymbol{q},-\omega)\Bigg]\,. \qquad (5.28)
\end{aligned}$$

We use here the $\boldsymbol{q} \neq 0$ version of (5.19),

$$P^0_{\mathrm{vc}}(\boldsymbol{k},\boldsymbol{k}+\boldsymbol{q},\omega) = \frac{1}{E_{\mathrm{c}}(\boldsymbol{k}+\boldsymbol{q}) - E_{\mathrm{v}}(\boldsymbol{k}) - \hbar\omega - \mathrm{i}\hbar\gamma_2}\,. \qquad (5.29)$$

Replacing the dipole matrix elements in (5.21) by plane-wave transition matrix elements and $\chi_{\mathrm{vcc'v'}}(\boldsymbol{k},\boldsymbol{k}',\omega)$ by the solution of (5.28) yields the following for the *susceptibility at finite wave vector*:

$$\begin{aligned}
\chi(\boldsymbol{q},\omega) = \frac{e^2}{\varepsilon_0 V}\sum_{\mathrm{vcv'c'}\boldsymbol{kk}'}&\Big(\langle \mathrm{v},\boldsymbol{k}|\,\mathrm{e}^{-\mathrm{i}\boldsymbol{qr}}\,|\mathrm{c},\boldsymbol{k}+\boldsymbol{q}\rangle\\
&\times\chi_{\mathrm{vcc'v'}}(\boldsymbol{k},\boldsymbol{k}+\boldsymbol{q},\boldsymbol{k}'+\boldsymbol{q},\boldsymbol{k}',\omega)\,\langle \mathrm{c'},\boldsymbol{k}'+\boldsymbol{q}|\,\mathrm{e}^{\mathrm{i}\boldsymbol{qr}}\,|\mathrm{v'},\boldsymbol{k}'\rangle\\
&+\langle \mathrm{c},\boldsymbol{k}|\,\mathrm{e}^{-\mathrm{i}\boldsymbol{qr}}\,|\mathrm{v},\boldsymbol{k}+\boldsymbol{q}\rangle\,\chi^*_{\mathrm{vcc'v'}}(\boldsymbol{k}+\boldsymbol{q},\boldsymbol{k},\boldsymbol{k}',\boldsymbol{k}'+\boldsymbol{q},-\omega)\\
&\times\langle \mathrm{v'},\boldsymbol{k}'+\boldsymbol{q}|\,\mathrm{e}^{\mathrm{i}\boldsymbol{qr}}\,|\mathrm{c'},\boldsymbol{k}'\rangle\Big)\,.
\end{aligned} \qquad (5.30)$$

The expansion of (5.30) for small wave vector reproduces immediately the susceptibility tensor (5.21), according to

$$\chi(\omega) \;=\; \boldsymbol{q}\,\underline{\chi}(\omega)\,\boldsymbol{q}\,. \tag{5.31}$$

Applying this projection onto the direction of the wave vector of the field to (5.24), we obtain

$$\chi(\omega) \;=\; P(\omega)\Big(1+\frac{1}{q^2}\,\boldsymbol{q}\,\underline{\chi}(\omega)\,\boldsymbol{q}\Big) - \varepsilon_0 V\,e^2 \sum_{\boldsymbol{G}\neq 0} P(0,\boldsymbol{G},\omega)v(\boldsymbol{G})\chi(\boldsymbol{G},0,\omega) \tag{5.32}$$

and, correspondingly, from (5.25)

$$\begin{aligned}\chi(\boldsymbol{G},0,\omega) &= P(\boldsymbol{G},0,\omega)\,\Big(1+\frac{1}{q^2}\,\boldsymbol{q}\,\underline{\chi}(\omega)\,\boldsymbol{q}\Big)\\ &\quad - \varepsilon_0 V\,e^2 \sum_{\boldsymbol{G}'\neq 0} P(\boldsymbol{G},\boldsymbol{G}',\omega)v(\boldsymbol{G}')\chi(\boldsymbol{G}',0,\omega)\,.\end{aligned} \tag{5.33}$$

We can now introduce the susceptibility $\tilde{\chi}(\omega)$ as defined in (2.167) and (2.173) and eliminate the macroscopic dielectric function resulting from the long-range contribution to the Coulomb potential in (5.21), (5.32) and (5.33). Without the projection (5.31), the elimination of this long-range interaction in (5.24) and (5.25) requires multiplication by the dielectric tensor. As a result, we find the following for the linear susceptibility and the macroscopic dielectric function:

$$\chi^{(1)}(\omega) \;=\; \frac{1}{q^2}\sum_{ij} q_i\,\tilde{\chi}_{ij}(\omega)\,q_j \;=\; \varepsilon(\omega)\;-\;1\,. \tag{5.34}$$

For cubic symmetry $\chi^{(1)}$, is again identical to the diagonal elements of the susceptibility tensor in (5.34).

Before we turn next to a more elaborate discussion of the effects of interaction on optical spectra, let us consider the electronic dielectric function due to noninteracting transitions. Using (5.19) in (5.30), we obtain the dielectric function as

$$\begin{aligned}\varepsilon(\boldsymbol{q},\omega) \;&=\; 1+e^2\,v(\boldsymbol{q})\sum_{\mathrm{vc}\boldsymbol{k}}\Bigg(\frac{\langle \mathrm{v},\boldsymbol{k}|\,\mathrm{e}^{\mathrm{i}\boldsymbol{qr}}\,|\mathrm{c},\boldsymbol{k}+\boldsymbol{q}\rangle\,\langle \mathrm{c},\boldsymbol{k}+\boldsymbol{q}|\,\mathrm{e}^{-\mathrm{i}\boldsymbol{qr}}\,|\mathrm{v},\boldsymbol{k}\rangle}{E_\mathrm{c}(\boldsymbol{k}+\boldsymbol{q})-E_\mathrm{v}(\boldsymbol{k})-\hbar\omega-\mathrm{i}\hbar\gamma_2}\\ &\quad+\;\frac{\langle \mathrm{v},\boldsymbol{k}+\boldsymbol{q}|\,\mathrm{e}^{-\mathrm{i}\boldsymbol{qr}}\,|\mathrm{c},\boldsymbol{k}\rangle\,\langle \mathrm{c},\boldsymbol{k}|\,\mathrm{e}^{\mathrm{i}\boldsymbol{qr}}\,|\mathrm{v},\boldsymbol{k}+\boldsymbol{q}\rangle}{E_\mathrm{c}(\boldsymbol{k})-E_\mathrm{v}(\boldsymbol{k}+\boldsymbol{q})+\hbar\omega+\mathrm{i}\hbar\gamma_2}\Bigg)\,.\end{aligned} \tag{5.35}$$

If we introduce the Fermi functions for $T=0$, we can write (5.35) in a more familiar notation [78]:

$$\begin{aligned}\varepsilon(\boldsymbol{q},\omega) = 1\;-\;e^2\,v(\boldsymbol{q})&\sum_{nm\boldsymbol{k}}\langle n,\boldsymbol{k}|\,\mathrm{e}^{\mathrm{i}\boldsymbol{qr}}\,|m,\boldsymbol{k}+\boldsymbol{q}\rangle\,\langle m,\boldsymbol{k}+\boldsymbol{q}|\,\mathrm{e}^{-\mathrm{i}\boldsymbol{qr}}\,|n,\boldsymbol{k}\rangle\\ &\times\,\frac{f_n(\boldsymbol{k}+\boldsymbol{q})\;-\;f_m(\boldsymbol{k})}{E_n(\boldsymbol{k}+\boldsymbol{q})-E_m(\boldsymbol{k})-\hbar\omega-\mathrm{i}\hbar\gamma_2}\,.\end{aligned} \tag{5.36}$$

Again, the expansion for small wave vector reproduces immediately our earlier result (4.24), found for noninteracting systems in Chap. 4. According to our derivation, this equation seems to be valid only for $f_n(\boldsymbol{k}) = \delta_{n\mathrm{v}}$. This restriction, however, is unnecessary. We can repeat our derivation immediately for arbitrary one-particle distributions, which may be determined by thermal excitation of the system, for example, i.e. our initial conditions are not those for zero temperature but those for finite temperature. The distribution may also result from an optical excitation of the system. In this case, however, we leave the linear regime.

The simplest description of nonlinear effects is obtained from the intraband transitions in (5.36), which are due to the generation of a nonequilibrium one-particle distribution by an external field. When one considers these contributions, a fundamental difference between inter- and intraband transitions becomes obvious. For the interband contributions $n \neq m$, the small-$\boldsymbol{q}$ expansion leads to dipole matrix elements and to a cancellation of the singularity of the Coulomb potential. In contrast, for intraband contributions $n = m$, the leading contribution to the matrix elements is 1, and thus the singularity of the Coulomb potential is not canceled. Obviously, this is the same behavior as in metallic systems, where intraband transitions are the leading contribution. The different properties of the interband and intraband contributions have important consequences, as (5.36) does not only describe optical properties but determines also the screening of the interaction. This will be discussed in detail in Chap. 10.

Let us come back to the strictly linear case, in which only interband contributions occur. The Bethe–Salpeter equation can be solved by means of various techniques. A transformation to a basis of localized orbitals is an appropriate way [74, 75] to take the short-range part of the Coulomb interaction into account. Long-range effects, however, have to be treated in $\boldsymbol{k}$-space [76]. Moreover, attention has been focused on modeling of dielectric functions [77], which are required to screen the exchange interaction in the calculation of quasiparticle energies (compare Sect. 10.6). Some mathematically more rigorous approaches are based on a plane-wave representation of Bloch functions and a treatment of the interaction contributions in $\boldsymbol{k}$ space [79, 80, 81].

An experimental spectrum showing the imaginary part of the dielectric function of silicon is compared in Fig. 5.1 with results calculated in various approximations [79]. The calculated curves show the $\boldsymbol{q} \to 0$ limit of (5.36), as well as the influence of local-field corrections and of the exchange contributions.[4] A systematic deviation between the experimental and theoretical data occurs without interaction. On the low-energy side the oscillator strength of the calculated spectrum is much lower than that of the experimental result, whereas an opposite deviation occurs on the high-energy side. In particular,

[4] Note that these calculations take dynamical screening of the electron–hole attraction into account, which is beyond the Hartree–Fock theory developed in this chapter (compare Chap. 10). Without screening, the exchange interactions are drastically overestimated.

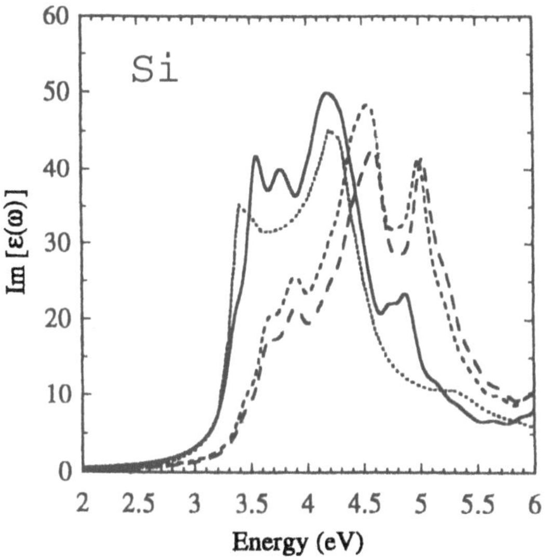

Fig. 5.1. Imaginary part of the dielectric function of Si versus energy in various approximations. The experimental values (*dotted line*) are compared with different theoretical approximations: no interaction (*short-dashed line*), local-field effects only (*long-dashed line*), local-field effects and screened exchange interaction (*full line*). Taken from [79]

this deviation becomes even worse on the low-energy side if local-field effects are taken additionally into account. Only when the screened exchange is included is there a pronounced transfer of oscillator strength to the low-energy side, which leads to excellent agreement with the experimental data. Similar results are found for other semiconductors and insulators [80, 81], where the influence of interaction effects is even more prominent. This is illustrated in Fig. 5.2, which shows the imaginary part of the dielectric function of MgO, again with and without interaction effects.

The essential physics of these ab initio calculations can be grasped by means of an approximate solution of the problem, which we discuss in the following.

5.3 Solution by Continued Fractions

We restrict ourselves to resonant contributions. In this case it is obvious from (5.18) that we have to solve a problem of the type

$$\chi_{\lambda\lambda'} = P^0_\lambda \delta_{\lambda\lambda'} + \sum_{\lambda''} P^0_\lambda V_{\lambda\lambda''} \chi_{\lambda''\lambda'} \,, \tag{5.37}$$

where we have introduced a multi-index $\lambda \equiv \{vc\boldsymbol{k}\}$ and an interaction matrix V, the definition of which is obvious from (5.18). We can solve this equation formally by iteration and represent the susceptibility in the form of a Neumann series

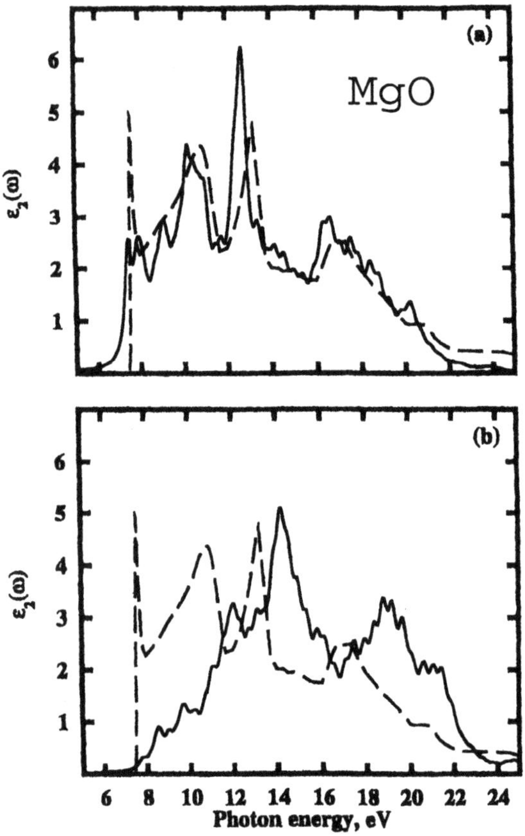

Fig. 5.2. Imaginary part of the dielectric function of MgO versus energy with (**a**) and without (**b**) electron–hole interaction in comparison with experiment (*dashed line*). Taken from [81]

$$\tilde{\chi} = \frac{1}{\varepsilon_0 V} \sum_{\lambda\lambda'} d_\lambda \chi_{\lambda\lambda'} d_{\lambda'}$$
$$= \sum_{\lambda\lambda'} w_\lambda \left(\delta_{\lambda\lambda'} + h_{\lambda\lambda'} + \sum_{\lambda''} h_{\lambda\lambda''} h_{\lambda''\lambda'} + \dots \right) w_{\lambda'} \tag{5.38}$$

where we have defined

$$h_{\lambda\lambda'} = \sqrt{P^0_\lambda} V_{\lambda\lambda'} \sqrt{P^0_{\lambda'}} \tag{5.39}$$

and

$$w_\lambda = \frac{\sqrt{P^0_\lambda} d_\lambda}{\sqrt{\varepsilon_0 V}} \; . \tag{5.40}$$

As the interaction matrix is by no means a weak perturbation, the series expansion will converge slowly only or even diverge. In this case a resummation as a *continued fraction* leads to a rapidly converging expansion. The method is quite general and is closely related to the Padè approximation technique

[82]. In the first step we transform the matrix $\underline{h}$ into a tridiagonal representation. This can be accomplished most easily if we consider $\underline{h}$ as a special representation of an operator h acting on a vector space spanned by states $|\lambda\rangle$. Now we construct a new basis such that $\langle 0|h|0\rangle = \sum w_\lambda \langle\lambda|h|\lambda'\rangle w_{\lambda'}$, which is realized if $|0\rangle = \sum w_\lambda|\lambda\rangle$. The next-higher state results from the action of h and explicit orthogonalization with respect to $|0\rangle$:

$$|1\rangle = h|0\rangle - a_0|0\rangle \ , \quad a_0 = \frac{\langle 0|h|0\rangle}{\langle 0|0\rangle} \ . \tag{5.41}$$

The next state has to be orthogonalized with respect to $|0\rangle$ as well as $|1\rangle$ and is obtained as

$$|2\rangle = h|1\rangle - a_1|1\rangle - b_1|0\rangle \ , \quad a_1 = \frac{\langle 1|h|1\rangle}{\langle 1|1\rangle} \ , \quad b_1 = \frac{\langle 1||1\rangle}{\langle 0|0\rangle} \ . \tag{5.42}$$

If $|3\rangle$ is orthogonal to both $|2\rangle$ and $|1\rangle$, it is necessarily orthogonal to $|0\rangle$. Thus we obtain the recursion

$$|n+1\rangle = h|n\rangle - a_n|n\rangle - b_n|n-1\rangle \ , \tag{5.43}$$

where

$$a_n = \frac{\langle n|h|n\rangle}{\langle n|n\rangle} \ , \quad b_n = \frac{\langle n|n\rangle}{\langle n-1|n-1\rangle} \ . \tag{5.44}$$

This recursion scheme can be easily mapped onto a matrix representation. Defining $K_{n\lambda} = \langle n|\lambda\rangle$, we obtain from (5.44)

$$a_n = \frac{\sum\limits_{\lambda\lambda'} K_{n\lambda} h_{\lambda\lambda'} K_{n\lambda'}}{\sum\limits_{\lambda} |K_{n\lambda}|^2} \ , \quad b_n = \frac{\sum\limits_{\lambda} |K_{n\lambda}|^2}{\sum\limits_{\lambda} |K_{n-1\lambda}|^2} \ , \tag{5.45}$$

with the recursion formula

$$K_{n+1\lambda} = \sum_{\lambda\lambda'} h_{\lambda\lambda'} K_{n\lambda'} - a_n K_{n\lambda} - b_n K_{n-1\lambda} \ . \tag{5.46}$$

In this new representation the Hamiltonian is obviously tridiagonal, with diagonal elements

$$h_{n,n} = a_n \tag{5.47}$$

and off-diagonal elements

$$h_{n,n-1} = h_{n-1,n} = \frac{\langle n|h|n-1\rangle}{\sqrt{\langle n-1\,|\,n-1\rangle\langle n\,|\,n\rangle}} = \sqrt{b_n} \ . \tag{5.48}$$

Now we come to the second step. According to (5.38), we have to calculate $\langle 0|(1-h)^{-1}|0\rangle$, i.e. the first diagonal matrix element of the inverse of a tridiagonal matrix. Using the standard rules of matrix inversion, we have the following for an $N \times N$ problem:

$$\langle 0|(1-h)^{-1}|0\rangle = \langle 0\,|\,0\rangle \frac{\det\{h^{(N-1)}\}}{\det\{h^{(N)}\}}$$
$$= \langle 0\,|\,0\rangle \frac{\det\{h^{(N-1)}\}}{(1-a_0)\det\{h^{(N-1)}\} - b_1 \det\{h^{(N-2)}\}} . \tag{5.49}$$

By iterating this procedure, we can express the result in the form of a continued fraction:

$$\langle 0|(1-h)^{-1}|0\rangle = \cfrac{\langle 0|0\rangle}{1-a_0-\cfrac{b_1}{1-a_1-\cfrac{b_2}{1-a_2-\dots}}} . \tag{5.50}$$

It is a straightforward exercise to show that the nth-order continued fraction reproduces the Neumann series up to order $(2n-1)$. In particular, if the Neumann series diverges, the continued fraction is truncated. To first order in the interaction, we obtain as an approximate solution

$$\tilde{\chi}(\omega) = \cfrac{\dfrac{1}{\varepsilon_0 V}\displaystyle\sum_\lambda d_\lambda P^0_\lambda d_\lambda}{1-\cfrac{\displaystyle\sum_{\lambda\lambda'} d_\lambda P^0_\lambda V_{\lambda\lambda'} P^0_{\lambda'} d_{\lambda'}}{\displaystyle\sum_\lambda d_\lambda P^0_\lambda d_\lambda}} . \tag{5.51}$$

In a more compact notation, this result takes the simple form

$$\tilde{\chi}(\omega) = \frac{P^0(\omega)}{1 - V(\omega)P^0(\omega)} , \tag{5.52}$$

where the definition of $V(\omega)$ is obtained from (5.51). If the interaction matrix $V_{\lambda\lambda'}$ depends only weakly on the wave vectors, the frequency dependence in $V(\omega)$ cancels and the effect of the Coulomb interaction can be described by a single parameter. This approximation describes, by definition, the interaction averaged over the whole Brillouin zone and thus describes essentially short-range interaction effects. Hence we cannot expect to obtain information about interaction effects which result from special regions, i.e. in the vicinity of band edges. Nevertheless, this simple model is well suited for demonstrating the basic features. It should be noted that an exact evaluation of the lowest-order continued fraction yields results which are not too far from exact solutions obtained by matrix inversion [76].

We illustrate the basic features of the simple model by the optical spectrum of ZnSe. Figure 5.3 shows the imaginary part of the dielectric function for a constant interaction parameter. We have assumed that the local-field corrections are overcompensated by the attractive exchange interaction. This implies $V > 0$. A general effect of this is that oscillator strength is transferred from the high- to the low-energy side of the spectrum, in accordance with the experimental data [83]. This is the same general trend as in the ab initio calculations discussed before.

The methods introduced so far are well suited to describing optical spectra over a wide range of energies, provided the range of the electron–hole interaction is restricted to the order of a few lattice constants. In the vicinity of band edges, however, many semiconductors exhibit Coulomb correlations which are much more long-ranged and lead to the formation of bound states with a correlation length large in comparison with the lattice constant. In order to take these effects into account, we have to look more carefully at the band edge region.

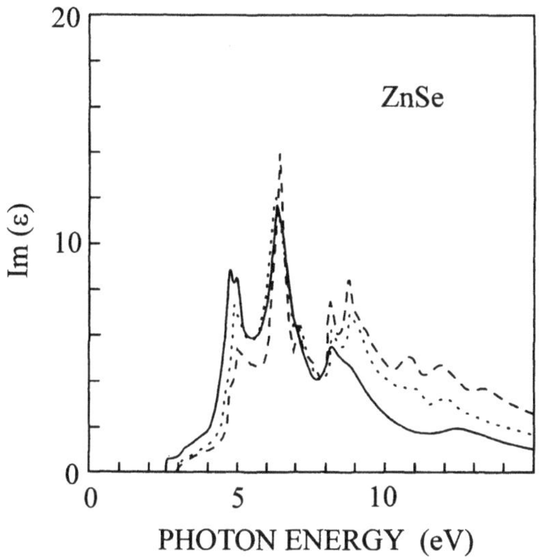

Fig. 5.3. Imaginary part of the dielectric function of ZnSe versus energy. The *full line* shows the experimental values, the *dashed line* results of the theory without interaction, and the *dotted line* results of the theory with a constant interaction parameter. The experimental data are taken from [83]

5.4 Problems

5.1. Derive the equations of motion for the fermion field operators $\psi_s^\dagger(\boldsymbol{r})$ and $\psi_s(\boldsymbol{r})$

5.2. Make use of the expansion (2.59), to show that the Wannier representation of the fermion field operators has the form

$$\psi_s(\boldsymbol{r}) = \frac{1}{\mathcal{N}} \sum_{n,\boldsymbol{L}} \eta_{s_n}(s) w_n(\boldsymbol{r} - \boldsymbol{L})\, c_{n,\boldsymbol{L}} \tag{5.53}$$

and, correspondingly, for $\psi_s(\boldsymbol{r})^\dagger$. Derive the commutation rules for the annilation and creation operators $c_{n,\boldsymbol{L}}$ and $c_{n,\boldsymbol{L}}^\dagger$.

5.3. Derive the Wannier representation of the equation of motion (5.7) by means of the Wannier representation (5.53). Discuss the various length scales, which occur in the problem with and without the effective-mass approximation. Hint: Distinguish between the dependence on lattice vectors, which varies on a scale of a few lattice constants, and the variations, which are related to large changes of the lattice vectors.

5.4. Consider, instead of the dipolar coupling $\boldsymbol{rE}$, the coupling to a vector potential $\boldsymbol{pA}$ and derive the pendant of (5.14) within the dipole approximation. Hint: Make use of the interaction Hamiltonian (2.103), where $\boldsymbol{A}$ can be treated as classical field within the dipole approximation. Note that the contributions, owing to the Coulomb interaction remain unchanged.

5.5. Consider, instead of the dipolar coupling $\boldsymbol{rE}$, the coupling to a scalar potential $\rho\Phi$ and derive the the pendant of (5.14) for finite photon wave vector $\boldsymbol{q}$.

5.6. Make use of the solution of Problem 5.3 to derive the generalization of (5.18) for finite photon wave vector $\boldsymbol{q}$. Show that the density–density correlation function is determined by (5.28) and discuss the limit $\boldsymbol{q} \to 0$. Hint: Note that the one-particle density matrix is not diagonal in the $\boldsymbol{k}$ vector, but has the form $p_{\rm cv}(\boldsymbol{k}, \boldsymbol{k}') = \delta_{\boldsymbol{k}',\boldsymbol{k}+\boldsymbol{q}}\, p_{\rm cv}(\boldsymbol{k}, \boldsymbol{k}+\boldsymbol{q})$.

5.7. Even rather crude approximations to the Coulomb matrix element (5.23) give some insight into the effect of the exchange interaction in the Bethe–Salpeter equation on optical spectra. Solve (5.22) for a parabolic two-band model and a factorized interaction matrix element of the form

$$V_{\rm ccvv}(\boldsymbol{k}, \boldsymbol{k}') = \frac{v_0}{k\, k'} \, . \tag{5.54}$$

Discuss the influence of the interaction on the dielectric function.

5.8. Any crossing of bands in the Brillouin zone leads to nonanalytic behavior of the Bloch functions in $\boldsymbol{k}$ space, and in turn, the Wannier functions of individual bands fail to be exponentially localized. This problem can be avoided if we consider a band complex, which is a group of bands isolated from all others above and below by a gap throughout the Brillouin zone.[5] The generalized Wannier functions of a band complex

$$\phi_i(\boldsymbol{r} - \boldsymbol{L}) = \sum_{n\boldsymbol{k}} c_{in}(\boldsymbol{k})\varphi_n(\boldsymbol{k}, \boldsymbol{r})\mathrm{e}^{-\mathrm{i}\boldsymbol{kL}} \tag{5.55}$$

are closely related to atomic orbitals but not identical. Use (5.55) to transform the Bethe–Salpeter equation to a local representation and reduce the solution to a matrix inversion.

[5] Unfortunately, single isolated bands do not exist in real systems

5.9. Start with the current–current correlation function for uncorrelated transitions in the long-wavelength limit and transform momentum matrix-elements into dipole matrix elements. Show that the Drude term, resulting from the $\boldsymbol{A}^2$ contribution, is exactly canceled. Hint: Make use of the results of Sect. 2.6.

5.10. Prove by means of complete induction that the nth-order continued fraction (5.50) reproduces the corresponding Neumann series up to order $2n-1$ exactly.

6. Correlated Transitions near the Band Edge

Let us now discuss the modifications of the optical Bloch equations (see Chap. 4) due to the Coulomb interaction among carriers. If optical transitions take place in the vicinity of the Γ point of the Brillouin zone, we can assert that the effective-mass approximation is valid, which allows us to analyze optical transitions in a small part of the Brillouin zone with very high accuracy. In real space, this implies that we ignore the lattice structure of the semiconductor and consider a scale which is large in comparison with one lattice constant.

6.1 The Semiconductor Bloch Equations

In order to derive the equations of motion, which have been obtained by means of various methods [84, 85, 86, 87], we go back to (5.12) and account for the leading long-range contributions only. Correspondingly, the short-range $\boldsymbol{G} \neq 0$ interaction resulting from the Hartree interaction does not contribute. The long-range Hartree interaction contributes exclusively for longitudinal fields and can be incorporated as before into the definition of the macroscopic susceptibility. The long-range Hartree interaction vanishes in the transverse case, which we consider in the following. A consistent treatment of the exchange interaction approximates matrix elements according to

$$\langle l\boldsymbol{k}|\mathrm{e}^{-\mathrm{i}\boldsymbol{k}'\boldsymbol{r}}|n\boldsymbol{k}+\boldsymbol{k}'\rangle = \delta_{ln} \ . \tag{6.1}$$

Within this approximation scheme, we obtain from (5.12)

$$\begin{aligned}
&\left(\mathrm{i}\hbar\frac{\partial}{\partial t} + E_i(\boldsymbol{k}) - E_j(\boldsymbol{k})\right) n_{ij}(\boldsymbol{k}) \\
&= \sum_l \Big(n_{lj}(\boldsymbol{k})\, \boldsymbol{d}_{li}(\boldsymbol{k})\boldsymbol{E} - n_{il}(\boldsymbol{k})\, \boldsymbol{d}_{jl}(\boldsymbol{k})\boldsymbol{E}\Big) \\
&+e^2 \sum_{l\boldsymbol{k}'} v(\boldsymbol{k}-\boldsymbol{k}')\Big[n_{lj}(\boldsymbol{k})\Big(n_{il}(\boldsymbol{k}') - \delta_{li}\delta_{lv}\Big) - n_{il}(\boldsymbol{k})\Big(n_{lj}(\boldsymbol{k}') - \delta_{lj}\delta_{lv}\Big)\Big] \ .
\end{aligned} \tag{6.2}$$

To arrange (6.2) into a more familiar form, we restrict ourselves to a two-band model and express the equations of motion for off-diagonal and diagonal elements separately. Using the same notation as that introduced for the optical Bloch equations in (4.20) and (4.21), we obtain the following for the transition amplitude:

$$\begin{aligned}&\left(\hbar\frac{\partial}{\partial t}+\mathrm{i}[E_{\mathrm{c}}(\boldsymbol{k})-E_{\mathrm{v}}(\boldsymbol{k})]+\hbar\gamma_2\right)p_{\mathrm{vc}}(\boldsymbol{k})\\&=\mathrm{i}\Big(\boldsymbol{d}_{\mathrm{cv}}\boldsymbol{E}(\boldsymbol{r},t)+e^2\sum_{\boldsymbol{k}'}v(\boldsymbol{k}-\boldsymbol{k}')p_{\mathrm{vc}}(\boldsymbol{k}')\Big)[f_{\mathrm{v}}(\boldsymbol{k})-f_{\mathrm{c}}(\boldsymbol{k})]\\&+\mathrm{i}e^2\sum_{\boldsymbol{k}'}v(\boldsymbol{k}-\boldsymbol{k}')[f_{\mathrm{v}}(\boldsymbol{k}')-f_{\mathrm{c}}(\boldsymbol{k}')-1]\,p_{\mathrm{vc}}(\boldsymbol{k}))\,.\end{aligned}\tag{6.3}$$

For the distribution function of the conduction band, we obtain

$$\begin{aligned}&\hbar\frac{\partial}{\partial t}f_{\mathrm{c}}(\boldsymbol{k})+\hbar\gamma_1^{\mathrm{c}}\,[f_{\mathrm{c}}(\boldsymbol{k})-f_{\mathrm{c}}^{(0)}(\boldsymbol{k})]\\&=-2\,\mathrm{Im}\Big[\Big(\boldsymbol{d}_{\mathrm{cv}}\boldsymbol{E}(\boldsymbol{r},t)+e^2\sum_{\boldsymbol{k}'}v(\boldsymbol{k}-\boldsymbol{k}')p_{\mathrm{vc}}(\boldsymbol{k}')\Big)p_{\mathrm{vc}}^*(\boldsymbol{k})\Big]\,.\end{aligned}\tag{6.4}$$

A corresponding equation holds for the valence band distribution function. It is easily obtained from (6.4) by interchanging the band indices. As described in Chap. 4, the phenomenological dephasing rate γ_2 subsumes all processes which lead to a decay of the polarization. Correspondingly, the energy relaxation rate γ_1 characterizes the timescale on which an optically generated one-particle distribution approaches a quasi-equilibrium distribution $f_{\mathrm{c,v}}^{(0)}$, which is given by a Fermi distribution. This distribution depends on the temperature and the chemical potential of the electrons or holes. The decay of the one-particle distributions due to radiative recombination requires an additional relaxation rate, which, however, is usually negligible on a femtosecond or picosecond timescale.

It is far from trivial to actually specify the conditions under which such phenomenological rates are physically meaningful. A thorough understanding requires a microscopic theory of these processes and their microscopic origin, which will be developed in Chap. 10. Some general criteria will however, be discussed at this point. We have to distinguish between intrinsic relaxation processes, which do not depend on the amount of carriers excited, and processes which result from mutual interaction among excited electrons and holes. The intrinsic processes may result from scattering by phonons, impurities, lattice imperfections, etc. These processes determine the linear and influence the nonlinear optical properties. The excitation-induced interaction processes become important in the nonlinear optical regime only, and change greatly with the excitation conditions. Depending on the intensity, duration, and center frequency of the excitation pulses, a variety of different nonequilibrium states of the semiconductor can be generated.

The pulse duration determines whether we are studying coherent or incoherent properties of the system or a mixture of both. If the pulse duration is short in comparison with all relevant relaxation processes, the system is driven by the external fields only, and dephasing and relaxation processes are negligible. This defines the coherent limit. In the opposite limit, the relaxation and dephasing processes are fast compared with the pulse duration and, correspondingly, the coherent properties cannot survive on the timescale of the pulse. This defines the incoherent regime. In reality we often find ourselves in between these two limits, which in a way is the most interesting case, as it allows us to study the temporal evolution of the coherently excited system into an incoherent, quasi-equilibrium state.

The second excitation parameter is the center frequency of the pulse, which determines the energy range in which the excitation takes place. On an ultrashort timescale, the excitation conditions are clearly related to the pulse duration via the Fourier theorem. With decreasing pulse duration, the spectral width of the pulse increases and, correspondingly, a broad energetic range of electrons is excited initially. The center frequency of the pulse can be above the band edge, in the vicinity of the band edge or well below the band edge. The latter case is called off-resonant excitation. Here no real carriers are excited and the dephasing is determined by the intrinsic processes only.

The excitation intensity determines the excited-carrier density and thus the basic physical properties of the nonequilibrium system. As far as the theoretical description of the system is concerned, the intensity determines whether or not we are in the regime in which perturbation theory is valid. It is intuitively clear that interaction processes between excited electrons and holes cannot be described by means of simple dephasing rates. Concerning intrinsic processes, constant rates may be accepted as a first, greatly simplified description. Hence one requirement for the application of the semiconductor Bloch equations is that the excitation-induced relaxation processes are negligible in comparison with the intrinsic ones. Strictly speaking, this criterion is only met for off-resonant excitation at low intensities. This is a severe restriction on the applicability of (6.3) and (6.4). Nevertheless, solutions of these equations can give an initial insight into many-particle effects in semiconductors.

Closely following Chap. 4, we transform the semiconductor Bloch equations into the electron–hole picture, obtaining

$$\begin{aligned}&\left(\hbar\frac{\partial}{\partial t}+\mathrm{i}[\tilde{E}_{\mathrm{e}}(\boldsymbol{k})+\tilde{E}_{\mathrm{h}}(\boldsymbol{k})]+\hbar\gamma_2\right)p_{\mathrm{vc}}(\boldsymbol{k})\\&=\mathrm{i}\,[1-f_{\mathrm{h}}(\boldsymbol{k})-f_{\mathrm{e}}(\boldsymbol{k})]\Big(\boldsymbol{d}_{\mathrm{cv}}\boldsymbol{E}(\boldsymbol{r},t)+e^2\sum_{\boldsymbol{k}'}v(\boldsymbol{k}-\boldsymbol{k}')p_{\mathrm{vc}}(\boldsymbol{k}')\Big)\end{aligned}\tag{6.5}$$

and

$$\hbar\frac{\partial}{\partial t}f_{\rm e}(\boldsymbol{k})+\hbar\gamma_1^{\rm e}[f_{\rm e}(\boldsymbol{k})-f_{\rm e}^{(0)}(\boldsymbol{k})]$$
$$=-2\,{\rm Im}\Big[\Big(\boldsymbol{d}_{\rm cv}\,\boldsymbol{E}(\boldsymbol{r},t)+e^2\sum_{\boldsymbol{k}'}v(\boldsymbol{k}-\boldsymbol{k}')\,p_{\rm vc}(\boldsymbol{k}')\Big)p_{\rm vc}^*(\boldsymbol{k})\Big]\,. \tag{6.6}$$

Here we have introduced the one-particle energies, renormalized by the static exchange interaction:

$$\tilde{E}_{\rm e}(\boldsymbol{k})=E_{\rm e}(\boldsymbol{k})-e^2\sum_{\boldsymbol{k}'}v(\boldsymbol{k}-\boldsymbol{k}')f_{\rm e}(\boldsymbol{k}') \tag{6.7}$$

and

$$\tilde{E}_{\rm h}(\boldsymbol{k})=E_{\rm h}(\boldsymbol{k})-e^2\sum_{\boldsymbol{k}'}v(\boldsymbol{k}-\boldsymbol{k}')f_{\rm h}(\boldsymbol{k}')\,. \tag{6.8}$$

In general, the hole distribution can differ from the electron distribution owing to different relaxation rates. For simplicity these are often set equal, and in this case we have $f_{\rm h}=f_{\rm e}$. Equations (6.5) and (6.6) represent the desired generalization of the optical Bloch equations (4.11) and (4.12).

In contrast to the description in terms of noninteracting electron–hole pairs, the static exchange interaction has dramatic consequences. As can be seen from (6.7) and (6.8), the Hartree–Fock one-particle energies are renormalized by the static exchange interactions. Owing to the excitation of the system, the charge density of the ground state is modified. In principle, this requires a new calculation of the band structure of the excited system. The exchange energies are nothing but the corresponding corrections in the effective-mass approximation. We have discussed similar effects within the temperature-dependent Hartree–Fock theory, where the renormalization of the bands was caused by the increasing occupation of conduction states with increasing temperature. In contrast to that case, however, the one-particle distributions are now nonequilibrium ones, which are generated by the optical excitation. An even more significant effect of the static Coulomb interaction occurs in the properties of electron–hole pairs, which are no longer independent; owing to contributions of the type $\sum v(\boldsymbol{k}-\boldsymbol{k}')p_{vc}(\boldsymbol{k}')$, pairs at different $\boldsymbol{k}$ vectors are coupled by the Coulomb potential. This coupling gives rise to drastic modifications of both the linear and the nonlinear optical properties. In particular, in the vicinity of the band edge, the linear optical spectrum exhibits completely new features due to the formation of bound states. The resulting linear optical properties are the subject of the next sections.

6.2 Linear Optical Properties: Bound Electron-Hole Pairs

An appropriate starting point for the discussion of optical properties in the vicinity of the band edge is either the semiconductor Bloch equations in the linear limit or the effective-mass approximation to the Bethe–Salpeter equation (5.22). As we have discussed the transition to the linear limit in Sect. 5.2 in general, we choose the second possibility. In (5.22), we neglect the off-resonant contribution and replace the plane wave matrix elements according to (6.1) again. Introducing the abbreviation $P_{\mathrm{vccv}}(\boldsymbol{k},\boldsymbol{k}',\omega) \equiv P(\boldsymbol{k},\boldsymbol{k}',\omega)$, we have

$$\begin{aligned}&\Big(E_{\mathrm{c}}(\boldsymbol{k}) - E_{\mathrm{v}}(\boldsymbol{k}) - \hbar\omega\mathrm{i}\hbar\gamma_2\Big)\, P(\boldsymbol{k},\boldsymbol{k}',\omega)\\ &= \delta_{\boldsymbol{k}\,\boldsymbol{k}'} + e^2 \sum_{\boldsymbol{k}''} v(\boldsymbol{k}-\boldsymbol{k}'')\, P(\boldsymbol{k}'',\boldsymbol{k}',\omega)\,. \end{aligned} \tag{6.9}$$

Within the effective mass approximation, we have $E_{\mathrm{c}}(\boldsymbol{k}) - E_{\mathrm{v}}(\boldsymbol{k}) = E_{\mathrm{g}} + (\hbar^2\boldsymbol{k}^2)/(2m_{\mathrm{r}})$. Performing the Fourier transformation into real space, we obtain a more familiar representation of the left-hand side:

$$\left(-\frac{\hbar^2\Delta}{2m_{\mathrm{r}}} + E_{\mathrm{g}} - \hbar\omega - \mathrm{i}\hbar\gamma_2 - e^2 v(\boldsymbol{r})\right) P(\boldsymbol{r},\boldsymbol{r}',\omega) = \delta(\boldsymbol{r}-\boldsymbol{r}')\,. \tag{6.10}$$

Apart from the different material parameters, the Hamiltonian is the same as that of the hydrogen problem. In complete analogy to an electron and a proton, an electron in the conduction band and a hole in the valence band can form bound states, which are called *excitons*. Thus the eigenfunctions, defined by

$$\left(-\frac{\hbar^2\Delta}{2m_{\mathrm{r}}} - e^2 v(\boldsymbol{r})\right)\psi_n(\boldsymbol{r}) = E_n\psi_n(\boldsymbol{r})\,, \tag{6.11}$$

are well known and allow an expansion according to

$$P(\boldsymbol{r},\boldsymbol{r}',\omega) = \sum_{nm} P_{nm}(\omega)\psi_n(\boldsymbol{r})\psi_m^*(\boldsymbol{r}')\,. \tag{6.12}$$

Inserting this expansion into (6.10), we find the following for the expansion coefficients:

$$P_{nm}(\omega) = \frac{\delta_{nm}}{E_n + E_{\mathrm{g}} - \hbar\omega - \mathrm{i}\hbar\gamma_2}\,. \tag{6.13}$$

If we identify the Coulomb Green's function by means of (6.12) as

$$P(\boldsymbol{r},\boldsymbol{r}',\omega) = G(\boldsymbol{r},\boldsymbol{r}',\omega - E_{\mathrm{g}} + \mathrm{i}\gamma_2)\,, \tag{6.14}$$

the result finally obtained is the well-known representation

$$G(\boldsymbol{r},\boldsymbol{r}',\omega) = \sum_n \frac{\psi_n(\boldsymbol{r})\psi_n^*(\boldsymbol{r}')}{E_n - \hbar\omega}\,. \tag{6.15}$$

Thus the linear optical properties in the vicinity of the band edge can be calculated from the real-space representation of (5.26) according to[1]

[1] Here and in the following we assume cubic symmetry.

$$P(\omega) = \frac{1}{\varepsilon_0} \int \mathrm{d}^3\boldsymbol{r}\, \mathrm{d}^3\boldsymbol{r}' d_{cv}(\boldsymbol{r})\, P(\boldsymbol{r},\boldsymbol{r}',\omega)\, d_{cv}(\boldsymbol{r}')\,. \tag{6.16}$$

Within the effective-mass approximation, the linear optical properties are completely determined if we know all eigenfunctions and eigenvalues of (6.11). The solution of the hydrogen problem can be found in many textbooks. Instead of reiterating these calculations, we shall proceed here in a different way and follow another approach. The Coulomb Green's function can be derived in a closed form, in real space [88, 89] and in Fourier space [90], without the detour of an eigenfunction expansion.

6.2.1 The Coulomb Green's Function

Using (6.11) in (6.15) and the completeness of the set of eigenfunctions, we find that the Coulomb Green's function obeys the equation

$$\left(-\frac{\hbar^2 \Delta}{2m_\mathrm{r}} - e^2 v(\boldsymbol{r}) - \hbar\omega\right) G(\boldsymbol{r},\boldsymbol{r}',\omega) = \delta(\boldsymbol{r}-\boldsymbol{r}')\,. \tag{6.17}$$

To get rid of the various constants, we make some scale changes, and measure $\boldsymbol{r}$ in units of the excitonic Bohr radius

$$r_\mathrm{B} = \frac{4\pi\hbar^2\varepsilon_0\varepsilon_\infty}{m_\mathrm{r}e^2} \tag{6.18}$$

and energies in units of the excitonic Rydberg energy

$$E_\mathrm{R} = \frac{m_\mathrm{r}e^4}{32\pi^2\hbar^2\varepsilon_0^2\varepsilon_\infty^2}\,. \tag{6.19}$$

A few typical numbers are given in Table 6.1.

By introducing these excitonic units, we have already anticipated that the electron–hole interaction is screened. For simplicity we have assumed that this screening can be described by the electronic dielectric constant. In reduced units, (6.17) takes the form

Table 6.1. Excitonic Rydberg energies E_R of several semiconductor materials at $T \leq 4.2\,\mathrm{K}$. Taken from [2]

Material	E_R/meV
Si	14.3
Ge	2.1–5.1
GaAs	4.2
AlAs	10–20
ZnSe	19
CdS	28
CuCl	190 (Z_3)

$$\left(\Delta_{\boldsymbol{r}} + \frac{2}{|\boldsymbol{r}|} + E\right) G(\boldsymbol{r}, \boldsymbol{r}', E) = -\delta(\boldsymbol{r} - \boldsymbol{r}') . \tag{6.20}$$

In a first step, we can eliminate the δ function on the right-hand side by using the fact that $-1/(4\pi|\boldsymbol{r} - \boldsymbol{r}'|)$ is the Green's function of the Laplace operator. Using

$$G(\boldsymbol{r}, \boldsymbol{r}') = \frac{g(\boldsymbol{r}, \boldsymbol{r}')}{4\pi|\boldsymbol{r} - \boldsymbol{r}'|} , \tag{6.21}$$

we obtain

$$\left(\Delta_{\boldsymbol{r}} - 2\frac{(\boldsymbol{r} - \boldsymbol{r}')}{|\boldsymbol{r} - \boldsymbol{r}'|^3} \nabla_{\boldsymbol{r}} + \frac{2}{|\boldsymbol{r}|} + E\right) g(\boldsymbol{r}, \boldsymbol{r}') = 0 , \tag{6.22}$$

where we have set

$$\delta(\boldsymbol{r} - \boldsymbol{r}')[1 - g(\boldsymbol{r}, \boldsymbol{r}')] = 0 . \tag{6.23}$$

This relation implies the normalization condition

$$g(\boldsymbol{r}, \boldsymbol{r}) = 1 . \tag{6.24}$$

It is a particularity of the Coulomb problem that the Hamiltonian is more than invariant under rotations in three dimensions. The corresponding rotation group is merely a subgroup of the true symmetry group, which is actually the group of rotations in four dimensions. A manifestation of this invariance is the degeneracy of the eigenvalues of (6.11), which depend neither on the angular momentum nor on the magnetic quantum number.[2] We shall not go into details of a group-theoretical analysis of the Coulomb problem, but note only that the hidden symmetry implies that, in the present problem, $g(\boldsymbol{r}, \boldsymbol{r}')$ depends on two variables only. We shall show this directly in two steps. We try first the new variables $u = |\boldsymbol{r}| + |\boldsymbol{r}'| = r + r'$ and $v = |\boldsymbol{r} - \boldsymbol{r}'|$. The transformation of (6.22), with the replacement $g(\boldsymbol{r}, \boldsymbol{r}') \to g(u, v, r')$, is lengthy but straightforward and is most easily carried out in Cartesian coordinates; the result is [88]

$$\left(\frac{\partial^2}{\partial u^2} + \frac{\partial^2}{\partial v^2} + \frac{v^2 + u^2 - 2ur'}{v(u - r')} \frac{\partial^2}{\partial v \partial u} \right.$$
$$\left. + \frac{v^2 - u^2 + 2ur'}{v^2(u - r')} \frac{\partial}{\partial u} + \frac{2}{u - r'} + E\right) g(u, v, r') = 0 .$$

Now, we can express the coefficient of the linear derivative by means of the derivative of the coefficient of the mixed derivative with respect to v. This allows us to eliminate the linear derivative. Defining

$$g(u, v, r') = \frac{\partial}{\partial v} \tilde{g}(u, v, r') , \tag{6.25}$$

[2] As a concomitant of this higher symmetry, there exists an additional constant of motion, i.e. the Runge–Lenz vector.

we obtain the following from (6.25):

$$\frac{\partial}{\partial v}\left(T(u,v,r') + \frac{2}{u-r'} + E\right)\tilde{g}(u,v,r') = 0\,, \tag{6.26}$$

where

$$T(u,v,r') = \frac{\partial^2}{\partial u^2} + \frac{\partial^2}{\partial v^2} + \frac{v^2+u^2-2ur'}{v(u-r')}\frac{\partial^2}{\partial v \partial u}\,. \tag{6.27}$$

Obviously, (6.26) still depends on r' as a parameter. This dependence can be eliminated completely by introducing, in the next step, the variables

$$x = u + v = r + r' + |\boldsymbol{r} - \boldsymbol{r}'| \tag{6.28}$$

and

$$y = u - v = r + r' - |\boldsymbol{r} - \boldsymbol{r}'|\,. \tag{6.29}$$

In these variables, our problem becomes separable. The kinetic energy takes the form

$$T(u,v,r') = \frac{1}{v(u-r')}\left(a(x,r')\frac{\partial^2}{\partial x^2} - a(y,r')\frac{\partial^2}{\partial y^2}\right)\,, \tag{6.30}$$

where

$$a(x,r') = x^2 - 2xr'\,. \tag{6.31}$$

Furthermore, we have the identities

$$\frac{2}{u-r'} = \frac{1}{v(u-r')}\left(\frac{a(x,r')}{x} - \frac{a(y,r')}{y}\right) \tag{6.32}$$

and

$$\frac{a(x,r') - a(y,r')}{v(u-r')} = \frac{1}{4}\,. \tag{6.33}$$

Lumping these results together, we find that $\tilde{g}$ obeys the equation

$$\frac{1}{v(u-r')}[a(x,r')\,O(x) - a(y,r')\,O(y)]\,\tilde{g}(x,y) = 0\,, \tag{6.34}$$

where we have defined the operator

$$O(x) = \frac{\partial^2}{\partial x^2} + \frac{1}{x} + \frac{E}{4}\,. \tag{6.35}$$

The representation (6.34) demonstrates that the Coulomb problem is separable in the variables x and y. Thus the solution is of the form [88] $\tilde{g}(x,y) = f(x)f(y)$, where f is a solution of $O(x)f(x) = 0$. This is equivalent to the equation

$$\left(\frac{\partial^2}{\partial z^2} + \frac{\lambda}{z} - \frac{1}{4}\right) f(z) = 0\,, \tag{6.36}$$

where we have substituted $z = -\mathrm{i}xk$, where $k = \sqrt{E}$, and introduced $\lambda = \mathrm{i}/k$. The solutions of (6.36) are the $\mu = 1/2$ Whittaker functions. The appropriate choice of independent solutions is determined by the boundary conditions at the origin and at infinity. The final solution must be regular at $r = 0$, be an outgoing spherical wave for $E \geq 0$, $r \to \infty$, and decay exponentially for $E \leq 0$, $r \to \infty$. This is guaranteed if we choose [88, 89]

$$\tilde{g}(x, y) = W_{\mathrm{i}/k,1/2}(-\mathrm{i}kx) M_{\mathrm{i}/k,1/2}(-\mathrm{i}ky) \, , \tag{6.37}$$

where $M_{\mathrm{i}/k,1/2}$ and $W_{\mathrm{i}/k,1/2}$ are the Whittaker functions of the first and second type, respectively [91, 140]. The Whittaker functions of the first type are given by

$$M_{\lambda,1/2}(z) = z\mathrm{e}^{-z/2} F(1-\lambda, 2, z) \, , \tag{6.38}$$

where $F(1-\lambda, 2, z)$ is the degenerate hypergeometric function. The Whittaker functions of the second type exhibit a logarithmic singularity and can be represented as

$$\begin{aligned} W_{\lambda,1/2}(z) = \frac{1}{\Gamma(1-\lambda)} \Big(& \mathrm{e}^{-z/2} + \lambda \ln(z) M_{\lambda,1/2}(z) \\ & + \sum_{n=0}^{\infty} \mathrm{e}^{-z/2} z^{n+1} \frac{\Gamma(1+n-\lambda)}{\Gamma(-\lambda) n! (n+1)!} [\Psi(1+n-\lambda) - \Psi(n+1) - \Psi(n+2)] \Big) \, , \end{aligned} \tag{6.39}$$

where Γ denotes the Γ function and Ψ is the logarithmic derivative of the Γ function, referred to as the Ψ function or digamma function. Transforming (6.25) into the x, y representation, we obtain using (6.37),

$$g(x, y) = \left(\frac{\partial}{\partial x} - \frac{\partial}{\partial y} \right) W_{\mathrm{i}/k,1/2}(-\mathrm{i}kx) M_{\mathrm{i}/k,1/2}(-\mathrm{i}ky) \, . \tag{6.40}$$

Finally, we have to take the normalization condition (6.24) into account, which requires, according to (6.40), that

$$\begin{aligned} M_{\mathrm{i}/k,1/2}(-\mathrm{i}kx) \frac{\partial}{\partial x} W_{\mathrm{i}/k,1/2}(-\mathrm{i}kx) - W_{\mathrm{i}/k,1/2}(-\mathrm{i}kx) \frac{\partial}{\partial x} M_{\mathrm{i}/k,1/2}(-\mathrm{i}kx) \\ = \mathcal{W} = 1 \, . \end{aligned} \tag{6.41}$$

This is just the Wronskian $\mathcal{W}$, which is independent of x, as can be easily checked by considering

$$\begin{aligned} \frac{\partial \mathcal{W}}{\partial x} = & M_{\mathrm{i}/k,1/2}(-\mathrm{i}kx) \frac{\partial^2}{\partial x^2} W_{\mathrm{i}/k,1/2}(-\mathrm{i}kx) \\ & - W_{\mathrm{i}/k,1/2}(-\mathrm{i}kx) \frac{\partial^2}{\partial x^2} M_{\mathrm{i}/k,1/2}(-\mathrm{i}kx) \, . \end{aligned} \tag{6.42}$$

Obviously, we have $\partial \mathcal{W}/\partial x = 0$, because W and M are solutions of (6.36). Thus we can evaluate the Wronskian for arbitrary arguments. For simplicity, we choose $x = 0$. As $M(0) = 0$ and $M'(0) = 1$, we obtain

$$\mathcal{W} = W_{\mathrm{i}/k,1/2}(0) = \frac{-\mathrm{i}k}{\Gamma(1-\mathrm{i}/k)} , \tag{6.43}$$

which yields the final result for the Green's function (6.21) in closed form

$$\begin{aligned} G(\boldsymbol{r},\boldsymbol{r}',E) &= -\frac{\Gamma(1-\mathrm{i}/k)}{4\pi|\boldsymbol{r}-\boldsymbol{r}'|}\frac{\mathrm{i}}{k} \\ &\times \left(\frac{\partial}{\partial x}-\frac{\partial}{\partial y}\right) W_{\mathrm{i}/k,1/2}(-\mathrm{i}kx)\, M_{\mathrm{i}/k,1/2}(-\mathrm{i}ky) . \end{aligned} \tag{6.44}$$

With exception of the simple poles of the Γ function, this expression is an analytic function of E in the complex plane with a branch cut along the positive real axis. If we approach this cut from above, the Green's function is retarded and causality is guaranteed. While the cut represents the continuous spectrum, the poles at $\mathrm{i}/k = n$, $n = 1, 2, 3, ..$, represent the bound states, which thus occur at energies $E = -E_{\mathrm{R}}/n^2$. This will be seen in more detail in the next section.

6.2.2 Optical Properties due to Bound Electron–Hole Pairs

The calculation of linear optical spectra according to (6.14) and (6.16) from the Coulomb Green's function can be considerably simplified if we assume that the dipole transition matrix elements are isotropic, i.e. they depend only on $|\boldsymbol{k}|$ or $|\boldsymbol{r}|$. In this case only excitons with s symmetry contribute. A complete neglect of the $\boldsymbol{k}$ dependence is not possible, as without any cutoff in $\boldsymbol{k}$ space the continuous part of the spectrum leads to divergences. This is displayed in the singular behavior of $G(0,0,E)$. For isotropic transition matrix elements, we can take the average of (6.44) with respect to the angle between $\boldsymbol{r}$ and $\boldsymbol{r}'$. With the substitution $|\boldsymbol{r}-\boldsymbol{r}'| = \sqrt{r^2+r'^2-2rr'z}$ in (6.28) and (6.29), the z integration can be performed immediately as it just cancels the derivatives, and yields, for $r \geq r'$,

$$\begin{aligned} G_{\mathrm{s}}(r,r',E) &= \int_{-1}^{1} \mathrm{d}z\, G(\boldsymbol{r},\boldsymbol{r}',E) \\ &= \frac{\mathrm{i}\,\Gamma(1-\mathrm{i}/k)}{4\pi k r r'} W_{\mathrm{i}/k,1/2}(-2\mathrm{i}kr) M_{\mathrm{i}/k,1/2}(-2\mathrm{i}kr') , \end{aligned} \tag{6.45}$$

where we have used again the fact that $M(0) = 0$. The result in the case $r' \geq r$ is obtained from (6.45) by interchanging r and r'. In the following we make use of the property that the Whittaker functions are analytic with respect to the index λ and that pole contributions arise from the Γ-function only. This can be immediately seen if we consider the function (compare (6.39))

$$g(\lambda) = \Gamma(1-\lambda) \sum_{n=0}^{\infty} \mathrm{e}^{\mathrm{i}kr}(-\mathrm{i}kr)^{n+1} \frac{\Gamma(1+n-\lambda)}{\Gamma(-\lambda)\Gamma(1-\lambda)n!(n+1)!}$$

$$\times \left[\Psi(1+n-\lambda) - \Psi(n+1) - \Psi(n+2)\right] M_{\lambda,1/2}(-2\mathrm{i}kr') \,, \tag{6.46}$$

where, at the moment we are taking into account only the contributions to the Whittaker function (6.39) in which the prefactor $\Gamma(1-\lambda)$ in (6.45) is not canceled. As all pole contributions from the Γ and Ψ functions in the sum cancel, $g(\lambda)$, like the Γ-function itself, is a meromorphic function. This allows us to expand (6.46) as a series of rational fractions of the form [92]

$$g(\lambda) = g_0 + \sum_{n=1}^{\infty} w_n \left(\frac{1}{n} + \frac{1}{\lambda - n} \right) , \tag{6.47}$$

where the w_n are obtained from the residues of $g(\lambda)$. Since the residues of $\Gamma(z)$ are $(-1)^m/m!$ at $z=-m$, we have to evaluate the sum in (6.46) in the limit $\lambda \to n$, where n is an arbitrary integer. By means of a rational-fraction expansion of the Γ and Ψ functions, we obtain

$$\lim_{\lambda \to n} \frac{\Gamma(1-\lambda+m)\Psi(1-\lambda+m)}{\Gamma(1-\lambda)\Gamma(-\lambda)}$$

$$= (-1)^n n! [\delta_{m,0} + (1-n)(2-n)....(m-n)] \,. \tag{6.48}$$

As in this limit the other contributions in the square bracket in (6.46) vanish, the residues are given by

$$w_n = n \sum_{m=0}^{\infty} \mathrm{e}^{\mathrm{i}kr}(-\mathrm{i}kr)^{m+1}$$

$$\times \frac{\delta_{m,0} + (1-n)(2-n)....(m-n)}{(m+1)!m!} M_{n,1/2}(-2\mathrm{i}kr') \,. \tag{6.49}$$

The sum over m represents nothing but the series expansion of $M_{n,1/2}(-2\mathrm{i}kr)$ (compare (6.38)). Using further the fact that $g_0 = g(0) = 0$, the rational-fraction expansion of $g(\lambda)$ takes the form

$$g(\lambda) = \sum_{n=1}^{\infty} \frac{\lambda}{\lambda - n} M_{n,1/2}(-2\mathrm{i}kr)\, M_{n,1/2}(-2\mathrm{i}kr'). \tag{6.50}$$

Together with the first contribution in (6.39) this representation allows to write (6.45) for $r \geq r'$, for example, in the form

$$G_{\mathrm{s}}(r, r', E)$$

$$= \frac{\mathrm{e}^{\mathrm{i}k(r+r')}}{\pi} \left\{ \left[\frac{1}{2r} - \ln(-2\mathrm{i}kr)\, F\!\left(1 - \frac{\mathrm{i}}{k}, 2, -2\mathrm{i}kr\right) \right] F\!\left(1 - \frac{\mathrm{i}}{k}, 2, -2\mathrm{i}kr'\right) \right.$$

$$\left. - \sum_{n=1}^{\infty} \frac{k}{\mathrm{i} - kn} F(1-n, 2, -2\mathrm{i}kr)\, F(1-n, 2, -2\mathrm{i}kr') \right\} . \tag{6.51}$$

This result displays explicitly the origin of the divergent contributions to $G(0,0,E)$. The first term, which reduces in the limit of large energies, i.e. $\mathrm{i}/k \to 0$, to the free-particle Green's function, is integrable, as is the logarithmic singularity. The integration in (6.16) requires a model for the spatial dependence of the transition matrix elements. From the wave vector dependence of the dipole matrix element (3.131), that we obtained from $\boldsymbol{kp}$ perturbation theory, the matrix elements in real space are obtained as

$$d_{\mathrm{vc}}(r) = d^0_{\mathrm{vc}} \frac{\alpha^2}{4\pi} \frac{\mathrm{e}^{-\alpha r}}{r} , \tag{6.52}$$

where $\alpha = \sqrt{E_{\mathrm{g}}}$ in excitonic units. As the gap energy is typically several orders of magnitude larger than the energy scale under consideration, i.e. $E \ll E_{\mathrm{g}}$, it is sufficient to use an expansion of (6.51) for kr and $kr' \ll 1$. Equation (6.52) can be used to show that corrections to the leading contributions are $\propto E/E_{\mathrm{g}}$ and can be safely neglected with respect to 1. This is consistent with the effective-mass approximation, which restricts us in most cases to a considerably smaller energy range. To obtain the leading contributions we use

$$F\Big(1 - \frac{\mathrm{i}}{k}, 2, -2\mathrm{i}kr'\Big) = 1 - \mathrm{i}kr' + O(r'^2) \tag{6.53}$$

and a corresponding expansion for the exponential function. The imaginary part of the logarithmic function must be treated with care. We consider a retarded Green's function, i.e. $E = (\hbar\omega - E_{\mathrm{g}} + \mathrm{i}\hbar\gamma_2)/E_{\mathrm{R}}$, and consider the limit $\gamma_2 \to 0$. Thus (6.51) yields the following for $r \geq r'$:

$$G_{\mathrm{s}}(r,r',E) = \frac{1}{\pi}\left(\frac{1}{2r} - \ln|2kr| + \frac{\mathrm{i}}{2}[k + \theta(k^2)\pi] + \sum_{n=1}^{\infty}\frac{k^2 n + \mathrm{i}k}{1 + k^2 n^2}\right) . \tag{6.54}$$

The contribution proportional to the θ function results from the imaginary part of the logarithmic function in the limit $\gamma_2 \to 0$. Now we are ready to calculate optical spectra according to (6.14) and (6.16). As we are considering the transverse case, the long-range Hartree interaction does not contribute and $\chi(\omega)$ is identical with $\tilde{\chi}(\omega)$. Also, we have already neglected $\boldsymbol{G} \neq 0$ contributions to the Coulomb potential within the effective-mass approximation. Therefore local-field corrections should be neglected too, and $P(\omega)$ can be identified with the linear susceptibility $\chi^{(1)}(\omega)$, which is obtained as

$$\begin{aligned}\chi^{(1)}(\omega) &= \frac{1}{\varepsilon_0}\int \mathrm{d}^3\boldsymbol{r}\,\mathrm{d}^3\boldsymbol{r}'\, d_{\mathrm{vc}}(r)\, G(\boldsymbol{r},\boldsymbol{r}',E)\, d_{\mathrm{vc}}(r') \\ &= \frac{8\pi^2}{\varepsilon_0}\int \mathrm{d}r\,\mathrm{d}r'\, r^2 r'^2 d_{\mathrm{vc}}(r)\, G_{\mathrm{s}}(r,r',E)\, d_{\mathrm{vc}}(r') .\end{aligned} \tag{6.55}$$

Using (6.2.2) and the corresponding result for $r' \geq r$ in (6.55), we obtain

$$\chi^{(1)}(\omega) = \frac{|d_{\mathrm{vc}}^0|^2}{4\pi\varepsilon_0}\left(\mathrm{i}[\sqrt{E}+\pi\theta(E)] + 2\sum_{n=1}^{\infty}\frac{nE+\mathrm{i}\sqrt{E}}{1+n^2E}\right). \tag{6.56}$$

The tedious constants which result from the integration of the $1/r$ and $\ln|2kr|$ contributions have already been neglected. The first contribution $\propto \sqrt{E}$ is the free-particle contribution which is proportional to the density of states. In the following we consider the imaginary part of $\chi^{(1)}$, which determines the absorption, and consider first the bound-state contributions, i.e. $E \leq 0$. To be consistent, we consider again the limit $\gamma_2 \to 0$ which allows to make use of the Dirac identity. Furthermore, we transform back to absolute units with the result

$$\mathrm{Im}\left(\chi^{(1)}(\omega)\right) = \theta(-E)\,\frac{|d_{\mathrm{vc}}^0|^2}{\varepsilon_0 r_{\mathrm{B}}^3}\sum_{n=1}^{\infty}\frac{1}{n^3}\,\delta\left(\hbar\omega - E_{\mathrm{g}} - \frac{E_{\mathrm{R}}}{n^2}\right). \tag{6.57}$$

Thus, below the band edge, an infinite series of resonances with rapidly decreasing oscillator strength appears. For large n these discrete levels form a quasi-continuum with constant oscillator strength. This can be seen by transforming the sum over n for large n into an integral and substituting $x = 1/n^2$. This substitution leads to the cancellation of the $1/n^3$ dependence, and $\mathrm{Im}(\chi^{(1)})$ becomes $|d^0|^2/(2\varepsilon_0 r_{\mathrm{B}}^3)$. We expect that the continuous spectrum approaches the same value at its onset .

For $E \geq 0$, we have to consider the imaginary part of (6.56). The sum over n can be evaluated exactly by considering

$$S(E) = \sum_{n=1}^{\infty}\frac{1}{1+n^2E} = \sum_{n=1}^{\infty}\sum_{m=0}^{\infty}\frac{(-1)^m}{E^{m+1}}\frac{1}{n^{2m+2}}\,. \tag{6.58}$$

We have initially complicated the problem by expanding the denominator into a geometric series. Now we can perform the summation over n by means of the Euler relation, which leads to the occurrence of Bernoulli numbers [140]:

$$S(E) = \sum_{m=0}^{\infty}\frac{(-1)^m}{E^{m+1}}\frac{(2\pi)^{2m+2}}{2(2m+2)!}|B_{2m+2}| = \sum_{m=1}^{\infty}\frac{1}{E^m}\frac{(2\pi)^{2m}}{2(2m)!}B_{2m}\,. \tag{6.59}$$

We make use of the fact that all Bernoulli numbers with odd index vanish except for $B_1 = -1/2$. This allows us to write (6.59) in the form

$$S(E) = \frac{1}{2}\left(\sum_{m=0}^{\infty}\left(\frac{2\pi}{\sqrt{E}}\right)^m\frac{B_m}{m!} - 1 + \frac{\pi}{\sqrt{E}}\right), \tag{6.60}$$

which is just the expansion of the generating function for the Bernoulli numbers. Thus we end up with

$$S(E) = \frac{1}{2}\left(\frac{x}{\mathrm{e}^x - 1} + \frac{x}{2} - 1\right), \tag{6.61}$$

where $x = 2\pi/\sqrt{E}$. Using this result in (6.56), we see that the 1 cancels the free-particle contribution, and the final result can be written in the form

$$\mathrm{Im}\Big(\chi^{(1)}(\omega)\Big) = \frac{|d^0_{\mathrm{vc}}|^2}{2\varepsilon_0 r_{\mathrm{B}}^3}\,\theta(E)\,\frac{\mathrm{e}^{2\pi\sqrt{E_{\mathrm{R}}/E}}}{\mathrm{e}^{2\pi\sqrt{E_{\mathrm{R}}/E}}-1}\,. \tag{6.62}$$

Together with (6.57), this is the so-called *Elliott formula*, which describes the optical absorption of a direct-gap semiconductor in the vicinity of the band edge. As expected, we have a continuous transition from the quasi-continuum formed by high-lying bound states to the continuum described by (6.62). This can be seen in the theory (Fig. 6.1) and in the experiment (Fig. 6.2). In contrast to the square-root dependence of the continuum for free-particles, the correlated continuum differs for $E \to 0$ and for $E \to \infty$ by a factor, which is often referred to as Coulomb or *Sommerfeld enhancement*. We should,

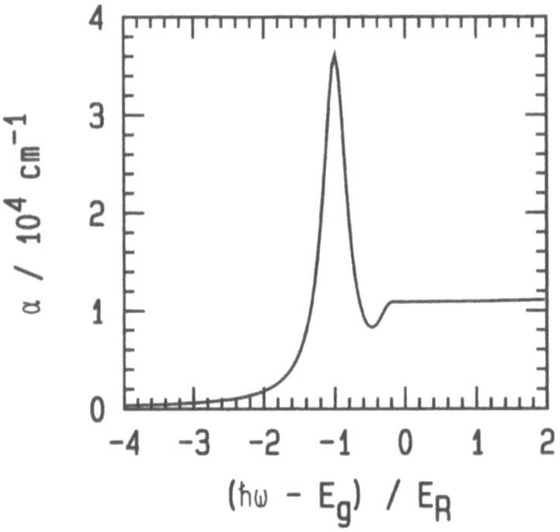

Fig. 6.1. Absorption spectrum of a three-dimensional semiconductor obtained from numerical inversion of the semiconductor Bloch equations; $\hbar\gamma_2 = 0.2\,E_{\mathrm{R}}$, GaAs parameters

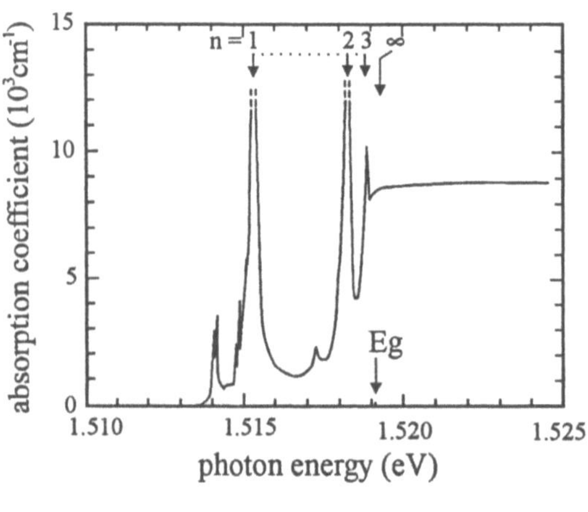

Fig. 6.2. Measured absorption spectrum of GaAs at a lattice temperature of 1.2 K. The lowest-energy peaks are due to impurity absorption. After [96]

however, keep in mind that, owing to the effective-mass approximation, our theory is limited to a small energy range in the vicinity of the band edge. Furthermore, our final result requires that the condition $kr \ll 1$ is fulfilled. As r is typically of the order of one or a few lattice constants a, our results are valid in the range $k \ll r_B/a$. In typical cases this corresponds to several Rydberg energies above the band edge.

The application of the Green's function method used in this section to nonlinear optical properties, described, for example, by the semiconductor Bloch equations, is straightforward under stationary excitation conditions only [93]. If the time evolution of the system comes into play, we need, instead of the Coulomb Green's function (6.44), the Coulomb propagator [94], i.e. the Fourier transform with respect to energy. A calculation of this quantity requires very lengthy expansions and has no real advantage over a direct numerical calculation, which we prefer in this case. The same is valid for the application of the method to semiconductors with reduced dimensionality, which can be easily performed along the lines discussed above in the pure two- and one-dimensional limits. More realistic models of semiconductor structures, however, require the application of numerical methods in any case; these will be described in the following.

6.2.3 Numerical Methods

Once more, we start from the exciton equation in $\boldsymbol{k}$ space (6.9) and consider the quantity

$$P_{\rm vc}(\boldsymbol{k},\omega) = \sum_{\boldsymbol{k}'} P_{\rm vccv}(\boldsymbol{k},\boldsymbol{k}',\omega)\, d_{vc}(\boldsymbol{k}')\,, \tag{6.63}$$

which is determined by

$$\left(\frac{\hbar^2\boldsymbol{k}^2}{2m_{\rm r}} - E\right) P_{\rm vc}(\boldsymbol{k}) - \frac{e^2 V}{(2\pi)^3}\int \mathrm{d}^3\boldsymbol{k}'\, v(\boldsymbol{k}-\boldsymbol{k}')P_{\rm vc}(\boldsymbol{k}') = d_{\rm vc}(\boldsymbol{k})\,. \tag{6.64}$$

But for the singularity of the Coulomb potential, the solution of this equation could be mapped onto a matrix inversion immediately. Fortunately, the singularity is integrable and can be removed. In the general case we can expand $P_{\rm vc}(\boldsymbol{k})$ by means of spherical harmonics, which leads to a projection of the Coulomb potential onto these functions. For simplicity we again restrict ourselves to the case of excitons with s symmetry, which implies that $P_{\rm vc}(\boldsymbol{k}) = P_{\rm vc}(k)$ and projection onto s states reduces to taking the angular average of the Coulomb potential. This results in the equation

$$\left(\frac{\hbar^2 k^2}{2m_{\rm r}} - E\right) P_{\rm vc}(k) - \int \mathrm{d}k'\, I(k,k')P_{\rm vc}(k') = d_{\rm vc}(k), \tag{6.65}$$

where we have introduced

$$I(k,k') = \frac{e^2}{4\pi^2\varepsilon_0\varepsilon_\infty}\frac{k'}{k}\ln\left|\frac{k+k'}{k-k'}\right| . \tag{6.66}$$

The integration can be performed by means of a numerical quadrature with abscissa k_i and weights w_i, $i = 1, \dots N$. To get rid of the singularity we add and subtract $P_{\mathrm{vc}}(k) \int \mathrm{d}k' I(k,k')$ and treat the subtracted contribution again by a numerical quadrature [95]. This way, the singular contributions cancel and we obtain

$$\left(\frac{\hbar^2 k_i^2}{2m_{\mathrm{r}}} - E\right) P_{\mathrm{vc}}(k_i) - \sum_{i \neq j} w_j \, I(k_i,k_j)[P_{\mathrm{vc}}(k_j) - a(k_i,k_j)P_{\mathrm{vc}}(k_i)]$$
$$+ P_{\mathrm{vc}}(k_i) \int dk' \, a(k_i,k') I(k_i,k') = d_{\mathrm{vc}}(k) , \tag{6.67}$$

where we have introduced a function $a(kk')$ such that $a(k,k) = 1$, which can be chosen to speed up the convergence rate for large k'. With $a(k,k') = 4k^4/(k^2+k'^2)^2$, the integral in (6.67) can be performed analytically and yields

$$\int \mathrm{d}k' \, a(k,k') I(kk') = \frac{e^2}{4\pi^2\varepsilon_0\varepsilon_\infty}\pi k . \tag{6.68}$$

Equation (6.67), together with (6.68), can be solved by matrix inversion. In practise, the k_i are defined on a finite interval, i.e. we introduce a cutoff k_{c} by replacing $d_{\mathrm{vc}}(k)$ by a θ function. To reproduce the analytical results within an accuracy of about 2 percent, it is sufficient to choose a mesh with $0 \leq k_i \leq 10 r_{\mathrm{B}}^{-1}$. Depending on the dephasing rate, the typical dimensions of the matrix range between 50 and 200. This dependence results from the property that the spectrum calculated from the matrix inversion is always discrete. A quasi-continuum results if the distance between neighboring energies is smaller than the dephasing rate, i.e. the spacing Δ_i between the points of support should fulfil $\Delta_i \leq \gamma_2/2k_i$.

In contrast to the analytical methods this scheme or similar ones can easily be applied to problems of nonlinear optics, such as the solution of the semiconductor Bloch equations and also to semiconductor heterostructures. In the following, as an example, we discuss the optical properties of quantum wells.

6.2.4 Excitons in Quantum Wells

As already discussed in Chap. 1, quantum well structures consist of alternating layers with very similar electronic and structural properties, in particular nearly the same lattice constant but different band-gap energies, which act as potential barriers for electrons and holes in the material with the smaller band gap. Actually, there is a continuous transition between the bands in the interface region. If this transition region, however, is narrow in comparison with the well width *and* we consider physics on a scale which is large in comparison with the lattice constant, we can neglect these microscopic details and describe the effect of different layers by an additional one-dimensional rectangular well potential. This already implies that effective-mass theory is valid and that we are restricting ourselves to a narrow region in $\boldsymbol{k}$ space. The corresponding one-particle wavefunctions have the form

$$\Psi_{in}(\boldsymbol{kr}) = \frac{1}{\sqrt{A}} g_{in}(z) \mathrm{e}^{\mathrm{i}\boldsymbol{k\rho}} u_n(0, \boldsymbol{r}) , \tag{6.69}$$

where $g_{in}(z)$ describes the ith eigenfunction of the well for valence and conduction states (n=\{v,c\}), $\boldsymbol{k}$ and $\boldsymbol{\rho}$ denote in-plane wave vectors and real-space vectors, and A is the area of the interface. In order to investigate the optical properties, we now have to represent the Bethe–Salpeter equation by means of the basis set (6.69). Starting with the linear version of the equation of motion (5.8), we obtain for the polarization propagator the general form

$$\Big(E_j - E_i - \hbar\omega - \mathrm{i}\hbar\gamma_2\Big) P_{ij} - e^2 \sum_{i'J'} \langle ji'|v|j'i\rangle P_{i'j'} = d_{ij} , \tag{6.70}$$

with quantum numbers that are arbitrary at this stage. Owing to the effective-mass approximation, we have again already neglected the Hartree contribution to the Coulomb interaction. In order to obtain the specific form of this equation for the wavefunctions (6.69), we have to replace a quantum number i, for example, according to $i \to \{i, n, \boldsymbol{k}\}$. Noting that $P_{ij}(\boldsymbol{k})$ is diagonal with respect to the in-plane wave vector, we can write the interaction matrix element as

$$\begin{aligned} \langle jc\boldsymbol{k}, i'v'\boldsymbol{k}'|\, v\, |j'c'\boldsymbol{k}', iv\boldsymbol{k}\rangle &= \frac{1}{\varepsilon_0\, \varepsilon_\infty V} \sum_{\boldsymbol{q}q_z} \frac{1}{\boldsymbol{q}^2 + q_z^2} \langle c\boldsymbol{k}|\mathrm{e}^{\mathrm{i}\boldsymbol{q\rho}}|c'\boldsymbol{k}'\rangle \\ \times \langle v'\boldsymbol{k}'|\mathrm{e}^{-\mathrm{i}\boldsymbol{q\rho}}|v\boldsymbol{k}\rangle & \int \mathrm{d}z\, \mathrm{d}z'\, g^*_{jc}(z) g_{j'c'}(z) \mathrm{e}^{\mathrm{i}q_z(z-z')} g^*_{i'v'}(z') g_{iv}(z') . \end{aligned} \tag{6.71}$$

Within the effective mass approximation we have, for example,

$$\langle c\boldsymbol{k}|\mathrm{e}^{\mathrm{i}\boldsymbol{q\rho}}|c'\boldsymbol{k}'\rangle = \delta_{\mathrm{cc}'}\delta_{\boldsymbol{k}',\boldsymbol{k}-\boldsymbol{q}} . \tag{6.72}$$

This allows us also to simplify the notation. We drop the band indices and introduce the convention that j indices imply conduction states whereas i indices imply valence states. Further, the sum over q_z can be replaced by an integral according to $\sum_{q_z} \to (L/2\pi) \int \mathrm{d}q_z$. This allows us to perform the q_z integration by means of the residue theorem, which yields

$$\langle j\boldsymbol{k}, i'\boldsymbol{k}'|v|j'\boldsymbol{k}', i\boldsymbol{k}\rangle = \frac{1}{2\varepsilon_0\varepsilon_\infty A} \frac{M_{ji',j'i}(\boldsymbol{k}-\boldsymbol{k}')}{|\boldsymbol{k}-\boldsymbol{k}'|} , \tag{6.73}$$

where the interaction matrix element is given by

$$M_{ji',ji}(\boldsymbol{q}) = \int \mathrm{d}z\,\mathrm{d}z'\, g_j^*(z) g_{j'}(z) \mathrm{e}^{-|\boldsymbol{q}||z-z'|} g_{i'}^*(z') g_i(z') \,; \tag{6.74}$$

we have also used $V = AL$. The transition matrix element on the right-hand side of (6.70) can be evaluated by means of (3.134) and will be defined here as

$$d_{ij}(\boldsymbol{k}) \;=\; d_{vc}(\boldsymbol{k})\,\delta_{ij}\,. \tag{6.75}$$

Thus, instead of (6.64), now we have to deal with the equation

$$\begin{aligned}&[E_j(\boldsymbol{k}) - E_i(\boldsymbol{k}) - E]\,P_{ij}(\boldsymbol{k})\\&-\frac{e^2 A}{(2\pi)^2}\sum_{i'j'}\int \mathrm{d}^2\boldsymbol{k}'\,\langle j\boldsymbol{k}, i'\boldsymbol{k}'|v|j'\boldsymbol{k}', i\boldsymbol{k}\rangle P_{i'j'}(\boldsymbol{k}') = d_{ij}(\boldsymbol{k})\,.\end{aligned} \tag{6.76}$$

What remains is to consider the singular contribution in (6.73), which can be easily separated by introducing a regular contribution by means of

$$\tilde{M}_{ji',j',i}(\boldsymbol{q}) = \int \mathrm{d}z\,\mathrm{d}z'\, g_j^*(z) g_{j'}(z) \Big(\mathrm{e}^{-|\boldsymbol{q}||z-z'|} - 1\Big) g_{i'}^*(z') g_i(z') \tag{6.77}$$

and making use of the orthonormality of the well functions. This leads to

$$\langle j\boldsymbol{k}, i'\boldsymbol{k}'|v|j'\boldsymbol{k}', i\boldsymbol{k}\rangle = \frac{1}{2\varepsilon_0\varepsilon_\infty A}\left(\frac{\delta_{jj'}\delta_{ii'}}{|\boldsymbol{k}-\boldsymbol{k}'|} + \frac{\tilde{M}_{ji',j'i}(\boldsymbol{k}-\boldsymbol{k}')}{|\boldsymbol{k}-\boldsymbol{k}'|}\right). \tag{6.78}$$

The first term represents the two-dimensional limit of the Coulomb potential, which is the only contribution if the well functions become extremely localized. This limit, however, is only of academic interest, as if the interfaces move to close together our assumptions cease to be valid.

In what follows, we consider only s excitons (as in the last section). In contrast to the three-dimensional case, the angular average cannot be performed analytically but requires a numerical integration. As the second contribution to the interaction (6.78) is regular in the limit $\boldsymbol{k} \to \boldsymbol{k}'$, the integration can be easily performed using standard methods. We define

$$V_{ji',j'i}(kk') = \frac{e^2}{8\pi^2\varepsilon_0\varepsilon_\infty}\int_0^{2\pi} \mathrm{d}\varphi\, \frac{k'\,\tilde{M}_{ji',j'i}\Big(\sqrt{k^2+k'^2-2kk'\cos\varphi}\Big)}{\sqrt{k^2+k'^2-2kk'\cos\varphi}}\,. \tag{6.79}$$

Considering the two-dimensional Coulomb potential, we introduce

$$I(k,k') = \frac{e^2}{8\pi^2\varepsilon_0\varepsilon_\infty}\int_0^{2\pi} \mathrm{d}\varphi\, \frac{k'}{(k^2+k'^2-2kk'\cos\varphi)^{1/2}} \tag{6.80}$$

correspondingly. As before, the $k \to k'$ singularity is removed by adding and subtracting $P_{ij}(k)\int \mathrm{d}k'\, a(k,k') I(k,k')$ with the same convergence factor as in the three-dimensional case. What remains is to calculate the integral

$$\frac{e^2}{8\pi^2\varepsilon_0\varepsilon_\infty}\int \mathrm{d}^2\boldsymbol{k}'\,\frac{a(k,|\boldsymbol{k}'|)}{|\boldsymbol{k}-\boldsymbol{k}'|} = \frac{e^2}{8\pi^2\varepsilon_0\varepsilon_\infty}\int \mathrm{d}^2\boldsymbol{k}'\,\frac{1}{|\boldsymbol{k}'|}a(k,|\boldsymbol{k}'+\boldsymbol{k}|)$$
$$= \frac{e^2}{\pi^2\varepsilon_0\varepsilon_\infty}k\int_0^\infty \mathrm{d}x\int_0^{2\pi}\mathrm{d}\varphi\frac{1}{(x^2+2+2x\cos\varphi)^2} = k\mathcal{I}\,. \tag{6.81}$$

After the φ integration has been performed analytically, the x integration can be done numerically, with the result

$$\mathcal{I} = \frac{e^2}{\pi\varepsilon_0\varepsilon_\infty}\int_0^\infty \frac{2+x^2}{(4+x^4)^{3/2}} = \frac{e^2}{\pi\varepsilon_0\varepsilon_\infty}0.4774\,. \tag{6.82}$$

With these ingredients, we obtain from (6.76)

$$[E_j(k)-E_i(k)-E]\,P_{ij}(k)$$
$$-\sum_{i'j'}\int_0^\infty \mathrm{d}k'\,V_{ji',j'i}(k,k')P_{i'j'}(k') + k\mathcal{I}P_{ij}(k) \tag{6.83}$$
$$-\int_0^\infty \mathrm{d}k'\,I(k,k')[P_{ij}(k')-P_{ij}(k)] = d_{ij}(k)\,.$$

After the integrations have been performed by means of numerical quadrature, this equation can be solved by matrix inversion and can serve as a starting point for discussing excitons in quantum wells. Some further work is necessary to calculate the matrix elements $\tilde{M}$. As an example, we consider the simplest possibility and assume a single quantum well with infinite barrier height. We consider the lowest subbands only, with eigenfunctions which are the same for electrons and holes,

$$g_0(z) = \sqrt{\frac{2}{L}}\cos(k_0 z)\,, \tag{6.84}$$

where $k_0 = \pi/L$. Using these wavefunctions, we obtain, after a straightforward calculation [97],

$$\tilde{M}_{0000}(q) = \frac{3qL}{4\pi^2 x} - 2\,\frac{1-qLx-q^2L^2x^2/2-\mathrm{e}^{-qL}}{q^2L^2x^2}\,, \tag{6.85}$$

where we have used the abreviations $q = |\boldsymbol{k}-\boldsymbol{k}'|$ and $x = 1+[qL/(2\pi)]^2$. To lowest order, (6.85) is $\propto q$ and (6.79) is regular in the limit $k \to k'$. Within this approximation, there is obviously no way back to the 3D limit. This limit is obtained from our general treatment for the case of densly lying subbands which form a continuum and allow us to introduce a wave vector k_z. Therefore our treatment is restricted to narrow wells for which the separation of subbands is sufficiently large. On the other hand, the well width must be sufficiently large too, to guarantee the validity of our basic assumptions. For GaAs/GaAlAs quantum wells, for example, we can estimate that our approximation is not too bad in the range 5–20 nm. Results obtained from the numerical solution of (6.83) within this approximation are shown in Fig. 6.3.

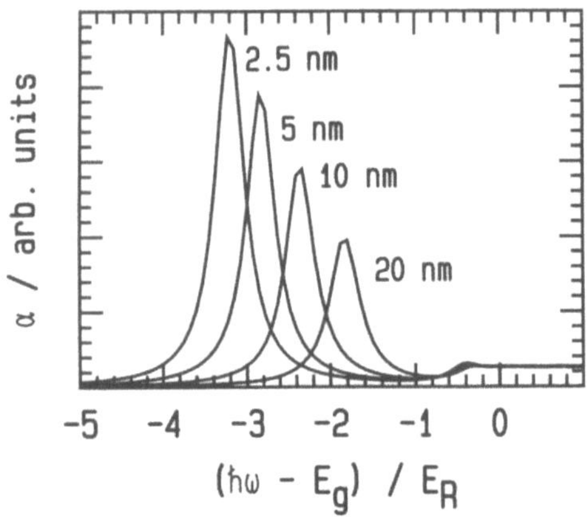

Fig. 6.3. Imaginary part of the dielectric function versus photon energy for different quantum well widths as indicated. The barriers are assumed to be infinitely high; GaAs parameters, damping fixed at $\hbar\gamma_2 = 0.17\,E_{\rm R}$. Note that the two-dimensional limit, where the binding energy would be $4E_{\rm R}$, is not reached even at quantum well widths as thin as 2.5 *nm*

Compared with excitons in bulk semiconductors, the binding energy is considerably enhanced and increases with decreasing well width. Correspondingly, the ratio of the oscillator strength of the 1s exciton to the continuum increases. This is an immediate consequence of the confinement of electrons and holes, which enhances their mutual interaction. The theoretical 2D-limit, in which the binding energy becomes identical to $4E_{\rm R}$, is an artefact and is never met with in real systems.

6.2.5 Propagation of Light: Polaritons and Cavity Polaritons

Let us focus on the 1s contribution to the susceptibility. According to (5.21), the sum of the resonant and off-resonant contributions can be obtained by means of (6.56) as

$$\chi^{(1)}(\omega) = \frac{2\,|d^0_{\rm vc}|^2\,\hbar\Omega_1}{\varepsilon_0 r_{\rm B}^3\,[(\hbar\Omega_1)^2 - (\hbar\omega)^2]}\,, \qquad (6.86)$$

where the 1s exciton energy $\hbar\Omega_1 := E_{\rm g} - E_{\rm R}$ and $\gamma_2 = 0$. The propagation of light is governed by the equation of motion for the photon Green's function (2.193).[3] For an infinite medium, this equation is readily solved by means of Fourier transformation, which yields[4]

$$\mathcal{D}^{\rm r}(\boldsymbol{q},\omega) = \frac{c_0^2}{\omega_{\boldsymbol{q}}^2 - \omega^2 - c_0^2\tilde{\chi}^{\rm t}(\boldsymbol{q},\omega)}\,. \qquad (6.87)$$

[3] Although we have derived this equation for classical fields only, it should be emphasized that the equation of motion for the photon Green's function is the same in a quantum mechanical treatment.

[4] For simplicity we restrict ourselves to cubic symmetry, in which case we have to consider scalar quantities only.

Making use of (2.199), we can reduce the transverse response function to (6.86). As before, all other transitions are lumped into the background dielectric function. With $\varepsilon(\omega) = \varepsilon_b\,[1+\chi^{(1)}(\omega)/\varepsilon_b]$, we introduce the renormalized light velocity and obtain

$$\mathcal{D}^r(\boldsymbol{q},\omega) = \frac{c^2}{c^2\boldsymbol{q}^2 - \omega^2[1+\chi^{(1)}(\omega)/\varepsilon_b]}\,. \tag{6.88}$$

The propagating modes are obtained from the two poles of the photon Green's function, which define the *polariton dispersion* [99] as solutions of

$$c_0^2 q^2 = \omega^2 n^2(\omega)\,, \tag{6.89}$$

where the complex refractive index is given by

$$n(\omega) = \sqrt{\varepsilon(\omega)} = \sqrt{\varepsilon_b}\sqrt{1+\frac{2\,\Omega_1\,\Delta_{LT}}{\hbar(\Omega_1^2-\omega^2)}}\,. \tag{6.90}$$

Here the *longitudinal–transverse splitting* is

$$\Delta_{LT} = \frac{|d_{vc}^0|^2}{\varepsilon_0\varepsilon_b r_B^3}\,. \tag{6.91}$$

The resulting dispersion is visualized in Fig. 6.4. Some examples of Δ_{LT} are given in Table 6.2. This quantity describes the energetic separation of the 1s energy from the pole of the longitudinal susceptibility. Owing to the $\boldsymbol{G}=0$ Hartree interaction, we have, according to (2.170),

$$\chi(\omega) = \frac{\chi^{(1)}(\omega)}{\varepsilon_b+\chi^{(1)}(\omega)}\,, \tag{6.92}$$

which has a pole at

$$\hbar\omega = \hbar\Omega_1 + \Delta_{LT}\,. \tag{6.93}$$

Correspondingly, the dielectric function is zero, i.e. $\varepsilon(\omega = \Omega_1 + \Delta_{LT}) = 0$, which allows the unusual case of electric fields that have longitudinal instead of transverse polarization because the Maxwell equation $\boldsymbol{qE} = \varepsilon\,\boldsymbol{qE}_{ext} = 0$ is fulfilled trivially. Thus for $\boldsymbol{q}=0$ the upper polariton branch becomes degenerate with the longitudinal exciton. In the energetic interval $[\hbar\Omega_1, \hbar(\Omega_1+\Delta_{LT})]$ the refractive index and the wave vector are purely imaginary. Hence no propagating electric field exists within this region.

The behavior depicted in Fig. 6.4 is closely similar to that of two coupled harmonic oscillators, one being the photon, the other being the 1s exciton. Their mutual coupling lifts the degeneracy and leads to two different eigenfrequencies of the coupled system. As a result of this, we find an *avoided crossing*. The flat dispersion relation close to $\hbar\Omega_1$ and $\hbar(\Omega_1+\Delta_{LT})$ leads to a pronounced *group velocity dispersion* $\mathrm{d}\omega/\mathrm{d}q$, which is depicted in the right part of Fig. 6.4. A detailed understanding of experimental data frequently requires the consideration of light propagation in a finite medium and thus

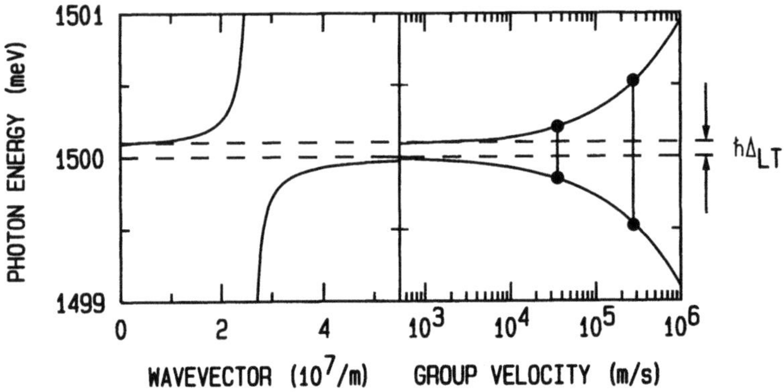

Fig. 6.4. *Left diagram*: dispersion of light in the vicinity of the 1s exciton resonance. *Right diagram*: energy versus group velocity. The parameters are $\hbar\Omega_1 = 1500\,\text{meV}$, $\hbar\gamma_2 = 0$, $n_b = 3.4$ and the longitudinal transverse splitting is $\hbar\Delta_{LT} = 0.1\,\text{meV}$

the coupling of the polariton modes to external photons. As the excitons exchange momentum with the propagating light field, we have to take into account also the excitonic center-of-mass motion. In this case, keeping a local model for the susceptibility such as (6.86) with a wave-vector-dependent exciton energy makes the problem troublesome. For each frequency ω we obtain two wave vectors $\boldsymbol{q}_1$ and $\boldsymbol{q}_2$ on the polariton branches, and Maxwell's boundary conditions are no longer sufficient to determine the fields. Attempts to overcome this difficulty by additional boundary conditions on the polarization (e.g. [100]) are not really satisfactory, as the problem is an immediate artefact of the assumed locality of the response function. Clearly this assumption fails in a half-space geometry. A systematic treatment requires the solution of Maxwell's equations together with the calculation of the nonlocal excitonic polarization, which for propagation in, for example, z direction depends on z and z' separately. Numerical investigations along these lines give a detailed microscopic explanation of the polariton effects found in absorption spectra without reference to additonal boundary conditions [101].

Table 6.2. Longitudinal–transverse splitting $\hbar\Delta_{LT}$ of the lowest-energy 1s exciton polariton of a few semiconductors, $T \leq 4.2\,K$. Taken from [2]

Material	$\hbar\Delta_{LT}$ / meV
GaAs	≈ 0.1
CdTe	≈ 0.4
ZnSe	≈ 1.3
CdS	≈ 2 (A)
CuCl	5.4–5.7 (Z_3)

Another striking effect of polariton propagation can be understood qualitatively without considering the nonlocality of the response function. Let the electric field at $z = 0$ be $\boldsymbol{E}(z = 0, t)$, with Fourier transform $\boldsymbol{E}(z = 0, \omega)$. After propagation to $z = L$, the electric field acquires a certain frequency-dependent phase (see Chap. 2) and becomes

$$\boldsymbol{E}(z = L, \omega) = \boldsymbol{E}(z = 0, \omega)\, \mathrm{e}^{\mathrm{i}\, q(\omega)\, L} , \tag{6.94}$$

where $q(\omega)$ is given by the polariton dispersion (6.89). An example of the corresponding intensity in the time domain $\propto |\boldsymbol{E}(z = L, t)|^2$ is shown in Fig. 6.5.

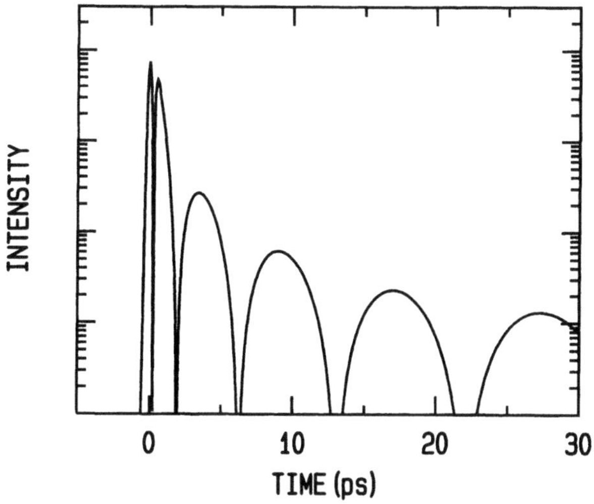

Fig. 6.5. Intensity of a short, Gaussian-shaped pulse that has propagated over a distance $L = 1\,\mu m$ in a semiconductor. The center frequency of the incident pulse is resonant with the 1s exciton. Other parameters are $\hbar\Omega_1 = 1500\,\mathrm{meV}$, $\hbar\gamma_2 = 0$, $n_\mathrm{b} = 3.4$, and $\hbar\Delta_\mathrm{LT} = 0.1\,\mathrm{meV}$. This phenomenon was first reported in [102]

The interference of light on the two polariton branches leads to the pronounced, nonperiodic oscillatory decay of the intensity. The distance between adjacent maxima scales inversely with both Δ_LT and L. The nonperiodic beating behavior can be understood with the help of Fig. 6.4. A short pulse excites a broad range of points on the polariton dispersion. Two components on the polariton dispersion, with a large and a small group velocity arrive early and late, respectively, and exhibit a large and a small energy difference, respectively (filled circles in Fig. 6.4). This large or small energy difference translates into a rapid or slow oscillation, respectively, in time. Consequently, the oscillation becomes slower at later times.

The avoided crossing of the exciton and the photon can be seen even more directly in the presence of a cavity. The intensity transmission coefficient $T(\omega)$

through a platelet of thickness L with mirrors of intensity reflectivity R on both sides is given by the Fabry–Perot formula, which can easily be derived by summation of all partial waves in the cavity. This formula is

$$T(\omega) = \frac{1}{1 + F \sin^2 \left[n(\omega)\, n_b \omega L / c_0 \right]}, \tag{6.95}$$

where we have introduced $F = 4R/(1-R)^2$, which is related to the *finesse* $\mathcal{F}$ of the cavity [107], the ratio of the energetic difference between adjacent transmission maxima to their half width, via $\mathcal{F} = \pi\sqrt{F}/2$. In Fig. 6.6 we again find the avoided crossing scenario. However, the effective coupling and hence also the observed *vacuum Rabi splitting* are not given simply by Δ_{LT} but also depend on the cavity length. This behavior is closely similar to the polariton propagation beats discussed above, which also depend on the interaction length. One can think of the two resonances in Fig. 6.6 as two points on the polariton dispersion Fig. 6.4, quantized owing to the presence of the cavity. Such quantized polariton modes are often referred to as cavity polaritons.

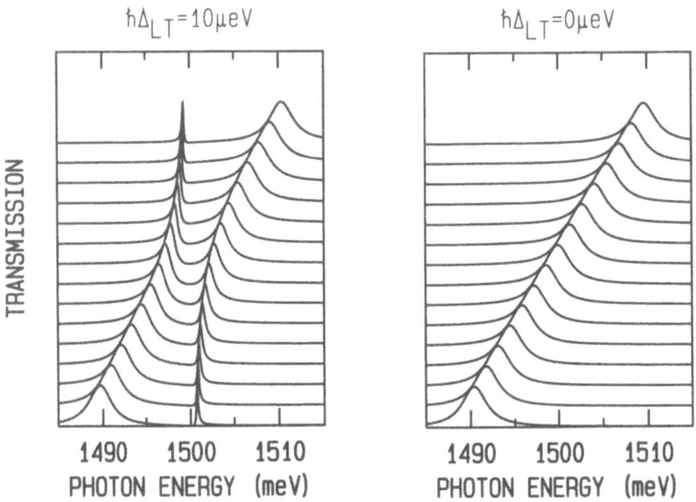

Fig. 6.6. Transmission of a semiconductor platelet with highly reflective mirrors on both sides. Parameters are $\hbar\Omega_1 = 1500\,\text{meV}$, $\hbar\gamma_2 = 0$, $n_b = 3.4$ and $F = 1000$. The thickness L of the platelet increases from *top* to *bottom* in steps of 1 nm; the central thickness L is 1.094 μm. This phenomenon was first reported in [103]. The authors used an epitaxially grown, slightly wedge-shaped microcavity at low temperature, containing several semiconductor quantum wells

6.3 Nonlinear Optical Properties

The semiconductor Bloch equations (6.3) and (6.4) are visualized in Fig. 6.7. An analysis of these equations corresponding to a realistic experimental situation in the regime of nonlinear optics, i.e. for finite occupation numbers, usually requires numerical solutions with considerable numerical effort. Hence, one would like to simplify the mathematics of the semiconductor Bloch equations and arrive at a more intuitive understanding. This is in fact possible in the case of resonant excitation of the 1s exciton [98], which has been studied in a large number of experiments. After describing this simplification, we discuss a number of numerical solutions.

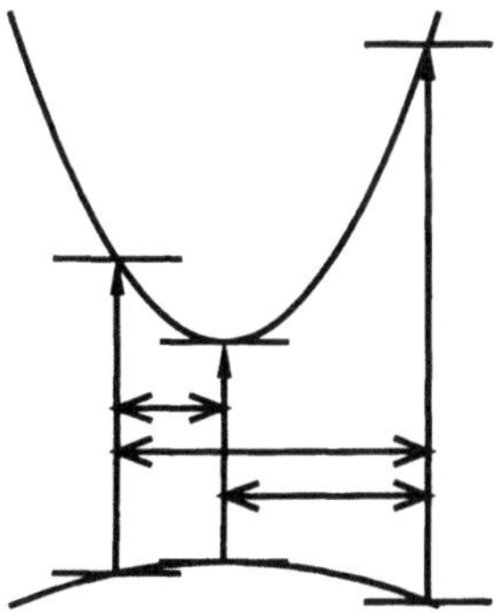

Fig. 6.7. Illustration of the semiconductor Bloch equations in the single particle band diagram

6.3.1 The Local-Field Approximation

We start from the semiconductor Bloch equations (6.3) and (6.4), neglect damping for the moment ($\gamma_1 = \gamma_2 = 0$). From (6.3), we obtain an equation for $|p_{\mathrm{vc}}(\boldsymbol{k})|^2$. Subtracting (6.4) leads to the relation

$$\frac{\partial}{\partial t}|p_{\mathrm{vc}}(\boldsymbol{k})|^2 = \frac{\partial}{\partial t}\{f_{\mathrm{e}}(\boldsymbol{k}) - [f_{\mathrm{e}}(\boldsymbol{k})]^2\} \, . \tag{6.96}$$

The solution is given by

$$f_{\mathrm{e}}(\boldsymbol{k}) = f_{\mathrm{h}}(\boldsymbol{k}) = \frac{1}{2}\left(1 \pm \sqrt{1 - 4|p_{\mathrm{vc}}(\boldsymbol{k})|^2}\right) , \tag{6.97}$$

which, for small excitation densities, owing to $|p_{\mathrm{vc}}(\boldsymbol{k})|^2 \ll 1$ (and hence $f_{\mathrm{e}}(\boldsymbol{k}) \ll 1$), is equivalent to

$$f_{\mathrm{e}}(\boldsymbol{k}) = f_{\mathrm{h}}(\boldsymbol{k}) = |p_{\mathrm{vc}}(\boldsymbol{k})|^2 \, . \tag{6.98}$$

This result is a simple consequence of the conservation of the length of the Bloch vector, and can be substituted in the equation of motion of the transition amplitude. We are interested in exclusive excitation of the 1s exciton.

Hence it is natural to expand the transition amplitude in the complete set of excitonic eigenfunctions with quantum number n, i.e.

$$p_{\mathrm{vc}}(\boldsymbol{k}) = \sum_n \phi_n(\boldsymbol{k}) p_n . \tag{6.99}$$

We introduce this expansion into the equation of motion of the transition amplitude in the semiconductor Bloch equations and consider the projection of the equation onto state m (we multiply by $\phi_m^*(\boldsymbol{k})$ and sum over $\boldsymbol{k}$). Taking advantage of the fact that the ϕ_n form an orthonormal set of states, we arrive at

$$\begin{aligned}
&\left(\hbar\frac{\partial}{\partial t} + \mathrm{i}\hbar\Omega_m\right)p_m \\
&= \mathrm{i}\phi_m^*(\boldsymbol{r}=0)\,\boldsymbol{d}_{\mathrm{cv}}\boldsymbol{E}(\boldsymbol{r},t) - 2\mathrm{i}\boldsymbol{d}_{\mathrm{cv}}\boldsymbol{E}(\boldsymbol{r},t)\sum_{\boldsymbol{k},n,n'} \phi_m^*(\boldsymbol{k})\phi_n(\boldsymbol{k})\phi_{n'}^*(\boldsymbol{k})p_n p_{n'}^* \\
&+2\mathrm{i}e^2 \sum_{\substack{\boldsymbol{k}\ \boldsymbol{k}',\\ n,\ n',\ n''}} v(\boldsymbol{k}-\boldsymbol{k}')\Big\{\phi_m^*(\boldsymbol{k})\phi_n(\boldsymbol{k})\phi_{n'}(\boldsymbol{k}')[\phi_{n''}^*(\boldsymbol{k}') - \phi_{n''}^*(\boldsymbol{k})]p_n p_{n'} p_{n''}^*\Big\} .
\end{aligned} \tag{6.100}$$

The Fourier transform of the mth excitonic eigenfunction is $\phi_m^*(\boldsymbol{r}=0) = \sum_{\boldsymbol{k}} \phi_m^*(\boldsymbol{k})$ and the excitonic eigenenergies from the Fourier transform of (6.11) are given by

$$[E_{\mathrm{e}}(\boldsymbol{k}) + E_{\mathrm{h}}(\boldsymbol{k})]\phi_m(\boldsymbol{k}) - e^2 \sum_{\boldsymbol{k}'} v(\boldsymbol{k}-\boldsymbol{k}')\phi_m(\boldsymbol{k}') = \hbar\Omega_m\phi_m(\boldsymbol{k}) \ , \tag{6.101}$$

which is nothing but the Schrödinger equation of the hydrogen problem in $\boldsymbol{k}$ space. This transformation is illustrated in Fig. 6.8. Restricting ourselves to the case of selective excitation of the 1s–exciton, i.e.

$$p_m = p_1\delta_{m,1} \ , \tag{6.102}$$

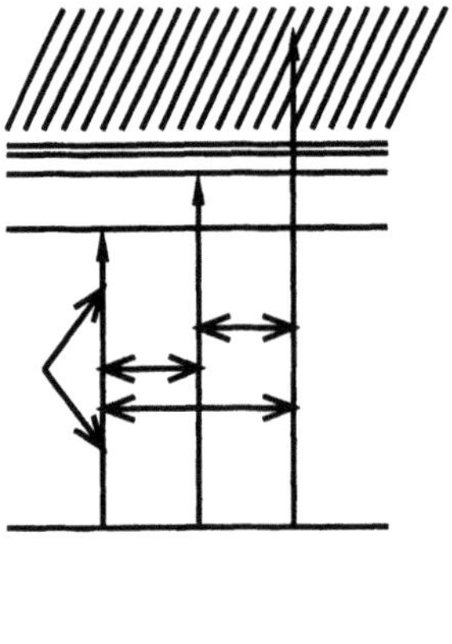

Fig. 6.8. Illustration of the semiconductor Bloch equations in the excitonic basis, schematically showing the Rydberg series and the excitonic continuum. This diagram must not be interpreted as an energy dispersion

the interaction with all excitonic levels other than the 1s exciton vanishes for $|p_1|^2 \ll 1$ and we arrive at the simple form [104]

$$\left(\hbar\frac{\partial}{\partial t} + i\hbar\bar{\Omega}_1 + \hbar\gamma_2\right) p_1 = i[\boldsymbol{d}_1 \boldsymbol{E}(\boldsymbol{r},t) + \bar{v}_{1,1}\, p_1]\left(1 - \bar{b}_1\, |p_1|^2\right) . \tag{6.109}$$

The Coulomb-enhanced transition matrix element is

$$\boldsymbol{d}_1 = \phi_1^*(\boldsymbol{r}=0)\, \boldsymbol{d}_{\mathrm{cv}} , \tag{6.104}$$

the phase-space-filling parameter is

$$b_1 = 2\sum_{\boldsymbol{k}} \frac{|\phi_1(\boldsymbol{k})|^2 \phi_1^*(\boldsymbol{k})}{\phi_1^*(\boldsymbol{r}=0)} , \tag{6.105}$$

and the interaction parameter is

$$v_{1,1} = 2\, e^2 \sum_{\boldsymbol{k},\, \boldsymbol{k}'} v(\boldsymbol{k}-\boldsymbol{k}')\, |\phi_1(\boldsymbol{k})|^2 \left(|\phi_1(\boldsymbol{k}')|^2 - \phi_1^*(\boldsymbol{k})\phi_1(\boldsymbol{k}')\right) . \tag{6.106}$$

We have also reintroduced the phenomenological damping γ_2 of the transition amplitude. This result can be cast into a form which has a closer analogy to a two-level description. Introducing the abbreviations

$$\bar{v}_{1,1} = v_{1,1} \,/\, b_1 \tag{6.107}$$

and

$$\bar{\Omega}_1 = \Omega_1 - \bar{v}_{1,1} , \tag{6.108}$$

we obtain the following from (6.103):

$$\left(\hbar\frac{\partial}{\partial t} + i\hbar\Omega_1 + \hbar\gamma_2\right) p_1 = i\boldsymbol{d}_1 \boldsymbol{E}(\boldsymbol{r},t)\left(1 - b_1\, |p_1|^2\right) + i\, v_{1,1}\, p_1\, |p_1|^2 . \tag{6.103}$$

This approximation is illustrated in Fig. 6.9. The linear-optical solution of (6.109) is clearly nothing but the 1s contribution to the Elliott formula (6.57). The factor $1-\bar{b}_1|p_1|^2$ describes saturation of the 1s exciton due to phase space filling and arises from the fact that excitons are neither bosons (for which $b_1 = 0$ would hold) nor fermions (with $b_1 = 2$) but, rather, composite particles that feel the Fermi statistics of the underlying constituents. The factor $\boldsymbol{d}_1\boldsymbol{E}(\boldsymbol{r},t) + \bar{v}_{1,1}\, p_1$, where $\bar{v}_{1,1}$ is the *exchange energy* between 1s excitons, can be interpreted as the renormalized Rabi frequency or as a local field[5] generated by the interaction of a 1s exciton with the surrounding excitons [98].

[5] The "local field" was introduced in the original publications because the mathematical form is closely similar to that of the actual local fields discussed in many elementary textbooks. This term is, however, somewhat misleading as the exchange interaction is due to the long-range Coulomb interaction, which is anything but local. In particular, this model should not be confused with the local-field corrections due to the short-range Hartree interaction.

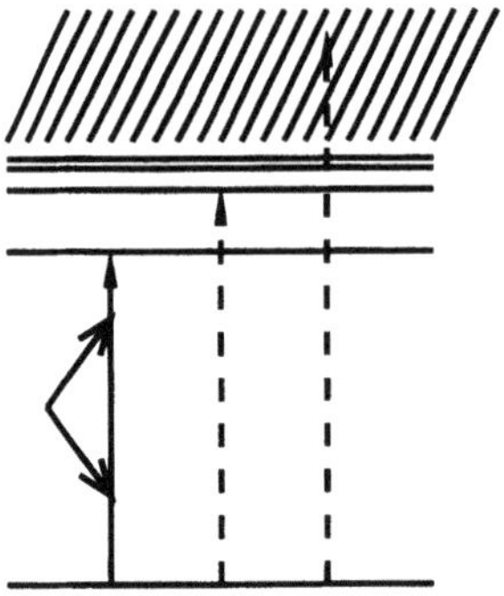

Fig. 6.9. Illustration of the local-field approximation in the excitonic basis

It is this factor that makes the dynamics of the 1s exciton qualitatively different from that of a two-level system. More generally, even our simple model demonstrates a basic relationship between linear and nonlinear properties. Any interaction which leads at the level of linear optics, to the formation of new quasiparticles – excitons in our case – generates interactions between these quasiparticles in the nonlinear regime. Let us illustrate this important finding with two examples.

Example (Four-Wave Mixing). We consider the four-wave-mixing geometry discussed in Chap. 4 and use the rotating wave approximation (see Sect. 4.3.2)

$$p_1 = \tilde{p}_1 \mathrm{e}^{-\mathrm{i}\omega t} , \tag{6.110}$$

which leads to

$$\left(\hbar\frac{\partial}{\partial t} + \mathrm{i}\hbar\Delta_1 + \hbar\gamma_2\right)\tilde{p}_1 = \mathrm{i}\left(\boldsymbol{d}_1\frac{\tilde{\boldsymbol{E}}(\boldsymbol{r},t)}{2} + \bar{v}_{1,1}\tilde{p}_1\right)\left(1 - b_1|\tilde{p}_1|^2\right) , \tag{6.111}$$

where $\hbar\Delta_1 = \hbar\bar{\Omega}_1 - \hbar\omega$ is the detuning of the photon energy $\hbar\omega$ with respect to $\hbar\bar{\Omega}_1$. The response to δ-shaped optical pulses

$$\frac{\tilde{\boldsymbol{E}}(\boldsymbol{r},t)}{2} = \tilde{\boldsymbol{E}}_1\,\mathrm{e}^{\mathrm{i}\boldsymbol{q}_1\boldsymbol{r}}\,\delta(t) + \tilde{\boldsymbol{E}}_2\,\mathrm{e}^{\mathrm{i}\boldsymbol{q}_2\boldsymbol{r}}\,\delta(t-T) \tag{6.112}$$

within the perturbation approach (see Sect. 4.3) up to third order, for resonant excitation ($\hbar\Omega_1 - \hbar\omega = 0$), is given by

$$\begin{aligned} \tilde{p}_1 &= -\frac{b_1}{2}\boldsymbol{d}_1|\boldsymbol{d}_1|^2\hbar^{-3}\tilde{\boldsymbol{E}}_2^2\tilde{\boldsymbol{E}}_1\mathrm{e}^{\mathrm{i}(2\boldsymbol{q}_2-\boldsymbol{q}_1)\boldsymbol{r}} \\ &\times\Big(\mathrm{i}\,\Theta(T)\Theta(t-T)\mathrm{e}^{-\gamma_2 t} + \frac{\bar{v}_{1,1}}{\hbar\gamma_2}\mathrm{e}^{-\gamma_2(t-2T)} \\ &\times\left[\Theta(T)\Theta(t-T)(\mathrm{e}^{-2\gamma_2 t} - \mathrm{e}^{-2\gamma_2 T}) + \Theta(-T)\Theta(t)(\mathrm{e}^{-2\gamma_2 t} - 1)\right]\Big) , \end{aligned} \tag{6.113}$$

which consists of two contributions. The first contribution is instantaneous and is proportional to the phase space parameter b_1; it yields the *free polarization decay.* The second contribution, proportional to $\bar{v}_{11}$, describes an *interaction-induced signal* [105], which always exhibits a delayed maximum. The polarization

$$\boldsymbol{P}(\boldsymbol{r},t) = \frac{1}{V}\boldsymbol{d}_1^* p_1 + \text{c.c.} \tag{6.114}$$

(derived from (4.9)) is obtained for $\bar{v}_{1,1}/(\hbar\gamma_2) \gg 1$ as

$$\begin{aligned}\boldsymbol{P}(\boldsymbol{r},t) = &-b_1|\boldsymbol{d}_1|^4\hbar^{-3}\tilde{\boldsymbol{E}}_2^2\tilde{\boldsymbol{E}}_1\frac{v_{1,1}}{\hbar\gamma_2}\cos[(2\boldsymbol{q}_2-\boldsymbol{q}_1)\boldsymbol{r}-\Omega_1 t]\,\mathrm{e}^{-\gamma_2(t-2T)}\\ &\times\left[\Theta(T)\Theta(t-T)\left(\mathrm{e}^{-2\gamma_2 t}-\mathrm{e}^{-2\gamma_2 T}\right)+\Theta(-T)\Theta(t)\left(\mathrm{e}^{-2\gamma_2 t}-1\right)\right].\end{aligned} \tag{6.115}$$

Two aspects of this result are important:

- A polarization is observed for both negative and positive time delays.
- The polarization exhibits a slow rise after whichever pulse is the second in time.

Both aspects are illustrated in Fig. 6.10. This observation has given rise to the notion that *the coherent radiation of excitons is distinctly different from that of atoms.* Finally, the energy, emitted in the $2\boldsymbol{q}_2-\boldsymbol{q}_1$ direction, is proportional to time integrated signal

$$\int \mathrm{d}t\,|\boldsymbol{P}|^2 \propto \Theta(T)\mathrm{e}^{-2T/T_2} + \Theta(-T)\mathrm{e}^{+4T/T_2}\,. \tag{6.116}$$

Even though the assumption of exclusively resonant excitation of the 1s exciton is not strictly valid in many experimental situations where short pulses (often with a bandwidth exceeding the Rydberg energy) are used, the local-field model still delivers almost quantitative results (when compared with numerical solutions of the semiconductor Bloch equations). A corresponding experimental result on GaAs quantum wells is shown in Fig. 6.11, and a selection of measured dephasing times is given in Table 6.3.

Table 6.3. Measured dephasing times T_2 of the direct-gap 1s exciton of several bulk semiconductors and semiconductor quantum wells at low temperature ($T \leq 4.2\,\mathrm{K}$) and low excitation density

Material	T_2
Ge, bulk	200–300 fs
GaAs, bulk	7 ps
GaAs/AlGaAs, 17 nm QW	2.2 ps
InGaAs, bulk	600 fs
InGaAs/InAlAs, 10 nm QW	4 ps
CdSSe, bulk	Up to 1 ns, depending on composition

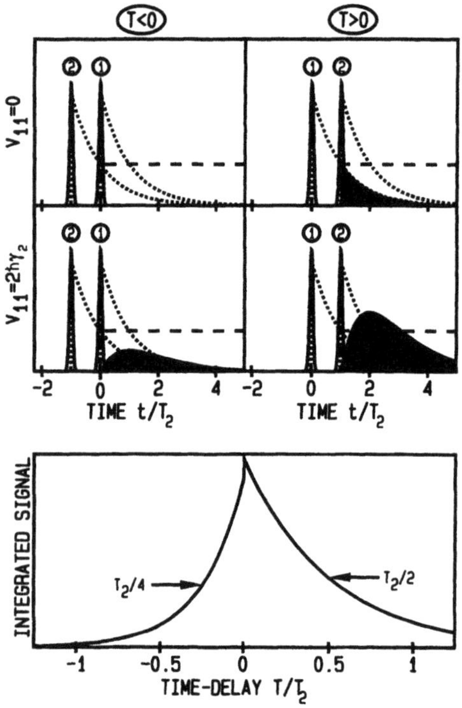

Fig. 6.10. Transient four-wave-mixing within the local-field approximation. The top demonstrates the influence and importance of the 1s exciton exchange interaction $v_{1,1}$. The *filled areas* are the third-order polarizations, the incident pulses are marked by the labels 1 and 2, the *dashed* step functions are the second-order occupation functions, and the *dotted lines* are the first-order polarizations. The *bottom* shows the temporally integrated intensity versus the time delay T for $\bar{v}_{1,1} = 2\hbar\gamma_2$

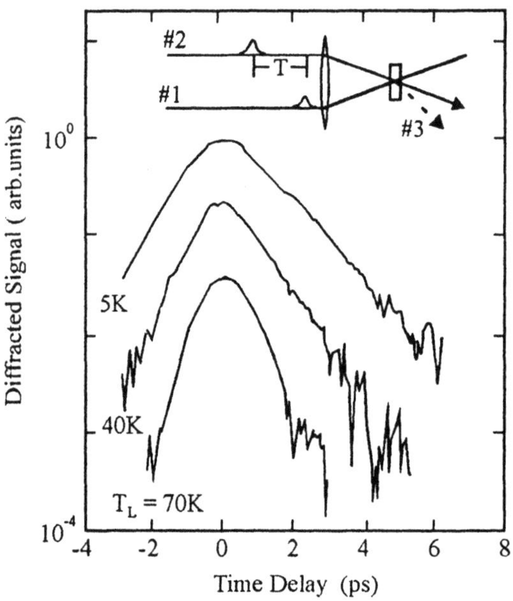

Fig. 6.11. Four-wave-mixing experiment resonant with the 1s heavy-hole exciton in a GaAs multiple-quantum-well structure at three different lattice temperatures T_L. Note that both the rise and the decay time change. Taken from [106]

Example (Optical Stark Effect). The concomitants of any optical excitation are the so-called *light-induced states* [108, 109, 110, 111], which are present on the timescale of the light field. Although this effect is always present in principle, it depends crucially on the excitation conditions whether it is of any relevance or not. For continuum excitation on an ultrashort timescale, the spectral width of the pulse is usually much larger than the light-induced gaps. When one approaches stationary excitation conditions, the physics is dominated by phase-space filling and relaxation processes, and light-induced effects are of minor importance in this limit also. What remains at high excitation intensities is a slight modification of the density of states. In the case of off-resonant excitation, however, light-induced states become important and are constitutive of the optical Stark effect. Let us examine the basic physics within the local-field model.[6] We consider a pump–probe configuration with a strong excitation beam $\boldsymbol{E}_{\mathrm{p}}$ and a weak test beam $\boldsymbol{E}_{\mathrm{t}}$ (similar to that in Sect. 4.3). As for large detuning all excitonic states contribute with the same order of magnitude, we assume that the center frequency of the steady-state pump ω_{p} is tuned sufficiently close to the low-energy side of the 1s exciton that the coupling of higher exciton states can be neglected. Under these conditions (6.103) allows us to compute the linear response (of the excited system) with respect to the weak test beam. Making the replacement $\boldsymbol{E} = \boldsymbol{E}_{\mathrm{p}} + \boldsymbol{E}_{\mathrm{t}}$ and, correspondingly, $p = p_{\mathrm{p}} + p_{\mathrm{t}}$ in (6.103) and considering only terms linear in the test beam leads us to

$$\begin{aligned}(\hbar\frac{\partial}{\partial t} + \mathrm{i}\hbar\Omega + \hbar\gamma_2)p_{\mathrm{t}} = \mathrm{i}\boldsymbol{d}\boldsymbol{E}_{\mathrm{t}}(1 - b|p_{\mathrm{p}}|^2) - \mathrm{i}\left(b\,\boldsymbol{d}\,\boldsymbol{E}_{\mathrm{p}}\,p_{\mathrm{p}}^* \,+\, 2v\,|p_{\mathrm{p}}|^2\right)p_{\mathrm{t}} \\ -\,\mathrm{i}\left(b\,\boldsymbol{d}\,\boldsymbol{E}_{\mathrm{p}}\,p_{\mathrm{p}} + v\,p_{\mathrm{p}}^2\right)p_{\mathrm{t}}^* \,. \qquad (6.117)\end{aligned}$$

By substituting $\boldsymbol{E}_{\mathrm{p}} = \tilde{\boldsymbol{E}}_{\mathrm{p}}\mathrm{e}^{-\mathrm{i}\omega_{\mathrm{p}}t}$, $\boldsymbol{E}_{\mathrm{t}} = \tilde{\boldsymbol{E}}_{\mathrm{t}}\mathrm{e}^{-\mathrm{i}\omega t}$, and $p_{\mathrm{p}} = \tilde{p}_{\mathrm{p}}\mathrm{e}^{-\mathrm{i}\omega_{\mathrm{p}}t}$, we see that p_{t} contains two contributions which oscillate with frequencies ω and $2\omega_{\mathrm{p}} - \omega$. This suggests the ansatz

$$p_{\mathrm{t}} = p_+\mathrm{e}^{-\mathrm{i}\omega t} + p_-\mathrm{e}^{-\mathrm{i}(2\omega_{\mathrm{p}}-\omega)t} \qquad (6.118)$$

and, after Fourier decomposition, yields the coupled set of equations

$$\begin{aligned}[\hbar\,(\Omega - \omega - \mathrm{i}\gamma_2) + b\,\boldsymbol{d}\,\tilde{\boldsymbol{E}}_{\mathrm{p}}\tilde{p}_{\mathrm{p}} + 2v\,\tilde{p}_{\mathrm{p}}^2]\,p_+ \;=\; \boldsymbol{d}\,\tilde{\boldsymbol{E}}_{\mathrm{t}}(1 - b\,\tilde{p}_{\mathrm{p}}^2) \\ -\,(b\,\boldsymbol{d}\,\tilde{\boldsymbol{E}}_{\mathrm{p}} + v\,\tilde{p}_{\mathrm{p}}^2)\,p_-^* \,, \qquad (6.119)\end{aligned}$$

$$[\hbar\,(\Omega + \omega - 2\omega_{\mathrm{p}} - \mathrm{i}\gamma_2) + b\,\boldsymbol{d}\,\tilde{\boldsymbol{E}}_{\mathrm{p}}\,\tilde{p}_{\mathrm{p}} + 2v\,\tilde{p}_{\mathrm{p}}^2]\,p_- \;=\; -(b\,\boldsymbol{d}\tilde{\boldsymbol{E}}_{\mathrm{p}} + v\,\tilde{p}_{\mathrm{p}}^2)\,p_+^* \,. \qquad (6.120)$$

[6] For clarity, we drop the index 1 on the quantities in (6.103).

Here we have used the fact that, under the assumed excitation conditions, $\tilde{p}_\mathrm{p}$ is real (neglecting the damping γ_2). The solution of (6.119) and (6.120) can be written in the form

$$p_+ = \frac{\boldsymbol{d}\tilde{\boldsymbol{E}}_\mathrm{t}(1 - b\tilde{p}_\mathrm{p}^2)[\mathcal{A} - \hbar(\omega - \omega_\mathrm{p} + \mathrm{i}\gamma_2)][\mathcal{A} + \hbar(\omega - \omega_\mathrm{p} + \mathrm{i}\gamma_2)]}{[\mathcal{A} - \hbar(\omega - \omega_\mathrm{p} + \mathrm{i}\gamma_2)][\mathcal{A} + \hbar(\omega - \omega_\mathrm{p} + \mathrm{i}\gamma_2)] - \mathcal{B}^2} \tag{6.121}$$

where we have used the abbreviations $\mathcal{A} = \hbar(\Omega - \omega_\mathrm{p}) + b\,\boldsymbol{d}\,\tilde{\boldsymbol{E}}_\mathrm{p}\,\tilde{p}_\mathrm{p} + 2\,v\,\tilde{p}_\mathrm{p}^2$ and $\mathcal{B} = b\,\boldsymbol{d}\,\tilde{\boldsymbol{E}}_\mathrm{p}\,\tilde{p}_\mathrm{p} + v\,\tilde{p}_\mathrm{p}^2$.

This expression for p_+, which is proportional to the linear susceptibility of the excited system, has poles at

$$\hbar\omega_\pm = \hbar\omega_\mathrm{p} - \mathrm{i}\hbar\gamma_2 \pm (\mathcal{A}^2 - \mathcal{B}^2)^{1/2}\,. \tag{6.122}$$

In Fig. 6.12, these poles are plotted as a function of pump intensity with (solid lines) and without (dashed lines) the Coulomb exchange interaction. The first pole (plussign in (6.122)) corresponds to the usual exciton contribution which is, however, shifted towards higher energies, owing to the action of the pump. The second pole (minus sign in (6.122)) corresponds to the excitation energy of a bound electron–hole pair, which is built up from valence and conduction states shifted up and down, respectively, by $\hbar\omega_\mathrm{p}$. These are the light-induced states, which, owing to their coupling to the quasiparticle states, are shifted to lower energies. As can easily be shown by Taylor expansion of (6.122) for small pump intensities, the energy shifts are proportional to the pump intensity and (asymptotically) inversely proportional to the pump detuning $\hbar(\Omega_1 - \omega_\mathrm{p})$. These dependences have actually been observed in experiments [112, 113] on semiconductor quantum wells (Fig. 6.13). The shifts are drastically enhanced by the exchange interaction $v_{1,1}$ among 1s excitons (compare the solid and dashed lines in Fig. 6.12).

The effect of the Coulomb exchange interaction becomes particularly clear in the case of resonant excitation ($\hbar\omega_\mathrm{p} = \hbar\Omega_1$). Without interaction (Fig. 6.12b, dashed line, $v_{1,1} = 0$), no shift is expected (which is immediately obvious from (6.122)), while the exciton exhibits a blue shift for $v_{1,1} > 0$ (Fig. 6.12b, solid lines, $v_{1,1} > 0$) according to (6.122). This blue shift has been observed in experiments on semiconductor quantum wells [114]. Intuitively, it originates from the repulsive exchange interaction ($v_{1,1} > 0$) of excitons with equal spin, which simply increases the energy of the system as the number of excitons increases. In contrast to this, excitons with opposite spin experience an attractive interaction, the bound states of which are called *biexcitons*. Biexcitons will be discussed in some detail in Chap. 9.

6.3.2 Numerical Solutions

The description of four-wave-mixing experiments or of the optical Stark effect within the local-field model covers merely the qualitative features. A more realistic and quantitative description requires numerical solutions – at least at the level of the semiconductor Bloch equations.

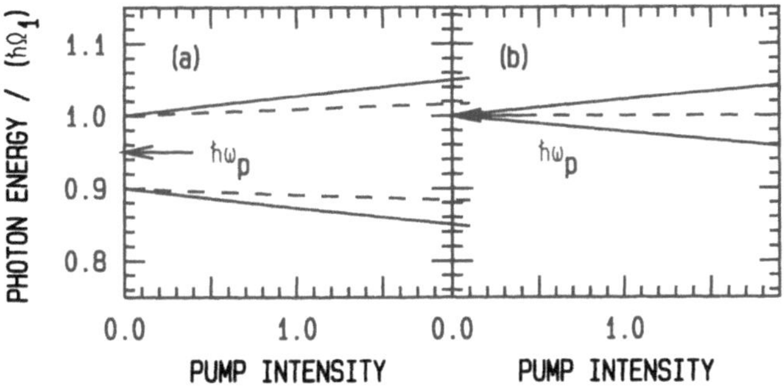

Fig. 6.12. Schematic illustration of optical transition energies, as observed in a pump–probe experiment, within the local-field model under steady-state excitation. *Dashed lines* $v_{1,1} = 0$; *solid lines* $v_{1,1} > 0$, $\hbar\gamma_2 = 0$. (*a*) Optical Stark effect, that is off-resonant excitation at $\hbar\omega_\mathrm{p} = 0.95\,\hbar\Omega_1$ (*arrow*), (*b*) resonant excitation of the 1s exciton, i.e. $\hbar\omega_\mathrm{p} = \hbar\Omega_1$. Clearly, this picture becomes unrealistic for energy shifts exceeding the exciton binding energy

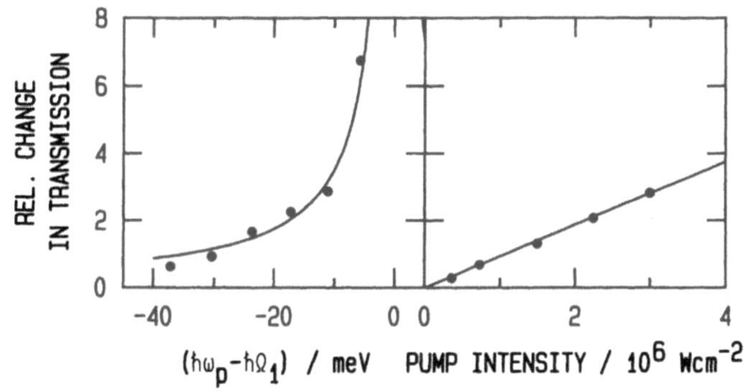

Fig. 6.13. Experimental results: the blue shift of the 1s exciton in a GaAs/AlGaAs semiconductor quantum well is proportional to the measured relative change in transmission (*filled circles*, in arbitrary units) observed in a pump–probe experiment using rather long (5 ps) pulses. The data are plotted versus detuning for fixed pump intensity (*left*) and versus pump intensity for fixed detuning (*right*). The *solid lines* correspond to a blue shift which is inversely proportional to the detuning (*left*) and proportional to the pump intensity (*right*), respectively. Data taken from [112]

Example (Excitonic Wave Packets). Figure 6.14 shows an example of some numerical solutions of the semiconductor Bloch equations. The center frequency of the Gaussian 107 fs (FWHM of the intensity) pulse is tuned to the 1s exciton resonance of bulk GaAs; $\hbar\gamma_1 = 0$, $\hbar\gamma_2 = 0.2E_\mathrm{R}$. The Rabi energy of each individual pulse is $\hbar\Omega_\mathrm{R} = 0.001E_\mathrm{R}$, which is well within the third-order perturbational limit. The diagram on the right schematically illustrates the excitation conditions. The 1s exciton and higher energy states are excited simultaneously, which creates an *excitonic wave packet*. The resulting oscillation, a *quantum beat*, has a period close to the inverse exciton binding energy, as indicated by the bar with length $\hbar E_R^{-1}$. Consequently, the spectrally resolved four-wave-mixing signal shows one peak at the 1s exciton energy and another one in the vicinity of the 2s resonance (see inset). As expected intuitively, the oscillation period of the excitonic wave packet corresponds to one orbital period of the exciton in classical physics. The slow rise (with respect to the 107 fs optical pulse) of the envelope of the four-wave-mixing signal in real time is again due to the exchange interaction among 1s excitons, an effect which has already been discussed within the local-field model.

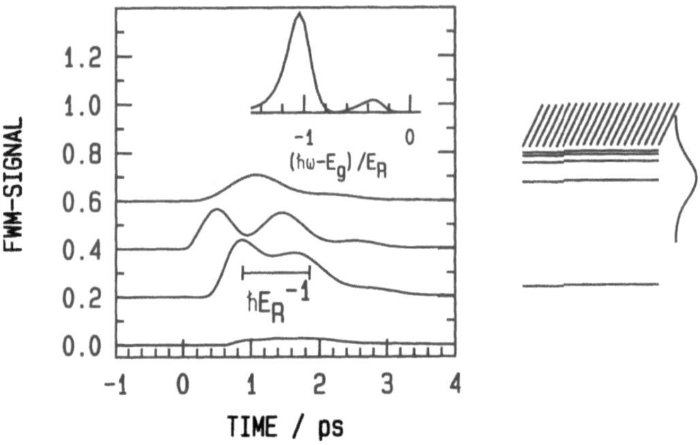

Fig. 6.14. Transient four-wave-mixing signal as a function of time. The time delay increases from *top* to *bottom* (-300 fs, 0 fs, 300 fs, 600 fs), and the traces are vertically displaced for clarity. The *inset* shows the spectrum of the diffracted beam for zero time delay. For details, see text

Experiments of this type [115] are useful in the presence of inhomogeneous broadening, in which case the exciton binding energy can no longer be determined unambiguously from the linear absorption spectrum. The oscillation in the time-delay domain, however, survives in the presence of inhomogeneous broadening, while the real-time response becomes a photon echo for large inhomogeneous widths [116]. This fact makes *quantum beat spectroscopy* an important tool for observing intrinsic structures underneath spectra which are smeared out by extrinsic effects. This idea is quite general and also works for luminescence [117] and pump–probe techniques.

Quantum beats – in the strict sense – refer to a situation in which two quantum mechanical states, related to one common ground state, interfere. Two states related to two different ground states also lead to an oscillation, a *polarization interference*, which, however, vanishes in the presence of inhomogeneous broadening (Fig. 6.15). In the case of a quantum beat, the quantum mechanical (electronic) wavefunctions interfere, leading to an oscillation of the emitted light field, while for a polarization interference one merely has interference of the light fields. The two distinct scenarios can be separated experimentally by detailed analysis of time resolved four-wave-mixing data [118].

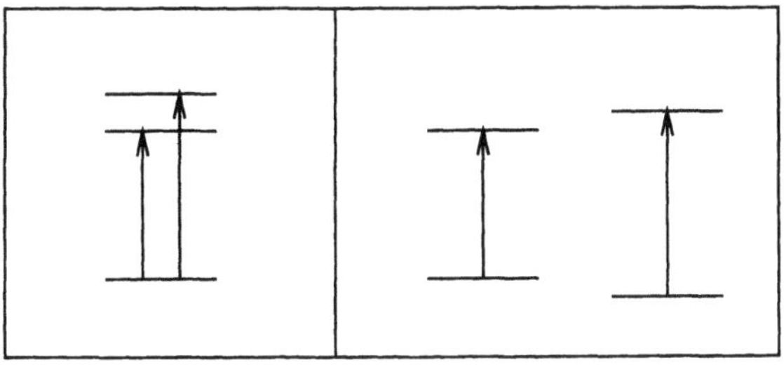

Fig. 6.15. Schematic illustration of the difference between a quantum beat (*left*) and a polarization interference (*right*)

Example (Optical Stark Effect). Experiments on the optical Stark effect usually employ pulsed excitation in order to achieve pump field strengths sufficient to lead to observable shifts of the resonances (see Fig. 6.16). Hence, a quantitative theory has to account for the time dependence of the field amplitudes.

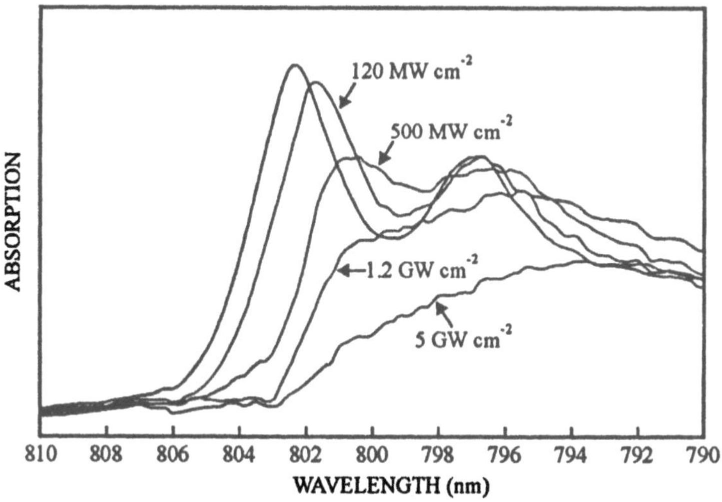

Fig. 6.16. Experimental results: optical Stark effect, as observed in a pump–probe experiment using optical pulses 400 fs in duration. The pump intensity is the parameter. After [113]

The finite spectral width of the pump (see Figs. 6.17 and 6.18) modifies the results discussed within the local-field model under steady-state conditions in that real excitation of quasiparticle states also comes into play. In the inset of Fig. 6.17, this shows up as a contribution to the carrier density which persists after the pump pulse has gone. Consequently, some bleaching of the exciton remains for a positive time delay.

For longer pulses and/or larger detunings, this persistent contribution loses relative weight with respect to the contribution adiabatically following the pump pulse [119]. In addition, one has not only four-wave-mixing contri-

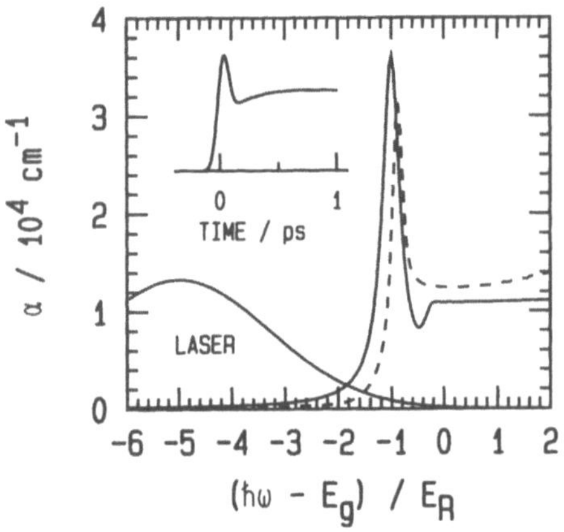

Fig. 6.17. Optical Stark effect. The 107 fs Gaussian-shaped optical pump pulse with Rabi energy $\hbar\Omega_R = 0.5\,E_R$ is tuned below the band edge as indicated. The probe pulse is weak. The *solid line* corresponds to the linear absorption spectrum, the *dashed line* to the absorption spectrum under excitation. The *inset* shows the carrier density versus time. The parameters are $\hbar\gamma_1 = 0$, $\hbar\gamma_2 = 0.2E_R$, zero time delay, and bulk GaAs parameters

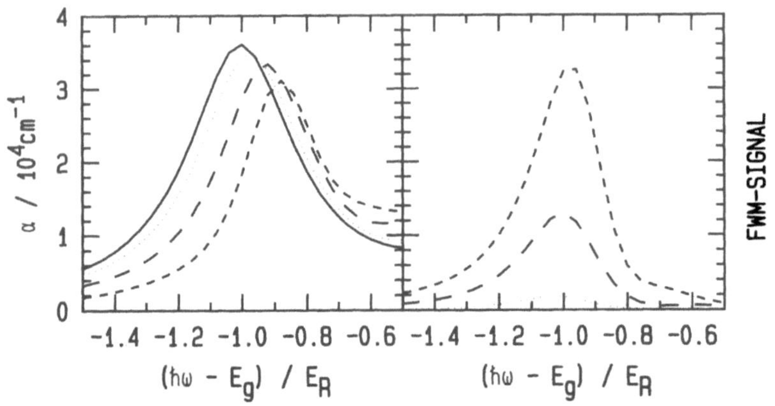

Fig. 6.18. As Fig. 6.17, but for the following pump Rabi energies (in units of E_R): 0.5 (*dashed*), 0.35 (*long-dashed*), 0.2 (*dotted*) and 0 (*solid*). The corresponding four-wave-mixing spectra are shown on the *right*

butions in the direction $2\boldsymbol{q}_2 - \boldsymbol{q}_1$, shown in Fig. 6.18, but also in the direction of the probe (see the discussion of pump–probe experiments in Sect. 4.3.3), which significantly modify the actual spectra. In particular, these contributions show up as oscillations in the spectra when the probe preceedes the pump. For a thorough discussion of this aspect, see [120, 97] and references therein. Furthermore, substantial modifications also arise from the off-resonantly excited continuum of states. In particular, oscillator strength is transferred from the 1s state to the continuum, as can be seen from Fig. 6.17. Finally, the light-induced states become spectrally broadened, corresponding to the finite pulse duration. Owing to the admixture of light-induced states in the exciton, the latter is not only shifted but also broadened, an effect which becomes dominant at large excitation densities (see Figs. 6.16 and 6.18).

There are only a few applications for which a description at the Hartree–Fock level yields nearly quantitative results. The optical Stark effect is one of these examples. Owing to the off-resonant excitation, in the ideal case, no real occupation of electron–hole pairs is present and, correspondingly, scattering processes are absent. Thus the dominating processes are phase space filling, the static exchange interaction,[7] and the dynamics of the excitation process. The semiconductor Bloch equations include this basic physics. In contrast to the *optical* Bloch equations discussed in Chap. 4, the various optical transitions are now no longer uncoupled, but interact via renormalized transition energies and via the renormalized Rabi frequency, which gives rise to the formation of excitons. On the level of nonlinear optics, this results, in the simplest case, in a local field, due to the Hartree–Fock exchange interaction among excitons. This interaction is similar to the molecular-field approximation for ferromagnets, where the origin of the local field (which also appears to be classical at first) is the quantum mechanical exchange interaction among spins.

From a pragmatical point of view, there is an intermediate time regime, in which the semiconductor Bloch equations have proven to be extremely successful in describing a large variety of nonlinear optical experiments close to the band edge. On a timescale shorter than the loss of coherence and longer than the duration of the relevant scattering processes, phenomenological constants can yield an appropriate approximation for phase and energy relaxation. Nevertheless, this is the Achilles heel of the semiconductor Bloch equations, which are, unable to describe the physics of relaxation, scattering, screening, or four-particle correlations, such as biexcitons. Before we turn to these aspects in Chaps. 9 and 12, we shall retain the Hartree–Fock approximation and consider the effect of static external fields.

---

[7] In certain semiconductor structures and materials higher-order correlation effects, i.e. the biexcitons already mentioned become important too, if the excitation takes place in the vicinity of these resonances (compare Chap. 9).

6.4 Problems

6.1. Derive Schrödinger equation for the longitudinal exciton. Show that the shift with respect to the transverse exciton is equal to the longitudinal–transverse splitting of the polariton dispersion (6.91). Hint: Make use of the results in Sect. 5.2.

6.2. Consider the reflection of light from a semiconductor surface at normal incidence. The photon energy is in the vicinity of the 1s exciton. Show that the intensity reflection coefficent $R(\omega)$ is given by

$$R(\omega) = \left| \frac{n(\omega) - 1}{n(\omega) + 1} \right|^2 .$$

An example of a reflection coefficient calculated using (6.90) is depicted in Fig. 6.19. Give an intuitive explanation of why the reflection coefficient is equal to one in the energy interval $[\hbar\Omega_1, \hbar(\Omega_1 + \Delta_{\mathrm{LT}})]$.

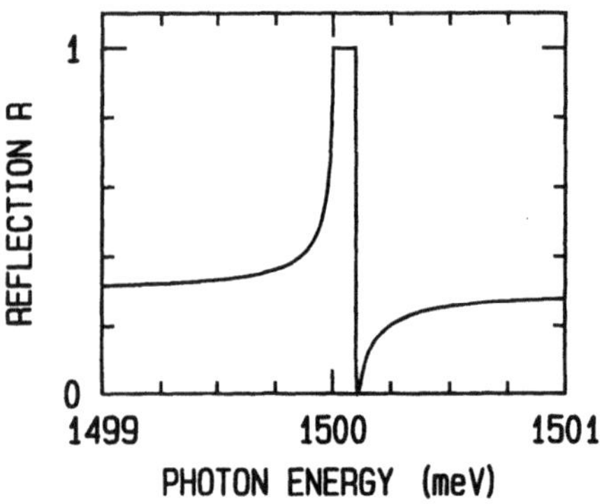

Fig. 6.19. Reflection coefficient in the vicinity of the 1s exciton resonance. Parameters as Fig. 6.4

6.3. The spectral shift linear in the exciton density due to the static exchange interaction (sometimes referred to as the static exciton self-energy) is obtained from (6.106) as

$$\Delta = 2e^2 \sum_{\boldsymbol{k},\boldsymbol{k}'} v(\boldsymbol{k} - \boldsymbol{k}') \left(|\phi_1(\boldsymbol{k})|^2 |\phi_1(\boldsymbol{k}')|^2 - |\phi_1(\boldsymbol{k})|^2 \phi_1^*(\boldsymbol{k}) \phi_1(\boldsymbol{k}') \right) n_{\mathrm{ex}} .$$

Show that for 1s exciton functions

$$\phi_1(\boldsymbol{k}) = \frac{8\sqrt{\pi}}{(1 + k^2)^2}$$

the shift obtained is $(13\pi/3)n_{\mathrm{ex}}$.

6.4. Derive an expression for the electron–hole interaction in a single quantum well analogous to (6.85) for the case of finite barrier height.

6.5. How can one measure γ_1 and γ_2 with the three-beam four-wave-mixing geometry depicted in Fig. 6.20? Compute the third-order response of the exciton in the direction $\boldsymbol{q}_3 + \boldsymbol{q}_2 - \boldsymbol{q}_1$ within the local-field model for δ pulses. What is different with respect to the optical Bloch equations?

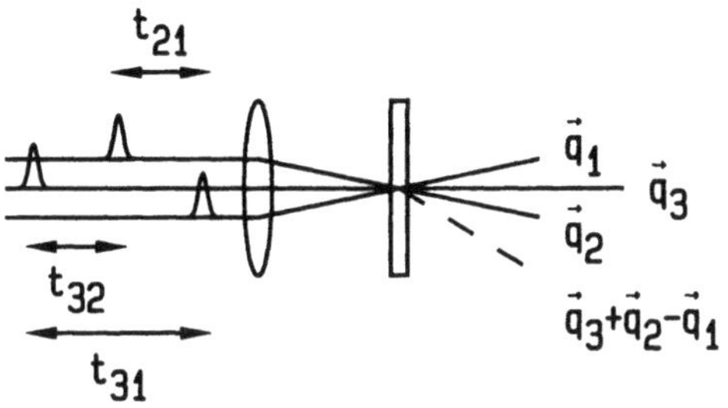

Fig. 6.20. Schematic of a three-beam four-wave-mixing experiment

6.6. Repeat Problem 4.6 for the local-field model.

6.7. What would be the response time constant of a semiconductor photodetector (see Sect. 4.4) operating close to the exciton resonance? Discuss the behavior qualitatively first and then analyze the problem using the local-field model.

7. Influence of Static Magnetic Fields

In semiconductor physics, a wealth of new phenomena are obtained from the application of static external fields. Whereas the most striking phenomena arise in transport, the optical properties are also radically modified by static fields. As in the case of optical fields, the properties of a system subject to a static magnetic field are determined by the interaction of the particles with this field. This interaction gives rise to the formation of new eigenstates at the one- or two-particle level, which in turn results in a direct response to the magnetic field as well as in a modified response to other external perturbations, such as optical fields. We shall not discuss the first aspect, which concerns the magnetic properties of the system, but instead shall focus on the second aspect in this chapter. Classically, charged particles orbit around the magnetic-field axis, in the case of free electrons with the *cyclotron frequency* $\omega_c^0 = eB/m$.[1] The corresponding confinement of the motion in quantum mechanics leads asymptotically (for large fields) to one-dimensional behavior in bulk semiconductors and is zero-dimensional for quantum wells in a perpendicular field. Both cases share many features with one- and zero-dimensional material systems, which can be realized in quantum wires and quantum dots, respectively. The analogy is indeed very close as long as we consider properties for which the center-of-mass motion of electron–hole pairs is negligible, an assumption usually valid for optical transitions.

The transition from two-dimensional to zero-dimensional behavior is illustrated in Fig. 7.1, which shows the dependence of the linear absorption of a GaAs/GaAlAs multiple-quantum-well sample with 8.5 nm well width on the magnetic field [122]. (Even more highly structured data can be found, for example, in [121].) At low fields only two discrete resonances, due to the heavy- and light-hole excitons, are observed , and at higher energies the spectrum is continuous. When the magnetic field is increased, this region of continuum absorption evolves into a series of additional discrete resonances. As will be shown below, the most prominent deviations from an equidistant spacing corresponding to the cyclotron frequency are due to the Coulomb interaction. The existence of well-separated discrete transition energies, which also occur in a similar form in magnetically confined three-dimensional systems, has an

[1] Note that the cyclotron frequency ω_c^0 is defined with respect to the free electron mass.

important implication. If we apply radiation in the microwave or far infrared range rather than optical fields, the frequency of the electromagnetic field can become resonant with the spacing of the energy levels and can thus induce transitions between them. The absorption of the microwave field gives valuable information about transition frequencies and, in particular, it allows us to measure effective masses in semiconductors with high accuracy. The theoretical description of the material system necessary in order to extract correct information from the experimental data is the same as in the case of optical experiments. In the following section, we focus our interest first on the one-particle properties, i.e. the eigenstates of Bloch electrons in a magnetic field, which will be used later to expand the semiconductor Bloch equations.

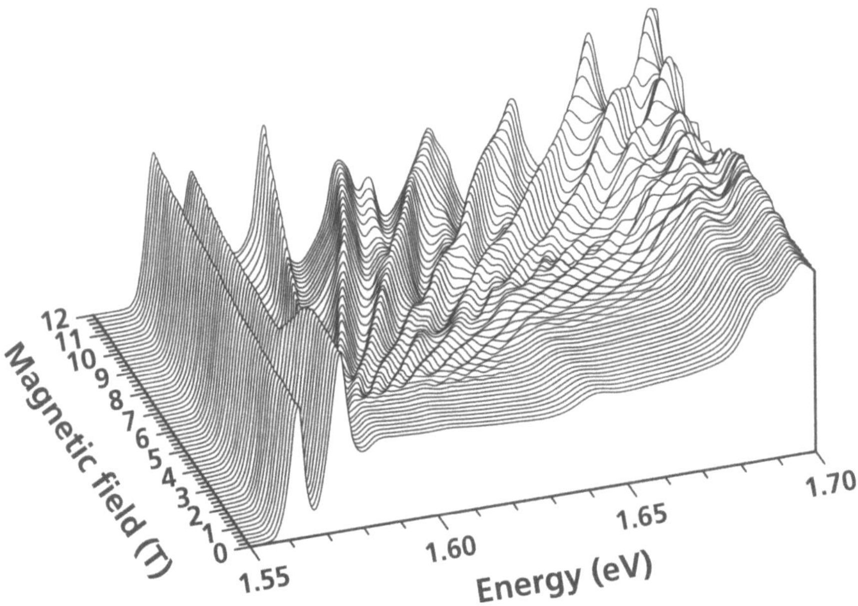

Fig. 7.1. Measured absorption spectra of an 8.5 nm wide GaAs quantum well sample subject to a perpendicular magnetic field. Taken from [122]

7.1 One-Particle Properties

Let us start with the one-particle Hamiltonian describing Bloch electrons under the action of a magnetic field,

$$\frac{\hbar^2}{2m}(\boldsymbol{p}+e\boldsymbol{A})^2+V(\boldsymbol{r})=h_0+h_{\text{int}}\ , \tag{7.1}$$

where h_0 is the unperturbed one-particle Hamiltonian and the perturbation h_{int} contains the contribution due to the magnetic field. The corresponding Schrödinger equation may serve as a general starting point for studying the eigenstates of electrons in a lattice-periodic potential and a magnetic field. However, for general energy bands, the solution of this problem may be hopelessly difficult. If we anticipate that we shall be mainly interested in optical transitions in the vicinity of the absorption edge, the problem can be considerably simplified by means of the effective mass-approximation. For states high up in a band, we need to take the effect of lattice periodicity, which produces a broadening of magnetic levels, more seriously. In order to derive the effective-mass equation, we make use of the expansion

$$\psi_m(\boldsymbol{r}) = \frac{1}{\sqrt{V}} \sum_{n\boldsymbol{k}} c_{n,m}(\boldsymbol{k}) \mathrm{e}^{i\boldsymbol{k}\boldsymbol{r}} u_n(0,\boldsymbol{r}) \tag{7.2}$$

which differs from (3.82) by the additional sum over $\boldsymbol{k}$ because of the broken lattice symmetry. By using the Schrödinger equation and the property that Bloch functions in effective-mass approximation fulfill the usual orthogonality relation, we find (similarly to Sect. 3.4) the eigenvalue equation

$$\begin{aligned}\left(\frac{\hbar^2\boldsymbol{k}^2}{2m} + E_{n,m}\right) c_{n,m}(\boldsymbol{k}) + \sum_{n'\neq n,} \frac{\hbar}{m}\boldsymbol{k}\,\boldsymbol{p}_{nn'} c_{n',m}(\boldsymbol{k})& \\ + \sum_{n'\neq n,\boldsymbol{k}'} \langle n\boldsymbol{k}|h_{int}|n'\boldsymbol{k}'\rangle c_{n',m}(\boldsymbol{k}') = 0\,,&\end{aligned} \tag{7.3}$$

where the interaction matrix element is defined using the wavefunctions (7.2) and we have set $E_{n,m} = E_n(0) - E_m$. Choosing the magnetic field to be along the z direction, such that the vector potential in the Landau gauge is given by $\boldsymbol{A} = (0, Bx, 0)$, we obtain for the interaction Hamiltonian

$$h_{\text{int}}(\boldsymbol{r}) = \omega_{\text{c}}^0 x p_y + \frac{m(\omega_{\text{c}}^0)^2}{2} x^2 = \frac{\hbar}{ml^2} x p_y + \frac{\hbar^2}{2m} \frac{x^2}{l^4}\,, \tag{7.4}$$

where we have introduced the *magnetic length* $l = \sqrt{\hbar/(m\omega_{\text{c}}^0)} = \sqrt{\hbar/(eB)}$, which is nothing but the radius of the classical cyclotron orbit. Even for high magnetic fields, this quantity is of the order of 10 to 100 nm and thus much larger than the lattice constant. This is the reason why the effective-mass approximation works and lattice periodicity plays a role for high-lying states only. With (7.4), the interaction matrix element takes the form

$$\begin{aligned}&\langle n\boldsymbol{k}|h_{\text{int}}|n'k'\rangle \\ &= \int \mathrm{d}^3\boldsymbol{r}\, \mathrm{e}^{\mathrm{i}(\boldsymbol{k}'-\boldsymbol{k})\boldsymbol{r}} u_n^*(0,\boldsymbol{r}) \left(\frac{\hbar}{m\,l^2}(p_y + \hbar k_y) + \frac{\hbar^2}{2m\,l^4}x^2\right) u_n(0,\boldsymbol{r})\,,\end{aligned} \tag{7.5}$$

where we have commuted the operator p_y with the plane wave. The dependence on x can be eliminated by using the replacement $x \to i\partial/\partial k_x$, and the integration yields

$$\langle n\boldsymbol{k}|h_{\rm int}|n'\boldsymbol{k}'\rangle = \delta(\boldsymbol{k}-\boldsymbol{k}')\left(\mathrm{i}\frac{\hbar}{ml^2}\left(p^y_{nn'}+\hbar k_y\delta_{nn'}\right)\frac{\partial}{\partial k_x}-\frac{\hbar^2}{2ml^4}\frac{\partial^2}{\partial k_x^2}\right). \tag{7.6}$$

Inserting this interaction into (7.3), we obtain

$$\begin{aligned}&\left(\frac{\hbar^2}{2m}(k_x^2+\mathcal{K}_y^2+k_z^2)+E_n(0)-E_m\right)c_{n,m}(\boldsymbol{k})\\&+\sum_{n'\neq n}\frac{\hbar}{m}(k_x\,p^x_{nn'}+\mathcal{K}_y\,p^y_{nn'}+k_z\,p^z_{nn'})c_{n',m}(\boldsymbol{k})=0\,,\end{aligned} \tag{7.7}$$

where

$$\mathcal{K}_y = k_y+\frac{\mathrm{i}}{l^2}\frac{\partial}{\partial k_x}\,. \tag{7.8}$$

Comparison of this result with (3.73) shows that the magnetic field enters the effective-mass equation only by means of the replacement (7.8). This is an important result [63, 64, 123], which allows one immediately to apply the results of the effective-mass theory worked out in Sects. 3.4 and 3.5 to the case of a magnetic field.

7.1.1 Effective Mass Theory for Isolated Bands

We start with isotropic, nondegenerate bands and ignore band-mixing effects. The ***kp*** term is treated perturbationally and leads, as in Sect. 3.4 to an effective mass $m_{\rm e}$. The resulting form of (7.7) suggests that we should transform it back to real space according to

$$c(\boldsymbol{r}) = \frac{1}{\sqrt{V}}\sum_{\boldsymbol{k}}\mathrm{e}^{\mathrm{i}\boldsymbol{k}\boldsymbol{r}}c(\boldsymbol{k})\,. \tag{7.9}$$

This transformation results in the equation

$$\left(-\frac{\hbar^2}{2m_{\rm e}}\Delta-\mathrm{i}\hbar\omega_{\rm c}x\frac{\partial}{\partial y}+\frac{m_{\rm e}\omega_{\rm c}^2}{2}x^2-E\right)c(\boldsymbol{r})=0\,, \tag{7.10}$$

where we have introduced the *cyclotron frequency* $\omega_{\rm c}=eB/m_{\rm e}$ with respect to the effective mass $m_{\rm e}$. This is the only reminder of the lattice-periodic potential and the solutions are the same as those for free electrons in a magnetic field, first derived by Landau [124]. Equation (7.10) can be easily reduced to the Schrödinger equation of a one-dimensional displaced harmonic oscillator. If we separate the dependence on the y and z coordinates by means of plane waves and drop band indices, we have

$$c(\boldsymbol{r}) = (L_yL_z)^{-1/2}\exp(\mathrm{i}k_yy+\mathrm{i}k_zz)\phi(x)\,, \tag{7.11}$$

which yields by means of (7.10) the following equation for $\phi(x)$:

$$\left[-\frac{\hbar^2}{2m_e}\frac{\partial^2}{\partial x^2}\frac{m_e\omega_c^2}{2}\left(x+\frac{\hbar}{m_e\omega_c}k_y\right)^2\right]\phi(x)=\left(E-\frac{\hbar^2}{2m_e}k_z^2\right)\phi(x)\,. \tag{7.12}$$

The solution of this equation can be immediately obtained from the solution of the undisplaced problem by the replacement $x \rightarrow x - l^2 k_y$. Thus the resulting *Landau levels* are given by

$$E_n(k_z)=\left(n+\frac{1}{2}\right)\hbar\omega_c+\frac{\hbar^2}{2m_e}k_z^2\,, \qquad n=0,1,2,\ldots \tag{7.13}$$

and the eigenfunctions can be written in terms of Hermite polynomials H_n as

$$\phi_{nk_y}(x)=\frac{1}{\sqrt{\sqrt{\pi}\,l\,2^n n!}}\exp\left[\frac{1}{2}\left(\frac{x}{l}+lk_y\right)^2\right]H_n\left(\frac{x}{l}+lk_y\right)\,. \tag{7.14}$$

A useful relation beween these eigenfunctions results from the recursion relations of the Hermite polynomials, which may be expressed by means of raising and lowering operators for the quantum number n:

$$\alpha\phi_n(x)=\frac{1}{\sqrt{2}}\left(l\frac{\partial}{\partial x}+\frac{x}{l}+lk_y\right)\phi_n(x)=\sqrt{n}\,\phi_{n-1}(x) \tag{7.15}$$

and

$$\alpha^\dagger\phi_n(x)=-\frac{1}{\sqrt{2}}\left(l\frac{\partial}{\partial x}-\frac{x}{l}-lk_y\right)\phi_n(x)=\sqrt{n+1}\,\phi_{n+1}(x)\,. \tag{7.16}$$

Without referring to the solutions of the harmonic oscillator, we could use the properties of these operators to represent the Hamiltonian in (7.12) and to derive eigenvalues and eigenfunctions. Since the energy does not depend on k_y, the eigenenergies are highly degenerate. The degeneracy can be obtained from the requirement that the displaced wavefunctions must have their centers within the crystal. Thus, approximately, the condition $-L_x/2 \le l^2 k_y \le L_x/2$ is met, which allows us to calculate the degeneracy of any energy level in (7.13) for given values of n and k_z according to

$$\sum_{k_y}\rightarrow L_y\int_{-L_x/(2\,l^2)}^{L_x/(2\,l^2)}\frac{\mathrm{d}k_y}{2\pi}=\frac{L_xL_y}{2\pi\,l^2}=L_xL_y\frac{e\,B}{h}\,. \tag{7.17}$$

Classically, this degeneracy corresponds to the different center-of-mass positions of the cyclotron orbits. The degeneracy per unit area and spin is given by the fraction eB/h. This suggests that we should introduce the dimensionless *filling factor* ν, corresponding to an electron sheet density n_e^{2d} according to

$$\nu=n_e^{2d}\frac{h}{e\,B}\,, \tag{7.18}$$

a result that will be important for the discussion of the quantum Hall effect (see Chap. 16). This degeneracy also occurs as a prefactor in the density of states, which, after the $k_z$integration has been performed, is obtained as

$$D_{\mathrm{e}}(E) = \frac{L_x L_y L_z}{4\pi^2 l^2} \sqrt{\frac{m_{\mathrm{e}}}{2\hbar}} \sum_n \theta(E - E_n)(E - E_n)^{-1/2} , \tag{7.19}$$

where

$$E_n = \hbar\omega_{\mathrm{c}} \left(n + \frac{1}{2} \right) . \tag{7.20}$$

The density of states of each Landau level exhibits the typical $1/\sqrt{E}$ singularity, characteristic of one-dimensional behavior. The structure of peaks in optical spectra and also many transport properties in the presence of a magnetic field can be traced back to this type of singularity in the density of states. In the limit of small magnetic field, the sum in (7.19) can be replaced by an integral and we recover the usual square-root dependence of the density of states of a parabolic band in three dimensions. In real systems the Landau levels are broadened. If we consider a Lorentzian broadening, we have to convolute with a Lorentzian of width γ. Instead of (7.19), we obtain

$$D_{\mathrm{e}}(E) = \frac{L_x L_y L_z}{8\pi^2 l^2} \sqrt{\frac{m_{\mathrm{e}}}{\hbar}} \sum_n \frac{\left(\sqrt{(E - E_n)^2 + \hbar\gamma^2} + E - E_n \right)^{1/2}}{\sqrt{(E - E_n)^2 + \hbar\gamma^2}} . \tag{7.21}$$

Figure. 7.2 illustrates the evolution of the density of states with increasing magnetic field. At sufficiently low fields the separation of the Landau levels is of the order of the broadening. Thus we see a typical square-root dependence,

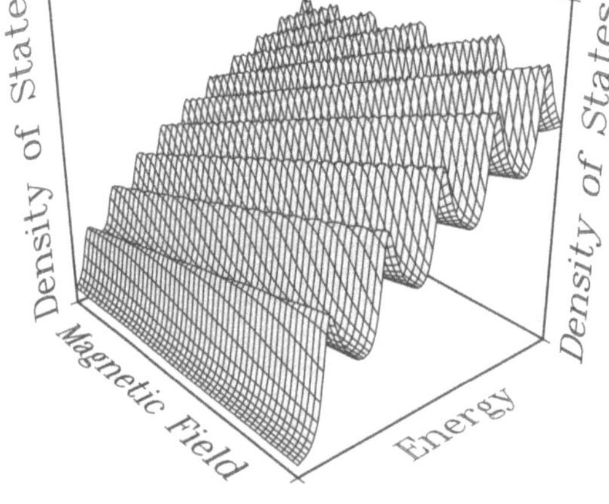

Fig. 7.2. Density of states of a three-dimensional semiconductor subject to a perpendicular magnetic field

which, however, is modulated by resonance structures. The latter vanish if the broadening becomes larger than the level spacing. With increasing field, the resonances evolve into a fan of distinct ridges. In cyclotron resonance experiments, one usually works at fixed energy and varies the strength of the magnetic field. This situation is considered in Fig. 7.3, where we show the conduction band density of states at several fixed energies as a function of the cyclotron frequency. The spacing of the resonances increases with increasing magnetic field but decreases with increasing energy, as expected. Thus far, our results are valid for states in the vicinity of the bottom of an isotropic, nondegenerate band, which may be, for example, the lowest conduction band of a typical semiconductor. If we consider valence bands, their usual degeneracy makes it necessary to treat (7.7) more carefully.

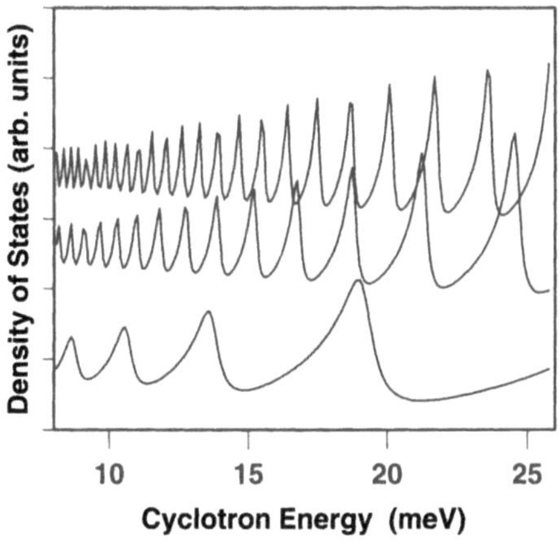

Fig. 7.3. As Fig. 7.2, but plotted as a function of magnetic field for the following energies: 270 meV, 160 meV, 50 meV (*top* to *bottom*)

7.1.2 Degenerate Bloch Electrons in a Magnetic Field

In Sect. 3.5, we have introduced band-mixing effects resulting from the $\boldsymbol{kp}$ contribution. Band mixing resulting from the magnetic field leads to effects of the same order of magnitude and cannot be ignored. We consider again the case of zinc blende semiconductors and restrict ourselves to the highest four valence bands, i.e. we neglect the coupling to the $J = 1/2$ bands, which are split off by spin–orbit interaction. For completeness, we also account for the spin Zeeman contribution h_s, which describes the coupling of the total angular-momentum operator to the external field. The expectation value of this operator for eigenstates of J_z is given by

$$\langle h_\mathrm{s} \rangle = g\mu_\mathrm{B}\, \kappa B \frac{\langle J_z \rangle}{\hbar} = \hbar\omega_\mathrm{c}^0 \kappa\, m_J \,, \tag{7.22}$$

where $\mu_\mathrm{B} = e\hbar/2m$ is the Bohr magneton and $g = 2$ is the g-factor of free electrons. The additional dimensionless material parameter κ takes into account the fact that, owing to spin–orbit interaction, the effective g-factor of band electrons deviates from that of free electrons. This is the only point at which spin–orbit interaction enters into our analysis. All other results that we need can immediately be taken over from (3.116), by means of the replacement (7.8). Performing the Fourier transformation with respect to k_x, we obtain the kinetic energy of the valence bands again in the form

$$\underline{h} = E_v(0)\,\underline{1} + \begin{pmatrix} a_{-3/2} & b^\dagger & c^\dagger & 0 \\ b & a_{-1/2} & 0 & c^\dagger \\ c & 0 & a_{1/2} & -b^\dagger \\ 0 & c & -b & a_{3/2} \end{pmatrix} . \tag{7.23}$$

The operators a_{m_J} are defined as

$$a_{m_J} = -\hbar\omega_c^0\left[\frac{1}{2}(\gamma_1 \pm \gamma_2)\left(-l^2\frac{\partial^2}{\partial x^2} + \frac{1}{l^2}(x + l^2k_y)^2\right) + \kappa m_J\right] + \frac{\hbar^2 k_z}{2m}(\gamma_1 \mp 2\gamma_2)k_z^2 . \tag{7.24}$$

The upper sign is valid for states with $m_J = \pm 3/2$ and the lower one for $m_J = \pm 1/2$. The operators b and c are obtained as follows:

$$b = \mathrm{i}\hbar\omega_c^0\sqrt{3}\gamma_3\left(x + l^2k_y - l^2\frac{\partial}{\partial x}\right)k_z \tag{7.25}$$

and

$$c = \hbar\omega_c^0\frac{\sqrt{3}}{2}\left[\gamma_2\left(l^2\frac{\partial^2}{\partial x^2} + \frac{1}{l^2}(x + l^2k_y)^2\right) - \gamma_3\left(\frac{\partial}{\partial x}(x + l^2k_y) + (x + l^2k_y)\frac{\partial}{\partial x}\right)\right] . \tag{7.26}$$

In the last line, according to a general rule of quantum mechanics, we have introduced the symmetrized product rather than the product of noncommuting factors. It is convenient to write these results in terms of the lowering and raising operators defined by (7.15) and (7.16), which fulfill the relations

$$x + l^2k_y = \frac{l}{\sqrt{2}}(\alpha + \alpha^\dagger) , \tag{7.27}$$

$$\frac{\partial}{\partial x} = \frac{1}{l\sqrt{2}}(\alpha - \alpha^\dagger) , \tag{7.28}$$

and allow us the replacements

$$a_{m_J} = -\hbar\omega_c^0\left[(\gamma_1 \pm \gamma_2)\left(\alpha^\dagger\alpha + \frac{1}{2}\right) + \kappa m_J\right] - \frac{\hbar^2k_z^2}{2m}(\gamma_1 \mp 2\gamma_2) , \tag{7.29}$$

$$b = \mathrm{i}\hbar\omega_c^0\sqrt{6}\gamma_3\, l\, k_z\alpha^\dagger , \tag{7.30}$$

$$c = \hbar\omega_c^0 \frac{\sqrt{3}}{2}[\gamma_2(\alpha\alpha + \alpha^\dagger\alpha^\dagger) - \gamma_3(\alpha\alpha - \alpha^\dagger\alpha^\dagger)] \,. \tag{7.31}$$

For the further analysis, we neglect warping, i.e. we set $\gamma_3 = \gamma_2$ and expand the effective-mass equation by means of the Landau functions (7.14). For the diagonal matrix elements in (7.23), this leads to

$$a_{m_J}^{nn'} = -\delta_{nn'}\left\{\hbar\omega_c^0\left[(\gamma_1 \pm \gamma_2)\left(n + \frac{1}{2}\right) + \kappa m_J\right] + \frac{\hbar^2 k_z^2}{2m}(\gamma_1 \mp 2\gamma_2)\right\} , \tag{7.32}$$

where the upper sign refers to $m_j = \pm 3/2$ states and the lower one to $m_j = \pm 1/2$ states. The off-diagonal elements are obtained as

$$b_{nn'} = \mathrm{i}\hbar\omega_c^0\sqrt{6}\gamma_2\, l\, k_z\sqrt{n}\; \delta_{n',n-1} \,, \tag{7.33}$$

$$b_{nn'}^\dagger = -\mathrm{i}\hbar\omega_c^0\sqrt{6}\gamma_2\, l\, k_z\sqrt{n+1}\; \delta_{n',n+1} \,, \tag{7.34}$$

$$c_{nn'} = \hbar\omega_c^0\sqrt{3}\gamma_2\sqrt{n(n-1)}\; \delta_{n',n-2} \,, \tag{7.35}$$

$$c_{nn'}^\dagger = \hbar\omega_c^0\sqrt{3}\gamma_2\sqrt{(n+1)(n+2)}\; \delta_{n',n+2} \,. \tag{7.36}$$

It is an instructive exercise to consider the bottom of the Landau bands only, i.e. $k_z = 0$. In this case only Landau levels with indices n and $n-2$ are coupled, Hence the valence band matrix splits into 2×2 submatrices of two different types, which are obtained as

$$\underline{h}_1 = E_\mathrm{v}(0)\,\underline{1} + \begin{pmatrix} -\hbar\omega_c^0\left[(\gamma_1+\gamma_2)\left(n-\frac{3}{2}\right)+\frac{3}{2}\kappa\right] & \hbar\omega_c^0\sqrt{3}\gamma_2\sqrt{n(n-1)} \\ \hbar\omega_c^0\sqrt{3}\gamma_2\sqrt{n(n-1)} & -\hbar\omega_c^0\left[(\gamma_1-\gamma_2)\left(n+\frac{1}{2}\right)-\frac{1}{2}\kappa\right] \end{pmatrix} \tag{7.37}$$

and

$$\underline{h}_2 = E_\mathrm{v}(0)\,\underline{1} + \begin{pmatrix} -\hbar\omega_c^0\left[(\gamma_1-\gamma_2)\left(n-\frac{3}{2}\right)+\frac{1}{2}\kappa\right] & \hbar\omega_c^0\sqrt{3}\gamma_2\sqrt{n(n-1)} \\ \hbar\omega_c^0\sqrt{3}\gamma_2\sqrt{n(n-1)} & -\hbar\omega_c^0\left[(\gamma_1+\gamma_2\left(n+\frac{1}{2}\right)-\frac{3}{2}\kappa\right] \end{pmatrix} . \tag{7.38}$$

This form is valid for $n \geq 2$; for $n = 0, 1$ we obtain the diagonal solutions

$$h_1 = E_\mathrm{v}(0) - \hbar\omega_c^0\left[(\gamma_1+\gamma_2)\left(n+\frac{1}{2}\right)-\frac{3}{2}\kappa\right] \tag{7.39}$$

and

$$h_2 = E_v(0) - \hbar\omega_c^0\left[(\gamma_1 - \gamma_2)\left(n + \frac{1}{2}\right) - \frac{1}{2}\kappa\right]. \tag{7.40}$$

The eigenvalues of (7.37) and (7.38) are easily found as

$$\begin{aligned} E_1^{\pm}(n) = E_v(0) - \hbar\omega_c^0\Bigg(\gamma_1\left(n - \frac{1}{2}\right) \pm \left(\gamma_2 - \frac{\kappa}{2}\right) \\ + \left\{\left[\gamma_1 - \kappa \mp \gamma_2\left(n - \frac{1}{2}\right)\right]^2 + 3\gamma_2^2 n(n-1)\right\}^{1/2}\Bigg), \end{aligned} \tag{7.41}$$

This result is valid for $n \geq 0$ and includes for $n = 0, 1$ the solutions (7.39) and (7.40) The other two solutions $E_2^{\pm}(n)$, valid for $n \geq 2$, are obtained by changing the sign of the square root. As an intrinsic effect of the magnetic field, the Kramers degeneracy arising from time-reversal symmetry is obviously lifted, and we obtain four separate eigenstates. To illustrate the properties of this level scheme, we consider in Fig. 7.4 the spectral function corresponding to (7.41) for the example of GaAs. Levels near the bottom of the Landau ladders have a rather irregular spacing. On the other hand, for high quantum numbers a regime with groups of level pairs appears. In this limit the levels reduce to equally spaced but split Landau levels centered around the energies $E_1 \simeq E_v(0) - \hbar\omega_c^0(\gamma_1 + 2\gamma_2)n$ and $E_2 \simeq E_v(0) - \hbar\omega_c^0(\gamma_1 - 2\gamma_2)n$. This allows

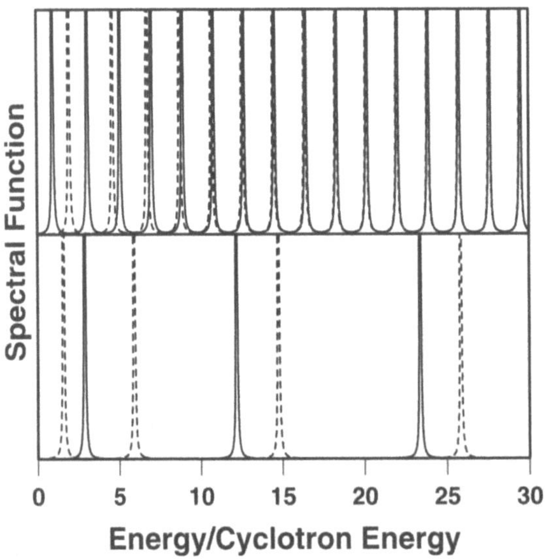

Fig. 7.4. Spectral function of valence band states at $k_z = 0$ versus energy in units of $\hbar\omega_c^0$. The upper series corresponds to the heavy-hole Landau levels, and the lower series to the light-hole Landau levels. Owing to the action of the magnetic field the spin degeneracy is lifted (*full* and *dashed* lines). Parameters correspond to GaAs

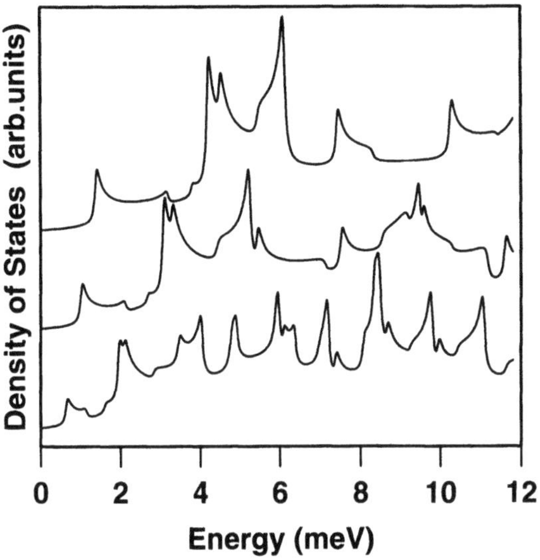

Fig. 7.5. Valence band density of states at magnetic fields of 10.5 T, 7.7 T, and 4.9 T (*top* to *bottom*).Parameters correspond to GaAs, where a field of ≈7 T corresponds to a free cyclotron energy of $\hbar\omega_c^0 = 1\,\mathrm{meV}$

us to identify $E_1^{\pm}$ and $E_2^{\pm}$ with the heavy- and light-hole Landau ladders, respectively.

The finite wave vector k_z has a profound influence on the density of states as it leads (compare Fig. 7.5) to shifts as well as to broadening of the levels and cannot be neglected. There are several extensions of the approximation scheme described thus far. Whereas anisotropic effects, i.e. $\gamma_2 \neq \gamma_3$, can be treated along the same lines as above but with slightly different definitions of parameters [64, 125], we have to enlarge the subspace of states if there is a strong admixture of the split-off band. In narrow-gap semiconductors there is even a strong interaction between the conduction and valence bands, which introduces a high degree of nonparabolicity. All these effects can be treated straightforwardly on the same footing as discussed above. Explicit solutions, however, can only be obtained numerically.

An elegant possible way to obtain decoupled valence bands again is the application of uniaxial stress, which enables one to obtain either more precise or, in some cases, even new information. Owing to symmetry breaking, a splitting of heavy- and light-hole states exists even for vanishing magnetic field. For a finite field, one obtains well separated groups of Landau ladders. This method has been widely used in practice to determine Luttinger parameters from cyclotron resonance data [126, 129].

In order to minimize broadening effects due to scattering by phonons and impurities, cyclotron resonance experiments have to be performed at low temperatures and on very pure samples. Under these conditions, the numbers of electrons and holes available in thermal equilibrium to participate in transitions within the Landau ladders of electrons or holes can be extremely small. Thus carriers are often created by other means, such as photoexcitation. As the excited carriers relax partly radiatively, one can detect the radiation emitted at the cyclotron frequency. This technique has also been applied to study band structure effects, as well as energy relaxation and the different scattering mechanisms involved [130]. The interpretation of the transition energies corresponding to transitions between Landau states within the same band in terms of one-particle energies is rather intuitive, but there is a pitfall. For both thermally and photoexcited carriers, the transitions take place between excited states of a many-particle system, which are strongly influenced by Coulomb effects. If we consider sufficiently high lying states the Coulomb interaction yields a shift and a broadening of one-particle resonances and may indeed be negligible in the analysis of this type of cyclotron resonance experiment. For states near the bandedge, however, the Coulomb interaction plays a dominant role and should not be neglected. The variety of experimental conditions, as well as the variety of different models and approximation schemes which have been used to analyze experimental data, led to a tremendous scatter in band structure parameters in the early work of this kind. However, careful theoretical analysis of cyclotron resonance data, and also of optical spectra [131, 132], has resulted in a well-established and consistent picture of the band structure of many semiconductors within the effective-mass approximation.

First-principle band structure calculations and the resulting information about the Bloch functions and energies allow one to perform the general expansion (7.1) up to convergence and thus to examine whether the restriction to a low-dimensional subspace of eigenstates can be justified or not. Such investigations have been performed for GaAs [134]. For states near the zone center, it is found that an eight-band model, consisting of the fourfold degenerate six uppermost valence bands and the two lowest conduction bands, reproduces the results of a pseudopotential calculation with good accuracy.

7.1.3 One-Particle States in Quantum Wells

The description of bulk one-particle states in a magnetic field which we have developed so far can easily be extended to the case of quantum wells. If we neglect band-bending effects in the interface region between the barrier and the well, the quantum well can be characterized by an additional confinement potential which enters into the effective-mass equation (7.10). In the simplest case, the magnetic field is perpendicular to the layers. In this case we have to replace the plane wave in the z direction in (7.11) by the well eigenfunctions introduced in Sect. 6.4. Thus instead of the eigenenergies (7.13), we obtain

$$E_{ni} = \left(n + \frac{1}{2}\right)\hbar\omega_c + E_i \,, \tag{7.42}$$

where E_i denotes the energy of the ith subband in the quantum well. We shall not go through the tedious derivation again, as – roughly speaking – all results remain valid if we replace the wave vector k_z by the quantum number labeling the eigenstates in the well. Thus the continuum of states due to the continous wave vector vanishes in favor of a set of of discrete levels. The complete lack of any continuum of states is particular to zero-dimensional systems. For each subband we have a set of Landau levels. For wide quantum wells, for which the subband splitting is still rather weak, this can lead to an involved energy-level structure. Even if the subband splitting is large compared with the separation of the Landau levels, the Landau fan, including band-mixing effects [133, 135, 136], is still very complex. As an example, Fig. 7.6 shows the lowest heavy- and light-hole Landau fans calculated for a 7.8 nm GaAs/AlGaAs quantum well sample [135].

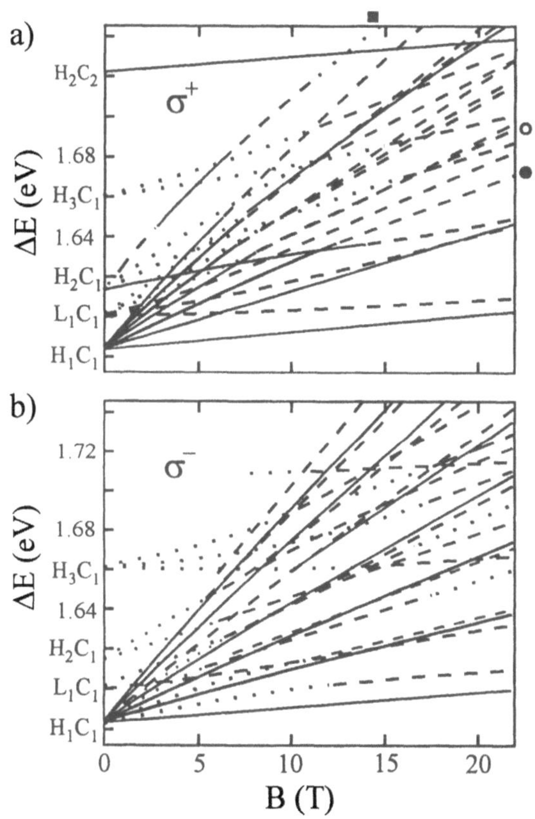

Fig. 7.6a,b. Calculated transition energies of a GaAs/AlGaAs quantum well sample for the two circular polarizations. The oscillator strength is coded by the line type and increases from *dotted* to *dashed* to *solid*. Taken from [135]

7.2 Optical Properties of Magneto-Excitons

The properties of excitons under the action of a magnetic field have been investigated with a number of different theoretical methods, which range from various perturbational approaches through elaborate variational techniques [131, 137, 138, 139] in bulk materials, to systematic expansion schemes which have been applied in confined systems [121, 136]. If we anticipate that we shall also want to investigate nonlinear optical properties, a systematic expansion is mandatory. We shall use the one-particle states introduced in the last section as the most natural basis set to expand the semiconductor Bloch equations. In principle we should start from the real-space representation of the density matrix equations (5.7) within the Hartree–Fock approximation, add the contributions due to the magnetic field to the one-particle Hamiltonian and use the wavefunctions defined in (7.2) as the basis set. This procedure can be shortened considerably by noting that the general representation of the density matrix equations (5.8) is also valid in the case with a magnetic field if we apply the following notation. We replace a single quantum number, e.g. j, by the set of quantum numbers $\underline{j} = \{\nu_j, j, k_{yj}, k_{zj}\}$, which labels the band index, the Landau quantum number, and the y and z components of the wave vector. The eigenenergies $E_{\underline{j}}$ for conduction band states on the left-hand side of (7.12) are given by the sum of the Landau levels (7.13) and the spin Zeeman contribution. Concerning valence band states, we also allow for off-diagonal elements according to our treatment of degenerate bands in the last section (compare (7.32)–(7.36)).

In the next step, we consider the optical transition matrix elements, which are given by

$$\begin{aligned} \boldsymbol{r}_{\underline{j}\underline{l}} &= \int \mathrm{d}^3\boldsymbol{r}\, u^*_{\nu_j}(0,\boldsymbol{r}) c^*_{j,k_{yj},k_{zj}}(\boldsymbol{r}) \boldsymbol{r} u_{\nu_l}(0,\boldsymbol{r}) c_{l,k_{yl},k_{zl}}(\boldsymbol{r}) \\ &= \sum_{\boldsymbol{L}} \int_{\Omega} \mathrm{d}^3\boldsymbol{r} u^*_{\nu_j}(0,\boldsymbol{r}) c^*_{j,k_{yj},k_{zj}}(\boldsymbol{r}+\boldsymbol{L})\,(\boldsymbol{r}+\boldsymbol{L})\, u_{\nu_l}(0,\boldsymbol{r}) c_{l,k_{yl},k_{zl}}(\boldsymbol{r}+\boldsymbol{L})\,, \end{aligned} \tag{7.43}$$

where $\boldsymbol{r}$ is restricted to the unit cell Ω. Now we make use of the fact that on the scale of one lattice constant, Landau states are slowly varying functions (for the experimentally accessible magnetic fields of a few tens of tesla). This allows us to neglect $\boldsymbol{r}$ in the argument as compared with $\boldsymbol{L}$ and to treat $\boldsymbol{L}$ as a continuous variable (compare Sect. 3.6). Using (7.11), we obtain for the dipole matrix element $\boldsymbol{d}_{\underline{j}\underline{l}} = -e\boldsymbol{r}_{\underline{j}\underline{l}}$ as a result

$$\boldsymbol{d}_{\underline{j}\underline{l}} = \delta_{k_{yj},k_{yl}} \delta_{k_{zj},k_{zl}} \delta_{jl} \boldsymbol{d}_{\nu_j\nu_l} + \delta_{\nu_j,\nu_l} \langle j k_{yj}, k_{zj} | \boldsymbol{d} | l k_{yl}, k_{zl} \rangle\,. \tag{7.44}$$

The first contribution describes the interband optical transitions, which are diagonal with respect to the Landau quantum numbers. The second contribution is responsible for intraband transitions and couples different Landau levels. By means of (7.27), one can easily show that the selection rules for these transitions, which may be induced, for example, by a microwave field in a cyclotron resonance experiment, are $\Delta n = \pm 1$. In the following we shall focus on optical transitions and neglect the intraband contributions in (7.44). What remains is the calculation of the Coulomb matrix elements, which is rather lengthy but straightforward. The most important steps of the derivation will be summarized in the following.

7.2.1 Evaluation of the Coulomb Matrix Element

We use a representation of the Coulomb matrix element corresponding to (5.10). By means of the states defined by (7.2) and (7.11) this yields

$$\begin{aligned}
&\langle \underline{lm}|v|\underline{ni}\rangle \\
&= e^2 \sum_{\boldsymbol{q}} v(\boldsymbol{q}) \int \mathrm{d}^3\boldsymbol{r}\mathrm{d}^3\boldsymbol{r}' u^*_{\nu_l}(0,\boldsymbol{r}) c^*_{l,k_{yl},k_{zl}}(\boldsymbol{r}) \mathrm{e}^{\mathrm{i}\boldsymbol{q}\boldsymbol{r}} u_{\nu_n}(0,\boldsymbol{r}) c_{n,k_{yn},k_{zn}}(\boldsymbol{r}) \\
&\times u^*_{\nu_m}(0,\boldsymbol{r}') c^*_{m,k_{ym},k_{zm}}(\boldsymbol{r}') \mathrm{e}^{-\mathrm{i}\boldsymbol{q}\boldsymbol{r}'} u_{\nu_i}(0,\boldsymbol{r}') c_{i,k_{yi},k_{zi}}(\boldsymbol{r}') \,.
\end{aligned} \tag{7.45}$$

As in the case of the optical transition matrix elements, we make use of the separability of scales on which lattice-periodic functions and Landau functions vary. This results in

$$\begin{aligned}
\langle \underline{lm}|v|\underline{ni}\rangle &= e^2 \sum_{\boldsymbol{q}} v(\boldsymbol{q}) F(k_{zl}k_{zn}q_z) F^*(k_{zi}k_{zm}q_z) \\
&\times J(lk_{yl}, nk_{yn}, q_x q_y) J^*(ik_{yi}, mk_{ym}, q_x q_y) \\
&\times (\delta_{\nu_l\nu_n} + \mathrm{i}\boldsymbol{q}\boldsymbol{r}_{\nu_l\nu_n})(\delta_{\nu_m\nu_i} - \mathrm{i}\boldsymbol{q}\boldsymbol{r}_{\nu_m\nu_i})\,,
\end{aligned} \tag{7.46}$$

where

$$J(ik_{yi}, mk_{ym}, q_x q_y) = \langle ik_{yi}|\mathrm{e}^{\mathrm{i}q_x x}|mk_{ym}\rangle F(k_{yi}k_{ym}q_y) \tag{7.47}$$

and the form factors are

$$F(k_{zi}k_{zm}q_z) = \langle k_{zi}|\mathrm{e}^{\mathrm{i}q_z z}|k_{zm}\rangle \,. \tag{7.48}$$

The wavefunctions are given by (7.14). Performing the y integration and making the substitution $x \to xl + (k_y - q_y/2 + iq_x/2)\,l^2$, we obtain the integral

$$\begin{aligned}
J(ik_{yi}, mk_{ym}, q_x, q_y) &= \delta_{k_{yi}+q_y,k_{ym}} \exp\left(-\frac{l^2}{2}(q_x^2+q_y^2) + iq_x l^2\left(k_{yi}+\frac{q_y}{2}\right)\right) \\
&\times \frac{1}{\sqrt{\pi}} \int_{-\infty}^{\infty} dx\, \mathrm{e}^{-x^2} H_i\left(\frac{x}{l} - \frac{l}{2}(q_y + iq_x)\right) H_m\left(\frac{x}{l} + \frac{l}{2}(q_y - iq_x)\right).
\end{aligned} \tag{7.49}$$

The x integration can be performed analytically and the result can be expressed by means of Laguerre polynomials (see [140]). For $m \geq i$, this leads to

$$J(ik_{yi}, mk_{ym}, q_x, q_y) = \delta_{k_{yi}+q_y, k_{ym}} \left(\frac{i!}{m!}\right)^{1/2} \left(-\frac{\mathrm{i}l}{\sqrt{2}}(q_y - \mathrm{i}q_x)\right)^{m-i}$$
$$\times \exp\left(-\frac{l^2}{4}(q_x^2 + q_y^2) + \mathrm{i}q_x l^2 \left(k_{yi} + \frac{q_y}{2}\right)\right) L_i^{m-i}\left(\frac{l^2}{2}(q_x^2 + q_y^2)\right) . \quad (7.50)$$

For $i \geq m$, the Landau indices have to be interchanged. Next we insert this result into (7.46) and the resulting Coulomb matrix elements into (5.10). The density matrix elements are diagonal with respect to the wave vectors k_y. Moreover, they do not depend on k_y. This is the same as in the case without interaction, and is again due to the degeneracy of the Landau levels. Hence, the density-matrix equations do not depend on k_y at all, and this formal variable can be dropped completely. This is also expected for general reasons, as the Coulomb interaction does not reduce the symmetry of the problem and thus cannot remove any degeneracy. The Hartree contributions in (5.10) are proportional to the dipole contributions in (7.44) and give rise to a longitudinal–transverse splitting (see Chap. 5). This effect will not be considered in the following and we shall focus on the evaluation of the exchange contributions. We take the continuum limit with respect to the $\boldsymbol{q}$ integration, transform to polar coordinates $q_z, q_\|$, and φ, and integrate with respect to φ. This integration yields a δ function with respect to the Landau quantum numbers. As a result, the exchange interaction matrix elements which enter into the density-matrix equations can be written in the form

$$\langle \nu_j j k_{zj}, \nu_m m k_{zm} | v | \nu_n n k_{zn}, \nu_i i k_{zi} \rangle$$
$$= \delta_{\nu_j \nu_n} \delta_{\nu_m \nu_i} \delta_{|n-j|,|m-i|} \langle j k_{zj}, m k_{zm} | v | n k_{zn}, i k_{zi} \rangle . \quad (7.51)$$

where

$$\langle j k_{zj}, m k_{zm} | v | n k_{zn}, i k_{zi} \rangle$$
$$= \frac{e^2}{4\varepsilon_0 \pi^2} (\zeta_{nj}\zeta_{im})^{1/2} \int_{-\infty}^{\infty} \mathrm{d}q_z \int_0^{\infty} \mathrm{d}q_\| \frac{q_\|}{q_z^2 + q_\|^2} F(k_{zj} k_{zn}, q_z) F^*(k_{zi} k_{zn}, q_z)$$
$$\times \exp\left(-\frac{l^2}{2} q_\|^2\right) \left(\frac{l^2}{2} q_\|^2\right)^{|i-m|} L_{\mathrm{Min}(i,m)}^{|i-m|}\left(\frac{l^2}{2} q_\|^2\right) L_{\mathrm{Min}(j,n)}^{|j-n|}\left(\frac{l^2}{2} q_\|^2\right) . \quad (7.52)$$

Here we have defined $\zeta_{nj} = \mathrm{Min}(j,n)!/\mathrm{Max}(j,n)!$. For later reference we point out that we have not yet utilized the properties of the form factors

F, which in the present three-dimensional case immediately allow us to perform the q_z integration. With this representation of the Coulomb interaction we obtain, using (5.8), the still rather general representation of the density matrix equation[2]

$$
\begin{aligned}
\mathrm{i}\hbar\frac{\partial}{\partial t}n_{\nu i,\nu' j}(k_z) &= \sum_{\nu''}[\boldsymbol{d}_{\nu''\nu'}\boldsymbol{E}n_{\nu i,\nu'' j}(k_z) - \boldsymbol{d}_{\nu\nu''}\boldsymbol{E}n_{\nu'' i,\nu' j}(k_z)] \\
&+ \sum_{\nu'' m}[h_{\nu'' m,\nu' j}(k_z)n_{\nu i,\nu'' m}(k_z) - h_{\nu' i,\nu'' m}(k_z)n_{\nu'' m,\nu' j}(k_z)] \\
&+ \sum_{\nu'' k_z' lmn}\{\langle lk_z, mk_z'|v|nk_z', ik_z\rangle[n_{\nu'' n,\nu m}(k_z') - \delta_{\nu''\nu}\delta_{nm}\delta_{\nu \mathrm{v}}]n_{\nu'' l,\nu' j}(k_z) \\
&- \langle lk_z', mk_z|v|jk_z, nk_z'\rangle[n_{\nu'' n,\nu' l}(k_z') - \delta_{\nu''\nu'}\delta_{ln}\delta_{\nu' \mathrm{v}}]n_{\nu i,\nu'' m}(k_z)\}\,. \qquad (7.53)
\end{aligned}
$$

This result becomes more familiar if we consider diagonal and off-diagonal elements of the density matrix separately. The resulting equations are still rather complicated owing to the existence of off-diagonal elements in the kinetic energy, which forces us to take off-diagonal elements of the density matrix with respect to Landau levels into account, too. We shall not discuss these contributions in general, but do so only for the linear case.

7.2.2 Linear Optical Properties

The linear equation for the transition amplitude p_{vc} (in the notation introduced in Sect. 5.2) is immediately obtained from (7.53) with $\nu = \mathrm{v}$ and $\nu' = \mathrm{c}$ if the diagonal conduction band distribution function vanishes, $f_{\mathrm{c}} = 0$, and the valence band distribution functions are given by $f_{\mathrm{vv}'}^{nm} = \delta_{\mathrm{vv}'}\delta_{mn}$. Evaluating also the δ function of the Coulomb matrix element (7.51), we obtain the following in linear approximation:

$$
\begin{aligned}
&\left(\mathrm{i}\hbar\frac{\partial}{\partial t} + E_{cj}(k_z) + \mathrm{i}\hbar\gamma_2\right)p_{\mathrm{vc}}^{ij}(k_z) - \sum_{v'm} h_{\mathrm{vv}'}^{im}(k_z)p_{\mathrm{v'c}}^{mj}(k_z) \\
&= \delta_{ij}\boldsymbol{d}_{\mathrm{cv}}\boldsymbol{E} + \sum_{k_z', n\geq M_{ij}} \langle n+j-i\,k_z', ik_z|v|jk_z, nk_z'\rangle\, p_{\mathrm{vc}}^{n,n+j-i}(k_z') \qquad (7.54) \\
&+ \sum_{k_z', n<i+j} \langle j+i-n\,k_z', ik_z|v|jk_z, nk_z'\rangle\, p_{\mathrm{vc}}^{n,j+i-n}(k_z')\,,
\end{aligned}
$$

---

[2] In the presence of a static magnetic field, the dipolar coupling $\boldsymbol{dE}$ can be derived again by means of the gauge transformation (2.110), where the phase in (2.111) is defined with respect to the the optical field. Now, the transformation gives rise to additional contributions, which are proportional to the product of the vector potentials of the optical field and of the magnetic field (see Problem 7.4). In most applications, these contributions are negligible and can be omitted.

where we have used the abbreviation $\mathcal{M}_{ij} = \text{Max}(0, i-j)$. Replacing the formfactors (7.48) in (7.52) by δ functions, the Coulomb interaction in a bulk system is obtained as

$$\begin{aligned}\langle mk_z', ik_z|v|jk_z, nk_z'\rangle &= \frac{e^2}{2\varepsilon_0\pi L_z}(\zeta_{jm}\zeta_{ni})^{1/2}\\ &\times \int_0^\infty \mathrm{d}x \frac{x^{|n-i|}}{(k_z-k_z')^2 l^2 + 2x}\,\mathrm{e}^{-x} L^{|n-i|}_{\mathrm{Min}(i,n)}(x) L^{|n-i|}_{\mathrm{Min}(j,m)}(x)\,.\end{aligned} \tag{7.55}$$

Before further discussion of (7.54), it is instructive to consider the physical meaning of the off-diagonal elements with respect to the Landau indices.

Let us consider the macroscopic polarization which is obtained as in (4.9) from the solutions of (7.54) by using

$$\boldsymbol{P}(t) = \frac{1}{2\pi l^2 L_z}\sum_{vcjk_z} \boldsymbol{d}_{\mathrm{cv}} p^{jj}_{\mathrm{vc}}(k_z,t)\,, \tag{7.56}$$

where the prefactor is obtained from the k_y summation, performed according to (7.17). Again, this stems from the degeneracy. Owing to the selection rule (7.43), only contributions which are diagonal with respect to the Landau quantum number occur, and for the same reason, there is no source of off-diagonal contributions in (7.54). This is an immediate consequence of the fact that without a magnetic field only s excitons contribute to the macroscopic polarization. The relation between the Landau quantum numbers and angular momentum-quantum numbers can easily be established by considering the energy of a free electron–hole pair.[3] From (7.13), we have

$$E_{\mathrm{c}n}(k_z) - E_{\mathrm{v}m}(k_z) = \left(n+\frac{1}{2}\right)\frac{\hbar eB}{m_{\mathrm{c}}} + \frac{\hbar^2k_z^2}{2m_{\mathrm{c}}} - \left(m+\frac{1}{2}\right)\frac{\hbar eB}{m_{\mathrm{v}}} - \frac{\hbar^2k_z^2}{2m_{\mathrm{v}}}\,. \tag{7.57}$$

We can introduce new quantum numbers $N = n$ and $M = N - m$ and rewrite this equation as

$$E_{\mathrm{c}n}(k_z) - E_{\mathrm{v}m}(k_z) = \left(N+\frac{1}{2}\right)\frac{\hbar eB}{m_{\mathrm{r}}} + M\frac{\hbar eB}{m_{\mathrm{v}}} + \frac{\hbar^2k_z^2}{2m_{\mathrm{r}}}\,, \tag{7.58}$$

where m_{r} is the reduced effective mass. This is just the representation of eigenstates of free electron–hole pairs that we would have obtained by working in the symmetrical gauge, i.e. $\boldsymbol{A} = (-By, Bx, 0)$, which introduces the L_z component of the angular momenta in (7.9) rather than the xp_y contribution, and x^2+y^2 instead of x^2. Since L_z is a constant of motion, it may be replaced by its eigenvalue M. A transformation to cylindrical coordinates with the axis in the z direction allows us to map the problem onto a two-dimensional harmonic oscillator with the eigenvalues given in (7.58). Thus contributions

[3] For simplicity, we consider a diagonal valence band energy only.

with $M \neq 0$ in the symmetrical gauge correspond to the existence of off-diagonal elements in the Landau gauge. In both formulations the sources of off-diagonal elements of the transition amplitude are the off-diagonal elements of the kinetic energy (7.33)–(7.36), which couple to diagonal elements of the transition amplitude. As in the case of one-particle states, this coupling leads to a splitting of exciton states – needless to say that a quantitative description of this splitting based on solutions of (7.54) or equivalent formulations can be obtained only by means of numerical methods. In optical spectra, band mixing leads, even in the linear case, to a pronounced dependence on the polarization of the incident light. Owing to the selection rules, different polarizations of the external source allow one to switch certain transitions on or off; these transitions contribute in different ways.

Example (Optical Selection Rules). The transition matrix elements for certain polarization geometries can be obtained easily from the symmetry properties of the Bloch functions. Here we assume zinc blende symmetry. The corresponding transformation properties have been described in (3.100)–(3.103). For circularly polarized light, for example, we have to consider the dipole matrix elements

$$\langle c|\boldsymbol{d}(\boldsymbol{e}_x \pm \mathrm{i}\boldsymbol{e}_y)/\sqrt{2}|v\rangle = -e\langle c|x \pm \mathrm{i}y|v\rangle/\sqrt{2}\,. \tag{7.59}$$

Corresponding matrix elements of x or y occur for linear polarization. By means of (3.100), and for s-like conduction band states $|c, j_z = \pm 1/2\rangle = |s\rangle s_\pm$ this leads to

$$\mathrm{e}\langle c, j_z = -1/2|x + \mathrm{i}y|v, j_z = -3/2\rangle/\sqrt{2} = \mathrm{e}\langle x|x|s\rangle = -d\,. \tag{7.60}$$

We can treat other transition matrix elements in a similar way; these are summarized in Table 7.1. All matrix elements not included in this table vanish. In the Hartree–Fock approximation, the transition amplitudes to the two different conduction bands in (7.54) are strictly separated and do not mix. Both, however, contribute to the macroscopic polarization (7.56). Circularly polarized light allows one to excite either heavy- or light-hole excitons in each subspace of transitions.

For linear polarization, both types are excited in each subspace. Owing to the removal of the Kramers degeneracy in a magnetic field, all four transi-

Table 7.1. Valence band (v) to conduction band (c) transition matrix elements for different polarizations

Polarization			$x+\mathrm{i}y$	$x-\mathrm{i}y$	x	y
$\|c, 1/2\rangle$	$\rightarrow$	$\|v, 3/2\rangle$	0	d	$d/\sqrt{2}$	$\mathrm{i}d/\sqrt{2}$
$\|c, -1/2\rangle$	$\rightarrow$	$\|v, 1/2\rangle$	0	$d/\sqrt{3}$	$d/\sqrt{6}$	$\mathrm{i}d/\sqrt{6}$
$\|c, 1/2\rangle$	$\rightarrow$	$\|v, -1/2\rangle$	$d/\sqrt{3}$	0	$d/\sqrt{6}$	$-\mathrm{i}d/\sqrt{6}$
$\|c, -1/2\rangle$	$\rightarrow$	$\|v, -3/2\rangle$	d	0	$d/\sqrt{2}$	$-\mathrm{i}d/\sqrt{2}$

tions have different energies. Even for linear spectra, this results in a strong dependence on the polarization.

Let us now turn to the discussion of the Coulomb interaction. It can easily be shown by partial integration of (7.55) that the Coulomb matrix element has a logarithmic singularity, which is typical of systems confined to one dimension. This singularity can be removed by the methods discussed in Sect. 6.2.3, and we end up with a problem which can again be solved by standard numerical methods. Compared with the case without a magnetic field, the only additional complication arises from the integral in (7.55), which, however, can be evaluated numerically too. Further, we should emphasize that it is advantageous to solve (7.54) in the time domain and then calculate the optical spectrum from the Fourier transform of (7.56). This avoids the inversion of large matrices, which becomes unavoidable if (7.54) is solved after Fourier transformation – a method which becomes impracticable if more than only a very few Landau levels are involved. For high magnetic fields, one obtains rapidly converging results. The parameter which governs the rate of convergence is the ratio of the cyclotron energy $\hbar\omega_c^r = \hbar eB/m_r$ to the excitonic Rydberg energy, i.e. $\lambda = \hbar\omega_c^r/2E_R$. For $\lambda \geq 2$, about ten Landau levels are required to achieve convergent results for the low-lying exciton states. If $\lambda \leq 1$, the number of Landau levels necessary for convergence increases rapidly and other expansion schemes or perturbational methods may be advantageous.

As an example, Fig. 7.7 illustrates the basic features of excitons in a magnetic field for GaAs parameters and $\lambda = 1.8$, corresponding to a field of 6 T. The absorption spectra for positive and negative circular polarizations are compared with the corresponding spectra resulting from uncorrelated transitions between Landau levels. As expected intuitively, for each transition between the Coulomb correlated Landau levels, we obtain a sharp resonance. In comparison with the transitions without interaction, the peaks are shifted to lower energies, owing to the binding energy of the magnetoexcitons. The shift and the increase in oscillator strength are most pronounced for the lowest exciton states and become nearly constant for higher Landau levels. The variety of small resonances in between the dominant transitions results from band-mixing effects. As the shifts of these resonances are different for the different one-particle transitions, the Coulomb interaction also effects an increase of the splitting of neighboring one-particle transition energies. Thus, even for high-lying states, the differences of transition energies between uncorrelated Landau levels are only approximately the same as those for correlated transitions.

If we are mainly interested in the vicinity of the absorption edge and details of the excitonic resonance are of minor importance, as, for example, in a number of nonlinear optical experiments, we may neglect band-mixing effects and $p_{vc}^{nn'}$ is diagonal with respect to the Landau indices. Then (7.54) can be written in the much simpler form

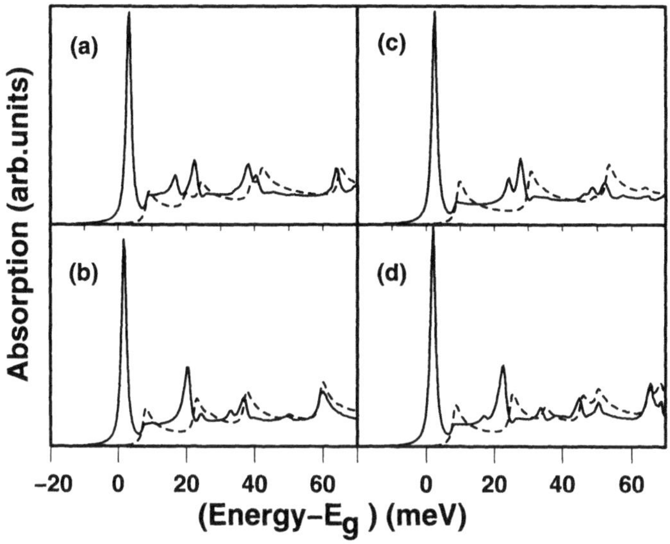

Fig. 7.7. Linear absorption of bulk GaAs at a magnetic field of 6 T, with (*full lines*) and without (*dashed lines*) Coulomb-interaction, decomposed with respect to the various transitions: (**a**) $|c, 1/2\rangle \to |v, 3/2\rangle$, (**b**) $|c, -1/2\rangle \to |v, 1/2\rangle$, (**c**) $|c, -1/2\rangle \to |v, -3/2\rangle$, (**d**) $|c, 1/2\rangle \to |v, -1/2\rangle$. The sum of the contributions (**a**) and (**b**), with weights 1 and 1/3, respectively, yield the absorption for light with negative circular polarization. The sum of the contributions (**c**) and (**d**), with weights 1 and 1/3, respectively, yield the absorption for light with positive circular polarization (see Table 7.1)

$$\left(\mathrm{i}\hbar\frac{\partial}{\partial t} + E_{\mathrm{c}}(\eta) - E_{\mathrm{v}}(\eta)\right)p_{\mathrm{vc}}(\eta) = \boldsymbol{d}_{\mathrm{cv}}\boldsymbol{E} + \sum_{\eta'} V(\eta,\eta')p_{\mathrm{vc}}(\eta') \,, \tag{7.61}$$

where we have introduced the shorthand notation $\eta = \{n, k_z\}$ and the abbreviation

$$V(\eta,\eta') = V(n,k_z,n',k_z') = \langle n'k_z', nk_z|v|nk_z, n'k_z'\rangle \,. \tag{7.62}$$

The simple structure obtained in the linear case is also conserved in the semiconductor Bloch equations if we continue to describe the valence bands in the diagonal approximation. It should be mentioned that the method of solution discussed above can also be readily applied in the nonlinear case with band mixing. It is straightforward to modify the general results of the last subsections for the case of a quantum well in a perpendicular magnetic field. In the limit of a very narrow well, we obtain approximately the idealized theory of magnetoexcitons in two dimensions [143, 144, 145].

7.2.3 Semiconductor Bloch Equations in Two and Three Dimensions

In the following we neglect the off-diagonal elements in the kinetic energy (see (7.53)), but still account for the twofold degeneracy of the valence bands. Using the notation of the linear case, we obtain for the transition amplitude

$$\begin{aligned}&\left(\hbar\frac{\partial}{\partial t}+\mathrm{i}[E_{\mathrm{c}}(\eta)-E_{\mathrm{v}}(\eta)+\hbar\gamma_2]\right)p_{\mathrm{vc}}(\eta)\\&=\mathrm{i}\sum_{\mathrm{v}'}[f_{\mathrm{vv}'}(\eta)-\delta_{\mathrm{vv}'}f_{\mathrm{c}}(\eta)]\left(\boldsymbol{d}_{\mathrm{v}'\mathrm{c}}\boldsymbol{E}+\sum_{\eta'}V(\eta,\eta')p_{\mathrm{v}'\mathrm{c}}(\eta')\right)\\&+\mathrm{i}\sum_{\eta'\mathrm{v}'}p_{\mathrm{v}'\mathrm{c}}(\eta)V(\eta,\eta')[f_{\mathrm{vv}'}(\eta')-\delta_{\mathrm{vv}'}f_{\mathrm{c}}(\eta')-\delta_{\mathrm{vv}'}]\,,\end{aligned}\tag{7.63}$$

and for the distribution function of the conduction band

$$\begin{aligned}&\hbar\frac{\partial}{\partial t}f_{\mathrm{c}}(\eta)+\hbar\gamma_1^{\mathrm{c}}[f_{\mathrm{c}}(\eta)-f_{\mathrm{c}}^0(\eta)]\\&=-2\,\mathrm{Im}\left[\sum_{\mathrm{v}}\left(\boldsymbol{d}_{\mathrm{cv}}\boldsymbol{E}+\sum_{\eta'}V(\eta,\eta')p_{\mathrm{vc}}(\eta')\right)p_{\mathrm{vc}}^*(\eta)\right].\end{aligned}\tag{7.64}$$

The diagonal and off-diagonal distributions of valence band electrons are determined by

$$\begin{aligned}&\hbar\frac{\partial}{\partial t}f_{\mathrm{vv}'}(\eta)+\hbar\gamma_1^{\mathrm{v}}[f_{\mathrm{vv}'}(\eta)-\delta_{\mathrm{vv}'}f_{\mathrm{vv}}^0(\eta)]\\&=-\mathrm{i}[\boldsymbol{d}_{\mathrm{cv}'}\boldsymbol{E}p_{\mathrm{vc}}^*(\eta)-\boldsymbol{d}_{\mathrm{cv}}\boldsymbol{E}p_{\mathrm{v}'\mathrm{c}}(\eta)]\\&-\mathrm{i}\sum_{\eta'}V(\eta,\eta')[p_{\mathrm{v}'\mathrm{c}}(\eta')p_{\mathrm{vc}}^*(\eta)-p_{\mathrm{v}'\mathrm{c}}(\eta)p_{\mathrm{vc}}^*(\eta')]\\&+\mathrm{i}\sum_{\eta'\mathrm{v}''}V(\eta,\eta')[f_{\mathrm{vv}''}(\eta')f_{\mathrm{v}''\mathrm{v}'}^*(\eta)-f_{\mathrm{vv}''}(\eta)f_{\mathrm{v}''\mathrm{v}'}^*(\eta')]\,.\end{aligned}\tag{7.65}$$

If we restrict these equations to a two-band model, they obviously exhibit exactly the same structure as the semiconductor Bloch equations without a magnetic field (see (6.5) and (6.6)). Replacing our abbreviation η by the wave vector of the bulk system and taking the corresponding limit in the Coulomb matrix elements, we indeed obtain the semiconductor Bloch equations for $\boldsymbol{B}=0$. The same structure is also obtained if we consider quantum wells or the two-dimensional limit [141]. We shall focus our interest on the latter in the following.

In order to modify our derivation in the last section for the case of idealized quantum wells, we have to replace the plane waves by the eigenfunctions of the quantum well sublevels. In a very narrow quantum well, only the lowest electron and hole sublevels contribute and we can suppress the corresponding dummy index in the following. In the same limit, the form factors in (7.52) become diagonal and independent of q_z. This allows us to perform the q_z integration analytically. Further, we utilize the δ function in (7.51) and use the fact that without band mixing only diagonal elements of the transition amplitude enter the semiconductor Bloch equations. As a result, the Coulomb interaction in the two-dimensional limit can be written as

$$V(n,n') = \frac{e^2}{4\pi\sqrt{2}\varepsilon_0 l}\,\zeta_{nn'}\,I_{nn'}\ , \tag{7.66}$$

where

$$I_{nn'} = \int \mathrm{d}x\,\mathrm{e}^{-x}\,\frac{x^{|n-n'|}}{\sqrt{x}}\left(L^{|n-n'|}_{\mathrm{Min}(nn')}(x)\right)^2 . \tag{7.67}$$

Inserting the definition of the Laguerre polynomials into this integral and integrating term by term, the resulting sums can be identified with certain hypergeometric functions, which finally allow us to write the integral for $n' \geq n$, for example, in the form [142]

$$I_{nn'} = \frac{n'!\Gamma(n'-n+1/2)\Gamma(n+1/2)}{\sqrt{\pi}n!n!(n'-n)!}g_{nn'}\ , \tag{7.68}$$

where the generalized hypergeometric function is given by

$$g_{nn'} = {}_3F_2\left(-n, n'-n+\frac{1}{2}, \frac{1}{2}; n'-n+1, \frac{1}{2}-n; 1\right) . \tag{7.69}$$

Finally we introduce excitonic units by means of (6.18) and (6.19) and obtain the Coulomb matrix elements as

$$V(n,n') = \left(\frac{2\lambda}{\pi}\right)^{1/2}\frac{\Gamma\big(n'-n+1/2\big)\Gamma\big(n+1/2\big)}{n!\,(n'-n)!}g_{nn'}\ . \tag{7.70}$$

As only finite sums are involved, the interaction matrix element can be easily calculated numerically.

With the derivation of the Coulomb interaction, we have completed the derivation of the semiconductor Bloch equations in the two-dimensional limit. Restricting ourselves to a two-band model and switching to the electron-hole picture (as in Sect. 4.1 and 6.1) with $1 - f_\mathrm{v} = f_\mathrm{h} = f_\mathrm{c} = f$, we find the following for the transition amplitude corresponding to (7.63)

$$\begin{aligned}&\left(\hbar\frac{\partial}{\partial t} + \mathrm{i}\lambda(2n+1) + \hbar\gamma_2\right)p_{\mathrm{vc}}(n) = \mathrm{i}[1-2f(n)]\boldsymbol{d}_{\mathrm{vc}}\boldsymbol{E}\\ &+\mathrm{i}\sum_{n'} V(n,n')\{[1-2f(n)]p_{\mathrm{vc}}(n') + 2p_{\mathrm{vc}}(n)f(n')\}\ .\end{aligned} \tag{7.71}$$

For the one-particle distribution function, we find

$$\hbar\frac{\partial}{\partial t}f(n) + \hbar\gamma_1[f(n) - f^0(n)] \\ = -2\,\mathrm{Im}\Big[\sum_{\mathrm{v}}\Big(\boldsymbol{d}_{\mathrm{cv}}\boldsymbol{E} + \sum_{n'} V(n,n')p_{\mathrm{vc}}(n')\Big)p^*_{\mathrm{vc}}(n)\Big] . \tag{7.72}$$

Compared with the semiconductor Bloch equations for a bulk system (6.5) and (6.6), the wave vectors have been replaced by Landau-level indices and no dependence on a continuous variable is left. Correspondingly, the kinetic energy of the relative motion of electron–hole pairs is completely quenched owing the combined action of the barriers and the magnetic field. We shall dicuss the consequences of this property in the following. The compensatory effect between different types of exchange interactions,[4] which is always present, yields a complete cancellation, as a consequence of the completely discrete spectrum resulting from quantum confinement. As can be immediately seen from (7.71), the static exciton–exciton interaction contributions, proportional to the occupation, vanish for $n' = n$. In the limit of high fields the coupling between the lowest and higher Landau levels becomes negligible. This limit requires a binding energy small compared with the cyclotron energy, a condition already met for $\lambda \geq 3$. In this case the only remaining Coulomb interaction is the intraband interaction matrix element. According to (7.66), this turns out to be

$$V(0,0) = e^2/(4\sqrt{2\pi}\varepsilon_0 l) , \tag{7.73}$$

giving just the binding energy of the lowest magnetoexcitons. Any mutual interaction between them, however, is missing. This complete lack of interaction is not an artefact of the Hartree–Fock approximation but is exactly valid [146, 147, 148], provided the system is spin-polarized, i.e. only one spin species exists in the system. This rare case of an exactly solvable many particle problem deserves some further attention.

7.2.4 Bose Condensation of Magnetoexcitons in Two Dimensions

Let us consider the idealized case of two-dimensional, interacting electron–hole pairs in the lowest Landau level. The many-particle Hamiltonian (2.54) is easily specified for this system and (apart from the gap energy) consists of interaction contributions only. By means of (7.46) and the form factors (7.50), we find

$$\hat{H} = \sum_{q,k,k',i,j} v(q,k-k')\, c^\dagger_{i,k-q/2}\, c^\dagger_{j,k'+q/2}\, c_{j,k'-q/2}\, c_{i,k+q/2} , \tag{7.74}$$

where $c^\dagger_{i,k}$ for $i \in \{\mathrm{e},\mathrm{h}\}$ creates electrons or holes in the lowest Landau state (7.14), and the interaction matrix element is given by[5]

[4] Technically speaking, these are self-energy and vertex corrections.

[5] Note that for holes, the displacement in (7.14) changes its sign.

$$v(q,k) = \sum_{q'} \frac{\mp e^2}{2\pi L^2 \varepsilon_0 \sqrt{q^2+q'^2}} \exp\Big(l^2 \frac{q^2+q'^2}{2} + iq'l^2(k)\Big) . \tag{7.75}$$

Apart from the different signs (the upper sign is valid for an electron–hole interaction), the interaction matrix element is the same for electron–electron, hole–hole and electron–hole interactions. The ground state of this system can be obtained by considering the commutator of the Hamiltonian with the electron–hole pair operator

$$d^\dagger = \frac{1}{\sqrt{N}} \sum_k c^\dagger_{\mathrm{e},k}\, c^\dagger_{\mathrm{h},-k} \, . \tag{7.76}$$

A lengthy but straightforward calculation yields

$$[H, d^\dagger] = -\sum_q v(q,0) d^\dagger = E_0 d^\dagger \, . \tag{7.77}$$

As can easily be seen, the energy E_0 is identical to the binding energy of the exciton we obtained in (7.73) within the Hartree–Fock approximation. Iterating (7.77), we find

$$[\hat{H}, (d^\dagger)^n] = nE_0 (d^\dagger)^n \, , \tag{7.78}$$

which implies, after acting on the vacuum state (in which neither electrons nor holes are occupied), that

$$\hat{H}(d^\dagger)^n|0\rangle = nE_0 (d^\dagger)^n |0\rangle \, . \tag{7.79}$$

Hence $(d^\dagger)^n|0\rangle$ is an exact eigenstate of the Hamiltonian, which is is formed by n noninteracting excitons with energy nE_0. If all available states in the lowest Landau level become occupied, we can choose $n = N$ and $(d^\dagger)^N|0\rangle$ is just the ground state. For more general conditions, we can construct the ground state from a coherent superposition of states with n pairs, which is analogous to the construction of coherent states of ideal bosons, introduced in Sect. 2.4. In the present case the construction of coherent pair states is closely related to the unitary transformation

$$\hat{U} = \mathrm{e}^{\alpha(d^\dagger - d)} \, , \tag{7.80}$$

which generates a special *Bogolyubov transformation* [149, 150] of fermion operators according to

$$\bar{c}_{i,k}(\alpha) = \hat{U}\, c_{i,k}\, \hat{U}^\dagger \, . \tag{7.81}$$

This transformation can easily be evaluated by differentiating with respect to the parameter α, which yields for the electron operator, for example,

$$\frac{\partial}{\partial \alpha} \bar{c}_{\mathrm{e},k}(\alpha) = \hat{U}\, [(d^\dagger - d), c_{\mathrm{e},k}]\, \hat{U}^\dagger = \frac{1}{\sqrt{N}} \hat{U}\, c^\dagger_{\mathrm{h},-k}\, \hat{U}^\dagger = \frac{1}{\sqrt{N}} \bar{c}^\dagger_{\mathrm{h},-k}(\alpha) \, . \tag{7.82}$$

The corresponding result for the hole operator reads

$$\frac{\partial}{\partial\alpha}\,\bar{c}^{\dagger}_{\mathrm{h},-k}(\alpha) = -\frac{1}{\sqrt{N}}\,\bar{c}_{\mathrm{e},k}(\alpha)\,. \tag{7.83}$$

Differentiating (7.82) once again and inserting (7.83), we see that the transformed operator is determined by the equation of motion of a classical harmonic oscillator. Together with the initial conditions, i.e. (7.81) and (7.82) for $\alpha = 0$, we obtain as the solution

$$\hat{U} c_{\mathrm{e},k} \hat{U}^{\dagger} = u\, c_{\mathrm{e},k} - v\, c^{\dagger}_{\mathrm{h},-k}\,, \tag{7.84}$$

$$\hat{U} c_{\mathrm{h},k} \hat{U}^{\dagger} = u\, c_{\mathrm{h},k} + v\, c^{\dagger}_{\mathrm{e},-k}\,, \tag{7.85}$$

where $u = \cos(\alpha/\sqrt{N})$, $v = \sin(\alpha/\sqrt{N})$, and $u^2 + v^2 = 1$. In the limit of large N, we have $\alpha = \sqrt{N}v/u$. The Bogolyubov transformation represents a rotation in electron–hole space, which is particularly simple in our case. Owing to the complete lack of kinetic energy, the coefficients u and v are independent of the wave vector. By means of the Bogolyubov transformation, it is straightforward to establish the invariance property of the Hamiltonian

$$\hat{U}\left[\hat{H} - \mu(\hat{N}_{\mathrm{e}} + \hat{N}_{\mathrm{h}})\right]\hat{U}^{\dagger} = \hat{H} - \mu(\hat{N}_{\mathrm{e}} + \hat{N}_{\mathrm{h}})\,, \tag{7.86}$$

which is fulfilled [147] provided we choose the chemical potential as $\mu = E_0/2$. Acting on the vacuum state (7.86) yields

$$\hat{H}\hat{U}^{\dagger}|0\rangle = \frac{E_0}{2}(\hat{N}_{\mathrm{e}} + \hat{N}_{\mathrm{h}})\,\hat{U}^{\dagger}|0\rangle\,. \tag{7.87}$$

Thus, apart from a normalization constant, $\hat{U}^{\dagger}|0\rangle$ is the eigenstate of the Hamiltonian $\hat{H} - E_0/2(\hat{N}_{\mathrm{e}} + \hat{N}_{\mathrm{h}})$ with vanishing eigenenergy. We can write this state in the more familar form of a BCS state [151],

$$|\Psi\rangle = u^N \exp\left(\frac{u}{v}\sum_k c^{\dagger}_{\mathrm{e},k} c^{\dagger}_{\mathrm{h},-k}\right)|0\rangle = \prod_k \left(u + v\, c^{\dagger}_{\mathrm{e},k} c^{\dagger}_{\mathrm{h},-k}\right)|0\rangle\,, \tag{7.88}$$

encountered in the study of superconductivity and also extensively used in the theory of Bose condensation of dense exciton systems [152, 153]. No higher-order terms contribute to the expansion, because of the exclusion principle. Generally, the BCS state serves as a variational state. As in the present case the action of the Hamiltonian (7.74) is proportional to the action of $d^{\dagger}$, no excited states, i.e. states outside the condensate, are admixed. Thus the BCS state represents the true ground state. From (7.87), we obtain for the ground state energy

$$\langle\Psi|\hat{H}|\Psi\rangle = \frac{E_0}{2}\langle\Psi|\hat{N}_{\mathrm{e}} + \hat{N}_{\mathrm{h}}|\Psi\rangle = v^2 N E_0\,. \tag{7.89}$$

Obviously, v^2 describes the fraction of occupied states, i.e. just the filling factor. For $v = 1$ we find again our earlier result (7.79) with $n = N$, and the BCS state reduces to the Slater determinant describing completely occupied electron and hole Landau levels. Interacting electron–hole pairs in the lowest Landau level in quasi-equilibrium thus lead to a Bose-condensed state for which the energy per pair is given by the Hartree–Fock single-exciton binding energy multiplied by the filling factor $\nu = v^2$. This is a surprising result. The Hartree–Fock approximation becomes exact in our idealized system. This is an immediate consequence of the electron–hole symmetry in the interaction-matrix elements. Excitons formed by these pairs are completely neutral objects, which cannot interact with each other. In quantum wells of finite width, distortions arise owing to the difference between the electron and hole wavefunctions. A similar effect results from the coupling to higher Landau levels at intermediate magnetic field. In both cases a more realistic description results in a system of weakly repulsively interacting magnetoexcitons [148]. As intuitively expected, the corrections are proportional to the filling factor ν.

Coherent pair states, which we have obtained as quasi-equilibrium states of electron–hole pairs in the lowest Landau level, can be prepared directly by optical excitation. Provided the optical field propagates perpendicular to the quantum well, no in-plane center-of-mass momentum results from the photon wave vector. In the low density limit, this allows one to excite magnetoexcitons with vanishing center-of-mass momentum and the deviations from our idealized assumptions are negligible. Correspondingly, the Hartree–Fock theory comprises the essential physics. With increasing density, i.e. increasing filling factor, the corrections can become large and the optically prepared state may be far from the quasi-equilibrium ground state of the system. In this regime, our assumptions are no longer valid and, correspondingly we have to consider processes beyond the Hartree–Fock approximation. The corresponding dynamic exciton–exciton interaction processes drive the system towards its quasi-equilibrium ground state. Whether this state can be reached on the timescale of ultrafast spectroscopy is a different question. At this point, we postpone further theoretical considerations to later chapters. In the following, experimental data are compared with results obtained from the semiconductor Bloch equations.

7.2.5 Nonlinear Absorption of Magnetoexcitons in Quantum Wells

A further complication in real quantum wells arises from the existence of light- and heavy-hole excitons, which yields contributions from both spin components. In general this leads to interactions beyond the theory of the last section. Owing to the splitting of heavy- and light-hole bands in quantum wells, however, it is possible to excite heavy- or light-hole excitons selectively with circularly polarized light (see Table 7.1), thus allowing approximately the preparation of a spin-polarized sytem. Figure 7.8 shows the linear and differential absorption spectra of a 70 nm GaAs/AlGaAs quantum well sample at a magnetic field of 12 T [154]. In the linear spectrum, the lowest three resonances can be attributed to the 1s heavy- and light-hole excitons and the

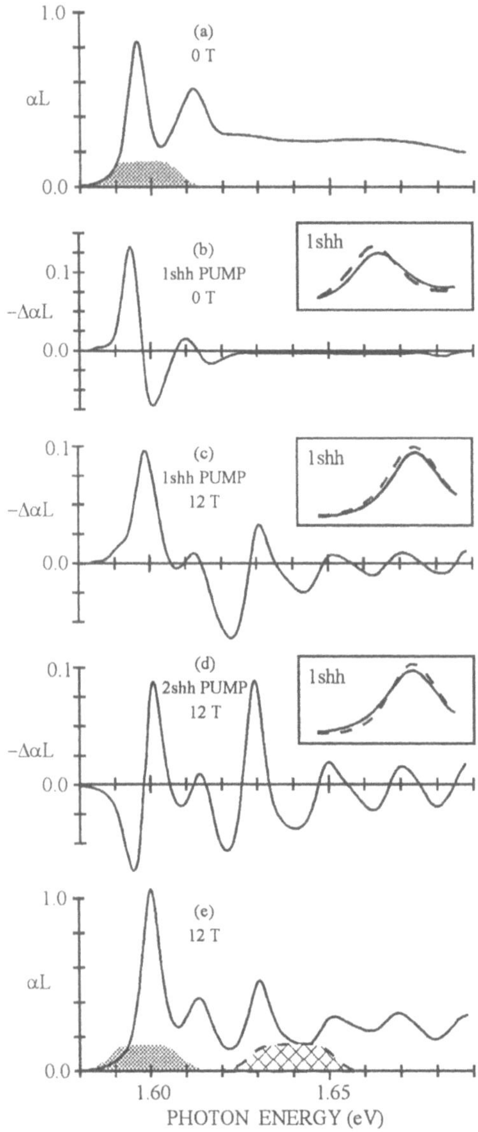

Fig. 7.8. Experimental results: linear and differential transmission spectra of a GaAs/GaAlAs quantum well. (**a**) Linear absorption at $B = 0\,\mathrm{T}$, (**b**) differential absorption at $B = 0\,\mathrm{T}$, (**c**) differential absorption at a magnetic field of 12 T for 1s heavy-hole exciton pumping and (**d**) for 2s heavy-hole exciton pumping; (**e**) shows the linear absorption spectrum at $B{=}12\,\mathrm{T}$. *Shaded* and *hatched areas* indicate the pump spectra for 1s and 2s heavy-hole exciton pumping, respectively. Insets: The corresponding 1s heavy-hole absorption 600 fs before (*dashed line*) and 600 fs after (*solid line*) the arrival of the pump. Taken from [154]

2s heavy-hole exciton. The first differential spectrum illustrates the resonant exitation of 1s heavy-hole excitons by a 100 fs pump pulse. In contrast to the case without a magnetic field,[6] no blueshift of the excitonic resonance occurs, but only a bleaching is observed, resulting from phase-space filling. This is due to the first term on the left-hand side of (7.71). This indicates that at high fields the exciton–exciton interaction in the lowest Landau level indeed

[6] Compare our discussion of the optical Stark effect in Sect. 6.3.

becomes quenched, in agreement with our theoretical results. The nonnegligible influence of the 1s heavy-hole population on higher states demonstrates, however, that the coupling to higher Landau levels is not completely negligible. The 2s heavy-hole exciton exhibits a redshift and higher states become broadened owing to their interaction with the 1s population. If a population is generated in the 2s heavy-hole exciton, this resonance becomes bleached with otherwise unchanged energetic position. As expected, the effect of the 2s heavy-hole population on the 1s heavy-hole exciton is a shift to lower energies. Apart from broadening effects, these basic features are in agreement with results obtained from numerical solution of the semiconductor Bloch equations. This is illustrated in Fig. 7.9, where the linear absorption is compared with the absorption spectra for 1s heavy-hole and 2s heavy-hole exciton pumping [154]. As in the experiments, we find that the pumped exciton resonance is essentially bleached only, whereas the other resonance is redshifted. All shift effects result from the weak coupling between different Landau levels and decrease further with increasing magnetic field. This example and more detailed studies [122] illustrate that magnetoexcitons within the lowest Landau level at high fields approach the state of a Bose-condensed system of non-interacting particles. On the other hand, if the Bose-condensed state is not

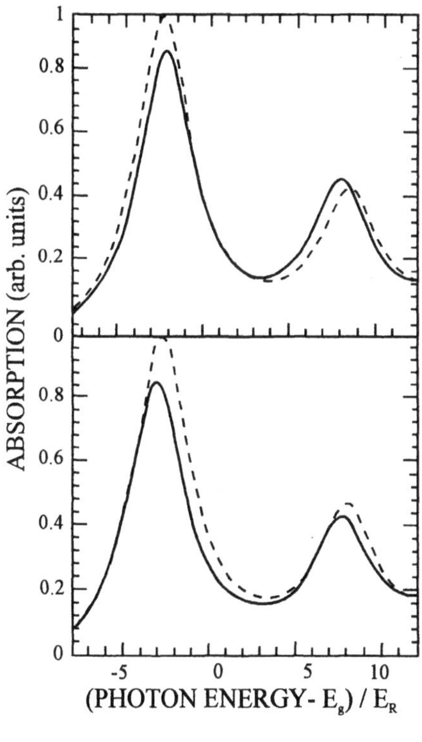

Fig. 7.9. Calculated linear (*dashed*) and nonlinear (*solid*) absorption spectra for 1s heavy-hole (*top*) and 2s heavy-hole (*bottom*) exciton pumping. Taken from [154]

prepared directly by means of the experimental excitation conditions but has to be reached by relaxation processes, we expect that the magnetoexcitons arriving in the lowest Landau level have a finite center-of-mass momentum. This excess energy can only be transferred to the lattice on a rather long timescale. In particular, for steady-state excitation, the continuous generation of highly energetic magnetoexcitons effectively prevents the system from cooling down to a pure Bose-condensed state. In this case, the repulsive interaction due to deviations from the ideal electron–hole symmetry can become large, as has been demonstrated for InGaAs/InAlAs quantum wells for *CW* excitation high above the absorption edge [155]. Owing to the small reduced effective mass in this material system, a magnetic field of 7 T corresponds to $\lambda = 5$, that is, the high-field regime. A pronounced blueshift of the heavy-hole exciton resonance persists even in this limit. A detailed understanding also requires the inclusion of scattering and relaxation processes. We shall come back to these aspects in Chaps. 10 and 11.

7.3 Problems

7.1. Show that the operator

$$(\hat{x}_0, \hat{y}_0) = \left(x + \mathrm{i}\frac{\hbar}{eB}\frac{\partial}{\partial y}\, ,\, y + \mathrm{i}\frac{\hbar}{eB}\frac{\partial}{\partial x} \right) \tag{7.90}$$

commutes with the one-particle Hamiltonian in (7.10). Give an interpretation of this quantity in terms of classical physics.

7.2. Instead of the Landau gauge, we can also use the symmetric gauge, where the vector potential is given by $\boldsymbol{A} = (-By, Bx, 0)$. Apply this gauge to derive the one-particle states of electrons in a homogeneous magnetic field. Calculate the interband transition energies and show that off-diagonal transitions in the Langau gauge correspond to transitions with $M = \langle L_z \rangle \neq 0$ in the symmetrical gauge.

7.3. Consider an electron in a quantum well with infinite barriers under the action of a homogeneous magnetic field perpendicular to the growth direction. Calculate the energy levels and discuss the limiting cases where the magnetic length is much smaller or larger than the well.

7.4. Perform the gauge transformation (2.110) of the Hamiltonian (2.23) in the presence of a static magnetic field. Discuss the additional contributions and estimate their order of magnitude. Hint: Define the gauge transformation with respect to the optical field.

7.5. Calculate the energy levels of an exciton in a strong magnetic field under the condition that the magnetic length is small compared with the excitonic Bohr radius.

7.6. Apply the semiconductor Bloch equations (7.71) and (7.72) to calculate the four-wave-mixing signal in the case of δ-pulse excitation. Assume a sufficiently strong magnetic field such that only the two lowest Landau levels, need to be taken into account. Hint: Calculate at first the required Coulomb matrix elements (7.70).

7.7. Prove the invariance property (7.86) of the Hamiltonian (7.74) and calculate the standard deviation of the particle number in a BCS state.

8. Influence of Static Electric Fields

8.1 Introduction

If a static electric field is applied to a hydrogen atom, the well-known Stark effect shifts its energy levels towards lower energy and lifts degeneracies. For atoms, the Rydberg energy is typically much larger than the induced level shifts (accessible electric fields are usually smaller than a few 10^5 V/cm). The situation is reversed for excitons in semiconductors. From perturbation theory, one obtains the result that the electric-field induced shifts are expected to be comparable to or even larger than the exciton Rydberg energy. But perturbation theory fails because electrons and holes are ripped apart by the field and true bound states do not exist. Owing to this *field ionization*, the exciton line broadens considerably and exhibits an absorption tail well below the excitonic resonance. With increasing fields, the resonance disappears before any appreciable shift is observed. Above the band edge, the electric field brings about an oscillatory behavior of the absorption coefficient. Both features, the oscillations and the absorption tail, were predicted long ago and constitute the *Franz–Keldysh effect* [156, 157, 158]. The corresponding theory, however, is insufficient to describe *electroabsorption* in bulk semiconductors quantitatively.

Yet more interesting are semiconductor heterostructures. If an electric field is applied perpendicular to the plane of the quantum well, the potential walls prevent the separation of electrons and holes. Thus energy shifts even larger than the exciton binding energy can be achieved. This effect, called the *quantum-confined Stark effect* [159], has proved to be useful in electro-optic modulators, in which an applied (*DC*) electric field modulates the optical properties – in the ideal case from perfect transparency to complete opaqueness.

Two other interesting effects of the electric field were originally predicted for bulk materials. As the width of an electronic band is determined by the overlap between the wavefunctions of neighboring atoms, intuitively one expects an energetic separation of the atomic states contributing to the band for large electric fields. In particular, if the potential drop over one unit cell is larger than the width of the band, one expects a strong reduction of the overlap, which is accompanied by an increasing localization of electron states. This effect has been called *Wannier–Stark localization* [161, 162]. Closely related is the existence of the so-called *Bloch oscillations* [59]. So far, these two effects have not been observed in bulk semiconductors, for various reasons.

The question, of how to describe the properties of Bloch electrons in the presence of a homogeneous electric field is nearly as old as solid-state physics. The problem has its origin in the unbounded perturbation $h_F = eFz$, where F is the electric field in the z direction, which becomes singular for extended systems.[1] The difficulty in applying standard perturbational techniques to such systems has been the reason for a long-term controversy about the existence of Wannier–Stark states in semiconductors [163, 164].

A single isolated band with energy $E_n(\boldsymbol{k})$ and periodicity $2\pi/a$ in $\boldsymbol{k}$ space, evolves under the action of a uniform electric field into a ladder spectrum with equal spacing eFa [165]. This spectrum, however, is not stable, as the perturbation leads to a coupling to other bands, which is of the same order of magnitude as the intraband processes. This interband coupling, together with other broadening mechanisms, transforms the Wannier–Stark energies into strongly broadened resonances. In bulk systems, this broadening turns out to be of the same order of magnitude as or even larger than the spacing. The evenly spaced ladder spectrum corresponds to an oscillatory motion in the time domain. Physically, these oscillations result from the periodicity of the energy band in $\boldsymbol{k}$ space, which gives rise to a periodic change of the direction of acceleration under the action of the electric field. These *Bloch oscillations* can only be observed if the period $2\pi/eFa$ is small compared with the inverse broadening, i.e. small compared with the lifetime of Wannier–Stark states. Thus, the conditions for observing Wannier–Stark ladders or Bloch oscillations are hardly met in bulk materials. In contrast to this, the material parameters in superlattices are much more favorable for observation of the two effects. In these artificial structures, the periodicity intervals are much larger than the unit cell in a bulk semiconductor and the width of the band results from the overlap between the wavefunctions of neighboring wells. Indeed, Wannier–Stark localization [166, 167], and Bloch oscillations [168, 169, 170] have been observed in experiments on semiconductor heterostructure superlattices.

The various phenomena which occur in semiconductor systems in uniform electric fields have been studied by a variety of different methods. We shall again apply the Hartree–Fock approximation and follow the concepts introduced in the last chapters. First, we consider the properties of one-particle states within the effective-mass approximation. We could then proceed as previously and make use of these states to expand the equation for the optical transition amplitude or, in the nonlinear case, the semiconductor Bloch equations. In contrast to quantum well systems or bulk semiconductors in magnetic fields, however, the resulting equations are even more involved and numerical solutions are more difficult. Therefore, we shall also discuss sim-

[1] For the application of perturbational techniques to singular perturbations and a rigorous mathematical treatment of this and related problems, see [160].

plified approaches for investigating the optical properties under the influence of electric fields.

8.2 Uncorrelated Optical Transitions in Uniform Electric Fields

In order to introduce the one-particle states of electrons in a semiconductor under the influence of a static, homogeneous electric field, we proceed similarly to the case of a static magnetic field in Sect. 7.1. Choosing the electric field F to be in the z direction, the wavefunction, obtained in the in effective-mass approximation is given by

$$c(\boldsymbol{r}) = \frac{1}{\sqrt{L_x L_y}} \exp(\mathrm{i}\, k_x x + \mathrm{i}\, k_y y)\phi(z) \, . \tag{8.1}$$

The pendant to (7.13) defines $\phi(z)$ as the solution of

$$\left(-\frac{\hbar^2}{2m_\mathrm{e}} \frac{\partial^2}{\partial z^2} - e\,F\,z \right) \phi(z) = \left(E - \frac{\hbar^2 k^2}{2m_\mathrm{e}} \right) \phi(z) \, , \tag{8.2}$$

where $\boldsymbol{k}$ denotes the wave vector perpendicular to the field. Defining the length parameter

$$l_\mathrm{e} = \left(\frac{\hbar^2}{2m_\mathrm{e} e\, F} \right)^{1/3} , \tag{8.3}$$

we can introduce the dimensionless variables $\zeta = z/l_\mathrm{e}$ and $\mathcal{E} = E/(eFl_\mathrm{e}) - l_\mathrm{e}^2\, k^2$. With this transformation, (8.2) can be cast into the form

$$\left(\frac{\partial^2}{\partial \zeta^2} - (\zeta - \mathcal{E}\,) \right) \phi(l_\mathrm{e}\zeta) = 0 \, , \tag{8.4}$$

which is equivalent to the differential equation

$$\left(\frac{\partial^2}{\partial s^2} - s \right) w(s) = 0 \, . \tag{8.5}$$

The solutions of (8.5) are the Airy functions Ai(s) and Bi(s) [171, 172], which are regular and irregular, respectively, at infinity. In this section, we can restrict ourselves to Ai(s), as only this function obeys the necessary boundary conditions. A useful integral representation of Ai(s) is given by

$$\mathrm{Ai}(s) = \frac{1}{2\pi} \int_{-\infty}^{\infty} \mathrm{d}t \exp\left[\mathrm{i} \left(\frac{t^3}{3} + st \right) \right] . \tag{8.6}$$

By means of (8.6), we obtain the orthogonality relation[2]

[2] Sometimes a different normalization, in which π is replaced by $\sqrt{\pi}$ in (8.6), is used in the literature.

$$\int_{-\infty}^{\infty} dt\,\mathrm{Ai}(t+\mathcal{E})\mathrm{Ai}(t+\mathcal{E}') = \delta(\mathcal{E}-\mathcal{E}') . \tag{8.7}$$

A generalization of this relation, which we require in the following, can be derived in the same way and yields

$$\int_{-\infty}^{\infty} dt\,\mathrm{Ai}[\alpha(t+\mathcal{E})]\mathrm{Ai}[\beta(t-\mathcal{E}')] = \frac{1}{(\alpha^3-\beta^3)^{1/3}}\mathrm{Ai}\left(\frac{\alpha\beta(\mathcal{E}+\mathcal{E}')}{(\alpha^3-\beta^3)^{1/3}}\right) . \tag{8.8}$$

From the representation (8.6), one can also obtain the asymptotic expansion of the Airy function. For large positive argument we have approximately [172]

$$\mathrm{Ai}(s) = \frac{1}{2\,s^{1/4}\,\pi^{1/2}}\exp\left(-\frac{2}{3}\,s^{3/2}\right) , \tag{8.9}$$

whereas for large negative argument the Airy function behaves asymptotically like

$$\mathrm{Ai}(s) = \frac{1}{|s|^{1/4}\,\pi^{1/2}}\sin\left(\frac{2}{3}\,|s|^{3/2}+\frac{\pi}{4}\right) . \tag{8.10}$$

For later reference, we state also the corresponding expansions for the Bi functions, which diverge for positive argument according to

$$\mathrm{Bi}(s) = \frac{1}{2\,s^{1/4}\,\pi^{1/2}}\exp\left(\frac{2}{3}\,s^{3/2}\right) \tag{8.11}$$

and show an oscillatory behavior at negative arguments

$$\mathrm{Bi}(s) = \frac{1}{|s|^{1/4}\,\pi^{1/2}}\cos\left(\frac{2}{3}\,|s|^{3/2}+\frac{\pi}{4}\right) . \tag{8.12}$$

In terms of Airy functions, the solution of (8.4) can be written as

$$\phi(z) = \frac{1}{\sqrt{C_\mathrm{e}}}\,\mathrm{Ai}\left(\frac{z}{l_\mathrm{e}}-\frac{E}{eFl_\mathrm{e}}+l_\mathrm{e}^2k^2\right) . \tag{8.13}$$

The constant C_e is determined by the usual normalization condition on δ functions. We obtain, by means of (8.7),

$$C_\mathrm{e} = eFl_\mathrm{e}^2 = \left(\frac{\hbar^4 eF}{4m_\mathrm{e}^2}\right)^{1/3} . \tag{8.14}$$

The corresponding spectrum is completely continuous and does not depend on any parameter which labels eigenenergies. Obviously, this property results immediately from the monotonically increasing potential energy of an electron in a uniform electric field. Corresponding results are obtained for holes.

8.2.1 Optical Absorption

Let us now turn to the calculation of uncorrelated optical transitions in the vicinity of the band edge. An appropriate starting point is the long-wavelength limit of (5.35), which has to be evaluated with the eigenfunctions (8.1) and the corresponding energies for electrons and holes in the effective-mass approximation. Owing to the monotonic increase of the potential, the thermodynamic limit requires some care. In order to have a well-defined dependence of the lower boundaries of the energies on the system size, we assume that our system is reframed in the z direction by infinite walls at $\pm L_z/2$. A general treatment of this problem will be given in Sect. 8.4. For the moment we are interested in the limit $L_z \to \infty$ only. If the potential drop is large compared with the spacing of energy levels in the quantum well, the ground state is obviously given by $eFL_z/2 + \Delta$, where Δ does not depend on the system size and does not contribute in the thermodynamic limit. Thus, to evaluate (5.35) we can use $eFL_z/2$ as the lower bound of the spectrum. Considering the imaginary part, we have

$$\begin{aligned}\mathrm{Im}\Big(\chi^{(1)}(\omega)\Big) &= \frac{\pi}{\varepsilon_0 V}\sum_{\boldsymbol{k}}\int_{-eFL_z/2}^{\infty}\mathrm{d}E\,\mathrm{d}E'\,|d_{\mathrm{vc}}(E,E')|^2\\ &\times\delta\Big(E_{\mathrm{g}}+\frac{\hbar^2k^2}{2m_{\mathrm{r}}}+E+E'-\hbar\omega\Big)\,. \end{aligned}\tag{8.15}$$

Assuming the separability of length scales (compare Sect. 3.6), the transition matrix element is given by

$$d_{\mathrm{vc}}(E,E') = \frac{d_{\mathrm{vc}}^0}{\sqrt{C_{\mathrm{e}}C_{\mathrm{h}}}}\int_{-\infty}^{\infty}\mathrm{d}z\,\mathrm{Ai}\Big(-\frac{z}{l_{\mathrm{h}}}-\frac{E}{eFl_{\mathrm{h}}}\Big)\,\mathrm{Ai}\Big(\frac{z}{l_{\mathrm{e}}}-\frac{E'}{eFl_{\mathrm{e}}}\Big)\,,\tag{8.16}$$

where we have used (8.9) and the corresponding result for holes. In accordance with (8.8), we have $d_{\mathrm{vc}}(E,E') = d_{\mathrm{vc}}(E+E')$. Thus, we can perform the E and E' integrations first and take the thermodynamic limit, with the result

$$\mathrm{Im}\Big(\chi^{(1)}(\omega)\Big) = \frac{1}{4\pi\varepsilon_0}\int d^2\boldsymbol{k}\,\Big|\mathrm{d}_{\mathrm{vc}}(\hbar\omega - E_{\mathrm{g}} - \frac{\hbar^2k^2}{2m_{\mathrm{r}}})\Big|^2\,.\tag{8.17}$$

Inserting the result of integration of the transition matrix element by means of (8.8), we finally obtain

$$\mathrm{Im}\Big(\chi^{(1)}(\omega)\Big) = \frac{|d_{\mathrm{vc}}^0|^2}{2\varepsilon_0\,|e|Fl_{\mathrm{r}}^2}\int_0^{\infty}\mathrm{d}k\,k\,\Big|\mathrm{Ai}\Big(\frac{E_{\mathrm{g}}+\hbar k^2/2m_{\mathrm{r}}-\hbar\omega}{eF\,l_{\mathrm{r}}}\Big)\Big|^2\,.\tag{8.18}$$

As an effect of the relation (8.8), the electron and hole masses have vanished and are replaced by the reduced mass. Correspondingly, l_{r} is defined analogously to (8.3), but with m_e replaced by m_r. Indeed, (8.14) can also be obtained by applying the pair picture (6.16) or (6.55) directly [173] and leaving the transition matrix elements unchanged. This treatment corresponds to a picture in which electrons and holes behave like correlated pairs, owing to their acceleration in opposite directions. Thus we can use, instead of (8.2), the corresponding equation with m_e replaced by m_r. The solution describes the electron–hole pair envelope functions, which are again given by an Airy function. Assuming, as in the excitonic case, that the transition dipole matrix elements $d_{vc}(z)$ are strongly localized on the scale on which the Airy functions vary significantly, we can set $z = 0$ in the Airy functions and obtain (8.14) again. The equivalence of the two derivations, however, is nontrivial and is only due to the relation (8.8). Now that we have introduced the pair picture, it is more convenient to use excitonic units. We introduce a dimensionless field parameter, which is determined by the ratio of the potential drop across a Bohr radius to the excitonic Rydberg energy, i.e. by $f = eFr_B/E_R$. In (8.18), this replaces the length parameter according to $l_r = f^{-1/3}\, r_B$.

The final k integration can be performed by means of (8.5), which yields, after multiplication by $\partial w/\partial s$, the relation

$$\frac{\partial}{\partial s}\left(\frac{\partial w}{\partial s}\right)^2 + s\frac{\partial}{\partial s}w^2 = \frac{\partial}{\partial s}\left[\left(\frac{\partial w}{\partial s}\right)^2 - sw^2\right] - w^2 = 0\,. \tag{8.19}$$

Integration gives the desired result,

$$\int \mathrm{d}s\, w^2(s) = \left(\frac{\partial w}{\partial s}\right)^2 - sw^2\,. \tag{8.20}$$

Making use of this relation in (8.18), we obtain the final expression

$$\mathrm{Im}\Big(\chi^{(1)}(\omega)\Big) = \frac{|d_{vc}^0|^2\, m_r\, f^{1/3}}{2\,\varepsilon_0 \hbar^2 r_B}\,\{[\mathrm{Ai}'(s)]^2 - s\,[\mathrm{Ai}(s)]^2\}\,, \tag{8.21}$$

where the argument s is given by

$$s = -\frac{\hbar\omega - E_g}{E_R\, f^{2/3}} \equiv -\frac{\Omega}{f^{2/3}}\,. \tag{8.22}$$

The properties of the optical transitions can be discussed right away using the asymptotic behavior of the Airy functions. For large positive arguments, we make use of (8.9). Obviously, the electric field induces a finite absorption below the band edge. As this range of energies is classically forbidden, for optical transition amplitudes as well as for the motion of an electron in a uniform electric field, the absorption tail represents a typical quantum mechanical effect, which can be also interpreted as the photon-assisted tunneling of electrons between valence and conduction band states [156]. For large negative arguments, the asymptotic behavior is given by (8.10). Thus, above the band edge (8.21) predicts an oscillatory behavior, with a frequency proportional to $1/F$. These oscillations, together with the absorption tail below the band edge, represent the basic features of the Franz–Keldysh effect. Compared with the experimental data, however, the predicted electroabsorption turns out to be much too weak. As in the case without an electric field, this shortcoming is due to the fact that we have neglected the Coulomb interaction.

8.3 Correlated Optical Transitions in Uniform Electric Fields

In order to account for the electron–hole interaction, we apply the pair picture and include the electric field in the exciton equation (6.11). Thus we have to solve

$$\left(-\frac{\hbar^2}{2m_r}\Delta - e^2 v(\boldsymbol{r}) - eFz \right) \psi(\boldsymbol{r}) = E\,\psi(\boldsymbol{r})\,, \tag{8.23}$$

which yields the optical spectra according to (6.15) and (6.55). Unfortunately, no analytical solutions of this equation are available. In the case of the hydrogen atom, perturbation theory is at least asymptotically valid in a large range of field strengths and yields the linear and quadratic Stark shifts. In the case of excitons, the allowed fields are orders of magnitude smaller, as is obvious from the effective field. In absolute units, this material-dependent parameter is related to the electric field, measured in units of V/m, by

$$f = 3.9 \times 10^{-8}\,\varepsilon_\infty^3\,\frac{m^2}{m_r^2}\,F\,. \tag{8.24}$$

Thus, one expects that in most semiconductors perturbation theory will fail even for extremely small fields, and a more rigorous treatment is necessary.

Introducing excitonic units in (8.23), we obtain

$$\left(-\Delta - \frac{2}{r} + fz \right) \psi(\boldsymbol{r}) = E\psi(\boldsymbol{r})\,. \tag{8.25}$$

As in the case without Coulomb interaction, the spectrum of this Hamiltonian is completely continuous and the question arises: What happens to the bound states? There are various possible ways to obtain an answer. For example, by making use of the Airy functions to expand the wavefunctions, the field can be treated exactly. If we consider the Green's function belonging to (8.25), we can replace the continuum by a discrete set of energies combined with a sufficiently large dephasing constant (compare the treatment of the purely excitonic continuum in Sect. 6.2). To achieve convergent results, unfortunately, the Coulomb matrix needs to be rather large and the solution of the problem requires time-consuming numerical work. We can, however, apply this type of expansion to produce a simplified model, which gives an initial idea of the physics of excitons in a uniform electric field.

8.3.1 An Analytical Model

Let us consider, instead of (8.23), the corresponding equation for the Green's function,

$$\left(-\Delta - \frac{2}{r} + fz - \Omega - \Gamma\right) G(\boldsymbol{r},\boldsymbol{r}',\omega) = \delta(\boldsymbol{r}-\boldsymbol{r}')\,, \tag{8.26}$$

where Ω is defined in (8.22) and $\Gamma = \hbar\gamma_2/E_{\mathrm{R}}$. In the following, we shall treat the lowest bound state $|\psi_1\rangle$ explicitly, and higher states approximately by an expansion in terms of Airy functions and plane waves. Analogously to (8.1) together with (8.13) for the pair states, these wavefunctions are given by

$$\langle \boldsymbol{r}|E\boldsymbol{q}\rangle_0 = \frac{1}{2\pi f^{1/6}} \exp(\mathrm{i}k_x x + \mathrm{i}k_y y)\,\mathrm{Ai}\left(-zf^{1/3} - \frac{E-\boldsymbol{k}^2}{f^{2/3}}\right), \tag{8.27}$$

where the plane waves have been normalized to δ functions. In principle, these states should be orthogonalized to the lowest state according to

$$|E\boldsymbol{q}\rangle = |E\boldsymbol{q}\rangle_0 - S^{\mathrm{e}}_{1,E\boldsymbol{q}}\,|\psi_1\rangle\,, \tag{8.28}$$

where $S^{\mathrm{e}}_{1,E\boldsymbol{q}} = \langle\psi_1|E\boldsymbol{q}\rangle_0$. However, the corrections, which are proportional to this overlap of the excitonic ground state with the Airy functions, contribute significantly for large fields only and will be neglected in the following. By means of (8.28), an expansion of the Green's function can be obtained in the form

$$\begin{aligned} G = {} & |\psi_1\rangle\, G_{1,1}\,\langle\psi_1| \\ & + \int \frac{\mathrm{d}^2\boldsymbol{q}}{(2\pi)^2}\mathrm{d}E\Big(|\psi_1\rangle\, G_{1,E\boldsymbol{q}}\,\langle E\boldsymbol{q}| + |E\boldsymbol{q}\rangle\, G_{E\boldsymbol{q},1}\,\langle\psi_1|\Big) \\ & + \int \frac{\mathrm{d}^2\boldsymbol{q}}{(2\pi)^2}\frac{\mathrm{d}^2\boldsymbol{q}'}{(2\pi)^2}\mathrm{d}E\,\mathrm{d}E'\,|E\boldsymbol{q}\rangle\, G_{E\boldsymbol{q},E'\boldsymbol{q}'}\,\langle E'\boldsymbol{q}'|\,. \end{aligned} \tag{8.29}$$

Inserting this expansion into (8.26) and projecting either onto $|\psi_1\rangle$ or onto $|E\boldsymbol{q}\rangle$, we obtain a set of four coupled equations, which can be simplified by the following approximation. Since the excitonic ground state is treated exactly in the limit of vanishing field, the residual Coulomb interaction leads mainly to a correlation of the continuum states. Focusing on the behavior of the excitonic resonance, we can neglect this correlation. Within this approximation, the Green's function can be calculated analytically, and we obtain as the final results

$$G_{1,1} = \left(E_1 - \Omega - \mathrm{i}\Gamma - \int \frac{\mathrm{d}^2\boldsymbol{q}}{(2\pi)^2}\mathrm{d}E\frac{Z_{1,E\boldsymbol{q}}Z_{E\boldsymbol{q},1}}{E+\boldsymbol{q}^2-\Omega-\mathrm{i}\Gamma}\right)^{-1}, \tag{8.30}$$

$$G_{1,E\boldsymbol{q}} = -G_{1,1}\,\frac{Z_{1,E\boldsymbol{q}}}{E+\boldsymbol{q}^2-\Omega-\mathrm{i}\Gamma}\,, \tag{8.31}$$

$$G_{E\boldsymbol{q},1} = -G_{1,1}\,\frac{Z_{E\boldsymbol{q},1}}{E+\boldsymbol{q}^2-\Omega-\mathrm{i}\Gamma}\,, \tag{8.32}$$

$$G_{E\boldsymbol{q},E'\boldsymbol{q}'} = \frac{\delta(E-E')\,\delta_{\boldsymbol{q},\boldsymbol{q}'}}{E+\boldsymbol{q}^2-\Omega-\mathrm{i}\Gamma} + \frac{Z_{E\boldsymbol{q},1}}{E+\boldsymbol{q}^2-\Omega-\mathrm{i}\Gamma}\,G_{1,1}\,\frac{Z_{1,E'\boldsymbol{q}'}}{E'+\boldsymbol{q}'^2-\Omega-\mathrm{i}\Gamma}\,, \tag{8.33}$$

where the transition matrix element is defined as

$$Z_{1,E\boldsymbol{q}} = f\,\langle\psi_1|z|E\boldsymbol{q}\rangle = f\,\langle\psi_1|z|E\boldsymbol{q}\rangle_0\,. \tag{8.34}$$

For intermediate fields, the numerical evaluation of these equations results in optical spectra which are very close to exact numerical solutions of (8.26). Our purpose in this section, however, is the derivation of an analytical expression for the absorption edge, which requires some approximations. In the vicinity of the excitonic resonance, the spectrum is dominated by the imaginary part of $G_{1,1}$, which is given by

$$\mathrm{Im}[G_{1,1}(\Omega)] = \frac{\Gamma+\Gamma_F}{(E_1+\Delta_F-\Omega)^2+(\Gamma+\Gamma_F)^2}\,, \tag{8.35}$$

where, in the limit $\Gamma=0$, the field-induced broadening is given by

$$\Gamma_F\,(\Omega) = \pi\int\frac{\mathrm{d}^2\boldsymbol{q}}{(2\pi)^2}\,\mathrm{d}E\,|Z_{1,E\boldsymbol{q}}|^2\,\delta(E-\Omega) \tag{8.36}$$

and Δ_F is the corresponding shift of the exciton line. Mathematically, the broadening is nothing but the immediate consequence of a completely continuous spectrum. The physical interpretation of this effect is, however, more appealing. Considering the effective potential in (8.26), we see that we have to distinguish between three regions for $z \le 0$. Whereas for small $|z|$ the potential is dominated by the Coulomb potential, for large $|z|$ it approaches the potential due to the field. These regions are separated by a potential barrier, which increases with increasing field. Owing to the finite tunneling probability, any excitonic bound state has an admixture of continuum states. This makes the spectrum continuous and as a remnant of the true bound states, we have broadened resonances only. In the time domain this may be interpreted as a finite lifetime of bound exciton states due to their ionization, resulting from the finite tunneling probability [177]. Our present treatment implies that, apart from the lowest bound state, all excited excitonic states are ionized. To discuss these effects in more detail, we now turn to the calculation of the transition matrix element. We make use of (8.27) and the Fourier representation of the excitonic ground state in the form

$$\psi_1(\boldsymbol{k},k_z) = \frac{8\sqrt{\pi}}{(\boldsymbol{k}^2+k_z^2+1)^2}\,. \tag{8.37}$$

Integrating the plane waves in the x and y directions, as well as the resulting δ function, we find that the matrix element is given by

$$Z_{1,Eq} = \frac{4f^{5/6}}{\sqrt{\pi}} \int \mathrm{d}z\,\mathrm{d}k_z \frac{z\exp(-\mathrm{i}k_z z)}{(q^2+1+k_z^2)^2} \mathrm{Ai}\Big(-zf^{1/3} - \frac{E-q^2}{f^{2/3}}\Big)$$
$$= f^{5/6}\sqrt{\pi}\int \mathrm{d}z \left(\frac{z\,|z|}{q^2+1} + \frac{z}{(q^2+1)^{3/2}}\right)$$
$$\times \exp\Big(-\sqrt{q^2+1}\,|z|\Big)\,\mathrm{Ai}\Big(-zf^{1/3} - \frac{E-q^2}{f^{2/3}}\Big)\,. \qquad (8.38)$$

Owing to the localization of the exponential function, the Airy function can be expanded in powers of z. We restrict ourselves to the leading order and obtain, by means of the asymptotic representation (8.9),

$$Z_{1,Eq} = 8f\,\frac{(q^2-E)^{1/4}}{(q^2+1)^3}\exp\Big(-\frac{2(q^2-E)^{3/2}}{3f}\Big) + O\left(\frac{1}{(q^2+1)^4}\right)\,. \qquad (8.39)$$

Now, we can again turn to the calculation of the induced broadening, for which our approximate calculation of the transition matrix element is valid. Well below the excitonic resonance, the dominant contributions to the q integral arise from the region $|q| \ll 1$. If we neglect the dependence on q in the denominator of $Z_{1,Eq}$, the integration can be performed analytically and yields the final result

$$\Gamma_F(\Omega) = 16f^3\exp\Big(-\frac{4|\Omega|^{3/2}}{3f}\Big)\,. \qquad (8.40)$$

Despite some rather drastic approximations, (8.40) predicts the correct order of magnitude for field strengths in the range of the classical ionization threshold, i.e. $f = 1$. More importantly, it also shows the essential features of the excitonic resonance under the action of the field. The broadening, as well as the shift Δ_F, which can be calculated along the same lines, follows an exponential law. The shift turns out to be much smaller than the broadening and approaches a maximum value of about 10% of the binding energy for $f \geq 1$. At even higher fields, the broadening becomes so strong that the excitonic resonance can no longer be resolved.

According to (8.35) and (8.40), the absorption tail for $\Omega \ll -1$ is determined by

$$\mathrm{Im}[G_{1,1}(\Omega)] = \frac{16f^3}{(1+\Omega)^2}\exp\Big(-\frac{4|\Omega|^{3/2}}{3f}\Big)\,, \qquad (8.41)$$

where we have neglected the shift. The lineshape is determined by two factors: the electroabsorption due to uncorrelated transitions, and an enhancement factor, resulting from the electron–hole interaction. An important consequence of this excitonic enhancement is the transformation of the Franz–Keldysh lineshape into an exponential absorption tail. Surprisingly, this behavior is predicted even by our approximate treatment.

8.3.2 Representation in Parabolic Coordinates

A method which is better suited for rigorous numerical solutions of the problem is based on the separability of the hydrogen problem in parabolic coordinates [172]. By means of these coordinates, which are defined as $\xi = r + z$, $\eta = r - z$, $\phi = \text{arctg}(y/x)$, (8.25) is transformed into the form

$$\left[-\frac{\partial}{\partial \xi}\left(\xi \frac{\partial}{\partial \xi}\right) - \frac{\partial}{\partial \eta}\left(\eta \frac{\partial}{\partial \eta}\right) - \frac{\xi + \eta}{4\xi\eta}\frac{\partial^2}{\partial \varphi^2} - 1 + \frac{\xi^2 - \eta^2}{8} f \right] \psi = E \frac{\xi + \eta}{4} \psi \, . \tag{8.42}$$

With the ansatz

$$\psi(\boldsymbol{r}) = \psi_1(\xi)\, \psi_2(\eta)\, \mathrm{e}^{\mathrm{i}m\varphi} \, , \tag{8.43}$$

we can separate this equation into

$$\left[-\frac{\partial}{\partial \xi}\left(\xi \frac{\partial}{\partial \xi}\right) + \frac{m^2}{4\xi} + \frac{\xi^2}{8} f - E\frac{\xi}{4} \right] \psi_1(\xi) = p\, \psi_1(\xi) \tag{8.44}$$

and

$$\left[-\frac{\partial}{\partial \eta}\left(\eta \frac{\partial}{\partial \eta}\right) + \frac{m^2}{4\eta} - \frac{\eta^2}{8} f - E\frac{\eta}{4} \right] \psi_2(\eta) = (1-p)\, \psi_2(\eta) \, . \tag{8.45}$$

The separation parameter p is also referred to as the parabolic quantum number. Eliminating the first-order derivatives by substituting $\psi_1(\xi) = b(\xi)/\sqrt{\xi}$ and $\psi_2)\eta = u(\xi)/\sqrt{\eta}$, we obtain two one-dimensional Schrödinger equations

$$\left(-\frac{\partial^2}{\partial \xi^2} + \frac{m^2 - 1}{4\xi^2} + \frac{\xi}{8} f - \frac{E}{4} - \frac{p}{\xi} \right) b(\xi) = 0 \tag{8.46}$$

and

$$\left(-\frac{\partial^2}{\partial \eta^2} + \frac{m^2 - 1}{4\eta^2} - \frac{\eta}{8} f - \frac{E}{4} - \frac{1-p}{\eta} \right) u(\eta) = 0 \, . \tag{8.47}$$

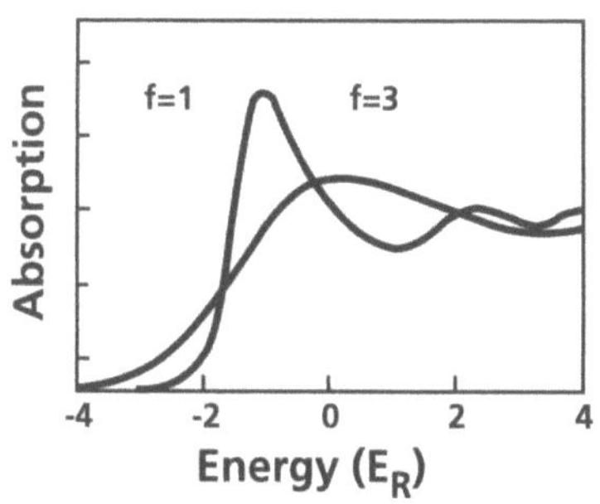

Fig. 8.1. Absorption spectra in two different strong electric fields. Taken from [176]

For direct, allowed optical transitions, only the case $m = 0$ has to be considered.

As the effective potential in (8.46) is positive for $\xi \to \infty$, the eigenstates are bound and we obtain a discrete set for the parabolic quantum numbers p for any energy E. The numerical determination of these quantum numbers requires a trial and error procedure, which starts with a guessed value and improves this value iteratively. The boundary conditions are determined by the behavior of $b(\xi)$ at $\xi = 0$ and $\xi \to \infty$. In order to take both conditions into account, we have to perform the integration inwards and outwards, starting at large ξ and at $\xi = 0$, respectively. The two solutions are matched at the outermost maximum. In this way, we can calculate the solutions of (8.46). In the second step we make use of the parabolic eigenvalues to calculate the eigenfunctions of (8.47). With these results, we again obtain an eigenfunction representation (6.15) of the Green's function (8.24) and the optical spectra according to (6.55). Numerical results obtained along these lines [174, 175, 176] are shown in Fig. 8.1 for two different fields and vanishing dephasing rate γ_2. As we have already seen from our analytical model, there is a strong broadening of the exciton line, which is accompanied by a small shift. Above the band edge, we obtain again an oscillatory behavior. Compared with the Franz–Keldysh effect, however, the magnitude of the absorption is much larger as a consequence of the Coulomb enhancement. To illustrate the striking consequences of this excitonic enhancement, the absorption spectrum is depicted in Fig. 8.2 on a logarithmic scale, with and without Coulomb interaction, for various electric fields.

As already predicted approximately by (8.41), the numerical results show a strictly exponential decay below the excitonic resonance, which results from the Coulomb enhancement of the Franz–Keldysh absorption tail. Thus, the probability of finding the electron and hole at the same place decreases exponentially owing to the field ionization induced by the electric field. There are far-reaching consequences of this feature of the absorption tail [178], as exponential behavior is also found in cases without an external electric field. Indeed, at high temperatures, the interaction of excitons with phonons of a certain type also brings about an exponential tail, which covers several orders of magnitude and which is referred to as the *Urbach tail* [179, 180]. We shall discuss this universal phenomenon in more detail in Sect. 11.4.

8.4 Quantum Wells in Electric Fields

For quantum well systems under the influence of an electric field perpendicular to the layers, field ionization, which governs the optical properties in bulk semiconductors, is strongly suppressed by the potential barriers. Correspondingly, the electroabsorption is completely different from the bulk case. Instead of strong broadening and almost negligible shifts, one obtains substantial shifts much larger than the binding energy, which are accompanied

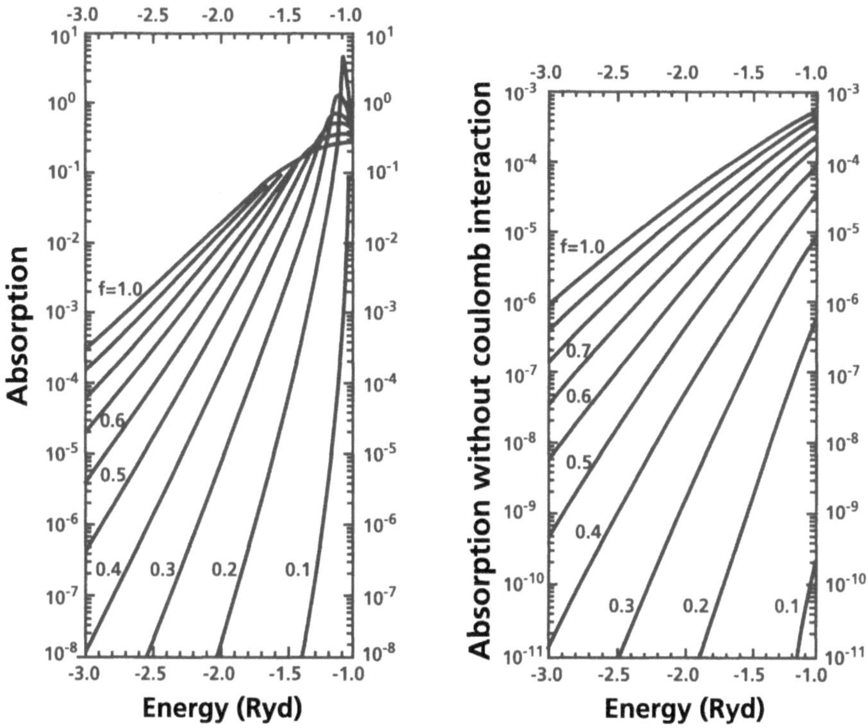

Fig. 8.2. Exponential absorption tails of excitons in various electric fields. Taken from [174]

by little broadening. An experimental example of such changes in the optical absorption spectrum is shown in Fig. 8.3, where the electric field was realized by the use of a reverse-biased p–i–n diode in which the quantum wells were embedded in the intrinsic (i) region. Notice that shifts as large as 2.5 times the zero-field exciton binding energy occur at 50 times the classical ionization field.

Intuitively, the reason for the *quantum-confined Stark effect* is obvious. Electrons and holes are pulled to opposite sides of the well. Thus the exciton is strongly polarized but remains a stable bound state as field ionization is prevented by the walls. To see this effect, however, it is necessary that the well thickness is much smaller than the excitonic Bohr radius, as otherwise the ionization process is no longer suppressed. For the same reason, the barriers need to be sufficiently high and thick to prevent tunneling through the barrier. If tunneling is strictly suppressed, there is no field-induced broadening and the change in the spectrum is determined by the shift. The position of the exciton line results from two contributions: the shift of the one-particle energies and the change in the excitonic binding energy. The influence of the field on the binding energy, however, is rather weak, in particular for narrow

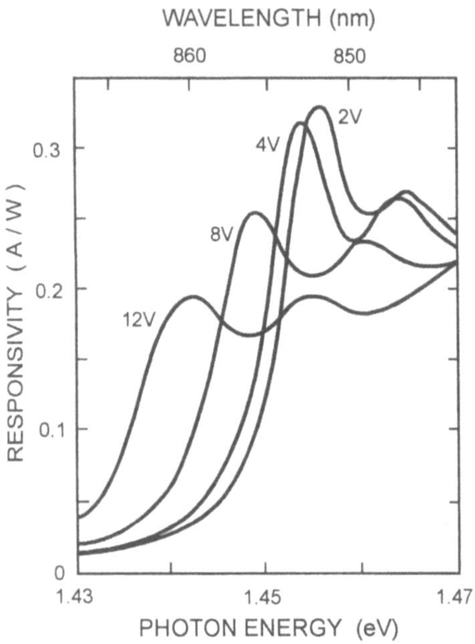

Fig. 8.3. Quantum confined Stark effect as observed in a p–i–n diode containing GaAs/AlGaAs multiple quantum wells in the intrinsic region. The responsivity (the measured photocurrent divided by the incident optical power) is approximately proportional to the optical absorption coefficient. The reverse-bias voltage is the parameter. Taken from [181]

wells. Thus, the first step in analyzing the quantum-confined Stark effect is again the calculation of the one-particle states.

Compared with the bulk case, the only difference is the confinement potential $U_e(z)$, which we have to add in (8.2). The solutions are again Airy functions. For simplicity, we consider a single quantum well in the range $-L_z/2 \leq z \leq L_z/2$ with infinite potential barriers in the following. Thus, the solution of (8.2) obeys the boundary conditions $\phi(-L_z/2) = \phi_n(L_z/2) = 0$ and takes the form

$$\phi(z) = a\,\mathrm{Ai}(\zeta) + b\,\mathrm{Bi}(\zeta)\,, \tag{8.48}$$

where

$$\zeta(z) = -\frac{z}{l_e} + \frac{E}{eFl_e} + l_e^2 k^2\,. \tag{8.49}$$

Inserting the boundary conditions, we obtain two coupled equations, which have nontrivial solutions if

$$\mathrm{Ai}\left[\zeta\left(\frac{L_z}{2}\right)\right]\mathrm{Bi}\left[\zeta\left(-\frac{L_z}{2}\right)\right] - \mathrm{Bi}\left[\zeta\left(\frac{L_z}{2}\right)\right]\mathrm{Ai}\left[\zeta\left(-\frac{L_z}{2}\right)\right] = 0 \tag{8.50}$$

is fulfilled. The solutions of (8.50) determine a set of discrete energy levels E_n. The corresponding expansion coefficients are given by

$$a_n\,\mathrm{Ai}\left[\zeta_n\left(\frac{L_z}{2}\right)\right] = -b_n\,\mathrm{Bi}\left[\zeta_n\left(\frac{L_z}{2}\right)\right], \tag{8.51}$$

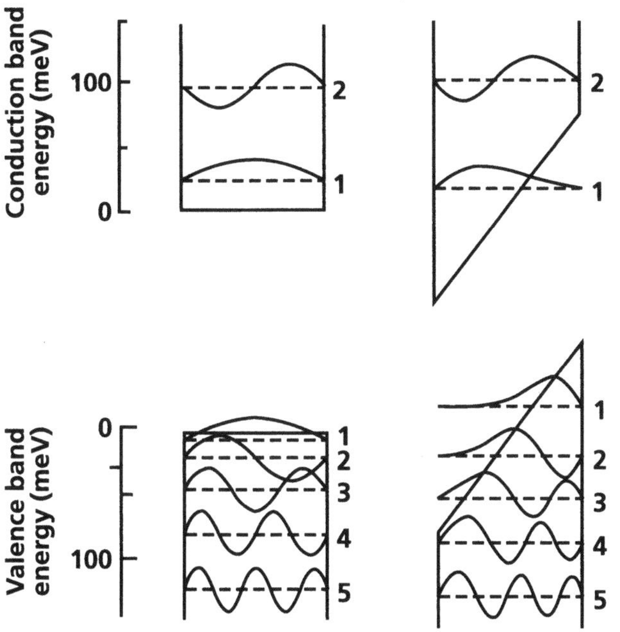

Fig. 8.4. Conduction and valence band energy levels and the corresponding wavefunctions at 0 and 10^5 V/cm in a 15 nm GaAs slab between infinite potential barriers. Taken from [181]

and the normalization condition on the wavefunction $\phi_n(z)$. Figure 8.4 shows solutions at zero electric field and at 10^5V/cm, with material parameters corresponding to electrons and heavy holes in a GaAs slab [181].

Owing to their lower confinement energy, heavier particles such as electrons in the valence bands are more easily perturbed by the electric field than are conduction band electrons. For the lowest transition, the shift is about 25 meV. In comparable systems, the shift of the excitonic resonance found experimentally is of the same order of magnitude. As expected, the shift of the one-particle energies accounts for the lion's share of the redshift in the absorption spectrum. This is confirmed by calculations of the binding energy.

Of course, one would like to expand (8.26), with the confinement potential $U_e(z) + U_h(z)$ added, in terms of the one-particle states (8.48). In principle, this approach is practicable, but it is rather involved. In order to model real systems, the minimum we have to do is to account for the finite height of the potential barriers, in order to allow for tunneling and, thus, for field ionization. In this case, the expansion in terms of one-particle states becomes unsuitable. The change of the binding energy with the applied field has been studied by variational techniques [182, 183, 184]. The results of these studies show the in-plane kinetic energy and the Coulomb interaction decrease, because electrons and holes are pulled apart by the field. The net effect obtained is a nearly insignificant decrease of the binding energy. In a GaAs quantum well of 15 nm width, the reduction at 10^5 V/cm is about 2.5 meV, i.e. 10% of the one-particle shift [184].

The quantum-confined Stark effect opens the door to an interesting modern device, the self-electrooptic-effect device (SEED). Assume that we irradiate the p–i–n diode with monochromatic light which is roughly resonant with the exciton line at zero applied field (see Fig. 8.5). Assume further that this diode is connected to a voltage source with a bias voltage U_0 via an additional ohmic series resistance R. For low incident optical power, most of the voltage drop occurs across the diode because the photocurrent is low. This voltage shifts the exciton absorption to lower photon energies; hence the absorption is low (point A in Fig. 8.5) and the transmission large. Increasing the incident optical power increases the photocurrent, which reduces the voltage drop across the diode. As a consequence of this, the exciton line shifts back to larger photon energy and the absorption coefficient rises, which in turn increases the photocurrent even more, and so forth. This positive feedback mechanism can lead to optical bistability and to switching into a strongly absorbing state (point B in Fig. 8.5) if the exciton absorption line remains sufficiently narrow under the influence of the electric field. Such SEED operation was first demonstrated in [185, 186].

8.5 Superlattices in Electric Fields

When the potential drop across one lattice constant becomes larger than the width of a band, the effective-mass approximation, which we have applied up to now, breaks down. This condition is hardly possible to meet in bulk semiconductors and structures consisting of only a few quantum wells. In a structure with a large number of wells which are strongly coupled owing to the overlap of neighboring wavefunctions of individual wells, however, minibands are built up. Their bandwidth ΔE can be tailored in the range of a few tens of meV. Thus the condition $eFa \geq \Delta E$, where a is the width of one period of the superlattice, can be realized with moderate fields. Ultimately, it is this property which gives rise to the most striking and novel electrooptic effects in superlattices. Qualitatively, many features can be understood without considering Coulomb correlation. Thus, we again start our considerations with the one-particle properties.

8.5.1 One-Particle States in Superlattices

In a superlattice consisting of $2N + 1$ wells, the one-particle equation (8.2) has to be replaced by

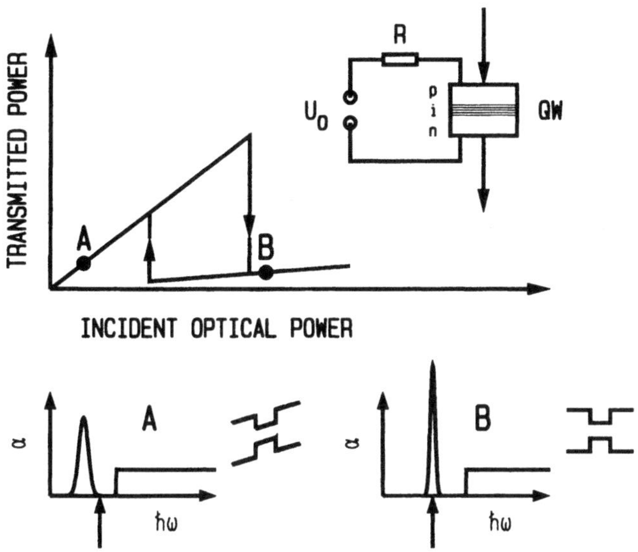

Fig. 8.5. Schematic illustration of an optically bistable self-electrooptic-effect device (SEED). At point A, the voltage across the p–i–n diode is large; at point B it is small. The *arrows* in the absorption spectra in the *lower part* of the figure indicate the operating photon energy

$$\left(-\frac{\hbar^2}{2m_\mathrm{e}}\frac{\partial^2}{\partial z^2} - eFz + \sum_{l=-N}^{N} U_\mathrm{e}(z-al)\right)\phi_n^\mathrm{e}(z) = \left(E_n^\mathrm{e} - \frac{\hbar^2 k^2}{2m_\mathrm{e}}\right)\phi_n^\mathrm{e}(z)\,. \tag{8.52}$$

As the superlattice potential consists of a series of rectangular wells and barriers, the exact solution could again be constructed by means of Airy functions. The evaluation of the boundary conditions, however, becomes extremely complicated. Therefore, we use in the following a different expansion scheme, which is particularly useful for intermediate and high fields. The potential $U_\mathrm{e}(z)$ is defined as

$$U_\mathrm{e}(z) = \begin{cases} U_\mathrm{e}^\mathrm{t}, & -\frac{a}{2} \leq z \leq -\frac{L_z}{2} \\ U_\mathrm{e}^\mathrm{b}, & -\frac{L_z}{2} \leq z \leq \frac{L_z}{2} \\ U_\mathrm{e}^\mathrm{t}, & \frac{L_z}{2} \leq z \leq \frac{a}{2} \end{cases}\,. \tag{8.53}$$

Additionally, we introduce an auxiliary potential

$$\tilde{U}_{\mathrm{e}}(z) = \begin{cases} U_{\mathrm{e}}^{\mathrm{t}}, & -\infty \leq z \leq -\dfrac{L_z}{2} \\ U_{\mathrm{e}}^{\mathrm{b}}, & -\dfrac{L_z}{2} \leq z \leq \dfrac{L_z}{2} \\ U_{\mathrm{e}}^{\mathrm{t}}, & \dfrac{L_z}{2} \leq z \leq \infty \end{cases} . \tag{8.54}$$

In the interval $-a/2 \leq z \leq a/2$, the two potentials are identical. We make use of (8.54) to introduce a complete set of orthogonal eigenfunctions and eigenenergies according to

$$\left(-\frac{\hbar^2}{2m_{\mathrm{e}}} \frac{\partial^2}{\partial z^2} + \tilde{U}_{\mathrm{e}}(z) \right) \varphi_j^{\mathrm{e}}(z) = E_j^{0\,\mathrm{e}} \varphi_j^{\mathrm{e}}(z) . \tag{8.55}$$

The expansion of the one-particle states by means of these functions takes the form

$$\phi_n^{\mathrm{e}}(z) = \sum_j \sum_{l=-N}^{N} c_{n,\,jl}^{\mathrm{e}} \varphi_j^{\mathrm{e}}(z - la) . \tag{8.56}$$

Together with (8.52) and (8.55), this representation yields the algebraic equation

$$\sum_{jl} \left[\left(E_j^{0\,\mathrm{e}} - E_n^{\mathrm{e}} \right) S_{im,\,jl}^{\mathrm{e}} + U_{im,jl}^{\mathrm{e}} + eFD_{im,jl}^{\mathrm{e}} \right] c_{n,\,jl}^{\mathrm{e}} = 0 . \tag{8.57}$$

Here we have introduced the following definitions:

$$S_{im,\,jl}^{\mathrm{e}} = \begin{cases} \delta_{ij} & l = m \\ s_{im,\,jl} & l \neq m \end{cases} , \tag{8.58}$$

where the overlap matrix is

$$s_{im,\,jl} = \int_{-\infty}^{\infty} \mathrm{d}z\, \varphi_i^{\mathrm{e}\,*}(z - am)\, \varphi_j^{\mathrm{e}}(z - al) . \tag{8.59}$$

This results from the nonorthogonality of the individual well wavefunctions at different sites. The matrix elements of the potential are given by

$$U_{im,\,jl}^{\mathrm{e}} = \sum_{l' \neq l} \int_{-\infty}^{\infty} \mathrm{d}z\, \varphi_i^{\mathrm{e}\,*}(z - am)\, U_{\mathrm{e}}(z - al')\, \varphi_j^{\mathrm{e}}(z - al) . \tag{8.60}$$

Note that we have already used the $l = l'$ contribution to obtain the eigenenergies $E_j^{0\,\mathrm{e}}$. Finally, we have the transition matrix element

$$D_{im,\,jl}^{\mathrm{e}} = \begin{cases} al\delta_{ij} + d_{ij}^{(0)} & l = m \\ ams_{im,\,jl} + d_{ij}^{(l-m)} & l \neq m \end{cases} , \tag{8.61}$$

where

$$d_{ij}^{(l)} = \int_{-\infty}^{\infty} \mathrm{d}z\, \varphi_i^{\mathrm{e}\,*}(z-am)\, z\, \varphi_j^{\mathrm{e}}(z-al)\,. \tag{8.62}$$

This set of equations can easily be solved numerically. Analytical one-particle states, however, are more appropriate for including interaction effects. Therefore, in (8.57) we separate the dominant contributions, which can be solved analytically. The remaining perturbations are easily taken into account when we expand the semiconductor Bloch equations and the application of numerical methods becomes unavoidable. Let us first discuss the physical conditions under which this procedure is meaningful.

We consider wells sufficiently narrow that the spacing of the energy levels is large enough to restrict ourselves to few low-lying electron and hole levels. Furthermore, the coupling between neighboring wells should be small enough to allow a restriction to next-nearest-neighbor integrals. If these conditions are satisfied, (8.58)–(8.61) can be approximated by

$$S_{im,\,jl}^{\mathrm{e}} = \delta_{ij}\,\delta_{ml}\,, \tag{8.63}$$

$$U_{im,\,jl}^{\mathrm{e}} = T_{ii}^{\mathrm{e}}\,\delta_{ij}(\delta_{m+1,\,l} + \delta_{m-1,\,l})\,, \tag{8.64}$$

where the transfer integral is

$$T_{ii}^{\mathrm{e}} = \int_{-\infty}^{\infty} \mathrm{d}z\, \varphi_i^{\mathrm{e}\,*}(z)\, U_{\mathrm{e}}(z-a)\, \varphi_i^{\mathrm{e}}(z-a)\,. \tag{8.65}$$

In addition

$$D_{im,\,jl}^{\mathrm{e}} = am\delta_{ij}\delta_{ml}\,. \tag{8.66}$$

Making use of these approximations in (8.57), we obtain

$$\left(E_i^{0\,\mathrm{e}} - E_{n,i}^{\mathrm{e}} + eFam\right) c_{n,i,m}^{\mathrm{e}} + T_{ii}^{\mathrm{e}}\left(c_{n,i,m+1}^{\mathrm{e}} + c_{n,i,m-1}^{\mathrm{e}}\right) = 0\,. \tag{8.67}$$

All other couplings between different energy levels and between more distant neighbors, as well as contributions proportional to the nearest-neighbor dipole and overlap matrix elements, can be treated as minor perturbations. The latter become especially small for high fields. Provided that boundary effects are negligible, we can map (8.67) for any quantum well level onto the recurrence relation for Bessel functions

$$2n\mathcal{J}_n(x) = x\Big[\mathcal{J}_{n+1}(x) + \mathcal{J}_{n-1}(x)\Big]\,. \tag{8.68}$$

This has the spectrum

$$E_{n,i}^{\mathrm{e}} = E_i^{0\,\mathrm{e}} + eFan\,, \tag{8.69}$$

and the eigenvectors are given by

$$c_{n,i,m}^{\mathrm{e}} = \mathcal{J}_{n-m}\left(\frac{2\,T_{ii}^{\mathrm{e}}}{eFa}\right) \equiv \mathcal{J}_{n-m}(x_i^{\mathrm{e}})\,. \tag{8.70}$$

By means of (8.56), the eigenstates are obtained as

$$\phi_n^{\mathrm{e}}(z) = \sum_j \sum_{m=-N}^{N} \mathcal{J}_{n-m}(x_j^{\mathrm{e}})\, \varphi_j^{\mathrm{e}}(z-am)\,. \tag{8.71}$$

The evenly spaced levels of the spectrum are referred to as the *Wannier–Stark ladder*; they differ from the exact solutions by edge effects only. In a finite superlattice, these boundary effects can be safely ignored for most states if the energy of high-lying states is large compared with the transfer integral. Thus, the condition $eFaN \gg T_{ii}^{\mathrm{e}}$ must be fulfilled, which requires either a high field or a thick superlattice. From (8.71), we see also formally that the Wannier–Stark spectrum is accompanied by a strong localization of the ladder states, which is an immediate consequence of the properties of Bessel functions for small arguments, i.e.

$$\mathcal{J}_n(x) = \frac{1}{|n|!}\, x^{|n|}\,. \tag{8.72}$$

Together with (8.56), this indicates that the admixture of wavefunctions of neighboring wells $l \neq n$ in the nth well decreases rapidly with increasing field. Physically, this behavior is not surprising at all. For vanishing field, the states of different wells are coupled by resonant tunneling. With increasing field, the resonance condition is switched off, owing to the energy difference between neighboring wells, which is just given by eFa. This results in a drastic decrease of tunneling rates and a decreasing coupling of the wells at high fields. Consequently, we see that, in this limit, the conditions required by our approximation scheme are met even at moderate fields of the order of a few tens of kV/cm. This is an immediate consequence of the period of the superlattice, which is typically a factor of 30 larger than the lattice constant in a bulk semiconductor. For thick superlattices it is useful to introduce reciprocal space also. If we consider the case $F = 0$ in (8.67), we have to search for plane-wave solutions; these result in the spectrum

$$E_{n,i}^{\mathrm{e}} = E_0^{i\,\mathrm{e}} - 2\,T_{ii}^{\mathrm{e}} \cos(k_n a)\,, \tag{8.73}$$

where $k_n = n\pi/[a(2N+1)]$ and $-N \leq n \leq N$, and the eigenvectors are

$$c_{n,i,m}^{\mathrm{e}} = \frac{1}{\sqrt{N+1}} \begin{cases} \sin(k_n ma), & \text{if m even} \\ \cos(k_n ma), & \text{if m odd} \end{cases}\,. \tag{8.74}$$

Owing to the resonant coupling of the wells, the lowest quantum well states evolve into *minibands* of widths $4\,T_{00}^{\mathrm{e}}$ and $4\,T_{00}^{\mathrm{h}}$ for electrons and holes, respectively. Because of the large lattice constant, however, the Brillouin zone is much smaller than in the bulk case. To illustrate the magnitude of the miniband width, we consider a GaAs/AlGaAs superlattice with an aluminum content of 0.3. Linear interpolation between the GaAs ($E_{\mathrm{g}} = 1.52\,\mathrm{eV}$) and AlAs ($E_{\mathrm{g}} = 3.14\,\mathrm{eV}$) band gaps yields a gap of 2.01 eV, which leads to shifts of the conduction and valence bands of 65% and 35%, respectively, of this amount. Thus, we have barrier heights $U^{\mathrm{e}} = 316\,\mathrm{meV}$ for electrons and $U^{\mathrm{h}} = 170\,\mathrm{meV}$ for holes. The effective masses for the barrier material can also be obtained by linear interpolation (see Table 3.2). With the solution of (8.55), we can calculate the transfer integrals, which determine, within the approximation (8.73), the width of the minibands. Results for various well widths and a barrier width of 2 nm are summarized in Table 8.1. With in-

Table 8.1. Miniband width of electron and hole minibands in units of meV for a GaAs/AlGaAs superlattice for various well widths.

Well width	4 nm	5 nm	6 nm	7 nm	8 nm	9 nm	10 nm
Electron	80.38	47.76	29.25	18.48	12.02	8.01	5.46
Heavy hole	31.32	14.93	7.68	4.15	2.34	1.37	0.82
Light hole	76.43	43.90	25.92	15.87	10.04	6.50	4.32

creasing well width, the transfer integrals decrease rapidly for both electrons and holes. Owing to the larger mass of holes, the width of the hole band is much smaller. We should, however, point out that the transfer integral is only an approximate measure of the miniband width, as (8.67) is valid for high fields only. In particular, for wide wells (≥ 9 nm) the band width is underestimated by nearly a factor of two. This results mainly from the admixture of higher quantum well states, which grows with the well width owing to the decreasing energetic spacing of the subbands. Thus, in general, a solution of (8.57) is required for zero field. In any case, in this regime of parameters, the miniband width is small enough to allow the formation of localized states at moderate fields, but also larger than the broadening due to scattering and tunneling processes. For this reason, Wannier–Stark ladders become observable even in linear optical spectra [167]. In the time domain, the electrons can execute Bloch oscillations, i.e. a periodic motion in $\boldsymbol{k}$ space with period $T_{\text{Bloch}} = 2\pi\hbar/eFa$. In agreement with the above conditions, this time has to be sufficiently small compared with typical timescales of scattering and tunneling. Otherwise, any oscillation is suppressed and the ladder structure is washed out. The close relation between Bloch oscillations and Wannier–Stark states will be discussed in more detail in the following example.

Example (Bloch Oscillations). We consider an electron in a lattice periodic potential $V(z)$ under the action of a uniform electric field. The Hamiltonian is given by

$$H = -\frac{\hbar^2}{2m_\text{e}}\frac{\partial^2}{\partial z^2} + V(z) - eFz\,. \tag{8.75}$$

For $F = 0$, the Hamiltonian is lattice-periodic and Bloch's theorem tells us that the wavefunctions take the form $\psi_n(k,z) = \text{e}^{\text{i}kz}u_n(k,z)$ with eigenvalues $E(k)$. For $F \neq 0$, we consider the time-dependent problem and apply the gauge transformation

$$\bar{\phi}(z,t) = \mathrm{e}^{-\mathrm{i}eFzt/\hbar}\phi(z,t) \tag{8.76}$$

to the wavefunctions. This allows us to transform the time-dependent Schrödinger equation into the representation

$$\mathrm{i}\hbar\frac{\partial}{\partial t}\bar{\phi}(z,t) = \left[-\frac{\hbar^2}{2m_\mathrm{e}}\left(\frac{\partial}{\partial z}+\mathrm{i}\frac{eFt}{\hbar}\right)^2 + V(z)\right]\bar{\phi}(z,t)\,. \tag{8.77}$$

Obviously, the Hamiltonian in (8.76) is now lattice-periodic for $F \neq 0$ also and the wavefunction should again factorize into a plane wave and a lattice-periodic function according to $\bar{\phi}(z,t) = \mathrm{e}^{\mathrm{i}kz}\bar{u}(z,t)$. From (8.77), we find

$$\mathrm{i}\hbar\frac{\partial}{\partial t}\bar{u}(z,t) = \left\{-\frac{\hbar^2}{2m_\mathrm{e}}\left[\frac{\partial}{\partial z}-\mathrm{i}\left(k+\frac{eFt}{\hbar}\right)\right]^2 + V(z)\right\}\bar{u}(z,t)\,. \tag{8.78}$$

The eigenvalues and eigenfunctions of the Hamiltonian on the right-hand side can evidently be obtained from those for $F = 0$ and are given by $E_n(k + eFt/\hbar)\bar{u}(z)$ and $u_n(k + eFt/\hbar, z)$, respectively. We make use of these eigenstates to represent $\bar{u}(z,t)$ according to

$$\bar{u}(z,t) = \sum_m c_m(t)u_m[k(t),z]\,, \tag{8.79}$$

where $k(t) = k + eFt/\hbar$. Using this expansion in (8.78) results in

$$\mathrm{i}\hbar\frac{\partial}{\partial t}c_n(t) - E_n[k(t)]c_n(t) + \sum_m \varphi_{nm}(t)c_m = 0\,, \tag{8.80}$$

where we have introduced

$$\varphi_{nm}(t) = \mathrm{i}\hbar\langle u_n[k(t)]|\frac{\partial}{\partial t}|u_m[k(t)]\rangle\,. \tag{8.81}$$

Taking the lowest subband only into account, (8.80) is easily integrated and yields, after reversal of the gauge transformation, the final result

$$\phi(z,t) = \exp\left(-\mathrm{i}\int_0^t \mathrm{d}t'\{E_0[k(t')] - \varphi_{00}(t')\}\right)\psi_0[k(t),z]\,, \tag{8.82}$$

where the so-called accelerated Bloch states are given by [187]

$$\psi_n[k(t),z] = \exp\{\mathrm{i}[k(t)z]\}\,u_n[k(t),z]\,. \tag{8.83}$$

The states (8.82) are referred to as *Houston states* [188, 189] and describe the dynamics of an electron in a periodic potential under the action of a uniform electric field in the time domain. The critical point in this approach is the neglect of the coupling to higher bands, resulting from (8.81). In performing the derivation, we see that the coupling is linear in the field and, thus, of the same order as the spacing of the Wannier–Stark levels. The magnitude of this coupling can be calculated by means of $\boldsymbol{kp}$ perturbation theory (see (3.87)), which allows us to relate (8.81) to momentum or dipole transition matrix elements. In general, the restriction to a single, isolated band is insufficient, as the interband coupling forces the decay of the Houston states. Whether the Wannier–Stark splitting or the interband coupling dominates, however, is a problem of material properties and requires realistic calculations. From a pragmatic point of view we can claim that in superlattices the lifetime broadening of Wannier–Stark levels due to interband transitions is small compared with the level spacing. Correspondingly, Bloch oscillations are not damped out.

From (8.82) the phenomenon of Bloch oscillations can be understood rather intuitively. As $k(t)$ and $k(t + T_{\text{Bloch}})$ are equivalent points in $\boldsymbol{k}$ space, $E[k(t)]$ and $E[k(t + T_{\text{Bloch}})]$ are the same and the result is a periodic behavior of the wavefunction. A wave packet constructed by superposition of Houston states obviously has the same periodicity. Thus, the electron executes a periodic motion. The same interpretation can also be obtained from the *acceleration theorem* [190], which is obtained by taking the derivative

$$\frac{\partial}{\partial t}k(t) = \frac{eF}{\hbar}\,. \tag{8.84}$$

The wavenumber increases with time until the zone boundary is reached. Then, the electron undergoes a Bragg reflection and the wavenumber evolves, until the other zone boundary is reached, and so forth. These oscillations in $\boldsymbol{k}$ space are accompanied by an oscillatory behavior in real space. The two types of oscillations, allow various experimental tests. Time-resolved optical experiments, such as four-wave mixing are best suited to demonstrating the existence of Wannier–Stark ladders and Bloch oscillations by the observation of quantum beats [168, 169, 196]. Figure 8.6 shows a time-integrated FWM signal for various electric fields. The linear decrease of the quantum-beat period with increasing static electric field shows clearly the existence of Wannier–Stark levels. In this range of field strengths, the decay times depend only weakly on the field. Increasing the field even more, tunneling to higher subbands and to continuum states comes into play. These field-induced interband processes are the pendant to Zener tunneling in bulk semiconductors [191], and lead to a delocalization of Wannier–Stark states. For high fields ($\geq 50\,\text{kV/cm}$) the delocalization is associated with a rapidly decreasing lifetime of these states due to Zener breakdown [192]. The corresponding damping of Bloch oscillations can be observed in time resolved pump-probe experiments and four-wave-mixing signals [193].

In real space the transitions between Wannier–Stark states can be interpreted as oscillation of a wave packet, which gives rise to the emission of terahertz radiation [170, 194]. Even a qualitatively correct description of these intraband transitions, however, requires the consideration of electron–hole correlation [195]. Before taking these excitonic effects into account, we shall first discuss electroabsorption as a further example for the application of the one-particle picture.

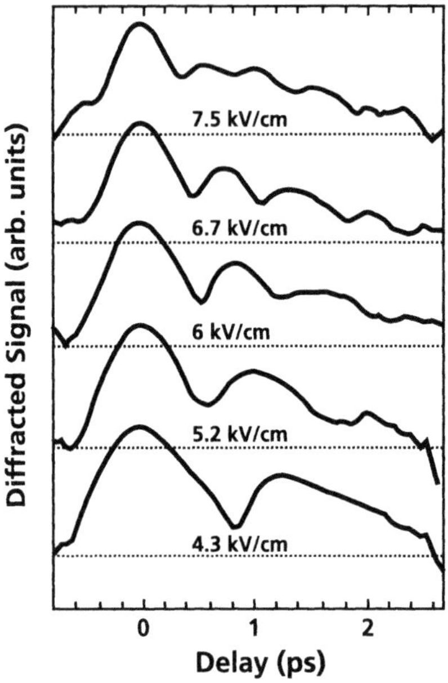

Fig. 8.6. Time-integrated four-wave-mixing signal as a function of the delay time for various electric fields. The sample contains 40 periods of 10 nm thick GaAs and 1.7 nm thick $Al_{0.3}Ga_{0.7}As$ layers. The bandwidth of the laser is 17meV and covers several transitions of the Wannier–Stark ladder. Taken from [169]

Example (Electroabsorption). Let us consider the optical spectrum of a superlattice resulting from uncorrelated transitions. We proceed as in Sect. 8.2.1 and calculate the imaginary part of the susceptibility from the lowest electron and hole quantum well levels, according to

$$\begin{aligned} \mathrm{Im}\Big(\chi^{(1)}(\omega)\Big) &= \frac{\pi}{\varepsilon_0 V} \sum_{nm\boldsymbol{k}} \left| d_{\mathrm{vc}}^0 \int \mathrm{d}z\, \phi_n^{\mathrm{e}\,*}(z) \phi_m^{\mathrm{h}}(z) \right|^2 \\ &\quad \times \delta\left(\tilde{E}_{\mathrm{g}} + \frac{\hbar^2 k^2}{2m_{\mathrm{r}}} + eFa(n-m) - \hbar\omega \right), \end{aligned} \tag{8.85}$$

where $\tilde{E}_{\mathrm{g}} = E_{\mathrm{g}} + E_0^{0\,\mathrm{e}} + E_0^{0\,h}$ and $V = L_x L_y\, a(2N+1)$. Neglecting overlap contributions, we obtain, using (8.71),

$$\begin{aligned} \int \mathrm{d}z\, \phi_n^{\mathrm{e}\,*}(z) \phi_m^{\mathrm{h}}(z) &= \sum_l \mathcal{J}_{n-m+l}(x_0^{\mathrm{e}})\, \mathcal{J}_l(-x_0^{\mathrm{h}}) \\ &= \mathcal{J}_{n-m}(x_0^{\mathrm{e}} + x_0^{\mathrm{h}}), \end{aligned} \tag{8.86}$$

where, in the last step, we have assumed $N \to \infty$ and have made use of Neumann's addition theorem for Bessel functions.[3] The final $\boldsymbol{k}$ integration results in a series of step functions

[3] Note that this is just a pendant to the relation (8.8), which again allows us to relate the one-particle picture to the pair representation.

$$\mathrm{Im}\Big(\chi^{(1)}(\omega)\Big)$$
$$= \frac{|\mathrm{d}^0_{\mathrm{vc}}|^2\, m_r}{2\,\varepsilon_0\,\hbar^2\, a} \sum_n \mathcal{J}_n^2(x_0^{\mathrm{e}} + x_0^{h})\,\theta(\hbar\omega - E_{\mathrm{g}} - E_0^{0\,e} - , E_0^{0\,h} - \mathrm{e}Fan) \tag{8.87}$$

for the optical absorption. The specific features resulting from the electric field become obvious if we compare this result with the limit of vanishing field. Owing to the formation of minibands (compare (8.73)), the onset of the absorption is lower by $2\,(T^{\mathrm{e}}_{0\,0} + T^{\mathrm{h}}_{0\,0})$ than for an isolated well. If the potential drop eFa is large compared with the miniband width, (8.87) shows that we obtain a weak absorption tail followed by an effective absorption edge, which is determined by the $n = 0$ contribution. Thus, with increasing field, this edge is shifted to higher energy and approaches the lowest transition energy of an isolated well. This again indicates the decline of the resonant coupling and the resulting localization of Wannier–Stark states. Indeed, such a blueshift is also observed in experimental absorption spectra [167]. Needless to say that the analysis of real systems require at least the inclusion of the electron–hole interaction.

8.5.2 Semiconductor Bloch Equations

Let us now turn to the expansion of the *semiconductor Bloch equations* in the case of a superlattice. An appropriate basis set is given by the one-particle states

$$\psi_n(\boldsymbol{k}, \boldsymbol{r}) = \frac{1}{\sqrt{L_x L_y}} \exp(\mathrm{i}k_x x + \mathrm{i}k_y y)\,\phi^{\mathrm{e}}_n(z)\,u_{\mathrm{c}}(0, \boldsymbol{r})\,, \tag{8.88}$$

where $\phi^{\mathrm{e}}_n(z)$ is defined in (8.71), and correspondingly for holes. Owing to the nonorthogonality of these wavefunctions on different sites in general, we should start from the real-space representation (5.7) of the density-matrix equations, rather than applying (5.8) directly, as in earlier applications. If we represent the fermion field operators by means of (8.77), corrections accounting for the difference between (8.57) and (8.67) can be incorporated systematically. For the sake of simplicity, however, we neglect the corresponding contributions in the following and restrict ourselves to the high-field approximation. Under these conditions, (8.67) is valid. In this case, the main effort is in the evaluation of the Coulomb matrix element, which is performed in the same way as in Sect. 7.2 (compare (7.45)). Again, we make use of the separability of length scales and obtain, for example, for the electron–hole interaction matrix element:

$$\begin{aligned}
&\langle c\,n\boldsymbol{k}_n, v\,n'\boldsymbol{k}_{n'}|v|c'\,m\boldsymbol{k}_m, v'\,m'\boldsymbol{k}_{m'}\rangle \\
&= e^2 \sum_{\boldsymbol{q}\,q_z} v\Big(\sqrt{|\boldsymbol{q}|^2+q_z^2}\Big)\delta_{c,c'}\,\delta_{v,v'}\,\delta_{\boldsymbol{q},\boldsymbol{k}_n-\boldsymbol{k}_m}\,\delta_{\boldsymbol{k}_n-\boldsymbol{k}_m,\boldsymbol{k}_{n'}-\boldsymbol{k}_{m'}} \\
&\times \int \mathrm{d}z\,\mathrm{d}z'\phi_n^{\mathrm{e}*}(z)\phi_{n'}^{\mathrm{h}*}(z')\exp[\mathrm{i}q_z(z-z')]\phi_m^{\mathrm{e}}(z)\phi_{m'}^{\mathrm{h}}(z')\,. \qquad (8.89)
\end{aligned}$$

Performing the $\boldsymbol{q}$ integration and inserting (8.71) with the lowest quantum well level only yields

$$\begin{aligned}
&\langle c\,n\boldsymbol{k}_n, v\,n'\boldsymbol{k}_{n'}|v|c'\,m\boldsymbol{k}_m, v'\,m'\boldsymbol{k}_{m'}\rangle \\
&= \delta_{c,c'}\,\delta_{v,v'}\,\delta_{\boldsymbol{k}_n-\boldsymbol{k}_m,\boldsymbol{k}_{n'}-\boldsymbol{k}_{m'}}\,\mathcal{I}_{n,n',m,m'}(\boldsymbol{k}_n-\boldsymbol{k}_m)\,, \qquad (8.90)
\end{aligned}$$

where we have introduced the abbreviation

$$\begin{aligned}
&\mathcal{I}^{\mathrm{eheh}}_{n,n',m,m'}(\boldsymbol{q}) \\
&= \sum_{\nu\nu'\mu\mu'} I^{\mathrm{eheh}}_{\nu,\nu',\mu,\mu'}(\boldsymbol{q})\,\mathcal{J}_{n-\nu}(x_0^{\mathrm{e}})\,\mathcal{J}_{n'-\nu'}(-x_0^{\mathrm{h}})\,\mathcal{J}_{m-\mu}(x_0^{\mathrm{e}})\,\mathcal{J}_{m'-\mu'}(-x_0^{\mathrm{h}})\,. \qquad (8.91)
\end{aligned}$$

Here we have defined the interaction matrix element (compare Sect. 6.2.4)

$$\begin{aligned}
&I^{\mathrm{eheh}}_{\nu\nu'\mu\mu'}(\boldsymbol{q}) \\
&= \int \mathrm{d}z\,\mathrm{d}z'\varphi_0^{\mathrm{e}*}(z-\nu a)\,\varphi_0^{\mathrm{h}*}(z'-\nu' a)v(\boldsymbol{q},z-z')\,\varphi_0^{\mathrm{e}}(z-\mu a)\,\varphi_0^{\mathrm{h}}(z'-\mu' a) \qquad (8.92)
\end{aligned}$$

and the interaction potential is given by

$$v(\boldsymbol{q},z-z') = \frac{e^2}{2\varepsilon_0\varepsilon_\infty L_x L_y}\frac{\mathrm{e}^{-|\boldsymbol{q}||z|}}{|\boldsymbol{q}|}\,. \qquad (8.93)$$

Next, we shall consider the case of a thick superlattice in which edge effects are negligible and well indices may be arbitrarily shifted. In this limit, we have already seen, for uncorrelated optical transitions (see (8.73) and (8.74)), that the transition matrix elements and, correspondingly, the polarization depend on the difference between the ladder indices only. Thus, the Coulomb matrix-element (8.79) couples to a transition amplitude with ladder index $m-m'$. Anticipating this coupling, we substitute $m \to m+m'$. With a few further shifts of indices, we can transform (8.80) into

$$\begin{aligned}
\mathcal{I}^{\mathrm{eheh}}_{n,n',m+m',m'}(\boldsymbol{q}) &= \sum_{\nu\nu'\mu\mu'} I^{\mathrm{eheh}}_{\nu,\nu'+\mu,0,\mu}(\boldsymbol{q})\,\mathcal{J}_{n-n'-\nu+\nu'+\mu+m'}(x_0^{\mathrm{e}})\mathcal{J}_{m'}(-x_0^{\mathrm{h}}) \\
&\times \mathcal{J}_{m+\mu+\mu'}(x_0^{\mathrm{e}})\,\mathcal{J}_{\mu'}(-x_0^{\mathrm{h}})\,. \qquad (8.94)
\end{aligned}$$

Finally, performing the m' and μ' summation, we can again make use of Neumann's addition theorem. For the electron–hole interaction which enters the semiconductor Bloch equations, we find the result

$$
\begin{aligned}
V(nm,\boldsymbol{q}) &= \sum_{m'} \mathcal{I}^{\text{eheh}}_{n+n',n',m+m',m'}(\boldsymbol{q}) \\
&= \sum_{\nu\nu'\mu} I^{\text{eheh}}_{\nu,\nu'+\mu,0,\mu}(\boldsymbol{q})\, \mathcal{J}_{n-\nu+\nu'+\mu}(x_0^{\text{e}}+x_0^{\text{h}})\mathcal{J}_{m+\mu}(x_0^{\text{e}}+x_0^{\text{h}})\,.
\end{aligned}
\tag{8.95}
$$

As expected, this result depends on two ladder indices only. The remaining part of the derivation is straightforward and is essentially the same as in the magnetic-field case (Sect. 7.2). We obtain the following final result for the transition amplitude:

$$
\begin{aligned}
&\left(\hbar\frac{\partial}{\partial t} + \mathrm{i}\,E(n,\boldsymbol{k}) + \hbar\gamma_2\right)p_{\text{vc}}(n,\boldsymbol{k}) \\
&\qquad = \mathrm{i}\,[1 - f_{\text{e}}(n,\boldsymbol{k}) - f_{\text{h}}(n,\boldsymbol{k})] \\
&\qquad \times \left(\boldsymbol{d}_{\text{vc}}(n)\boldsymbol{E} + \sum_{n',\boldsymbol{k}'} V(nn',\boldsymbol{k}-\boldsymbol{k}')\,p_{\text{vc}}(n',\boldsymbol{k}')\right),
\end{aligned}
\tag{8.96}
$$

where the renormalized transition energies are

$$
\begin{aligned}
E(n,\boldsymbol{k}) &= E_0^{0\,\text{e}} + E_0^{0\,\text{h}} + \frac{\hbar\boldsymbol{k}^2}{2m_{\text{r}}} + eFan \\
&\quad - \sum_{n',\boldsymbol{k}'} V(nn',\boldsymbol{k}-\boldsymbol{k}')[f_{\text{e}}(n',\boldsymbol{k}') + f_{\text{h}}(n',\boldsymbol{k}')]
\end{aligned}
\tag{8.97}
$$

and the transition matrix elements (compare (8.86))

$$
\boldsymbol{d}_{\text{vc}}(n) = \boldsymbol{d}^0_{\text{vc}}\,\mathcal{J}_n(x_0^{\text{e}} + x_0^{h})\,. \tag{8.98}
$$

The electron distribution function is determined by

$$
\begin{aligned}
&\hbar\frac{\partial}{\partial t} f_{\text{e}}(n,\boldsymbol{k}) + \hbar\gamma_1[f_{\text{e}}(n,\boldsymbol{k}) - f_{\text{e}}^0(n,\boldsymbol{k})] \\
&= -2\,\mathrm{Im}\sum_{\text{v}}\left(\boldsymbol{d}_{\text{cv}}(n)\boldsymbol{E} + \sum_{n'} V(nn',\boldsymbol{k}-\boldsymbol{k}')p_{\text{vc}}(n'\boldsymbol{k}')\right)p^*_{\text{vc}}(n\boldsymbol{k})\,.
\end{aligned}
\tag{8.99}
$$

The result for the holes is analogous.

By applying similar methods to those discussed in Sect. 6.2.3, the numerical solution of these equations can be performed straightforwardly. Linear electroabsorption spectra are shown in Fig. 8.7 for a 12-period GaAs/AlGaAs superlattice. Here we have assumed the same material parameters as in the calculation of the miniband widths and we have chosen a well width of 90 nm. In contrast to the result (8.87) for free transitions, any staircase is replaced by a strong excitonic resonance, which is most pronounced for the $n = 0$

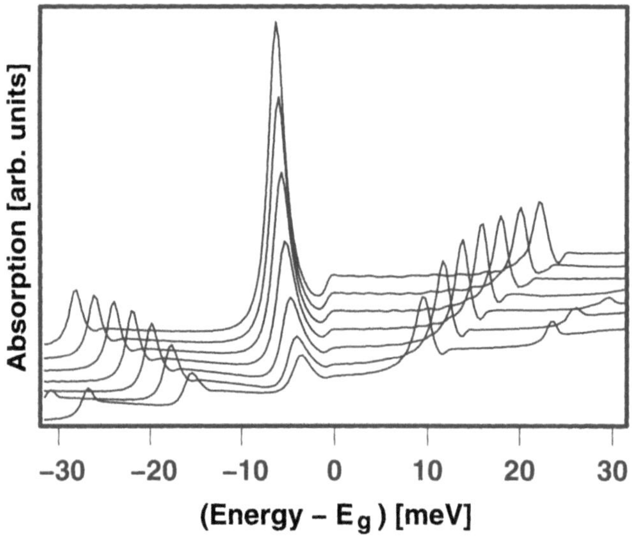

Fig. 8.7. Linear electroabsorption spectra of a 12-period GaAs/AlGaAs superlattice for various field strengths

transition. Since the field determines the localization of electron–hole pairs within the well, the binding energy depends sensitively on the field strength. This dependence leaves its mark also on time-resolved four-wave-mixing signals. As we have seen in Sect. 6.3, even the simple local-field approximation predicts the occurrence of an instantaneous and an *interaction-induced signal*, which is always delayed. Whether this signal is resolved or not depends crucially on the ratio of interaction strength to dephasing rate. One therefore expects that the ratio of the instantaneous to the delayed signal changes sensitively when the field is varied. Indeed, this behavior is obtained from solutions of the semiconductor Bloch equations, which are shown for various fields in Fig. 8.8. In a range from 5.6 to 8.5 kV/cm, the ratio of the instantaneous to the delayed signal varies by more than an order of magnitude. The corresponding experimental signals exhibit exactly the same field dependence [197]. Without the strong excitonic enhancement of the Wannier–Stark states (see Fig. 8.7), the pronounced quantum beats shown in Fig. 8.8 are replaced by an extremely weak modulation. Thus, the distinct signature of Bloch oscillations (see Fig. 8.6) would vanish without electron–hole correlation. Correspondingly, also the emission of terahertz radiation would be orders of magnitude weaker.

This is again one of many examples of the outstanding role of the Coulomb interaction in the optical properties of semiconductors and semiconductor heterostrucures. Surprisingly, a lot of experimental findings can be understood within the Hartree–Fock approximation, which we have applied up to this point – at least qualitatively. There are, however, many features which are

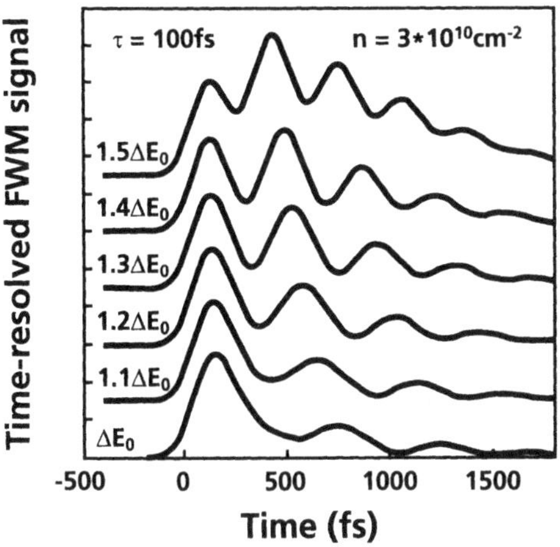

Fig. 8.8. Calculated field dependence of the time-resolved four-wave-mixing signal in a GaAs/GaAlAs superlattice with a well width of 9 nm and a barrier width of 2.5 nm. The potential drop $\Delta E_0 = 6.5$ meV corresponds to a field strength of 5.6 kVcm. Taken from [197]

based on correlation effects beyond this approximation scheme. Such effects and various approaches to their analysis are the subject of the next chapters.

8.6 Problems

8.1. Prove the generalized closure relation for Airy functions (8.8) by means of the representation (8.6).

8.2. Calculate the tunneling probability of the lowest excitonic state for small electric fields, which is given by

$$P = \lim_{z \to -\infty} 2\pi \int \mathrm{d}\rho\, \rho |v_z \psi(\rho, z)|^2 , \tag{8.100}$$

where v_z denotes the velocity. P is the probability current through a plane perpendicular to the field for z far outside the potential barrier. Hint: Make use of the representation of the wave equation in parabolic coordinates and seek a solution of (8.47) in the limit $\eta \gg 1$.

8.3. Apply (8.50) to derive the the Franz–Keldysh effect in a bulk semiconductor. Hint: Consider the limit $L_z \to \infty$ and make use of the asymptotic representation of the Airy functions (8.8)–(8.12).

8.4. Calculate the optical spectrum of a superlattice in the case of vanishing electric field. Hint: Consider the tight-binding limit, i.e. apply (8.73) and (8.74).

8.5. Discuss the motion of an electron in a superlattice under the action of a uniform electric field in real space. Hint: Make use of the dispersion (8.76) and the theorem (3.87).

8.6. Generalize the semiconductor Bloch equations (8.96) and (8.99) to the case in which the coupling to higher quantum well subbands needs to be taken into account.

9. Biexcitons

In close analogy to the hydrogen molecule, formed by the binding of two hydrogen atoms, two excitons of opposite spin can form a bound state, called a *biexciton*. It is clear that the reduction of the energy of two excitons by the additional biexciton binding energy E_{bi} is yet smaller than the exciton binding energy E_R itself. As a rule of thumb, *Hayne's rule* [198] tells us that the biexciton binding energy in three dimensions is 10% of the exciton binding energy. In quantum wells, the experimental results indicate a value of 20%. For two excitons of equal spin, on the other hand, the interaction is effectively repulsive. This behaviour is obtained even within the Hartree–Fock approximation, where the effective exciton–exciton exchange interaction arising from the Pauli exclusion principle leads to a blueshift of the exciton as discussed in Sect. 6.3. The effective force remains repulsive if correlation effects are included.

Biexcitons only show up in nonlinear optical experiments, where two different types of biexciton formation are possible. In the course of an optical excitation, biexcitons can be generated *coherently*. This direct excitation should be distinguished from the formation of *incoherent biexcitons* as a result of the relaxation of hot carriers or incoherent excitons. If, for example, photons generate 1s excitons resonantly, absorption into the biexciton state via a photon of opposite spin becomes possible and leads to *induced absorption* at a photon energy given by the 1s exciton energy minus the biexciton binding energy; see Fig. 9.1. Alternatively, the biexciton can be excited by *off-resonant excitation* by two-photon absorption, that is, via a virtual intermediate state [205, 206, 207]. For two photons of equal energy, this resonance occurs at the exciton energy minus half the biexciton binding energy. If, on the other hand, electron–hole pairs are generated far above the absorption threshold, they cool down by giving their excess energy to the lattice (see Chap. 12). These relaxation processes can be fast, in particular, fast in comparison with the radiative recombination. In that case, formation of excitons and even of biexcitons becomes possible. In the ideal case, at intermediate densities, a dilute gas of biexcitons is the most stable quasi-thermal final state. As a biexciton can decay radiatively into one photon and one 1s exciton, luminescence can also serve as a spectroscopic tool. The different recombination lines can be assigned via their power dependence. While the exciton recombination is

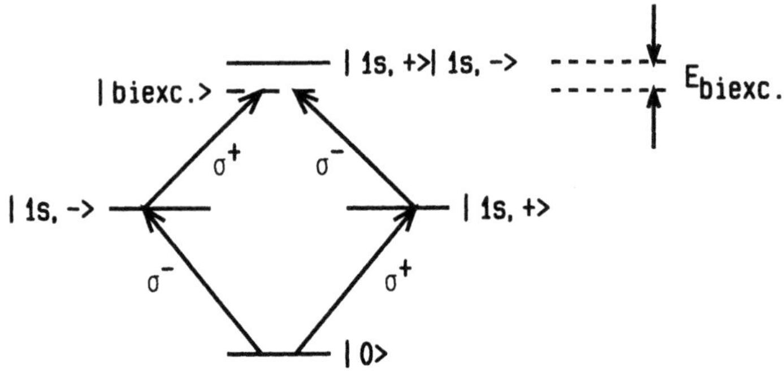

Fig. 9.1. The energies of the ground state $|0\rangle$, the two 1s exciton states $|1s,+\rangle$ and $|1s,-\rangle$ (+ corresponds to spin $+\hbar$, − to spin $-\hbar$), the two-exciton product state $|1s,+\rangle|1s,-\rangle$, and the biexciton state $|\text{biexc}\rangle$ lowered by the biexciton binding energy E_{bi}. The possible optical transitions with light of circular polarizations σ_+ and σ_-, are indicated by the *arrows*

linear in the excitation power, the biexciton luminescence is, ideally, quadratic in the incident fluence. In many experiments, however, the exponent of the biexciton contribution is between one and two.

Spectroscopic evidence for biexcitons was first reported in luminescence experiments on CuCl [199], and later also in a variety of other bulk crystals (see, for example, [200] and references therein). Biexciton effects are comparatively pronounced in CuCl (see Fig. 9.2), as the exciton binding energy is huge, with $E_R = 190\,\text{meV}$ (see Table 6.1). A large amount of work has also been published on semiconductor quantum wells [201]. Many experiments have employed (transient) four-wave-mixing techniques (see Sects. 4.3 and 6.3), which, as far as biexcitons are concerned, were pioneered in [202]. The rich variety of different polarizations of the two incoming beams, as well as of the analyzer for the diffracted beam, allows to perform a clear assignment of biexciton effects. The biexciton binding energy can be determined directly from the inverse of the period of the quantum beating between the exciton energy and half the biexciton energy (see also the discussion on quantum beating in Sect. 6.3.2).

In semiconductor quantum wells, the reported biexciton binding energies show a substantial scatter, which has various origins. From theoretical calculations, one expects a pronounced dependence of the binding energy on confinement [208, 209, 210]. Although the magnitude of this effect depends on the assumptions made in the model, it is quite clear that for a given material system, the binding energy is still a complicated function of the depth and width of the wells and barriers. Additionally, most results are greatly influenced by inhomogeneity of the samples caused by disorder. Such disorder effects can be due to alloy fluctuations or fluctuations in the quantum well width. In this case biexcitons tend to form in local potential minima, which leads to a substantially enhanced binding energy due to the additional, effectively zero-dimensional, confinement.

In thick samples, the effects of light propagation come into play also, similarly to the case of the excitonic resonance. Strictly speaking, the mixed state of light and the biexciton, the biexciton polariton (bipolariton), has to be discussed as a fusion of two exciton polaritons (see Sect. 6.2.5) rather than two excitons [203]. This theoretical approach has been used to explain time-resolved luminescence experiments [204], for example, Also, in addition to these briefly mentioned complications, the simple picture suggested in Fig. 9.1 needs a caveat. As we have seen in Sect. 6.3, exciton–exciton interaction is present even at the Hartree–Fock level. A priori, it is not obvious whether these interaction contributions and others beyond the Hartree–Fock approximation are incorporated into the biexcitonic bound states or not. As we shall see in this chapter, bound biexcitons are indeed not the whole story, and we cannot avoid performing a careful analysis of the various interaction contributions. In the following we shall focus on a systematic treatment of an interacting electron–hole system in terms of powers of the incident light field, rather than on performing any sort of perturbation theory treatment in the Coulomb potential. It is important to note that, besides the immediate applications to biexcitons, such a procedure may also be used to take account of higher correlations in a many-body system in a more general context.

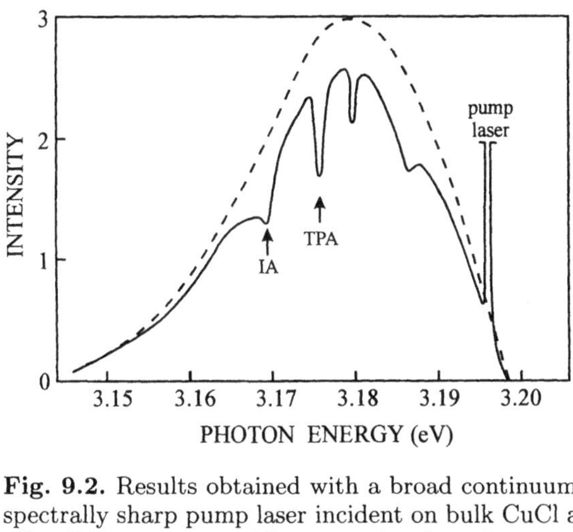

Fig. 9.2. Results obtained with a broad continuum probe together with a strong, spectrally sharp pump laser incident on bulk CuCl at T = 4.2 K. *Dashed line* pump off; *solid line* pump on. The transmitted continuum probe shows the induced absorption (IA) at the biexciton resonance as well as the two-photon absorption (TPA). The exciton resonance itself is not within the energy regime depicted; the other resonances are not attributed to biexcitons. Taken from [200]

9.1 Truncation of the Many-Particle Problem in Coherently Driven Systems

We start again with the Hamiltonian of a many-particle system coupled to an external electromagnetic field (5.1). In a plane-wave representation, we have $H = H_0 + H_{\text{ext}}$ within the effective-mass approximation, where

$$H_0 = \sum_{\boldsymbol{k}e} E_e(\boldsymbol{k})\, c^\dagger_{e,\boldsymbol{k}} c_{e,\boldsymbol{k}} + \sum_{\boldsymbol{k}h} E_h(\boldsymbol{k})\, c^\dagger_{h,-\boldsymbol{k}} c_{h,-\boldsymbol{k}} + \frac{e^2}{2} \sum_{\boldsymbol{k}\boldsymbol{k}'\boldsymbol{q}} v(\boldsymbol{q})$$
$$\times \Bigg(\sum_{ee'} c^\dagger_{e,\boldsymbol{k}+\boldsymbol{q}} c^\dagger_{e',\boldsymbol{k}'-\boldsymbol{q}} c_{e',\boldsymbol{k}'} c_{e,\boldsymbol{k}} + \sum_{hh'} c^\dagger_{h,\boldsymbol{k}+\boldsymbol{q}} c^\dagger_{h',\boldsymbol{k}'-\boldsymbol{q}} c_{h',\boldsymbol{k}'} c_{h,\boldsymbol{k}} \tag{9.1}$$
$$- \sum_{eh} \left(c^\dagger_{e,\boldsymbol{k}+\boldsymbol{q}} c^\dagger_{h,\boldsymbol{k}'-\boldsymbol{q}} c_{h,\boldsymbol{k}'} c_{e,\boldsymbol{k}} + c^\dagger_{h,\boldsymbol{k}-\boldsymbol{q}} c^\dagger_{e,\boldsymbol{k}'+\boldsymbol{q}} c_{e,\boldsymbol{k}'} c_{h,\boldsymbol{k}} \right) \Bigg)$$

and

$$H_{\text{ext}} = \sum_{\boldsymbol{k}eh} \left(\boldsymbol{d}_{eh}\, \boldsymbol{E}\, c^\dagger_{e,\boldsymbol{k}} c^\dagger_{h,-\boldsymbol{k}} + \boldsymbol{d}_{he}\, \boldsymbol{E}\, c_{h,-\boldsymbol{k}} c_{e,\boldsymbol{k}} \right) . \tag{9.2}$$

Here we have already switched to the electron–hole picture, where

$$c_{v,\boldsymbol{k}} = c^\dagger_{h,-\boldsymbol{k}} , \quad c_{c,\boldsymbol{k}} = c_{e,\boldsymbol{k}} , \tag{9.3}$$

and restricted ourselves to the dominant contributions to the electron–hole interaction, in which pairs of electrons and holes are involved. The band indices e, h include the spin quantum number. Let us consider first the equation of motion for the transition amplitude. In contrast to our treatment in Sect. 5.1, we do not use the Hartree–Fock approximation. In this case, we have by means of the Heisenberg equations (5.3) and (5.4)

$$\left(-\mathrm{i}\hbar\frac{\partial}{\partial t} - \mathrm{i}\hbar\gamma_2 - E_e(\boldsymbol{k}) - E_h(\boldsymbol{k}) \right) p_{eh}(\boldsymbol{k}) + e^2 \sum_{\boldsymbol{q}} v(\boldsymbol{q}) p_{eh}(\boldsymbol{k}+\boldsymbol{q})$$
$$= \sum_{h'} \boldsymbol{d}_{h'e} \boldsymbol{E} [\delta_{hh'} - f_{hh'}(\boldsymbol{k})] - \sum_{e'} \boldsymbol{d}_{he'} \boldsymbol{E}\, f_{e'e}(\boldsymbol{k}) + \frac{\partial}{\partial t} p_{eh}(\boldsymbol{k}) \bigg|_{\text{int}} , \tag{9.4}$$

where the interaction contribution is given by

$$\frac{\partial}{\partial t} p_{eh}(\boldsymbol{k}) \bigg|_{\text{int}} = e^2 \sum_{\boldsymbol{k}'\boldsymbol{q}} v(\boldsymbol{q}) \Bigg(\sum_{e'} \Big(\langle c^\dagger_{e,\boldsymbol{k}+\boldsymbol{q}}\, c^\dagger_{h,-\boldsymbol{k}}\, c^\dagger_{e',\boldsymbol{k}'}\, c_{e',\boldsymbol{k}'+\boldsymbol{q}} \rangle$$
$$- \langle c^\dagger_{e,\boldsymbol{k}}\, c^\dagger_{h,-\boldsymbol{k}-\boldsymbol{q}}\, c^\dagger_{e',\boldsymbol{k}'+\boldsymbol{q}}\, c_{e',\boldsymbol{k}'} \rangle \Big) + \sum_{h'} \Big(\langle c^\dagger_{e,\boldsymbol{k}}\, c^\dagger_{h,-\boldsymbol{k}-\boldsymbol{q}}\, c^\dagger_{h',-\boldsymbol{k}'}\, c_{h',-\boldsymbol{k}'-\boldsymbol{q}} \rangle$$
$$- \langle c^\dagger_{e,\boldsymbol{k}+\boldsymbol{q}}\, c^\dagger_{h,-\boldsymbol{k}}\, c^\dagger_{h',-\boldsymbol{k}'-\boldsymbol{q}}\, c_{h',-\boldsymbol{k}'} \rangle \Big) \Bigg) . \tag{9.5}$$

In order to go beyond the Hartree–Fock approximation, we could write down the equations of motion for the four-point functions and introduce a factorization into the next higher order in the resulting hierarchy of the equations of motion. In semiconductors, there may be conditions under which such a truncation procedure can be performed exactly [211, 212, 213, 214]. This procedure is based on the observation that in the ground state of a semiconductor the electron and hole distribution functions vanish and any finite electron or hole distribution is a result of the action of external fields. If there is no loss of memory with respect to the excitation process, all excited particles exist in the form of coherently driven electron–hole pairs. Under these conditions, the exact truncation is obtained by means of perturbation theory with respect to the external fields. In the following we shall discuss this procedure in some detail.

9.1.1 Decomposition of Expectation Values

In the first step we transform all expectation values in (9.5) into a representation in which electron and hole operators occur as pairs only. This is accomplished by making use of particle number conservation. Even in the presence of external fields, the total Hamiltonian commutes with the particle number operator, which, in the electron–hole picture, is given by

$$\hat{N} = \sum_{ie} c^\dagger_{ei} c_{ei} + \sum_{ih} c_{hi} c^\dagger_{hi} = \sum_{ie} c^\dagger_{ei} c_{ei} - \sum_{ih} (c^\dagger_{hi} c_{hi} - 1) , \tag{9.6}$$

where we have used the abbreviation $i \equiv \boldsymbol{k}_i$. Thus we can expand a four-point function according to

$$\begin{aligned} N\langle c^\dagger_{e1} c^\dagger_{h2} c^\dagger_{e3} c_{e4}\rangle &= \langle c^\dagger_{e1} c^\dagger_{h2} c^\dagger_{e3} c_{e4} \hat{N}\rangle \\ &= \sum_{ie'} \langle c^\dagger_{e1} c^\dagger_{h2} c^\dagger_{e3} c_{e4} c^\dagger_{e'i} c_{e'i}\rangle - \sum_{ih'} \langle c^\dagger_{e1} c^\dagger_{h2} c^\dagger_{e3} c_{e4} (c^\dagger_{h'i} c_{h'i} - 1)\rangle . \end{aligned} \tag{9.7}$$

After putting the right-hand side into normal order, this results in

$$\langle c^\dagger_{e1} c^\dagger_{h2} c^\dagger_{e3} c_{e4}\rangle = \sum_{ih'} \langle c^\dagger_{e1} c^\dagger_{h2} c^\dagger_{e3} c^\dagger_{h'i} c_{h'i} c_{e4}\rangle - \sum_{ie'} \langle c^\dagger_{e1} c^\dagger_{h2} c^\dagger_{e3} c^\dagger_{e'i} c_{e'i} c_{e4}\rangle . \tag{9.8}$$

The first term on the right-hand side already has the desired form. The second term can be expanded once more, by iterating the insertion of the particle number operator twice. This expansion leads to one contribution that includes five electron–hole pair operators. The additional contributions can be mapped, by further expansion steps, onto a series of expectation values of an increasing number of electron–hole pair operators. The purpose of this procedure is to obtain a representation which allows a classification with respect to the order of the external field. That this is indeed possible can be immediately seen after transformation into the interaction picture and making use of the time evolution under the action of H_{ext} given by (9.2). For the first contribution in (9.8), we obtain

$$\langle c^\dagger_{e1} c^\dagger_{h2} c^\dagger_{e3} c^\dagger_{h'i} c_{h'i} c_{e4} \rangle = \langle S^\dagger(t,t_0) \bar{c}^\dagger_{e1} \bar{c}^\dagger_{h2} \bar{c}^\dagger_{e3} \bar{c}^\dagger_{h'i} \bar{c}_{h'i} \bar{c}_{e4} S(t,t_0) \rangle \, , \tag{9.9}$$

where the time evolution operator $S(t, t_0)$ results from the iteration of the integral equation (2.150) (see also the next chapter). In order to compensate the action of the electron–hole pair operators $\bar{h}_i \bar{e}_j$ and $\bar{e}^\dagger_i \bar{h}^\dagger_j$, we have to expand S up to at least the first order and $S^\dagger$ up to the second order in H_{ext}. Thus the leading contribution in (9.9) is of third order in the field. In general, the leading order is simply given by the number of electron–hole pair operators. After expansion of the second contribution in (9.8), we find, using the same argument,

$$\begin{aligned} &\langle c^\dagger_{e1} c^\dagger_{h2} c^\dagger_{e3} c^\dagger_{e'i} c_{e'i} c_{e4} \rangle \\ &= \frac{1}{2} \sum_{jkh'h''} \langle c^\dagger_{e1} c^\dagger_{h2} c^\dagger_{e3} c^\dagger_{h'j} c^\dagger_{e'i} c^\dagger_{h''k} c_{h''k} c_{e'i} c_{h'j} c_{e4} \rangle + \mathcal{O}(E^7) \, . \end{aligned} \tag{9.10}$$

Normally ordered expectation values of this type, which consist of a sequence of electron–hole pair annilation operators followed by another sequence of electron–hole pair creation operators, factorize to leading order in the field. This can easily be shown using, again, the expansion of the time evolution operator. We see, for example, from (9.9) that the action of $\bar{c}_{ni} \bar{c}_{e4} \bar{H}_{\text{ext}}$ on the ground state brings us back to the groundstate. Thus, if we insert a complete set of Fock states, only the groundstate contributes and any additional contributions are at least of fifth order in the field. This yields the factorization

$$\langle c^\dagger_{e1} c^\dagger_{h2} c^\dagger_{e3} c^\dagger_{h'i} c_{h'i} c_{e4} \rangle = \langle c^\dagger_{e1} c^\dagger_{h2} c^\dagger_{e3} c^\dagger_{h'i} \rangle \langle c_{h'i} c_{e4} \rangle + \mathcal{O}(E^5) \, . \tag{9.11}$$

In general, we have to leading order,

$$\begin{aligned} &\langle c^\dagger_{e1} c^\dagger_{h1'} c^\dagger_{e2} c^\dagger_{h2'} \dots c^\dagger_{en} c^\dagger_{hn'} c_{e\bar{1}} c_{h\bar{1}'} c_{e\bar{2}} c_{h\bar{2}'} \dots c_{em} c_{hm'} \rangle \\ &= \langle c^\dagger_{e1} c^\dagger_{h1'} c^\dagger_{e2} c^\dagger_{h2'} \dots c^\dagger_{en} c^\dagger_{hn'} \rangle \langle c_{e\bar{1}} c_{h\bar{1}'} c_{e\bar{2}} c_{h\bar{2}'} \dots c_{em} c_{hm'} \rangle + \mathcal{O}(E^{n+m+2}) \, . \end{aligned} \tag{9.12}$$

Beyond the leading order in the field, the factorization of expressions such as that in (9.11) is more elaborate. As an example, we obtain the following up to fifth order, together with the leading-order factorization of (9.10):

$$\begin{aligned} &\langle c^\dagger_{e1} c^\dagger_{h2} c^\dagger_{e3} c_{e4} \rangle = \sum_{ih'} \langle c^\dagger_{e1} c^\dagger_{h2} c^\dagger_{e3} c^\dagger_{h'i} \rangle \langle c_{h'i} c_{e4} \rangle \\ &- \sum_{ijke'h'h''} \Big[\Big(\langle c_{h'j} c_{e'i} c_{h''k} c_{e4} \rangle - \langle c_{h'j} c_{e'i} \rangle \langle c_{h''k} c_{e4} \rangle \Big) \\ &\times \langle c^\dagger_{e1} c^\dagger_{h2} c^\dagger_{e3} c^\dagger_{h'j} \rangle \langle c^\dagger_{e'i} c^\dagger_{h''k} \rangle + \Big(\langle c_{h'j} c_{e'i} \rangle \langle c_{h''k} c_{e4} \rangle - \frac{1}{2} \langle c_{h'j} c_{e'i} c_{h''k} c_{e4} \rangle \Big) \\ &\times \langle c^\dagger_{e1} c^\dagger_{h2} c^\dagger_{e3} c^\dagger_{h'j} c^\dagger_{e'i} c^\dagger_{h''k} \rangle \Big] + \mathcal{O}(E^7) \, . \end{aligned} \tag{9.13}$$

To any given order in the field, we could now write down the equations of motion for the higher-order expectation values and truncate these equations to the same order in the field. This leads to closed systems of equations, without further approximations. The resulting problem is still a challenge, as even to third order in the field, we have to solve the quantum mechanical four-particle problem. To fifth order, as can be seen from (9.13), we have to deal with the six-particle problem and so forth. Therefore an expansion to arbitrary order is not really helpful. Applications of this procedure have been restricted to the consideration of four-particle correlations only [216, 217, 218, 219, 220] or to a treatment of six-particle correlations in terms of products of two-particle and four-particle correlation functions [221, 222]. Before we continue to discuss the problem within the $\chi^{(3)}$ approximation in some detail, we should pause for a moment and emphasize the conditions under which our considerations are valid.

The expansion scheme is rather general and requires nothing but particle number conservation. In contrast, the classification of expectation values with respect to the power of the field, and the factorization rely on the assumption of coherence. Up to this point, we have not built in any microscopic mechanism, which is responsible for the loss of coherence, but have introduced the dephasing constant, for example in (9.4), by hand. In principle, any process which gives rise to dephasing in the equations of motion has its origin in the Hamiltonian. If we take such contributions, such as the coupling to lattice vibrations (see Chap. 11), seriously, the above arguments are no longer valid. Also, without going into formal details, it is obvious that dephasing processes which destroy the coherence of electron–hole pairs are simultaneously the source of incoherent contributions to the one-particle and pair distribution functions. This can already be seen on the phenomenological level.

Considering the Bloch vector in the presence of dephasing, we have, at low densities, instead of the conservation law, the relation

$$f_{ee'}(\boldsymbol{k},t) = f^{\mathrm{c}}_{ee'}(\boldsymbol{k},t) + (\gamma_1 - 2\gamma_2)\int_{-\infty}^{t} \mathrm{d}t'\, f^{\mathrm{c}}_{ee'}(\boldsymbol{k},t') \,, \tag{9.14}$$

where the coherent contribution is

$$f^{\mathrm{c}}_{ee'}(\boldsymbol{k},t) = \sum_{h} p_{eh}(\boldsymbol{k},t)p^{*}_{e'h}(\boldsymbol{k},t) \,, \tag{9.15}$$

and there is also an incoherent contribution to the one-particle distribution function. The incoherent part vanishes only if $\gamma_1 = \gamma_2 = 0$, namely, if there is no dephasing at all. The other formal possibility, $\gamma_1 = 2\gamma_2$, is never met with in real systems. Strictly speaking, the above considerations are rigorously valid only in the coherent limit, i.e. when there is no dephasing at all. Otherwise, any incoherent contribution to the distribution functions prevents an exact factorization. But even if dephasing is present, we can consider a timescale on which the incoherent contributions are small in comparison with the dominating coherent ones. In this case, which is realized under the experimental conditions of four-wave mixing, for example, the results obtained in the coherent limit are expected to be a meaningful approximation. If incoherent contributions become important, we need to take them into account in a better way. In that case we again rely on approximate treatments of the many-particle problem. Nevertheless, the coherent approach may help to construct such approximate treatments. In the following, we continue our discussion of the coherent limit at the $\chi^{(3)}$ level.

9.2 Equations of Motion in the Coherent Limit

We make use of the relation (9.11) to factorize the four-point functions in (9.5) within the $\chi^{(3)}$ approximation. The resulting expectation values of two pair-creation operators describe, as will be seen below, the correlated states of two electron–hole pairs, including both scattering and bound-state contributions. Rather than deal with the bare biexcitonic amplitude, we introduce a biexcitonic correlation function as follows:

$$b_{eh}^{e'h'}(\boldsymbol{k}_1,\boldsymbol{k}_2,\boldsymbol{k}_3,\boldsymbol{k}_4) = \langle c^\dagger_{e,\boldsymbol{k}_1} c^\dagger_{h,-\boldsymbol{k}_2} c^\dagger_{e',\boldsymbol{k}_3} c^\dagger_{h',-\boldsymbol{k}_4}\rangle$$
$$- \langle c^\dagger_{e,\boldsymbol{k}_1} c^\dagger_{h,-\boldsymbol{k}_2}\rangle\langle c^\dagger_{e',\boldsymbol{k}_3} c^\dagger_{h',-\boldsymbol{k}_4}\rangle + \langle c^\dagger_{e,\boldsymbol{k}_1} c^\dagger_{h',-\boldsymbol{k}_4}\rangle\langle c^\dagger_{e',\boldsymbol{k}_3} c^\dagger_{h,-\boldsymbol{k}_2}\rangle \,. \qquad (9.16)$$

The advantage of this decomposition is that the static exciton–exciton interaction contributions that are obtained in the Hartree–Fock approximation occur explicitly. Thus we are able to write the interaction part as a sum of two terms,

$$\left.\frac{\partial}{\partial t}p_{eh}(\boldsymbol{k})\right|_{\text{int}} = \left.\frac{\partial}{\partial t}p_{eh}(\boldsymbol{k})\right|_{\text{HF}} + \left.\frac{\partial}{\partial t}p_{eh}(\boldsymbol{k})\right|_{\text{corr}} , \qquad (9.17)$$

where the Hartree–Fock contribution contains, of course, coherent one-particle distributions f^{c} only and is given by

$$\left.\frac{\partial}{\partial t}p_{eh}(\boldsymbol{k})\right|_{\text{HF}} = \sum_{\boldsymbol{q},e'h'} e^2 v(\boldsymbol{k}-\boldsymbol{q})[p_{eh'}(\boldsymbol{q})f^{\text{c}}_{h'h}(\boldsymbol{k})\,\delta_{ee'}$$
$$+ p_{e'h}(\boldsymbol{q})f^{\text{c}}_{ee'}(\boldsymbol{k})\,\delta_{hh'} - p_{e'h}(\boldsymbol{k})f^{\text{c}}_{ee'}(\boldsymbol{q})\,\delta_{hh'} - p_{eh'}(\boldsymbol{k})f^{\text{c}}_{h'h}(\boldsymbol{q})\,\delta_{ee'}] \,. \qquad (9.18)$$

The correlational part is obtained from

$$\left.\frac{\partial}{\partial t}p_{eh}(\boldsymbol{k})\right|_{\text{corr}} = \sum_{\boldsymbol{k}'\boldsymbol{q},e'h'} e^2 v(\boldsymbol{q})\Big\{[p^*_{e'h'}(\boldsymbol{k}') - p^*_{e'h'}(\boldsymbol{k}'+\boldsymbol{q})]$$
$$\times[b_{eh}^{e'h'}(\boldsymbol{k}+\boldsymbol{q},\boldsymbol{k},\boldsymbol{k}',\boldsymbol{k}'+\boldsymbol{q}) + b_{eh}^{e'h'}(\boldsymbol{k},\boldsymbol{k}+\boldsymbol{q},\boldsymbol{k}'+\boldsymbol{q},\boldsymbol{k}')]\Big\} \,. \qquad (9.19)$$

In order to close the system of equations, we have, finally, to derive the equation for the biexcitonic correlation function. Making again use of the Heisenberg equations (5.3) and (5.4), this is a straightforward but lengthy calculation, which leads us to six-point functions in the interaction contributions. Treating these six-point functions as described in Sect. 9.1.1 at the $\chi^{(3)}$ level this results in

$$
\begin{aligned}
&\left(\mathrm{i}\hbar\frac{\partial}{\partial t} + \mathrm{i}\gamma_\mathrm{b} + \Delta_{eh}^{e'h'}(\boldsymbol{k}+\boldsymbol{q},\boldsymbol{k},\boldsymbol{k}',\boldsymbol{k}'+\boldsymbol{q})\right) b_{eh}^{e'h'}(\boldsymbol{k}+\boldsymbol{q},\boldsymbol{k},\boldsymbol{k}',\boldsymbol{k}'+\boldsymbol{q})\\
&+\sum_{\boldsymbol{q}'} e^2 v(\boldsymbol{q}')[b_{eh}^{e'h'}(\boldsymbol{k}+\boldsymbol{q}+\boldsymbol{q}',\boldsymbol{k},\boldsymbol{k}'-\boldsymbol{q}',\boldsymbol{k}'+\boldsymbol{q})\\
&+b_{eh}^{e'h'}(\boldsymbol{k}+\boldsymbol{q},\boldsymbol{k}-\boldsymbol{q}',\boldsymbol{k}',\boldsymbol{k}'+\boldsymbol{q}+\boldsymbol{q}') - b_{eh}^{e'h'}(\boldsymbol{k}+\boldsymbol{q}+\boldsymbol{q}',\boldsymbol{k},\boldsymbol{k}',\boldsymbol{k}'+\boldsymbol{q}+\boldsymbol{q}')\\
&-b_{eh}^{e'h'}(\boldsymbol{k}+\boldsymbol{q},\boldsymbol{k}+\boldsymbol{q}',\boldsymbol{k}'+\boldsymbol{q}',\boldsymbol{k}'+\boldsymbol{q}) - b_{eh}^{e'h'}(\boldsymbol{k}+\boldsymbol{q}+\boldsymbol{q}',\boldsymbol{k}+\boldsymbol{q}',\boldsymbol{k}',\boldsymbol{k}'+\boldsymbol{q})\\
&-b_{eh}^{e'h'}(\boldsymbol{k}+\boldsymbol{q},\boldsymbol{k},\boldsymbol{k}'+\boldsymbol{q}',\boldsymbol{k}'+\boldsymbol{q}+\boldsymbol{q}')]\\
&= e^2 v(\boldsymbol{q})[p_{eh}(\boldsymbol{k}) - p_{eh}(\boldsymbol{k}+\boldsymbol{q})][p_{e'h'}(\boldsymbol{k}'+\boldsymbol{q}) - p_{e'h'}(\boldsymbol{k}')]\\
&- e^2 v(\boldsymbol{k}-\boldsymbol{k}')[p_{e'h}(\boldsymbol{k}) - p_{e'h}(\boldsymbol{k}')][p_{eh'}(\boldsymbol{k}'+\boldsymbol{q}) - p_{eh'}(\boldsymbol{k}+\boldsymbol{q})]\,. \qquad (9.20)
\end{aligned}
$$

Here we have defined $\Delta_{eh}^{e'h'}(\boldsymbol{k}+\boldsymbol{q},\boldsymbol{k},\boldsymbol{k}',\boldsymbol{k}'+\boldsymbol{q}) = E_e(\boldsymbol{k}+\boldsymbol{q})+E_h(\boldsymbol{k})+E_{h'}(\boldsymbol{k}'+\boldsymbol{q}) + E_{e'}(\boldsymbol{k}')$. The left-hand side is nothing but the $\boldsymbol{k}$ space representation of a Schrödinger equation of two electrons and two holes interacting with each other. Apart from the different masses and interaction strength, this system is equivalent to a hydrogen molecule and has been extensively studied. In particular, the ground state of the real-space representation of this Hamiltonian has been used to investigate the biexcitonic binding energy in various semiconductors [224, 225, 226] and also in quantum wells [208, 227] by means of variational methods.

9.2.1 Variational Methods

In the real-space representation, the Hamiltonian corresponding to the left-hand side of (9.20) has the form

$$
\begin{aligned}
H_\mathrm{b} =& -\frac{\hbar^2}{2m_\mathrm{e}}(\Delta_e + \Delta_{e'}) - \frac{\hbar^2}{2m_\mathrm{h}}(\Delta_h + \Delta_{h'})\\
&+ e^2\Big(v(\boldsymbol{r}_e - \boldsymbol{r}_e') + v(\boldsymbol{r}_h - \boldsymbol{r}_h') - v(\boldsymbol{r}_e - \boldsymbol{r}_h) \qquad (9.21)\\
&- v(\boldsymbol{r}_e - \boldsymbol{r}_h') - v(\boldsymbol{r}_e' - \boldsymbol{r}_h) - v(\boldsymbol{r}_e' - \boldsymbol{r}_h')\Big).
\end{aligned}
$$

The most familiar approach to this type of problem is the Heitler–London method as applied to the hydrogen molecule [228], which can be immediately transferred to excitons if $m_{\mathrm{h}} \gg m_{\mathrm{e}}$. In this approach, the trial function is an antisymmetrized product of excitonic ground-state functions. As the product of the spin states of electrons and holes can be represented by either symmetric (triplet) or antisymmetric (singlet) states, the position-space contribution in these states has just the opposite symmetry. As shown in

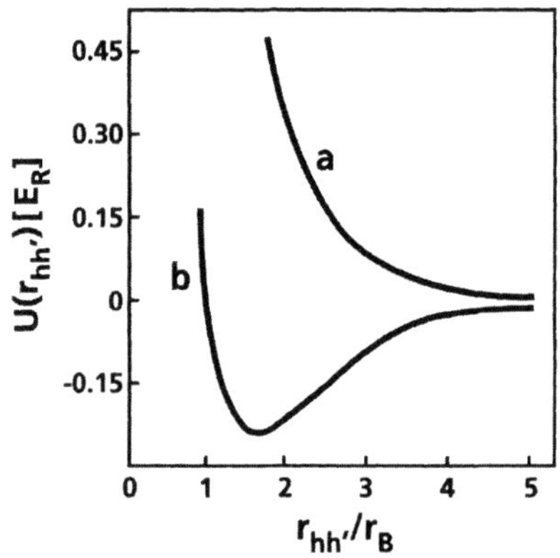

Fig. 9.3. Exciton–exciton interaction potential as a function of the separation of the holes in the Heitler–London approximation. *(a)* triplet state, *(b)* singlet state

Fig. 9.3, in the singlet state the expectation value of the Hamiltonian (9.21) exhibits a pronounced minimum as a function of the separation of the holes. The corresponding attractive interaction may be traced back to the maximum of charge density between the holes that results from the symmetric position-space function. Formally, it is the exchange interaction due to the antisymmetrization of the trial wavefunction which is responsible for the attractive interaction and thus for the existence of bound states. Considering the simplicity of this method, the results are not too bad. The potential exhibits a minimum of $-0.23\,E_{\mathrm{R}}$ at a separation of $1.64\,r_{\mathrm{B}}$, compared with the experimental values of $-0.34\,E_{\mathrm{R}}$ and $1.4\,r_{\mathrm{B}}$. In the triplet state, the interaction becomes repulsive. For a finite hole mass the problem is more involved, as now we have also a contribution from the kinetic energy. Indeed, the binding energy of a biexciton depends sensitively on the ratio of the electron to the hole mass and can be shown to be a monotonically decreasing function of the mass ratio [223]. Higher accuracy can be achieved by means of more refined trial functions, such as a function of the following type [224]:

$$\begin{aligned}\varphi_{\mathrm{bi}}(\boldsymbol{r}_e\boldsymbol{r}_h\boldsymbol{r}_{e'}\boldsymbol{r}_{h'}) &= \mathcal{N}\{|\boldsymbol{r}_{hh'}|^{\gamma}\exp[-\alpha(|\boldsymbol{r}_{eh}|+|\boldsymbol{r}_{e'h'}|+|\boldsymbol{r}_{eh'}|+|\boldsymbol{r}_{e'h}|)] \\ &\times \exp(-\delta|\boldsymbol{r}_{hh'}|)\cosh[-\beta(|\boldsymbol{r}_{e'h}|+|\boldsymbol{r}_{eh'}|-|\boldsymbol{r}_{eh}|-|\boldsymbol{r}_{e'h'}|)]\}\,, \qquad (9.22)\end{aligned}$$

where $\boldsymbol{r}_{eh} = \boldsymbol{r}_e - \boldsymbol{r}_h$, $\mathcal{N}$ is the normalization constant, and α, β, γ, and δ serve as variational parameters. The resulting dependence of the binding energy on the ratio of the electron to the hole mass is shown in Fig. 9.4. Although one

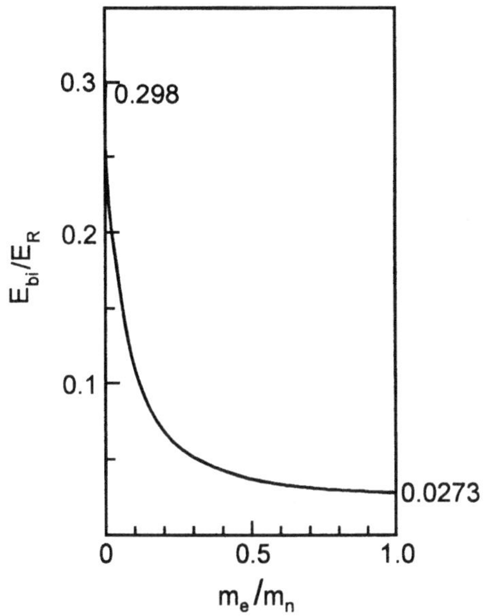

Fig. 9.4. Dependence of the biexciton binding energy on the mass ratio of electrons and holes. Taken from [224]

obtains a bound state for all mass ratios, in the limit of equal masses the binding energy is reduced by more than an order of magnitude to a value of $0.027E_{\mathrm{R}}$.

The existence of bound states is not the only peculiarity of (9.20), but is the most important one. Other remarkable features result from the existence of two types of scattering continua. The first continuum appears because of the excitonic continuum in the source terms and exhibits a pronounced dependence on excitation conditions. The second continuum is generated by the finite center-of-mass motion of the excitons and may be regarded as a biexcitonic scattering continuum. Even to obtain approximate solutions of (9.20), which afford a reasonable description of bound states and scattering continua is a difficult task. The most systematic approach, which has the advantage of allowing also a separate treatment of the different contributions, is an expansion in terms of excitonic eigenfunctions.

9.2.2 Eigenfunction Expansion

From its definition, it is obvious that the biexcitonic correlation function has the following symmetry property (due to anticommutation of Fermi operators) with respect to exchange of electrons:

$$b_{eh}^{e'h'}(\boldsymbol{k}+\boldsymbol{q},\boldsymbol{k},\boldsymbol{k}',\boldsymbol{k}'+\boldsymbol{q}) = -\,b_{e'h}^{eh'}(\boldsymbol{k}',\boldsymbol{k},\boldsymbol{k}+\boldsymbol{q},\boldsymbol{k}'+\boldsymbol{q})\,. \tag{9.23}$$

The corresponding property with respect to exchange of holes is

$$b_{eh}^{e'h'}(\boldsymbol{k}+\boldsymbol{q},\boldsymbol{k},\boldsymbol{k}',\boldsymbol{k}'+\boldsymbol{q}) = -\,b_{eh'}^{e'h}(\boldsymbol{k}+\boldsymbol{q},\boldsymbol{k}'+\boldsymbol{q},\boldsymbol{k}',\boldsymbol{k})\,. \tag{9.24}$$

In principle, this property can be fulfilled by any expansion by means of a complete set of eigenfunctions. From a practical point of view, however, it is favorable to guarantee the symmetry properties of each individual term of an expansion. Thus we consider a sum of four auxiliary quantities as follows:

$$\begin{aligned} b_{eh}^{e'h'}(\boldsymbol{k}+\boldsymbol{q},\boldsymbol{k},\boldsymbol{k}',\boldsymbol{k}'+\boldsymbol{q}) &= \tilde{b}_{eh}^{e'h'}(\boldsymbol{k}+\boldsymbol{q},\boldsymbol{k},\boldsymbol{k}',\boldsymbol{k}'+\boldsymbol{q}) \\ -\tilde{b}_{e'h}^{eh'}(\boldsymbol{k}',\boldsymbol{k},\boldsymbol{k}+\boldsymbol{q},\boldsymbol{k}'+\boldsymbol{q}) &- \tilde{b}_{eh'}^{e'h}(\boldsymbol{k}+\boldsymbol{q},\boldsymbol{k}'+\boldsymbol{q},\boldsymbol{k}',\boldsymbol{k}) \\ +\tilde{b}_{e'h'}^{eh}(\boldsymbol{k}',\boldsymbol{k}'+\boldsymbol{q},\boldsymbol{k}+\boldsymbol{q},\boldsymbol{k}) \,. \end{aligned} \tag{9.25}$$

The auxiliary quantities may be expanded by means of products of excitonic eigenfunctions. If we consider one spin-degenerate conduction band, the band index e labels the spin quantum number only and the exciton wavefunctions can be classified by the hole index alone. In this case the expansion has the form

$$\begin{aligned} &\tilde{b}_{eh}^{e'h'}(\boldsymbol{k}_1,\boldsymbol{k}_2,\boldsymbol{k}_3,\boldsymbol{k}_4) \\ &= \sum_{nm} \phi_n^h(\alpha\boldsymbol{k}_1+\beta\boldsymbol{k}_2)\phi_{nm}^{h'}(\alpha\boldsymbol{k}_3+\beta\boldsymbol{k}_4)B_{eh,nm}^{e'h'}(\boldsymbol{k}_1-\boldsymbol{k}_2) \,, \end{aligned} \tag{9.26}$$

where the eigenfunctions are obtained as solutions of

$$\begin{aligned} &\left(\frac{\hbar^2(\boldsymbol{k}+\boldsymbol{q})^2}{2m_{\rm e}}+\frac{\hbar^2\boldsymbol{k}^2}{2m_{\rm h}}+E_e(0)+E_h(0)\right)\phi_n^h(\boldsymbol{k}+\alpha\boldsymbol{q}) \\ &-\sum_{k'} e^2 v(\boldsymbol{k}-\boldsymbol{k}')\,\phi_n^h(\boldsymbol{k}'+\alpha\boldsymbol{q}) = E_n^h(\boldsymbol{q})\phi_n^h(\boldsymbol{k}+\alpha\boldsymbol{q}) \,. \end{aligned} \tag{9.27}$$

Here $\alpha = m_{\rm h}/M$, $\beta = m_{\rm e}/M$ and $M = m_{\rm e}+m_{\rm h}$. By means of (9.25), we are thus led to the representation

$$\begin{aligned} b_{eh}^{e'h'}(\boldsymbol{k}+\boldsymbol{q},\boldsymbol{k},\boldsymbol{k}',\boldsymbol{k}'+\boldsymbol{q}) &= \sum_{nm}\Big(\phi_n^h(\boldsymbol{k}+\alpha\boldsymbol{q})\phi_m^{h'}(\boldsymbol{k}'+\beta\boldsymbol{q})b_{eh,nm}^{e'h'}(\boldsymbol{q}) \\ -\phi_n^h(\alpha\boldsymbol{k}'+\beta\boldsymbol{k})\phi_m^{h'}(\alpha\boldsymbol{k}+\beta\boldsymbol{k}'+\boldsymbol{q})&b_{e'h,nm}^{eh'}(\boldsymbol{k}'-\boldsymbol{k})\Big) \,, \end{aligned} \tag{9.28}$$

where we have introduced a new expansion coefficient

$$b_{eh,nm}^{e'h'}(\boldsymbol{q}) = B_{eh,nm}^{e'h'}(\boldsymbol{q}) + B_{e'h',mn}^{eh}(-\boldsymbol{q}) \,. \tag{9.29}$$

The two types of contributions in the expansion (9.28) result from the fact that there are two ways to form excitons pairs from two electrons and two holes. In order to have expansion coefficients which are odd or even with respect to the interchange of spin quantum numbers alone, it is convenient at this point to form linear combinations. As the case of equal spin indices is trivial, it is sufficient to consider explicitly two cases, which correspond to a spin singlet (b^-) and the $m_j = 0$ triplet state (b^+) of two electrons. Using

$$b^{e'h'\pm}_{eh}(\boldsymbol{k}+\boldsymbol{q},\boldsymbol{k},\boldsymbol{k}',\boldsymbol{k}'+\boldsymbol{q}) = \frac{1}{2}\left(b^{e'h'}_{eh}(\boldsymbol{k}+\boldsymbol{q},\boldsymbol{k},\boldsymbol{k}',\boldsymbol{k}'+\boldsymbol{q}) \pm b^{eh'}_{e'h}(\boldsymbol{k}+\boldsymbol{q},\boldsymbol{k},\boldsymbol{k}',\boldsymbol{k}'+\boldsymbol{q})\right) \tag{9.30}$$

in (9.28), we arrive at

$$b^{e'h'\pm}_{eh}(\boldsymbol{k}+\boldsymbol{q},\boldsymbol{k},\boldsymbol{k}',\boldsymbol{k}'+\boldsymbol{q}) = \sum_{nm}\Big(\phi^h_n(\boldsymbol{k}+\alpha\boldsymbol{q})\phi^{h'}_m(\boldsymbol{k}'+\beta\boldsymbol{q})b^{\pm}_{nm}(\boldsymbol{q}) \mp\phi^h_n(\alpha\boldsymbol{k}'+\beta\boldsymbol{k})\phi^{h'}_m(\alpha\boldsymbol{k}+\beta\boldsymbol{k}'+\boldsymbol{q})b^{\pm}_{nm}(\boldsymbol{k}'-\boldsymbol{k})\Big)\,, \tag{9.31}$$

where the expansion coefficients are

$$b^{\pm}_{nm}(\boldsymbol{q}) = \frac{1}{2}\left(b^{e'h'}_{eh,nm}(\boldsymbol{q}) \pm b^{eh'}_{e'h,nm}(\boldsymbol{q})\right)\,. \tag{9.32}$$

All other transformation properties are determined by the equation of motion for the expansion coefficients. Inserting the expansion (9.31) together with (9.30) into the biexciton equation (9.20) and using an excitonic representation for the transition amplitude, we obtain the following for the coefficients:

$$\mathrm{i}\hbar\frac{\partial}{\partial t}b^{\pm}_{nm}(\boldsymbol{q}) + \sum_{n'm'q'} H^{\pm}_{nm,n'm'}(\boldsymbol{q},\boldsymbol{q}')b^{\pm}_{n'm'}(\boldsymbol{q}') = \sum_{n'm'} C^{\pm}_{nmn'm'}(\boldsymbol{q})\left(p^{eh}_{n'}p^{e'h'}_{m'} \pm p^{e'h}_{n'}p^{eh'}_{m'}\right)\,. \tag{9.33}$$

The quantities introduced in this equation will be discussed in the following. In order not to overburden the notation, hole-band indices are suppressed in various abbreviations used below. The source of the biexcitonic correlation is proportional to a product of excitonic transition amplitudes, rather than to a product of a transition amplitude and an external field, as is suggested by phenomenological models:

$$C^{\pm}_{nmn'm'}(\boldsymbol{q}) = \frac{1}{2}\sum_{lkq'}(\hat{1}\mp\hat{S})^{-1}_{nm,lk}(\boldsymbol{q},\boldsymbol{q}')\left[V^{\mathrm{c}}_{lk,n'm'}(\boldsymbol{q}')\mp V^{\mathrm{xc}}_{lk,n'm'}(\boldsymbol{q}')\right]\,. \tag{9.34}$$

The coupling to the driving amplitudes consists of a direct Coulomb interaction matrix element

$$V^{\mathrm{c}}_{lk,n'm'}(\boldsymbol{q}) = e^2 v(\boldsymbol{q})M^h_{ln'}(\boldsymbol{q})M^{h'}_{km'}(-\boldsymbol{q})\,, \tag{9.35}$$

which is expressed by means of excitonic transition matrix elements

$$M^h_{nn'}(\boldsymbol{q}) = \langle\phi^h_n|\exp(i\alpha\boldsymbol{q}\boldsymbol{r}) - \exp(-i\beta\boldsymbol{q}\boldsymbol{r})|\phi^h_{n'}\rangle\,, \tag{9.36}$$

and a corresponding exchange matrix element given by

$$V^{\mathrm{xc}}_{nm,n'm'}(\boldsymbol{q}) = \sum_{kk'} e^2 v(\boldsymbol{k}-\boldsymbol{k}')\phi^{h*}_n(\boldsymbol{k}+\alpha\boldsymbol{q})\phi^{h'*}_m(\boldsymbol{k}'+\beta\boldsymbol{q})$$
$$\times\left(\phi^h_{n'}(\boldsymbol{k})-\phi^h_{n'}(\boldsymbol{k}')\right)\left(\phi^{h'}_{m'}(\boldsymbol{k}'+\boldsymbol{q})-\phi^{h'}_{m'}(\boldsymbol{k}+\boldsymbol{q})\right). \quad (9.37)$$

As expected, the source vanishes in the singlet case for equal spin indices of the electrons. Otherwise, it contributes according to the selection rules for the transition amplitude. Further, we have defined the overlap matrix

$$S_{nm\,n'm'}(\boldsymbol{q},\boldsymbol{q}')$$
$$=\sum_k \phi^{h*}_n(\boldsymbol{k}+\alpha\boldsymbol{q})\phi^{h'*}_m(\boldsymbol{k}+\boldsymbol{q}'+\beta\boldsymbol{q})\phi^h_{n'}(\boldsymbol{k}+\alpha\boldsymbol{q}')\phi^{h'}_{m'}(\boldsymbol{k}+\boldsymbol{q}+\beta\boldsymbol{q}'), \quad (9.38)$$

which results from the nonorthogonality of the excitonic wavefunctions associated with the two combinations of electrons and holes. Finally, the Hamiltonian matrix is given by

$$H^{\pm}_{nm,n'm'}(\boldsymbol{q},\boldsymbol{q}') = \left(E^h_n(\boldsymbol{q})+E^{h'}_m(\boldsymbol{q})+i\hbar\gamma_{\mathrm{b}}\right)\delta_{nn'}\delta_{mm'}\delta_{\boldsymbol{q},\boldsymbol{q}'}$$
$$+\sum_{lk\boldsymbol{k}}\left(\hat{1}\mp\hat{S}\right)^{-1}_{nm,lk}(\boldsymbol{q},\boldsymbol{k})W^{\pm}_{lk,n'm'}(\boldsymbol{k},\boldsymbol{q}'), \quad (9.39)$$

where the interaction matrix is

$$W^{\pm}_{nm,n'm'}(\boldsymbol{q},\boldsymbol{q}') = e^2 v(\boldsymbol{q}-\boldsymbol{q}')M^h_{nn'}(\boldsymbol{q}'-\boldsymbol{q})M^{h'}_{mm'}(\boldsymbol{q}-\boldsymbol{q}')$$
$$\pm e^2\sum_{kk'} v(\boldsymbol{k}-\boldsymbol{k}')\phi^{h*}_n[\boldsymbol{k}+\alpha(\boldsymbol{q}-\boldsymbol{q}')]\,\phi^{h'*}_m[\boldsymbol{k}'+\beta(\boldsymbol{q}+\boldsymbol{q}')]$$
$$\times\left(\phi^h_{m'}[\boldsymbol{k}'+\alpha(\boldsymbol{q}-\boldsymbol{q}')+\beta(\boldsymbol{q}+\boldsymbol{q}')]-\phi^h_{m'}[\boldsymbol{k}+\alpha(\boldsymbol{q}-\boldsymbol{q}')+\beta(\boldsymbol{q}+\boldsymbol{q}')]\right)$$
$$\times\left(\phi^{h'}_{n'}(\boldsymbol{k}')-\phi^{h'}_{n'}(\boldsymbol{k})\right). \quad (9.40)$$

The result of the eigenfunction expansion becomes more familiar if we consider the limit of infinite hole mass ($\alpha=1$, $\beta=0$) and restrict the expansion to 1s functions only. In this case, the kinetic energy vanishes and the interaction and the overlap matrices depend on $\boldsymbol{q}-\boldsymbol{q}'$ only. The transformation to real space yields a local potential depending on the distance between the holes. By construction, the result yields again just the Heitler–London theory. In the singlet state, i.e. for the lower sign in (9.40), the interaction becomes attractive owing to the exchange interaction. For a finite hole mass, the interaction remains nonlocal both in $\boldsymbol{k}$ space and in position space.

To complete the excitonic representation of the equations of motion we perform the same transformation as in (9.4) in the coherent limit, i.e. the one-particle distribution functions are replaced by their coherent contribution (9.15). The result of this expansion leads to a generalization of the local field model introduced in Chap. 6. Apart from the inclusion of more than two bands, the important new ingredient is the correlational contribution. Using the static Hartree–Fock contribution (9.18), we can now obtain the transition amplitude from

$$\begin{aligned}
&\left(- \mathrm{i}\hbar \frac{\partial}{\partial t} - \mathrm{i}\hbar\gamma_2 - E_n^h \right) p_n^{eh} \\
&= \boldsymbol{d}_{eh} \boldsymbol{E} \tilde{\phi}_n^h(0) - \sum_{h'ml} \boldsymbol{d}_{eh'} \boldsymbol{E}\, b_{nml}^{hh'h} p_n^{eh'} p_l^{eh*} + \boldsymbol{d}_{eh} \boldsymbol{E}\, b_{nml}^{hh'h'} p_n^{eh'} p_l^{eh'*} \\
&+ \sum_{h'mlk} V_{nmlk}^{st} p_m^{eh'*} p_k^{eh'} p_l^{eh} + \left. \frac{\partial}{\partial t} p_n^{eh} \right|_{\mathrm{corr}} ,
\end{aligned} \tag{9.41}$$

where the phase-space parameter is defined as

$$b_{nml}^{hh'h} = \sum_{\boldsymbol{k}} \phi_n^h(\boldsymbol{k}) \phi_m^{h'}(\boldsymbol{k}) \phi_l^{h*}(\boldsymbol{k}) \tag{9.42}$$

and the static interaction contribution is given by

$$\begin{aligned}
V_{nmlk}^{\mathrm{st}} &= \sum_{\boldsymbol{kq}} e^2 v(\boldsymbol{k} - \boldsymbol{q}) \Big(\phi_n^{h*}(\boldsymbol{k}) \phi_m^{h'*}(\boldsymbol{k}) - \phi_n^{h*}(\boldsymbol{k}) \phi_m^{h'*}(\boldsymbol{q}) \Big) \\
&\quad \times \Big(\phi_l^h(\boldsymbol{k}) \phi_k^{h'}(\boldsymbol{q}) + \phi_l^h(\boldsymbol{q}) \phi_k^{h'}(\boldsymbol{k}) \Big) .
\end{aligned} \tag{9.43}$$

The correlational contribution may be expressed as sum of a singlet and a triplet contribution,

$$\left. \frac{\partial}{\partial t} p_n^{eh} \right|_{\mathrm{corr}} = \left. \frac{\partial}{\partial t} p_n^{eh} \right|_{+} + \left. \frac{\partial}{\partial t} p_n^{eh} \right|_{-} , \tag{9.44}$$

which can be cast into the form

$$\left. \frac{\partial}{\partial t} p_n^{eh} \right|_{\pm} = \sum_{\boldsymbol{q}, e'h'mlk} \Big(V_{lk,nm}^{c}(\boldsymbol{q}) \mp V_{lk,nm}^{xc}(\boldsymbol{q}) \Big)^* p_m^{e'h'*} b_{lk}^{\pm}(\boldsymbol{q}) . \tag{9.45}$$

The band indices of $b^{\pm}$ are defined according to (9.32). The expansion derived so far respresents a systematic method for dealing with four-particle correlations in optical spectra. In contrast to variational approaches, it yields not only the lowest bound state but also excited states; in particular, the biexcitonic scattering continuum is obtained from the solution of (9.33). It should, however, be noted that the expansion becomes impracticable if more than a few of the lowest excitonic eigenstates contribute to the source, i.e. the excitonic transition amplitudes in (9.34). This restriction is, however, consistent with the present approach, as excitation conditions under which unbound electron–hole pairs dominate require different methods. One reason is the interaction with free carriers, in particular screening effects (see the next chapter), which, at least at sufficiently high densities, destroys the correlation effects, as well as the coherence discussed in this chapter. As expected, our results derived so far include the coherent contributions to these processes in lowest order only. An appropriate extension of the present approach, including the incoherent interactions to arbitrary order is possible but awkward.

9.3 Bound-State and Scattering Contributions

In the following we focus on the excitation conditions that we considered earlier in the discussion of the local-field approximation in Sect. 6.3. If only the lowest exciton state is excited, the sums over intermediate states in (9.33) and (9.34) converge rapidly and the dimensions of the problem remain manageable by numerical methods. As biexcitonic bound and scattering states exhibit different features in nonlinear optical emission, however, it is instructive to separate the lowest bound state in the equations of motion and treat the remaining contributions perturbationally.

9.3.1 Separation of Bound States

Although the Hamiltonian matrix (9.39) is certainly non-Hermitian because of the overlap matrix, it is readily transformed into a Hermitian representention by means of the substitution

$$b_\lambda^\pm = \sum_{\lambda'} (\hat{1} \mp \hat{S})^{-1/2}_{\lambda\lambda'} \, \tilde{b}^\pm_{\lambda'} \,, \tag{9.46}$$

where we have used the multiple index $\lambda = \{nm\boldsymbol{q}\}$, and by corresponding transformations for the other quantities. We consider the transformed equation (9.33) for the singlet state in the compact notation

$$\mathrm{i}\hbar \frac{\partial}{\partial t} \tilde{b}_\lambda^- + \sum_{\lambda'} \tilde{H}^-_{\lambda\lambda'} \tilde{b}^-_{\lambda'} = \tilde{S}^-_\lambda \,, \tag{9.47}$$

where $\tilde{S}^-_\lambda$ results from the source in (9.33). The Hamiltonian may be diagonalized by means of a unitary transformation, defined by the eigenvalue equation

$$\sum_{\lambda'} \tilde{H}^-_{\lambda\lambda'} \tilde{U}_{\lambda'\mu} = E_\mu \tilde{U}_{\lambda\mu} \,. \tag{9.48}$$

By substituting

$$\tilde{b}^-_\lambda = \sum_\mu \tilde{U}_{\lambda\mu} b_\mu \tag{9.49}$$

in (9.47), we can separate the equation of motion for the lowest eigenvalue, with $\mu = 0$, and introduce a remaining contribution which results from projecting out the lowest eigenstate. Reversing the symmetrization in the resulting equations, we obtain the following for the bound state contribution:

$$\mathrm{i}\hbar\frac{\partial}{\partial t}b_0 + E_0 b_0 = \sum_{\lambda\lambda'} U^{\dagger}_{0\lambda}(\hat{1}+\hat{S})_{\lambda\lambda'}\mathcal{S}^{-}_{\lambda'} \ . \tag{9.50}$$

The remaining contribution δb^{-}_{λ} is determined by

$$\mathrm{i}\hbar\frac{\partial}{\partial t}\delta b^{-}_{\lambda} + \sum_{\lambda'} H^{-}_{\lambda\lambda'}\delta b^{-}_{\lambda'} = \sum_{\lambda'} \mathcal{P}_{\lambda,\lambda'}\mathcal{S}^{-}_{\lambda'} = \delta\mathcal{S}^{-}_{\lambda} \ , \tag{9.51}$$

where we have defined the projector

$$\mathcal{P}_{\lambda,\lambda'} = \delta_{\lambda,\lambda'} - \sum_{\lambda''} U_{\lambda 0}U^{\dagger}_{0\lambda''}(\hat{1}+\hat{S})_{\lambda''\lambda'} \ . \tag{9.52}$$

The two contributions yield the total amplitude as follows:

$$b^{-}_{\lambda} = U_{\lambda 0}b_0 + \delta b^{-}_{\lambda} = b^{0}_{\lambda} + \delta b^{-}_{\lambda} \ . \tag{9.53}$$

The last relation allows us to rewrite (9.50) as

$$\mathrm{i}\hbar\frac{\partial}{\partial t}b^{0}_{\lambda} + E_0 b^{0}_{\lambda} = \sum_{\lambda}(\delta_{\lambda\lambda'} - \mathcal{P}_{\lambda\lambda'})\mathcal{S}^{-}_{\lambda'} \ . \tag{9.54}$$

Now we can again make use of our expansion and project the lowest eigenvector of (9.48) back onto a $\boldsymbol{k}$ space representation according to

$$\begin{aligned}\varphi^{0}_{\mathrm{b}}(\boldsymbol{k}+\boldsymbol{q},\boldsymbol{k},\boldsymbol{k}',\boldsymbol{k}'+\boldsymbol{q}) = \sum_{nm}\Big\{&U_{nm\boldsymbol{q},0}\,\phi^{h}_{n}(\boldsymbol{k}+\alpha\boldsymbol{q})\phi^{h'}_{m}(\boldsymbol{k}'+\beta\boldsymbol{q})\\ +U_{nm,\boldsymbol{k}'-\boldsymbol{k},0}\,\phi^{h}_{n}(\alpha\boldsymbol{k}'+\beta\boldsymbol{k})&\phi^{h'}_{m}(\alpha\boldsymbol{k}+\beta\boldsymbol{k}'+\boldsymbol{q})\Big\} \ . \end{aligned} \tag{9.55}$$

The result is just the lowest biexciton eigenfunction in $\boldsymbol{k}$ space. On the other hand, if we have found this eigenstate by some other method, for example, with the help of a variational function such as (9.22) or by means of more refined methods [226], we can apply the inverse of the transformation (9.55), which yields the lowest eigenvector as

$$U_{nm\boldsymbol{q},0} = \sum_{n'm'\boldsymbol{q}'}(\hat{1}+\hat{S})^{-1}_{nm,n'm'}(\boldsymbol{q},\boldsymbol{q}')s_{n'm'}(\boldsymbol{q}') \ , \tag{9.56}$$

where the matrix element is

$$s_{nm}(\boldsymbol{q}) = \sum_{\boldsymbol{k}\boldsymbol{k}'}\varphi^{0}_{\mathrm{bi}}(\boldsymbol{k}+\boldsymbol{q},\boldsymbol{k},\boldsymbol{k}',\boldsymbol{k}'+\boldsymbol{q})\phi^{h*}_{n}(\boldsymbol{k}+\alpha\boldsymbol{q})\phi^{h'*}_{m}(\boldsymbol{k}'+\beta\boldsymbol{q}) \ . \tag{9.57}$$

In this way, the source in (9.50) and the projector (9.52) can be expressed by means of the biexcitonic ground state and the problems of calculating the lowest bound state and the higher-lying states are completely decoupled. Each of these may be treated by the method which is best suited. Indeed, if we require high accuracy for the bound state, variational calculations are superior. The contributions arising additional to the bound-state, can be discussed by means of perturbation theory with respect to the Coulomb potential.

9.3.2 Biexcitonic Scattering Contributions

In order to obtain results which are correct up to second order in the Coulomb potential, it is sufficient to approximate the Hamiltonian (9.36) by the diagonal part, i.e. by the sum of the two exciton energies. With respect to the calculation of the transition amplitude, this corresponds to a second order Born approximation. The solution for the biexcitonic amplitude can then be written in the form

$$\delta b^{-}_{nm}(\boldsymbol{q},t) = -\frac{\mathrm{i}}{\hbar}\sum_{n'm'}\int_{-\infty}^{t}\mathrm{d}t'\,\mathrm{e}^{-\mathrm{i}[E^{h}_{n}(\boldsymbol{q})+E^{h'}_{m}(\boldsymbol{q})+\mathrm{i}\hbar\gamma_{\mathrm{b}}](t-t')/\hbar}$$
$$\times\,\delta C^{-}_{nmn'm'}(\boldsymbol{q})\left(p^{eh}_{n'}(t')p^{e'h'}_{m'}(t') - p^{e'h}_{n'}(t')p^{eh'}_{m'}(t')\right)\,, \qquad (9.58)$$

where δC results from the action of the projector (9.52) on the right hand side of (9.34). A corresponding result holds for the triplet contribution b^{+}, but with the projection removed. The physical meaning of this contribution becomes more obvious if we consider the Fourier transform, which takes the form

$$\delta b^{-}_{nm}(\boldsymbol{q},\omega)$$
$$= \sum_{n'm'}\delta C^{-}_{nmn'm'}(\boldsymbol{q})\int\frac{\mathrm{d}\omega'}{2\pi}\,\frac{p^{eh}_{n'}(\omega-\omega')p^{e'h'}_{m'}(\omega') - p^{e'h}_{n'}(\omega-\omega')p^{eh'}_{m'}(\omega')}{\hbar\omega - E^{h}_{n}(\boldsymbol{q}) - E^{h'}_{m}(\boldsymbol{q}) + \mathrm{i}\hbar\gamma_{\mathrm{b}}}\,, \qquad (9.59)$$

where to lowest order in the field, the transition amplitude is given by

$$p^{eh}_{n}(\omega) = \frac{\boldsymbol{d}_{eh}\boldsymbol{E}(\omega)\tilde{\phi}(0)}{\hbar\omega - E^{h}_{n}(0) + \mathrm{i}\hbar\gamma_{2}}\,. \qquad (9.60)$$

Depending on the excitation conditions, the correlation function may exhibit resonances in the vicinity of

$$\hbar\omega = E^{h}_{n'}(0) + E^{h'}_{m'}(0)\,. \qquad (9.61)$$

These resonances correspond to an exciton–exciton interaction process, where two coherently excited excitons with negligible wave vector scatter into final states $E^{h}_{n}(\boldsymbol{q})$ and $E^{h'}_{m}(\boldsymbol{q})$. Thus $b_{nm}(\boldsymbol{q},\omega)$ has an imaginary part, which acts in (9.41) as a coherently induced dephasing of the transition amplitude [229, 230]. If we consider the scattering of free electron–hole pairs only, the corresponding process is referred to as polarization scattering. This is obtained either as a special case of (9.58) or, more directly, from (9.20) with the interaction on the left-hand side neglected.

Often, the frequency dependence is weak and one can choose a static approximation, i.e. we can set $\hbar\omega = E^{h}_{n'}(0) + E^{h'}_{m'}(0)$ in (9.59). In the time domain, the pendant of this treatment is the *Markov approximation*, which can be used in (9.58). If we separate the rapidly oscillating phase factors of the transition amplitudes, their remaining part can be assumed to be slowly varying in comparison with the phase factors in the integrand and approximated by its value at the upper limit. In this way, we neglect *memory effects* in the correlation function with respect to the history of the transition amplitude. If we do this, performing the time integration with respect to the phase factor only yields the same result as the static limit in the frequency domain. If, on the other hand, the correlation function exhibits sharp structures in the frequency domain, the Markov approximation necessarily fails and the correlation function (9.57) represents an independent dynamical degree of freedom, additional to the bound-state contribution. This can occur in confined systems, as, for example, under the action of a strong magnetic field [219, 220, 231], and in quantum well sytems [232]. In a four-wave-mixing signal, the non-Markovian coupling of the four-particle correlation function to the transition amplitude shows up as a slow rise time for negative time delay [219]. If the dephasing of the biexcitonic correlations is sufficiently small, the same non-Markovian effect contributes to a redshift of the excitonic resonance for off-resonant excitation [232]. It is worth emphasizing that both biexcitonic bound states and scattering states contribute to this effect.

Let us focus first on the Markov limit, which gives the result

$$\delta b^-_{nm}(\boldsymbol{q},t) = -\sum_{n'm'} \frac{\delta C^-_{nmn'm'}(\boldsymbol{q}) \left(p^{eh}_{n'}(t)p^{e'h'}_{m'}(t) - p^{e'h}_{n'}(t)p^{eh'}_{m'}(t)\right)}{E^h_n(\boldsymbol{q}) + E^{h'}_m(\boldsymbol{q}) - E^h_{n'}(0) - E^{h'}_{m'}(0) - i\hbar\gamma_{\mathrm{b}}} . \tag{9.62}$$

This approximation yields a structure very similar to that of the static interaction contributions. The weights and the allowed combinations of band indices, however, are completely different. This leads to a coupling of the two spin subspaces. As we shall see in the next section, the consequences of this feature are directly measurable in experiments.

The correlational contribution (9.44) can now be written as the sum of a bound and a scattering part. Owing to the finite overlap matrix and the projection of the bound-state contribution, the general results of our treatment require an intricate notation. In order to survey the structure of the theory, we shall introduce in the following the simplifying, although not completely realistic, assumption that only the lowest exciton state contributes. Correspondingly, we shall omit all exciton quantum numbers. Under these assumptions, the scattering contribution, obtained from (9.45) together with (9.62), takes the form

$$\left.\frac{\partial}{\partial t}p^{eh}\right|_{\mathrm{scatt}} = \sum_{e'h'} \left(K^{hh'}_+ p^{eh} p^{e'h'} p^{e'h'*} + K^{hh'}_- p^{e'h} p^{eh'} p^{e'h'*} \right) , \tag{9.63}$$

where we have introduced

$$\begin{aligned} K^{hh'}_\pm = \sum_{\boldsymbol{q}\boldsymbol{q}'} \Big(& V^{\mathrm{c}*}(\boldsymbol{q}) T^\pm(\boldsymbol{q}\boldsymbol{q}') V^{\mathrm{c}}(\boldsymbol{q}') + V^{\mathrm{xc}*}(\boldsymbol{q}) T^\pm(\boldsymbol{q}\boldsymbol{q}') V^{\mathrm{xc}}(\boldsymbol{q}') \\ & - V^{\mathrm{xc}*}(\boldsymbol{q}) T^\mp(\boldsymbol{q}\boldsymbol{q}') V^{\mathrm{c}}(\boldsymbol{q}') - V^{\mathrm{c}*}(\boldsymbol{q}) T^\mp(\boldsymbol{q}\boldsymbol{q}') V^{\mathrm{xc}}(\boldsymbol{q}') \Big) \end{aligned}$$

$$\times \left(E_n^h(\boldsymbol{q}) + E_m^{h'}(\boldsymbol{q}) - E_{n'}^h(0) - E_{m'}^{h'}(0) - i\hbar\gamma_\mathrm{b}\right)^{-1} . \tag{9.64}$$

Here

$$T^{\pm}(\boldsymbol{q}\boldsymbol{q}') = \frac{1}{2}\left((\hat{1} - \hat{S})^{-1}(\boldsymbol{q}\boldsymbol{q}') \pm \sum_{\boldsymbol{k}} \mathcal{P}(\boldsymbol{q}\boldsymbol{k})(\hat{1} + \hat{S})^{-1}(\boldsymbol{k}\boldsymbol{q}')\right) . \tag{9.65}$$

If $S = 0$ and $\mathcal{P}(\boldsymbol{q}\boldsymbol{k}) = \delta_{\boldsymbol{q},\boldsymbol{k}}$, this expression reduces to $T^+ = \delta_{\boldsymbol{q},\boldsymbol{q}'}$ and $T^- = 0$. In this limit one can show easily that $K^+ \gg K^-$, as the Coulomb matrix element is usually larger than the exchange contribution. For finite overlap matrix elements and a finite projection, this relation is valid also. Finally, by means of (9.54), the bound-state contribution can be cast into the form

$$\left.\frac{\partial}{\partial t} p^{eh}\right|_{\text{bound}} = \sum_{e'h'} \mathcal{C}_\mathrm{b}^{hh'} (1 - \delta_{ee'}) b^{hh'} p^{e'h'*} , \tag{9.66}$$

where the biexcitonic coupling coefficient is

$$\begin{aligned} \mathcal{C}_\mathrm{b}^{hh'} &= \sum_{\boldsymbol{q}\boldsymbol{q}'} [V^{\mathrm{c}*}(\boldsymbol{q}) + V^{\mathrm{xc}*}(\boldsymbol{q})][V^{\mathrm{c}}(\boldsymbol{q}') + V^{\mathrm{xc}}(\boldsymbol{q}')] \\ &\times \left((\hat{1} + \hat{S})^{-1}(\boldsymbol{q},\boldsymbol{q}') - \sum_{\boldsymbol{k}} \mathcal{P}(\boldsymbol{q},\boldsymbol{k})(\hat{1} + \hat{S})^{-1}(\boldsymbol{k},\boldsymbol{q}')\right) , \end{aligned} \tag{9.67}$$

and the amplitude is the solution of

$$\left(\mathrm{i}\hbar\frac{\partial}{\partial t} + E_0^{hh'} + \mathrm{i}\hbar\gamma_\mathrm{b}\right) b^{hh'} = (1 - \delta_{ee'}) p^{eh} p^{e'h'} . \tag{9.68}$$

The correlational contributions introduced in this way generalize the local-field approximation introduced in Sect. 6.3.1 with respect to both, the bound biexcitons and the coherent exciton–exciton scattering. In spite of the simplifications, which may make it impossible to obtain really quantitative results, this model includes the essential physics under the excitation conditions assumed [233].

9.4 Signatures of Biexcitonic Bound States

A striking feature of the interactions between coherently excited excitons is their pronounced dependence on the polarization of the external fields [216, 234, 235]. This dependence can only partly be traced back to valence band mixing [236], as discussed in Sects. 3.5. Within the present model, polarization dependences result from both the scattering and the bound-state contributions and exhibit different features in nonlinear absorption (as measured in a pump–probe experiment) and four-wave-mixing signals. Both cases will be discussed in the following.

9.4.1 Nonlinear Absorption

The nonlinear absorption can be discussed for cases in which the test and pump pulses propagate either in the same direction or in different directions. Here we choose the first possibility. As in Sect. 6.3, the transition amplitude is given by the sum $p = p_{\mathrm{t}} + p_{\mathrm{p}}$ (where the subscripts "t" and "p" denote the test pulse and the probe pulse, respectively). Considering first the scattering contribution, we note that the second term in (9.63) contributes for equal electron spins only and can be lumped together with the static exciton–exciton interaction. Only a marginal polarization dependence results from these contributions from the simultaneous excitation of light and heavy holes. We omit this tedious contribution in the following and obtain the scattering contribution for the transition amplitude driven by the test pulse as

$$\left.\frac{\partial}{\partial t} p_{\mathrm{t}}^{eh}\right|_{\mathrm{scatt}} = \sum_{e'h'} K_{+}^{hh'} \left[\left(p_{\mathrm{p}}^{eh} p_{\mathrm{t}}^{e'h'} + p_{\mathrm{t}}^{eh} p_{\mathrm{p}}^{e'h'} \right) p_{\mathrm{p}}^{e'h'*} + p_{\mathrm{p}}^{eh} p_{\mathrm{p}}^{e'h'} p_{\mathrm{t}}^{e'h'*} \right] . \tag{9.69}$$

This term includes transitions in both subspaces, which can be evaluated by means of the selection rules summarized in Table 7.1. Let us consider the four geometries in which the pump and test beams are either circularly polarized (in the same or the opposite direction) or have linear polarizations (parallel or orthogonal). For equal circular polarizations we have only transitions in one subspace, but both combinations of the test and pump pulses in (9.69) contribute. If the circular polarizations have opposite directions, one of these combinations is switched off. If the beams have the same linear polarization, both combinations of test and pump pulses and both subspaces contribute. The products of the transition amplitudes in different subspaces cancel for orthogonal linear polarizations. Thus the weights of the scattering contribution are different in all four cases. These differences become even more pronounced for the bound-state contribution, which takes the form

$$\left.\frac{\partial}{\partial t} p_{\mathrm{t}}^{eh}\right|_{\mathrm{bound}} = \sum_{e'h'} C_{\mathrm{b}}^{hh'} (1 - \delta_{ee'}) \left[\left(b_{\mathrm{tp}}^{hh'} + b_{pt}^{hh'} \right) p_{\mathrm{p}}^{e'h'*} + b_{\mathrm{pp}}^{hh'} p_{\mathrm{t}}^{e'h'*} \right] . \tag{9.70}$$

The biexciton amplitude b_{tp} is driven by a product of the pump and probe transition amplitudes. If both beams have the same circular polarization, the dominant bound-state contribution of heavy holes is switched off, and bound states formed from light-hole and heavy-hole excitons occur. Their influence, however, is negligible because of their weak oscillator strength. Biexcitons from heavy-hole excitons contribute only in the case of opposite circular polarizations and of equal linear polarizations. For crossed linear polarizations the bound-state amplitudes in (9.70) cancel. The biexciton amplitude b_{pp} does not contribute to the nonlinear absorption at the $\chi^{(3)}$ level.

Owing to the scattering contributions, we have quantitative differences in the nonlinear-absorption lineshape. In contrast, the bound-state contributions can be switched on or off by varying the polarization. Corresponding signatures of biexcitons are found under the conditions of the optical Stark effect. The Hartree–Fock theory as discussed in Sect. 6.3, predicts a blueshift of the absorption lineshape for excitation below the excitonic resonance. If there is a biexcitonic bound-state contribution, however, this blueshift is converted into a redshift if a spectrally narrow pump pulse crosses the biexcitonic resonance from above. This effect has been observed in CuCl for collinear polarizations [237]. For crossed linear polarizations, it vanishes. Let us discuss this effect in some detail.

Example (Optical Stark Effect). We start with the same assumptions as in Sect. 6.3.1 and calculate the linear response of a stationary excited system with collinearly polarized pump and test pulses propagating in the same direction. Using the same simplified notation, we have, instead of (6.117),

$$
\begin{aligned}
&\left(\hbar\frac{\partial}{\partial t} + \mathrm{i}\hbar\Omega_1 + \hbar\gamma_2\right)p_\mathrm{t} \\
&= \mathrm{i}\boldsymbol{dE}_\mathrm{t}(1 - b|p_\mathrm{p}|^2) + \mathrm{i}\,\mathcal{C}_\mathrm{b}\left(2b_\mathrm{tp}p_\mathrm{p}^* + b_\mathrm{pp}p_\mathrm{t}^*\right) \\
&-\mathrm{i}\left(b\boldsymbol{dE}_\mathrm{p}p_\mathrm{p}^* + 2(v + K_- + 2K_+)\,|p_\mathrm{p}|^2\right)p_\mathrm{t} \\
&-\mathrm{i}\,\left(b\boldsymbol{dE}_\mathrm{p}p_\mathrm{p} + (v + K_- + 2K_+)p_\mathrm{p}^2\right)p_\mathrm{t}^* \,.
\end{aligned} \tag{9.71}
$$

By means of the ansatz (6.118), b_tp and b_pp can be calculated from (9.68). Fourier decomposition again yields a coupled set of equations for p_+ and p_-. If we restrict ourselves to the third order in the fields, both equations are decoupled, and we only need to solve

$$
\left(\hbar(\Omega_1 - \omega - \mathrm{i}\gamma_2) + S - \frac{2\mathcal{C}_\mathrm{bi}\tilde{p}_\mathrm{p}^2}{E_0 - \hbar\omega_\mathrm{p} - \hbar\omega - \mathrm{i}\hbar\gamma_\mathrm{b}}\right)p_+ = \boldsymbol{d}\tilde{\boldsymbol{E}}_\mathrm{t}(1 - b\tilde{p}_\mathrm{p}^2)\,, \tag{9.72}
$$

where the RPA contribution to the shift is

$$
S = b\boldsymbol{d}\tilde{\boldsymbol{E}}_\mathrm{p}\tilde{p}_\mathrm{p} + 2(v + K_- + 2K_+)\tilde{p}_\mathrm{p}^2\,. \tag{9.73}
$$

Obviously, if the central frequency of the pump pulse is well above the biexcitonic resonance, i.e. $\hbar\omega_\mathrm{p} < E_0 - \hbar\omega = 2\hbar\Omega_1 - E_\mathrm{bi} - \hbar\omega$, where E_bi is again the biexcitonic binding energy, the bound-state contribution increases the blueshift. This can change if the pump frequency crosses the biexcitonic resonance. More precisely, we have the excitonic resonance at $\omega = \Omega_1$ and thus a vanishing shift, if

$$
\hbar\omega_\mathrm{p} = \hbar\Omega_1 - E_\mathrm{bi} - \frac{\mathcal{C}_\mathrm{bi}\tilde{p}_\mathrm{p}^2}{S} + \left(\frac{\mathcal{C}_\mathrm{bi}^2\tilde{p}_\mathrm{p}^4}{S^2} - \gamma_\mathrm{b}^2\right)^{1/2}\,. \tag{9.74}
$$

This relation implies a condition on the ratio of the biexcitonic coupling coefficient to the biexcitonic dephasing. In order to convert the blueshift into a redshift, we have to fulfill the relation

$$\frac{C_{\mathrm{bi}}}{\hbar\gamma_{\mathrm{b}}} \geq \frac{S}{\tilde{p}_{\mathrm{p}}^{2}} \,. \tag{9.75}$$

From this relation it can be understood why this redshift, which was predicted theoretically [238], was first observed in CuCl (see Fig. 9.5). Here the biexcitonic oscillator strength is much larger than in many other semiconductors. In most other material systems the biexciton coupling coefficient is too weak to overcompensate the RPA shift. Thus, from (9.75), one expects only a reduction of the blueshift when a biexcitonic resonance is crossed. A non-Markovian treatment of the coupling to the biexcitonic correlation shows, however, that the correlational shift to lower energies becomes enhanced with decreasing dephasing of the biexcitonic bound and scattering states – an effect which is completely missing in the Markov limit. In Fig. 9.6 the nonlinear absorption for off-resonant excitation of InGaAs quantum wells is shown for different pump–probe polarization configurations [232]. Although the observed redshift becomes pronounced for oppositely circularly polarized pulses, this is definitely a rather indirect manifestation of bound biexcitons.

9.4.2 Four-Wave Mixing

Much more direct signatures of bound biexcitons than those obtained in nonlinear absorption are obtained in four-wave-mixing (FWM) signals. We proceed as in Sect. 4.3 and decompose the equations with respect to the propagation direction of the various fields. Calculating the signal in the $2\boldsymbol{q}_2 - \boldsymbol{q}_1$ direction, we have, instead of (9.69)

$$\left.\frac{\partial}{\partial t} p^{eh}_{2\boldsymbol{q}_2-\boldsymbol{q}_1}\right|_{\mathrm{scatt}} = \sum_{e'h'} K^{hh'}_{+}\, p^{eh}_{\boldsymbol{q}_2} p^{e'h'}_{\boldsymbol{q}_2} p^{e'h'*}_{\boldsymbol{q}_1} \,, \tag{9.76}$$

(9.70) has to be replaced by

$$\left.\frac{\partial}{\partial t} p^{eh}_{2\boldsymbol{q}_2-\boldsymbol{q}_1}\right|_{\mathrm{bound}} = \sum_{e'h'} C^{hh'}_{\mathrm{b}} (1-\delta_{ee'}) b^{hh'}_{\boldsymbol{q}_2\boldsymbol{q}_2} p^{e'h'*}_{\boldsymbol{q}_1} \,, \tag{9.77}$$

and the other contributions to the signal are the same as already dicussed in Sect. 6.3. Again we start by discussing the polarization dependences. For different circular polarizations, the FWM signal is zero, as any transition amplitude that contributes for σ_+ light vanishes for σ_- light, and vice versa. This statement no longer holds exactly if off-diagonal elements of the kinetic energy are taken into account, which become the source of a finite FWM signal. Nevertheless, this signal is almost two orders of magnitude smaller than the signal obtained when the two pulses have the same circular polarization. In this geometry, however, the dominant heavy-hole bound-state

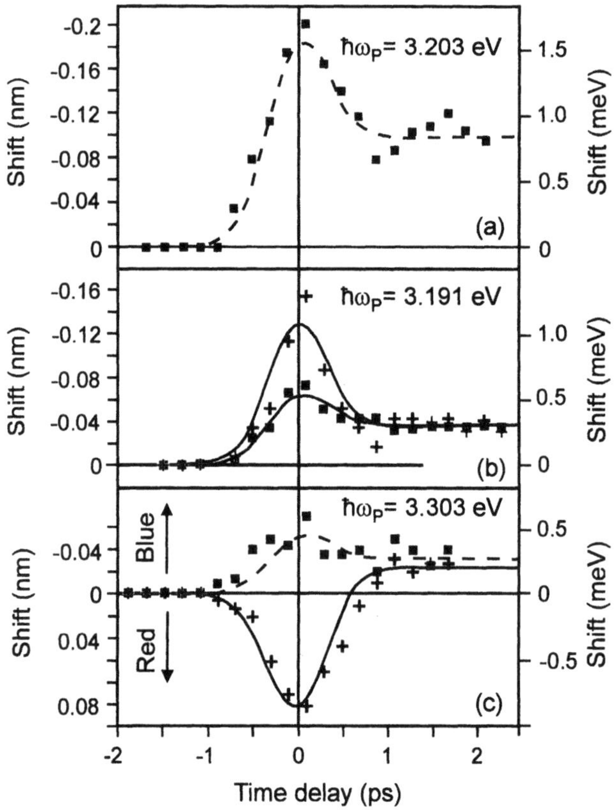

Fig. 9.5a–c. Shift of the excitonic resonance in CuCl as a function of the delay between the pump and test pulses for various pump frequencies; the biexcitonic resonance is at 3.185 eV. *Crosses* and *squares* denote parallel and perpendicular polarizations, respectively, of the pump and test pulses. The lines are guides for the eye. Taken from [237]

contribution vanishes and only bound states formed from heavy- and light-hole excitons are possible. Heavy-hole biexcitons contribute in the cases of parallel and crossed linear polarizations. The scattering contribution (9.76), however, vanishes in the latter case and the magnitude of the FWM signal is considerably reduced. Moreover, for parallel polarizations effects of bound biexcitons occur on the background of the RPA contribution. In the crossed linear case these effects may become more dominant.

Provided the biexcitonic coupling is sufficiently strong, there are three distinct signatures of bound states in FWM signals. The most obvious one is found in spectrally resolved FWM signals, which exhibit an additional resonance centered at E_{bi} below the exciton. In time-resolved and time-integrated signals this resonance gives rise to pronounced quantum beats, with period $\tau = h/E_{\mathrm{bi}}$, which, however, at the $\chi^{(3)}$ level occur at negative time delay

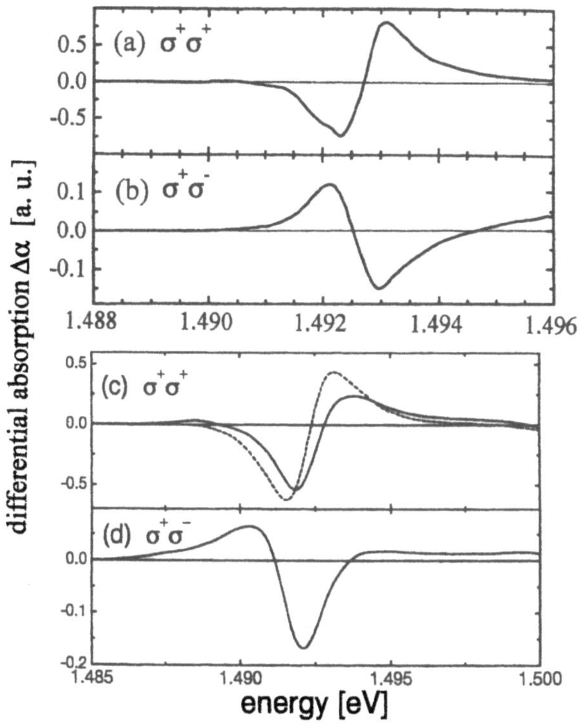

Fig. 9.6. Differential absorption spectra of InGaAs quantum wells at zero delay for excitation 4.5 meV below the heavy-hole exciton. (**a**), (**b**) Experimental data for equal and opposite circular polarizations, respectively, of pump and probe. (**c**), (**d**) show corresponding theoretical results obtained from numerical solution of (9.4) together with (9.17)–(9.20) for a two-dimensional quantum well model. The *dashed line* displays the Hartree–Fock result. Data taken from [232]

only. Depending on the excitation conditions, this beating can be superimposed on heavy-hole/light-hole quantum beats. A less obvious signature is the rise time of the time-integrated FWM signal. As we have seen in Sect. 6.3, the signal grows in proportion to $\exp(-4\,\gamma_2\,T)$ with a time delay T, owing to static exciton–exciton interaction. The same behavior is obtained from all interaction contributions which can be treated in the Markov limit. For bound biexcitons, which represent an independent dynamic degree of freedom, this statement is not true. If the dephasing of bound biexcitons fulfills the relation $\gamma_b < 2\gamma_2$,[1] then the signal at negative time delay exhibits a slower compo-

[1] The validity of such an inequality can be proved from a microscopic theory of dephasing processes such as that discussed in Chaps. 10 and 11. The inequality results from the fact that additionally to the sum of the dephasing rates of the two excitons there are correlational contributions. Technically, these additional contributions are vertex corrections.

nent, which results from biexcitonic transitions. Moreover, this component can become dominant if the $\boldsymbol{E}_2$ pulse is tuned selectively to the biexcitonic resonance. These signatures, as well as their polarization dependences, have been studied in AlGaAs quantum well systems [216, 239, 240, 241, 242].

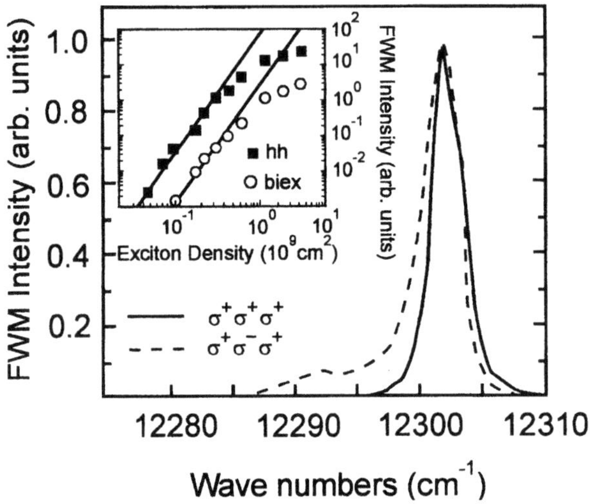

Fig. 9.7. Spectrally resolved FWM signal for two different combinations of the polarizations of three 1.0 ps pulses. The *inset* shows the growth of the signal at the exciton and biexciton resonance as a function of the exciton density and indicates the range in which a $\chi^{(3)}$ approximation (*solid line*) is valid. Taken from [239]

Even more selective studies of the various interaction contributions become possible if the four-wave mixing is performed with three external pulses. In this case the scattering contribution for the signal propagating in the direction $\boldsymbol{q}_1 + \boldsymbol{q}_2 - \boldsymbol{q}_3$ takes the form

$$\left.\frac{\partial}{\partial t} p^{eh}_{\boldsymbol{q}_1+\boldsymbol{q}_2-\boldsymbol{q}_3}\right|_{\text{scatt}} = \sum_{e'h'} K^{hh'}_{+} \left(p^{eh}_{\boldsymbol{q}_1} p^{e'h'}_{\boldsymbol{q}_2} + p^{eh}_{\boldsymbol{q}_2} p^{e'h'}_{\boldsymbol{q}_1} \right) p^{e'h'*}_{\boldsymbol{q}_3} , \tag{9.78}$$

and the bound-state contribution is given by

$$\left.\frac{\partial}{\partial t} p^{eh}_{\boldsymbol{q}_1+\boldsymbol{q}_2-\boldsymbol{q}_3}\right|_{\text{bound}} = \sum_{e'h'} C^{hh'}_{\text{b}} (1-\delta_{ee'}) \left(b^{hh'}_{\boldsymbol{q}_1\boldsymbol{q}_2} + b^{hh'}_{\boldsymbol{q}_2\boldsymbol{q}_1} \right) p^{e'h'*}_{\boldsymbol{q}_3} . \tag{9.79}$$

Now it is obvious that if the $\boldsymbol{E}_1$,$\boldsymbol{E}_2$, and $\boldsymbol{E}_3$ pulses have polarizations σ_+, σ_-, and σ_+, respectively, or σ_-, σ_+, and σ_+, all phase-space and static interaction contributions vanish. Only (9.78) and (9.79), which couple the two transition

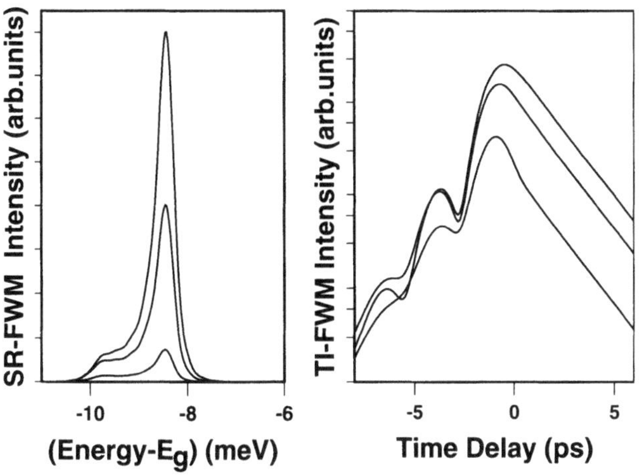

Fig. 9.8. The *right-hand side* shows spectrally resolved FWM signals calculated for three 1.0 ps pulses and the $\sigma_+\sigma_-\sigma_+$ polarization geometry. The excitation energy is tuned to the exciton resonance at -8.4 meV, between the resonances at -9,2 meV, and to the biexciton resonance at -9.9 meV (*top* to *bottom*). The *left-hand side* shows the corresponding time-integrated FWM signal as a function of the time delay T_{23} for $T_{12} = 0$. Parameters correspond to a 15 nm GaAs quantum well

subspaces, contribute. Corresponding experimental results [239] are shown in Fig. 9.7. The spectrally resolved FWM signal shows clearly the biexcitonic resonance below the heavy-hole exciton. If all pulses have the same circular polarization this contribution vanishes, as expected. Using the dephasing rates of excitons and biexcitons as adjustable parameters, it not difficult to reproduce these experimental data. Corresponding results for spectrally resolved and time-integrated FWM signals and their dependence on the excitation energy are shown in Fig. 9.8. Owing to the similtaneous excitation of the excitonic and the biexcitonic resonance, we obtain pronounced quantum beats for negative time delay T_{23}, in agreement with experimantal data.

If the excitation intensity is increased, higher order contributions come into play at even rather moderate exciton densities (see inset of Fig. 9.7). At the $\chi^{(5)}$ level, we obtain also quantum beats for positive time delay, and higher order diffracted signals, for example, in $3\boldsymbol{q}_1 - 2\boldsymbol{q}_2$ direction, can be detected [222].

Increasing the density further even more, perturbation theory breaks down. Owing to the interaction of biexcitons, excitons, and free carriers the dephasing rates of biexcitons and excitons increase with the excitation density and give rise to a rapid decay of coherence. Correspondingly the coherently generated correlations are destroyed. In this regime, the coherently controlled truncation of the many-particle problem is inappropriate, and we prefer a different approach, which can also be applied to derive the results of this chapter [231, 243, 244].

9.5 Problems

9.1. Derive the equation of motion for the density–density correlation function

$$n_{eh}^{e'h'}(\boldsymbol{k}_1, \boldsymbol{k}_2, \boldsymbol{k}_3, \boldsymbol{k}_4)$$
$$= \langle c^\dagger_{e,\boldsymbol{k}_1} c^\dagger_{h,-\boldsymbol{k}_2} c_{e',\boldsymbol{k}_3} c_{h',-\boldsymbol{k}_4}\rangle - \langle c^\dagger_{e,\boldsymbol{k}_1} c^\dagger_{h,-\boldsymbol{k}_2}\rangle\langle c_{e',\boldsymbol{k}_3} c_{h',-\boldsymbol{k}_4}\rangle \tag{9.80}$$

and show that the source vanishes in the coherent limit. Hint: Make use of the Heisenberg equations (5.3) and (5.4).

9.2. Prove the relation (9.14).

9.3. Derive an expression for the renormalization of the gap energy due to the resonant excitation of 1s excitons.

9.4. Calculate the four-wave-mixing signal for crossed linear and collinear polarizations for δ pulse excitation. Show that the simultaneous excitation of excitons and biexcitons leads to quantum beats for positive time delay.

9.5. Consider the polarization dependences of four-wave-mixing signals in the case of three external pulses. Discuss which interaction contributions can be switched off in which polarization geometry.

9.6. Repeat Problem 9.5 for the nonlinear absorption.

10. Nonequilibrium Green's Functions

As we have already seen in the previous chapters, the dynamics of electrons in a semiconductor are the result of the solution of a many-body problem. This is due to the direct Coulomb interaction among carriers and the mediated interaction via phonons. So far, the theoretical approaches that we have developed to tackle this difficult problem have all been based on the *density matrix* approach and – apart from the $\chi^{(n)}$ expansion in the last chapter – on the Hartree–Fock approximation. This *mean-field* approach, however, neglects important physics such as *static* and *dynamic screening*, as well as all kinds of *relaxation*, *correlation*, and *scattering* processes. Before reviewing various aspects of relaxation in Chap. 12, we have to get ready for some more sophisticated mathematics.

Whereas the Hartree–Fock approximation can be obtained in any attempt to solve the many-particle problem, beyond this approximation, different methods can yield different approximation schemes and hence different results. The preferred method depends not only on the problem but also on the taste of the physicist investigating it.

From the example of the $\chi^{(n)}$ expansion, we have seen that the density matrix approach leads to a hierarchy of kinetic equations, which requires a systematic *decoupling scheme*. While this approach is best suited for obtaining results (even rigorous ones) in the coherent regime, the coherent truncation scheme becomes meaningless if incoherent occupations of excited electrons and holes come into play. Unfortunately, this is often the case for electronic excitations in semiconductors. Nevertheless, we could still apply the density matrix method in the incoherent case to develop approximation schemes beyond the Hartree–Fock level. In order to close the system of equations, for example, on the one-particle level, a reduction of the sources of higher-order correlations to one-particle quantities is required. In many cases such a procedure is rather arbitrary and is guided by a priori knowlegde of the desired result. Although it is possible to derive systematic approximation schemes within the density matrix approach, there are other methods which approach the problem in a more elegant manner.

We can distinguish between two classes of theories. One approach, frequently applied in nonlinear optics and quantum optics, describes the time evolution of observables by means of the time evolution of the *statistical ope-*

rator, reduced to variables of a *relevant subsystem*, which usually has a few degrees of freedom only. This procedure constitutes the Nakajima–Zwanzig approach [245, 246, 247]. The properties of this subsystem are, however influenced by the rest of the world; this has many degrees of freedom, which act as a bath or *reservoir.* Often one is interested only in the dynamics of the small subsystem and can neglect its influence on the reservoir. Thus the latter can be assumed to be in thermodynamic equilibrium. In the simplest case the action of the reservoir may be described in terms of damping, which leads to irreversibility of the time evolution of the subsystem. Whenever a separation of the entire problem into a relevant subsystem and a reservoir (with a necessarily unknown dynamics) is meaningful, this is a powerful approach. However, in solids, in particular when we anticipate nonequilibrium conditions, we prefer a different method which treats all constituents of the system on the same footing (such as the density matrix formalism).

Up to now we have investigated *one-particle density matrices* of the form $\langle c_n^\dagger(t)c_{n'}(t)\rangle$ or expressed in terms of field operators $\langle\psi^\dagger(\boldsymbol{r}t)\psi(\boldsymbol{r}'t)\rangle$. The introduction of two separate times for each operator, i.e. consideration of expectation values such as $\langle c_n^\dagger(t)c_{n'}(t')\rangle$ or $\langle\psi^\dagger(\boldsymbol{r}t)\psi(\boldsymbol{r}'t')\rangle$, leads us naturally to the concept of *Green's functions* [248, 249, 250]. This allows the construction of consistent approximation methods on the one-, two-, ..., N-particle levels. As a penalty, we have the complication of more than one time variable.

The theory of many-particle Green's functions and, in particular, of their application to nonequilibrium systems has been developed by many authors [251, 252, 253, 254] and reviewed numerous times (see e.g. [255]). In the following we shall give a compact introduction into the theory of nonequilibrium Green's functions, which is oriented around the structure introduced for the equilibrium case [256]. We start our considerations with an extension of linear response theory, introduced in Sect. 2.6.3, to the case of arbitrary external perturbations.

10.1 Time Evolution under the Action of External Fields

Let us consider the expectation value of an operator $\hat{O}$ for a system which starts from its ground state at zero temperature and evolves in time under the action of the Hamiltonian $H + H_{\text{ext}}(t)$. According to (2.146) and (2.147), we obtain

$$\begin{aligned}\langle\Psi(t)|\hat{O}|\Psi(t)\rangle &= \langle\Psi_0|\hat{O}(t)|\Psi_0\rangle = \langle\Psi_0|U^\dagger(t,t_0)\hat{O}U(t,t_0)|\Psi_0\rangle \\ &= \langle\Psi_0|S^\dagger(t,t_0)\bar{O}(t)S(t,t_0)|\Psi_0\rangle\,, \end{aligned} \tag{10.1}$$

where we have set $|\Psi_0\rangle = |\Psi(t_0)\rangle$. In order to keep our notation compact, we shall replace the expectation value $\langle\Psi_0|\ldots|\Psi_0\rangle$ by $\langle\ldots\rangle$ from now on.[1] In

[1] It should be noted that we can use instead of the expectation value $\langle\Psi_0|\ldots|\Psi_0\rangle$ also the statistical average $\mathrm{Tr}\{\hat{\varrho}\ldots\}$, where $\hat{\varrho}$ is the statistical operator of the

what follows, our aim is to derive a formally exact expression for the *time-evolution operator*, which so far has been treated in first-order perturbation theory only. By iteration, we obtain from the integral equation (2.150) an infinite expansion of the form

$$S(t,t_0) = 1 - \frac{\mathrm{i}}{\hbar}\int_{t_0}^{t} \mathrm{d}t' \bar{H}_{\mathrm{ext}}(t') + \left(\frac{\mathrm{i}}{\hbar}\right)^2 \int_{t_0}^{t} \mathrm{d}t' \bar{H}_{\mathrm{ext}}(t') \int_{t_0}^{t'} \mathrm{d}t'' \bar{H}_{\mathrm{ext}}(t'') - \ldots \tag{10.2}$$

As the simplest example, we consider the quadratic contribution. There are two different ways to write this term such that both integrals have the same upper limit. We can either introduce $\theta(t'-t'')$ in the second integral or first interchange the variables t' and t'' and introduce $\theta(t''-t')$ in the first integral. This allows us to write the quadratic contribution in the form

$$\int_{t_0}^{t} \mathrm{d}t' \bar{H}_{\mathrm{ext}}(t') \int_{t_0}^{t'} \mathrm{d}t'' \bar{H}_{\mathrm{ext}}(t'') = \frac{1}{2}\int_{t_0}^{t} \mathrm{d}t'\mathrm{d}t'' \left[\bar{H}_{\mathrm{ext}}(t'')\bar{H}_{\mathrm{ext}}(t')\theta(t''-t') + \bar{H}_{\mathrm{ext}}(t')\bar{H}_{\mathrm{ext}}(t'')\theta(t'-t'')\right] . \tag{10.3}$$

The expression in square brackets is just the definition of a *time-ordered product* for the simplest case of two operators, which we can abbreviate by acting with a positive *time-ordering operator*, denoted by $\mathbf{T}_+$, on the product of the operators. By definition, this results in a sequence of operators which is ordered from left to right with decreasing time arguments. It is straightforward to show that for the nth order contribution in the expansion of (10.2), there are $n!$ possible ways to arrange the product of n operators. The corresponding $n!$ different sequences of operators can again be written in compact form by means of $\mathbf{T}_+$. Obviously, the resulting expansion has the structure of the series expansion of an exponential function. This allows the formal summation of the series, which results in the representation

$$S(t,t_0) = \mathbf{T}_+ \exp\left(-\frac{\mathrm{i}}{\hbar}\int_{t_0}^{t} \mathrm{d}t' \bar{H}_{\mathrm{ext}}(t')\right) . \tag{10.4}$$

To calculate the expectation value (10.1), we also need the adjoint operator $S^\dagger(t,t_0) = S(t_0,t)$. Going again through the derivation of (10.4), we obtain the same structure as before. Owing to the interchange of t and t_0, however, we now have a negative time order. This results in

system in equilibrium (see Sect. 3.2). The formalism, which we develop in the following, is the same.

$$S(t_0,t) = \mathbf{T}_- \exp\left(-\frac{\mathrm{i}}{\hbar}\int_t^{t_0} \mathrm{d}t' \bar{H}_{\mathrm{ext}}(t')\right) , \tag{10.5}$$

where the action of $\mathbf{T}_-$ generates a sequence of operators which is ordered from left to right with increasing time arguments. The occurence of both *positive* and *negative* time ordered contributions in the expectation value has important consequences. Let assume us for the moment that $S(t,t_0)$ does not describe the time evolution under the action of H_{ext} but evolves the system under the action of an internal interaction which can be switched on adiabatically. In such a case, the action of $S(t_0,t)$ on the ground state is again proportional to the ground state, and no admixture of excited states can occur. Thus we would obtain

$$\langle S(t_0,t)\bar{O}(t)S(t,t_0)\rangle = \langle S(t_0,t)\rangle\langle\bar{O}(t)S(t,t_0)\rangle . \tag{10.6}$$

Using this factorization again, we have

$$\langle S(t_0,t)S(t,t_0)\rangle = \langle S(t_0,t)\rangle\langle S(t,t_0)\rangle = 1 , \tag{10.7}$$

which, together with (10.6), yields

$$\langle S(t_0,t)\bar{O}(t)S(t,t_0)\rangle = \frac{\langle\bar{O}(t)S(t,t_0)\rangle}{\langle S(t,t_0)\rangle} . \tag{10.8}$$

Thus, under *equilibrium conditions*, only *positive time ordering* occurs. This factorization is obviously not possible if excited states come into play owing to time-dependent external fields. In nonequilibrium situations this requires us to account for both types of time ordering. Formally, it is most convenient to introduce one single time ordering along a time contour C, which extends from t_0 to t and then goes back to t_0. We can further utilize the relation $S(t,\infty)S(\infty,t) = 1$ and set $t_0 = -\infty$ in order to extend the time contour from $-\infty$ to ∞ and then back to $-\infty$. This allows us to write the expectation value in several different but equivalent forms:

$$\begin{aligned}
\langle\hat{O}(t)\rangle &= \langle S(-\infty,t)\bar{O}(t)S(t,-\infty)\rangle \\
&= \langle S(-\infty,t)S(t,\infty)S(\infty,t)\bar{O}(t)S(t,-\infty)\rangle \\
&= \langle S(-\infty,\infty)[\mathbf{T}_+ S(\infty,-\infty)\bar{O}(t)]\rangle \\
&= \langle[\mathbf{T}_- S(-\infty,\infty)\bar{O}(t)]S(\infty,-\infty)\rangle \\
&= \langle\mathbf{T}_{\mathrm{c}} S^\dagger(\infty,-\infty)S(\infty,-\infty)\bar{O}(t)\rangle .
\end{aligned} \tag{10.9}$$

In the last step we have introduced a time ordering along a contour C, which is illustrated in Fig. 10.1. Like the original definition of the expectation value, this representation describes nothing but the time evolution of the system from $-\infty$ to t under the action of an external perturbation and the evolution back from t to $-\infty$. Let us assume that we could distinguish between the perturbation which evolves the system forward in time and that which evolves it backward, i.e. between perturbations lying on the upper and on the lower branch of the time contour C. This suggests that we should

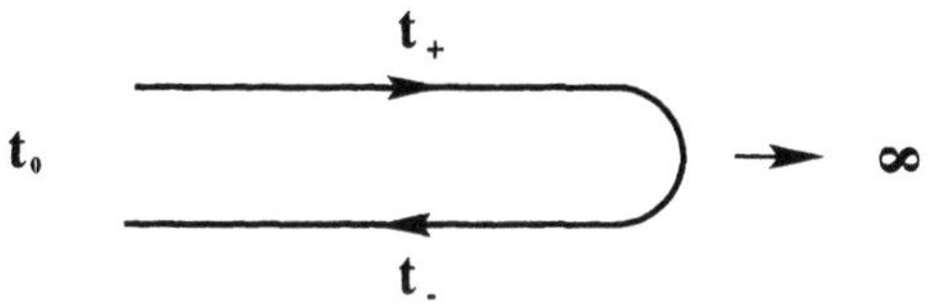

Fig. 10.1. Time contour for time ordering in nonequilibrium

define

$$S_+ = S_+(\infty, -\infty) = \mathbf{T}_+ \exp\left(-\frac{\mathrm{i}}{\hbar}\int_{-\infty}^{\infty} \mathrm{d}t' \bar{H}^+_{\mathrm{ext}}(t')\right) \tag{10.10}$$

and

$$S_-^\dagger = S_-(-\infty, \infty) = \mathbf{T}_- \exp\left(\frac{\mathrm{i}}{\hbar}\int_{-\infty}^{\infty} \mathrm{d}t' \bar{H}^-_{\mathrm{ext}}(t')\right) . \tag{10.11}$$

Correspondingly, the alternative formulations in (10.9) are no longer equivalent and we have two types of expectation values. If we consider an operator which evolves under the action of $H^+_{\mathrm{ext}}(t)$, we label its time argument by a branch index and denote it as $(t+)$. Otherwise, for the action of $H^-_{\mathrm{ext}}(t)$, we denote the argument as $(t-)$. Depending on wether the time arguments lie on the upper branch or on the lower branch of the time contour, we shall obtain different results. As $S_-^\dagger S_+ \neq 1$, we have to normalize these expectation values, and obtain, as a generalization of (10.9),

$$\langle \hat{O}(t+)\rangle = \frac{\langle S_-^\dagger [\mathbf{T}_+ S_+ \bar{O}(t)]\rangle}{\langle S_-^\dagger S_+\rangle} \tag{10.12}$$

and, correspondingly, for the case of a time argument on the lower branch

$$\langle \hat{O}(t-)\rangle = \frac{\langle [\mathbf{T}_- S_-^\dagger \bar{O}(t)] S_+\rangle}{\langle S_-^\dagger S_+\rangle} . \tag{10.13}$$

Physically this distinction is completely meaningless, since we have nothing but $H_{\mathrm{ext}}(t)$ and thus we should set $H^+_{\mathrm{ext}}(t) = H^-_{\mathrm{ext}}(t)$ and, correspondingly, $S_- = S_+$ in any final result. Mathematically, however, it is a powerful trick which will allow us to derive the equations of motion of two-time functions in compact form.

10.2 Definitions of One-Particle Green's Functions

We start our discussion with the definition of the one-particle Green's functions, which represent generalizations of the one-particle density matrices to the case of two time arguments. These definitions are suggested by the straightforward generalization of the expectation values (10.12) and (10.13), which for two-time functions obviously leads to four different combinations of time arguments and thus to four different types of Green's functions. In order to keep our notation compact we combine the spatial, temporal, and spin variables into a single argument according to $1 = \{\boldsymbol{r}_1, t_1, \sigma_1\}$ and in general terms $n = \{\boldsymbol{r}_n, t_n, \sigma_n\}$. Later on, we shall include also the branch index in the definition of the argument. At the moment we still treat the branch index explicitly and define the following four Green's functions. If both time arguments are on the upper branch of the time contour, we define

$$\begin{aligned} G^{++}(12) &= -\frac{\mathrm{i}}{\hbar} \frac{\langle \mathbf{T}_{\mathrm{c}} \psi(1+) \psi^\dagger(2+) \rangle}{\langle S_-^\dagger S_+ \rangle} \\ &= -\frac{\mathrm{i}}{\hbar} \frac{\langle S_-^\dagger [\mathbf{T}_+ S_+ \bar{\psi}(1) \bar{\psi}^\dagger(2)] \rangle}{\langle S_-^\dagger S_+ \rangle} . \end{aligned} \tag{10.14}$$

The time ordering also implies a change of sign due to the anticommuting Fermi operators, i.e.

$$\mathbf{T}_+ \psi(1) \psi^\dagger(2) = \theta(t_1 - t_2) \psi(1) \psi^\dagger(2) - \theta(t_2 - t_1) \psi^\dagger(2) \psi(1) . \tag{10.15}$$

In the limit $H^{+}_{\mathrm{ext}}(t) = H^{-}_{\mathrm{ext}}(t)$, (10.14) corresponds formally to the usual definition of the time-ordered one-particle Green's function at zero temperature. Owing to the external fields, however, again no factorization like (10.8) is possible and, also, three other types of Green's functions come into play. As the corresponding definition for the case in which both time arguments are on the lower branch, we have

$$\begin{aligned} G^{--}(12) &= \frac{\mathrm{i}}{\hbar} \frac{\langle \mathbf{T}_{\mathrm{c}} \psi(1-) \psi^\dagger(2-) \rangle}{\langle S_-^\dagger S_+ \rangle} \\ &= -\frac{\mathrm{i}}{\hbar} \frac{\langle [\mathbf{T}_- S_-^\dagger \bar{\psi}(1) \bar{\psi}^\dagger(2)] S_+ \rangle}{\langle S_-^\dagger S_+ \rangle} . \end{aligned} \tag{10.16}$$

If the two time arguments are on different branches, we have to introduce

$$\begin{aligned} G^{+-}(12) &= \frac{\mathrm{i}}{\hbar} \frac{\langle \mathbf{T}_{\mathrm{c}} \psi(1+) \psi^\dagger(2-) \rangle}{\langle S_-^\dagger S_+ \rangle} \\ &= -\frac{\mathrm{i}}{\hbar} \frac{\langle [\mathbf{T}_- S_-^\dagger \bar{\psi}^\dagger(2)] [\mathbf{T}_+ S_+ \bar{\psi}(1)] \rangle}{\langle S_-^\dagger S_+ \rangle} . \end{aligned} \tag{10.17}$$

This propagator has a simple physical interpretation. It is just the transition amplitude from a state where a particle is removed from the system at a point $\boldsymbol{r}_1$ and at a time t_1 to a state where the particle is removed at a point $\boldsymbol{r}_2$ and at a time t_2. In other words (10.17) is the probability amplitude for the *propagation of a hole* from $\boldsymbol{r}_1, t_1$ to $\boldsymbol{r}_2, t_2$. Correspondingly, the propagator

$$G^{-+}(12) = -\frac{\mathrm{i}}{\hbar}\frac{\langle \mathbf{T}_{\mathrm{c}}\psi(1-)\psi^{\dagger}(2+)\rangle}{\langle S_{-}^{\dagger}S_{+}\rangle}$$
$$= -\frac{\mathrm{i}}{\hbar}\frac{\langle [\mathbf{T}_{-}S_{-}^{\dagger}\bar{\psi}(1)][\mathbf{T}_{+}S_{+}\bar{\psi}^{\dagger}(2)]\rangle}{\langle S_{-}^{\dagger}S_{+}\rangle} \tag{10.18}$$

describes the probability amplitude for the *propagation of a particle* which is added to the system. Transforming back to the Heisenberg picture, we can write the four Green's functions (10.14) and (10.16)–(10.18) in the more compact form

$$G(\bar{1}\bar{2}) = -\frac{\mathrm{i}}{\hbar}\, b_2\, \frac{\langle \mathbf{T}_{\mathrm{c}}\psi(\bar{1})\psi^{\dagger}(\bar{2})\rangle}{\langle S_{-}^{\dagger}S_{+}\rangle}\,, \tag{10.19}$$

where the branch index $b_n = \pm$ is included in the definition of the argument according to $\bar{n} = \{\boldsymbol{r}_n, t_n, \sigma_n, b_n\}$. As is obvious from their definitions, the four functions are not independent but fulfill the relation

$$G^{++}(12) + G^{+-}(12) = G^{--}(12) + G^{-+}(12)$$
$$= -\frac{\mathrm{i}}{\hbar}\,\theta(t_1 - t_2)\,\langle \psi(1)\psi^{\dagger}(2) + \psi^{\dagger}(2)\psi(1)\rangle$$
$$= \theta(t_1 - t_2)[G^{-+}(12) + G^{+-}(12)] \equiv G^{\mathrm{r}}(12)\,, \tag{10.20}$$

which defines the *retarded Green's function* G^{r}. The *advanced Green's function* G^{a} is obtained from

$$G^{++}(12) - G^{-+}(12) = G^{--}(12) - G^{+-}(12)$$
$$= \frac{\mathrm{i}}{\hbar}\,\theta(t_2 - t_1)\,\langle \psi(1)\psi^{\dagger}(2) + \psi^{\dagger}(2)\psi(1)\rangle$$
$$= -\theta(t_2 - t_1)[G^{-+}(12) + G^{+-}(12)) \equiv G^{\mathrm{a}}(12]\,. \tag{10.21}$$

A knowledge of the *propagators* G^{+-} and G^{-+} is obviously sufficient to determine the two other types of Green's functions, in particular the *spectral functions* G^{r} and G^{a}. In an equilibrium situation even one function would be sufficient to completely characterize the system. In this case, we merely have to determine the spectral properties, as the information about the occupation can be obtained from quantum statistics (see Sect. 3.2). In this case, it is a simple exercise to analyze the physical meaning of the above definitions.

Example (Representation in the Quasiparticle Approximation). Let us consider for example the G^{+-} propagator. According to (2.59) and (2.60), we can expand the field operators by means of a complete set of one-particle states, which yields, if $S_{-}^{\dagger} = S_{+} = 1$,

$$G^{+-}(12) = \sum_{jj'} \varphi_j(\boldsymbol{r}_1\sigma_1)\,\varphi_{j'}^{*}(\boldsymbol{r}_2\sigma_2)\,G_{jj'}^{+-}(t_1t_2)\,. \tag{10.22}$$

The projection onto the one-particle states is defined as

$$G^{+-}_{jj'}(t_1t_2) = -\frac{\mathrm{i}}{\hbar}\langle c^\dagger_{j'}(t_2)c_j(t_1)\rangle$$
$$= -\frac{\mathrm{i}}{\hbar}\sum_n \langle \Phi_0^{(N)}|c^\dagger_{j'}(t_2)|\Phi_n^{(N-1)}\rangle\langle\Phi_n^{(N-1)}|c_j(t_1)|\Phi_0^{(N)}\rangle\,. \quad (10.23)$$

In the last step, we have inserted a complete set of $(N-1)$-particle states which are eigenstates of the many-particle Hamiltonian H. This property, together with

$$c_j(t_1) = \mathrm{e}^{-\mathrm{i}Ht_1/\hbar}c_j\mathrm{e}^{\mathrm{i}Ht_1/\hbar}\,, \quad (10.24)$$

allows us to write

$$G^{+-}_{jj'}(t_1t_2)$$
$$= -\frac{\mathrm{i}}{\hbar}\sum_n \mathrm{e}^{\mathrm{i}\,(E_n^{(N-1)}-E_0^{(N)})(t_2-t_1)/\hbar}\langle\Phi_0^{(N)}|c^\dagger_{j'}|\Phi_n^{(N-1)}\rangle\langle\Phi_n^{(N-1)}|c_j|\Phi_0^{(N)}\rangle\,. \quad (10.25)$$

Here $E_n^{(N-1)} - E_0^{(N)} = E_n^{(N-1)} - E_0^{(N-1)} - (E_0^{(N)} - E_0^{(N-1)}) = E_n$ is the difference between the *exact excitation energy* and the chemical potential of the $(N-1)$-particle system, which we introduced earlier within the Hartree–Fock approximation in Sect. 3.1. Within this approximation, the ground state is given by a single Slater determinant and we can apply (2.41) and (2.42) to obtain

$$G^{+-}(12) = -\frac{\mathrm{i}}{\hbar}\sum_j \varphi_j(\boldsymbol{r}_1\sigma_1)\varphi_j^*(\boldsymbol{r}_2\sigma_2)\mathrm{e}^{\mathrm{i}E_j(t_2-t_1)/\hbar}\,n_j\,. \quad (10.26)$$

In the same way, the G^{-+} function is obtained as

$$G^{-+}(12) = -\frac{\mathrm{i}}{\hbar}\sum_j \varphi_j(\boldsymbol{r}_1\sigma_1)\varphi_j^*(\boldsymbol{r}_2\sigma_2)\mathrm{e}^{\mathrm{i}E_j(t_2-t_1)/\hbar}(1-n_j)\,. \quad (10.27)$$

Thus the G^{+-}- and G^{-+} propagators are related to the properties of occupied and unoccupied states, respectively. This is obviously generally valid, as (10.25) in the equal-time limit reduces to the density matrix and diagonal contributions determine the one-particle distribution functions, whereas off-diagonal elements determine the transition amplitudes (compare Sect. 10.5). In the sum of G^{+-} and G^{-+}, the information about the occupation cancels. Thus the retarded and advanced Green's functions contain directly only the information about the one-particle spectrum. Taking the Fourier transform of (10.26) and (10.27), we obtain a spectrum which consists of a series of δ singularities at the Hartree–Fock energies weighted with the occupation number. Correspondingly, the retarded and advanced Green's functions are meromorphic functions in the complex frequency plane, which have simple poles slightly below and slightly above the real axis, respectively.

Beyond the Hartree–Fock theory, the operators in (10.25) connect the ground state with many excited states of the $(N-1)$-particle system. Thus (10.25) remains off-diagonal even without external fields. The time dependence is no longer determined by single plane waves, but rather by a complicated superposition of oscillations, which correspond in the frequency domain, to a superposition of δ functions. As the level spacing vanishes in the thermodynamic limit, the discrete poles in the complex frequency plane merge to form a branch line. If the deviations from the mean-field theory are not too large, we expect that the sharp singularities will evolve into broadened structures, reflecting the *finite lifetime* of one-particle states in an interacting many-particle system. The corresponding *poles in the complex frequency plane* are constitutive of a general definition of *quasiparticle* energies, in particular of the *band structure.* A representation such as (10.25) is, however, not well suited for investigations beyond the mean-field theory and other methods which allow systematic approximation schemes are required. In particular, nonequilibrium situations require us – as in the special case of the semiconductor Bloch equations – to determine the information about the one-particle distributions and transition amplitudes from quantum kinetic equations.

10.3 Equations of Motion of One-Particle Green's Functions

We consider the time evolution of the Green's functions under the action of the Hamiltonian (5.1), which has the real-space representation (compare (2.63))

$$
\begin{aligned}
H = &\sum_{\sigma} \int \mathrm{d}^3\boldsymbol{r}\, \psi_\sigma^\dagger(\boldsymbol{r}t)[h(\boldsymbol{r}t) + e\phi_{\mathrm{ext}}(\boldsymbol{r}t)]\psi_\sigma(\boldsymbol{r}t) \\
&+ \frac{e^2}{2}\sum_{\sigma\sigma'} \int \mathrm{d}^3\boldsymbol{r}\mathrm{d}^3\boldsymbol{r}'\psi_\sigma^\dagger(\boldsymbol{r}t)\psi_{\sigma'}^\dagger(\boldsymbol{r}'t)v(\boldsymbol{r}-\boldsymbol{r}')\psi_{\sigma'}(\boldsymbol{r}'t)\psi_\sigma(\boldsymbol{r}t)\,. \qquad (10.28)
\end{aligned}
$$

The one-particle Hamiltonian is given by

$$
h(\boldsymbol{r}t) = -\frac{\hbar^2}{2m}\Delta + e\,U(\boldsymbol{r}) + \boldsymbol{d}\boldsymbol{E}(\boldsymbol{r},t)\,, \qquad (10.29)
$$

further, we have added $\phi_{\mathrm{ext}}(\boldsymbol{r}t)$ as an additional, time-dependent external potential. From (10.19), we obtain the equation of motion

$$
\begin{aligned}
\mathrm{i}\hbar\frac{\partial}{\partial t_1}G(\bar{1}\bar{2}) = &\frac{\mathrm{i}}{\hbar}\, b_2\langle \mathbf{T}_{\mathrm{c}}[H,\psi(\bar{1})]\psi^\dagger(\bar{2})\rangle \\
&+ \delta_{b_1,b_2}\delta(t_1-t_2)\langle\psi(\bar{1})\psi^\dagger(\bar{2}) + \psi^\dagger(\bar{2})\psi(\bar{1})\rangle\,. \qquad (10.30)
\end{aligned}
$$

The second contribution on the right-hand side results from the derivative of the θ functions and because of the anticommutation of the field operators, can be reduced to

$$\delta(\bar{1}-\bar{2}) = \delta(\boldsymbol{r}_1 - \boldsymbol{r}_2)\delta(t_1 - t_2)\delta_{\sigma_1,\sigma_2}\delta_{b_1,b_2} \, . \tag{10.31}$$

Inserting the commutator with the Hamiltonian (10.28) and introducing the compact notations

$$v(\bar{1}-\bar{2}) = v(\boldsymbol{r}_1 - \boldsymbol{r}_2)\delta(t_1 - t_2)\delta_{b_1,b_2} \tag{10.32}$$

and

$$\int \mathrm{d}\bar{1} = \sum_{b_1\sigma_1} \int \mathrm{d}^3\boldsymbol{r}_1 \, \mathrm{d}t_1 \, , \tag{10.33}$$

we obtain

$$\begin{aligned} \mathrm{i}\hbar \frac{\partial}{\partial t_1} G(\bar{1}\bar{2}) &= \delta(\bar{1}-\bar{2}) + [h(\bar{1}) + e\,\phi_{\mathrm{ext}}(\bar{1})]\, G(\bar{1}\bar{2}) \\ &\quad - \frac{\mathrm{i}}{\hbar} b_2 e^2 \int \mathrm{d}\bar{3}\, v(\bar{1}^+ - \bar{3}) \langle \mathbf{T}_{\mathrm{c}} \psi^\dagger(\bar{3}^+)\psi(\bar{3})\psi(\bar{1})\psi^\dagger(\bar{2})\rangle \, . \end{aligned} \tag{10.34}$$

The notation $\bar{1}^+$ means that the time argument t_1 is replaced by $t_1 + b_1\delta$, where the shift $\delta \to +0$ guarantees the correct sequence of operators under the action of $\mathbf{T}_{\mathrm{c}}$. As in the case of the equation of motion for the one-particle density matrix (5.7), we have to face the problem that a two-particle function enters the equation of the one-particle Green's function. If we were to apply the Hartree–Fock factorization as before, we could easily recover all earlier results obtained within the density-matrix formalism. Our aim in the following, however, is to derive an exact equation for the one-particle Green's functions. This can be accomplished by means of the Green's functions defined in the last section, which may be utilized as generating functionals, i.e. the two-particle functions can be expressed by means of functional derivatives of the one-particle functions with respect to the external potential in (10.28). Let us consider, for example, the functional derivative of (10.14) with respect to $\phi_{\mathrm{ext}}(3+)$, which, by definition, acts on the S_+ operator only:

$$\begin{aligned} \frac{\delta G^{++}(12)}{\delta \phi_{\mathrm{ext}}(3+)} &= -\frac{\mathrm{i}}{\hbar \langle S_-^\dagger S_+\rangle} \left\langle S_-^\dagger \left(\mathbf{T}_+ \frac{\delta S_+}{\delta\phi_{\mathrm{ext}}(3+)} \bar{\psi}(1)\bar{\psi}^\dagger(2) \right) \right\rangle \\ &\quad - \frac{G^{++}(12)}{\langle S_-^\dagger S_+\rangle} \left\langle S_-^\dagger \frac{\delta S_+}{\delta\phi_{\mathrm{ext}}(3+)} \right\rangle . \end{aligned} \tag{10.35}$$

Making use of the functional derivative of (10.10),

$$\frac{\delta S_+}{\delta\phi_{\mathrm{ext}}(3+)} = -\frac{\mathrm{i}\,e}{\hbar} \mathbf{T}_+ S_+ \bar{\psi}^\dagger(3)\bar{\psi}(3) \, , \tag{10.36}$$

we find

$$\begin{aligned} \frac{\delta G^{++}(12)}{\delta\phi_{\mathrm{ext}}(3+)} &= -\frac{e}{\hbar^2} \frac{\langle S_-^\dagger [\mathbf{T}_+ S_+ \bar{\psi}^\dagger(3)\bar{\psi}(3)\bar{\psi}(1)\bar{\psi}^\dagger(2)]\rangle}{\langle S_-^\dagger S_+\rangle} \\ &+ \frac{\mathrm{i}\,e}{\hbar} G^{++}(12) \frac{\langle S_-^\dagger [\mathbf{T}_+ S_+ \bar{\psi}^\dagger(3)\bar{\psi}(3)]\rangle}{\langle S_-^\dagger S_+\rangle} \end{aligned}$$

$$= e\,G^{++}(12)\,G^{++}(33^+) - \frac{e}{\hbar^2}\,\langle \mathbf{T}_c\psi^\dagger(3^+ +)\psi(3+)\psi(1+)\psi^\dagger(2+)\rangle\,. \quad (10.37)$$

Results corresponding to (10.37) for the other types of Green's functions are readily obtained by means of

$$\frac{\delta S_-^\dagger}{\delta\phi_{\text{ext}}(3-)} = \frac{\mathrm{i}\,e}{\hbar}\,\mathbf{T}_- S_-^\dagger \bar{\psi}^\dagger(3)\bar{\psi}(3)\,. \quad (10.38)$$

Thus we find in general that

$$\frac{\delta G(\bar{1}\bar{2})}{\delta\phi_{\text{ext}}(\bar{3})} = -b_2\,b_3\frac{e}{\hbar^2}\,\langle \mathbf{T}_c\psi^\dagger(\bar{3}^+)\psi(\bar{3})\psi(\bar{1})\psi^\dagger(\bar{2})\rangle + e\,G(\bar{1}\bar{2})\,G(\bar{3}\bar{3}^+) \quad (10.39)$$

relates two-particle functions to functional derivatives of one-particle functions. This result allows us to write the equation of motion (10.34) as

$$\begin{aligned}&\mathrm{i}\hbar\frac{\partial}{\partial t_1}G(\bar{1}\bar{2})\\ &= \delta(\bar{1}-\bar{2}) + (h(\bar{1}) + \mathrm{e}\,\phi_{\text{eff}}(\bar{1}))\,G(\bar{1}\bar{2}) + \mathrm{i}\hbar\,\mathrm{e}\,b_1\int \mathrm{d}\bar{3}\,v(\bar{1}^+ - \bar{3})\frac{\delta G(\bar{1}\bar{2})}{\delta\phi_{\text{ext}}(\bar{3})}\,,\end{aligned} \quad (10.40)$$

where we have already used the fact that the Coulomb interaction is diagonal with respect to the branch indices and have defined

$$\phi_{\text{eff}}(\bar{1}) = \phi_{\text{ext}}(\bar{1}) - \mathrm{i}\hbar\,e\,b_1\int \mathrm{d}\bar{3}\,v(\bar{1}-\bar{3})G(\bar{3}\bar{3}^+)\,. \quad (10.41)$$

As $-\mathrm{i}\hbar e b_1 G(\bar{1}\bar{1}^+)$ is nothing but the charge density, the second term in (10.41) can easily be identified with the sum of the Hartree potential (3.2), which we introduced in Sect. 3.1, and the induced contributions discussed in Sect. 2.6.4.

Exchange and correlation effects are described by the functional-derivative contribution, which still requires the calculation of a two-particle Green's function. In order to decouple the hierarchy formally, we introduce the one-particle self-energy.[2] This is accomplished by the trivial identity

$$G(\bar{1}\bar{2}) = \int \mathrm{d}\bar{4}\bar{5}\,G(\bar{1}\bar{4})\,G^{-1}(\bar{4}\bar{5})\,G(\bar{5}\bar{2})\,, \quad (10.42)$$

which allows us to express the functional derivative of G by means of the functional derivative of G^{-1}. Differentiating with respect to G, we obtain

$$\frac{\delta G(\bar{1}\bar{2})}{\delta\phi_{\text{ext}}(\bar{3})} = -\int \mathrm{d}\bar{4}\bar{5}\,G(\bar{1}\bar{4})\frac{\delta G^{-1}(\bar{4}\bar{5})}{\delta\phi_{\text{ext}}(\bar{3})}G(\bar{5}\bar{2})\,. \quad (10.43)$$

Now, the equation of motion can be cast into a closed form. We define the self-energy as

[2] In constructing approximation schemes (see Sect. 10.6), we shall meet the hierarchy problem again.

$$\Sigma(\bar{1}\bar{2}) = -\mathrm{i}\hbar\, e\, b_1 \int \mathrm{d}\bar{3}\bar{4}\, v(\bar{1}^+ - \bar{3}) G(\bar{1}\bar{4}) \frac{\delta G^{-1}(\bar{4}\bar{2})}{\delta \phi_{\mathrm{ext}}(\bar{3})}, \tag{10.44}$$

and obtain the following from (10.40), using (10.43):

$$\left(\mathrm{i}\hbar\frac{\partial}{\partial t_1} - h(\bar{1}) - e\,\phi_{\mathrm{eff}}(\bar{1})\right) G(\bar{1}\bar{2}) - \int \mathrm{d}\bar{3}\, \Sigma(\bar{1}\bar{3})\, G(\bar{3}\bar{2}) = \delta(\bar{1} - \bar{2})\,. \tag{10.45}$$

An equivalent integral representation, the *Dyson equation* [260], is obtained by acting on (10.45) with G_0 from the left-hand side:

$$G(\bar{1}\bar{2}) = G_0(\bar{1}\bar{2}) + \int \mathrm{d}\bar{3}\bar{4}\, G_0(\bar{1}\bar{3}) \Sigma(\bar{3}\bar{4}) G(\bar{4}\bar{2})\,, \tag{10.46}$$

where G_0 is defined as the solution of (10.45) with the self-energy set equal to zero and represents just the Hartree approximation. The Dyson equation determines the renormalized one-particle Green's functions and thus the energy and lifetime of quasiparticles exactly. Unfortunately, this claim is correct only in principle, as we have to rely on approximations for the self-energy. The equations determining the Green's function and other propagators can be visualized by means of *Feynman diagrams* [261], which represent a precise transcription of analytical expressions into a graphical language. The Dyson equation is represented in Fig. 10.2. Iteration of (10.45) yields a series expansion with the structure

$$G = G_0 + G_0 \Sigma G_0 + G_0 \Sigma G_0 \Sigma G_0 + \dots\,, \tag{10.47}$$

which can easily be summed to obtain another, equivalent Dyson equation

$$G(\bar{1}\bar{2}) = G_0(\bar{1}\bar{2}) + \int \mathrm{d}\bar{3}\bar{4}\, G(\bar{1}\bar{3}) \Sigma(\bar{3}\bar{4}) G_0(\bar{4}\bar{2})\,. \tag{10.48}$$

Acting on this equation with G_0^{-1} from the right-hand side yields the equation of motion in the form

$$\left(-\mathrm{i}\hbar\frac{\partial}{\partial t_2} - h(\bar{2}) - e\,\phi_{\mathrm{eff}}(\bar{2})\right) G(\bar{1}\bar{2}) - \int \mathrm{d}\bar{3}\, G(\bar{1}\bar{3})\, \Sigma(\bar{3}\bar{2}) = \delta(\bar{1} - \bar{2})\,. \tag{10.49}$$

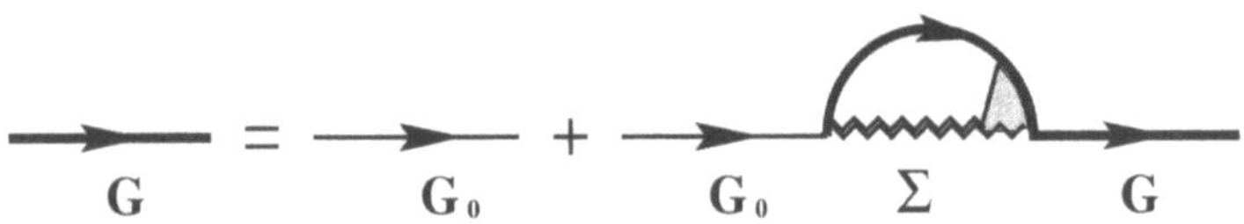

Fig. 10.2. Diagrammatic representation of the Dyson equation. The *thick wavy line* corresponds to the screened Coulomb interaction and the *shaded triangle* represents the vertex function

The self-energy describes exactly the *renormalization of one-particle states* due to the interaction with the surrounding many-particle system. Of course, an exact evaluation is possible only for some rather pathological models, and for real systems we have to rely on approximation schemes. These depend on the contributions to the inverse Green's function which we take into account in the evaluation of the functional derivative. From (10.45), we find the following for G^{-1}:

$$G^{-1}(\bar{1}\bar{2}) = \left(\mathrm{i}\hbar\frac{\partial}{\partial t_1} - h(\bar{1}) - e\,\phi_{\mathrm{eff}}(\bar{1})\right)\delta(\bar{1}-\bar{2}) - \Sigma(\bar{1}\bar{2})\,. \tag{10.50}$$

The lowest-order contribution to the functional derivative of G^{-1} is obtained from $\phi_{\mathrm{ext}}(\boldsymbol{r}t)$ itself and yields

$$\frac{\delta G^{-1}(\bar{1}\bar{2})}{\delta\phi_{\mathrm{ext}}(\bar{3})} = -e\,\delta(\bar{1}\bar{2})\delta(\bar{1}\bar{3}) \tag{10.51}$$

and thus a static self-energy contribution

$$\Sigma^{\mathrm{st}}(\bar{1}\bar{2}) = \mathrm{i}\hbar\, e^2\, b_1\, v(\bar{1}^+ - \bar{2})\, G(\bar{1}\bar{2})\,. \tag{10.52}$$

A brief calculation shows that this approximation together with an eigenfunction expansion such as (10.22), yields the Hartree–Fock equations introduced in Sect. 3.1. From the definition (10.44), it is straightforward to show that the self-energies obey the same linear dependences as the Green's functions, i.e.

$$\Sigma^{++}(12) + \Sigma^{+-}(12) = \Sigma^{--}(12) + \Sigma^{-+}(12) \equiv \Sigma^{\mathrm{r}}(12) \tag{10.53}$$

and

$$\Sigma^{++}(12) - \Sigma^{-+}(12) = \Sigma^{--}(12) - \Sigma^{+-}(12) \equiv \Sigma^{\mathrm{a}}(12)\,. \tag{10.54}$$

The advanced and retarded self-energies are, however, not simply given by the sum of Σ^{+-} and Σ^{-+} times a θ function, because they contain the instantaneous Hartree–Fock contributions. Thus we have

$$\Sigma^{\mathrm{r}}(12) = \Sigma^{\mathrm{st}}(12) + \theta(t_1 - t_2)[\Sigma^{+-}(12) + \Sigma^{-+}(12)] \tag{10.55}$$

and

$$\Sigma^{\mathrm{a}}(12) = \Sigma^{\mathrm{st}}(12) - \theta(t_2 - t_1)[\Sigma^{+-}(12) + \Sigma^{-+}(12)]\,. \tag{10.56}$$

Any approximation scheme beyond the static one requires the consideration of a more complex system of equations [258, 259], which we derive below in a form which is known from equilibrium theories [256].

10.4 Screened Interaction, Polarization, and Vertex Function

The bare Coulomb interaction is a quantity that is rather badly suited for an expansion with respect to the interaction strength. It is therefore advantageous, and this is also suggested by the structure of (10.44), to introduce a screened Coulomb interaction. By applying the chain rule for functional derivatives, we can introduce the derivative with respect to the effective potential. This allows us to write the self-energy as

$$\Sigma(\bar{1}\bar{2}) = -\mathrm{i}\hbar\, e\, b_1 \int \mathrm{d}\bar{3}\bar{4}\, w(\bar{3}\bar{1}^+)\, G(\bar{1}\bar{4})\, \frac{\delta G^{-1}(\bar{4}\bar{2})}{\delta \phi_{\mathrm{eff}}(\bar{3})}\,, \tag{10.57}$$

where the *screened interaction* is

$$w(\bar{1}\bar{2}) = \int \mathrm{d}\bar{3}\, v(\bar{2}-\bar{3})\, \frac{\delta \phi_{\mathrm{eff}}(\bar{1})}{\delta \phi_{\mathrm{ext}}(\bar{3})} = \int \mathrm{d}\bar{3}\, v(\bar{2}-\bar{3})\, \epsilon^{-1}(\bar{1}\bar{3})\,. \tag{10.58}$$

Screening is described by the inverse dielectric function, which we have already introduced in (2.162), where it was the starting point for calculating the macroscopic response of the system. Using the definition of the effective potential (10.41) together with (10.43) and the chain rule, we obtain

$$w(\bar{1}\bar{2}) = v(\bar{2}-\bar{1}) + \frac{\varepsilon_0}{V} \int \mathrm{d}\bar{3}\bar{4}\, v(\bar{1}-\bar{3})\, P(\bar{3}\bar{4})\, w(\bar{4}\bar{2})\,, \tag{10.59}$$

where we have defined, in accordance with (2.160), the *polarization propagator*

$$\begin{aligned} P(\bar{1}\bar{2}) &= -\mathrm{i}\hbar\, V\, b_1 \frac{e}{\varepsilon_0} \frac{\delta G(\bar{1}\bar{1}^+)}{\delta \phi_{\mathrm{eff}}(\bar{2})} \\ &= \mathrm{i}\hbar\, V\, b_1 \frac{e}{\varepsilon_0} \int \mathrm{d}\bar{3}\bar{4}\, G(\bar{1}\bar{3}) \frac{\delta G^{-1}(\bar{3}\bar{4})}{\delta \phi_{\mathrm{eff}}(\bar{2})} G(\bar{4}\bar{1})\,. \end{aligned} \tag{10.60}$$

We note without explicit proof that the screened interaction and the polarization function fulfill the same linear dependences as the Green's functions and self-energies, i.e. relations of the form

$$a^{++} + a^{+-} = a^{--} + a^{-+} = a^{\mathrm{r}} \tag{10.61}$$

and

$$a^{++} - a^{-+} = a^{--} - a^{+-} = a^{\mathrm{a}} \tag{10.62}$$

are valid for both w and P.

Our system of equations defining the self-energy is closed by the equation for the vertex functions, which also enters the polarization propagator. From (10.50), we obtain (by steps similar to those used before)

$$\begin{aligned}\Gamma(\bar{1}\bar{2}\bar{3}) &= \frac{\delta G^{-1}(\bar{1}\bar{2})}{\delta \phi_{\text{eff}}(\bar{3})} = -e\,\delta(\bar{1}-\bar{2})\,\delta(\bar{1}-\bar{3}) - \frac{\delta \Sigma(\bar{1}\bar{2})}{\delta \phi_{\text{eff}}(\bar{3})} \\ &= -e\,\delta(\bar{1}-\bar{2})\,\delta(\bar{1}-\bar{3}) - \int \mathrm{d}\bar{4}\bar{5}\,\frac{\delta \Sigma(\bar{1}\bar{2})}{\delta G(\bar{4}\bar{5})}\,\frac{\delta G(\bar{4}\bar{5})}{\delta \phi_{\text{eff}}(\bar{3})} \\ &= -e\,\delta(\bar{1}-\bar{2})\,\delta(\bar{1}-\bar{3}) - \int \mathrm{d}\bar{4}\bar{5}\bar{6}\bar{7}\,\frac{\delta \Sigma(\bar{1}\bar{2})}{\delta G(\bar{4}\bar{5})}\,G(\bar{4}\bar{6})\,G(\bar{7}\bar{5})\Gamma(\bar{6}\bar{7}\bar{3})\,.\end{aligned} \tag{10.63}$$

Contributions proportional to $\delta\Sigma/\delta G$ are referred to as *vertex corrections* and describe interaction processes at the two-particle level. The system of equations introduced in this section has the same structure as in equilibrium theories [256] and can serve as a general starting point for constructing systematic approximations to the self-energy. Depending on the problem, we can either attempt a summation of a selected class of dominant contributions or perform an expansion with respect to the screened interaction. Starting such an iterative solution, we can first neglect vertex corrections in (10.63) and obtain by means of (10.57) together with (10.59) and (10.60) an approximation for the self-energy and hence a set of equations, which determines the Green's function. Making use of this approximation, we calculate $\delta\Sigma/\delta G$, and include vertex corrections in the next step (see Sect. 10.6). The iteration of such a procedure generates an expansion in terms of the screened interaction and the Green's function defined as self-consistent solution of the Dyson equation. We can associate with each such term a self-energy diagram (see Fig. 10.2). It is easily seen that an nth-order self-energy diagram, for example, consists of n interaction lines which are connected to $2n-1$ Green's functions at $2n-2$ internal vertices and 2 external vertices. At each internal vertex, we have an incoming and an outgoing Green's function. Making use of our analytical expressions, it is an easy task to establish the rules, which determine the calculation of such self-energy diagrams. Thus it is also possible to construct approximations to the self-energy by means of a diagrammatic language [261, 29].

The polarization propagator has a fundamental significance and can be expressed as a contraction, i.e. the limit for $\bar{1} \to \bar{1}'$ and $\bar{2} \to \bar{2}'$, of a slightly more general function $P(\bar{1}\bar{1}'\bar{2}\bar{2}')$. The latter obeys a *Bethe–Salpeter equation* [262] which is obtained after insertion of (10.63) into (10.60), as

$$\begin{aligned}P(\bar{1}\bar{1}'\bar{2}\bar{2}') = &-\mathrm{i}\hbar V b_1 \frac{e^2}{\varepsilon_0}\,G(\bar{1}\bar{2})\,G(\bar{2}'\bar{1}') \\ &- \int \mathrm{d}\bar{3}\bar{4}\bar{5}\bar{6}\,b_1 b_5\,G(\bar{1}\bar{3})\,G(\bar{4}\bar{1}')\,\frac{\delta\Sigma(\bar{3}\bar{4})}{\delta G(\bar{5}\bar{6})}\,P(\bar{5}\bar{6}\bar{2}\bar{2}')\,.\end{aligned} \tag{10.64}$$

A graphical representation of this equation in the so-called *screened ladder approximation*, in which $\delta\Sigma/\delta G$ is approximated by the screened Coulomb interaction, is shown in Fig. 10.3. As is obvious from its definition (10.60),

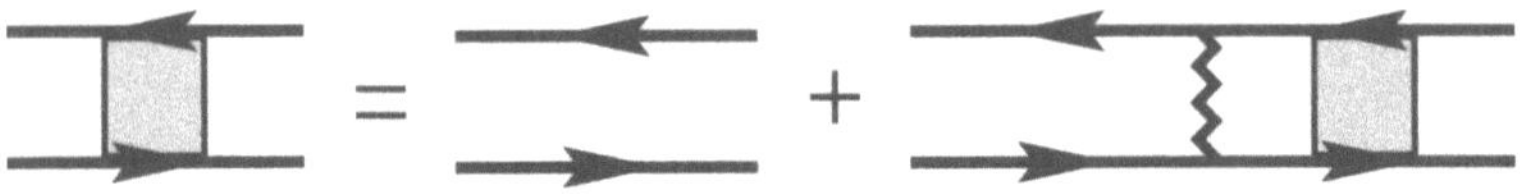

Fig. 10.3. Graphical representation of the Bethe–Salpeter equation for the polarization propagator in the ladder approximation. The *thick wavy line* represents the screened Coulomb interaction

the polarization functions determine not only the screened interaction, but also the response of the system to external perturbations. The bubble diagram, without any interaction, leads to a dielectric function, which in the long-wavelength limit reduces to (5.36). The sum of bubble diagrams (see Fig. 2.1 and (2.161)) already introduced in Sect. 2.6 corresponds to the iterative solution of the integral equation (10.59) and, in turn, is nothing but the inversion of the dielectric matrix (see Sect. 5.2). This Hartree approximation, linearized in the external field, is referred to as the *random-phase approximation* (RPA) [263].

In Chap. 5, we derived the response from the off-diagonal elements of the one-particle density matrix with respect to band indices.[3] and derived the Bethe–Salpeter equation (5.22) within the Hartree–Fock approximation; thus these are results of a one-particle theory. In the present formulation the polarization propagator is introduced as a two-particle quantity from the very beginning. As approximations are involved, the two formulations are in general not equivalent. Exchange and correlation effects in one-particle properties are determined by the self-energy, whereas the effective electron–hole interaction entering the Bethe–Salpeter equation (10.64) is the functional derivative of the self-energy with respect to the Green's function. Only within the Hartree–Fock approximation, in which this derivative is nothing but the bare Coulomb potential, does one obtain the same Bethe–Salpeter equation as for the off-diagonal elements of the one-particle Green's function at the two-particle level. In this limit (10.64) reduces to (5.22) and represents again the simplest approximation in which one can introduce electron–hole bound states, which result in this case from the summation of the infinite series of ladder diagrams in Fig. 10.3. Beyond the Hartree–Fock level, more or less important additional contributions can arise at the two-particle level if we start with an approximation for the self-energy. Before we introduce different approximation schemes in detail, we turn first to a general discussion of the equations of motion.

[3] We could clearly perform the corresponding derivation starting from off-diagonal elements of the one-particle Green's function.

10.5 Quantum Kinetic Equations

The equations of motion (10.45) and (10.49) include not only the complete many-particle problem in equilibrium but also all additional complications arising from the presence of external fields. Therefore a variety of steps is necessary to reduce these equations to a more convinient form. Let us consider first the dependence on spatial coordinates. We perform the expansion (10.22) by means of eigenfunctions of the form (3.95), which anticipates the effective-mass approximation. Thus we represent the Green's function as

$$G(\bar{1}\bar{2}) = \frac{1}{V} \sum_{nn'\boldsymbol{k}\boldsymbol{k}'} \mathrm{e}^{-\mathrm{i}\boldsymbol{k}\boldsymbol{r}_1} u_n^*(0,\boldsymbol{r}_1) G_{nn'}(\boldsymbol{k}\boldsymbol{k}', \bar{t}_1\bar{t}_2) \mathrm{e}^{\mathrm{i}\boldsymbol{k}'\boldsymbol{r}_2} u_{n'}(0,\boldsymbol{r}_2)\,, \quad (10.65)$$

where $\bar{t}_1 \equiv t_1 b_1$. In a homogeneous system we would have a dependence on one $\boldsymbol{k}$ vector only, which would occur as the Fourier variable of the *microscopic variable* $\Delta\boldsymbol{r} = \boldsymbol{r}_1 - \boldsymbol{r}_2$. We can separate this dependence by $\boldsymbol{k} \to \boldsymbol{k} + \alpha_{nn'}\boldsymbol{q}$ and $\boldsymbol{k}' \to \boldsymbol{k} - \beta_{nn'}\boldsymbol{q}$, where $\alpha_{nn'} = |m_n|/(|m_n| + |m_{n'}|)$, $\beta_{nn'} = |m_{n'}|/(|m_n| + |m_{n'}|)$, and $\boldsymbol{q}$ is the center-of-mass momentum. Performing the $\boldsymbol{q}$ summation, we introduce

$$G_{nn'}(\boldsymbol{k}, \boldsymbol{r}, \bar{t}_1\bar{t}_2) = \sum_{\boldsymbol{q}} G_{nn'}(\boldsymbol{k} + \alpha_{nn'}\boldsymbol{q}, \boldsymbol{k} - \beta_{nn'}\boldsymbol{q}, \bar{t}_1\bar{t}_2) \mathrm{e}^{-\mathrm{i}\boldsymbol{q}\boldsymbol{r}} \quad (10.66)$$

and write the Green's function in the form

$$G(\bar{1}\bar{2}) = \frac{1}{V} \sum_{nn'\boldsymbol{k}} \mathrm{e}^{-\mathrm{i}\boldsymbol{k}\Delta\boldsymbol{r}} u_n^*(0,\boldsymbol{r}_1)\, G_{nn'}(\boldsymbol{k}, \boldsymbol{r}, \bar{t}_1\bar{t}_2)\, u_{n'}(0,\boldsymbol{r}_2)\,. \quad (10.67)$$

The variable $\boldsymbol{r} = \alpha_{nn'}\boldsymbol{r}_1 + \beta_{nn'}\boldsymbol{r}_2$ is usually referred to as the *macroscopic variable* and accounts for those inhomogeneities of the system which occur in addition to the inhomogeneity described by lattice-periodic functions. These inhomogeneities may have various physical origins, depending on the problem. In optics, inhomogeneities result from the spatial dependence of the external fields, inhomogeneous excitation conditions due to the finite penetration depth of the field, propagation effects, etc. In the case of transport phenomena, inhomogeneous carrier distributions are constitutive.

The transformation of the equations of motion (10.45) and (10.49) using the representation (10.67), is performed by projecting onto Bloch states, where the integrations are taken over a single unit cell. The contributions of the optical field $\boldsymbol{E}(\boldsymbol{r}_1)$ and $\boldsymbol{E}(\boldsymbol{r}_2)$, are treated in the same manner, as that used in deriving the transition matrix element (3.134). This leads to a dependence of the field on the center-of-mass momentum. As we have shown in Sect. 4.1, we obtain in real space, within the dipole approximation, a dependence on the macroscopic variable $\boldsymbol{r}$. We shall not repeat this derivation, but turn to the interaction contributions.

We decompose, similar as in Sect. 5.1, the self-energy and the Hartree potential into those contributions which are already present in the system without external fields and those which take the action of the external fields into account. The first set of contributions define the one-particle Hamiltonian of the unperturbed system, and have been already treated in the band structure and thus are already incorporated into the effective-mass approximation. For simplicity, we restrict ourselves to the nondegenerate case, which yields one-particle energies proportional to $(\boldsymbol{k} + \alpha_{nn'}\boldsymbol{q})^2$ and $(\boldsymbol{k} - \beta_{nn'}\boldsymbol{q})^2$. The dependence on $\boldsymbol{q}$ is transformed to derivatives with respect to $\boldsymbol{r}$. For the sum of the external and induced potentials, which is denoted by V_{eff} in the following, we shall use a Fourier representation. This allows us to describe rapidly varying potentials also.[4]

For the field-dependent self-energy we shall not introduce a new notation, but shall arrange that any contribution independent of external perturbations will be removed in the following (compare Sect. 5.1). We represent the self-energy in the same form as the Green's function in (10.66) and (10.67). We use these representations in (10.45), project onto Bloch states, and Fourier transform with respect to the center-of-mass momentum. With these ingredients, we can cast the equation of motion (10.45) into the form

$$
\begin{aligned}
&\left(\mathrm{i}\hbar\frac{\partial}{\partial t_1} - \frac{\hbar^2}{2m_n}(\boldsymbol{k}^2 + 2\mathrm{i}\alpha_{nn'}\boldsymbol{k}\boldsymbol{\nabla}_{\boldsymbol{r}} - \alpha_{nn'}^2\Delta_{\boldsymbol{r}})\right)G_{nn'}(\boldsymbol{k},\boldsymbol{r},\bar{t}_1\bar{t}_2)\\
&-e\sum_{\boldsymbol{q}} \mathrm{e}^{\mathrm{i}\boldsymbol{q}\boldsymbol{r}}\, V_{\text{eff}}(\boldsymbol{q},t_1)\, G_{nn'}(\boldsymbol{k}-\beta_{nn'}\boldsymbol{q},\boldsymbol{r},\bar{t}_1\bar{t}_2)\\
&= \delta(\bar{t}_1-\bar{t}_2)\,\delta_{nn'} - \frac{1}{2}\sum_m \boldsymbol{d}_{nm}\boldsymbol{E}(\boldsymbol{r},t_1)\, G_{mn'}(\boldsymbol{k},\boldsymbol{r},\bar{t}_1\bar{t}_2)\\
&+\frac{1}{V^2}\sum_{mqq'}\int \mathrm{d}\bar{t}_3\,\mathrm{d}^3\boldsymbol{r}_1\,\mathrm{d}^1\boldsymbol{r}_2\,\mathrm{e}^{\mathrm{i}(\boldsymbol{k}-\boldsymbol{q}')\boldsymbol{r}_1}\,\mathrm{e}^{-\mathrm{i}(\boldsymbol{k}-\boldsymbol{q})\boldsymbol{r}_2}\\
&\times \Sigma_{nm}(\boldsymbol{q},\boldsymbol{r}+\beta_{nn'}\boldsymbol{r}_1 + (\beta_{nm}-\beta_{nn'})\,\boldsymbol{r}_2,\bar{t}_1\bar{t}_3)\\
&\times G_{mn'}(\boldsymbol{q}',\boldsymbol{r}+\alpha_{nn'}\boldsymbol{r}_2 + (\alpha_{mn'}-\alpha_{nn'})\,\boldsymbol{r}_1,\bar{t}_3\bar{t}_2)\,.
\end{aligned}
\tag{10.68}
$$

The *nonlocal spatial coupling* between the self-energy and the Green's function can be simplified considerably if the microscopic and macroscopic scales become separable. As $\boldsymbol{r}_1$ and $\boldsymbol{r}_2$ in (10.64) occur as Fourier variables of the wave vectors $\boldsymbol{q}'$ and $\boldsymbol{q}$, respectively, they obviously vary on a microscopic scale. In both optics and in transport, this scale is typically orders of magnitude smaller than the macroscopic scale characterized by $\boldsymbol{r}$. Owing to the slow variation on the macroscopic scale, we can enforce spatial locality by means of a gradient expansion of the self-energy and the Green's function around the macroscopic variable $\boldsymbol{r}$ in powers of the variables $\boldsymbol{r}_1$ and $\boldsymbol{r}_2$. Representing these powers as derivatives with respect to the wave vectors $\boldsymbol{q}'$ and $\boldsymbol{q}$, we can perform the spatial integrations and only $\boldsymbol{q} = \boldsymbol{q}' = \boldsymbol{k}$ contributes.

[4] With this representation, we anticipate applications in transport theory, where, for example, the barriers in a quantum well system can be included in this potential.

This procedure yields a *Wigner–Weyl expansion* [264, 265, 266], which allows us a local representation in a closed form. Here, we restrict ourselves to the linear gradient contributions.

With respect to time, we also introduce the macroscopic and microscopic time variables $t = (t_1 + t_2)/2$ and $\Delta t = t_1 - t_2$, respectively. Subtracting from (10.68) the corresponding adjoint equation, which we obtain from (10.49), the derivatives with respect to the microscopic time cancel, and the quantum kinetic equation can be cast into the the form

$$\begin{aligned}
&\left[\mathrm{i}\hbar\frac{\partial}{\partial t} - \frac{\hbar^2\boldsymbol{k}^2}{2m_n} + \frac{\hbar^2\boldsymbol{k}^2}{2m_{n'}} - \mathrm{i}\left(\frac{\hbar^2\alpha_{nn'}}{m_n} + \frac{\hbar^2\beta_{nn'}}{m_{n'}}\right)\boldsymbol{k}\boldsymbol{\nabla}_{\boldsymbol{r}}\right] G_{nn'}(\boldsymbol{k},\boldsymbol{r},\bar{t}_1\bar{t}_2) \\
&+\left(\frac{\hbar^2\alpha_{nn'}^2}{2m_n} - \frac{\hbar^2\beta_{nn'}^2}{2m_{n'}}\right)\Delta_{\boldsymbol{r}} G_{nn'}(\boldsymbol{k},\boldsymbol{r},\bar{t}_1\bar{t}_2) - e\sum_{\boldsymbol{q}} \mathrm{e}^{\mathrm{i}\boldsymbol{q}\boldsymbol{r}}\, V_{\mathrm{eff}}(\boldsymbol{q},t_1) \\
&\times[G_{nn'}(\boldsymbol{k} - \beta_{nn'}\boldsymbol{q},\boldsymbol{r},\bar{t}_1\bar{t}_2) - G_{nn'}(\boldsymbol{k} + \alpha_{nn'}\boldsymbol{q},\boldsymbol{r},\bar{t}_1\bar{t}_2)] \\
&= -\sum_m \Big[\boldsymbol{d}_{nm}\boldsymbol{E}(\boldsymbol{r},t_1)\, G_{mn'}(\boldsymbol{k},\boldsymbol{r},\bar{t}_1\bar{t}_2) - G_{nm}(\boldsymbol{k},\boldsymbol{r},\bar{t}_1\bar{t}_2)\, \boldsymbol{d}_{mn'}\boldsymbol{E}(\boldsymbol{r},t_2) \\
&-\int \mathrm{d}\bar{t}_3 \Big(\Sigma_{nm}(\boldsymbol{k},\boldsymbol{r},\bar{t}_1\bar{t}_3) G_{mn'}(\boldsymbol{k},\boldsymbol{r},\bar{t}_3\bar{t}_2) - G_{nm}(\boldsymbol{k},\boldsymbol{r},\bar{t}_1\bar{t}_3)\Sigma_{mn'}(\boldsymbol{k},\boldsymbol{r},\bar{t}_3\bar{t}_2) \\
&+\mathrm{i}\,[\alpha_{nn'}\boldsymbol{\nabla}_{\boldsymbol{r}'}\boldsymbol{\nabla}_{\boldsymbol{k}} - \beta_{nn'}\boldsymbol{\nabla}_{\boldsymbol{r}}\boldsymbol{\nabla}_{\boldsymbol{k}'} + (\beta_{nm} - \beta_{nn'})\boldsymbol{\nabla}_{\boldsymbol{r}}\boldsymbol{\nabla}_{\boldsymbol{k}} \\
&-(\alpha_{mn'} - \alpha_{nn'})\boldsymbol{\nabla}_{\boldsymbol{r}'}\boldsymbol{\nabla}_{\boldsymbol{k}'}][\Sigma_{nm}(\boldsymbol{k},\boldsymbol{r},\bar{t}_1\bar{t}_3)\, G_{mn'}(\boldsymbol{k}',\boldsymbol{r}',\bar{t}_3\bar{t}_2) \\
&+ G_{nm}(\boldsymbol{k},\boldsymbol{r},\bar{t}_1\bar{t}_3)\, \Sigma_{mn'}(\boldsymbol{k}',\boldsymbol{r}',\bar{t}_3\bar{t}_2)]\Big|_{\substack{\boldsymbol{r}'=\boldsymbol{r}\\ \boldsymbol{k}'=\boldsymbol{k}}}\Big)\Big]\,.
\end{aligned} \tag{10.69}$$

Without using an explicit notation, we have implied that the time variables are expressed by the corresponding microscopic and macroscopic variables, i.e. $t_1 = t + \Delta t/2$ and $t_2 = t - \Delta t/2$. The result (10.69) can serve as a general starting point for discussing quantum kinetics in inhomogeneous systems, when interband and intraband processes are involved. We shall come back to the discussion of spatial inhomogeneities in connection with transport properties in Chap. 15. At the moment we consider interband excitations only and set $V_{\mathrm{eff}} = 0$. Assuming homogeneous excitation conditions, we neglect all derivatives with respect to $\boldsymbol{r}$. Thus inhomogeneities are taken into account, if at all, by a parametric dependence on the global macroscopic variable $\boldsymbol{r}$ only, which is dropped in the following.

In the next step we consider the decomposition with respect to the different types of Green's functions. First we focus on the G^{+-} function, for which (by means of the relations (10.61) and (10.62)) the self-energy contributions take the form

$$\int \mathrm{d}3\,[\Sigma^{+-}(13)\,G^{--}(32) + \Sigma^{++}(13)\,G^{+-}(32)]$$
$$= \int \mathrm{d}3\,[\Sigma^{+-}(13)\,G^{a}(32) + \Sigma^{r}(13)\,G^{+-}(32)]\,. \tag{10.70}$$

Thus, in the limit of weak inhomogeneity, we obtain as a special case of (10.69)

$$\left(\mathrm{i}\hbar\frac{\partial}{\partial t} - E_n(\boldsymbol{k}) + E_{n'}(\boldsymbol{k})\right) G^{+-}_{nn'}(\boldsymbol{k}, t_1t_2)$$
$$= -\sum_m [\boldsymbol{d}_{nm}\boldsymbol{E}(\boldsymbol{r}, t_1)\,G^{+-}_{mn'}(\boldsymbol{k}, t_1, t_2) \;-\; G^{+-}_{nm}(\boldsymbol{k}, t_1, t_2)\,\boldsymbol{d}_{mn'}\boldsymbol{E}(\boldsymbol{r}, t_2)]$$
$$+\sum_m \int \mathrm{d}\bar{t}_3[\Sigma^{\mathrm{r}}_{nm}(\boldsymbol{k}, t_1t_3)\,G^{+-}_{mn'}(\boldsymbol{k}, t_3t_2) + \Sigma^{+-}_{nm}(\boldsymbol{k}, t_1t_3)\,G^{\mathrm{a}}_{mn'}(\boldsymbol{k}, t_3t_2)$$
$$-G^{\mathrm{r}}_{nm}(\boldsymbol{k}, t_1t_3)\,\Sigma^{+-}_{mn'}(\boldsymbol{k}, t_3t_2) - G^{+-}_{nm}(\boldsymbol{k}, t_1t_3)\,\Sigma^{\mathrm{a}}_{mn'}(\boldsymbol{k}, t_3t_2)]\,, \tag{10.71}$$

where $E_n(\boldsymbol{k})$ is the effective-mass one-particle energy. Concerning the corresponding equation for the G^{-+} function, we have to consider the relation

$$\int \mathrm{d}3\,[\Sigma^{-+}(13)\,G^{++}(32) + \Sigma^{--}(13)\,G^{-+}(32)]$$
$$= \int \mathrm{d}3\,[\Sigma^{-+}(13)\,G^{a}(32) + \Sigma^{r}(13)\,G^{-+}(32)]\,, \tag{10.72}$$

which shows that we only have to interchange $+-$ with $-+$ in (10.71). The relations (10.70) and (10.72) reflect a typical feature of the coupling with respect to branch indices. Retarded functions occur on the left-hand side of $+-$ functions and advanced ones on the right-hand side. This guarantees that causality is fulfilled in any order of perturbation theory with respect to external fields.

Having discussed the *nonlocal spatial coupling* in the quantum kinetic equations in some detail, we turn now to the corresponding features in the time domain. Whereas the macroscopic and microscopic timescales are also separable with respect to transport problems, this is not valid in many applications in optics.

10.5.1 The Two-Time Formalism

Formally, we could proceed as in the case of the spatial variables and assume that the system changes much more slowly on the macroscopic timescale than on the microscopic one. The weak temporal nonlocality would again allow a Wigner–Weyl expansion and a restriction to linear derivatives.[5] Thus the macroscopic time variable t occurs as a global variable and we can perform the Fourier transformation with respect to the microscopic time $\Delta t = t_1 - t_2$. In

[5] An application of this approach will be given in Sect. 13.3.

optical experiments on an ultrashort timescale, however, the time evolutions of the system on the microscopic and macroscopic scales are of the same order of magnitude and the assumption of separability fails. Further, there are other reasons, which will become evident in the following, to take the *nonlocal temporal coupling* seriously and to treat it in more detail.

Let us consider the time evolution of the system on the macroscopic timescale. In order to obtain a conveniently defined initial-value problem for the two-time functions, we use a slightly different definition of the macroscopic variable. Defining $t_1 = t$ and $t_2 = t - \Delta t$ yields again the quantum kinetic equation (10.71). We separate the static Hartree–Fock contribution (10.52) according to (10.55) and (10.56), with the result

$$\Sigma^{\mathrm{st}}_{nm}(\boldsymbol{k},t) = \mathrm{i}e^2 \sum_{\boldsymbol{k}'} v(\boldsymbol{k}-\boldsymbol{k}')[\hbar G^{+-}_{nm}(\boldsymbol{k}',t,t) + \mathrm{i}\delta_{vn}\delta_{nm}] \, . \tag{10.73}$$

Finally, we use again the relations (10.20) and (10.21) and split the integration for the retarded quantities according to

$$\int_{t_0}^{t} = \int_{t_0}^{t-\Delta t} + \int_{t-\Delta t}^{t} , \tag{10.74}$$

where t_0 is a time sufficiently long before the excitation. With these modifications, we obtain the following from (10.71):

$$\begin{aligned}
&\left(\mathrm{i}\hbar\frac{\partial}{\partial t} - E_n(\boldsymbol{k}) + E_{n'}(\boldsymbol{k})\right) G^{+-}_{nn'}(\boldsymbol{k},t,\tau) \\
&= -\sum_m \Bigg(\left[\boldsymbol{d}_{nm}\boldsymbol{E}(\boldsymbol{r},t) + \Sigma^{\mathrm{st}}_{nm}(\boldsymbol{k},t)\right] G^{+-}_{mn'}(\boldsymbol{k},t,\tau) \\
&- G^{+-}_{nm}(\boldsymbol{k},t,\tau)\left[\boldsymbol{d}_{mn'}\boldsymbol{E}(\boldsymbol{r},\tau) + \Sigma^{\mathrm{st}}_{mn'}(\boldsymbol{k},\tau)\right] \\
&- \int_{t_0}^{\tau} \mathrm{d}t' \, [\Sigma^{+-}_{nm}(\boldsymbol{k},t,t')\, G^{-+}_{mn'}(\boldsymbol{k},t',\tau) - G^{+-}_{nm}(\boldsymbol{k},t,t')\, \Sigma^{-+}_{mn'}(\boldsymbol{k},t',\tau) \\
&- \Sigma^{-+}_{nm}(\boldsymbol{k},t,t')\, G^{+-}_{mn'}(\boldsymbol{k},t',\tau) + G^{-+}_{nm}(\boldsymbol{k},t,t')\, \Sigma^{+-}_{mn'}(\boldsymbol{k},t',\tau)] \\
&- \int_{\tau}^{t} \mathrm{d}t' \, [\Sigma^{+-}_{nm}(\boldsymbol{k},t,t')\, G^{+-}_{mn'}(\boldsymbol{k},t',\tau) - G^{+-}_{nm}(\boldsymbol{k},t,t')\, \Sigma^{+-}_{mn'}(\boldsymbol{k},t',\tau) \\
&- \Sigma^{-+}_{nm}(\boldsymbol{k},t,t')\, G^{+-}_{mn'}(\boldsymbol{k},t',\tau) + G^{-+}_{nm}(\boldsymbol{k},t,t')\, \Sigma^{+-}_{mn'}(\boldsymbol{k},t',\tau)] \Bigg) ,
\end{aligned} \tag{10.75}$$

where we have used the abbreviation $t - \Delta t \equiv \tau$. Again, the corresponding equations for the G^{-+} propagator are obtained by interchanging $+-$ with $-+$. If all correlational contributions beyond the Hartree–Fock approximation are replaced by phenomenological relaxation rates we need to consider $\Delta t = 0$ only, and one easily recovers from (10.75) the *semiconductor Bloch equations*. According to the definitions, the one-particle distribution functions and transition amplitudes are related to the G^{+-} functions by

$$p_{nm}(\boldsymbol{k},\boldsymbol{r},t) = \mathrm{i}\hbar\, G^{+-}_{nm}(\boldsymbol{k},\boldsymbol{r},t,t) \tag{10.76}$$

and

$$f_n(\boldsymbol{k},\boldsymbol{r},t) = \mathrm{i}\hbar\, G^{+-}_{nn}(\boldsymbol{k},\boldsymbol{r},t,t)\,. \tag{10.77}$$

Also, beyond the Hartree–Fock approximation, we are only interested in the diagonal elements with respect to time. The evaluation of the correlational contributions requires, however, knowledge of the Green's functions for $\Delta t \neq 0$ also. There are several possible ways to proceed further. Let us first discuss a method which takes the dependence on two times into account without further assumptions.

In this method, we separate the rapidly oscillating phase factors in all Green's functions according to

$$G_{nn'}(\boldsymbol{k},t,t') = \mathrm{e}^{-\mathrm{i}[E_n(\boldsymbol{k})t - E_{n'}(\boldsymbol{k})t']/\hbar} g_{nn'}(\boldsymbol{k},t,t')\,. \tag{10.78}$$

For the self-energies, we set

$$\Sigma_{nn'}(\boldsymbol{k},t,t') = \mathrm{e}^{-\mathrm{i}[E_n(\boldsymbol{k})t/\hbar - E_{n'}(\boldsymbol{k})t']/\hbar} \sigma_{nn'}(\boldsymbol{k},t,t')\,, \tag{10.79}$$

correspondingly. Furthermore, we introduce the renormalized Rabi frequency

$$\Delta_{nm}(\boldsymbol{k},t) = \mathrm{e}^{-\mathrm{i}[E_n(\boldsymbol{k}) - E_m(\boldsymbol{k})]t/\hbar}[\boldsymbol{d}_{nm}\boldsymbol{E}(\boldsymbol{r},t) + \sigma^{\mathrm{st}}_{nm}(\boldsymbol{k},t)]\,, \tag{10.80}$$

where the static Hartree–Fock self-energies are

$$\begin{aligned}\sigma^{\mathrm{st}}_{nm}(\boldsymbol{k},t) &= -\mathrm{i}e^2 \sum_{\boldsymbol{k}'} v(\boldsymbol{k}')\, \mathrm{e}^{-\mathrm{i}[E_n(\boldsymbol{k}') - E_n(\boldsymbol{k}) - E_m(\boldsymbol{k}) + E_m(\boldsymbol{k}')]\,t/\hbar} \\ &\qquad \times [\hbar g^{+-}_{nm}(\boldsymbol{k}+\boldsymbol{k}',t) + \mathrm{i}\delta_{nv}\delta_{nm}]\,.\end{aligned} \tag{10.81}$$

With these ingredients, the kinetic equation (10.75) is reduced to

$$\begin{aligned}&\mathrm{i}\hbar\frac{\partial}{\partial t} g^{+-}_{nn'}(\boldsymbol{k},t,\tau) \\ &= -\sum_m \Bigg(\Delta_{nm}(\boldsymbol{k},t) g^{+-}_{mn'}(\boldsymbol{k},t,\tau) - g^{+-}_{nm}(\boldsymbol{k},t,\tau)\Delta_{mn'}(\boldsymbol{k},\tau) \\ &- \int_{t_0}^{\tau} \mathrm{d}t'\, [\sigma^{+-}_{nm}(\boldsymbol{k},t,t') g^{-+}_{mn'}(\boldsymbol{k},t',\tau) - g^{+-}_{nm}(\boldsymbol{k},t,t')\sigma^{-+}_{mn'}(\boldsymbol{k},t',\tau) \\ &- \sigma^{-+}_{nm}(\boldsymbol{k},t,t') g^{+-}_{mn'}(\boldsymbol{k},t',\tau) + g^{-+}_{nm}(\boldsymbol{k},t,t')\sigma^{+-}_{mn'}(\boldsymbol{k},t',\tau)] \\ &- \int_{\tau}^{t} \mathrm{d}t'\, [\sigma^{+-}_{nm}(\boldsymbol{k},t,t') g^{+-}_{mn'}(\boldsymbol{k},t',\tau) - g^{+-}_{nm}(\boldsymbol{k},t,t')\sigma^{+-}_{mn'}(\boldsymbol{k},t',\tau) \\ &- \sigma^{-+}_{nm}(\boldsymbol{k},t,t') g^{+-}_{mn'}(\boldsymbol{k},t',\tau) + g^{-+}_{nm}(\boldsymbol{k},t,t')\sigma^{+-}_{mn'}(\boldsymbol{k},t',\tau)] \Bigg)\,.\end{aligned} \tag{10.82}$$

In any approximation, the self-energy contributions can be expressed as functionals of G^{+-} and G^{-+}. Thus the two corresponding equations together form a closed set of equations, which can be solved as an initial-value problem. Since for $t = t_0 \to -\infty$, the propagators depend on the difference time only, a knowledge of $g_{nm}^{+-,-+}(\boldsymbol{k}, t_0, t_0 - \Delta t)$ is required over a sufficiently wide stripe. Considering the time evolution on the macroscopic timescale, we have to distinguish between those correlations, which are present even without external fields, and those which are switched on due to the excitation of the system. In many cases, we can assume that the system is uncorrelated at the initial time, and the values of the propagators $g_{nm}^{+-,-+}$ are either 0 or 1. There may, however, be contributions to the self-energy, such as in the case of electron–phonon interaction (see Sect. 11.4), which contribute at t_0, and cannot be incorporated into the effective mass approximation. Such correlations have an important influence on the temporal evolution of the system also after the field is switched on, and have to be included in the initial values.[6] When the external field is switched on, the system evolves in time along the macroscopic time axis and the microscopic scale Δt enters the equations merely as a parameter. In order to evaluate the correlational contributions, we have to perform the solution with a sufficiently large scale $\Delta t \geq 0$. This scale, however, is limited to a finite range, as the Green's functions are decaying functions with respect to the difference time.[7] As illustrated in Fig. 10.4, we can calculate the time evolution along a stripe for which $\Delta t \geq 0$; the width of this stripe depends on the *memory depth*

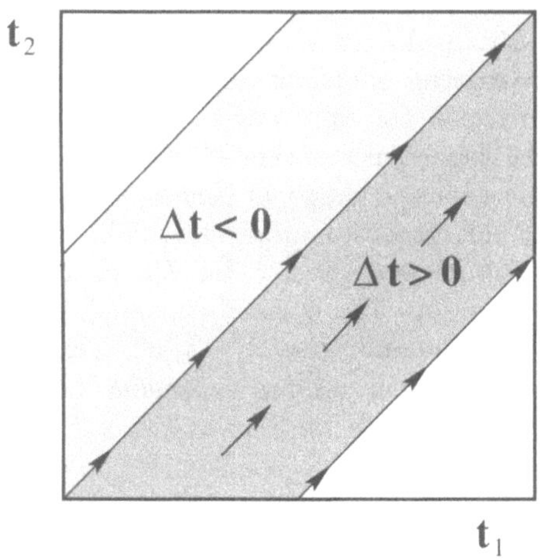

Fig. 10.4. Schematic illustration of the solution of the quantum kinetic equations in the two-time plane. The propagators are calculated along a stripe with $t_1 \geq t_2$ parallel to the $t_1 = t_2$ axis. The case $t_1 \leq t_2$ is obtained from symmetry

[6] Note that different but equivalent formulations of the two-time formalism calculate the time evolution of the two equations (10.45) and (10.49) in the two-time plane by time-stepping along the two time axes [278]. In this case, however, it is unsuitable to define the initial values by means of the propagators of the unexcited but correlated system.

[7] We shall see later that this property results from the imaginary part of the self-energy in the frequency domain.

and is determined by the properties of the self-energy itself. This yields the functions $G^{+-,-+}(t,t')$ for $t \geq t'$. If we need the opposite case, i.e. $t \leq t'$, we make use of the symmetry relation

$$G_{nm}^{+-,-+}(\boldsymbol{k},t,t') = -(G_{mn}^{+-,-+})^*(\boldsymbol{k},t',t)\,. \tag{10.83}$$

The scheme developed thus far is practical and can be implemented on a computer. This allows one to study correlation and scattering processes within a two-time formalism which treats the memory due to nonlocal temporal coupling and thus *non-Markovian* effects without further approximations. A further advantage of this approach is the exact treatment of spectral properties, which include, for example, additional structures due to coherently excited excitons and light-induced modifications of the spectra (compare Sect. 6.3). Only the two-time formalism takes these effects rigorously into account. Any other approximate treatment relies on a quasiparticle approximation.

The dimension of the set of equations increases linearly with the number of values of Δt required. Thus Green's functions that decay slowly with respect to the difference time are hard to tackle numerically. There have, nevertheless, been a number of applications of the two-time approach, for example to exciton–phonon interaction [270, 271], electron–electron correlations [272, 274, 273, 275, 276, 277] and nuclear collisions [278, 279, 280].

10.5.2 Reduction of Propagators to Single Time Functions

Another non-Markovian treatment is obtained from a weak-coupling expansion [267, 268] with respect to the scattering contributions. If we have to avoid an explicit treatment using two times, the only way to proceed is to construct an ansatz which expresses the dependence of the $G^{+-,-+}$ functions on two times in terms of spectral functions and diagonal elements, i.e. the one-particle distribution functions and the transition amplitudes. This basic idea is suggested by the dependences (10.20) and (10.21), but the choice of the time arguments remains an open question. The original choice [251], to take the distribution functions at the macroscopic time $(t_1+t_2)/2$, violates causality unless locality in time is enforced by an infinite expansion. Let us search for solutions of the form

$$G_{nn'}^{+-,-+}(\boldsymbol{k},t,t') = \sum_m \left[a_{nm}(\boldsymbol{k},t,t')\, G_{mn'}^{+-,-+}(\boldsymbol{k},t',t') + G_{nm}^{+-,-+}(\boldsymbol{k},t,t)\, b_{mn'}(\boldsymbol{k},t,t')\right], \tag{10.84}$$

with the obvious constraints that arise for $t = t'$,

$$a_{nm}(\boldsymbol{k},t,t) = c\,\delta_{nm} \tag{10.85}$$

and

$$b_{mn}(\boldsymbol{k},t,t) = (1-c)\delta_{nm}\,. \tag{10.86}$$

Using (10.20) in the form

$$G^{+-}_{nn'}(\boldsymbol{k},t,t) + G^{-+}_{nn'}(\boldsymbol{k},t,t) = -\frac{\mathrm{i}}{\hbar}\delta_{nn'}\,, \tag{10.87}$$

we find

$$\begin{aligned} G^{+-}_{nn'}(\boldsymbol{k},t,t') + G^{-+}_{nn'}(\boldsymbol{k},t,t') &= -\frac{\mathrm{i}}{\hbar}[a_{nn'}(\boldsymbol{k},t,t') + b_{nn'}(\boldsymbol{k},t,t')] \\ &= G^{r}_{nn'}(\boldsymbol{k},t,t') - G^{a}_{nn'}(\boldsymbol{k},t,t')\,. \end{aligned} \tag{10.88}$$

The only conclusion we can draw from this result is that both, a and b may be expressed as suitable linear combinations of retarded and advanced Green's functions, or may even vanish. A possible choice is obtained, if we also account for the causality requirement. In that case we should have a sequence of functions of the form (10.70) and are thus led to the ansatz

$$\begin{aligned} G^{+-,-+}_{nn'}(\boldsymbol{k},t,t') = \mathrm{i}\hbar\sum_m [&G^{\mathrm{r}}_{nm}(\boldsymbol{k},t,t')\,G^{+-,-+}_{mn'}(\boldsymbol{k},t',t') \\ &- G^{+-,-+}_{nm}(\boldsymbol{k},t,t)\,G^{\mathrm{a}}_{mn'}(\boldsymbol{k},t,t')]\,, \end{aligned} \tag{10.89}$$

where the retarded and advanced functions obey the equations

$$\begin{aligned} &\left(\mathrm{i}\hbar\frac{\partial}{\partial t_1} - E_n(\boldsymbol{k})\right) G^{\mathrm{r,a}}_{nn'}(\boldsymbol{k},t_1t_2) = \delta_{nn'}\delta(t_1-t_2) \\ &-\sum_m \left(\boldsymbol{d}_{nm}\boldsymbol{E}(\boldsymbol{r},t_1)\,G^{\mathrm{r,a}}_{mn'}(\boldsymbol{k},t_1,t_2) - \int \mathrm{d}t_3\,\Sigma^{\mathrm{r,a}}_{nm}(\boldsymbol{k},t_1t_3)\,G^{\mathrm{r,a}}_{mn'}(\boldsymbol{k},t_3t_2)\right). \end{aligned} \tag{10.90}$$

It is a simple exercise to show that at the Hartree–Fock level or for any other static approximation to the self-energy, (10.89) is indeed a solution of the quantum kinetic equations, but not a definite one. In (10.89), we can replace G^{r} and G^{a} by any linear combination of these spectral functions, which obeys (10.88), or even interchange them, and again we have a solution within the Hartree–Fock approximation. This is obvious from (10.90), as without dynamical self-energies, G^{r} and G^{a} obey the same equation. Thus the choice is somewhat arbitrary and guided by the idea of obtaining a maximum of retardation and thus of *memory effects*. Using the generalized Kadanoff–Baym ansatz to evaluate the correlational contributions in (10.82), we see that, by construction, we can restrict ourselves to the case $\Delta t = 0$. Instead of the macroscopic time, either the first or the second time argument of the G^{+-} function occurs in the distribution functions in such a way that the one-particle distribution functions and transition amplitudes originating from the Green's functions and from the self-energies always enter in retarded form. The opposite limit is obtained if we interchange G^{r} and G^{a} in (10.90). Memory effects vanish completely and the spectral functions only have to be integrated.

A less arbitrary answer can be found for the transition amplitudes in the low-density limit, which implies $g^{+-}_{\mathrm{vv}}(\boldsymbol{k},t,t-\Delta t) = g^{-+}_{\mathrm{cc}}(\boldsymbol{k},t,t-\Delta t) = -\mathrm{i}/\hbar$ and $g^{-+}_{\mathrm{vv}}(\boldsymbol{k},t,t-\Delta t) = g^{+-}_{\mathrm{cc}}(\boldsymbol{k},t,t-\Delta t) = 0$. Solving (10.82) without scattering contributions linear in the external field, we see immediately that, for example, $g^{+-}_{\mathrm{vc}}(\boldsymbol{k},t,t-\Delta t)$ depends indeed on the second time argument only. Whereas the same holds for $g^{-+}_{\mathrm{cv}}(\boldsymbol{k},t,t-\Delta t)$, we find a dependence on the first argument for $g^{-+}_{\mathrm{vc}}(\boldsymbol{k},t,t-\Delta t)$ and $g^{+-}_{\mathrm{cv}}(\boldsymbol{k},t,t-\Delta t)$. These considerations lead us in accordance with (10.88) to the ansatz

$$G^{+-}_{\mathrm{cv}}(\boldsymbol{k},t,t') = \mathrm{i}\hbar \sum_n G^{+-}_{cn}(\boldsymbol{k},t,t)[G^{\mathrm{r}}_{n\mathrm{v}}(\boldsymbol{k},t,t') - G^{\mathrm{a}}_{n\mathrm{v}}(\boldsymbol{k},t,t')] \quad (10.91)$$

and

$$G^{-+}_{\mathrm{vc}}(\boldsymbol{k},t,t') = \mathrm{i}\hbar \sum_n G^{-+}_{vn}(\boldsymbol{k},t,t)[G^{\mathrm{r}}_{n\mathrm{c}}(\boldsymbol{k},t,t') - G^{\mathrm{a}}_{n\mathrm{c}}(\boldsymbol{k},t,t')] \,. \quad (10.92)$$

Correspondingly, we have

$$G^{-+}_{\mathrm{cv}}(\boldsymbol{k},t,t') = \mathrm{i}\hbar \sum_n [G^{\mathrm{r}}_{cn}(\boldsymbol{k},t,t') - G^{\mathrm{a}}_{cn}(\boldsymbol{k},t,t')]G^{-+}_{nv}(\boldsymbol{k},t',t') \quad (10.93)$$

and

$$G^{+-}_{\mathrm{vc}}(\boldsymbol{k},t,t') = \mathrm{i}\hbar \sum_n [G^{\mathrm{r}}_{\mathrm{v}n}(\boldsymbol{k},t,t') - G^{\mathrm{a}}_{\mathrm{v}n}(\boldsymbol{k},t,t')]G^{+-}_{nc}(\boldsymbol{k},t',t') \,. \quad (10.94)$$

These relations reproduce the low-density limit correctly, and allow us to treat the coherent part of the diagonal propagators in a similar manner. In the high-density limit, when the occupation is inversed, the properties discussed before are interchanged. Correspondingly, we have to interchange t and t' in the time arguments of the G^{+-} and G^{-+} functions on the right-hand side. Thus we see that the type of a suitable linear combination depends crucially on the type of propagator under consideration.

Any approximate treatment of the two-time behavior should coincide with exact results of the density matrix theory, if we consider the same limiting case (compare Problem 10.10), i.e. the $\chi^{(3)}$ limit. In this limit the Green's function approach can be directly compared with the density matrix approach of Sect. 9.2 [231, 244, 243]. In contrast to a treatment by means of (10.89), the relations (10.91)–(10.94) yield the correct scattering contributions at the $\chi^{(3)}$ level. In particular, it follows exactly that both retarded and nonretarded couplings occur in the correlational contributions to the quantum kinetic equation for the transition amplitudes [243].[8] Furthermore, the off-diagonal elements of the spectral functions in (10.89) lead to additional contributions, which are not present in the exact results.

These artefacts can be avoided, if we restrict the ansatz (10.89) to diagonal elements of the spectral function. In this form, there are a number

[8] This is easily seen, if we insert the result of a formal integration of (9.20) into (9.19). Only the contributions resulting from the biexcitonic correlation function couple retarded in (9.19).

of applications, for example, electron–phonon interaction, where the results coincide with a weak coupling expansion in the framework of density matrix theory (compare Sect. 11.4). Beyond the weak coupling limit, the problem of calculating the two-time dependence from (10.82) is only transferred to the spectral functions (10.90). Thus the advantage of this approach disappears unless it is justifiable to assume a quasiparticle approximation for the retarded and advanced Green's functions.

If we separate the rapidly oscillating phase factors, the ansatz (10.89) takes the form

$$g_{nn'}^{+-,-+}(\boldsymbol{k},t,t') = \mathrm{i}\hbar \sum_m [g_{nm}^{\mathrm{r}}(\boldsymbol{k},t,t') g_{mn'}^{+-,-+}(\boldsymbol{k},t',t') - g_{nm}^{+-,-+}(\boldsymbol{k},t,t) g_{mn'}^{\mathrm{a}}(\boldsymbol{k},t,t')] \,. \tag{10.95}$$

If we now apply the undamped quasiparticle approximations $g_{nn'}^{\mathrm{r}}(\boldsymbol{k},t,t') = \theta(t-t')\,\delta_{nn'}$ and $g_{nn'}^{\mathrm{a}}(\boldsymbol{k},t,t') = -\theta(t'-t)\,\delta_{nn'}$, the problem is drastically simplified, as only single-time functions enter the kinetic equations. This undamped quasiparticle approximation for the retarded and advanced Green's functions in (10.95) is, however a pitfall. Whereas in the two-time formalism memory effects are limited by the finite range of the Green's functions or self-energies, the dependence on two times is transferred by the ansatz (10.95) to the retarded and advanced Green's functions. Consequently, an undamped quasiparticle approximation for $g^{\mathrm{r,a}}$ neglects the basic mechanism, which limits the range of the memory and cannot lead to meaningful results. The importance of a consistent treatment of finite-lifetime effects has been demonstrated in a number of numerical studies [269]. If the damping of quasiparticles resulting from the imaginary part of the self-energy becomes much larger than the inverse of typical times over which the distribution functions or transition amplitudes change significantly, we approach the Markovian limit. In this case, the time evolution due to scattering depends only on the actual state of the system and not on its past. We shall discuss this limit in more detail after we have obtained explicit representations for the self-energy.

10.6 The Self-Energy in Different Approximations

To construct consistent approximations beyond the Hartree–Fock approximation is far from trivial. The self-energy itself and the polarization propagator, which determines the screened interaction, depend sensitively on the vertex-corrections. Thus there is a complicated functional dependence of one-particle properties on two-particle properties and vice versa, and we have to consider the consistence of approximations at the one-particle and the two-particle levels. In principle, there are two,different approaches to attacking the tremendously coupled system of equations, which we derived in Sect. 10.4.

- *Iterative procedure.* The sequence of steps can be defined by the vertex-function (10.63), which again yields by means of the chain rule the recurrence

$$\begin{aligned}\Gamma_{n+1}(\bar{1}\bar{2}\bar{3}) &= -\delta(\bar{1}-\bar{2})\,\delta(\bar{1}-\bar{3}) - \frac{\delta\Sigma_n(\bar{1}\bar{2})}{\delta\phi_{\text{eff}}(\bar{3})} \\ &= -\delta(\bar{1}-\bar{2})\,\delta(\bar{1}-\bar{3}) - \int \mathrm{d}\bar{4}\bar{5}\bar{6}\bar{7}\,\frac{\delta\Sigma_n(\bar{1}\bar{2})}{\delta G_0(\bar{4}\bar{5})}\,G_0(\bar{4}\bar{6})\,G_0(\bar{7}\bar{5})\Gamma_1(\bar{6}\bar{7}\bar{3})\,.\end{aligned} \tag{10.96}$$

  We start with the Hartree–approximation, i.e. $\Sigma_0 = 0$, which delivers G_0, $\Gamma_1 = -\delta(\bar{1}-\bar{2})\,\delta(\bar{1}-\bar{3})$ and the screened interaction w_1. In the subsequent step we obtain Σ_1, G_1 and Γ_2 and so forth. The effect of this iteration is twofold. In the n-th step, the Green's functions contributing to $\Sigma^{(n)}[G^{(n-1)}]$ become dressed by an additional interaction line and additionally new types of diagrams are generated. Indeed this type of expansion in terms of G_0 seems to converge, at least for strongly simplified model systems [282].
- *Self-consistent approximations.* We select a certain class of self-energy diagrams $\Sigma[G]$. The Dyson equation becomes a nonlinear functional equation of the Green's functions, which thus has to be solved self-consistently. The solution corresponds, obviously, to the summation of a certain class of diagrams up to infinite order in the interaction, whereas others which contribute even in low order are neglected.[9] The difficulty is in finding the correct way to choose a subset of diagrams for each order. To avoid arbitrariness, additional criteria for the construction of an approximation scheme are required.

In order to deliver physically meaningful results, any approximation should guarantee certain macroscopic conservation laws. This condition can be imposed by the postulate that all diagrams contributing to the self-energy, are obtained from the functional derivative of a functional $\Phi[G]$ with respect to G. Solving the Dyson equation self-consistently with a *Φ-derivable self-energy* yields a Green's function which conserves particle number, energy, and momentum [283]. Physically, the functional Φ is nothing but the sum of all closed diagrams representing the *exchange* and *correlation* contributions to the *exact ground-state energy* as a functional of the fully renormalized one-particle Green's function [284]. Thus before we construct an approximation to the Φ functional, we must make a detour to the calculation of the ground-state energy.

[9] As the Green's function is fully dressed by interaction effects, we have to take care of self-energy insertions, in order to avoid double counting of diagrams.

10.6.1 Ground-State Energy

The ground-state energy results from the expectation value of (10.28) with the ground state wavefunction $|\Psi_0\rangle$, which we write in the form

$$E_0 = \langle\Psi_0|H_0|\Psi_0\rangle + \langle\Psi_0|H_{\mathrm{int}}|\Psi_0\rangle \equiv \langle H_0\rangle + \langle H_{\mathrm{int}}\rangle\,. \tag{10.97}$$

The interaction contribution can be readily expressed in terms of a two-particle Green's function as

$$\langle H_{\mathrm{int}}\rangle = \frac{e^2}{2}\int \mathrm{d}^3\boldsymbol{r}\,\mathrm{d}^3\boldsymbol{r}'\,v(\boldsymbol{r}-\boldsymbol{r}')\langle \mathbf{T}_+\psi_\sigma(\boldsymbol{r}t)\psi_{\sigma'}(\boldsymbol{r}'t)\psi^\dagger_{\sigma'}(\boldsymbol{r}'t^+)\psi^\dagger_\sigma(\boldsymbol{r}t^+)\rangle\,. \tag{10.98}$$

By making use of (10.39), the calculation of (10.98) can be reduced to

$$\langle H_{\mathrm{int}}\rangle = \frac{1}{2}\int \mathrm{d}^3\boldsymbol{r}\,\mathrm{d}^3\boldsymbol{r}\,v(\boldsymbol{r}-\boldsymbol{r}')\Big(\rho(\boldsymbol{r})\rho(\boldsymbol{r}') - \hbar\,\frac{\delta\rho(\boldsymbol{r}t)}{\delta\phi_{\mathrm{ext}}(\boldsymbol{r}'t)}\Big)\,, \tag{10.99}$$

where the first term is the purely electronic part of the electrostatic interaction, which remains finite when taken together with the electron–nuclei and nuclei–nuclei interaction. The second term describes exchange and correlation. Unfortunately $\langle H_{\mathrm{int}}\rangle$ is not the complete interaction contribution to the ground-state energy. As the *exact* wavefunctions of the interacting system are involved, the expectation value $\langle H_0\rangle$ contains interaction contributions too. The task of separating the exchange and correlation energy can be accomplished by means of the *Hellmann–Feynman theorem* [285]. This theorem states, for a Hamiltonian that depends on a continous parameter λ and has eigenfunctions $\Psi^{(\lambda)}$ and eigenvalues $E^{(\lambda)}$, that

$$\frac{\partial E^{(\lambda)}}{\partial\lambda} = \langle\Psi_0^{(\lambda)}|\frac{\partial H^{(\lambda)}}{\partial\lambda}|\Psi_0^{(\lambda)}\rangle\,. \tag{10.100}$$

Let us vary the coupling strength of the Coulomb interaction and replace the Hamiltonian (10.28) according to $H \to H^{(\lambda)} = H_0 + \lambda H_{\mathrm{int}}$. Applying (10.100), we find

$$E_0^{(1)} - E_0^{(0)} = \int_0^1 \mathrm{d}\lambda\,\frac{1}{\lambda}\langle\Psi_0^{(\lambda)}|\lambda H_{\mathrm{int}}|\Psi_0^{(\lambda)}\rangle\,, \tag{10.101}$$

which is just the desired difference of the ground-state energy between the interacting system $E_0^{(1)}$ and the noninteracting system $E_0^{(0)}$. As (10.99) can be immediately generalized to arbitrary values of λ, we make use of the corresponding result to represent the right-hand side of (10.101) by means of the density response. Neglecting the Hartree contribution, we can write the exchange and correlation contribution as

$$\Delta E = \frac{\hbar}{2}\int \mathrm{d}^3\boldsymbol{r}\,\mathrm{d}^3\boldsymbol{r}'\int_0^1 \mathrm{d}\lambda\,v(\boldsymbol{r}-\boldsymbol{r}')\frac{\delta\rho^{(\lambda)}(\boldsymbol{r}t)}{\delta\phi_{\mathrm{ext}}(\boldsymbol{r}'t)}\,. \tag{10.102}$$

Provided we know the *density–density correlation function* for the required range of coupling strengths, ΔE can be calculated. Fortunately, the procedure is rather simple.

10.6.2 The Screened Hartree–Fock Approximation

The density response in (10.102) can be unambigously represented in terms of *connected Feynman diagrams.* For a given subset of diagrams with n Coulomb lines, which is obviously of the order λ^{n-1}, we can perform the λ integration in (10.102). The result is simply a factor $1/n$. Thus the structure of the exchange and correlation energy or of the functional Φ, can be described by

$$\Phi[G] = -\sum_n \frac{1}{2n} \int \mathrm{d}\bar{1}\bar{2}\bar{3}\bar{4}\, G(\bar{1}\bar{2}) \mathcal{I}_n\left([G], \bar{1}\bar{2}\bar{3}\bar{4}\right) G(\bar{3}\bar{4})\,. \qquad (10.103)$$

The quantity $\mathcal{I}_n$ denotes an effective two-particle interaction, in which all connected Feynman diagrams with n Coulomb lines, which are coupled to $(2n-2)$ fully dressed Green's functions, can be taken into consideration. Allowed contributions, however, are *irreducible* diagrams only, i.e. they do not become disconnected into two parts when a single interaction or particle line is cut. Further, they have to obey the symmetry requirements

$$\mathcal{I}_n\left([G], \bar{1}\bar{2}\bar{3}\bar{4}\right) = \mathcal{I}_n\left([G], \bar{3}\bar{4}\bar{1}\bar{2}\right) = \left[\mathcal{I}_n\left([G^\dagger], \bar{2}\bar{1}\bar{4}\bar{3}\right)\right]^*\,, \qquad (10.104)$$

which guarantee the correct structure of the self-energy and of the effective interaction $\delta\Sigma/\delta G$ in the Bethe–Salpeter equation. Needless to say, that the selection of diagrams depends on the system under consideration.

The simplest functional beyond the Hartree–Fock theory which accounts for the screening of the bare Coulomb interaction is shown in Fig. 10.5. Performing the functional derivative with respect to G,[10] we obtain the *screened Hartree–Fock approximation* to the self-energy,

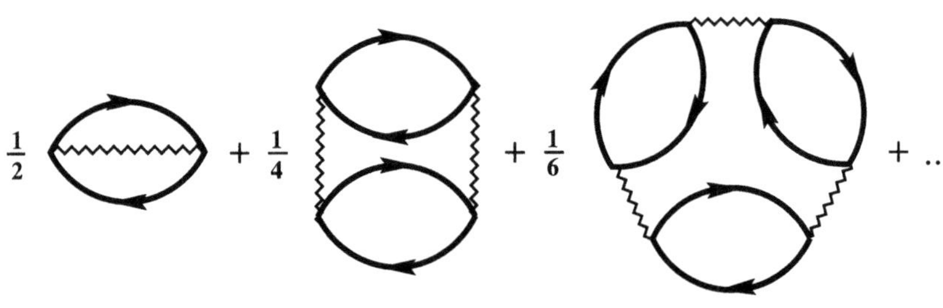

Fig. 10.5. Diagrammatic representation of the functional $\Phi[G]$ generating the GW approximation to the self-energy. The first contribution corresponds to the Hartree–Fock approximation

[10] Note, that functional differentiation with respect to G is nothing but plucking a line out of a diagram.

$$\Sigma(\bar{1}\bar{2}) = i\hbar e^2 b_1 w(\bar{2}\bar{1}^+) G(\bar{1}\bar{2}) . \qquad (10.105)$$

By construction, the screened interaction is defined within the RPA. If we neglect vertex-corrections in (10.57) and (10.64), we obtain the same result. This approximation, often referred to as the *GW approximation*, is frequently applied in solid-state physics, not only to describe the dynamics of excited systems, but also as a standard approximation in *quasiparticle band structure* calculations [54, 55, 56, 57, 58]. If the renormalized quantities are replaced by G_0 and w_1, (10.105) corresponds to Σ_1 of the iterative scheme.

Let us consider the consistency of the *GW* approximation. At the two-particle level an approximation is conserving, in particular the *f-sum rule* is obeyed (compare Problem 2.4), if the electron–hole interaction can be obtained from the second functional derivative of Φ or $\delta\Sigma/\delta G$. This leads in the present case to the electron–hole interaction

$$\frac{\delta\Sigma(\bar{1}\bar{2})}{\delta G(\bar{3}\bar{4})} = i\hbar e^2 b_1 [w(\bar{2}\bar{1})\,\delta(\bar{1}\bar{3})\,\delta(\bar{2}\bar{4}) + w(\bar{1}\bar{3})\,G(\bar{3}\bar{4})\,G(\bar{2}\bar{1})\,w(\bar{4}\bar{2})] , \qquad (10.106)$$

where the last contribution results from the derivative $\delta w/\delta G$. Thus we are already beyond the screened ladder approximation for the polarization function. In contrast, the screened interaction in the self-energy is calculated without any interaction, which in the self-consistent version necessarily violates the f-sum rule. In principle, this inconsistency between the treatment of one-particle and two-particle properties can only be removed if we take the same vertex corrections into account in the calculations of Σ and P [286]. This requirement is of particular importance, as there is always a strong cancellation between the two contributions. Any improvement beyond the screened Hartree–Fock approximation can be constructed by means of an expansion in powers of the screened interaction. Physically, the shortcomings of the self-consistent *GW* approximation show up as excessively large gaps in quasiparticle bandstructures [57, 58]. This problem is closely related to the failure of the RPA to describe optical spectra of semiconductors (compare Sect. 5.2). Also, for finite wave vectors the RPA underestimates screening in the vicinity of the band edge, owing to the neglect of the electron–hole attraction. This is precisely the reason for incorrect quasiparticle energies. Including screened ladder diagrams in the Φ functional leads to a variety of additional types of self-energies and electron–hole interactions. Whether such additional contributions beyond an improved *GW* approximation have to be considered or not depends ultimately on their magnitude.

Also, in the case of excited semiconductors, the *GW* approximation is not consistent and has to be improved. We postpone consideration of this for the moment, as we shall first elaborate on the properties of the screened interaction within the RPA.

10.7 The Screened Interaction

In semiconductors under the action of external fields, the excited electrons and holes give rise to intraband processes and thus to a new type of screening which is not present in the unexcited state. In the following we shall focus our interest on these excitation-induced screening processes. We start from (10.59) and split the polarization propagator into an interband and an intraband contribution. The solution with respect to the interband contributions replaces the bare Coulomb interaction by an interaction which is screened by all interband processes. These are nonresonant and can be described by a background dielectric function, which we incorporate as we did earlier in the definition of the bare Coulomb interaction. In the next step we perform the decomposition with respect to branch indices. The w^{+-} functions contain the information about the *occupation of longitudinal elementary excitations* of the system, e.g. *plasmons*, and describe the interaction mediated by them. By means of the relations (10.61)–(10.62) we obtain from (10.59)

$$w^{+-}(12) = \frac{\varepsilon_0}{V}\int \mathrm{d}34\, v(1-3)[P^{+-}(34)\, w^{\mathrm{a}}(42) + P^{r}(34)\, w^{+-}(42)] \,, \qquad (10.107)$$

and the *retarded interaction* obeys the equation

$$\begin{aligned} w^{\mathrm{r}}(12) &= v(1-2) + \frac{\varepsilon_0}{V}\int \mathrm{d}34\, v(1-3)P^{\mathrm{r}}(34)w^{\mathrm{r}}(42) \\ &= \int \mathrm{d}3\, \epsilon_{\mathrm{r}}^{-1}(13)v(2-3) \,, \end{aligned} \qquad (10.108)$$

where the *retarded dielectric function* is

$$\epsilon_{\mathrm{r}}(12) = \delta(1-2) - \frac{\varepsilon_0}{V}\int \mathrm{d}3\, v(1-3)\, P^{\mathrm{r}}(32) \,. \qquad (10.109)$$

Corresponding equations are obtained for w^{-+} and w^{a}. Formally solving (10.107) by means of (10.108), we find

$$w^{+-}(12) = \frac{\varepsilon_0}{V}\int \mathrm{d}34\, w^{\mathrm{r}}(13)\, P^{+-}(34)\, w^{\mathrm{a}}(42) \,. \qquad (10.110)$$

In an excited system the polarization propagator consists of the sum of of two contribution, i.e. $P = P_{\mathrm{g}} + P_{\mathrm{i}}$, where P_{g} determines the linear response of the unexcited system, and P_{i} denotes the excitation-induced contributions. The corresponding decomposition of the dielectric function is less trivial, and will be discussed in the following

10.7.1 Separation of the Intraband and the Interband Susceptibility

Let us consider the total susceptibility, which has to be calculated according to (compare (2.161))

$$\chi_{\text{tot}} = \left(1 - \frac{\varepsilon_0}{V}(P_{\text{g}} + P_{\text{i}})v\right)^{-1} (P_{\text{g}} + P_{\text{i}}) \,. \tag{10.111}$$

For convinience, we have used an obvious matrix notation. We can easily separate the two contributions to the polarization function with respect to the susceptibility, and obtain as a result

$$\chi_{\text{tot}} = \chi + \delta\chi \,, \tag{10.112}$$

where

$$\delta\chi = \left(1 + \frac{\varepsilon_0}{V}\chi v\right) \chi_{\text{i}} \left(1 + \frac{\varepsilon_0}{V}\chi v\right) \tag{10.113}$$

and the intraband susceptibility is

$$\chi_{\text{i}} = \left(1 - \frac{\varepsilon_0}{V} P_{\text{i}}\, w\right)^{-1} P_{\text{i}} \,. \tag{10.114}$$

Here χ and w denote the susceptibility and screened interaction, respectively, of the unexcited semiconductor. If we are interested in an energy range, which is small compared with the gap energy, these quantities can be treated statically. Correspondingly, the excited carriers occupy only a small fraction of the Brillouin zone, and can be treated in effective-mass approximation. Thus, we only need to consider the long-range contributions to the quantities in (10.114), and obtain after spatial Fourier transformation the integral equation

$$\chi_{\text{i}}(\boldsymbol{q}, \bar{t}_1\bar{t}_2) = P_{\text{i}}(\boldsymbol{q}, \bar{t}_1\bar{t}_2) + \varepsilon_0 V \int \mathrm{d}\bar{t}_3\, P_{\text{i}}(\boldsymbol{q}, \bar{t}_1\bar{t}_3) w(\boldsymbol{q})\, \chi_{\text{i}}(\boldsymbol{q}, \bar{t}_3\bar{t}_2) \,, \tag{10.115}$$

where $w(\boldsymbol{q}) = w(\boldsymbol{q}, \omega = 0) = v(\boldsymbol{q})/\varepsilon(\boldsymbol{q}, \omega = 0) \simeq v(\boldsymbol{q})/\varepsilon_\infty$. In order to see through the various relations most easily, we assume separability of the macroscopic and microscopic timescales, defined as earlier by $t = (t_1 + t_2)/2$ and $\Delta t = t_1 - t_2$. We then obtain the intraband susceptibility in a closed form. From (10.115), we have

$$\begin{aligned} \chi_{\text{i}}^{+-}(\boldsymbol{q}, \omega, t) &= P_{\text{i}}^{+-}(\boldsymbol{q}, \omega, t) + \varepsilon_0 V\, w(\boldsymbol{q})\, P_{\text{i}}^{+-}(\boldsymbol{q}, \omega, t)\, \chi_{\text{i}}^{\text{a}}(\boldsymbol{q}, \omega, t) \\ &\quad + \varepsilon_0 V\, w(\boldsymbol{q})\, P_{\text{i}}^{\text{r}}(\boldsymbol{q}, \omega, t)\, \chi_{\text{i}}^{+-}(\boldsymbol{q}, \omega, t) \,. \end{aligned} \tag{10.116}$$

The retarded function is determined by

$$\chi_{\text{i}}^{\text{r}}(\boldsymbol{q}, \omega, t) = P_{\text{i}}^{\text{r}}(\boldsymbol{q}, \omega, t) + \varepsilon_0 V\, w(\boldsymbol{q})\, P_{\text{i}}^{\text{r}}(\boldsymbol{q}, \omega, t)\, \chi_{\text{i}}^{\text{r}}(\boldsymbol{q}, \omega, t) \,, \tag{10.117}$$

and a corresponding equation holds for $\chi_{\rm i}^{\rm a}$. Eliminating $\chi_{\rm i}^{\rm a}$ in (10.116), we can write the solution as

$$\begin{aligned}\chi_{\rm i}^{+-}(\boldsymbol{q},\omega,t) &= \frac{P_{\rm i}^{+-}(\boldsymbol{q},\omega,t)}{[1-\varepsilon_0 V\,w(\boldsymbol{q})P_{\rm i}^{\rm r}(\boldsymbol{q},\omega,t)]\,[1-\varepsilon_0 V\,w(\boldsymbol{q})P_{\rm i}^{\rm a}(\boldsymbol{q},\omega,t)]}\\ &= \frac{P_{\rm i}^{+-}(\boldsymbol{q},\omega,t)}{\varepsilon_{\rm i}^{\rm r}(\boldsymbol{q},\omega,t)\varepsilon_{\rm i}^{\rm a}(\boldsymbol{q},\omega,t)}\,,\end{aligned} \tag{10.118}$$

where the intraband dielectric function is denoted by $\varepsilon_{\rm i}$. In the same way, we find the following for the total electronically screened interaction from (10.59):

$$w^{\rm r}(\boldsymbol{q},\omega,t) = \frac{w(\boldsymbol{q})}{(1-\varepsilon_0 V\,w(\boldsymbol{q})P_{\rm i}^{\rm r}(\boldsymbol{q},\omega,t))} = \frac{v(\boldsymbol{q})}{\varepsilon_\infty\varepsilon_{\rm i}^{\rm r}(\boldsymbol{q},\omega,t)}\,. \tag{10.119}$$

Clearly, if our assumption of separability of macroscopic and microscopic timescales is not valid, we have to deal with integral equations of the type (10.115). Relations of the type (10.118) have been introduced by intuitive arguments in Sect. 2.7, when we separated the resonant contributions from the background contributions of the dielectric function. We shall make extensive use of these results, when we introduce the elctron–phonon interaction in an excited system in Sect. 11.5

10.7.2 The Screened Interaction in Random Phase Appproximation

We now turn to the calculation of the polarization propagator, which contains intraband transitions between noninteracting particles only. This means that we neglect vertex corrections in (10.60). After decomposition with respect to branch indices and by means of an eigenfunction expansion, we obtain

$$\begin{aligned}P^{\rm r}(\boldsymbol{q},t,t') = \frac{{\rm i}\hbar e^2}{V\varepsilon_0}\sum_{nn'mm'\boldsymbol{k}} &\Big(G^{\rm r}_{nn'}(\boldsymbol{k}+\boldsymbol{q},t,t')G^{+-}_{mm'}(\boldsymbol{k},t',t)\\ &+G^{+-}_{nn'}(\boldsymbol{k}+\boldsymbol{q},t,t')G^{\rm a}_{mm'}(\boldsymbol{k},t',t)\Big)\\ &\times\langle n,\boldsymbol{k}+\boldsymbol{q}|\,{\rm e}^{-{\rm i}\boldsymbol{q}\boldsymbol{r}}\,|m',\boldsymbol{k}\rangle\langle m,\boldsymbol{k}|\,{\rm e}^{{\rm i}\boldsymbol{q}\boldsymbol{r}}\,|n',\boldsymbol{k}+\boldsymbol{q}\rangle\,.\end{aligned} \tag{10.120}$$

This representation is still rather general, as it contains both the polarization function of the unexcited system and the contributions induced by the excitation. For vanishing external field, only interband transitions are present, which yield the dielectric function introduced in Sect. 5.2. In the following, we shall focus on the excitation-induced intraband contributions, i.e. the long-wavelength limit for $m'=n$ and $n'=m$, which we have denoted as $P_{\rm i}^{\rm r}$.

From (10.113) and (10.115) it follows that within the static approximation the interband contributions can be lumped together into the dielectric function ε_∞, which screens the free electron charge. Formally, we can account for this effect by the replacement $v(\boldsymbol{q}) \to w(\boldsymbol{q})$.

If we express the retarded and advanced Green's functions by the G^{+-} and G^{-+} functions, respectively, we can treat this representation within the two-time formalism and solve (10.108) in the time domain. In order to proceed analytically, we insert the ansatz (10.89) with diagonal spectral functions, which yields

$$P_{\mathrm{i}}^{\mathrm{r}}(\boldsymbol{q},t,t') = -\frac{\hbar^2 e^2}{V\varepsilon_0} \sum_{nm\boldsymbol{k}} [G_{nm}^{\mathrm{r}}(\boldsymbol{k}+\boldsymbol{q},t,t')\, G_{nn}^{\mathrm{a}}(\boldsymbol{k},t',t)\, G_{mn}^{+-}(\boldsymbol{k}+\boldsymbol{q},t',t') \\ - G_{nn}^{\mathrm{r}}(\boldsymbol{k}+\boldsymbol{q},t,t')\, G_{mn}^{\mathrm{a}}(\boldsymbol{k},t',t)\, G_{nm}^{+-}(\boldsymbol{k}+\boldsymbol{q},t',t')] \,. \quad (10.121)$$

Since the off-diagonal elements of G^{r} and G^{a} depend separately on both times, a further evaluation requires again an ansatz for these functions, which may be constructed from the case of noninteracting particles [258, 259]. We shall not go into details of the lengthy analysis here. We note merely that these off-diagonal elements lead to coherently induced intraband transitions on the timescale over which the optical polarization is present. Thus these contributions are of importance for times short compared with the dephasing time – in particular, the corresponding screening mechanism is similar to that resulting from the diagonal elements of the Green's functions in (10.121), which we consider in the following.

Using the quasiparticle approximation for the retarded and advanced functions (compare (10.78)), we obtain from (10.121)

$$P_{\mathrm{i}}^{\mathrm{r}}(\boldsymbol{q},t_1,t_2) = \mathrm{i}\,\theta(t_1-t_2)\,\frac{e^2}{\hbar V\varepsilon_0}\sum_{n\boldsymbol{k}} \mathrm{e}^{-\mathrm{i}[E_n(\boldsymbol{k}+\boldsymbol{q})-E_n(\boldsymbol{k})](t_1-t_2)/\hbar} \\ \times [f_n(\boldsymbol{k}+\boldsymbol{q},t_2) - f_n(\boldsymbol{k},t_2)] \,, \quad (10.122)$$

which, together with the spatial Fourier transform[11] of (10.108)

$$w^{\mathrm{r}}(\boldsymbol{q},t_1,t_2) = w(\boldsymbol{q})\,\delta(t_1-t_2) + \varepsilon_0 V \int_{t_2}^{t_1} \mathrm{d}t_3\, w(\boldsymbol{q}) P_{\mathrm{i}}^{\mathrm{r}}(\boldsymbol{q},t_1,t_3) w^{\mathrm{r}}(\boldsymbol{q},t_3,t_2) \,, \quad (10.123)$$

determines the screened interaction in the time domain. It is convenient to separate the instantaneous and noninstantaneous parts of the interaction by means of the intraband contribution to the *density–density correlation function* χ, introduced in Sect. 2.6.4. As an abbreviation, we introduce $D^{\mathrm{r}}(\boldsymbol{q},t_1,t_2) = \varepsilon_0 V w(\boldsymbol{q})\chi(\boldsymbol{q},t_1,t_2)$ which allows us to write (10.123) as

$$w^{\mathrm{r}}(\boldsymbol{q},t_1,t_2) = w(\boldsymbol{q})[\delta(t_1-t_2) + D^{\mathrm{r}}(\boldsymbol{q},t_1,t_2)] \,, \quad (10.124)$$

[11] Note that we have assumed spatial homogeneity, otherwise we would obtain a coupling similar to that in (10.68).

where D^{r} obeys the equation

$$D^{\mathrm{r}}(\boldsymbol{q},t_1,t_2) = D_0^{\mathrm{r}}(\boldsymbol{q},t_1,t_2) + \int_{t_2}^{t_1} \mathrm{d}t_3\, D_0^{\mathrm{r}}(\boldsymbol{q},t_1,t_3)\, D^{\mathrm{r}}(\boldsymbol{q},t_3,t_2)\,. \quad (10.125)$$

This integral equation describes the summation of the intraband contribution of the bubble diagrams (see Fig. 2.1).

$$D_0^{\mathrm{r}}(\boldsymbol{q},t_1,t_2) = \varepsilon_0 V w(\boldsymbol{q})\, P_{\mathrm{i}}^{\mathrm{r}}(\boldsymbol{q},t_1,t_2) = \frac{1}{|\boldsymbol{q}|^2\varepsilon_\infty} P_{\mathrm{i}}^{r}(\boldsymbol{q},t_1,t_2)\,. \quad (10.126)$$

An immediate conclusion obtainable from (10.125) is the retardation of screening effects. As D^{r} vanishes in the limit $t_1 \to t_2$, there is no instantaneous action, but screening is built up on the timescale of excitation with increasing time interval $t_1 - t_2$. This effect is further enhanced by the non-Markovian coupling of the distribution functions, i.e. the occurence of the time argument t_2, which is an immediate consequence of the generalized Kadanoff–Baym ansatz. Compared with the two-time formalism, memory effects are maximized by this ansatz and are thus expected to be overestimated. A general calculation of the screened interaction requires numerical methods, in both quasi-equilibrium and in the general nonequilibrium case. On the other hand, approximations are possible. Analytical results can be obtained within the *plasmon–pole approximation*, which is guided by the long-wavelength limit, i.e. $\boldsymbol{q} \to 0$. In the following discussion we start with the quasi-equilibrium limit.

Example (Plasmon-Pole Approximation in Quasi-Equilibrium). Provided the time dependence of the one-particle distribution functions becomes negligible, we can solve (10.123) or (10.125) by means of Fourier transformation with respect to difference variables. In this fashion, we obtain an algebraic equation with the solution

$$w^{\mathrm{r}}(\boldsymbol{q},\omega) = \frac{w(\boldsymbol{q})}{1-\varepsilon_0 V w(\boldsymbol{q}) P_{\mathrm{i}}^{r}(\boldsymbol{q},\omega)}\,, \quad (10.127)$$

where the polarization propagator is

$$P_{\mathrm{i}}^{\mathrm{r}}(\boldsymbol{q},\omega) = -\frac{e^2}{\varepsilon_0 V}\sum_{n\boldsymbol{k}} \frac{f_n(\boldsymbol{k}+\boldsymbol{q}) - f_n(\boldsymbol{k})}{E_n(\boldsymbol{k}+\boldsymbol{q}) - E_n(\boldsymbol{k}) - \hbar\omega - \mathrm{i}\delta}\,, \quad (10.128)$$

which is often referred to as *Lindhard's formula* [287] and is nothing but the intraband contribution to the dielectric function, we have already introduced by different means in Sect. 5.2. Let us consider two limiting cases.

- *Limit* $\omega = 0$ *and* $\boldsymbol{q} \to 0$. Setting $\omega = 0$ and taking the long-wavelength limit, we have

$$\begin{aligned} P_{\mathrm{i}}^{\mathrm{r}}(\boldsymbol{q},0) &= -\frac{e^2}{\varepsilon_0 V}\sum_{n\boldsymbol{k}} m_n \frac{\boldsymbol{q}\nabla_{\boldsymbol{k}} f_n(\boldsymbol{k})}{\hbar^2 \boldsymbol{q}\boldsymbol{k}} \\ &= -\frac{e^2}{\varepsilon_0 V}\sum_{n\boldsymbol{k}} \frac{m_n}{\hbar^2|\boldsymbol{k}|}\frac{\partial}{\partial|\boldsymbol{k}|} f_n(\boldsymbol{k})\,, \end{aligned} \quad (10.129)$$

where in the last step we have assumed an isotropic distribution function. Thus the static polarization approaches a finite value in the limit $\boldsymbol{q} \to 0$, which is an immediate consequence of the vanishing energy transfer for intraband transitions. Defining the *screening momentum* as

$$\kappa^2 = -\boldsymbol{q}^2 \varepsilon_0 V w(\boldsymbol{q}) P_{\mathrm{i}}^{\mathrm{r}}(\boldsymbol{q}, 0) = -\frac{P_{\mathrm{i}}^{\mathrm{r}}(\boldsymbol{q}, 0)}{\varepsilon_\infty} , \tag{10.130}$$

we can write the statically screened interaction as

$$w^{\mathrm{r}}(\boldsymbol{q}, 0) = \frac{1}{\varepsilon_0 V} \frac{1}{\boldsymbol{q}^2 + \kappa^2} . \tag{10.131}$$

The cancellation of the $1/\boldsymbol{q}^2$ singularity of the bare Coulomb interaction leads to an exponential decay of the interaction, i.e. to a *Yukawa potential*, in real space.

- *Limit* $\omega \neq 0$ *and* $\boldsymbol{q} \to 0$. Taking the long-wavelength limit but keeping ω finite, on the other hand, yields a different result. We transform (10.128) into the form

$$P_{\mathrm{i}}^{\mathrm{r}}(\boldsymbol{q}, \omega) = \frac{e^2}{\varepsilon_0 V} \sum_{n\boldsymbol{k}} \frac{2[E_n(\boldsymbol{k}+\boldsymbol{q}) - E_n(\boldsymbol{k})]}{(\hbar\omega + \mathrm{i}\delta)^2 - [E_n(\boldsymbol{k}+\boldsymbol{q}) - E_n(\boldsymbol{k})]^2} f_n(\boldsymbol{k}) \tag{10.132}$$

and take the long-wavelength limit, with the result

$$P_{\mathrm{i}}^{\mathrm{r}}(\boldsymbol{q}, \omega) = \frac{e^2}{\varepsilon_0 V} \frac{\boldsymbol{q}^2}{(\omega + \mathrm{i}\delta)^2} \sum_{n\boldsymbol{k}} \frac{f_n(\boldsymbol{k})}{m_n} . \tag{10.133}$$

This allows us to write (10.126) in the form

$$D_0^{\mathrm{r}}(\boldsymbol{q}, \omega) = \frac{\omega_{\mathrm{pl}}^2}{(\omega + \mathrm{i}\delta)^2} , \tag{10.134}$$

where the *plasma frequency* is

$$\omega_{\mathrm{pl}}^2 = \frac{e^2}{\varepsilon_\infty \varepsilon_0 V} \sum_{n\boldsymbol{k}} \frac{f_n(\boldsymbol{k})}{m_n} = \frac{e^2}{\varepsilon_\infty \varepsilon_0 m_{\mathrm{r}}} \frac{N}{V} . \tag{10.135}$$

In the last step, we have assumed a two-band model.

At the plasma frequency, the dielectric function becomes zero. Correspondingly, the inverse dielectric function or the density–density correlation function has a pole, as can be immediately seen from (10.125), which yields, after Fourier transformation in the present limit,

$$D^{\mathrm{r}}(0, \omega) = \frac{\omega_{\mathrm{pl}}^2}{(\omega + \mathrm{i}\delta)^2 - \omega_{\mathrm{pl}}^2} . \tag{10.136}$$

The plasma frequency is thus a collective excitation of the electron-hole system in the vicinity of vanishing wave vector, but is obviously not the only one. For finite wave vector we also have the continuum of pair excitations, which lead to a very effective damping of the plasmon excitation (see Problem 10.8). The plasmon-pole approximation now asserts that the effect of the continuum of pair excitations can be simulated by an extension of the plasmon pole into this pair continuum [288]. At the lowest level, this requires us to introduce a $\boldsymbol{q}$ dependence which reproduces the static limit, thus leading to $\omega_{\boldsymbol{q}}^2 = \omega_{\mathrm{pl}}^2(1 + \boldsymbol{q}^2/\kappa^2)$. Furthermore, we have the condition that screening has to disappear on small spatial scales. In Fourier space, this translates into large wave vectors. Thus we can introduce an effective screening momentum which decreases with increasing wave vector. A frequently used form for the plasmon-pole dispersion is [289, 290]

$$\omega_{\boldsymbol{q}}^2 = \omega_{\mathrm{pl}}^2(1 + \boldsymbol{q}^2/\kappa^2) + \alpha \boldsymbol{q}^4 , \tag{10.137}$$

where α is a numerical constant. Unfortunately α, however, cannot be determined unambigously. Within this approximation, the density–density correlation function takes the form

$$D^{\mathrm{r}}(\boldsymbol{q}, \omega) = \frac{\omega_{\mathrm{pl}}^2}{(\omega + \mathrm{i}\delta)^2 - \omega_{\boldsymbol{q}}^2} . \tag{10.138}$$

This kind of single-pole approximation can be generalized to a rational-function expansion with parameters adjusted to reproduce the exact density–density correlation function [291]. This allows one to represent the screened interaction in terms of a few single-pole contributions. Thus dynamical screening effects, for example in the Bethe–Salpeter equation, can be treated beyond the plasmon-pole approximation.

If we consider nonequilibrium situations, the time dependence of the distribution functions does not allow a solution of (10.125) by means of Fourier transformation. Only if the macroscopic and microscopic timescales are separable can we apply a Markovian approximation and introduce a global macroscopic time. In this limit the dependence of the distribution functions on this macroscopic time is the only difference with respect to the quasi-equilibrium case. For optical excitation on an ultrashort timescale, however, this procedure becomes questionable and we have to a solve the problem in the time domain. In order to make contact with the solutions in quasi-equilibrium situations, we can Fourier transform the solution $D^{\mathrm{r}}(\boldsymbol{q}, t, t - \Delta t)$ with respect to Δt. After the pulse has passed the system, corresponding results are very similar to those obtained in the Markovian approximation or even in the quasi-equilibrium limit. As an example, the inverse dielectric function $\varepsilon^{-1}(\boldsymbol{q}, t, \omega)$ of an electron–hole system, after the excitation by an 50 fs pulse, is shown in Fig. 10.6. Qualitatively, the spectrum displays the same features as in quasi-equilibrium. For small wave vectors, we obtain a pronounced pole at the plasma frequency, which merges with increasing wave vector in the intraband continuum of electrons. Owing to the much larger hole mass, the continuum of hole excitations is separated. This behavior allows us again an approximate description by means of a plasmon-pole approximation.

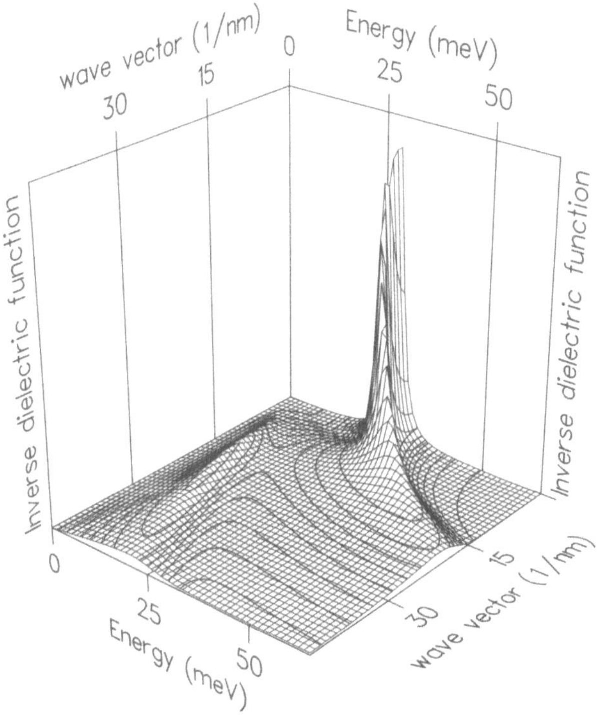

Fig. 10.6. Imaginary part of the inverse dielectric function (RPA) of an electron–hole system, at t=50 fs after excitation with a pulse of 50 fs duration and a central frequency of $\hbar\omega_0 = E_g$. The displayed structures correspond to the plasmon pole and the intraband continua of electrons and holes at a density of $5 \times 10^{17}\,\mathrm{cm}^{-3}$. GaAs parameters

Example (Plasmon-Pole Approximation in the Time Domain). As a guideline for constructing a time-dependent plasmon-pole approximation, we can refer to the Fourier transform of (10.138), which reads

$$D^{\mathrm{r}}(\boldsymbol{q}, t_1 - t_2) = -\theta(t_1 - t_2)\frac{\omega_{\mathrm{pl}}^2}{\omega_{\boldsymbol{q}}} \sin[\omega_{\boldsymbol{q}}(t_1 - t_2)] \,. \tag{10.139}$$

Let us consider first the long-wavelength limit again. From (10.122), we obtain for the leading contribution

$$D_0^{\mathrm{r}}(\boldsymbol{q}, t_1 - t_2) = -\theta(t_1 - t_2)(t_1 - t_2)\,\omega_{\mathrm{pl}}^2(t_2)\,. \tag{10.140}$$

The second derivative yields

$$\frac{\partial^2}{\partial t_1^2} D_0^{\mathrm{r}}(\boldsymbol{q}, t_1 - t_2) = -\delta(t_1 - t_2)\,\omega_{\mathrm{pl}}^2(t_1)\,, \tag{10.141}$$

which is nothing but the Fourier transform of (10.134), but with a time-dependent plasma frequency. Differentiating (10.125) with respect to t_1 and inserting (10.139) shows, as is already suggested by (10.139), that the density–density correlation obeys the equation of a classical harmonic oscillator,

$$\left(\frac{\partial^2}{\partial t_1^2} + \omega_{\rm pl}^2(t_1)\right) D^{\rm r}(\boldsymbol{q}, t_1, t_2) = 0\,. \tag{10.142}$$

The initial conditions

$$D^{\rm r}(\boldsymbol{q}, t_1, t_1) = 0 \tag{10.143}$$

and

$$\left.\left(\frac{\partial}{\partial t_1} D^{\rm r}(\boldsymbol{q}, t_1, t_2)\right)\right|_{t_1=t_2} = -\omega_{\rm pl}^2(t_1) \tag{10.144}$$

follow immediately from (10.125) together with (10.140). If the plasma frequency varies sufficiently slowly with the period of oscillation itself, we can search for an oscillatory solution, which should reduce to (10.139) for a constant plasma frequency. This requires us to replace $\omega_{\rm pl}(t_1)$ by $\omega_{\boldsymbol{q}}(t_1)$ in (10.142) and keep the initial conditions unchanged. An approximate solution, which is correct up to contributions of the order $\dot{\omega}_{\rm pl}^2$ and $\ddot{\omega}_{\rm pl}$, is easily found as [292]

$$D^{\rm r}(\boldsymbol{q}, t_1, t_2) = -\theta(t_1 - t_2) \frac{\omega_{\rm pl}^2(t_2)}{\sqrt{\omega_{\boldsymbol{q}}(t_1)\omega_{\boldsymbol{q}}(t_2)}} \sin\left(\int_{t_2}^{t_1} {\rm d}t\, \omega_{\boldsymbol{q}}(t)\right). \tag{10.145}$$

The plasma frequency depends on the total density of excited carriers, rather than on the shape of the distribution functions. As the density is not changed by scattering processes (except radiative recombination) but only by the excitation process, the time dependence of the plasma frequency comes into play only in a very short initial period after the excitation process. In this coherent regime, however, the screening is effectively suppressed – an effect which is expected to have consequences which can be observed in experiments.

Considering (10.145) in the form $D^{\rm r}(\boldsymbol{q}, t, t - \Delta t)$, we can take the Fourier transform again with respect to Δt. This allows us, to define a *dynamically screened and time-dependent interaction.* Within a certain approximation scheme, which will be discussed in the next section, this type of interaction occurs in the calculation of the scattering contributions. More generally, the result (10.145) or any other reasonable approximation, together with (10.124), can be applied to evaluate the w^{+-} functions given in (10.110) and thus the self-energies entering into the quantum kinetic equations.

10.8 The Second-Order Born Approximation

An obvious shortcoming of the screened Hartree–Fock approximation arises from the fact that according to (10.110), the w^{+-} functions are of second order in the screened interaction $w^{\mathrm{r,a}}$. In order to improve this lack of consistency we have to include vertex corrections that are at least linear in w. Alternatively, we can consider additional types of diagrams in the generating functional Φ. From (10.57) and (10.63), we have in first order in $\delta\Sigma/\delta G$,

$$\begin{aligned}
&\Sigma(\bar{1}\bar{2})\\
&= \mathrm{i}\hbar e^2 b_1 \left(w(\bar{2}\bar{1}^+)\, G(\bar{1}\bar{2}) + \int \mathrm{d}\bar{3}\bar{4}\bar{5}\bar{6}\, w(\bar{3}\bar{1}^+)\, G(\bar{1}\bar{4}) \frac{\delta\Sigma(\bar{4}\bar{2})}{\delta G(\bar{5}\bar{6})}\, G(\bar{5}\bar{3})\, G(\bar{3}\bar{6}) \right)\\
&= \mathrm{i}\hbar e^2 b_1 \left(w(\bar{2}\bar{1}^+)\, G(\bar{1}\bar{2}) \right.\\
&\left. + \mathrm{i}\hbar e^2 \int \mathrm{d}\bar{3}\bar{4}\, b_4 w(\bar{3}\bar{1}^+)\, G(\bar{1}\bar{4})\, w(\bar{2}\bar{4})\, G(\bar{4}\bar{3})\, G(\bar{3}\bar{2}) \right) . \qquad (10.146)
\end{aligned}$$

A diagrammatic representation of this second-order approximation is shown in Fig 10.7. Vertex corrections which are obtained from the derivatives of the screened interaction $\delta w/\delta G$, yield by means of (10.63) higher-order contributions of the ladder type. These can be summarized as additional T-matrix contributions to the self-energy, which are also shown in Fig. 10.7, but have already been omitted in (10.146). We perform a decomposition with respect to branch-indices and make use of (10.110) and (10.60), where we neglect vertex corrections to the polarization propagator. Keeping contributions of second order in $w^{\mathrm{r,a}}$ only, we find

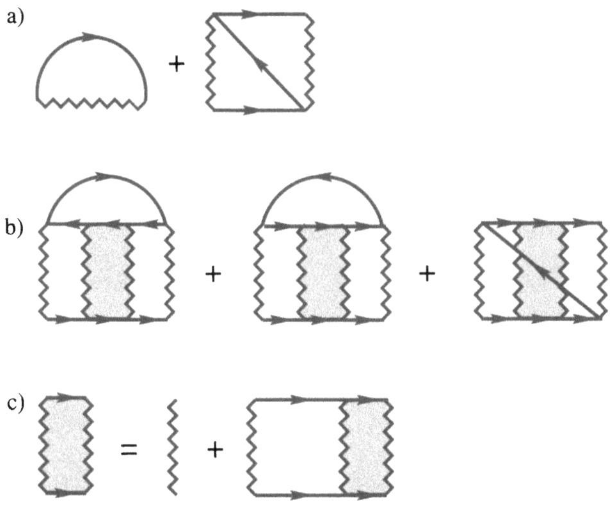

Fig. 10.7. Diagrams contributing to the self-energy beyond the screened Hartree–Fock approximation: *a)* to the second-order Born approximation, *b)* additional contributions within a T-matrix approximation, *c)* diagrammatic representation of the T-matrix equation

$$\Sigma^{+-}(12) = (\mathrm{i}\hbar e^2)^2 \Big(\int \mathrm{d}34\, w^{\mathrm{a}}(31) G^{+-}(34) G^{-+}(43) w^{\mathrm{r}}(24) G^{+-}(12)) - \int \mathrm{d}34\, w^{\mathrm{a}}(31) G^{+-}(14) G^{-+}(43) w^{\mathrm{r}}(24) G^{+-}(32) \Big) . \quad (10.147)$$

The first contribution corresponds to the screened Hartree–Fock approximation, whereas the second one is the corresponding exchange contribution. An eigenfunction expansion of the self-energy can again be performed by means of (10.65). We treat the interaction matrix element as in the derivation of the semiconductor Bloch equations within the effective-mass approximation and obtain the result

$$\begin{aligned}
\Sigma^{+-}_{nm}(\boldsymbol{k}, t_1, t_2) &= (\mathrm{i}\hbar e^2)^2 \sum_{\boldsymbol{q}\boldsymbol{k}'n'm'} \Big(\int \mathrm{d}t_3\, \mathrm{d}t_4\, w^{\mathrm{a}}(\boldsymbol{q}, t_3, t_1)\, w^{\mathrm{r}}(\boldsymbol{q}, t_2, t_4) \\
&\times G^{+-}_{nm}(\boldsymbol{k}+\boldsymbol{q}, t_1, t_2)\, G^{-+}_{n'm'}(\boldsymbol{k}'+\boldsymbol{q}, t_4, t_3)\, G^{+-}_{m'n'}(\boldsymbol{k}', t_3, t_4) \\
&- \int \mathrm{d}t_3\, \mathrm{d}t_4\, w^{\mathrm{a}}(\boldsymbol{q}, t_3, t_1)\, w^{\mathrm{r}}(\boldsymbol{k}-\boldsymbol{k}', t_2, t_4) \\
&\times G^{+-}_{nn'}(\boldsymbol{k}+\boldsymbol{q}, t_1, t_4)\, G^{-+}_{n'm'}(\boldsymbol{k}'+\boldsymbol{q}, t_4, t_3)\, G^{+-}_{m'm}(\boldsymbol{k}', t_3, t_2) \Big) . \quad (10.148)
\end{aligned}$$

Owing to the existence of off-diagonal elements of the Green's function, the self-energy, too, is off-diagonal with respect to band indices. In order to deal with the case that we need in the kinetic equation (10.82), we apply (10.78) and (10.79) and separate the rapidly oscillating phase factors in (10.148) also. As a final result, we obtain

$$\begin{aligned}
\sigma^{+-}_{nm}(\boldsymbol{k}, t_1, t_2) &= (\mathrm{i}\hbar e^2)^2 \sum_{\boldsymbol{q}\boldsymbol{k}'n'm'} \Big(\int \mathrm{d}t_3\, \mathrm{d}t_4\, w^{\mathrm{a}}(\boldsymbol{q}, t_3, t_1)\, w^{\mathrm{r}}(\boldsymbol{q}, t_2, t_4) \\
&\times \mathrm{e}^{\mathrm{i}[E_n(\boldsymbol{k})-E_n(\boldsymbol{k}+\boldsymbol{q})]t_1/\hbar - \mathrm{i}[E_m(\boldsymbol{k})-E_m(\boldsymbol{k}+\boldsymbol{q})]t_2/\hbar} \\
&\times \mathrm{e}^{\mathrm{i}[E_{n'}(\boldsymbol{k}')-E_{n'}(\boldsymbol{k}'+\boldsymbol{q})]t_4/\hbar - \mathrm{i}[E_{m'}(\boldsymbol{k}')-E_{m'}(\boldsymbol{k}'+\boldsymbol{q})]t_3/\hbar} \\
&\times g^{+-}_{nm}(\boldsymbol{k}+\boldsymbol{q}, t_1, t_2)\, g^{-+}_{n'm'}(\boldsymbol{k}'+\boldsymbol{q}, t_4, t_3)\, g^{+-}_{m'n'}(\boldsymbol{k}', t_3, t_4) \\
&- \int \mathrm{d}t_3\, \mathrm{d}t_4\, w^{\mathrm{a}}(\boldsymbol{q}, t_3, t_1)\, w^{\mathrm{r}}(\boldsymbol{k}-\boldsymbol{k}', t_2, t_4) \\
&\times \mathrm{e}^{\mathrm{i}[E_n(\boldsymbol{k})-E_n(\boldsymbol{k}+\boldsymbol{q})]t_1/\hbar - \mathrm{i}[E_m(\boldsymbol{k})-E_m(\boldsymbol{k}')]t_2/\hbar} \\
&\times \mathrm{e}^{\mathrm{i}[E_{n'}(\boldsymbol{k}+\boldsymbol{q})-E_{n'}(\boldsymbol{k}'+\boldsymbol{q})]t_4/\hbar - \mathrm{i}[E_{m'}(\boldsymbol{k}')-E_{m'}(\boldsymbol{k}'+\boldsymbol{q})]t_3/\hbar} \\
&\times g^{+-}_{nn'}(\boldsymbol{k}+\boldsymbol{q}, t_1, t_4)\, g^{-+}_{n'm'}(\boldsymbol{k}'+\boldsymbol{q}, t_4, t_3)\, g^{+-}_{m'm}(\boldsymbol{k}', t_3, t_2) \Big) . \quad (10.149)
\end{aligned}$$

Replacing the indices $+-$ with $-+$ and vice versa, we obtain the corresponding equation for σ^{-+}. Within a restriction to binary collisions, we have now closed our system of quantum kinetic equations which determines dephasing and relaxation processes. To solve these equations without further assumptions is still a challenge. In particular, the exchange contribution requires a much larger computational effort than does the screened Hartree–Fock contribution. Several approximation schemes can be derived from (10.149) and (10.82). These approximations concern two different aspects of the scattering process. The first one is the noninstantaneous interaction due to the time dependence of the screened interaction; the second one results from the temporal convolution of self-energies and Green's functions.

Apart from the very initial stage of the excitation process, the dependence of the screened interaction on two separate times becomes rather smooth. This can already be seen from the simple plasmon-pole approximation (10.141), in which $\omega_{\rm pl}$ becomes completely time independent after the excitation, and the screening momentum varies very smoothly with time. This time dependence can be taken into account in a simple way if the ansatz (10.95) is applied. As a result, one obtains a modulation of the dephasing rates by plasmon oscillations [281], which has also been observed experimentally [293]. More rigorous numerical studies not making use of (10.95) have been performed for instantaneous interactions only, up to now [272, 273, 275, 276].

In order to make contact with kinetic equations of the *Boltzmann* type [251], we neglect this non-Markovian coupling. In principle, this requires that the Green's functions vary slowly compared with the rapidly oscillating phase factors. Then we can approximate the Green's functions in (10.149) and (10.82) at the upper limit of the time integrals. Formally, we also obtain this Markovian limit, if we apply (10.95), but with the retarded and advanced Green's functions interchanged. For notational simplicity we restrict ourselves to a statically screened interaction $w(\boldsymbol{q})$. Finally, reversing the separation of phase factors and switching to the electron–hole picture, we obtain for the correlational contributions to (10.75) within the Markovian limit [294, 295]:

$$
\begin{aligned}
\mathrm{i}\hbar\frac{\partial}{\partial t}p_{\rm vc}(\boldsymbol{k})\Big|_{\rm corr} &= \sum_{\boldsymbol{k}'\boldsymbol{q},ab=\{e,h\}} \{2[w(\boldsymbol{q})]^2 - \delta_{ab}w(\boldsymbol{q})w(\boldsymbol{k}-\boldsymbol{k}')\} \\
&\times \Big[\Big(g[E_a(\boldsymbol{k}) - E_a(\boldsymbol{k}+\boldsymbol{q}) + E_b(\boldsymbol{k}'+\boldsymbol{q}) - E_b(\boldsymbol{k}')] \\
&\times p_{\rm vc}(\boldsymbol{k})\,\{[1-f_a(\boldsymbol{k}+\boldsymbol{q})]f_b(\boldsymbol{k}'+\boldsymbol{q})[1-f_b(\boldsymbol{k}')] \\
&+ f_a(\boldsymbol{k}+\boldsymbol{q})[1-f_b(\boldsymbol{k}'+\boldsymbol{q})]f_b(\boldsymbol{k}') - p^*_{\rm vc}(\boldsymbol{k}')p_{\rm vc}(\boldsymbol{k}'+\boldsymbol{q})\}\Big) \\
&- \Big(\boldsymbol{k}\rightleftharpoons(\boldsymbol{k}+\boldsymbol{q}), \boldsymbol{k}'\rightleftharpoons(\boldsymbol{k}'+\boldsymbol{q})\Big)\Big] \\
&- \sum_{\boldsymbol{k}'\boldsymbol{q},a,b=\{e,h\}} (1-\delta_{ab})\,w(\boldsymbol{q})w(\boldsymbol{k}-\boldsymbol{k}')
\end{aligned}
$$

$$
\begin{aligned}
&\times \Big[g[E_a(\boldsymbol{k}) - E_a(\boldsymbol{k}+\boldsymbol{q}) - E_b(\boldsymbol{k}'+\boldsymbol{q}) + E_b(\boldsymbol{k}')]\\
&\times \Big(p_{\rm vc}(\boldsymbol{k}+\boldsymbol{q})\{[1-f_a(\boldsymbol{k})]f_b(\boldsymbol{k}'+\boldsymbol{q})[1-f_b(\boldsymbol{k}')]\\
&+ f_a(\boldsymbol{k})[1-f_b(\boldsymbol{k}'+\boldsymbol{q})]f_b(\boldsymbol{k}')\} - p_{\rm vc}(\boldsymbol{k}'+\boldsymbol{q})\{[1-f_a(\boldsymbol{k})]\\
&\times f_a(\boldsymbol{k}+\boldsymbol{q})[1-f_b(\boldsymbol{k}')] + f_a(\boldsymbol{k})[1-f_a(\boldsymbol{k}+\boldsymbol{q})]f_b(\boldsymbol{k}')\}\Big)\\
&- g[-E_a(\boldsymbol{k}) + E_a(\boldsymbol{k}+\boldsymbol{q}) + E_b(\boldsymbol{k}'+\boldsymbol{q}) - E_b(\boldsymbol{k}')]\\
&\times [p_{\rm vc}(\boldsymbol{k})p^*_{\rm vc}(\boldsymbol{k}+\boldsymbol{q})p_{\rm vc}(\boldsymbol{k}') - p_{\rm vc}(\boldsymbol{k})p^*_{\rm vc}(\boldsymbol{k}'+\boldsymbol{q})p_{\rm vc}(\boldsymbol{k}')]\Big]\,,
\end{aligned}
\tag{10.150}
$$

which determines the dephasing and spectral shifts of the transition amplitude (10.76). Without the restriction to static screening, we could have performed the time integrations of the screened interaction in (10.149). This results in dynamically screened and time-dependent interactions

$$w^{\rm r}(\boldsymbol{q},t,\omega) = \int {\rm d}\Delta t\, w^{\rm r}(\boldsymbol{q},t,t-\Delta t)\,{\rm e}^{-{\rm i}\omega\Delta t} \tag{10.151}$$

and

$$w^{\rm a}(\boldsymbol{q},\omega,t) = \int {\rm d}\Delta t\, w^{\rm a}(\boldsymbol{q},t-\Delta t,t)\,{\rm e}^{{\rm i}\omega\Delta t}\,, \tag{10.152}$$

where the arguments are of the form $\hbar\omega = E_n(\boldsymbol{k}) - E_n(\boldsymbol{k}+\boldsymbol{q})$. We have introduced this form of dynamical and time-dependent screening already in Sect. 10.6.2 to illustrate the features of the solution of (10.125). Since a static approximation assesses screening effects too heavily and is unsuited to give a reasonable description of the buildup of screening, the inclusion of these dynamical aspects is required. The corresponding improvement of (10.150) and (10.153) is straightforward, but the representation of the results, however, is rather lengthy and left as an excercise for the reader (see Problem 10.11). The same is valid for the treatment of the scattering processes, which govern the relaxation of the electron distribution function (10.77), and are described by

$$
\begin{aligned}
{\rm i}\hbar\frac{\partial}{\partial t}f_{\rm e}(\boldsymbol{k})\Big|_{\rm corr} &= \sum_{\boldsymbol{k}'\boldsymbol{q},b=\{e,h\}} \{2[w(\boldsymbol{q})]^2 - \delta_{ab}w(\boldsymbol{q})w(\boldsymbol{k}-\boldsymbol{k}')\}\\
&\times \Big(2\pi{\rm i}\delta[E_{\rm e}(\boldsymbol{k}) - E_{\rm e}(\boldsymbol{k}+\boldsymbol{q}) + E_b(\boldsymbol{k}'+\boldsymbol{q}) - E_b(\boldsymbol{k}')]\\
&\times \{[1-f_{\rm e}(\boldsymbol{k})]f_{\rm e}(\boldsymbol{k}+\boldsymbol{q})[1-f_b(\boldsymbol{k}'+\boldsymbol{q})]f_b(\boldsymbol{k}')\\
&- f_{\rm e}(\boldsymbol{k})[1-f_{\rm e}(\boldsymbol{k}+\boldsymbol{q})]f_b(\boldsymbol{k}'+\boldsymbol{q})[1-f_b(\boldsymbol{k}')]\}
\end{aligned}
$$

$$
\begin{aligned}
&+g[E_\mathrm{e}(\boldsymbol{k})-E_\mathrm{e}(\boldsymbol{k}+\boldsymbol{q})+E_b(\boldsymbol{k}'+\boldsymbol{q})-E_b(\boldsymbol{k}')]\\
&\times 2\,\mathrm{Re}[p_\mathrm{vc}(\boldsymbol{k}')p^*_\mathrm{vc}(\boldsymbol{k}'+\boldsymbol{q})]\,[f_\mathrm{e}(\boldsymbol{k})-f_\mathrm{e}(\boldsymbol{k}+\boldsymbol{q})]\\
&+g[E_\mathrm{h}(\boldsymbol{k})-E_\mathrm{h}(\boldsymbol{k}+\boldsymbol{q})+E_b(\boldsymbol{k}'+\boldsymbol{q})-E_b(\boldsymbol{k}')]\\
&\times 2\,\mathrm{Re}[p_\mathrm{vc}(\boldsymbol{k})p^*_\mathrm{vc}(\boldsymbol{k}+\boldsymbol{q})]\,[f_b(\boldsymbol{k}')-f_b(\boldsymbol{k}'+\boldsymbol{q})]\Big)\\
&+\sum_{\boldsymbol{k}'\boldsymbol{q}} w(\boldsymbol{q})w(\boldsymbol{k}-\boldsymbol{k}')\\
&\times\Big(g[-E_\mathrm{e}(\boldsymbol{k})+E_\mathrm{e}(\boldsymbol{k}+\boldsymbol{q})+E_\mathrm{h}(\boldsymbol{k}'+\boldsymbol{q})-E_\mathrm{h}(\boldsymbol{k}')]\\
&\times\{2\,\mathrm{Re}[p_\mathrm{vc}(\boldsymbol{k}')p^*_\mathrm{vc}(\boldsymbol{k}'+\boldsymbol{q})]\,[f_\mathrm{e}(\boldsymbol{k})-f_\mathrm{e}(\boldsymbol{k}+\boldsymbol{q})]\\
&-2\,\mathrm{Re}[p_\mathrm{vc}(\boldsymbol{k}')p^*_\mathrm{vc}(\boldsymbol{k}+\boldsymbol{q})]\,[f_\mathrm{e}(\boldsymbol{k})-f_\mathrm{h}(\boldsymbol{k}'+\boldsymbol{q})]\}\\
&+g[E_\mathrm{h}(\boldsymbol{k})-E_\mathrm{h}(\boldsymbol{k}+\boldsymbol{q})-E_\mathrm{e}(\boldsymbol{k}'+\boldsymbol{q})+E_\mathrm{e}(\boldsymbol{k}')]\\
&\times\{2\,\mathrm{Re}[p_\mathrm{vc}(\boldsymbol{k})p^*_\mathrm{vc}(\boldsymbol{k}+\boldsymbol{q})]\,[f_\mathrm{e}(\boldsymbol{k}')-f_\mathrm{e}(\boldsymbol{k}'+\boldsymbol{q})]\\
&-2\,\mathrm{Re}[p_\mathrm{vc}(\boldsymbol{k})p^*_\mathrm{vc}(\boldsymbol{k}'+\boldsymbol{q})]\,[f_\mathrm{e}(\boldsymbol{k}')-f_\mathrm{h}(\boldsymbol{k}+\boldsymbol{q})]\}\Big)\,.
\end{aligned}
\tag{10.153}
$$

The corresponding contribution for holes is obtained by interchanging the indices e and h. Further, we have performed the spin summation and defined $g(\omega) = -\mathcal{P}(1/\omega) + i\pi\delta(\omega)$, where $\mathcal{P}$ denotes the principal value.

Owing to the Markovian approximation, the spectral properties are reduced to the Fourier transform of the phase factors in (10.149). Thus our results imply a quasiparticle approximation. In both correlational contributions, two different types of processes occur. In (10.150) we have contributions which are linear and of third order in the transition amplitude. The sum of all linear contributions is always positive and contributes to the dephasing of the transition amplitude. In contrast, the *polarization scattering*, i.e. the third-order contributions, is responsible for a redistribution of transition amplitudes in $\boldsymbol{k}$ space without destroying the coherence. On the contrary, the decay of the transition amplitude can even be reduced. Corresponding contributions, which are of second order in the transition amplitude, occur in (10.153). Obviously, this polarization scattering dies out on the timescale of the dephasing and we are left with the usual binary-collision contributions, which drive the one-particle distributions towards quasi-equilibrium distribution functions. These scattering processes, which are left when all coherent contributions have decayed, govern transport phenomena also. Owing to the external fields the one-particle distribution functions in this case depend also on the macroscopic spatial variable, and according to (10.69), gradient contributions may come into play. If the distribution functions approach an quasi-equilibrium state, the condition of *detailed balance* is fulfilled. This requires that the contributions to incoherent scattering into a state and out of this state, on the right-hand side of (10.153) compensate one another exactly. As consequence (10.153) vanishes completely.

A non-Markovian version of (10.150) and (10.153) is obtained straightforwardly, if we evaluate (10.149) by means of the ansatz (10.95). As already discussed, this approach, however, fails to reproduce the exact results of the density matrix theory in the $\chi^{(3)}$ limit, and obviously, overestimates memory effects. In contrast to the two-time approach, not only the Markovian limit but also this non-Markovian version, neglects correlations in the spectral functions of excited carriers, and also light-induced effects. Moreover, if incoherent exciton distributions come into play, the second-order Born approximation fails to take *incoherent exciton–exciton scattering* into account. A description of these processes requires a summation of an infinite series of ladder diagrams (compare Fig. 10.7), i.e. the *T-matrix approximation* [255, 296]. This is similar to the case of the Bethe–Salpeter equation discussed above. In spite of these shortcomings, (10.150) and (10.153) have been applied successfully to a variety of problems in semiconductor optics. Before we, turn to applications of these equations and the phenomenology of relaxation processes, however, we have to consider the second important cause of dephasing and relaxation, the interaction with phonons.

10.9 Problems

10.1. Derive the representations of the Green's functions G^{++}, G^{--}, $G^{\rm r}$, and $G^{\rm a}$ in the frequency domain within the Hartree–Fock approximation. Discuss the position of poles in the complex frequency plane.

10.2. Make use of the approximation (10.52) for the self-energy in the Dyson equation and apply an eigenfunction expansion to derive the Hartree–Fock equations. What modifications of the eigenfunction expansions are necessary if you use a screened Hartree–Fock approximation for the self-energy?

10.3. Show that the generalized Kadanoff–Baym ansatz (10.89) solves the Hartree–Fock approximation to the quantum kinetic equations (10.45) and (10.49).

10.4. Verify the Hellmann–Feynman theorem (10.100).

10.5. Apply the iterative approach to the vertex function (10.96) to derive all self-energy and polarization diagrams to third order.

10.6. Express the exchange and correlation energy (10.102) in terms of the screened interaction and the polarization propagator or in terms of the self-energy and the Green's function.

10.7. Calculate the Φ-functional for the sum of bubble diagrams, shown in Fig. 10.5. Show, for a homogeneous electron gas, that the functional is given by

$$\Phi = -\frac{\mathrm{i}\hbar V}{2(2\pi)^4}\int \mathrm{d}\boldsymbol{q}^3\,\mathrm{d}\omega\,\ln[1-\varepsilon_0 V\,v(\boldsymbol{q})P_0(\boldsymbol{q},\omega)]\,, \tag{10.154}$$

where P_0 is the free polarization propagator.

10.8. Prove that plasma oscillations become damped for wave vectors above $q = q_\mathrm{c}$, where q_c is determined by the equation

$$q_\mathrm{c}^2 = \frac{k_\mathrm{F}}{\pi r_\mathrm{B}}\left[\left(2+\frac{q_\mathrm{c}}{k_\mathrm{F}}\right)\ln\left(1+2\frac{k_\mathrm{F}}{q_\mathrm{c}}\right)-2\right]\,, \tag{10.155}$$

where $k_\mathrm{F} = \left(3\pi^2 N/V\right)^{1/3}$ is the Fermi momentum.

10.9. Derive, for an electron–hole plasma, the "perfect screening" sum rule

$$\lim_{\boldsymbol{q}\to 0}\int_0^\infty \frac{\mathrm{d}\omega}{\omega}\frac{\mathrm{Im}[\epsilon(\boldsymbol{q},\omega)]}{|\epsilon(\boldsymbol{q},\omega)|^2} = \frac{\pi}{2}\,. \tag{10.156}$$

Discuss the physical meaning of this result.

10.10. Apply the coherent limit to calculate the correlational contribution to the quantum kinetic equation for the transition amplitude (10.82) in the second-order Born approximation (10.149) up to third order in the external field. Make use either of the Markovian approximation or of the generalized Kadanoff–Baym ansatz. Compare the results with the correlational contribution which is obtained in the second-order Born approximation from (9.19) together with (9.20). Discuss the differences.

10.11. Derive the generalization of (10.150) and (10.153) for the case of a dynamically screened interaction. Hint: Start from (10.82) together with (10.149); treat the propagators within the Markovian approximation, and perform then the temporal integrations.

11. The Electron–Phonon Interaction

As we have seen in Chap. 10, the Green's function theory offers a systematic way to describe many-particle systems under arbitrary nonequilibrium conditions. Up to now, however, we have restricted ourselves to a rigid lattice. Under these conditions, nonequilibrium concerns only the particle system. The time evolution into a true quasi-equilibrium state, however, requires *dissipative processes*, i.e. the *transfer of energy* from the particle system to the lattice, which results from interaction with lattice vibrations. Moreover, completely different nonequilibrium states can arise because of this interaction, in particular if the lattice itself is driven out of equilibrium by interaction with the particle system. Thus our theory of scattering processes developed so far needs improvement.

As is already obvious from classical considerations, any displacement of a nucleus or ion core from its equilibrium position induces a change of the electronic charge density, which gives rise to a field and thus to an *effective electron–electron interaction.* What follows now is nothing but the mathematical formulation of this simple picture within the framework of response theory. This approach allows us to treat *electron–phonon coupling* on a strictly microscopic footing. It emphasizes the dynamical aspects of this interaction from the very beginning. As a central result, we shall show that phonons mediate an additional contribution to the screened interaction, which was introduced in Sect. 10.4. The results derived there remain valid and we simply have to add the phonon-induced contribution.

The evaluation of phonon-induced interactions requires a brief review of some fundamentals of phonon physics. The microscopic theory of *interatomic force constants* shows that the properties of lattice vibrations themselves are determined by the *electronic response* of the system. This intimate coupling of electronic and phonon properties becomes especially obvious in nonequilibrium situations where the excited particle system acts back on the phonon system. These processes give rise to a *renormalization of the phonon spectra* and to *nonequilibrium phonon distribution functions.* The corresponding quantum kinetic equations will be the final result of this chapter.

11.1 The Phonon-Induced Interaction

We consider the complete Hamiltonian

$$H = H_{\rm el} + H_{\rm n} + H_{\rm ext} \,. \tag{11.1}$$

The electronic part $H_{\rm el}$ is given by (10.28), where $\phi_{\rm ext} = 0$ and the one-particle Hamiltonian is

$$h(\boldsymbol{r}t) = -\frac{\hbar^2}{2m}\Delta + eU(\boldsymbol{r}, \{\boldsymbol{R}(\boldsymbol{L\kappa})\}) + \boldsymbol{dE}(\boldsymbol{r}, t) \,. \tag{11.2}$$

In contrast to (10.29), the lattice potential U depends now on the actual positions of the nuclei $\boldsymbol{R}(\boldsymbol{L\kappa}) = \boldsymbol{L} + \boldsymbol{\kappa} + \boldsymbol{u}(\boldsymbol{L\kappa})$, defined in terms of the equilibrium lattice vector $\boldsymbol{L}$, the basis vector within the unit cell $\boldsymbol{\kappa}$, and the lattice displacements $u_\alpha(\boldsymbol{L\kappa})$. The contribution of the nuclei is given by

$$\begin{aligned} H_{\rm n} = &- \sum_{\boldsymbol{L\kappa}} \frac{\hbar^2}{2M_\kappa} \frac{\partial^2}{\partial[u_\alpha(\boldsymbol{L\kappa})]^2} \\ &+ \frac{1}{2} \sum_{\boldsymbol{L\kappa}, \boldsymbol{L}'\boldsymbol{\kappa}'} \int \mathrm{d}^3\boldsymbol{r}\, \mathrm{d}^3\boldsymbol{r}'\, \rho_{\boldsymbol{\kappa}}[\boldsymbol{r} - \boldsymbol{R}(\boldsymbol{L\kappa})] v(\boldsymbol{r} - \boldsymbol{r}') \rho_{\boldsymbol{\kappa}'}[\boldsymbol{r}' - \boldsymbol{R}(\boldsymbol{L}'\boldsymbol{\kappa}')] \,, \end{aligned} \tag{11.3}$$

The charge density of the nuclei $\langle \hat\rho_n(\boldsymbol{r})\rangle$ is represented as a sum of local charge densities $\rho_\kappa[\boldsymbol{r} - \boldsymbol{R}(\boldsymbol{L\kappa})]$, which would be δ functions in the case of bare nuclei. In applications, however, it is more convenient to consider *rigid ion cores* instead of bare nuclei. In this case, ρ_κ denotes the charge density of the ion cores. For our present purposes, this difference is not important. The coupling of the electrons and nuclei to external sources is summarized in $H_{\rm ext}$ and given by

$$H_{\rm ext} = \int \mathrm{d}^3\boldsymbol{r}\, \{[\hat\rho_{\rm el}(\boldsymbol{r}) + \hat\rho_{\rm n}(\boldsymbol{r})]\, \phi_{\rm ext}(\boldsymbol{r}, t) + \hat\rho_n(\boldsymbol{r}) J(\boldsymbol{r}, t)\} \,. \tag{11.4}$$

For reasons which will become clear below, we have added an additional external source $J(\boldsymbol{r}, t)$ [297], which couples to the charge density of the nuclei only.[1] Our aim is now to calculate the total linear response of the system, including the contribution from the nuclei, i.e. the variation of the total electrostatic potential with the external potential. We use again the abbreviated notation of Chap. 10, including the branch index. We have

$$\begin{aligned} \frac{\delta V(\bar 1)}{\delta\phi_{\rm ext}(\bar 2)} &= \delta(\bar 1 - \bar 2) + \int \mathrm{d}3\, v(\bar 1 - \bar 3) \left(\frac{\delta\langle\hat\rho_{\rm el}(\bar 3)\rangle}{\delta\phi_{\rm ext}(\bar 2)} + \frac{\delta\langle\hat\rho_{\rm n}(\bar 3)\rangle}{\delta\phi_{\rm ext}(\bar 2)} \right) \\ &= \delta(\bar 1 - \bar 2) + \int \mathrm{d}\bar 3\bar 4\, v(\bar 1 - \bar 3) \frac{\delta\langle\hat\rho_{\rm el}(\bar 3)\rangle}{\delta V(\bar 4)} \frac{\delta V(\bar 4)}{\delta\phi_{\rm ext}(\bar 2)} \end{aligned}$$

[1] If you worry about the physical realization of this kind of coupling, you can think about mechanical forces generating an elastic deformation of the crystal. Here, it is merely a mathematical trick which allows a simple derivation.

$$+ \int \mathrm{d}\bar{3}\, v(\bar{1}-\bar{3})\, \frac{\delta \langle \hat{\rho}_{\mathrm{n}}(\bar{3})\rangle}{\delta \phi_{\mathrm{ext}}(\bar{2})} \,. \tag{11.5}$$

Solving with respect to $\delta V / \delta \phi_{\mathrm{ext}}$, we obtain

$$\frac{\delta V(\bar{1})}{\delta \phi_{\mathrm{ext}}(\bar{2})} = \varepsilon^{-1}(\bar{1},\bar{2}) + \int \mathrm{d}\bar{3}\, w(\bar{1},\bar{3})\, \frac{\delta \langle \hat{\rho}_{\mathrm{n}}(\bar{3})\rangle}{\delta \phi_{\mathrm{ext}}(\bar{2})} \,, \tag{11.6}$$

where the *electronic dielectric function* is

$$\varepsilon(\bar{1},\bar{2}) = \delta(\bar{1}-\bar{2}) - \int \mathrm{d}\bar{3}\, v(\bar{1}-\bar{3})\, \frac{\delta \langle \hat{\rho}_{\mathrm{el}}(\bar{3})\rangle}{\delta V(\bar{2})} \tag{11.7}$$

and w is the screened interaction. The derivative $\delta \rho_{\mathrm{el}} / \delta V$ differs from the purely electronic polarization which we introduced earlier, owing to the phonon contribution to the total potential. Neglecting this phonon contribution to the polarization function is one of the ingredients of the *adiabatic approximation* [298, 299]. This approximation reduces the dielectric function and the screened interaction in (11.6) to the purely electronic quantities, which we have already introduced in (10.58) and (10.59). What remains is the calculation of the lattice contribution, which follows immediately by the techniques described in Sect. 10.3. We consider the expectation value of the density of nuclei, which, according to (10.12) and (10.13), is given by

$$\langle \hat{\rho}_n(\bar{1})\rangle = \frac{\langle T_{\mathrm{c}} S_{-}^{\dagger} S_{+} \bar{\rho}_n(\bar{1})\rangle}{\langle S_{-}^{\dagger} S_{+}\rangle} \,. \tag{11.8}$$

The density response of the nuclei under the action of H_{ext}, given by (11.4), can be calculated by means of the relations (10.36) and (10.38) as

$$\begin{aligned}
\frac{\delta \langle \hat{\rho}_{\mathrm{n}}(\bar{1})\rangle}{\delta \phi_{\mathrm{ext}}(\bar{2})} &= -\frac{\mathrm{i}}{\hbar} \frac{\langle T_{\mathrm{c}}[\hat{\rho}_{\mathrm{el}}(\bar{2}) + \hat{\rho}_n(\bar{2})]\hat{\rho}_n(\bar{1})\rangle}{\langle S_{-}^{\dagger} S_{+}\rangle} \\
&\quad + \frac{\mathrm{i}}{\hbar} \frac{\langle T_{\mathrm{c}} \hat{\rho}_n(\bar{1})\rangle \langle T_{\mathrm{c}}[\hat{\rho}_n(\bar{2}) + \hat{\rho}_{\mathrm{el}}(\bar{2})]\rangle}{\langle S_{-}^{\dagger} S_{+}\rangle^2} \\
&= -\frac{\mathrm{i}}{\hbar} \frac{\langle T_{\mathrm{c}}[\Delta\hat{\rho}_{\mathrm{el}}(\bar{2}) + \Delta\hat{\rho}_n(\bar{2})]\Delta\hat{\rho}_n(\bar{1})\rangle}{\langle S_{-}^{\dagger} S_{+}\rangle} \,.
\end{aligned} \tag{11.9}$$

In the last step, we have introduced the deviation operator $\Delta\hat{\rho} = \hat{\rho} - \langle \hat{\rho}\rangle$. Now the additional external field J comes into play, which allows us to eliminate the mixed electron–nuclei contribution. By steps completely analogous to those used before, we find

$$\frac{\delta \langle \hat{\rho}_{\mathrm{el}}(\bar{1})\rangle}{\delta J(\bar{2})} = -\frac{\mathrm{i}}{\hbar} \frac{\langle T_{\mathrm{c}}\, \Delta\hat{\rho}_{\mathrm{n}}(\hat{2})\, \Delta\hat{\rho}_{\mathrm{el}}(\bar{1})\rangle}{\langle S_{-}^{\dagger} S_{+}\rangle} \,, \tag{11.10}$$

which, together with (11.9), yields the result

$$\frac{\delta \langle \hat{\rho}_{\mathrm{n}}(\bar{1})\rangle}{\delta \phi_{\mathrm{ext}}(\bar{2})} = \frac{\delta \langle \hat{\rho}_{\mathrm{el}}(\bar{1}) + \hat{\rho}_{\mathrm{n}}(\bar{1})\rangle}{\delta J(\bar{2})} = D(\bar{1},\bar{2}) + \frac{\delta \langle \hat{\rho}_{\mathrm{el}}(\bar{1})\rangle}{\delta J(\bar{2})} \,, \tag{11.11}$$

where we have defined the *density–density correlation function* of the nuclei

$$D(\bar{1},\bar{2}) = -\frac{\mathrm{i}}{\hbar}\frac{\langle T_c\, \Delta\hat{\rho}_n(\bar{2})\, \Delta\hat{\rho}_n(\bar{1})\rangle}{\langle S_-^\dagger S_+\rangle} = \frac{\delta\langle\hat{\rho}_n(\bar{1})\rangle}{\delta J(\bar{2})}\,. \tag{11.12}$$

We can again apply the chain rule to (11.11) in order to eliminate the $\delta\langle\hat{\rho}_{\mathrm{el}}\rangle/\delta J$ contribution. This yields

$$\frac{\delta\langle\hat{\rho}_n(\bar{1})\rangle}{\delta\phi_{\mathrm{ext}}(\bar{2})} = D(\bar{1},\bar{2}) + \int \mathrm{d}\bar{3}\bar{4}\, \frac{\delta\langle\hat{\rho}_{\mathrm{el}}(\bar{2})\rangle}{\delta V(\bar{3})}\, v(\bar{3}-\bar{4})\, \frac{\delta\langle\hat{\rho}_{\mathrm{el}}(\bar{4}) + \hat{\rho}_n(\bar{4})\rangle}{\delta J(\bar{1})} \tag{11.13}$$

Making use oncemore of the relation (11.11), we solve the resulting equation with respect to $\delta\langle\hat{\rho}_n\rangle/\delta\phi_{\mathrm{ext}}$ and express the solution in terms of the electronic inverse dielectric function again. After insertion in (11.6), this yields the total dielectric screening function as

$$\frac{\delta V(\bar{1})}{\delta\phi_{\mathrm{ext}}(\bar{2})} = \varepsilon^{-1}(\bar{1},\bar{2}) + \int \mathrm{d}\bar{3}\bar{4}\, w(\bar{1},\bar{3})\, D(\bar{3},\bar{4})\, \varepsilon^{-1}(\bar{4},\bar{2})\,. \tag{11.14}$$

The desired *effective electron–electron interaction induced by lattice vibrations* is thus finally given by

$$w_{\mathrm{eff}}(\bar{1}\bar{2}) = w(\bar{1},\bar{2}) + \int \mathrm{d}\bar{3}\bar{4}\, w(\bar{1},\bar{3})\, D(\bar{3},\bar{4})\, w(\bar{4},\bar{2})\,. \tag{11.15}$$

This is an important result. We have reduced the problem of electron–phonon interaction to the replacement of the electronically screened interaction introduced in Sect. 10.3 by the effective interaction (11.15). It should be emphasized that this replacements concerns the screened interaction only, not the bare Coulomb interaction arising in or from Hartree contributions. In other words, there are no closed electron loops which are connected to the rest of a diagram by a phonon line.

If we had decomposed our system into valence electrons and ion cores rather than into electrons and bare nuclei, it is obvious that only valence electrons would have contributed to the electronic screening function. For later reference, we perform the decomposition with respect to branch indices and consider the spatial and temporal Fourier transform. Making use of the relation (10.61), we find, for example, for the retarded interaction

$$\begin{aligned} w^{\mathrm{r}}_{\mathrm{eff}}(\boldsymbol{G}+\boldsymbol{q},\boldsymbol{G}'+\boldsymbol{q},\omega) &= w^{\mathrm{r}}(\boldsymbol{G}+\boldsymbol{q},\boldsymbol{G}'+\boldsymbol{q},\omega) \\ &+ \sum_{\boldsymbol{K}\boldsymbol{K}'} w^{\mathrm{r}}(\boldsymbol{G}+\boldsymbol{q},\boldsymbol{K}+\boldsymbol{q},\omega) D^{\mathrm{r}}(\boldsymbol{K}+\boldsymbol{q},\boldsymbol{K}'+\boldsymbol{q},\omega) w^{\mathrm{r}}(\boldsymbol{K}'+\boldsymbol{q},\boldsymbol{G}+\boldsymbol{q},\omega)\,. \end{aligned} \tag{11.16}$$

The density–density correlation function of the nuclei will be reduced in the following to a quantity of more practical interest, i.e. the phonon Green's function within the *harmonic approximation*.

11.2 The Phonon Green's Function

The derivation of the phonon Green's function from the density–density correlation function requires the usual concepts and tools of phonon physics. We shall avoid any lengthy repetition and only briefly recall the basic terms which enter the derivation. Within the harmonic approximation, we expand the ionic charge density up to first order in the lattice displacements $u_\alpha(\boldsymbol{L\kappa})$ with respect to the equilibrium positions of ions:

$$\hat{\rho}_n(\boldsymbol{r},t) = \sum_{\boldsymbol{L\kappa}} \rho_{\boldsymbol{\kappa}}(\boldsymbol{r}-\boldsymbol{L}-\boldsymbol{\kappa}) + \sum_{\boldsymbol{L\kappa}\alpha} \nabla_\alpha \rho_{\boldsymbol{\kappa}}(\boldsymbol{r}-\boldsymbol{L}-\boldsymbol{\kappa})\, u_\alpha(\boldsymbol{L\kappa},t)\,, \quad (11.17)$$

where α denotes the Cartesian components. This expansion reduces the density–density correlation function (11.12) to

$$D(\boldsymbol{r},t,\boldsymbol{r}',t') = \sum_{\boldsymbol{L\kappa}\alpha,\boldsymbol{L}'\boldsymbol{\kappa}'\beta} \nabla_\alpha \rho_{\boldsymbol{\kappa}}(\boldsymbol{r}-\boldsymbol{L}-\boldsymbol{\kappa}) D_{\alpha\beta}(\boldsymbol{L\kappa},\boldsymbol{L}'\boldsymbol{\kappa}',t,t') \nabla_\beta \rho_{\boldsymbol{\kappa}'}(\boldsymbol{r}'-\boldsymbol{L}'-\boldsymbol{\kappa}')\,, \quad (11.18)$$

where the phonon Green's function in real space is

$$D_{\alpha\beta}(\boldsymbol{L\kappa},\boldsymbol{L}'\boldsymbol{\kappa}',\bar{t},\bar{t}') = -\frac{\mathrm{i}}{\hbar}\langle T_c u_\alpha(\boldsymbol{L\kappa},\bar{t}) u_\beta(\boldsymbol{L}'\boldsymbol{\kappa}',\bar{t}')\rangle\,. \quad (11.19)$$

In order to proceed further, we have to consider the lattice displacements in some detail.

11.2.1 Eigenmodes of Lattice Vibrations

Within the adiabatic and the harmonic approximation, we make use of (11.17), and separate in the electronic part of the Hamiltonian (11.1) the contributions up to the second order in the displacement. Together with the direct contribution of the nuclei (11.3), the lattice Hamiltonian can be cast into the form

$$H = -\sum_{\boldsymbol{L\kappa}\alpha} \frac{\hbar^2}{2M_{\boldsymbol{\kappa}}} \frac{\partial^2}{\partial[u_\alpha(\boldsymbol{L\kappa})]^2} + \frac{1}{2}\sum_{\boldsymbol{L\kappa}\alpha,\boldsymbol{L}'\boldsymbol{\kappa}'\beta} u_\alpha(\boldsymbol{L\kappa}) \Phi_{\alpha\beta}(\boldsymbol{L\kappa},\boldsymbol{L}'\boldsymbol{\kappa}') u_\beta(\boldsymbol{L}'\boldsymbol{\kappa}')\,. \quad (11.20)$$

The interatomic *force constants* are the sum of two contributions:

$$\Phi_{\alpha\beta}(\boldsymbol{L\kappa},\boldsymbol{L}'\boldsymbol{\kappa}') = \Phi^{\mathrm{ion}}_{\alpha\beta}(\boldsymbol{L\kappa},\boldsymbol{L}'\boldsymbol{\kappa}') + \Phi^{\mathrm{el}}_{\alpha\beta}(\boldsymbol{L\kappa},\boldsymbol{L}'\boldsymbol{\kappa}')\,. \quad (11.21)$$

Let us consider first the case $\boldsymbol{L} \neq \boldsymbol{L}'$, $\boldsymbol{\kappa} \neq \boldsymbol{\kappa}'$, which we label by $\tilde{\Phi} = \tilde{\Phi}^{\mathrm{ion}} + \tilde{\Phi}^{\mathrm{el}}$. The first contribution $\tilde{\Phi}^{\mathrm{ion}}$ describes the direct ion–ion interaction and is given, by means of (11.3), as

$$\tilde{\Phi}^{\mathrm{ion}}_{\alpha\beta}(\boldsymbol{L\kappa},\boldsymbol{L}'\boldsymbol{\kappa}')$$
$$= \int \mathrm{d}^3\boldsymbol{r}\,\mathrm{d}^3\boldsymbol{r}'\nabla_\alpha\rho_{\boldsymbol{\kappa}}(\boldsymbol{r}-\boldsymbol{L}-\boldsymbol{\kappa})v(\boldsymbol{r}-\boldsymbol{r}')\nabla_\beta\rho_{\boldsymbol{\kappa}'}(\boldsymbol{r}'-\boldsymbol{L}'-\boldsymbol{\kappa}') . \quad (11.22)$$

The second contribution, $\tilde{\Phi}^{\mathrm{el}}$, is mediated by electrons and has to be calculated from the second derivative of an eigenvalue of the electronic Hamiltonian (10.29) with respect to the displacement. We derive this contribution by making use of the Hellmann–Feynman theorem (10.100). Thus, in our case, the derivative of the electronic eigenenergy with respect to the lattice displacements which enters the lattice Hamiltonian is given by

$$\frac{\partial E_{\mathrm{el}}}{\partial u_\alpha(\boldsymbol{L\kappa})} = \int \mathrm{d}^3\boldsymbol{r}\frac{\partial U(\boldsymbol{r})}{\partial u_\alpha(\boldsymbol{L\kappa})}\langle\hat{\rho}_{\mathrm{el}}(\boldsymbol{r})\rangle \quad (11.23)$$

and is determined by the derivative of the electron–ion interaction U only. If we represent U as a sum of ionic potentials centered at the lattice sites,

$$U(\boldsymbol{r}) = \sum_{\boldsymbol{L\kappa}} U_{\boldsymbol{\kappa}}(\boldsymbol{r}-\boldsymbol{L}-\boldsymbol{\kappa}) , \quad (11.24)$$

the electronic contribution to the force constants for $\boldsymbol{L} \neq \boldsymbol{L}'$, $\boldsymbol{\kappa} \neq \boldsymbol{\kappa}'$ is obtained as

$$\tilde{\Phi}^{\mathrm{el}}_{\alpha\beta}(\boldsymbol{L\kappa},\boldsymbol{L}'\boldsymbol{\kappa}') = \int \mathrm{d}^3\boldsymbol{r}\,\nabla_\alpha U_{\boldsymbol{\kappa}}(\boldsymbol{r}-\boldsymbol{L}-\boldsymbol{\kappa})\frac{\delta\langle\hat{\rho}_{\mathrm{el}}(\boldsymbol{r})\rangle}{\delta u_\beta(\boldsymbol{L}'\boldsymbol{\kappa}')}$$
$$= \int \mathrm{d}^3\boldsymbol{r}\,\mathrm{d}^3\boldsymbol{r}'\,\nabla_\alpha U_{\boldsymbol{\kappa}}(\boldsymbol{r}-\boldsymbol{L}-\boldsymbol{\kappa})\frac{\delta\langle\hat{\rho}_{\mathrm{el}}(\boldsymbol{r})\rangle}{\delta U(\boldsymbol{r}')}\frac{\delta U(\boldsymbol{r}')}{\delta u_\beta(\boldsymbol{L}'\boldsymbol{\kappa}')}$$
$$= \frac{\varepsilon_0}{V}\int \mathrm{d}^3\boldsymbol{r}\,\mathrm{d}^3\boldsymbol{r}'\,\nabla_\alpha U_{\boldsymbol{\kappa}}(\boldsymbol{r}-\boldsymbol{L}-\boldsymbol{\kappa})\chi(\boldsymbol{r},\boldsymbol{r}')\nabla_\beta U_{\boldsymbol{\kappa}'}(\boldsymbol{r}'-\boldsymbol{L}'-\boldsymbol{\kappa}') . (11.25)$$

As U acts as an external potential with respect to the electronic density, we have identified the variational derivative $\delta\rho_{\mathrm{el}}(\boldsymbol{r})/\delta U(\boldsymbol{r}')$ with $\varepsilon_0\chi(\boldsymbol{r},\boldsymbol{r}')/V$, where $\chi(\boldsymbol{r},\boldsymbol{r}') \equiv \chi(\boldsymbol{r},\boldsymbol{r}',\omega=0)$ is the static electronic susceptibility. The case $\boldsymbol{L}=\boldsymbol{L}'$, $\boldsymbol{\kappa}=\boldsymbol{\kappa}'$ can be treated by means of the condition of *infinitesimal translational invariance*, i.e.

$$\sum_{\boldsymbol{L}'\boldsymbol{\kappa}'} \Phi_{\alpha\beta}(\boldsymbol{L\kappa},\boldsymbol{L}'\boldsymbol{\kappa}') = 0 . \quad (11.26)$$

Combining the two contributions, we can write the complete force constants as

$$\Phi_{\alpha\beta}(\boldsymbol{L\kappa},\boldsymbol{L}'\boldsymbol{\kappa}') = \tilde{\Phi}_{\alpha\beta}(\boldsymbol{L\kappa},\boldsymbol{L}'\boldsymbol{\kappa}') - \delta_{\boldsymbol{L},\boldsymbol{L}'}\delta_{\boldsymbol{\kappa},\boldsymbol{\kappa}'}\sum_{\boldsymbol{L}'\boldsymbol{\kappa}'}\tilde{\Phi}_{\alpha\beta}(\boldsymbol{L\kappa},\boldsymbol{L}'\boldsymbol{\kappa}') . \quad (11.27)$$

The introduction of the electronic susceptibility into (11.24) reduces the theory of phonon spectra to the theory of the dielectric response of our system. As we have seen earlier, the response of a semiconductor consists of two contributions. First we have the linear response to weak fields, which contains interband transitions only. The second contribution results from the excitation of the system by strong external fields, which induces intraband transitions too. Whereas the linear response determines the phonon spectrum of the unexcited system, the induced contributions give rise to a renormalization of phonon spectra and to nonequilibrium occupations of phonon modes. We shall consider first the unexcited system.

Using the lattice Hamiltonian (11.20), we can proceed to calculate the eigenmodes of the lattice vibrations. The equation of motion,

$$\frac{\partial}{\partial t}u_\alpha(\boldsymbol{L\kappa}) = \frac{\mathrm{i}}{\hbar}[H, u_\alpha(\boldsymbol{L\kappa})] = \frac{\mathrm{i}\hbar}{M_{\boldsymbol{\kappa}}}\frac{\partial}{\partial u_\alpha(\boldsymbol{L\kappa})}, \tag{11.28}$$

shows that $\partial u/\partial t$ and u are noncommuting variables (compare our treatment of the quantized photon field in Sect. 2.4), which obey the relation

$$\left[\frac{\partial}{\partial t}u_\alpha(\boldsymbol{L\kappa}), u_\beta(\boldsymbol{L'\kappa'})\right] = \frac{\mathrm{i}\hbar}{M_{\boldsymbol{\kappa}}}\delta_{\boldsymbol{L},\boldsymbol{L'}}\delta_{\boldsymbol{\kappa},\boldsymbol{\kappa'}}\delta_{\alpha,\beta}\,. \tag{11.29}$$

Iteration of (11.28) yields

$$\frac{\partial^2}{\partial t^2}u_\alpha(\boldsymbol{L\kappa}) = -\frac{1}{\hbar^2}[H,[H,u_\alpha(\boldsymbol{L\kappa})]], \tag{11.30}$$

which results in

$$M_{\boldsymbol{\kappa}}\frac{\partial^2}{\partial t^2}u_\alpha(\boldsymbol{L\kappa}) + \sum_{\boldsymbol{L'\kappa'}\beta}\Phi_{\alpha\beta}(\boldsymbol{L\kappa},\boldsymbol{L'\kappa'})\,u_\beta(\boldsymbol{L'\kappa'}) = 0\,. \tag{11.31}$$

We can solve the equation of motion (11.31) by making use of the lattice periodicity and representing the displacements by means of a Fourier expansion

$$u_\alpha(\boldsymbol{L\kappa},t) = \frac{1}{\sqrt{NM_{\boldsymbol{\kappa}}}}\sum_{\boldsymbol{q}}\mathrm{e}^{\mathrm{i}\boldsymbol{q}(\boldsymbol{L}+\boldsymbol{\kappa})}u_{\alpha\boldsymbol{\kappa}}(\boldsymbol{q},t)\,. \tag{11.32}$$

Owing to (11.29), the transformed displacements now fulfil the commutation relations

$$\left[\frac{\partial}{\partial t}u_{\alpha\boldsymbol{\kappa}}(\boldsymbol{q},t), u_{\beta\boldsymbol{\kappa'}}(\boldsymbol{q'},t)\right] = \mathrm{i}\hbar\,\delta_{\boldsymbol{q},\boldsymbol{q'}}\delta_{\boldsymbol{\kappa},\boldsymbol{\kappa'}}\delta_{\alpha,\beta}\,. \tag{11.33}$$

Utilizing the translational invariance

$$\Phi_{\alpha\beta}(\boldsymbol{L}+\boldsymbol{L'},\boldsymbol{\kappa},\boldsymbol{L'\kappa'}) = \Phi_{\alpha\beta}(\boldsymbol{L\kappa},0\boldsymbol{\kappa'})\,, \tag{11.34}$$

we obtain from (11.31)

$$\frac{\partial^2}{\partial t^2}u_{\alpha\boldsymbol{\kappa}}(\boldsymbol{q},t) + \sum_{\boldsymbol{\kappa'}\beta}M_{\alpha\boldsymbol{\kappa},\beta\boldsymbol{\kappa'}}(\boldsymbol{q})u_{\beta\boldsymbol{\kappa'}}(\boldsymbol{q},t) = 0\,, \tag{11.35}$$

where the *dynamical matrix* is obtained from the Fourier transformation of the force constants according to

$$M_{\alpha\kappa,\beta\kappa'}(\boldsymbol{q}) = \frac{1}{\sqrt{M_\kappa M_{\kappa'}}} \sum_{\boldsymbol{L}} \mathrm{e}^{-\mathrm{i}\boldsymbol{q}(\boldsymbol{L}+\boldsymbol{\kappa}-\boldsymbol{\kappa}')} \Phi_{\alpha\beta}(\boldsymbol{L}\boldsymbol{\kappa}, 0\boldsymbol{\kappa}') . \tag{11.36}$$

Equation (11.35) defines an eigenvalue problem,which determines the eigenmodes $e_{\alpha\kappa j}(\boldsymbol{q})$ and the eigenfrequencies $\omega_j(\boldsymbol{q})$ of the lattice vibrations according to

$$\omega_j^2(\boldsymbol{q})\, e_{\alpha\kappa j}(\boldsymbol{q}) - \sum_{\kappa'\beta} M_{\alpha\kappa,\beta\kappa'}(\boldsymbol{q})\, e_{\beta\kappa' j}(\boldsymbol{q}) = 0 . \tag{11.37}$$

We can use these eigenvectors to expand the Fourier components of the displacement in terms of *phonon operators*:

$$u_{\alpha\kappa}(\boldsymbol{q}, t) = \sum_j \frac{\hbar}{\sqrt{2\hbar\omega_j(\boldsymbol{q})}} e_{\alpha\kappa j}(\boldsymbol{q}) \left(b_{j,\boldsymbol{q}}(t) + b^\dagger_{j,-\boldsymbol{q}}(t) \right) . \tag{11.38}$$

These operators have the trivial time dependence

$$b_{j,\boldsymbol{q}}(t) = b_{j,\boldsymbol{q}}\, \mathrm{e}^{-\mathrm{i}\omega_j(\boldsymbol{q})t} . \tag{11.39}$$

From the commutator relation (11.33) together with the expansion (11.38), we find that the phonon operators fulfill the usual Boson commutation relations

$$\left[b_{j,\boldsymbol{q}}, b^\dagger_{j',\boldsymbol{q}'} \right] = \delta_{\boldsymbol{q},\boldsymbol{q}'} \delta_{j,j'} \tag{11.40}$$

and

$$\left[b_{j,\boldsymbol{q}}, b_{j',\boldsymbol{q}'} \right] = \left[b^\dagger_{j,\boldsymbol{q}}, b^\dagger_{j',\boldsymbol{q}'} \right] = 0 . \tag{11.41}$$

In terms of eigenfrequencies and phonon operators, the lattice Hamiltonian (11.20) takes the usual form

$$H = \sum_{j\boldsymbol{q}} \hbar\omega_j(\boldsymbol{q}) \left(b^\dagger_{j,\boldsymbol{q}}(t) b_{j,\boldsymbol{q}}(t) + \frac{1}{2} \right) . \tag{11.42}$$

Practical calculations of phonon spectra have been performed at various levels of sophistication [300], which range from more or less complicated phenomenological models of force constants up to evaluations of microscopic theories. The latter are mostly based on the calculation of the density response within the framework of density functional theory [301, 302]. As an example of calculated phonon spectra, Fig. 11.1 shows results for Si and GaAs (for details of the calculation see [301, 303]), which illustrate typical features of the phonon dispersion in diatomic semiconductors. Owing to the infinitesimal translational invariance (11.26), three acoustic branches exist, for which the phonon energies vanish in the limit $\boldsymbol{q} \to 0$. The other three branches correspond to one longitudinal optical (LO) and two transverse optical (TO) modes. In an elementary semiconductor such as silicon these modes are degenerate at $\boldsymbol{q} = 0$, but are split in polar semiconductors. As will be shown

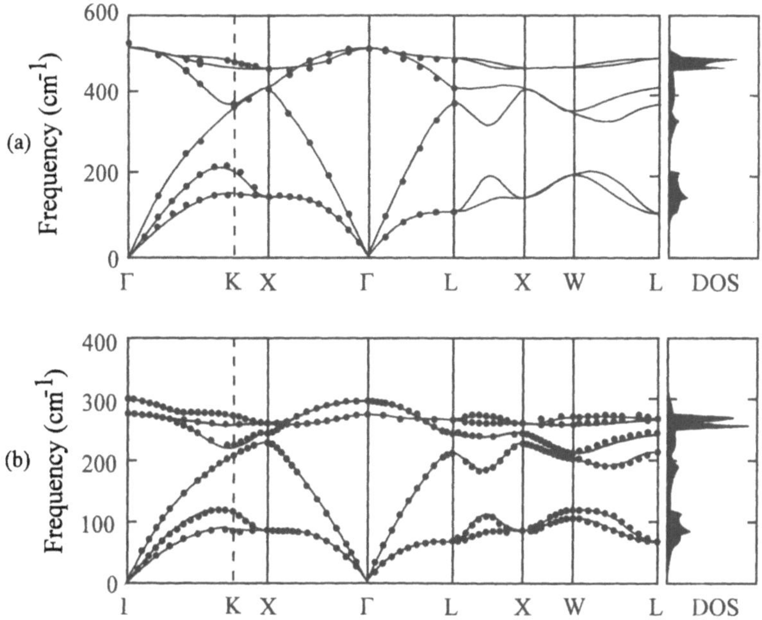

Fig. 11.1. Calculated phonon dispersion and densities of states of bulk Si (**a**) and bulk GaAs (**b**) in comparison with experimental data, denoted by *dots*. Taken from [303]

later, this longitudinal–transverse splitting results from long-range polarization fields which are generated by the effective charges centered on the two sublattices.

11.2.2 Green's Function Representation of the Density–Density Correlation Function

We have introduced now all ingredients necessary to transform the real-space representation of the density–density correlation function into a more convenient form. Owing to the lattice periodicity of the ionic charge densities, the spatial Fourier transform of (11.18) takes the form

$$D(\boldsymbol{G}+\boldsymbol{q},\boldsymbol{G}'+\boldsymbol{q},\bar{t}_1\bar{t}_2)=\sum_{\boldsymbol{\kappa}\boldsymbol{\kappa}'\alpha\beta j}(G_\alpha+q_\alpha)\rho_{\boldsymbol{\kappa}}(\boldsymbol{G}+\boldsymbol{q})\mathrm{e}^{\mathrm{i}\boldsymbol{G}\boldsymbol{\kappa}}$$

$$\times\frac{1}{\sqrt{M_{\boldsymbol{\kappa}}M_{\boldsymbol{\kappa}'}}}D_{\alpha\boldsymbol{\kappa},\beta\boldsymbol{\kappa}'}(\boldsymbol{q},\bar{t}_1\bar{t}_2)(G'_\beta+q_\beta)\rho_{\boldsymbol{\kappa}'}(\boldsymbol{G}'+\boldsymbol{q})\mathrm{e}^{\mathrm{i}\boldsymbol{G}'\boldsymbol{\kappa}'}\,, \tag{11.43}$$

where the $\boldsymbol{q}$-space representation of the *phonon Green's function* is

$$D_{\alpha\boldsymbol{\kappa},\beta\boldsymbol{\kappa}'}(\boldsymbol{q},\bar{t}_1\bar{t}_2) = \frac{\sqrt{M_{\boldsymbol{\kappa}}M_{\boldsymbol{\kappa}'}}}{N}\sum_{\boldsymbol{q}} \mathrm{e}^{-\mathrm{i}\boldsymbol{q}(\boldsymbol{L}-\boldsymbol{L}'+\boldsymbol{\kappa}-\boldsymbol{\kappa}')} D_{\alpha\beta}(\boldsymbol{L}\boldsymbol{\kappa},\boldsymbol{L}'\boldsymbol{\kappa}',\bar{t}_1\bar{t}_2)\,. \tag{11.44}$$

By means of the expansion (11.32), this transformation yields

$$D_{\alpha\boldsymbol{\kappa},\beta\boldsymbol{\kappa}'}(\boldsymbol{q},\bar{t}_1\bar{t}_2) = -\frac{\mathrm{i}}{\hbar}\langle T_{\mathrm{c}} u_{\alpha\boldsymbol{\kappa}}(\boldsymbol{q},\bar{t}_1) u_{\beta\boldsymbol{\kappa}'}(-\boldsymbol{q},\bar{t}_2)\rangle\,. \tag{11.45}$$

As before, we still restrict ourselves to the case in which external excitation fields are absent. In this case the force constants are independent of time, and we can derive the equations of motion for the unperturbed Green's function, making use of (11.35) and the commutator (11.33). We obtain as a result

$$\begin{aligned}\frac{\partial^2}{\partial t_1^2} D_{\alpha\boldsymbol{\kappa},\beta\boldsymbol{\kappa}'}(\boldsymbol{q},\bar{t}_1\bar{t}_2) + \sum_{\gamma\boldsymbol{\kappa}''} M_{\alpha\boldsymbol{\kappa},\gamma\boldsymbol{\kappa}''}(\boldsymbol{q}) D_{\gamma\boldsymbol{\kappa}'',\beta\boldsymbol{\kappa}'}(\boldsymbol{q},\bar{t}_1\bar{t}_2) \\ = b_1\,\delta_{b_1,b_2}\delta(t_1-t_2)\delta_{\boldsymbol{\kappa},\boldsymbol{\kappa}'}\delta_{\alpha,\beta}\,,\end{aligned} \tag{11.46}$$

which can be further reduced by means of the eigenfunction expansion (11.38). This expansion allows us to factorize (11.45) for each phonon mode according to

$$D_{\alpha\boldsymbol{\kappa},\beta\boldsymbol{\kappa}'}(\boldsymbol{q},\bar{t}_1\bar{t}_2) = \sum_j d_{\alpha\boldsymbol{\kappa},\beta\boldsymbol{\kappa}'}(\boldsymbol{q},j) D_j(\boldsymbol{q},\bar{t}_1\bar{t}_2)\,, \tag{11.47}$$

into a weight factor

$$d_{\alpha\boldsymbol{\kappa},\beta\boldsymbol{\kappa}'}(\boldsymbol{q},j) = \frac{e_{\alpha\boldsymbol{\kappa}j}(\boldsymbol{q}) e_{\beta\boldsymbol{\kappa}'j}(\boldsymbol{q})}{2\omega_j(\boldsymbol{q})} \tag{11.48}$$

and the phonon Green's functions

$$D_j(\boldsymbol{q},\bar{t}_1\bar{t}_2) = -\mathrm{i}b_2\Big\langle T_{\mathrm{c}}\Big(b_{j,\boldsymbol{q}}(\bar{t}_1)\, b^\dagger_{j,\boldsymbol{q}}(\bar{t}_2) + b^\dagger_{j,-\boldsymbol{q}}(\bar{t}_1)\, b_{j,-\boldsymbol{q}}(\bar{t}_2)\Big)\Big\rangle\,. \tag{11.49}$$

This reduces the equation of motion to

$$\left(\frac{\partial^2}{\partial t^2} + \omega_j^2(\boldsymbol{q})\right) D_{0j}(\boldsymbol{q},\bar{t}_1\bar{t}_2) = 2\omega_j(\boldsymbol{q})\, b_1\,\delta_{b_1,b_2}\delta(t_1-t_2)\,. \tag{11.50}$$

The four zeroth-order Green's functions can be calculated directly from the equation of motion or, alternatively, from (11.49), with the time dependence of phonon operators given by (11.39). The result is

$$D_{0j}^{++}(\boldsymbol{q},t_1t_2) = -\mathrm{i}\Big(2N_{0j}(\boldsymbol{q})\cos[\omega_j(\boldsymbol{q})(t_1-t_2)] + \mathrm{e}^{-\mathrm{i}\omega_j(\boldsymbol{q})|t_1-t_2|}\Big)\,, \tag{11.51}$$

$$D_{0j}^{+-}(\boldsymbol{q},t_1t_2) = \mathrm{i}\Big(2N_{0j}(\boldsymbol{q})\cos[\omega_j(\boldsymbol{q})(t_1-t_2)] + \mathrm{e}^{i\omega_j(\boldsymbol{q})(t_1-t_2)}\Big)\,, \tag{11.52}$$

$$D_{0j}^{-+}(\boldsymbol{q},t_1t_2) = -\mathrm{i}\Big(2N_{0j}(\boldsymbol{q})\cos[\omega_j(\boldsymbol{q})(t_1-t_2)] + \mathrm{e}^{-i\omega_j(\boldsymbol{q})(t_1-t_2)}\Big)\,, \tag{11.53}$$

$$D_{0j}^{--}(\boldsymbol{q}, t_1 t_2) = \mathrm{i}\Big(2N_{0j}(\boldsymbol{q})\cos[\omega_j(\boldsymbol{q})(t_1 - t_2)] + \mathrm{e}^{i\omega_j(\boldsymbol{q})|t_1 - t_2|}\Big) . \quad (11.54)$$

For finite temperature, we can calculate the expectation value $N_j(\boldsymbol{q})$ by means of a grand canonical average and obtain by steps similar to those used in the case of fermions (see Sect. 3.2) a *Bose distribution function*,

$$N_{0j}(\boldsymbol{q}) = \langle b_{j,\boldsymbol{q}}^{\dagger} b_{j,\boldsymbol{q}} \rangle = \frac{1}{\mathrm{e}^{\beta\hbar\omega_j(\boldsymbol{q})} - 1} . \quad (11.55)$$

Since the number of phonons is arbitrary, they have no chemical potential and the phonon distribution vanishes at zero temperature. Large phonon occupations can be generated by nonlinear optical effects. The time evolution of the Green's functions in (11.49) then is no longer simply determined by a phase factor but is determined by kinetic equations. Before we come back to this topic in Sect. 11.5, we shall at first simplify our still rather complicated description of electron–phonon coupling. Owing to the factorization in (11.49), we can evaluate the coupling for any combination of branch indices. Having calculated the coupling for one type of phonon-induced interaction, we immediately know the other ones. In the following we shall consider the retarded density–density correlation function (11.18), which, after spatial and temporal Fourier transformation, is given by

$$D^{\mathrm{r}}(\boldsymbol{G}+\boldsymbol{q}, \boldsymbol{G}'+\boldsymbol{q}, \omega) = \sum_{\boldsymbol{\kappa}\boldsymbol{\kappa}'\alpha\beta j} (G_\alpha + q_\alpha)\rho_{\boldsymbol{\kappa}}(\boldsymbol{G}+\boldsymbol{q})\,\mathrm{e}^{\mathrm{i}\boldsymbol{G}\boldsymbol{\kappa}} \frac{d_{\alpha\boldsymbol{\kappa},\beta\boldsymbol{\kappa}'}(\boldsymbol{q}, j)}{\sqrt{M_{\boldsymbol{\kappa}} M_{\boldsymbol{\kappa}'}}}$$
$$\times\, D_{0j}^{\mathrm{r}}(\boldsymbol{q}, \omega)\,(G'_\beta + q_\beta)\,\rho_{\boldsymbol{\kappa}'}(\boldsymbol{G}'+\boldsymbol{q})\,\mathrm{e}^{\mathrm{i}\boldsymbol{G}'\boldsymbol{\kappa}'} , \quad (11.56)$$

where the *retarded phonon Green's function* is

$$D_{0j}^{\mathrm{r}}(\boldsymbol{q}, \omega) = \frac{2\omega_j(\boldsymbol{q})}{(\omega + \mathrm{i}\delta)^2 - \omega_j^2(\boldsymbol{q})} , \quad (11.57)$$

which is obtained from the Fourier transformation of $D^{++} + D^{+-}$.

11.3 Electron–Phonon Coupling in the Long-Wavelength Limit

It is obvious from our general result for the effective interaction (11.16) that two properties determine the strength of the electron–phonon coupling. The first one is the magnitude of the electronic polarizability, and the second is the gradient of the ionic charge density. However, we can avoid working with these microscopic quantities even for arbitrary phonons throughout the entire Brillouin zone. If we again anticipate the effective-mass approximation for the treatment of scattering processes, we can restrict the considerations to phonons in the vicinity of the Γ point. This allows us to evaluate the results in the *long-wavelength limit*. In the following, we shall see that this limit allows us to express the electron–phonon coupling in terms of macroscopic quantities. Inserting (11.56) into (11.16), we obtain the following for the long-range contribution to the effective interaction:

$$w^{\mathrm{r}}_{\mathrm{eff}}(\boldsymbol{q},\omega) = w^{\mathrm{r}}(\boldsymbol{q},\omega) + \sum_{\alpha\boldsymbol{\kappa}\beta\boldsymbol{\kappa}'j} F_{\alpha\boldsymbol{\kappa}}(\boldsymbol{q}) \frac{d_{\alpha\boldsymbol{\kappa},\beta\boldsymbol{\kappa}'}(\boldsymbol{q},j)}{\sqrt{M_{\boldsymbol{\kappa}}M_{\boldsymbol{\kappa}'}}} D^{\mathrm{r}}_{0j}(\boldsymbol{q},\omega)\, F_{\beta\boldsymbol{\kappa}'}(\boldsymbol{q})\,, \tag{11.58}$$

where we have simplified the notation by setting $w^{\mathrm{r}}_{\mathrm{eff}}(\boldsymbol{q},\boldsymbol{q},\omega) \equiv w^{\mathrm{r}}_{\mathrm{eff}}(\boldsymbol{q},\omega)$. The long-ranged electronically screened interaction is given by

$$w^{\mathrm{r}}(\boldsymbol{q},\omega) = \frac{v(\boldsymbol{q})}{\varepsilon^{\mathrm{r}}(\boldsymbol{q},\omega)}\,, \tag{11.59}$$

and we have introduced the interaction coefficients in the form

$$F_{\alpha\boldsymbol{\kappa}}(\boldsymbol{q}) = \sum_{\boldsymbol{G}} w^{\mathrm{r}}(\boldsymbol{q},\boldsymbol{G}+\boldsymbol{q},\omega)(G_\alpha+q_\alpha)\rho_{\boldsymbol{\kappa}}(\boldsymbol{G}+\boldsymbol{q})\mathrm{e}^{\mathrm{i}\boldsymbol{G}\boldsymbol{\kappa}} = v(\boldsymbol{q})q_\alpha\rho_{\boldsymbol{\kappa}}(\boldsymbol{q})$$

$$+\varepsilon_0 V\sum_{\boldsymbol{G}} v(\boldsymbol{q})\chi^{\mathrm{r}}(\boldsymbol{q},\boldsymbol{G}+\boldsymbol{q},\omega)v(\boldsymbol{G}+\boldsymbol{q})(G_\alpha+q_\alpha)\rho_{\boldsymbol{\kappa}}(\boldsymbol{G}+\boldsymbol{q})\mathrm{e}^{\mathrm{i}\boldsymbol{G}\boldsymbol{\kappa}}\,. \tag{11.60}$$

With these ingredients, we can introduce a still rather general *phonon-induced interaction*, i.e. the second term in (11.58), which takes the form

$$w^{\mathrm{r}}_{\mathrm{ph}}(\boldsymbol{q},\omega) = \sum_j M_j^2(\boldsymbol{q})D^{\mathrm{r}}_j(\boldsymbol{q},\omega)\,, \tag{11.61}$$

where the coupling constants for the different branches are given by

$$M_j^2(\boldsymbol{q}) = \sum_{\alpha\beta\boldsymbol{\kappa}\boldsymbol{\kappa}'} F_{\alpha\boldsymbol{\kappa}}(\boldsymbol{q}) \frac{d_{\alpha\boldsymbol{\kappa},\beta\boldsymbol{\kappa}'}(\boldsymbol{q},j)}{\sqrt{M_{\boldsymbol{\kappa}}M_{\boldsymbol{\kappa}'}}} F_{\beta\boldsymbol{\kappa}'}(\boldsymbol{q})\,. \tag{11.62}$$

Although it is possible to evaluate this result quite generally, it can be considerably simplified in an expansion for small wave vector. Thus in the next step, we separate the long-range part, i.e. the $\boldsymbol{G}=0$ contribution to the sum and the long-range part of the susceptibility given by the relation (2.170), which yields

$$v(\boldsymbol{q})\chi^{\mathrm{r}}(\boldsymbol{q},\boldsymbol{G}+\boldsymbol{q},\omega) = w^{\mathrm{r}}(\boldsymbol{q},\omega)\,\tilde{\chi}^{\mathrm{r}}(\boldsymbol{q},\boldsymbol{G}+\boldsymbol{q},\omega)\,. \tag{11.63}$$

With this relations, we can write (11.60) as

$$F_{\alpha\boldsymbol{\kappa}}(\boldsymbol{q}) = w^{\mathrm{r}}(\boldsymbol{q},\omega)\,Z_{\alpha\boldsymbol{\kappa}}(\boldsymbol{q})\,. \tag{11.64}$$

The effective charge associated with a lattice wave of wave vector $\boldsymbol{q}$

$$Z_{\alpha\boldsymbol{\kappa}}(\boldsymbol{q})$$

$$= q_\alpha\rho_{\boldsymbol{\kappa}}(\boldsymbol{q}) + \varepsilon_0 V\sum_{\boldsymbol{G}\neq 0}\tilde{\chi}(\boldsymbol{q},\boldsymbol{G}+\boldsymbol{q})\,v(\boldsymbol{G}+\boldsymbol{q})\,(G_\alpha+q_\alpha)\rho_{\boldsymbol{\kappa}}(\boldsymbol{G}+\boldsymbol{q})\,\mathrm{e}^{\mathrm{i}\boldsymbol{G}\boldsymbol{\kappa}}\,, \tag{11.65}$$

is just the charge per unit displacement in the α direction of the sublattice $\boldsymbol{\kappa}$. We shall now consider the long-wavelength limit and restrict ourselves to the leading contribution to the susceptibility in an expansion for small wave vector. This is given by

$$\tilde{\chi}(\boldsymbol{q}, \boldsymbol{G}+\boldsymbol{q}) = \sum_{\beta} q_\beta \tilde{\chi}_\beta(0, \boldsymbol{G}) , \tag{11.66}$$

where $\tilde{\chi}_\beta$ denotes the q_β derivative of $\tilde{\chi}$ with respect to the first argument. Thus an expansion for small wave vector yields the representation

$$Z_{\alpha\boldsymbol{\kappa}}(\boldsymbol{q} \to 0) = \sum_{\beta} Z_{\alpha\beta}(\boldsymbol{\kappa}) q_\beta , \tag{11.67}$$

where the *dynamical charge tensor* is

$$Z_{\alpha\beta}(\boldsymbol{\kappa}) = Z_{\boldsymbol{\kappa}}\delta_{\alpha\beta} + \varepsilon_0 V \sum_{\boldsymbol{G}\neq 0} \tilde{\chi}_\beta(0, \boldsymbol{G})\, v(\boldsymbol{G})\, G_\alpha\, \rho_{\boldsymbol{\kappa}}(\boldsymbol{G})\, \mathrm{e}^{\mathrm{i}\boldsymbol{G}\boldsymbol{\kappa}} \tag{11.68}$$

and the ionic charge is $Z_{\boldsymbol{\kappa}} = \rho_{\boldsymbol{\kappa}}(0)$.

11.3.1 Coupling to Longitudinal Optical Phonons

Further evaluation of the effective interaction requires some more detailed specification of the problem. We shall restrict ourselves to semiconductors with cubic symmetry, for which the dynamical charge tensor becomes diagonal, i.e. $Z_{\alpha\beta}(\boldsymbol{\kappa}) = \delta_{\alpha\beta} z_{\boldsymbol{\kappa}}$. Furthermore, we consider diatomic crystals, for which a pure longitudinal mode exists. With these constraints, we can write the effective interaction (11.58), using (11.64) and (11.68), as

$$w^{\mathrm{r}}_{\mathrm{eff}}(\boldsymbol{q}, \omega) = w^{\mathrm{r}}(\boldsymbol{q}, \omega) + [w^{\mathrm{r}}(\boldsymbol{q}, \omega)]^2 \sum_{j} \frac{|d_j(\boldsymbol{q})|^2}{(\omega + \mathrm{i}\delta)^2 + \omega_j^2(\boldsymbol{q})} , \tag{11.69}$$

where we have introduced

$$d_j(\boldsymbol{q}) = \sum_{\boldsymbol{\kappa}\alpha} \frac{Z_{\boldsymbol{\kappa}}}{\sqrt{M_{\boldsymbol{\kappa}}}} q_\alpha\, e_{\alpha\boldsymbol{\kappa}j}(\boldsymbol{q}) . \tag{11.70}$$

From this result, we can immediately conclude that only *longitudinal phonon modes* contribute. *Transverse modes* are projected out in (11.70) owing to the inner product of a wave vector and an eigenvector. This corresponds to the rather intuitive picture that only longitudinal vibrations can build up a macroscopic field which contributes to the screening of macroscopic disturbances. The contribution of acoustic modes vanishes in the limit $\boldsymbol{q} \to 0$ as the eigenvectors are proportional to $\sqrt{M_{\boldsymbol{\kappa}}}$, and we are left with

$$\sum_{\boldsymbol{\kappa}} Z_{\boldsymbol{\kappa}} = 0 . \tag{11.71}$$

This condition of charge neutrality is a special case of the *acoustic sum rule* [304, 305] for the dynamical charge tensor,

$$\sum_{\kappa} Z_{\alpha\beta}(\kappa) = 0 \,, \tag{11.72}$$

and will be discussed in more detail in the next section. We can evaluate (11.69) further by inserting the eigenvectors for longitudinal modes, denoted by $j = \mathrm{LO}$, which because of the general structure of the eigenvalue problem for the dynamical matrix (11.36), are of the form

$$e_{\alpha 1 \mathrm{LO}}(\boldsymbol{q}) = \sqrt{\frac{M_2}{M_1 + M_2}} \frac{q_\alpha}{|\boldsymbol{q}|} \tag{11.73}$$

and

$$e_{\alpha 2 \mathrm{LO}}(\boldsymbol{q}) = -\sqrt{\frac{M_1}{M_1 + M_2}} \frac{q_\alpha}{|\boldsymbol{q}|} \,. \tag{11.74}$$

With the condition $Z_1 = -Z_2 = Z$, we obtain

$$d_{\mathrm{LO}}(\boldsymbol{q}) = |\boldsymbol{q}| \frac{Z}{\sqrt{M}} \,, \tag{11.75}$$

where M is the reduced mass of the ions. Up to this point our treatment of electron–phonon coupling has been strictly microscopic and has been reduced to the dynamical charge. It is now an easy task to express this quantity in terms of macroscopic dielectric functions. As above, we consider the case of intrinsic coupling only, i.e. no intraband transitions are present. As (apart from narrow-gap semiconductors) all electronic transition energies are large in comparison with phonon energies, we can neglect the dependence on frequency and the $\boldsymbol{q}$ in the electronic dielectric function and set $\varepsilon^{\mathrm{r}}(\boldsymbol{q}, \omega) = \varepsilon_\infty$ in (11.59). Thus, using (11.75), we can identify the effective dielectric function from (11.69) as

$$\frac{1}{\varepsilon_{\mathrm{eff}}(\omega)} = \frac{1}{\varepsilon_\infty} + \frac{Z^2 v(\boldsymbol{q}) \, |\boldsymbol{q}|^2}{M \varepsilon_\infty^2 \omega_{\mathrm{LO}}^2} \frac{\omega_{\mathrm{LO}}^2}{(\omega + \mathrm{i}\delta)^2 - \omega_{\mathrm{LO}}^2} \,, \tag{11.76}$$

where ω_{LO} is the *longitudinal optical-phonon frequency.* This result allows us to write the dynamical charge in terms of the static dielectric functions according to

$$\frac{Z^2}{M V \varepsilon_0 \varepsilon_\infty^2 \omega_{\mathrm{LO}}^2} = \frac{1}{\varepsilon_\infty} - \frac{1}{\varepsilon_{\mathrm{eff}}(0)} \,. \tag{11.77}$$

Obviously, the effective dielectric function becomes zero at the LO phonon frequency and has a pole at

$$\omega^2 = \omega_{\mathrm{LO}}^2 - \frac{Z^2}{M V \varepsilon_0 \varepsilon_\infty} \,, \tag{11.78}$$

which can be identified with the frequency of the *transverse phonon mode* ω_{TO}. Thus the *longitudinal–transverse splitting* is determined by the dynamical charge and may be expressed as

$$\omega_{LO}^2 - \omega_{TO}^2 = \omega_{LO}^2 \left(1 - \frac{\varepsilon_\infty}{\varepsilon_{\text{eff}}(0)}\right) , \tag{11.79}$$

which is equivalent to the *Lyddane–Sachs–Teller* relation [306]

$$\frac{\omega_{TO}^2}{\omega_{LO}^2} = \frac{\varepsilon_\infty}{\varepsilon_{\text{eff}}(0)} . \tag{11.80}$$

As we have seen, the longitudinal–transverse splitting (11.79) results from the fact that longitudinal oscillations of charged ions produce a macroscopic field, in contrast to the transverse modes. If we use the corresponding macroscopic polarization in Maxwell's equations, we can proceed as in our treatment of the exciton–polariton and find, by the same steps, the dispersion of the *phonon–polariton* (compare Sect. 6.2.5). The microscopic origin of the splitting, however, is completely different from the excitonic case, where it results from the coupling to the light field.

As a final result, we obtain the phonon-induced interaction, which, in the long-wavelength limit, results from the coupling of electrons and LO phonons according to (11.76) and (11.77):

$$w_{LO}^{r}(\boldsymbol{q},\omega) = M_{LO}^2(\boldsymbol{q}) \frac{2\omega_{LO}}{(\omega + i\delta)^2 - \omega_{LO}^2} , \tag{11.81}$$

where the coupling matrix element is

$$M_{LO}^2(\boldsymbol{q}) = v(\boldsymbol{q}) \frac{\omega_{LO}}{2} \left(\frac{1}{\varepsilon_\infty} - \frac{1}{\varepsilon_{\text{eff}}(0)}\right) \equiv v(\boldsymbol{q}) \frac{\omega_{LO}}{2\varepsilon^*} . \tag{11.82}$$

This result is a rather special case of (11.61) and is usually referred to as the *Fröhlich interaction.* Alternatively, we could have introduced this type of electron–phonon coupling by considering just the interaction energy of an electron with the macroscopic polarization due to LO phonons [307]. The Fröhlich interaction is *long-ranged,* as the effective dielectric function in (11.76) becomes independent of the wave vector and thus the phonon-induced interaction retains the singularity of the Coulomb interaction. A few selected material parameters, which determine the LO phonon coupling are listed in Table 11.1.

Table 11.1. A few selected material parameters determining the coupling strength of LO phonons. Taken from [2]

	GaAs	ZnO	CdS	CuCl
ε_∞	10.9	3.7	5.2	3.7
$\varepsilon_{\text{eff}}(0)$	12.5	8.7	8.4	7.4
$\hbar\omega_{LO}(meV)$	36.8	72	38	27.2

If we perform the expansion for small wave vectors to higher order, we obtain corrections resulting from higher moments of the dynamical charge (11.65), where the singularity is canceled. In this case, acoustic phonons contribute also.

11.3.2 Coupling to Acoustic Phonons

Let us consider (11.58) again, but now without separating the long-range part. For acoustic phonons, we can perform the sum with respect to $\boldsymbol{\kappa}$ in (11.60) and introduce the ionic potentials (11.24) via Poisson's equation. Further, we transform partially back to real space. Thus we obtain from (11.60) the representation

$$\begin{aligned}\sum_{\boldsymbol{\kappa}} F_{\alpha\boldsymbol{\kappa}}(\boldsymbol{q}) &= v(\boldsymbol{q})\Big(q_\alpha \rho_{\boldsymbol{\kappa}}(\boldsymbol{q}) + \varepsilon_0 V \int \mathrm{d}^3\boldsymbol{r}\, \chi^{\mathrm{r}}(\boldsymbol{q},\boldsymbol{r},\omega)\, \delta U_\alpha(\boldsymbol{r},\boldsymbol{q})\Big) \\ &= v(\boldsymbol{q})\Big(q_\alpha \rho_{\boldsymbol{\kappa}}(\boldsymbol{q}) + \delta\rho_\alpha(\boldsymbol{q})\Big)\,. \end{aligned} \tag{11.83}$$

We have introduced

$$\delta U_\alpha(\boldsymbol{r},\boldsymbol{q}) = \frac{\partial}{\partial r_\alpha} \sum_{\boldsymbol{L}\boldsymbol{\kappa}} U_{\boldsymbol{\kappa}}(\boldsymbol{r}-\boldsymbol{L}-\boldsymbol{\kappa})\mathrm{e}^{\mathrm{i}\boldsymbol{q}(\boldsymbol{L}+\boldsymbol{\kappa})}\,, \tag{11.84}$$

which is just the change in the ionic potential per unit displacement in the α-direction; $\delta\rho_\alpha$ is the corresponding induced change in the electronic charge density. Now we make use of the basic definition of the susceptibility (2.156), which, together with the identity (10.43) (compare (10.60) and (10.120)), yields

$$\chi^{\mathrm{r}}(\boldsymbol{q},\boldsymbol{r},\omega=0) = \int \mathrm{d}^3\boldsymbol{r}'\,\mathrm{d}^3\boldsymbol{r}''\, P_0^{\mathrm{r}}(\boldsymbol{q},\boldsymbol{r}',\boldsymbol{r}'')\frac{\delta G^{-1}(\boldsymbol{r}',\boldsymbol{r}'')}{\delta(eU(\boldsymbol{r}))}\,. \tag{11.85}$$

Using an eigenfunction expansion for the static RPA polarization propagator, we obtain the representation

$$\begin{aligned}&\delta\rho_\alpha(\boldsymbol{q}) \\ &= \frac{e}{V}\sum_{nn'\boldsymbol{k}} \langle n\boldsymbol{k}|\mathrm{e}^{-\mathrm{i}\boldsymbol{q}\boldsymbol{r}}|n'\boldsymbol{k}+\boldsymbol{q}\rangle \frac{f_n(\boldsymbol{k})-f_{n'}(\boldsymbol{k}+\boldsymbol{q})}{E_n(\boldsymbol{k})-E_{n'}(\boldsymbol{k}+\boldsymbol{q})}\langle n'\boldsymbol{k}+\boldsymbol{q}|\delta G_\alpha^{-1}|n\boldsymbol{k}\rangle\,,\end{aligned} \tag{11.86}$$

where δG^{-1} is the change in the one-particle Hamiltonian which is induced by δU. Let us consider first the limit $\boldsymbol{q}\to 0$, which we have already used in the last section. An acoustic mode in this limit corresponds to a small shift of the crystal as a whole, because the electronic density rigidly follows the shift of the ion cores. Thus at $\boldsymbol{q}=0$ the change of G^{-1}, which is induced by the derivative of the ionic potential, is just the derivative of G^{-1}, i.e. of the exact one-particle Hamiltonian h. Correspondingly, we write δG^{-1} in (11.86) as the commutator of h with the momentum operator and obtain

$$\begin{aligned}\delta\rho_\alpha(\boldsymbol{q}\to 0) &= \frac{\mathrm{i}\,e}{V\hbar}\sum_{nn'\boldsymbol{k}}\langle n\boldsymbol{k}|\boldsymbol{qr}|n'\boldsymbol{k}\rangle\frac{f_n(\boldsymbol{k})-f_{n'}(\boldsymbol{k})}{E_n(\boldsymbol{k})-E_{n'}(\boldsymbol{k})}\langle n'\boldsymbol{k}|[p_\alpha,h]|n\boldsymbol{k}\rangle \\ &= \frac{\mathrm{i}\,e}{V\hbar}\sum_{nn'\boldsymbol{k}}[f_n(\boldsymbol{k})-f_{n'}(\boldsymbol{k})]\langle n\boldsymbol{k}|\boldsymbol{qr}|n'\boldsymbol{k}\rangle\langle n'\boldsymbol{k}|p_\alpha|n\boldsymbol{k}\rangle\,. \end{aligned} \quad (11.87)$$

Now, the closure relation can be applied to yield

$$\delta\rho_\alpha(\boldsymbol{q}\to 0) = \frac{\mathrm{i}\,e}{V\hbar}\sum_{n\boldsymbol{k}} f_n(\boldsymbol{k})\langle n\boldsymbol{k}|[\boldsymbol{qr},p_\alpha]|n\boldsymbol{k}\rangle = q_\alpha\rho_{\mathrm{el}}(\boldsymbol{q}=0)\,. \quad (11.88)$$

As expected, the total electronic charge density compensates the ionic charge density, i.e. the $\boldsymbol{q}\to 0$ limit of the first term in (11.83). In this way, the acoustic sum rule (11.72) results from charge neutrality of the crystal.[2]

The tradition has been that higher-order contributions beyond the rigorous $\boldsymbol{q}\to 0$ limit are classified and parameterized as different types of electron–phonon interactions. Provided there is no center of inversion, the next contribution to an expansion of the susceptibility is of the order $|\boldsymbol{q}|^3$, which gives rise to the *piezoelectric coupling* of electrons and acoustic phonons [308]. This mechanism is a result of the action of electric fields which accompany the electronic density modulations induced by acoustic phonons. When this is taken together with the Coulomb interaction in (11.83) and the linear dispersion of acoustic phonons, the resulting dependence of the induced interaction on the wave vector is $\sim 1/|\boldsymbol{q}|$. In the next higher order, we obtain a $|\boldsymbol{q}|$ dependence. This is the most prominent coupling to acoustic phonons, and is known as *deformation-potential coupling* [310, 311, 312]. If we take only the ionic contribution into account, we obtain the common definition of the deformation-potential coupling constant, i.e. the long-wavelength limit of $D=\sum_\kappa v(\boldsymbol{q})[\rho_\kappa(\boldsymbol{q})-\rho_\kappa(0)]$. In principle, the induced electronic density also contributes in this order of an expansion in powers of $\boldsymbol{q}$. The resulting phonon-induced interaction can be written as

$$w_{\mathrm{A}}(\boldsymbol{q},\omega) = M_{\mathrm{A}}^2(\boldsymbol{q})\frac{2\omega_{\mathrm{A}}(\boldsymbol{q})}{(\omega+i\delta)^2-\omega_{\mathrm{A}}^2(\boldsymbol{q})}\,, \quad (11.89)$$

where the coupling matrix element

$$M_{\mathrm{A}}^2(\boldsymbol{q}) = D^2\frac{\hbar|\boldsymbol{q}|^2}{2\rho V\omega_{\mathrm{A}}(\boldsymbol{q})}\,; \quad (11.90)$$

ρ is the density of the crystal. The basic idea of the deformation-potential approach is to identify the effect of the local perturbation due to acoustic phonons with the shift of the band energies produced by a homogeneous strain of the same magnitude. In this case the deformation constant D measures

[2] Clearly, this result can be obtained also beyond the RPA approximation.

the rate of change of the band energies with pressure. The introduction of this analogy allows one to determine the coupling strength experimentally. In the case of degenerate bands, not only longitudinal but also transverse acoustic phonons contribute to the deformation-potential coupling [313]. Let us mention finally, that deformation potential coupling also exists for optical phonons [309].

11.4 The Phonon Self-Energy

In the preceeding sections we have derived helpful forms of the phonon-induced interaction which are valid in the long-wavelength limit. Thus these forms are consistent with the effective mass-approximation, which we have already anticipated for the treatment of electron–phonon scattering. Thus we can link up with the many-particle theory of the last chapter and introduce phonon contributions into the quantum kinetic equations. As we have shown in the derivation of the general result (11.16), we have to add the phonon-induced contributions to the electronically screened interaction (10.59). Together with (10.57), this defines the phonon self-energies which enter our general quantum kinetic equations (10.69) or (10.82). As the phonon-induced interaction is not a functional of the one-particle Green's function, the problem is slightly less complicated than the electron–electron interaction. As in the purely electronic case, we are dependent on approximation schemes. An expansion of the self-energy in powers of the phonon-induced interaction is easily generated by means of (10.57) using

$$w_{\mathrm{ph}}(\boldsymbol{q},\bar{t}_1\bar{t}_2) = \sum_j M_j^2(\boldsymbol{q}) D_j(\boldsymbol{q},\bar{t}_1\bar{t}_2) \,, \tag{11.91}$$

which is obtained in exactly the same way as the retarded interaction (11.61). The four types of phonon Green's functions entering (11.91) are given by (11.51)–(11.54).[3] The contribution linear in the phonon-induced interaction, i.e. the one-phonon self-energy, is given by

$$\Sigma_{nm}^{(1)}(\boldsymbol{k},\bar{t}_1\bar{t}_2) = -\mathrm{i}\hbar e^2 b_1 \sum_{\boldsymbol{q}j} M_j^2(\boldsymbol{q}) D_j(\boldsymbol{q},\bar{t}_2\bar{t}_1) G_{nm}(\boldsymbol{k}+\boldsymbol{q},\bar{t}_1\bar{t}_2) \,, \tag{11.92}$$

whereas the two-phonon contribution is obtained as

$$\begin{aligned}
&\Sigma_{nm}^{(2)}(\boldsymbol{k},\bar{t}_1\bar{t}_2) \\
&= (\hbar e^2)^2 \sum_{\boldsymbol{q}\boldsymbol{q}'jj'} \int \mathrm{d}\bar{t}_3\,\mathrm{d}\bar{t}_4\, b_1 b_4 M_j^2(\boldsymbol{q}) D_j(\boldsymbol{q},\bar{t}_3\bar{t}_1) G_{nn'}(\boldsymbol{k}+\boldsymbol{q},\bar{t}_1\bar{t}_4) \\
&\quad \times M_{j'}^2(\boldsymbol{q}') D_{j'}(\boldsymbol{q}',\bar{t}_2\bar{t}_4) G_{n'm'}(\boldsymbol{k}+\boldsymbol{q}+\boldsymbol{q}',\bar{t}_4\bar{t}_3) G_{m'm}(\boldsymbol{k}+\boldsymbol{q}',\bar{t}_3\bar{t}_2).
\end{aligned} \tag{11.93}$$

[3] The more general case of an excited system will be treated in Sect. 11.5

It is worth emphasizing that the particle propagators which enter the self-energies are completely renormalized with respect to the phonon-induced interaction. For excited systems a renormalization of the phonon Green's functions also occurs, owing to the excitation-induced response functions. Perturbational results may be obtained by replacing G by an approximate solution of the Dyson equation, e.g. $G_0 + G_0 \Sigma G_0$. In order to obtain an initial insight into the effect of phonon-induced interactions, we consider first a stationary situation with a single electron in the conduction band coupled to LO phonons at zero temperature.

11.4.1 The Polaron

The model of an electron coupled to a dispersionless phonon bath has been a long-standing problem of theoretical physics for more than four decades and has been attacked with a variety of different methods. We use this model as an instructive example of the application of Green's function techniques under equilibrium conditions and begin our discussion with the calculation of the one-phonon self-energy (11.92). This yields the following for the retarded function:

$$\begin{aligned}\Sigma_{\mathrm{cc}}^{\mathrm{r}\,(1)}(\boldsymbol{k},\omega) = {} & \mathrm{i}\hbar e^2 \sum_{\boldsymbol{q}} M_{\mathrm{LO}}^2(\boldsymbol{q}) \int \frac{\mathrm{d}\omega'}{2\pi} [G_{\mathrm{cc}}^{\mathrm{r}}(\boldsymbol{k}+\boldsymbol{q},\omega+\omega') D_{0L}^{+-}(\boldsymbol{q},\omega') \\ & - G_{\mathrm{cc}}^{+-}(\boldsymbol{k}+\boldsymbol{q},\omega+\omega') D_{0L}^{\mathrm{a}}(\boldsymbol{q},\omega')] \,. \end{aligned} \tag{11.94}$$

The retarded self-energy enters the solution of the Dyson equation

$$G_{\mathrm{cc}}^{\mathrm{r}}(\boldsymbol{k},\omega) = \frac{1}{\hbar\omega - E_{\mathrm{c}}^0(\boldsymbol{k}) - \Sigma_{\mathrm{cc}}^{r}(\boldsymbol{k},\omega)} \,, \tag{11.95}$$

the poles of which determine the renormalized quasiparticle spectrum. Owing to the frequency dependence of the self-energy, this usually leads to a non-linear problem. If we anticipate that an effective-mass representation of the one-particle energies will be retained, the effect of phonons is expected to result in a shift, a change of the effective mass, and a broadening of one-particle energies. The corresponding new quasiparticles are called *polarons*.

Example (One-Phonon Processes). There are two ways to proceed further. We can either approximate the electronic Green's functions by the zeroth-order function or use a suitable parameterization for the renormalized Green's function. Let us consider the simple perturbation theory first. After frequency integration, we obtain from (11.94)

$$\begin{aligned}\Sigma_{\mathrm{cc}}^{\mathrm{r}\,(1)}(\boldsymbol{k},\omega) = {} & -e^2 \sum_{\boldsymbol{q}} M_{\mathrm{LO}}^2(\boldsymbol{q}) \left(\frac{N_0(\boldsymbol{q}) + f_{\mathrm{c}}(\boldsymbol{k}+\boldsymbol{q})}{E_{\mathrm{c}}^0(\boldsymbol{k}+\boldsymbol{q}) - \hbar\omega - \hbar\omega_{\mathrm{LO}} - \mathrm{i}\delta} \right. \\ & \left. + \frac{1 + N_0(\boldsymbol{q}) - f_{\mathrm{c}}(\boldsymbol{k}+\boldsymbol{q})}{E_{\mathrm{c}}^0(\boldsymbol{k}+\boldsymbol{q}) - \hbar\omega + \hbar\omega_{\mathrm{LO}} - \mathrm{i}\delta} \right) . \end{aligned} \tag{11.96}$$

The first term corresponds to phonon absorption, and the second one to emission processes. For a single electron in the conduction band at zero temperature, we have $N_0(\boldsymbol{q}) = 0$ and $f_c(\boldsymbol{k}) = 0$ and only *emission processes* are possible. These, however, can take place only for frequencies above the threshold $\omega \geq \omega_{LO}$, as otherwise the imaginary part of the emission term vanishes. Consequently, *below threshold* there is *no broadening of the quasiparticle energies.* Using (11.82), the remaining integral can be written in the form

$$\mathrm{Re}\left(\Sigma_{cc}^{(1)}(\boldsymbol{k},\omega)\right) = -\frac{\alpha}{\pi}\left(\frac{2m_c}{\hbar}\right)^{1/2}(\hbar\omega_L)^{3/2} \times \int_0^\infty dq \int_{-1}^{1} dz \frac{1}{k^2+q^2+2kqz+(\Delta_c+\hbar\omega_{LO}-\hbar\omega)\,2m_c/\hbar^2} \,. \tag{11.97}$$

The integrations are easily performed and yield, by means of

$$\int_0^\infty dq \frac{1}{q^2-k^2}\ln(q^2+c) = \frac{\pi}{k}\arcsin\left(\frac{k^2}{k^2+c}\right)\,, \tag{11.98}$$

the final expression for the self-energy

$$\mathrm{Re}\left(\Sigma_{cc}^{(1)}(\boldsymbol{k},\omega)\right) = -\alpha(\hbar\omega_{LO})^{3/2}\left(\frac{1}{E_c^0(\boldsymbol{k})}\right)^{1/2}\arcsin\left(\frac{E_c^0(\boldsymbol{k})}{E_c^0(\boldsymbol{k})+\hbar\omega_L-\hbar\omega}\right)^{1/2} . \tag{11.99}$$

We have introduced in the above the dimensionless *polaron constant*

$$\alpha = \frac{e^2}{4\pi\varepsilon_0\hbar}\left(\frac{m_c}{2\hbar\omega_{LO}}\right)^{1/2}\left(\frac{1}{\varepsilon_\infty}-\frac{1}{\varepsilon_{\mathrm{eff}}(0)}\right), \tag{11.100}$$

which determines the shift of the band energy and the mass renormalization. To retain the effective-mass approximation, we can expand the arcsin function for small wavevector according to $\arcsin x = x + x^3/6$ and represent the pole of the Green's function (11.95) by $\hbar\omega = E_c(\boldsymbol{k})$, where the polaron energy is

$$E_c(\boldsymbol{k}) = \frac{\hbar^2\boldsymbol{k}^2}{2m_c^*} + \Delta_c \,. \tag{11.101}$$

Here the unknown quantities are the shift of the quasiparticle energy Δ_c and the renormalized mass m_c^*, which are obtained from

$$\mathrm{Re}[\Sigma_{cc}^{(1)}(\boldsymbol{k},E_c(\boldsymbol{k}))] = -\frac{\alpha\,(\hbar\omega_{LO})^{3/2}}{[E_c^0(\boldsymbol{k})-E_c(\boldsymbol{k})+\hbar\omega_{LO}]^{1/2}} - \frac{\alpha\,(\hbar\omega_{LO})^{3/2}E_c^0(\boldsymbol{k})}{6\,[E_c^0(\boldsymbol{k})-E_c(\boldsymbol{k})+\hbar\omega_{LO}]^{3/2}} \,. \tag{11.102}$$

By considering $\boldsymbol{k} = 0$, we obtain the following for the shift:

$$\Delta_c = -\frac{\alpha(\hbar\omega_L)^{3/2}}{(\hbar\omega_{LO} - \Delta_c)^{1/2}} = -\alpha\hbar\omega_{LO} + O(\alpha^2), \tag{11.103}$$

which, for weak coupling, reduces to just the polaron constant. The mass ratio is obtained from (11.102) by expanding the denominators up to order $\boldsymbol{k}^2$ and is given by

$$\frac{m_c^*}{m_c}\left(1 - \frac{3\Delta_c}{2\hbar\omega}\right) = 1 - \frac{4\Delta_c}{3\hbar\omega}. \tag{11.104}$$

The result linear in α is

$$m_c^* = m_c\left(1 + \frac{\alpha}{6}\right). \tag{11.105}$$

We see that we obtain a downward shift of the band energy combined with an increase of the effective mass. To leading order in α, this is the correct result. But our treatment becomes rather poor for intermediate values of the polaron constant. This is, partly, just as one might expect from the use of unrenormalized Green's functions in (11.92). Insertion of G^r from (11.95) shows that it is better to calculate

$$\Sigma_{cc}^{r\,(1)}(\boldsymbol{k},\omega) = -\alpha\hbar\omega_{LO}\left(\frac{\hbar^3\omega_{LO}}{2m_c}\right)^{1/2}$$
$$\times\frac{1}{2\pi^2}\int d^3\boldsymbol{q}\,\frac{1}{\boldsymbol{q}^2}\,\frac{1}{E_c^0(\boldsymbol{k}+\boldsymbol{q}) + \Sigma_{cc}^{r\,(1)}(\boldsymbol{k}+\boldsymbol{q},\hbar\omega - \hbar\omega_{LO}) - \hbar\omega + \hbar\omega_{LO}}. \tag{11.106}$$

The problem now is that the frequency dependence in the self-energy of the integrand is different from that on the left-hand side, i.e. it is just shifted by a phonon energy. This structure is due to the fact that at zero temperature and for an empty band, only combinations of retarded Green's functions and phonon propagators contribute – a feature which occurs to arbitrary order of phonon lines. If the frequency dependence were weak we could use a simple parameterization such as (11.101). Owing to the interaction with a single, sharp resonance, which gives rise to sidebands, this is not the case in the polaron problem. Fortunately, this property allows us to reduce the problem to a recursion. As a generalization of (11.101), we define

$$E_c(\boldsymbol{k},n) = E_c^0(\boldsymbol{k}) + \Sigma[\boldsymbol{k}, E_c(\boldsymbol{k}) - n\hbar\omega_{LO}] = \frac{\hbar^2\boldsymbol{k}^2}{2m_c^*(n)} + \Delta_c(n)\,. \tag{11.107}$$

According to (11.106), the calculation of the shift and mass to nth order requires quantities of the $(n+1)$th order. The integration in (11.106) may be performed as before. As a result, we obtain for the shift

$$\frac{\Delta_c(n)}{\hbar\omega_{LO}} = \Delta_n = -\alpha_{n+1}\,g_{n+1}\,, \tag{11.108}$$

where $g_n = n + \Delta_n - \Delta_0$ and $\alpha_n = \alpha/[g_n(g_n\mu_n)^{1/2}]$, and the mass ratio μ_n is obtained from

$$\frac{m_\mathrm{c}}{m_\mathrm{c}^*(n)} = \mu_n = 1 - \alpha_{n+1}\left(\frac{\mu_0}{2} - \frac{\mu_{n+1}}{3}\right) . \tag{11.109}$$

For sufficiently large n, we can use $\Delta_n = 0$ and $\mu_n = 1$ as starting values for the recursion. Finally, the values of the shift and mass ratio for $n = 0$ have to be determined self-consistently. The solution of the recursion implies the *summation of a certain class of diagrams up to arbitrary order.* Compared with the simple perturbational treatment, this gives a considerable improvement.

Example (Two-Phonon Processes). For intermediate values of the polaron constant, higher-order contributions are no longer negligible. For the two-phonon self-energy, we find from (11.93)

$$\begin{aligned}
&\Sigma_\mathrm{cc}^{\mathrm{r}\,(2)}(\boldsymbol{k}, \omega_n) \\
&= -\alpha^2(\hbar\omega_\mathrm{LO})^{3/2}\frac{\hbar^2}{\pi m_\mathrm{c}}\int \mathrm{d}^3\boldsymbol{q}\,\mathrm{d}^3\boldsymbol{q}'\left(\frac{1}{\boldsymbol{q}^2\boldsymbol{q}'^2[E_\mathrm{c}(\boldsymbol{k}+\boldsymbol{q}, n) + \hbar\omega_\mathrm{LO} - \hbar\omega_n]}\right. \\
&\times\left.\frac{1}{[E_\mathrm{c}(\boldsymbol{k}+\boldsymbol{q}+\boldsymbol{q}', n) + \hbar\omega_\mathrm{LO} - \hbar\omega_n][E_\mathrm{c}(\boldsymbol{k}+\boldsymbol{q}', n) + \hbar\omega_\mathrm{LO} - \hbar\omega_n]}\right) ,
\end{aligned} \tag{11.110}$$

where the energies are defined by (11.107) and $\hbar\omega_n = E_\mathrm{c}(\boldsymbol{k}) - n\hbar\omega_\mathrm{LO}$. To obtain the shift, we can set $\boldsymbol{k} = 0$. With respect to the mass correction, it is most convenient to use the following, according to (11.107):

$$\frac{1}{m_\mathrm{c}^*(n)} = \frac{1}{m_\mathrm{c}} + \frac{1}{3\hbar^2}\Delta_{\boldsymbol{k}}\Sigma_\mathrm{cc}^\mathrm{r}[\boldsymbol{k}, E_\mathrm{c}(\boldsymbol{k} - n\hbar\omega_\mathrm{LO})]\Big|_{\boldsymbol{k}=0} . \tag{11.111}$$

Thus we have to perform the integration of (11.110) and its derivatives for $\boldsymbol{k} = 0$ only. Nevertheless the calculation is lengthy, and we note only that all integrations may be reduced to the integral

$$\begin{aligned}
&\int \mathrm{d}q\,\mathrm{d}q'\frac{1}{2qq'(q^2+\alpha)(q'^2+\beta)}\ln\left(\frac{(q+q')^2+\gamma}{(q-q')^2+\gamma}\right) \\
&= \frac{\pi^2}{2\alpha\beta}[\ln(\sqrt{\gamma}+\sqrt{\alpha}) + \ln(\sqrt{\gamma}+\sqrt{\beta}) \\
&-\ln(\sqrt{\gamma}+\sqrt{\alpha}+\sqrt{\beta}) - \ln(\sqrt{\gamma}) + |\sqrt{\alpha}+\sqrt{\beta}|)] ,
\end{aligned} \tag{11.112}$$

and its derivatives with respect to the parameters α, β, and γ. As the final result for the shift, we have now, instead of (11.108),

$$\Delta_n = \Delta_n^{(1)} - \alpha_{n+1}^2\, g_{n+1}\ln\left(\frac{(r+1)^2}{r(r+2)}\right) \tag{11.113}$$

and the mass ratio is given by

$$\mu_n = \mu_n^{(1)} - \alpha_{n+1}^2(\mu_0 R - \mu_{n+1} K) \ , \tag{11.114}$$

with the definitions

$$R = 2\ln\left(\frac{(r+1)^2}{r(r+2)}\right) + \frac{1}{r+1} - \left(1 - \frac{1}{2r}\right)\left(\frac{2}{r+1} - \frac{1}{r+2} - \frac{1}{r}\right) , \tag{11.115}$$

$$K = \frac{2}{3}(1+r^2)\ln\left(\frac{(r+1)^2}{r(r+2)}\right) + \frac{1}{r+1} + \frac{2}{3}\frac{r^2}{r+1} - \frac{1}{2}$$
$$-\frac{1}{6}(r^2-1)\left(\frac{8}{r+1} - \frac{4}{r+2} - \frac{4}{r} - \frac{2}{(r+1)^2} + \frac{1}{r+2)^2} + \frac{1}{r^2}\right) , \tag{11.116}$$

and

$$r = \left(\frac{n+2+\Delta_{n+2}-\Delta_0}{n+1+\Delta_{n+1}-\Delta_0}\right)^{1/2} . \tag{11.117}$$

$\Delta_n^{(1)}$ and $\mu_n^{(1)}$ denote the right-hand sides of (11.108) and (11.109), respectively. Numerical results obtained from the simple perturbational treatment and the self-consistent one- and two-phonon approximations are given in Table 11.2 and denoted by $\Delta_c^{(0)}$, $\Delta_c^{(1)}$, and $\Delta_c^{(2)}$, respectively. The mass ratio, also shown, is obtained from the two-phonon calculation. Without calculating higher-order diagrams, the accuracy of our results can only be estimated. In contrast, variational methods at least give rigorous upper bounds for the polaron shift. The most accurate solution of the problem has been given by Feynman, who showed, how the solution of the polaron problem could be cast into a functional integral [314]. For a long period this approach led to the best available results over the entire range of polaron constants, which could be improved only slightly by laborious numerical methods [315]. For comparison the results obtained by Feynman's method and by numerical methods are shown in Table 11.2, too and are denoted by $\Delta_c^{(F)}$ and $\Delta_c^{(B)}$.The self-consistent one-phonon approximation is not too bad up to values of $\alpha = 1$.

Table 11.2. The polaron shift (in units of $\hbar\omega_{LO}$) and the renormalization of the electron mass of in various approximations

α	1	2	3	4	5	6
$\Delta_c^{(0)}$	-0.754	-1.314	-1.794	-2.227	-2.626	-3.000
$\Delta_c^{(1)}$	-0.909	-1.701	-2.421	-3.092	-3.726	-4.329
$\Delta_c^{(2)}$	-1.013	-1.964	-2.845	-3.671	-4.451	-5.195
$\Delta_c^{(F)}$	-1.013	-2.055	-3.133	-4.231	-5.407	-6.646
$\Delta_c^{(B)}$	-1.020	-2.080	-3.184	-4.314		
m_c/m_c^*	0.839	0.748	0.695	0.657	0.634	0.613

In the two-phonon approximation, the shift differs only very little from the best variational bounds up to polaron constants of $\alpha = 3$. This covers the range of coupling strengths in most semiconductors and insulators. Higher values of the polaron constant are found in certain alkali halides, such as KCl ($\alpha = 3.69$) and RbCl ($\alpha = 4.09$).

For large values of the polaron constant ($\alpha \approx 6$), certain variational calculations show a drastic, discontinuous change of the polaron mass (e.g. [316, 317]). This has led to the interpretation that the polaron becomes localized. However, if this process is considered exactly as a function of increasing α, it is – in contrast to widespread belief – not at all dramatic. As has been proved, the ground state wavefunction of the polaron is delocalized for all coupling strengths and the polaron mass is an analytic function of the polaron constant [318, 319, 320]. Nevertheless, the polaron wavefunction increases in the region of space around the center of an ion displacement. In real systems this may lead to large polaron masses. This case of strong coupling, however, is not suitable for description within the continuum approximation, which is only reasonable as long as all relevant length scales are large compared with the lattice constant. This is not the case in the strong-coupling limit, for which the continuum approximation is little more than an interesting mathematical model.

11.4.2 Dephasing Induced by Phonons

In the last section we considered the effect of phonons below the emission threshold at zero temperature, which results in a renormalization of band energies. For finite temperatures absorption of phonons also comes into play. Apart from a temperature dependence of the polaron shift and the mass renormalization, the most important consequence is the *finite lifetime* of polaron states. As all propagators depend on difference times only, we could tackle this problem in the frequency domain along lines similar to those used for the polaron problem at zero temperature. At the self-consistent level, however, this approach is cumbersome and a solution in the time domain, which applies (10.68) within our present approximations, has several advantages. As the most important advantage, the method guarantees, in arbitrary approximations to the self-energy, a self-consistent treatment of the particle propagators. As we have seen, this is essential for physically meaningful results beyond the perturbational regime. When the external fields are switched on, the solution of the polaron problem enters the time evolution of the propagators according to (10.82) as the initial value.

The simplest starting point is again the one-phonon self-energy (11.92). Even if this approximation is sufficient for weak coupling, the electron–hole interaction is included only partially. The off-diagonal elements of the particle Green's functions certainly take electron–hole correlations into account, but only for the optically excited states. The final states of phonon emission and absorption processes are uncorrelated. In order to account for the Coulomb

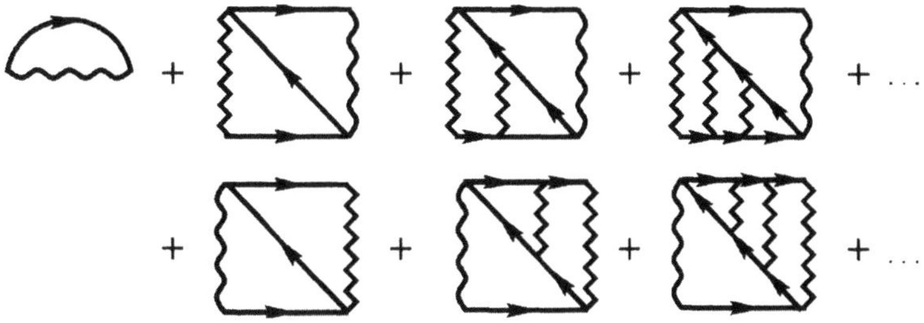

Fig. 11.2. Diagrammatic representation of the Coulomb renormalization of final states of phonon scattering within the ladder approximation

interaction between these final states, we have to include *vertex corrections*. A diagrammatic representation of these contributions within the ladder approximation is shown in Fig. 11.2. The summation of the ladder diagrams leads again to the formation of excitonic states. In this way, the final states of phonon scattering are treated on the same footing as the initial states. Analytically, the self-energy takes form

$$\begin{aligned}\Sigma(\bar{1}\bar{2}) = \mathrm{i}\hbar e^2 b_1 \Big[& w_{\mathrm{ph}}(\bar{2}\bar{1})\, G(\bar{1}\bar{2}) \\ & + \mathrm{i}\hbar e^2 \int \mathrm{d}\bar{3}\bar{4}\, b_4\, w_{\mathrm{ph}}(\bar{3}\bar{1})\, G(\bar{1}\bar{4})\, w(\bar{2}\bar{4}) \Big(G(\bar{4}\bar{3})\, G(\bar{3}\bar{2}) \\ & + \mathrm{i}\hbar e^2 \int \mathrm{d}\bar{5}\bar{6}\, b_6\, G(\bar{4}\bar{5}) G(\bar{3}\bar{6}) w(\bar{5}\bar{6}) G(\bar{5}\bar{3})\, G(\bar{6}\bar{2}) + \ldots \Big) \\ & + \mathrm{i}\hbar e^2 \int \mathrm{d}\bar{3}\bar{4}\, b_4 w_{\mathrm{ph}}(\bar{2}\bar{4})\, G(\bar{3}\bar{2})\, w(\bar{3}\bar{1}) \Big(G(\bar{1}\bar{4})\, G(\bar{4}\bar{3}) \\ & + \mathrm{i}\hbar e^2 \int \mathrm{d}\bar{5}\bar{6}\, b_6\, G(\bar{1}\bar{6})\, G(\bar{5}\bar{3}) w(\bar{5}\bar{6})\, G(\bar{6}\bar{4}) G(\bar{4}\bar{5}) + \ldots \Big) \Big] . \end{aligned} \tag{11.118}$$

We perform the eigenfunction expansion as earlier and obtain the representation

$$\begin{aligned}\Sigma_{nm}(\boldsymbol{k}, \bar{t}_1\bar{t}_2) = & -\mathrm{i}\hbar e^2 b_1 \sum_{\boldsymbol{q}j} \Big[M_j^2(\boldsymbol{q}) D_j(\boldsymbol{q}, \bar{t}_2\,\bar{t}_1) G_{nm}(\boldsymbol{k}-\boldsymbol{q}, \bar{t}_1\,\bar{t}_2) \\ +\mathrm{i}\hbar e^2 & \sum_{\boldsymbol{k}'n'm'} \int \mathrm{d}\bar{t}_3\, \mathrm{d}\bar{t}_4\, b_4 M_j^2(\boldsymbol{q}) D_j(\boldsymbol{q}, \bar{t}_3\,\bar{t}_1) G_{nn'}(\boldsymbol{k}-\boldsymbol{q}, \bar{t}_1\,\bar{t}_4) \\ \times w(\boldsymbol{k} & -\boldsymbol{k}', \bar{t}_2\,\bar{t}_4) \Big(G_{n'm'}(\boldsymbol{k}'-\boldsymbol{q}, \bar{t}_4\,\bar{t}_3)\, G_{m'm}(\boldsymbol{k}', \bar{t}_3\,\bar{t}_2) \\ +\mathrm{i}\hbar e^2 & \sum_{k''m''n''} \int \mathrm{d}\bar{t}_5\, \mathrm{d}\bar{t}_6\, b_6\, G_{n'm''}(\boldsymbol{k}'-\boldsymbol{q}, \bar{t}_4\,\bar{t}_5)\, G_{n''m}(\boldsymbol{k}', \bar{t}_6\,\bar{t}_2) \\ \times w(\boldsymbol{k}' & -\boldsymbol{k}'', \bar{t}_5\,\bar{t}_6)\, G_{m''m'}(\boldsymbol{k}''-\boldsymbol{q}, \bar{t}_5\,\bar{t}_3)\, G_{m'n''}(\boldsymbol{k}'', \bar{t}_3\bar{t}_6) + \ldots \Big) \end{aligned}$$

$$+\mathrm{i}\hbar e^2 \sum_{\boldsymbol{k}'n'm'} \int \mathrm{d}\bar{t}_3\, \mathrm{d}\bar{t}_4\, b_4 M_j^2(\boldsymbol{q}) D_j(\boldsymbol{q}, \bar{t}_2\, \bar{t}_4) G_{n'm}(\boldsymbol{k}+\boldsymbol{q}, \bar{t}_3\, \bar{t}_2)$$

$$\times w(\boldsymbol{k}-\boldsymbol{k}', \bar{t}_3\, \bar{t}_1) \Big(G_{nm'}(\boldsymbol{k}', \bar{t}_1\, \bar{t}_4)\, G_{m'n'}(\boldsymbol{k}'+\boldsymbol{q}, \bar{t}_4\, \bar{t}_3)$$

$$+\mathrm{i}\hbar e^2 \sum_{k''m''n''} \int \mathrm{d}\bar{t}_5\, \mathrm{d}\bar{t}_6\, b_6\, G_{nm''}(\boldsymbol{k}', \bar{t}_1\, \bar{t}_6)\, G_{n''n'}(\boldsymbol{k}'+\boldsymbol{q}, \bar{t}_5\, \bar{t}_3)$$

$$\times w(\boldsymbol{k}'-\boldsymbol{k}'', \bar{t}_5\, \bar{t}_6)\, G_{m''m'}(\boldsymbol{k}'', \bar{t}_6\, \bar{t}_4)\, G_{m'n''}(\boldsymbol{k}''+\boldsymbol{q}, \bar{t}_4 \bar{t}_5) + \ldots \Big) \Big] . \quad (11.119)$$

This result illustrates the complexity of a consistent theory of dephasing due to phonons for general excitation conditions. Let us consider first the uncorrelated contribution to (11.119). The representation which enters the quantum kinetic equations (10.82) is obtained after separation of the phase factors according to (10.78) and (10.79), and results in

$$\sigma_{nm}^{+-}(\boldsymbol{k}, t_1 t_2) = -\hbar e^2 \sum_{\boldsymbol{q}j} \Big(\{ M_j(\boldsymbol{q}) \left[1 + N_{0j}(\boldsymbol{q})\right]$$

$$\times U_{nn}^{j\,-}(\boldsymbol{k}-\boldsymbol{q}, \boldsymbol{k}, t_1)\, U_{mm}^{j\,-}(\boldsymbol{k}-\boldsymbol{q}, \boldsymbol{k}, -t_2)\, g_{nm}^{+-}(\boldsymbol{k}-\boldsymbol{q}, t_1\, t_2)\} \quad (11.120)$$

$$+ \{1 + N_{0j}(\boldsymbol{q}) \to N_{0j}(\boldsymbol{q}), \omega_j(\boldsymbol{q}) \to -\omega_j(\boldsymbol{q})\} \Big) ,$$

where we have defined

$$U_{nm}^{j\,\pm}(\boldsymbol{k}-\boldsymbol{q}, \boldsymbol{k}, t) = \mathrm{e}^{-\mathrm{i}\,[E_n(\boldsymbol{k}-\boldsymbol{q}) - E_m(\boldsymbol{k}) \pm \hbar\omega_j(\boldsymbol{q})]\, t/\hbar} . \quad (11.121)$$

Correspondingly, σ_{nm}^{-+} is obtained from (11.120) by the replacements $\omega_j(\boldsymbol{q}) \to -\omega_j(\boldsymbol{q})$ and $g_{nm}^{+-} \to g_{nm}^{-+}$. Within the two-time formalism, numerical studies have been performed using this approximation for scattering by LO phonons [270, 271]. If the LO phonon energy is large compared with the excitonic binding energy, the final states are high in the continuum of pair excitations. Under these conditions, the neglect of final-state correlations is no reason for pangs of conscience. Up to coupling constants α of the order 0.3, the results give quantitative descriptions of the temperature dependence of the optical absorption and, in particular, of its lineshape. If, however, phonon absorption takes place in the vicinity of the band edge, (11.120) has to be improved. Since the general evaluation of (11.119) is rather lengthy, we shall focuss in the following on the lineshape of linear optical spectra and, for notational simplicity, neglect the polaron renormalization of the ladder contributions. In this limit we have $g_{\mathrm{cc}}^{+-} = g_{\mathrm{vv}}^{-+} = 0$ and $g_{\mathrm{cc}}^{-+} = g_{\mathrm{vv}}^{+-} = -\mathrm{i}/\hbar$, and the problem is simplified considerably.[4] Only the electron–hole interaction contributes, and electron–electron and hole–hole interactions vanish. Further, the electronic screening reduces to a background dielectric function and we can replace the dynamically screened interaction by a static interaction. First, we consider

[4] The general evaluation is straightforward and left as an exercise for the reader.

$\sigma_{\rm vc}^{+-}$, to which – because $g_{\rm cc}^{+-} = 0$ – only the first group of ladders on the right-hand side of (11.119) contributes. We perform the decomposition with respect to branch indices, making use of the relations (10.20) and (10.21), separate the phase factors again, and obtain

$$\sigma_{\rm vc}^{+-}(\boldsymbol{k},t_1\,t_2) = -\hbar e^2 \sum_{\boldsymbol{q}\,j} \Bigg\{ \Bigg[M_j(\boldsymbol{q})\,[1+N_{0j}(\boldsymbol{q})]\, U_{\rm vv}^{j\,-}(\boldsymbol{k}-\boldsymbol{q},\boldsymbol{k},t_1)$$

$$\times \Bigg(U_{\rm cc}^{j\,-}(\boldsymbol{k}-\boldsymbol{q},\boldsymbol{k},-t_2)\, g_{\rm vc}^{+-}(\boldsymbol{k}-\boldsymbol{q},t_1\,t_2)$$

$$- \frac{\mathrm{i}e^2}{\hbar} \sum_{\boldsymbol{k}'} U_{\rm vc}^{j\,-}(\boldsymbol{k}-\boldsymbol{q},\boldsymbol{k},-t_2) v(\boldsymbol{k}-\boldsymbol{k}')\, P_{\rm vc}^{j\,-}(\boldsymbol{k}'-\boldsymbol{q},\boldsymbol{k}',t_2) \Bigg) \Bigg]$$

$$+ \Bigg[1+N_{0j}(\boldsymbol{q}) \to N_{0j}(\boldsymbol{q}),\ \omega_j(\boldsymbol{q}) \to -\omega_j(\boldsymbol{q}) \Bigg] \Bigg\} . \tag{11.122}$$

The summation of the ladder diagrams has been mapped onto the integral equation

$$P_{\rm vc}^{j\,\pm}(\boldsymbol{k}-\boldsymbol{q},\boldsymbol{k},t) = -\,U_{\rm vc}^{j\,\pm}(\boldsymbol{k}-\boldsymbol{q},\boldsymbol{k},t) \int_{-\infty}^{t} \mathrm{d}t'\, [U_{\rm cc}^{j\,\pm}(\boldsymbol{k}-\boldsymbol{q},\boldsymbol{k},-t')$$

$$\times\, g_{\rm vc}^{+-}(\boldsymbol{k}-\boldsymbol{q},t\,t') - U_{\rm vv}^{j\,\pm}(\boldsymbol{k}-\boldsymbol{q},\boldsymbol{k},-t')\, g_{\rm vc}^{-+}(\boldsymbol{k},t'\,t)]$$

$$+ \frac{\mathrm{i}e^2}{\hbar} \sum_{\boldsymbol{k}'} \int_{-\infty}^{t} \mathrm{d}t'\, U_{\rm vc}^{j\,\pm}(\boldsymbol{k}-\boldsymbol{q},\boldsymbol{k},t-t') v(\boldsymbol{k}-\boldsymbol{k}') P_{\rm vc}^{j\,\pm}(\boldsymbol{k}'-\boldsymbol{q},\boldsymbol{k}',t') , \tag{11.123}$$

which describes *phonon-assisted excitonic transitions.*

A numerical treatment of phonon scattering with correlated electron–hole pairs within the two-time formalism is still a challenge – even in the linear case. We shall restrict ourselves in the following onto the limit of equal times in the quantum kinetic equations. This allows us to summarize (11.122) and the diagonal element $\sigma_{\rm vv}^{+-}$ given by (11.120) by means of (11.123), with the result

$$\int_{-\infty}^{t_1} \mathrm{d}t'\, [\sigma_{\rm vc}^{+-}(\boldsymbol{k},t_1\,t')\, g_{\rm cc}^{-+}(\boldsymbol{k},t'\,t_1) + \sigma_{\rm vv}^{+-}(\boldsymbol{k},t_1\,t')\, g_{\rm vc}^{-+}(\boldsymbol{k},t'\,t_1)]$$

$$= -e^2 \sum_{\boldsymbol{q}\,j} \Bigg\{ M_j(\boldsymbol{q})\,[1+N_{0j}(\boldsymbol{q})]\, U_{\rm cv}^{j\,-}(\boldsymbol{k},\boldsymbol{k},t_1) P_{\rm vc}^{j\,-}(\boldsymbol{k}-\boldsymbol{q},\boldsymbol{k},t_1)\}$$

$$+ \{1+N_{0j}(\boldsymbol{q}) \to N_{0j}(\boldsymbol{q}),\ \omega_j(\boldsymbol{q}) \to -\omega_j(\boldsymbol{q})\} \Bigg) . \tag{11.124}$$

The second nonvanishing contribution to the self-energies $\sigma_{\rm vc}^{-+}$ is due to the second group of ladder contributions in (11.119) (see Fig. 11.2), together with the diagonal contribution $\sigma_{\rm cc}^{-+}$, and yields, correspondingly,

$$\int_{-\infty}^{t_1} \mathrm{d}t' \, [g^{+-}_{\mathrm{vv}}(\boldsymbol{k}, t_1\, t')\, \sigma^{-+}_{\mathrm{vc}}(\boldsymbol{k}, t'\, t_1) + g^{+-}_{\mathrm{vc}}(\boldsymbol{k}, t_1\, t')\, \sigma^{-+}_{\mathrm{cc}}(\boldsymbol{k}, t'\, t_1)]$$

$$= e^2 \sum_{\boldsymbol{q}\, j} \Big(\{ M_j(\boldsymbol{q})\, [1 + N_{0j}(\boldsymbol{q})]\, U^{j\,-}_{\mathrm{cv}}(\boldsymbol{k}, \boldsymbol{k}, t_1) P^{j\,-}_{\mathrm{vc}}(\boldsymbol{k}, \boldsymbol{k}+\boldsymbol{q}, t_1) \}$$

$$+ \{ 1 + N_{0j}(\boldsymbol{q}) \to N_{0j}(\boldsymbol{q}),\, \omega_j(\boldsymbol{q}) \to -\omega_j(\boldsymbol{q}) \} \Big) \, . \tag{11.125}$$

In the equal-time limit, we have to replace the two-time functions in (11.123) by single-time functions. Applying the generalized Kadanoff–Baym ansatz, allows us to describe the phonon-assisted transition by a quantum kinetic equation. We make use of (10.95) with diagonal spectral functions, apply (10.76), and reverse the separation of phase factors. We then take the time derivative of (11.123) and obtain

$$\begin{aligned} &\left(\mathrm{i}\hbar \frac{\partial}{\partial t} - E_{\mathrm{c}}(\boldsymbol{k}) + E_{\mathrm{v}}(\boldsymbol{k}-\boldsymbol{q}) \pm \hbar\omega_j(\boldsymbol{q}) \right) P^{j\,\pm}_{\mathrm{vc}}(\boldsymbol{k}-\boldsymbol{q}, \boldsymbol{k}, t) \\ &\qquad + e^2 \sum_{\boldsymbol{k}'} v(\boldsymbol{k}-\boldsymbol{k}') P^{j\,\pm}_{\mathrm{vc}}(\boldsymbol{k}'-\boldsymbol{q}, \boldsymbol{k}', t) \\ &\qquad = p_{\mathrm{vc}}(\boldsymbol{k}-\boldsymbol{q}, t) - p_{\mathrm{vc}}(\boldsymbol{k}, t) \, . \end{aligned} \tag{11.126}$$

Finally, we can summarize the various contributions obtained from (11.124) and (11.125) to the quantum kinetic equation for the transition amplitude, which reduces to

$$\left(\mathrm{i}\hbar \frac{\partial}{\partial t} - E_{\mathrm{c}}(\boldsymbol{k}) + E_{\mathrm{v}}(\boldsymbol{k}) \right) p_{\mathrm{vc}}(\boldsymbol{k}, t) + e^2 \sum_{\boldsymbol{k}'} v(\boldsymbol{k}-\boldsymbol{k}') p_{\mathrm{vc}}(\boldsymbol{k}', t)$$

$$= \boldsymbol{d}_{\mathrm{cv}} \boldsymbol{E}(t) + \mathrm{i}\hbar \frac{\partial}{\partial t} p_{\mathrm{vc}}(\boldsymbol{k}, t) \Big|_{\mathrm{phon}} \, , \tag{11.127}$$

where the phonon scattering contribution is given by

$$\mathrm{i}\hbar \frac{\partial}{\partial t} p_{\mathrm{vc}}(\boldsymbol{k}, t) \Big|_{\mathrm{phon}} = \hbar\, e^2 \sum_{\boldsymbol{q} j} \Big(\{ M_j(\boldsymbol{q}) [1 + N_{0j}(\boldsymbol{q})]\, [P^{j\,-}_{\mathrm{vc}}(\boldsymbol{k}-\boldsymbol{q}, \boldsymbol{k}, t)$$

$$-P^{j\,-}_{\mathrm{vc}}(\boldsymbol{k}, \boldsymbol{k}+\boldsymbol{q}, t)] \} + \{ 1 + N_{0j}(\boldsymbol{q}) \to N_{0j}(\boldsymbol{q}),\, \omega_j(\boldsymbol{q}) \to -\omega_j(\boldsymbol{q}) \} \Big) \, . \tag{11.128}$$

In this limit the Green's function theory, combined with the Kadanoff–Baym ansatz, coincides with the results obtained from density matrix theory [325]. This is consistent with the restriction to one-phonon processes, which implies the assumption of sufficiently weak coupling. In this case, our reduction of two-time propagators to single-time functions is justified.

A touchstone for any theory of electron–phonon coupling is the lineshape of the optical absorption, in particular the behavior below the exciton resonance. As already noted in Sect. 8.3, at high temperatures the absorption edge exhibits an exponential low-energy tail, which is referred to as the *Urbach tail.* A nonperturbative treatment of the electron–phonon interaction based on (11.120) within the two-time formalism gives rise to such an exponential lineshape in the weak-coupling regime [270]. The basic physical effect, however, can also be understood by means of a simplified treatment, which we consider in the following.

Example (The Urbach Tail). Let us summarize first the experimental facts. Absorption spectra at high temperatures not only exhibit an exponential tail over several orders of magnitude, but also show a typical temperature dependence. Extrapolating these tails to higher energies, one finds that all lines intersect in one point, which lies roughly at the energetic position of the zero-temperature exciton. Empirically, these features can be summarized by means of the rule [179, 180, 326]

$$\alpha(\omega) = \alpha_0\, \mathrm{e}^{-\sigma\,(\hbar\omega_0-\hbar\omega)/k_\mathrm{B}T}\,. \tag{11.129}$$

The material constants α_0, σ, and ω_0 depend only weakly on temperature. A classical experimental example for potassium iodide is shown in Fig. 11.3 [327]. Similar behavior is found in a variety of semiconductors and ionic crystals.

At low temperatures, in most semiconductors, one observes phonon-sidebands. There are, however, also materials, for example, alkali halides, in which the tail persists even at low temperatures.

On the basisi of the experimental data, LO phonons are expected to be responsible for this effect. We start our analysis of this effect with (11.126)–(11.128) and drop the phonon index in the following. First we transform into the frequency domain and into an excitonic representation, which is best suited for analytical calculations. Noting that

$$E_\mathrm{c}(\boldsymbol{k}+\beta\boldsymbol{q}) - E_\mathrm{v}(\boldsymbol{k}-\alpha\boldsymbol{q}) = \frac{\hbar^2\boldsymbol{k}^2}{2m_\mathrm{r}} + \frac{\hbar^2\boldsymbol{q}^2}{2M}\,, \tag{11.130}$$

where $\alpha = m_\mathrm{h}/M$, $\beta = m_\mathrm{e}/M$, and $M = m_\mathrm{e} + m_\mathrm{h}$, we have the expansion

$$P^\pm_\mathrm{vc}(\boldsymbol{k}-\boldsymbol{q},\boldsymbol{k},\omega) = \sum_n \phi_n(\boldsymbol{k}-\beta\boldsymbol{q})P^\pm_n(\omega)\,, \tag{11.131}$$

where ϕ_n are the excitonic eigenstates, and correspondingly,

$$p_\mathrm{vc}(\boldsymbol{k},\omega) = \sum_n \phi_n(\boldsymbol{k})p_n(\omega)\,. \tag{11.132}$$

This allows us to cast (11.126) into the form

$$[E_n(\boldsymbol{q}) \pm \hbar\omega_\mathrm{LO} - \hbar\omega]P^\pm_n(t) = -\sum_m \langle n|\varrho_{\alpha\beta}(\boldsymbol{r})|m\rangle p_m(\omega)\,, \tag{11.133}$$

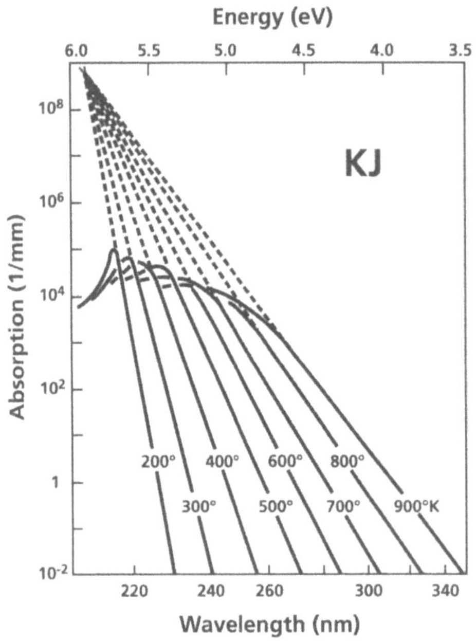

Fig. 11.3. Temperature dependence of the absorption edge of potassium iodide. Taken from [327]

where we have used the abbreviation

$$\varrho_{\alpha\beta}(\boldsymbol{r}) = \mathrm{e}^{-\mathrm{i}\alpha\boldsymbol{q}\boldsymbol{r}} - \mathrm{e}^{\mathrm{i}\beta\boldsymbol{q}\boldsymbol{r}} \tag{11.134}$$

and have neglected the wave vector dependence of the phonon energy. Inserting the solution into the corresponding expansion of (11.127), we find as a result

$$(E_n - \hbar\omega)\, p_n(\omega) - \sum_{n'} \Sigma_{nn'}(\omega)\, p_{n'}(\omega) = \boldsymbol{d}_{\mathrm{vc}}\boldsymbol{E}(\omega)\,, \tag{11.135}$$

where the *exciton–phonon self-energy* is

$$\begin{aligned}\Sigma_{nn'}(\omega) &= \hbar e^2 \sum_{m\boldsymbol{q}} M_{\mathrm{LO}}^2(\boldsymbol{q})\, \langle n|\varrho^*_{\alpha\beta}(\boldsymbol{r})|m\rangle\langle m|\varrho_{\alpha\beta}(\boldsymbol{r})|n'\rangle \\ &\times \left(\frac{N_{\mathrm{LO}}}{E_m(\boldsymbol{q}) - \hbar\omega_{\mathrm{LO}} - \hbar\omega + \mathrm{i}\delta} + \frac{1 + N_{\mathrm{LO}}}{E_m(\boldsymbol{q}) + \hbar\omega_{\mathrm{LO}} - \hbar\omega + \mathrm{i}\delta}\right).\end{aligned} \tag{11.136}$$

Since the self-energy depends on frequency, the resulting lineshape of the optical absorption is non-Lorentzian and can lead at $\hbar\omega = E_1 - \hbar\omega_{\mathrm{LO}}$ to a *phonon-sideband* at low temperatures. Owing to the important contributions of scattering states, (11.136) is not the best starting point for numerical investigations, as it is much easier to solve (11.126) together with (11.127) directly in $\boldsymbol{k}$ space. Here we proceed analytically and introduce next the real-space representation of the self-energy, according to

$$\Sigma_{nn'}(\omega) = \int \mathrm{d}^3\boldsymbol{r}\, \mathrm{d}^3\boldsymbol{r}'\, \phi_n^*(\boldsymbol{r})\Sigma(\boldsymbol{r}, \boldsymbol{r}', \omega)\phi_{n'}(\boldsymbol{r}')\,. \tag{11.137}$$

In order to perform the wave vector integration, we replace the Coulomb potential in the coupling matrix element (11.82) according to $1/\boldsymbol{q}^2 \rightarrow 1/(\boldsymbol{q}^2 + \lambda^2)$ and take the limit $\lambda \rightarrow 0$ at the end of the calculation. Using spherical coordinates, we perform the angle integrations first. The q integration is carried out by means of the residue theorem and yields, for $E_m \pm \hbar\omega_{\rm LO} - \hbar\omega \geq 0$,

$$\begin{aligned} \Sigma(\boldsymbol{r}, \boldsymbol{r}', \omega) &= \frac{e^2 \omega_{\rm LO} M}{8\pi\varepsilon_0 \varepsilon^* \hbar} \sum_m \phi_m^*(\boldsymbol{r}) \phi_m(\boldsymbol{r}') \\ &\quad \times \left[(1 + N_{\rm LO})\, \mathcal{F}(\Delta_m^+, \boldsymbol{r}, \boldsymbol{r}') + N_{\rm LO}\, \mathcal{F}(\Delta_m^-, \boldsymbol{r}, \boldsymbol{r}')\right] , \end{aligned} \tag{11.138}$$

where we have defined the quantity

$$\Delta_m^{\pm} = \left| \frac{2M}{\hbar^2} (E_m \pm \hbar\omega_{\rm LO} - \hbar\omega) \right|^{1/2} \tag{11.139}$$

and the function

$$\begin{aligned} \mathcal{F}(x, \boldsymbol{r}, \boldsymbol{r}') &= \frac{1}{x^2} \left(\frac{1 - g(x\alpha|\boldsymbol{r} - \boldsymbol{r}'|)}{\alpha|\boldsymbol{r} - \boldsymbol{r}'|} + \frac{1 - g(x\beta|\boldsymbol{r} - \boldsymbol{r}'|)}{\beta|\boldsymbol{r} - \boldsymbol{r}'|} \right. \\ &\quad \left. - \frac{1 - g(x|\alpha\boldsymbol{r} + \beta\boldsymbol{r}'|)}{|\alpha\boldsymbol{r} + \beta\boldsymbol{r}'|} - \frac{1 - g(x|\beta\boldsymbol{r} + \alpha\boldsymbol{r}'|)}{|\beta\boldsymbol{r} + \alpha\boldsymbol{r}'|} \right) , \end{aligned} \tag{11.140}$$

where

$$g(y) = \begin{cases} {\rm e}^{-y} & \text{if } E_m \pm \hbar\omega_{\rm LO} - \hbar\omega \geq 0 \\ \cos(y) & \text{if } E_m \pm \hbar\omega_{\rm LO} - \hbar\omega \leq 0 \end{cases} . \tag{11.141}$$

In general, (11.138) still represents a rather complicated nonlocal, energy-dependent potential. The dominant contributions are obtained, however, for small $\Delta_m^{\pm}$, which suggests an expansion of (11.141). For both functions, we obtain in leading order

$$\mathcal{F}(\Delta_m^{\pm}, \boldsymbol{r}, \boldsymbol{r}') \simeq \frac{1}{2}(|\alpha\boldsymbol{r} + \beta\boldsymbol{r}'| + |\beta\boldsymbol{r} + \alpha\boldsymbol{r}'| - |\boldsymbol{r} - \boldsymbol{r}'|) , \tag{11.142}$$

which is independent of the exciton energy. Thus we can perform the summation over excitonic quantum numbers in (11.138) and make use of the closure relation. As a result, the self-energy becomes local and takes the form

$$\Sigma(\boldsymbol{r}, \boldsymbol{r}') = \frac{e^2 \omega_{\rm LO} M}{8\pi\varepsilon_0 \varepsilon^* \hbar} \delta(\boldsymbol{r} - \boldsymbol{r}')\, |\boldsymbol{r}|\, (1 + 2N_{\rm LO}) . \tag{11.143}$$

Thus, in leading order, we rediscover a refined version of a phenomenological model [178].[5] The exciton–phonon self-energy is nothing but the action

[5] In this model the electron–phonon interaction is described by the action of a homogeneous electric field. Performing the corresponding expansion for small wave vector in the Hamiltonian, leads to ill-defined expressions which have to be repaired by additional assumptions. As we have seen, such a problem does not occur, if we take the long-wavelength limit in the equations of motion.

of a homogeneous electric field, which is determined by

$$F = \frac{e\,\omega_{\mathrm{LO}} M}{8\pi\varepsilon_0\varepsilon^*\hbar} \coth\left(\frac{\hbar\omega_{\mathrm{LO}}}{2k_{\mathrm{B}}T}\right) \simeq \frac{e\,M}{2\pi\varepsilon_0\varepsilon^*\hbar^2}\,k_{\mathrm{B}}T\,. \tag{11.144}$$

We have analyzed this problem in detail in Sect. 8.3, where we have found that the absorption tail is always exponential. According to our simple analytical result (8.41), for high temperatures we indeed have the temperature dependence given by (11.129). In addition, for alkali halides, for example, (11.129) is still valid at low temperatures if we replace $k_{\mathrm{B}}T$ by an effective temperature

$$k_{\mathrm{B}}T_{\mathrm{eff}} = \frac{\hbar\omega_{\mathrm{LO}}}{2} \coth\left(\frac{\hbar\omega_{\mathrm{LO}}}{2k_{\mathrm{B}}T}\right), \tag{11.145}$$

in accordance with (11.143). Also, the trends predicted by (11.144) for the material dependence of the steepness parameter σ agree with experimental data.

The fact that the leading contribution in our expansion does not depend on the intermediate states in (11.138) has important implications. Our actual treatment of the phonon-assisted transitions is obviously not completely consistent with our description of the transition amplitude, as the contributions from the exciton–phonon interaction are missing in (11.126). Consideration of these contributions requires the explicit inclusion of two-phonon processes. On a formal level this is straightforward, using the method described above, and even higher-order processes can be taken into account without problems. The resulting features show up only in the properties of the intermediate states in (11.138), which otherwise do not occur in the leading contribution to our expansion. Thus the range of validity of (11.143) is significantly more extensive than suggested by our actual derivation. However, the importance of the corrections which result from the sum of the higher-order contributions may vary. This is expected in particular for large electron–phonon coupling constants, where the cumulant expansion proposed in [328] has several advantages. In the nonlinear regime, however, the Green's function approach is best suited for including all relevant effects in a consistent manner; in particular the features, owing to the time evolution of non-equilibrium one-particle distributions. We shall consider the relaxation of the one-particle distribution function as a further example.

Example (Relaxation due to Electron–Phonon Scattering). As is obvious from the expression for the electron–phonon self–energy (11.120), the time evolution of optically generated carriers is strongly influenced by scattering from phonons. Let us discuss first a frequently used approximation. We consider again the quantum kinetic equation (10.82) and focus on the relaxation of the nonequilibrium carrier distribution. If the excitation process itself is fast compared with the typical timescale on which electron–phonon scattering becomes effective, we can restrict ourselves to the diagonal elements of the Green's function and self-energies in the scattering contributions. Physically this means that we neglect the polarization scattering, which we have discussed in Sect. 10.6 for scattering within the particle system. Anticipating further the Markovian limit, we have to consider

$$\mathrm{i}\hbar\frac{\partial}{\partial t}g^{+-}_{\mathrm{cc}}(\boldsymbol{k},t,t) = \Delta_{\mathrm{cv}}(\boldsymbol{k},t)g^{+-}_{\mathrm{vc}}(\boldsymbol{k},t,t) - g^{+-}_{\mathrm{cv}}(\boldsymbol{k},t,t)\Delta_{\mathrm{vc}}(\boldsymbol{k},t)$$
$$-\int_{-\infty}^{t}[\sigma^{+-}_{\mathrm{cc}}(\boldsymbol{k},t,t')g^{-+}_{\mathrm{cc}}(\boldsymbol{k},t',t) - g^{+-}_{\mathrm{cc}}(\boldsymbol{k},t,t')\sigma^{-+}_{\mathrm{cc}}(\boldsymbol{k},t',t)$$
$$-\sigma^{-+}_{\mathrm{cc}}(\boldsymbol{k},t,t')g^{+-}_{\mathrm{cc}}(\boldsymbol{k},t',t) + g^{-+}_{\mathrm{cc}}(\boldsymbol{k},t,t')\sigma^{+-}_{\mathrm{cc}}(\boldsymbol{k},t',t)]\,. \qquad (11.146)$$

After insertion of (11.120), the time integration reduces in the Markovian approximation to the integration of the phase factors. Thus the relaxation of the electron distribution function is determined by

$$\frac{\partial}{\partial t}f_{\mathrm{c}}(\boldsymbol{k}) = -\frac{2}{\hbar}\,\mathrm{Im}\Big[\Big(\boldsymbol{d}_{\mathrm{cv}}\,\boldsymbol{E}(\boldsymbol{r},t) + e^2\sum_{\boldsymbol{k}'}v(\boldsymbol{k}-\boldsymbol{k}')\,p_{\mathrm{vc}}(\boldsymbol{k}')\Big)p^{*}_{\mathrm{vc}}(\boldsymbol{k})\Big]$$
$$+2\pi\,e^2\sum_{\boldsymbol{q}\,j}M_j^2(\boldsymbol{q})\,\Big(\delta[E_{\mathrm{c}}(\boldsymbol{k}+\boldsymbol{q}) - E_{\mathrm{c}}(\boldsymbol{k}) - \hbar\omega_j(\boldsymbol{q})]$$
$$\times\{[1+N_{0j}(\boldsymbol{q})]f_{\mathrm{c}}(\boldsymbol{k}+\boldsymbol{q})[1-f_{\mathrm{c}}(\boldsymbol{k})] - N_{0j}(\boldsymbol{q})[1-f_{\mathrm{c}}(\boldsymbol{k}+\boldsymbol{q})]f_{\mathrm{c}}(\boldsymbol{k})\}$$
$$+\{1+N_{0j}(\boldsymbol{q}) \to N_{0j}(\boldsymbol{q}),\ \omega_j(\boldsymbol{q}) \to -\omega_j(\boldsymbol{q})\}\Big)\,, \qquad (11.147)$$

where the transition amplitude may be calculated within the Hartree–Fock approximation. There are four fundamental processes, which change the occupation of a state with wave vector $\boldsymbol{k}$: electrons are scattered from higher states with wavevectors $\boldsymbol{k}+\boldsymbol{q}$ into the state at $\boldsymbol{k}$ by *emission of phonons*, and electrons are scattered from the state with wavevector $\boldsymbol{k}$ into higher states at $\boldsymbol{k}+\boldsymbol{q}$ by *absorption of phonons*. Two corresponding processes occur if the state with wave vector $\boldsymbol{k}+\boldsymbol{q}$ is lower in energy and emission is interchanged with absorption. The same scattering processes determine the relaxation of the hole distribution function. With increasing density, the Coulomb interaction that enters the coupling matrix element becomes screened, as is already obvious from the general structure of the phonon-induced interaction (11.15). We shall come back to this point in the next section.

More rigorous results are obtained from solutions of (10.82) within the two–time formalism. As an example of LO phonon quantum kinetics by means of the one-phonon self-energy (11.92), the time evolution of the electron distribution is shown in Fig. 11.4 for bulk GaAs parameters at room temperature. The buildup of phonon sidebands is clearly displayed and leads to a stationary distribution about 3 ps after the excitation. As electron–electron scattering is not taken into account, the sidebands are much more pronounced than in

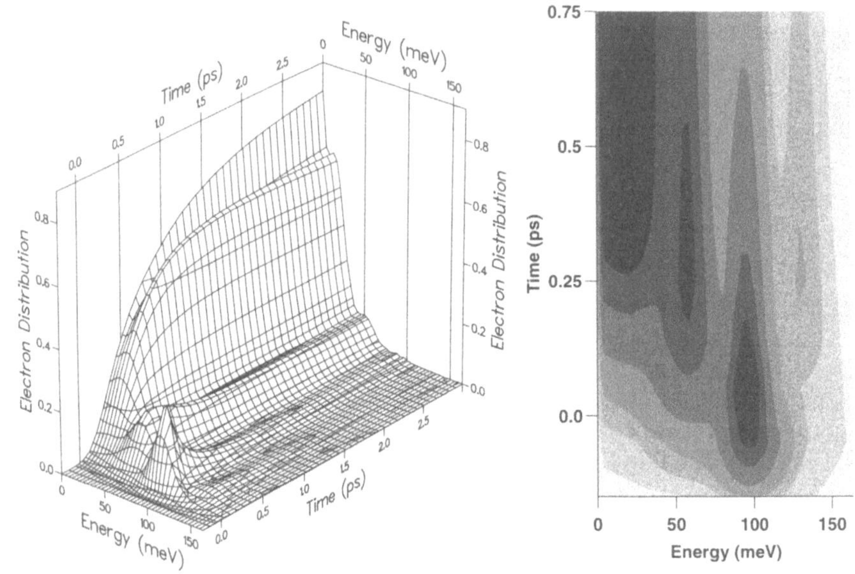

Fig. 11.4. Relaxation of the electron distribution in bulk GaAs due to LO phonon scattering (*left-hand side*). The distribution is generated 92 meV above the band edge by a 100 fs pulse centered at t = 0. The Rabi frequency is 2 meV. The *right-hand side* shows the behaviour for short times in more detail

real sytems and a quasi-equilibrium state cannot be reached. Compared with corresponding calculations within the Markovian approximation, the main effect is a slowing down of the relaxation rates by about 20%. This reduction of scattering rates is one of the typical differences between Boltzmann kinetics and quantum kinetics. In contrast to (11.147) the quantum kinetic equations do not conserve energy in a single scattering event. Thus the scattering becomes less effective. For coherent quantities, such as four-wave-mixing signals, the deviations between Boltzmann kinetics and quantum kinetics are more pronounced and lead to essential qualitative differences (see Sect. 12.1). In the case of weak coupling, the results obtained within the two-time formalism are very similar to those obtained by means of the ansatz (10.89). This correspondence, however, gets lost with increasing coupling strength. In any case, the Coulomb correlation of final states becomes important in the vicinity of the band edge, and it is better to apply (11.119).

The electron distribution shown in Fig. 11.5 obviously suffers a loss of energy, which is transferred to the lattice. Correspondingly, one might expect naively that the *lattice temperature* increases. In fact, this is the case for long times. The way in which this increase of temperature occurs, however, is a little more involved.

11.5 Nonequilibrium Phonons

The properties of phonons and their coupling to the particle system discussed in the preceding sections have been restricted to a semiconductor in its ground state. In the presence of excited carriers, the dispersion and occupation of phonons, and their coupling to the electronic system become modified. This results immediately, as already mentioned, from the electronic part of the force constants. In the presence of an optical excitation, the linear susceptibility in (11.24) exhibits *intraband transitions* in addition. The neglect of phonon energies in comparison with electronic transition energies, which was one ingredient of the adiabatic approximation, is not possible for these contributions. Thus we have to go beyond the static limit of the susceptibility in (11.24), which requires us to start from the complete Hamiltonian (11.3).

In the following we shall elaborate on the consequences of the excitation-induced contribution to the dielectric response, which we have separated from the contribution of the unexcited system in Sect. 10.7.1. We have to make use of these results with respect to the force constants (11.25) and, correspondingly, the phonon Green's function, and also in the coupling constant of the phonon-induced interaction (11.58). After these preliminary remarks, we turn first to the renormalization of the phonon Green's function.

11.5.1 Renormalization of Phonons

Let us again calculate the equation of motion for the lattice displacement, but now with the Hamiltonian (11.1). As (11.28) and 11.30) are still valid, we obtain

$$M_{\boldsymbol{\kappa}} \frac{\partial^2}{\partial t_1^2} u_\alpha(\boldsymbol{L\kappa}, \bar{t}_1)$$
$$= \int \mathrm{d}^3\boldsymbol{r}\, \mathrm{d}^3\boldsymbol{r}' \left(\frac{\partial \rho_\kappa[\boldsymbol{r} - \boldsymbol{R}(\boldsymbol{L\kappa})]}{\partial u_\alpha(\boldsymbol{L\kappa})} v(\boldsymbol{r} - \boldsymbol{r}')[\rho_n(\boldsymbol{r}', \bar{t}_1) + \rho_{el}(\boldsymbol{r}', \bar{t}_1)] \right) . \tag{11.148}$$

If we expand the charge densities with respect to the displacement. the leading contribution is linear in $u_\alpha(\boldsymbol{L\kappa})$ as there is no net force for vanishing displacement. Within the harmonic approximation, we keep the linear contributions only and introduce the force constants as in Sect. 11.2. The case $\boldsymbol{L} = \boldsymbol{L}'$, $\boldsymbol{\kappa} = \boldsymbol{\kappa}'$ is again treated by means of infinitesimal translational invariance. A new feature is obtained from the intraband contribution to the susceptibility, which turns up in the equation of motion and yields, instead of (11.31),

$$M_{\kappa}\frac{\partial^2}{\partial t_1^2}u_\alpha(\boldsymbol{L\kappa},\bar{t}_1)+\sum_{\boldsymbol{L}'\boldsymbol{\kappa}'\beta}\Big(\Phi_{\alpha\beta}(\boldsymbol{L\kappa},\boldsymbol{L}'\boldsymbol{\kappa}')u_\beta(\boldsymbol{L}'\boldsymbol{\kappa}',\bar{t}_1)$$
$$+\int \mathrm{d}\bar{t}_2\,\Phi^{(i)}_{\alpha\beta}(\boldsymbol{L\kappa},\boldsymbol{L}'\boldsymbol{\kappa}',\bar{t}_1,\bar{t}_2)u_\beta(\boldsymbol{L}'\boldsymbol{\kappa}',\bar{t}_2)\Big)=0\,. \tag{11.149}$$

The susceptibility of the unexcited system is again treated statically, whereas the intraband contribution gives rise to a noninstantaneous force, which together with (11.25) is obtained from the following:

$$\tilde{\Phi}^{(i)}_{\alpha\beta}(\boldsymbol{L\kappa},\boldsymbol{L}'\boldsymbol{\kappa}',\bar{t}_1,\bar{t}_2)$$
$$=\frac{\varepsilon_0}{V}\int \mathrm{d}^3\boldsymbol{r}\,\mathrm{d}^3\boldsymbol{r}'\,\nabla_\alpha U_{\boldsymbol{\kappa}}(\boldsymbol{r}-\boldsymbol{L}-\boldsymbol{\kappa})\delta\chi(\boldsymbol{r},\boldsymbol{r}',\bar{t}_1,\bar{t}_2)\nabla_\beta U_{\boldsymbol{\kappa}'}(\boldsymbol{r}'-\boldsymbol{L}'-\boldsymbol{\kappa}')\,. \tag{11.150}$$

We can now proceed as before, perform the Fourier expansion (11.32) and derive the equation of motion of the phonon Green's function (11.46), which has the structure

$$\frac{\partial^2}{\partial t_1^2}D_{\alpha\boldsymbol{\kappa},\beta\boldsymbol{\kappa}'}(\boldsymbol{q},\bar{t}_1\bar{t}_2)+\sum_{\gamma\boldsymbol{\kappa}''}M_{\alpha\boldsymbol{\kappa},\gamma\boldsymbol{\kappa}''}(\boldsymbol{q})D_{\gamma\boldsymbol{\kappa}'',\beta\boldsymbol{\kappa}'}(\boldsymbol{q},\bar{t}_1\bar{t}_2)$$
$$+\sum_{\gamma\boldsymbol{\kappa}''}\int \mathrm{d}\bar{t}_3\,M^{(i)}_{\alpha\boldsymbol{\kappa},\gamma\boldsymbol{\kappa}''}(\boldsymbol{q},\bar{t}_1,\bar{t}_3)D_{\gamma\boldsymbol{\kappa}'',\beta\boldsymbol{\kappa}'}(\boldsymbol{q},\bar{t}_3\bar{t}_2)$$
$$=b_1\,\delta_{b_1,b_2}\delta(t_1-t_2)\delta_{\boldsymbol{\kappa},\boldsymbol{\kappa}'}\delta_{\alpha,\beta}\,. \tag{11.151}$$

In order to calculate $M^{(i)}$ explicitly, we make use of (10.113). Noting that P_{i} is long-ranged only, i.e. $P_{\mathrm{i}}(\boldsymbol{G}+\boldsymbol{q})$ contributes for $\boldsymbol{G}=0$ only, the intraband contribution to the dynamical matrix is obtained as

$$M^{(i)}_{\alpha\boldsymbol{\kappa},\beta\boldsymbol{\kappa}'}(\boldsymbol{q},\bar{t}_1\bar{t}_2)=\frac{\varepsilon_0 V}{\sqrt{M_{\boldsymbol{\kappa}}M_{\boldsymbol{\kappa}'}}}F_{\alpha\boldsymbol{\kappa}}(\boldsymbol{q})\chi_{\mathrm{i}}(\boldsymbol{q},\bar{t}_1\bar{t}_2)F_{\beta\boldsymbol{\kappa}'}(\boldsymbol{q})\,, \tag{11.152}$$

where χ_{i} is the solution of (10.115). Not accidentally, we have found the same coupling coefficients $F_{\alpha\boldsymbol{\kappa}}(\boldsymbol{q})$ as those introduced in the effective interaction (11.61). Thus, by means of the eigenmode expansion (11.47), we can again introduce the general electron–phonon coupling constant defined in (11.62) and obtain, finally,

$$\left(\frac{\partial^2}{\partial t_1^2}+\omega_j^2(\boldsymbol{q})\right)D_j(\boldsymbol{q},\bar{t}_1\bar{t}_2)$$
$$=2\omega_j(\boldsymbol{q})\,b_1\,\delta_{b_1,b_2}\delta_{t_1-t_2}-2\omega_j(\boldsymbol{q})M_j^2(\boldsymbol{q})\varepsilon_0 V\int \mathrm{d}\bar{t}_3\,\chi_{\mathrm{i}}(\boldsymbol{q},\bar{t}_1\bar{t}_3)\,D_j(\boldsymbol{q},\bar{t}_3\bar{t}_2). \tag{11.153}$$

Alternatively, this result may be cast into the form of a *Dyson equation* using the equation of motion of the zeroth-order Green's function (11.50):

$$D_j(\boldsymbol{q},\bar{t}_1\bar{t}_2) = D_{0j}(\boldsymbol{q},\bar{t}_1\bar{t}_2) - \varepsilon_0 V\int \mathrm{d}\bar{t}_3\mathrm{d}\bar{t}_4\, b_3\, D_{0j}(\boldsymbol{q},\bar{t}_1\bar{t}_3)\, M_j^2(\boldsymbol{q})\, \chi_\mathrm{i}(\boldsymbol{q},\bar{t}_3\bar{t}_4)\, D_j(\boldsymbol{q},\bar{t}_4\bar{t}_2), \tag{11.154}$$

in which $M_j^2\chi_\mathrm{i}$ acts as the self-energy of the phonons. As a consequence, we obtain a renormalization of the phonon dispersion and the generation of nonequilibrium phonon distributions. The latter are obtained from quantum kinetic equations which may be derived from (11.150) along similar lines to the discussion in Sect. 10.5. First, we restrict ourselves to a Markovian treatment of the electron–phonon scattering, which allows us to reduce the Dyson equation to a *Boltzmann equation* for the phonon distribution function. As (11.154) is invariant with respect to an interchange of D_j and D_{0j} in the self-energy contribution, we find the following as an equivalent representation of (11.153):

$$\left(\frac{\partial^2}{\partial t_2^2} + \omega_j^2(\boldsymbol{q})\right) D_j(\boldsymbol{q},\bar{t}_1\bar{t}_2) = 2\omega_j(\boldsymbol{q})\, b_1\, \delta_{b_1,b_2}\delta_{t_1-t_2} - 2\omega_j(\boldsymbol{q}) M_j^2(\boldsymbol{q})\varepsilon_0 V\int \mathrm{d}\bar{t}_3\, D_j(\boldsymbol{q},\bar{t}_1\bar{t}_3)\, \chi_\mathrm{i}(\boldsymbol{q},\bar{t}_3\bar{t}_2)\,. \tag{11.155}$$

In order to derive a quantum kinetic equation, we proceed in the same way as in the case of the kinetic equations (10.72) and (10.74).

11.5.2 Kinetic Equation for the Phonon Green's Function

We shall consider first the simplest case only, in which the macroscopic and microscopic timescales are separable. This allows an expansion of the scattering contributions, which to leading order, results in replacements of the form

$$\begin{aligned} D_j^{+-}\left(\boldsymbol{q}, t_3 - t_2, \frac{t_3+t_2}{2}\right) &= D_j^{+-}\left(\boldsymbol{q}, t_3 - t_2, \frac{t_1+t_2}{2} + \frac{t_3-t_1}{2}\right) \\ &\simeq D_j^{+-}\left(\boldsymbol{q}, t_3 - t_2, \frac{t_1+t_2}{2}\right), \end{aligned} \tag{11.156}$$

and correspondingly for χ_i. Making use of

$$\frac{\partial^2}{\partial t_1^2} - \frac{\partial^2}{\partial t_2^2} = 2\frac{\partial}{\partial t}\frac{\partial}{\partial \Delta t} \tag{11.157}$$

and performing Fourier transformation with respect to the difference variables, we obtain from (11.153) and (11.155) the kinetic equation

$$\frac{\partial}{\partial t} D_j^{+-}(\boldsymbol{q},\omega,t) = \frac{\omega_j(\boldsymbol{q})}{\omega} M_j^2(\boldsymbol{q})\varepsilon_0 V[D_j^{+-}(\boldsymbol{q},\omega,t)\,\chi_{\mathrm{i}}^{-+}(\boldsymbol{q},\omega,t) - \chi_{\mathrm{i}}^{+-}(\boldsymbol{q},\omega,t)\,D_j^{-+}(\boldsymbol{q},\omega,t)] \,. \tag{11.158}$$

From this result, it is a trifling step to a Boltzmann equation for the nonequilibrium phonon distribution function. We make use of (10.118), and represent the phonon Green's functions in a form corresponding to (11.52) and (11.53). The intraband polarizability is treated as in Sect. 10.6 without vertex corrections (compare (10.64) and the derivation of (10.121) and (10.128)) and is obtained, within the quasiparticle approximation, as

$$P_{\mathrm{i}}^{+-}(\boldsymbol{q},\omega,t) = \frac{2\pi \mathrm{i} e^2}{\varepsilon_0 V} \sum_{\boldsymbol{k},c=e,h} f_{\mathrm{c}}(\boldsymbol{k},t)[1 - f_{\mathrm{c}}(\boldsymbol{k}+\boldsymbol{q},t)]\,\delta[E_{\mathrm{c}}(\boldsymbol{k}+\boldsymbol{q}) - E_{\mathrm{c}}(\boldsymbol{k}) - \hbar\omega] \,, \tag{11.159}$$

Correspondingly, we obtain $P_{\mathrm{i}}^{-+}(\boldsymbol{q},\omega,t) = -P_{\mathrm{i}}^{+-}(\boldsymbol{q},-\omega,t)$. Finally, we perform the frequency integration, which yields $2N_j(\boldsymbol{q},t)$ on the left-hand side, and the Boltzmann equation takes the form

$$\frac{\partial}{\partial t} N_j(\boldsymbol{q}) = 2\pi e^2 \sum_{\boldsymbol{k},c=e,h} \frac{M_j^2(\boldsymbol{q})}{\varepsilon^{\mathrm{r}}[\boldsymbol{q},\omega_j(\boldsymbol{q})]\varepsilon^{\mathrm{a}}[\boldsymbol{q},\omega_j(\boldsymbol{q})]} \times \delta[E_{\mathrm{c}}(\boldsymbol{k}+\boldsymbol{q}) - E_{\mathrm{c}}(\boldsymbol{k}) - \hbar\omega_j(\boldsymbol{q})]\{[1+N_j(\boldsymbol{q})] \times [1 - f_{\mathrm{c}}(\boldsymbol{k})] f_{\mathrm{c}}(\boldsymbol{k}+\boldsymbol{q}) - N_j(\boldsymbol{q}) f_{\mathrm{c}}(\boldsymbol{k})[1 - f_{\mathrm{c}}(\boldsymbol{k}+\boldsymbol{q})]\} \,. \tag{11.160}$$

This result exhibits some remarkable features, which are worth discussing in some detail. Let us consider first the stationary case. This is realized if the right-hand side vanishes, and constitutes a condition of detailed balance. It is an elementary exercise to show that this condition is fulfilled if the particle distributions and the phonon distribution are given by Fermi functions and a Bose function (11.55), respectively, at the same temperature. Thus, not only any deviation of the particle distribution functions from equilibrium but also any temperature difference between the carriers and the lattice is a source of nonequilibrium phonons. In particular, this mechanism is effective if carriers are excited far above the band edge [321]. Their relaxation towards a quasi-equilibrium distribution (compare Fig. 11.5) is accompanied by the emission of phonons. On an ultrashort timescale in the range of a few hundred femtoseconds, this process is dominated by LO phonons, provided there are carriers above the emission threshold. In this way, the large excess energy of the carriers is transferred to the lattice, which becomes heated owing to generation of nonequilibrium LO phonons. The buildup of nonequilibrium LO phonons is illustrated in Fig. 11.5 by results obtained within the two-time formalism for the same excitation conditions as in Fig. 11.4.

At low lattice temperatures, the carriers may reach a quasi-equilibrium distribution, but their temperature is still higher. Their temperature is brought to the same level as that of the lattice by the generation of acoustic phonons. This process typically takes place on a timescale of several hundred picoseconds. In the course of the cooling down of the carrier distribution, the nonequilibrium phonons generated decay owing to *anharmonic interactions* [322, 323, 324]. The final state, if reached before radiative recombination, is characterized by quasi-equilibrium distributions of carriers and phonons at the same temperature. Further details will be discussed in the next chapter.

As in the case of electron–electron scattering, the Markovian theory of electron–phonon scattering is questionable. For scattering from LO phonons the situation is even worse, as the self-energy has sharp structures in the frequency domain. As higher-order contributions to the expansion in (11.156) are determined by derivatives of self-energies and Green's functions with respect to the frequency, the corrections to the leading contribution are expected to be large. Therefore a non-Markovian treatment which avoids this expansion from the very beginning is preferable. A corresponding generalization of (11.160) can be derived either within the two-time formalism or by means of a generalized Kadanoff–Baym ansatz. Here, we leave the two-time

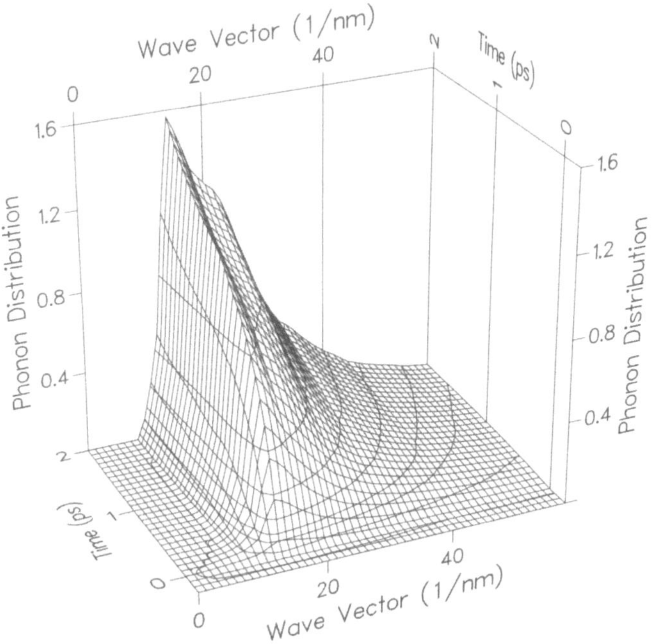

Fig. 11.5. Generation of nonequilibrium LO phonons in bulk GaAs due to relaxation of carriers. The carrier distribution is generated at 92 meV above the band edge by a 100 fs pulse, which is centered at t=0. The Rabi energy is 2 meV

formulation as an exercise for the reader and consider the latter case as an example. The intraband polarization function is calculated by means of the ansatz (10.89). Correspondingly, we represent the phonon Green's function according to

$$\begin{aligned} &D_j^{+-}(\boldsymbol{q}, t_1 t_2) \\ &= \mathrm{i}\Big(2\cos[\omega_j(\boldsymbol{q})(t_1 - t_2)][\theta(t_1 - t_2)N_j(\boldsymbol{q}, t_2) + \theta(t_2 - t_1)N_j(\boldsymbol{q}, t_1)] \\ &+ \mathrm{e}^{i\omega_j(\boldsymbol{q})(t_1 - t_2)}\Big) \end{aligned} \tag{11.161}$$

and

$$\begin{aligned} &D_j^{-+}(\boldsymbol{q}, t_1 t_2) \\ &= -\mathrm{i}\Big(2\cos[\omega_j(\boldsymbol{q})(t_1 - t_2)][\theta(t_1 - t_2)N_j(\boldsymbol{q}, t_2) + \theta(t_2 - t_1)N_j(\boldsymbol{q}, t_1)] \\ &+ \mathrm{e}^{-i\omega_j(\boldsymbol{q})(t_1 - t_2)}\Big) . \end{aligned} \tag{11.162}$$

Making use of these representations in (11.155), we perform the Δt integration which results from (11.157), and obtain as the final result

$$\begin{aligned} &\frac{\partial}{\partial t}N_j(\boldsymbol{q}, t) \\ &= \frac{2e^2}{\hbar}\sum_{\boldsymbol{k}, c=e,h}\frac{M_j^2(\boldsymbol{q})}{\varepsilon^{\mathrm{r}}[\boldsymbol{q}, \omega_j(\boldsymbol{q})]\varepsilon^{\mathrm{a}}[\boldsymbol{q}, \omega_j(\boldsymbol{q})]}\frac{2\hbar\omega_j(\boldsymbol{q})}{E_{\mathrm{c}}(\boldsymbol{k}+\boldsymbol{q}) - E_{\mathrm{c}}(\boldsymbol{k}) + \hbar\omega_j(\boldsymbol{q})} \\ &\times\int_{-\infty}^{t}\mathrm{d}t'\cos\{[E_{\mathrm{c}}(\boldsymbol{k}+\boldsymbol{q}) - E_{\mathrm{c}}(\boldsymbol{k}) - \hbar\omega_j(\boldsymbol{q})](t-t')/\hbar\} \\ &\times\{[1 + N_j(\boldsymbol{q}, t')][1 - f_{\mathrm{c}}(\boldsymbol{k}, t')]f_{\mathrm{c}}(\boldsymbol{k}+\boldsymbol{q}, t') \\ &- N_j(\boldsymbol{q}, t')f_{\mathrm{c}}(\boldsymbol{k}, t')[1 - f_{\mathrm{c}}(\boldsymbol{k}+\boldsymbol{q}, t')]\} . \end{aligned} \tag{11.163}$$

For weak coupling the results obtained from (11.163) and within the two-time formalism are nearly identical. As in the case of electron–electron scattering, the non-Markovian dynamics have the effect of reducing the scattering rates and thus of slowing down the generation of nonequilibrium phonons. In the next chapter, we shall come back to the physical implications of these *hot-phonon distributions* in more detail. However, we should not conceal the fact that there is still a blemish in the present result, as we have made use again of the Markovian solution of (10.115). This approximation, however, is of a different quality from a Markovian approximation of (11.163) and may be an excusable inconsistency, which in principle can be corrected.

Owing to (10.119), the intraband transitions of excited carriers lead to a screening of the electron–phonon coupling, This screening enters in to different ways. Repeating the derivation of (11.61) from (11.16) via (11.58), and by means of (10.119), we see that the phonon-induced interaction is screened by the square of the intraband dielectric function. Replacing the unrenormalized phonon Green's function by the renormalized one, we obtain instead of (11.61) for the effective interaction

$$w_{\text{eff}}^{\text{r}}(\boldsymbol{q},\omega) = \frac{w(\boldsymbol{q})}{\varepsilon_{\text{i}}^{\text{r}}(\boldsymbol{q},\omega)} + \frac{M_j^2(\boldsymbol{q})}{[\varepsilon_{\text{i}}^{\text{r}}(\boldsymbol{q},\omega)]^2} D_j^{\text{r}}(\boldsymbol{q},\omega) . \tag{11.164}$$

This contrasts with the screening of the electron–phonon coupling in the Dyson equation for the retarded phonon Green's function. As follows immediately from (10.117) together with the retarded solution of (11.154), this screening is only linear:

$$D_j^{\text{r}}(\boldsymbol{q},\omega) = D_{0j}^{\text{r}}(\boldsymbol{q},\omega) + \varepsilon_0 V D_{0j}^{\text{r}}(\boldsymbol{q},\omega) \frac{M_j^2(\boldsymbol{q})}{\varepsilon_{\text{i}}^{\text{r}}(\boldsymbol{q},\omega)} P_{\text{i}}^{\text{r}}(\boldsymbol{q},\omega)\, D_j^{\text{r}}(\boldsymbol{q},\omega) , \tag{11.165}$$

where $D_{0j}^{\text{r}}(\boldsymbol{q},\omega)$ is given by (11.57). The intimate coupling of the particle and phonon systems which is expressed in both of these equations can yield drastic modifications of spectral properties, which in return require further improvement of the quantum kinetic equations.

Example (Plasmon–Phonon Coupling). If the plasmon energy and the LO phonon energy are of the same order of magnitude, one expects a hybridization of the two modes in the energetic region where the level crossing occurs. The simplest version of this effect is obtained if we treat the intraband dielectric function within the plasmon-pole approximation. First, we consider the solution of (11.165). Suppressing for convenience the $\boldsymbol{q}$ and ω dependence, this can be written as

$$D_{\text{LO}}^{\text{r}} = \frac{D_{0\text{LO}}\varepsilon_{\text{i}}^{\text{r}}}{\varepsilon_{\text{i}}^{\text{r}} - \varepsilon_0 V D_{0\text{LO}}^{\text{r}} M_{\text{LO}}^2 P_{\text{i}}} . \tag{11.166}$$

Inserting this result into (11.164), we obtain the following for the effective interaction:

$$\begin{aligned} w_{\text{eff}} &= \frac{w\varepsilon_{\text{i}}^{\text{r}} + (1 - \varepsilon_0 V w P_{\text{i}}^{\text{r}}) D_{0\text{LO}}^{\text{r}} M_{\text{LO}}^2}{\varepsilon_{\text{i}}^{\text{r}}(\varepsilon_{\text{i}}^{\text{r}} - \varepsilon_0 V D_{0\text{L}} M_{\text{LO}}^2 P_{\text{i}})} \\ &= \frac{w + M_{\text{LO}}^2 D_{0\text{LO}}}{1 - \varepsilon_0 V P_{\text{i}}(w + M_{\text{LO}}^2 D_{0\text{LO}})} . \end{aligned} \tag{11.167}$$

By means of (11.79), (11.81), and $w = v/\varepsilon_\infty$, we obtain

$$1 + \frac{M_{\text{LO}}^2}{w} D_{0\text{LO}} = \frac{\omega^2 - \omega_{\text{TO}}^2}{(\omega + \text{i}\delta)^2 - \omega_{\text{LO}}^2} , \tag{11.168}$$

which allows us to write the effective dielectric function in the form

$$\varepsilon_{\text{eff}}^{\text{r}}(\boldsymbol{q},\omega) = \varepsilon_\infty \left(1 - \varepsilon_0 V w(\boldsymbol{q}) P_{\text{i}}^{\text{r}}(\boldsymbol{q},\omega) - \frac{\omega_{\text{LO}}^2 - \omega_{\text{TO}}^2}{(\omega + \text{i}\delta)^2 - \omega_{\text{TO}}^2} \right) . \tag{11.169}$$

The zeros of this dielectric function can be discussed analytically if we approximate the intraband polarization function within the plasmon-pole approximation. According to (10.110), (10.133), and (10.138) the electronic contribution is given by

$$\varepsilon_0 V w(\boldsymbol{q}) P_{\rm i}(\boldsymbol{q},\omega) = \frac{\omega_{\rm pl}^2}{(\omega+{\rm i}\delta)^2 + \omega_{\rm pl}^2 - \omega_{\boldsymbol{q}}^2}\,, \tag{11.170}$$

and we obtain the zeros from (11.169) as

$$\omega_{1,2}^2(\boldsymbol{q}) = \frac{1}{2}\left(\omega_{\rm LO}^2 + \omega_{\boldsymbol{q}}^2 \pm \left[\omega_{\rm LO}^2 - \omega_{\boldsymbol{q}}^2 + 4\,\omega_{\rm pl}^2\,(\omega_{\rm LO}^2 - \omega_{\rm TO}^2)\right]^{1/2}\right); \tag{11.171}$$

This allows us to write the effective inverse dielectric function as

$$\varepsilon_{\rm eff}^{-1}(\boldsymbol{q},\omega) = \frac{1}{\varepsilon_\infty}\left(1 + \frac{w_1(\boldsymbol{q})}{(\omega+{\rm i}\delta)^2 - \omega_1^2(\boldsymbol{q})} - \frac{w_2(\boldsymbol{q})}{(\omega+{\rm i}\delta)^2 - \omega_2^2(\boldsymbol{q})}\right), \tag{11.172}$$

where the spectral weights are

$$w_{1,2}(\boldsymbol{q}) = \frac{\omega_{\rm TO}^2(\omega_{\rm pl}^2 - \omega_{\boldsymbol{q}}^2) + \omega_1^2(\boldsymbol{q})\,\omega_2^2(\boldsymbol{q}) - \omega_{1,2}^2(\omega_{\rm pl}^2 + \omega_{\rm LO}^2 - \omega_{\rm TO}^2)}{\omega_1^2(\boldsymbol{q}) - \omega_2^2(\boldsymbol{q})}\,. \tag{11.173}$$

There are two limiting cases. If the plasma frequency is large in comparison with the frequency of LO phonons, the plasma contribution to the screening is essentially unchanged, whereas the phonons are renormalized. In the opposite limit the phonons remain unchanged and the plasma frequency is screened with $\varepsilon_{\rm eff}(0)$ rather than with ε_∞. More general results obtained by means of the RPA dielectric function of an optically generated electron–hole plasma are illustrated in Fig. 11.6. Although details are different from the treatment within the plasmon-pole approximation, the general features of the dispersion (11.173) are clearly displayed.

The effects of plasmon–phonon coupling in the quantum kinetic equations for the nonequilibrium phonon distribution are rigorously included if we solve (10.115) and (11.153) within the real-time formalism. Any further reduction of these equations, as performed in the last section, requires assumptions about the spectral properties of phonons and the intraband susceptibility. If we treat the latter within the plasmon-pole approximation and use instead of the ansatz (11.162) the renormalized phonon dispersion, we can include plasmon–phonon coupling in a single time equation of the type (11.163).

The plasmon–phonon coupling due to the renormalization of the phonon Green's function (11.154) also enters the electron–phonon self-energy and thus the quantum kinetic equations for the particles. The intuitive picture of relaxing electrons which transfer energy to the lattice by emission of phonons is no longer valid in this case. If we consider for the moment the change of the dispersion only, we find that, according to (11.171), there is a drastic

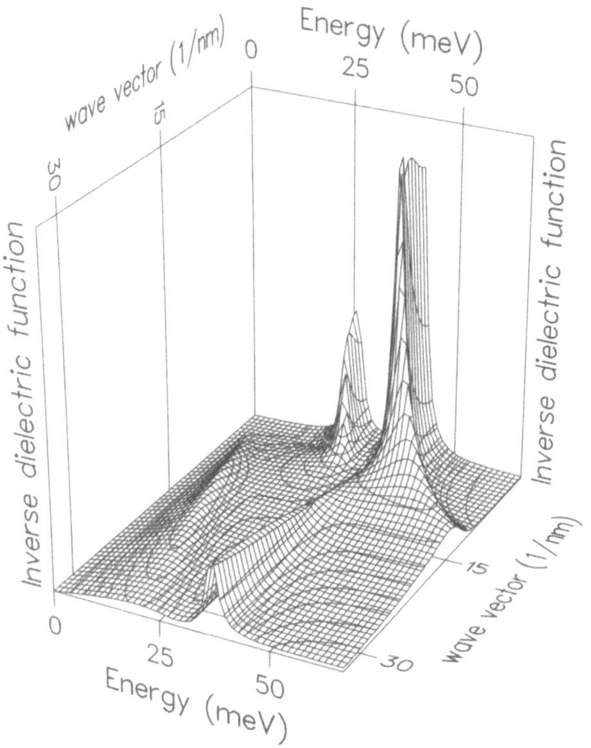

Fig. 11.6. Imaginary part of the inverse dielectric function (RPA) of an electron–hole plasma coupled to LO phonons at room temperature, and t=50 fs after excitation with a pulse of 50 fs duration and a central frequency of $\hbar\omega_0 = E_g$. The displayed structures correspond to the plasmon–phonon coupled mode and the intraband continua of electrons and holes at a density of $8 \times 10^{17}\,\mathrm{cm}^{-3}$. GaAs parameters

increase of the possible transition channels,[6] owing to the particle-like mode. Depending on the excess energy and the lattice temperature, this could result in an increase of the rate of energy transfer to the lattice by several orders of magnitude. This simple picture is, however, misleading. Not only the dispersion but also the distribution functions are renormalized owing to the mode coupling. As the particle-like mode has essentially the same excess energy as the particles, its contribution to the relaxation rate becomes nearly negligible. This is most easily seen if we assume quantum-statistical distribution functions for the phonons and the particles with different temperatures. Taking the renormalization due to the plasmon–phonon coupling into account, it is a straightforward exercise to show that the particle-like coupled mode is

[6] More formally, we have a much larger volume in $\boldsymbol{k}$ space for which the condition $E(\boldsymbol{k}+\boldsymbol{q}) - E(\boldsymbol{k}) - \hbar\omega_{1,2}(\boldsymbol{q})$ is fulfilled.

effectively at the particle temperature [329]. Thus the net effect of the mode coupling on the relaxation rates is not very impressive.

This is one of a wealth of examples to which we can apply the concepts and tools that we have developed in the last two chapters. The reward for our efforts is a unified picture, which allows the interpretation and in many cases also the quantitative analysis of a variety of phenomena in semiconductor systems. We come to this phenomenology in the next chapter.

11.6 Problems

11.1. Calculate the eigenmodes of a linear chain with next-nearest-neighbor interactions and two ions of different masses in the unit cell.

11.2. Consider a monatomic linear chain with next nearest neighbor interactions. One of the masses is replaced by a smaller one. Show that this impurity gives rise to an exponentially localized solution of the equations of motion and calculate the eigenfrequency of this mode.

11.3. Use the Hamiltonian (11.42) to investigate the thermodynamic properties of a free system of acoustic phonons within the Debye approximation [330]. The dispersion is assumed to be linear, with a cut-off frequency denoted as ω_D. Determine this frequency. Show that the chemical potential vanishes. Finally, calculate the specific heat and discuss the limiting cases $T \to \infty$ and $T \to 0$. Hint: Calculate the thermodynamic potential (see Sect. 3.2) and derive the energy according to the relation

$$E = -T^2 \frac{\partial}{\partial T}\left(\frac{\Omega}{T}\right) . \qquad (11.174)$$

11.4. Prove that the chemical potential of phonons vanishes in an interacting electron–phonon system.

11.5. Consider the phonon modes in a diatomic cubic lattice. Show that in the long wavelength limit the longitudinal eigenmodes are given by (11.73) and (11.74).

11.6. Derive the equations of motion for the transition amplitude $p_{\mathrm{vc}}(\boldsymbol{k})$ and the phonon–assisted transition amplitudes describing emission and absorption of phonons, i.e. $\langle b_{j,\boldsymbol{q}} c^\dagger_{v,\boldsymbol{k}+\boldsymbol{q}} c_{c,\boldsymbol{k}}\rangle$ and $\langle b^\dagger_{j,-\boldsymbol{q}} c^\dagger_{v,\boldsymbol{k}+\boldsymbol{q}} c_{c,\boldsymbol{k}}\rangle$, respectively. Show that for linear optical spectra, the lineshape is determined by (11.126) together with (11.127). Hint: Make use of the electron–phonon interaction Hamiltonian

$$H_{el,phon} = \sum_{\boldsymbol{q}\boldsymbol{k}j} M_j(\boldsymbol{q}) \left(b^\dagger_{j,-\boldsymbol{q}} + b_{j,\boldsymbol{q}}\right) \left(c^\dagger_{v,\boldsymbol{k}+\boldsymbol{q}} c_{v,\boldsymbol{k}} + c^\dagger_{c,\boldsymbol{k}+\boldsymbol{q}} c_{c,\boldsymbol{k}}\right) . \qquad (11.175)$$

The particle Hamiltonian and the phonon Hamiltonian are given by (9.1) and (11.42), respectively.

11.7. Prove that the scattering contribution in (11.147) conserves the particle number.

11.8. Show that the detailed-balance condition is fulfilled for (11.160), i.e. the left-hand side vanishes, if the distribution functions for particles and phonons are given by Fermi and Bose distribution functions, respectively, at equal temperatures. Calculate the energy loss rate in the case that the particles are in quasi-equilibrium but at a higher temperature than the phonon system.

11.9. Make use of (11.154) to estimate the renormalization of the dispersion and the lifetime of acoustic phonons in an excited semiconductor. Hint: Approximate the intraband susceptibility within the RPA and assume quasi-equilibrium for the particle distribution functions.

11.10. Derive the generalization of the kinetic equation (11.163) in the case that plasmon–phonon coupling is included.

12. Scattering and Screening Processes

As we have seen, the temporal evolution of electrons and holes in a semiconductor is influenced by a rich variety of possible scattering processes. In the simplest case, these scattering processes transfer energy or excitation from a part of the semiconductor – often referred to as "the system" – to the rest of the semiconductor – "the bath". This artificial distinction of the system from the bath is perfectly obvious in the example of electron–phonon scattering. In Chaps. 4 and 6, we have described these processes by simple phenomenological effective scattering rates $\gamma_1 = T_1^{-1}$ and $\gamma_2 = T_2^{-1}$ for the relaxation of the distribution functions and of the transition amplitude, respectively. This picture is – if at all – justified if a single transition such as the 1s exciton dominates. If continuum states come into play the effective scattering rates become dependent on the energy and on the wave vector. At this level, the decay of a transition amplitude or occupation is thought of as scattering *out of* a given state in terms of Fermi's golden rule:[1] for example, an electron in state $|a\rangle$ has *either* not emitted a phonon *or* has emitted a phonon and ends in state $|b\rangle$. The rates γ_1 and γ_2 are proportional to the transition probability from $|a\rangle$ to $|b\rangle$. Obviously, this simple picture fails – in particular, in the case of carrier–carrier scattering – if the scattering rates are dependent on the occupation of the final states,

An improvement takes into account the scattering back *into* a certain state also. If we insist as before on Fermi's golden rule, the dynamics of the system are Markovian, i.e. at any time t, the dynamics are given by the state of the system at that same time t and, again, the energy is conserved in an individual scattering event. This is the level of ordinary *Boltzmann kinetics*, which describe the dynamical effects of *collisions*, i.e. how particles are transferred from one configuration in $\boldsymbol{k}$ space to another. The Markovian approximation, however, neglects not only the history of the time evolution of the system, but also the renormalization of the spectral properties, i.e. the time and energy dependent damping and shift of quasiparticle states.[2] These *kinetic* effects require a non-Markovian approach, which can be elaborated on

[1] This implies the limit $t \to \infty$, in which a single scattering event conserves energy. Without this limit the uncertainty relation prevents energy conservation and arbitrary states can contribute to the final states.

[2] Note that any consistent inclusion of finite lifetime effects requires a non-Markovian treatment

various levels of sophistication. A rigorous description can only be achieved within the two-time formalism (see Sect. 10.5), which guarantees that all interaction effects are treated in a consistent manner. Despite its shortcomings, the reduction to single-time propagators, discussed in Sect. 10.5.2, represents in many cases the only practicable and physically meaningful way to account for the non-Markovian effects. The theory is often referred to as *quantum kinetics* at this level.

Intuitively, memory effects imply that the scattering process is not complete. In terms of our example above, this means the electron can be in a superposition of states $|a\rangle$ and $|b\rangle$ and the scattering is not irreversible. However, any additional damping mechanism which leads to a finite lifetime of quasiparticle states and thus limits the range of the memory in the quantum kinetic equations results in *irreversibility* of the scattering processes. Thus, in simple terms, one has two characteristic timescales which are involved: the scattering time and the duration of the interaction process (equivalent to the memory time in the language of non-Markovian relaxation). This is the main difference between non-Markovian kinetic equations and the Markovian Boltzmann kinetics. The corresponding underlying theory has already been discussed in Chaps. 10 and 11.

The rapid development of short-pulse lasers has motivated numerous experimental studies and has led to a wealth of information on scattering processes in semiconductors. An understanding of these processes is important for understanding optical and transport phenomena in semiconductors, and hence for any type of semiconductor device. There are only a few experimental situations in which a description in terms of relaxation rates is really meaningful. Most experiments require an analysis based on quantum kinetics to describe carrier–phonon scattering and carrier–carrier scattering properly. This concerns not only the quantitative understanding but also qualitative features. In doped or dirty materials, carrier–impurity scattering can also become important; in disordered semiconductors the dynamics are greatly influenced by the disorder potential.

12.1 Carrier–Phonon Scattering

Interestingly, one of the early experiments on ultrafast carrier dynamics was performed using excitation with long (5 ps) pulses and *time-integrated photoluminescence* detection [331]. Figure 12.1 shows the underlying idea.

A GaAs crystal was doped with an acceptor material (Mg at $7\times10^{16}\,\mathrm{cm}^{-3}$) at a rather low concentration, which introduces a sharp acceptor energy level, indicated by the dashed line[3] in Fig. 12.1. The resulting optical transitions from the conduction band to the acceptor level represent exclusively

[3] Such a picture has to be used with some caution. Strictly speaking, the translational invariance of the crystal is broken because of the randomly distributed doping atoms, and hence the wave vector is no longer a good quantum number.

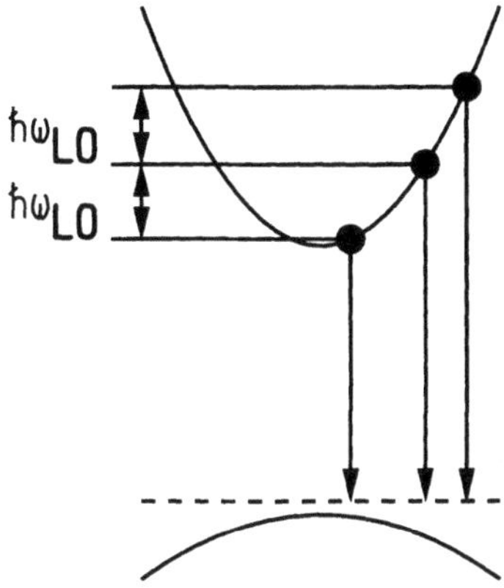

Fig. 12.1. Schematic illustration of optical transitions from the conduction band to the acceptor level in a p-doped semiconductor

the electron dynamics without any (direct) influence of the hole dynamics. The crystal was excited far above the band gap and the photoluminescence was detected for various excitation densities. Figure 12.2 shows the resulting spectra.

For low excitation density, one sees a cascade of LO phonon replicas as evidence for the dominance of electron–phonon scattering. As a rule of thumb, each such process takes a time of about 200 fs, and hence can be considered as an internal clock [331]. For higher excitation densities, the influence of electron–electron scattering increases and the phonon replicas are smeared out. It has been estimated, from the data, that electron–LO-phonon and electron–electron scattering in GaAs are of comparable strength at a density of $8 \times 10^{16}\,\text{cm}^{-3}$.[4]

The buildup of the phonon replicas in the luminescence reflects directly the phonon sidebands in the electron distribution function (compare Fig. 11.4). In order to understand this fundamental relationship between the measured luminescence signal and the microscopic properties of the sample in some detail, a detour to the calculation of luminescence signals is required.

12.1.1 Luminescence Spectra

The observed luminescence results from the radiative recombination of electron–hole pairs. Considering the detector as a photon counter, we have to consider the incoming photons per unit time, i.e. the photon flux in the stationary case. Thus we have

$$I_{\text{Lum}}(\boldsymbol{q},t) \propto \frac{\partial}{\partial t}\langle a^\dagger_{\boldsymbol{q}}(t) a_{\boldsymbol{q}}(t)\rangle = \frac{\partial}{\partial t} n_{\text{p}}(\boldsymbol{q},t)\,, \tag{12.1}$$

The picture becomes meaningful if we neglect any immediate influence of the doping on the dispersion of the bands and represent the spatially local states of the doping atoms as a superposition of many states in wave vector space.

[4] This estimate relates to the screened electron–phonon coupling and not the low-density coupling (see Sect. 11.4).

where the dispersion relation is $|\boldsymbol{q}| = \omega/c$ and the photon density is $n_p(\boldsymbol{q}, t)$ (compare Sect. 2.4), which is obtained from the photon Green's function. The retarded photon Green's function has already been introduced in Sect. 2.6 (see (2.194)). In the nonequilibrium case, we define the generalization according to

$$\underline{D}(\bar{1}\bar{2}) = -\frac{\mathrm{i}}{\hbar}[\langle \boldsymbol{A}(\bar{1})\boldsymbol{A}(\bar{2})\rangle - \langle \boldsymbol{A}(\bar{1})\rangle\langle \boldsymbol{A}(\bar{2})\rangle] . \tag{12.2}$$

The dynamics are determined by the interaction Hamiltonian (2.105) and yield a Dyson equation, which has the same general structure as in the classical case (2.193) (see also Problems 12.1 and 12.3):

$$\underline{D}(\bar{1}\bar{2}) = \underline{D}_0(\bar{1}\bar{2}) + \frac{1}{\mu_0 V} \int \mathrm{d}\bar{3}\bar{4}\, \underline{D}_0(\bar{1}\bar{3})\, \underline{P}^{\mathrm{t}}(\bar{3}\bar{4})\, \underline{D}(\bar{4}\bar{2}) . \tag{12.3}$$

Making use of the relation (2.85), we can transform (12.2) into a representation in $\boldsymbol{q}$ space,

$$\underline{D}(\boldsymbol{q}, \bar{t}_1\bar{t}_2) = -\mathrm{i}b_2 \sum_{\lambda\lambda'} \boldsymbol{e}_{\lambda\boldsymbol{q}}\boldsymbol{e}_{\lambda'\boldsymbol{q}} \langle T_c[a_{\boldsymbol{q}\lambda}(\bar{t}_1)\, a^\dagger_{\boldsymbol{q}\lambda'}(\bar{t}_2) + a^\dagger_{-\boldsymbol{q}\lambda}(\bar{t}_1)\, a_{-\boldsymbol{q}\lambda'}(\bar{t}_2)]\rangle , \tag{12.4}$$

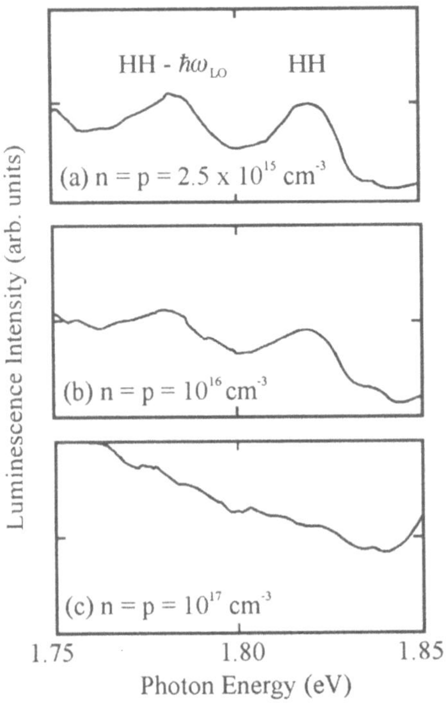

Fig. 12.2a–c. Temporally integrated photoluminescence from conduction band to acceptor transitions in p-doped GaAs. Taken from [331]

where we have omitted coherent contributions such as $\langle a^{\dagger}_{q\lambda}(\bar{t}_1)\rangle\langle a_{q\lambda'}(\bar{t}_2)\rangle$. For simplicity, we assume cubic symmetry, for which the transverse polarization propagator and, in turn, the photon Green's function can be reduced to scalar quantities $P^{\mathrm{t}}(\boldsymbol{q},t,t')$ and $\mathcal{D}(\boldsymbol{q},t,t')$. It is therefore sufficient to consider a single transverse mode. Omitting, correspondingly, the dependence on the mode index λ, (12.3) can be cast into the form

$$\begin{aligned}
&\mathcal{D}(\boldsymbol{q},\bar{t}_1\bar{t}_2)\\
&= \mathcal{D}_0(\boldsymbol{q},\bar{t}_1\bar{t}_2) - \int \mathrm{d}\bar{t}_3\mathrm{d}\bar{t}_4\, b_3\, \mathcal{D}_0(\boldsymbol{q},\bar{t}_1\bar{t}_3)\, P^{\mathrm{t}}(\boldsymbol{q},\bar{t}_3\,\bar{t}_4)\, \mathcal{D}(\boldsymbol{q},\bar{t}_4\bar{t}_2)\,.
\end{aligned} \tag{12.5}$$

The four zeroth-order Green's functions are formally the same as in (11.51)–(11.54), but with the replacements $N_{0j}(\boldsymbol{q}) \to n_{\mathrm{p}}(\boldsymbol{q})$ and $\omega_j(\boldsymbol{q}) \to \omega_{\boldsymbol{q}}$ and a prefactor $c_0^2/(2\omega_{\boldsymbol{q}})$. Indeed, from a formal point of view, the electron–phonon problem and the electron–photon problem are equivalent. The main difference results from the explicit form of the propagators and the coupling. The further evaluation of the Dyson equation is the same as in the derivation of (11.160) in the Boltzmann limit or (11.163) in the non-Markovian case. Making use of the long-wavelength limit for P^{t}, the transverse response can be expressed by means of the longitudinal response according to (2.199). In order to take the Coulomb interaction properly into account, the interband response function has to be calculated as the solution of a Bethe–Salpeter equation, at least within the screened Hartree–Fock approximation (see Sects. 10.6.1 and 5.2.1). Owing to the time evolution of the one-particle distribution functions and the screened interaction, this is a complicated dynamical problem. At low excitation densities, these dynamical aspects are of minor importance, and the problem can be considerablly simplified, without neglecting the Coulomb correlation completely [332]. If we consider processes sufficiently far above the absorption edge, the Coulomb correlation between electrons and holes leads merely to an enhancement, which depends only weakly on energy. In this case, we can treat the polarization propagator within the RPA by means of the interband contribution to (10.120) for $\boldsymbol{q} \to 0$. Further, we restrict ourselves in the following to the Markovian approximation. As a result, we obtain

$$\begin{aligned}
\hbar\frac{\partial}{\partial t}n_{\mathrm{p}}(\boldsymbol{q}) &= \frac{\pi\,\hbar\omega_{\boldsymbol{q}}\,d_{\mathrm{vc}}^2}{\varepsilon_0 V}\sum_{\boldsymbol{k}}\delta[E_{\mathrm{c}}(\boldsymbol{k}) - E_{\mathrm{v}}(\boldsymbol{k}) - \hbar\omega_{\boldsymbol{q}}]\\
&\times\{[1 - f_{\mathrm{v}}(\boldsymbol{k})]f_{\mathrm{c}}(\boldsymbol{k}) + n_{\mathrm{p}}(\boldsymbol{q})[f_{\mathrm{c}}(\boldsymbol{k}) - f_{\mathrm{v}}(\boldsymbol{k})]\}\,.
\end{aligned} \tag{12.6}$$

Performing the wave vector integration and switching to the electron–hole picture, we obtain

$$\hbar\frac{\partial}{\partial t}n_{\mathrm{p}}(\boldsymbol{q})$$
$$= \frac{2\,d_{\mathrm{vc}}^2}{\pi\varepsilon_0}\left(\frac{m_{\mathrm{r}}\hbar\omega_{\boldsymbol{q}}}{2\hbar^2}\right)^{3/2}\{f_{\mathrm{h}}(x)f_{\mathrm{e}}(x) - n_{\mathrm{p}}(\boldsymbol{q})[1 - f_{\mathrm{e}}(x) - f_{\mathrm{h}}(x)]\}\,, \quad (12.7)$$

where we have used the abbreviation $x = \sqrt{2m_{\mathrm{r}}\hbar\omega_{\boldsymbol{q}}}$. If the final states are not hole but impurity states as in the experiments discussed above, we have to replace the reduced mass by the electron mass. The simple relationship between the luminescence signal and the one-particle distribution function (12.7) has to be improved to obtain a more quantitative description. Unless we are very far above the band edge, the inclusion of Coulomb correlation between electrons and holes is required, as well as a non-Markovian treatment. An alternative method to the Greens's function treatment is a density-matrix approach, which introduces *photon-assisted transition amplitudes* [333]. Formally, this method is again closely related to the corresponding treatment of the electron–phonon problem (compare, for example, (11.126)).

The time evolution of the one-particle distribution function is determined by scattering within the particle system, scattering from phonons, and the radiative recombination itself. All processes which evolve faster in time than the radiative recombination leave their traces in the luminescence signal. On the time scale of LO phonon scattering, the change of the electron distribution due to radiative recombination is very small. Thus the time evolution, as displayed, for example, in Fig. 11.4, is directly reflected in the luminescence signal. Clearly, an instantaneous relationship is obtained only within the Markov approximation.

One tends to think of an electron–phonon scattering event as a process in which a phonon is absorbed or emitted (randomly) at some instant in time, the particle changes its state with a certain transition probability into all energetically allowed states (see (11.147), and the process is then complete. Such an intuitive picture corresponds precisely to the Markov approximation, combined with a view through a quick-motion camera, for which the continuous buildup of phonon sidebands in the one-particle distribution occurs instantaneously. When we consider the scattering over short timeintervals, however, the energy is not conserved in an elementary scattering event. All electronic transitions, for which $[E_{\mathrm{c}}(\boldsymbol{k}+\boldsymbol{q}) - E_{\mathrm{c}}(\boldsymbol{k}) - \hbar\omega_{\mathrm{LO}}]\,t > \hbar$ is fulfilled can contribute according to (11.163). This is a manifestation of the uncertainty principle and leads, in turn, to memory effects in the scattering processes. Only after the system has executed many oscillations do those transitions for which energy is not conserved, vanish owing to interference effects. In a very pictorial fashion, one can think of memory effects as an oscillation of an electron back and forth between some initial state with wave vector $\boldsymbol{k}$ and some final state with wave vector $\boldsymbol{k}+\boldsymbol{q}$. The states which finally dominate the optical transition are depicted in Fig. 12.3.

A corresponding experimental result [334] is shown in Fig. 12.4. Here the full circles are simple fits, delivering an oscillation period of 100 fs. As illu-

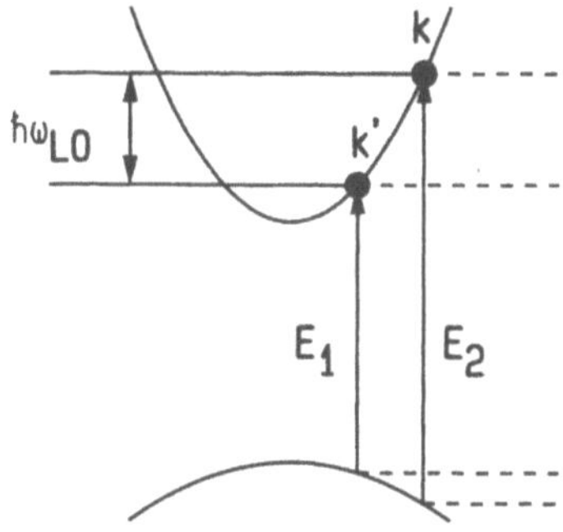

Fig. 12.3. Schematic illustration of phonon scattering in the conduction band and the resulting optical transition energies E_1 and E_2.

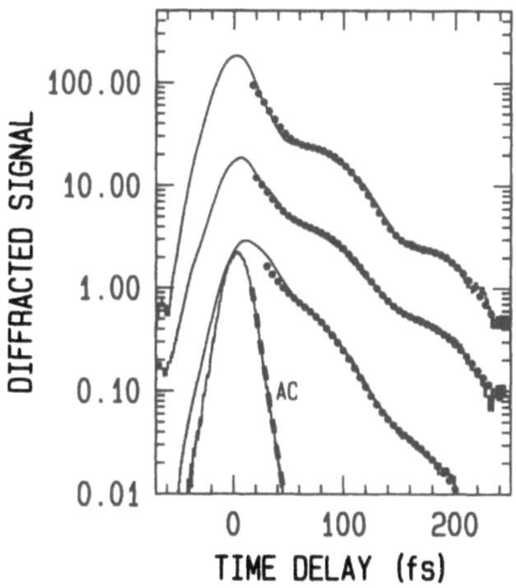

Fig. 12.4. Transient four-wave-mixing signal as a function of time delay. A layer of bulk GaAs at 77 K has been excited with 14 fs pulses, resonant with the band edge. The carrier density increases from *top* to *bottom*: $1.2 \times 10^{16}\,\mathrm{cm}^{-3}$, $1.9 \times 10^{16}\,\mathrm{cm}^{-3}$, $6.3 \times 10^{16}\,\mathrm{cm}^{-3}$. The curve labeled AC is the autocorrelation of the laser pulse. After [334]

strated in Fig. 12.3, the energy difference in the conduction band (the LO phonon energy $\hbar\omega_{\mathrm{LO}}$) translates, according to (12.7), into a larger energy difference in the optical transitions, and hence to a faster oscillation, owing to the finite curvature of the valence band. With elementary algebra, one immediately obtains $E_2 - E_1 = (1 + m_{\mathrm{e}}/m_{\mathrm{h}})\,\hbar\omega_{\mathrm{LO}}$.

Owing to the non-Markovian kinetics the threshold for LO phonon scattering is softened. Even though the system is excited below the one-LO phonon threshold, the absence of energy conservation brings about ultrafast relaxation. This channel becomes effective in the intermediate coupling regime ($\alpha \simeq 1$), and has been demonstrated by means of spectrally resolved transmission for the case of heavy holes in CdTe. The experimental results can be reproduced by two-time quantum kinetic calculations [335].

12.1.2 Four-Wave-Mixing Experiments

Four-wave-mixing experiments on an ultrashort timescale allow a temporal resolution of the oscillations that characterize the scattering by LO phonons.

Exciting the band gap of GaAs resonantly at low temperature and moderate density using optical pulses 14 fs in duration delivers signals (see Fig. 12.4) which exhibit such oscillations. Let us consider the origin of these memory effects in some detail.

Example (Memory Effects in Four-Wave-Mixing Signals). In what follows, we shall discuss the basic mechnism which gives rise to non-Markovian signatures in the diffracted signal. Although quantitative results require the inclusion of Coulomb correlation, the present excitation conditions allow us to neglect the electron–hole attraction without wiping out the principal effect. Starting from (11.127), we have

$$\Big(\mathrm{i}\hbar\frac{\partial}{\partial t} - E_{\mathrm{c}}(\boldsymbol{k}) + E_{\mathrm{v}}(\boldsymbol{k}) + \mathrm{i}\hbar\gamma_2\Big)p_{\mathrm{vc}}(\boldsymbol{k},t) = \boldsymbol{d}_{\mathrm{cv}}\boldsymbol{E}(t)[f_{\mathrm{v}}(\boldsymbol{k}) - f_{\mathrm{c}}(\boldsymbol{k})]$$
$$+\hbar e^2\sum_{\boldsymbol{q}}\Big\{M_{\mathrm{LO}}(\boldsymbol{q})N_{0\mathrm{LO}}[P^{+}_{\mathrm{vc}}(\boldsymbol{k},\boldsymbol{k}+\boldsymbol{q},t) - P^{+}_{\mathrm{vc}}(\boldsymbol{k}-\boldsymbol{q},\boldsymbol{k},t)]\Big\}$$
$$+\left\{N_{0\mathrm{LO}} \to 1 + N_{0\mathrm{LO}},\, \omega_{\mathrm{LO}} \to -\omega_{\mathrm{LO}}\right\}\Big)\,, \tag{12.8}$$

where the phonon-assisted transitions result from

$$\Big(\mathrm{i}\hbar\frac{\partial}{\partial t} - E_{\mathrm{c}}(\boldsymbol{k}+\boldsymbol{q}) + E_{\mathrm{v}}(\boldsymbol{k}) + \hbar\omega_{\mathrm{LO}} + \mathrm{i}\hbar\gamma_2\Big)P^{+}_{\mathrm{vc}}(\boldsymbol{k},\boldsymbol{k}+\boldsymbol{q},t)$$
$$= p_{\mathrm{vc}}(\boldsymbol{k},t) - p_{\mathrm{vc}}(\boldsymbol{k}+\boldsymbol{q},t) \tag{12.9}$$

and γ_2 accounts for all other dephasing processes. We consider the same four-wave-mixing geometry described in Sect. 4.3, which yields the following contribution in the $2\boldsymbol{q}_2 - \boldsymbol{q}_1$ direction for $T \geq 0$:[5]

$$\tilde{p}^{(3)}_{\mathrm{vc}}(\boldsymbol{k},t)$$
$$= -\mathrm{i}d_{\mathrm{cv}}|d_{\mathrm{cv}}|^2\hbar^{-3}\tilde{\boldsymbol{E}}_2^2\tilde{\boldsymbol{E}}_1\mathrm{e}^{\mathrm{i}(2\boldsymbol{q}_2-\boldsymbol{q}_1)\boldsymbol{r}}\Theta(T)\Theta(t-T)\mathrm{e}^{-\gamma_2 t}\,\mathrm{e}^{-\mathrm{i}\Delta(\boldsymbol{k})(t-2T)}$$
$$-\mathrm{i}e^2\sum_{\boldsymbol{q}}\Bigg[\Bigg(M^2_{\mathrm{LO}}(\boldsymbol{q})\,N_{0\mathrm{LO}}\int_{-\infty}^{t}\mathrm{d}t'\,\mathrm{e}^{-\mathrm{i}[\Delta(\boldsymbol{k})-\mathrm{i}\gamma_2](t-t')}$$
$$\times[P^{+(3)}_{\mathrm{vc}}(\boldsymbol{k},\boldsymbol{k}+\boldsymbol{q},t') - P^{+(3)}_{\mathrm{vc}}(\boldsymbol{k}-\boldsymbol{q},\boldsymbol{k},t')]\Bigg)$$
$$+\Big(N_{0\mathrm{LO}} \to 1 + N_{0\mathrm{LO}},\, \omega_{\mathrm{LO}} \to -\omega_{\mathrm{LO}}\Big)\Bigg]\,, \tag{12.10}$$

where the detuning is $\hbar\Delta(\boldsymbol{k}) = E_{\mathrm{c}}(\boldsymbol{k}) - E_{\mathrm{v}}(\boldsymbol{k}) - \hbar\omega_0$. In the following, we assume weak coupling and treat the phonon-assisted transitions perturbationally, i.e. the source in (12.9) is approximated by the first contribution on the right-hand side of (12.10). Thus we obtain from the integration of (12.9)

[5] Note that within the present approximation, the phonon contribution to the one-particle distributions in (12.8) yields no diffracted signal and only the free distributions contribute.

$$P_{\mathrm{vc}}^{+(3)}(\boldsymbol{k},\boldsymbol{k}+\boldsymbol{q},t) = \mathrm{i}d_{\mathrm{cv}}|d_{\mathrm{cv}}|^2\hbar^{-4}\tilde{\boldsymbol{E}}_2^2\tilde{\boldsymbol{E}}_1\mathrm{e}^{\mathrm{i}(2\boldsymbol{q}_2-\boldsymbol{q}_1)\boldsymbol{r}}\Theta(T)\Theta(t-T)$$

$$\times\mathrm{e}^{-\mathrm{i}[\Delta(\boldsymbol{k})-\mathrm{i}\gamma_2]t}\,\mathrm{e}^{-\mathrm{i}\Delta_{\mathrm{c}}(\boldsymbol{k}+\boldsymbol{q},\boldsymbol{k})t}\left(\frac{1}{\Delta_{\mathrm{c}}(\boldsymbol{k}+\boldsymbol{q},\boldsymbol{k})}\left(\mathrm{e}^{\mathrm{i}\Delta_{\mathrm{c}}(\boldsymbol{k}+\boldsymbol{q},\boldsymbol{k})t}-\mathrm{e}^{\mathrm{i}\Delta_{\mathrm{c}}(\boldsymbol{k}+\boldsymbol{q},\boldsymbol{k})T}\right)\right.$$

$$\left.-\frac{1}{\Delta_{\mathrm{v}}(\boldsymbol{k}+\boldsymbol{q},\boldsymbol{k})}\left(\mathrm{e}^{\mathrm{i}\Delta_{\mathrm{v}}(\boldsymbol{k}+\boldsymbol{q},\boldsymbol{k})t}-\mathrm{e}^{\mathrm{i}\Delta_{\mathrm{v}}(\boldsymbol{k}+\boldsymbol{q},\boldsymbol{k})T}\right)\right)\,, \tag{12.11}$$

where we have defined $\Delta_{\mathrm{a}}(\boldsymbol{k}+\boldsymbol{q},\boldsymbol{k}) = E_a(\boldsymbol{k}+\boldsymbol{q}) - E_a(\boldsymbol{k}) - \hbar\omega_{\mathrm{LO}}$. What remains is the calculation of the time-integrated four-wave-mixing signal, which is given within the RWA by

$$\mathcal{S}(T) = \frac{2|d_{\mathrm{vc}}|^2}{V^2}\sum_{\boldsymbol{k}\,\boldsymbol{k}'}\int_{-\infty}^{\infty}\mathrm{d}t\,p_{\mathrm{vc}}^{(3)}(\boldsymbol{k},t)p_{\mathrm{vc}}^{(3)\,*}(\boldsymbol{k}',t)$$

$$= \left(|d_{\mathrm{cv}}|^4\hbar^{-3}\tilde{\boldsymbol{E}}_2^2\tilde{\boldsymbol{E}}_1\right)^2\frac{1}{V^2}\sum_{\boldsymbol{k}\,\boldsymbol{k}'}[S_0(\boldsymbol{k},\boldsymbol{k}') + S_{\mathrm{phon}}(\boldsymbol{k},\boldsymbol{k}')]\,. \tag{12.12}$$

The contributions due to free transitions are given by

$$S_0(\boldsymbol{k},\boldsymbol{k}') = \frac{\mathrm{i}\mathrm{e}^{\mathrm{i}[\Delta_{\mathrm{r}}(\boldsymbol{k},\boldsymbol{k}')+2\mathrm{i}\gamma_2]T}}{\Delta_{\mathrm{r}}(\boldsymbol{k},\boldsymbol{k}')-2\mathrm{i}\gamma_2}\,, \tag{12.13}$$

where we have used the abbreviation

$$\Delta_{\mathrm{r}}(\boldsymbol{k},\boldsymbol{k}') = \frac{\hbar^2(\boldsymbol{k}^2-\boldsymbol{k}'^2)}{2m_{\mathrm{r}}}\,. \tag{12.14}$$

In accordance with our perturbational treatment, we restrict ourselves to the linear contribution from the phonon-assisted transitions and obtain, by means of (12.10) and (12.11)

$$S_{\mathrm{phon}}(\boldsymbol{k},\boldsymbol{k}') = \frac{\mathrm{i}e^2}{\hbar}\sum_{\boldsymbol{q}}\left\{\left[M_{\mathrm{LO}}^2(\boldsymbol{q})\,N_{0\mathrm{LO}}\right.\right.$$

$$\times\frac{1}{[\Delta_{\mathrm{c}}(\boldsymbol{k}+\boldsymbol{q},\boldsymbol{k}')-\Delta_{\mathrm{v}}(\boldsymbol{k},\boldsymbol{k}')-\omega_{\mathrm{LO}}-2\mathrm{i}\gamma_2][\Delta_{\mathrm{r}}(\boldsymbol{k},\boldsymbol{k}')-2\mathrm{i}\gamma_2]}$$

$$\left.\times\left(\frac{\mathrm{e}^{\mathrm{i}[\Delta_{\mathrm{r}}(\boldsymbol{k}+\boldsymbol{q},\boldsymbol{k}')+2\mathrm{i}\gamma_2]T}}{\Delta_{\mathrm{r}}(\boldsymbol{k}+\boldsymbol{q},\boldsymbol{k}')-2\mathrm{i}\gamma_2}-\frac{\mathrm{e}^{\mathrm{i}[\Delta_{\mathrm{r}}(\boldsymbol{k},\boldsymbol{k}')+2\mathrm{i}\gamma_2]T}}{\Delta_{\mathrm{r}}(\boldsymbol{k},\boldsymbol{k}')-2\mathrm{i}\gamma_2}\right)\right]$$

$$\left.+\left[N_{0\mathrm{LO}}\to 1+N_{0\mathrm{LO}},\,\omega_{\mathrm{LO}}\to-\omega_{\mathrm{LO}}\right]\right\}+\left\{\Delta_{\mathrm{c}}\rightleftharpoons-\Delta_{\mathrm{v}}\right\}+\mathrm{c.c.} \tag{12.15}$$

In the limit $\gamma_2 \to 0$, we can decompose the energy denominators into principal values and δ functions. A detailed inspection of the various terms shows that only the combination of the pole contributions from the first and second energy denominators on the right-hand side of (12.15) gives rise to an oscillatory behavior in the time-integrated signal. After integration of both δ functions, we obtain

$$\sum_{\boldsymbol{k},\boldsymbol{k}'} S_{\text{phon}}(\boldsymbol{k},\boldsymbol{k}') \propto -\,\mathrm{e}^{-2\gamma_2 T} \sin\left(\frac{m_\mathrm{e}}{m_\mathrm{r}}\,\omega_\mathrm{LO}\,T\right) + \dots\,, \tag{12.16}$$

where the dots indicate all other contributions. These oscillations, with frequency $m_\mathrm{e}/m_\mathrm{r}\omega_\mathrm{LO} = (1 + m_\mathrm{e}/m_\mathrm{h})\,\omega_\mathrm{LO}$ (see Fig. 12.3), modulate the decay of the signal with an amplitude depending on the excitation intensity and the electron–phonon coupling. They are an immediate consequence of memory effects and vanish if we treat (12.10) within the Markovian approximation. Beyond our present linear treatment of the phonon-assisted transitions, similar effects arise if we also include the density dependent contributions in the source of (12.9) (compare (11.120). These give rise to oscillations with twice the LO phonon frequency, in a similar manner to the explicit inclusion of two-phonon processes. Oscillations of this type have been observed experimentally in ZnSe [336].

12.1.3 Nonequilibrium Phonons

While carriers can rapidly thermalize among themselves via carrier–carrier scattering, this does not necessarily mean that they are in thermal equilibrium with the rest of the semiconductor crystal. Even though LO phonon scattering does represent a fast transfer of energy to the lattice, the resulting quantum-statistical distribution functions of the carriers are usually not in thermal equilibrium with the lattice (compare the time evolution of the electron distribution function in Fig. 11.4) As was shown in Sect. 11.5, any remaining difference between the lattice and carrier temperatures gives rise to subsequent energy relaxation to the lattice temperature, often referred to as *carrier cooling.* This cooling is brought about by the emission of nonequilibrium phonons (see (11.160) and (11.163)), and takes place on a timescale of tens of picoseconds at low temperatures (several picoseconds at room temperature). By employing a lineshape analysis of the high-energy tail of time-resolved photoluminescence spectra [337], one can derive the carrier temperature as a function of the time after impulsive excitation, an example of which is depicted in Fig. 12.5.

The cooling times become longer at higher excitation densities owing to two effects, screening of the Fröhlich interaction, and so-called *hot-phonons.* As large amounts of energy are transferred from the carrier system to the LO phonons, the latter can no longer be considered as a thermal bath and heatup, in turn, reabsorption of these nonequilibrium phonons transfers energy back to the charge carriers. Clearly, not all LO phonon states are involved, as only a small fraction of the entire Brillouin zone can couple to the electronic system (see Fig. 11.5). LO phonons subsequently relax via interaction with acoustic phonons on a timescale of several picoseconds prior to thermalization among LO phonons themselves. This hot-phonon scenario clearly slows down the cooling of the carriers [337]. In Chap. 13 we shall see that these nonequilibrium phonon effects directly influence the dynamics of light emission from semiconductor lasers. The cooling of carriers in quantum wells shows little difference from the bulk case [337], even though there has been some discussion about the influence of confined phonon modes. Carrier cooling has also been investigated in II–VI semiconductors [338] and in the nonpolar semiconductor germanium [339].

In indirect-gap materials, phonon scattering (via deformation potential coupling) can efficiently ($\sim$ 200 fs in germanium) transfer electrons from the center of the Brillouin zone to one of the lower energy side valleys [340]. This scattering process is comparably fast in direct gap materials at larger energies [341]. Interestingly, the reverse process is much slower [342]. Ultrafast intervalley scattering has also been investigated in type II quantum well structures [343].

12.2 Carrier–Carrier Scattering

Clearly, electron–LO-phonon scattering is inhibited for electron energies well below $E_c(\boldsymbol{k} = 0) + \hbar\omega_{LO}$. The remaining dominant channel is electron–electron scattering, which depends sensitively on the carrier density. One example using low carrier densities ($4 \times 10^{13}\,\mathrm{cm}^{-3}$), again employing p-doped

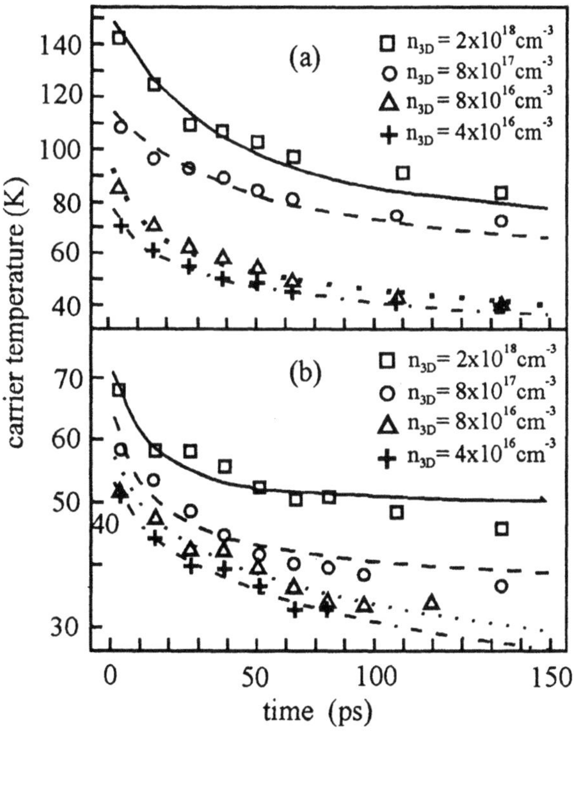

Fig. 12.5a,b. Carrier temperature as a function of time after impulsive excitation at t = 0 for a GaAs/AlGaAs quantum well sample. The data points were obtained from the analysis of transient photoluminescence spectra. Taken from [337]

GaAs (for the same reasons as discussed above), is depicted in Fig. 12.6 [344]. As we have seen, the time-resolved luminescence spectra directly reflect the change with time of the electron occupations functions. Initially, the occupation functions are nonthermal owing to the impulsive excitation with picosecond pulses, and they thermalize only on a timescale of several tens of picoseconds.

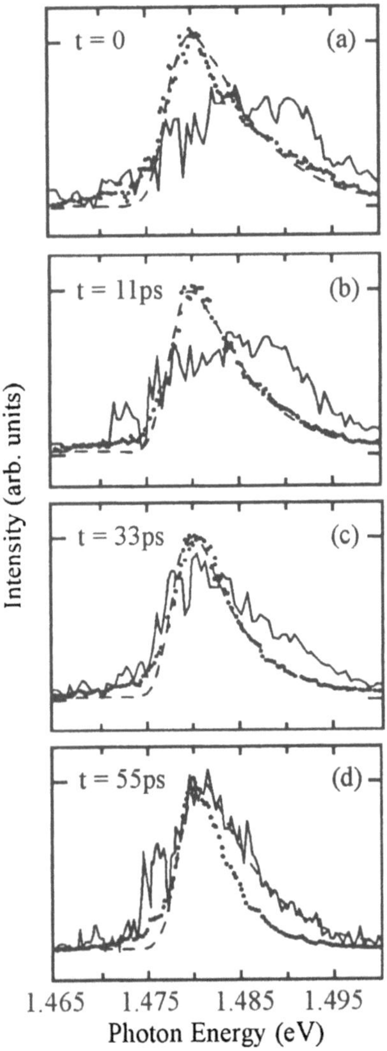

Fig. 12.6a–d. Time-resolved photoluminescence due to conduction-band-to-acceptor transitions in p-doped bulk GaAs. The excitation is below the threshold for LO phonon emission. Notice the slow relaxation of the initial, nonthermal electron distribution, which is directly proportional to the photoluminescence intensity. Taken from [344]

In Fig. 12.7 the result of another experiment, a femtosecond pump–probe experiment (see Sect. 4.3) on GaAs quantum wells, is shown [345]. The pump pulse excites electron–hole pairs which inhibit further transitions into occupied states when the probe pulse is applied. In the simple one-particle picture, the resulting differential transmission is directly proportional to the induced change in absorption, and hence proportional to the sum of the electron and hole occupation functions ($\sim (f_e + f_h)$), as the density of states is constant in two dimensions. Obviously, the initial spectral hole (Fig. 12.7) disappears on a timescale of about 100 fs – much faster than in Fig. 12.6 – owing to substantial carrier–carrier scattering.

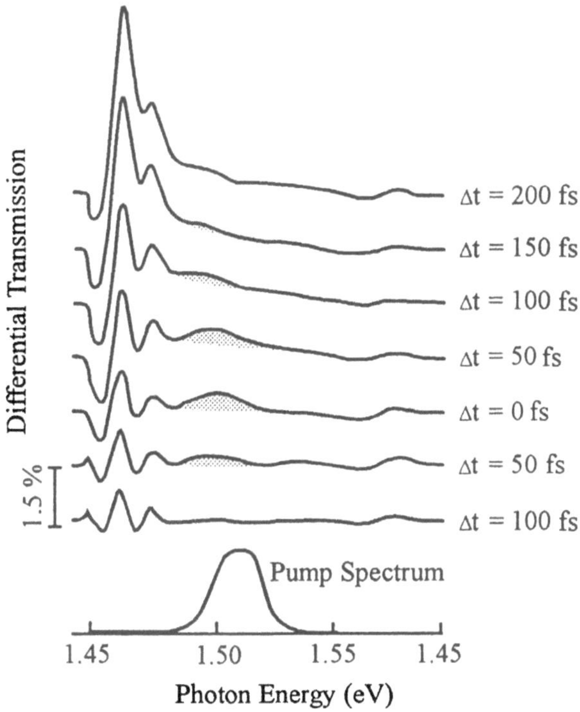

Fig. 12.7. Results of a femtosecond pump–probe experiment on GaAs quantum wells. The density of excited carriers is 2×10^{10} cm^{-2}. Taken from [345]

For electron energies more than one LO phonon energy above the band gap energy ($> E_c(\boldsymbol{k} = 0) + \hbar\omega_{LO}$), relaxation is generally fast because of LO-phonon–emission and the large number of final states for carrier–carrier scattering. In contrast to the luminescence experiments discussed above, in this case a very fast thermalization (energy relaxation) is observed [346]. This is illustrated by the transient photoluminescence spectra shown in Fig. 12.8. Owing to carrier–carrier scattering at large densities, the spectra are completely structureless even at times shortly after excitation.

Even though the dependence of carrier–carrier scattering on the carrier density has been investigated in many different experiments, only a few studies have shown a clear dependence over an appreciable range of carrier den-

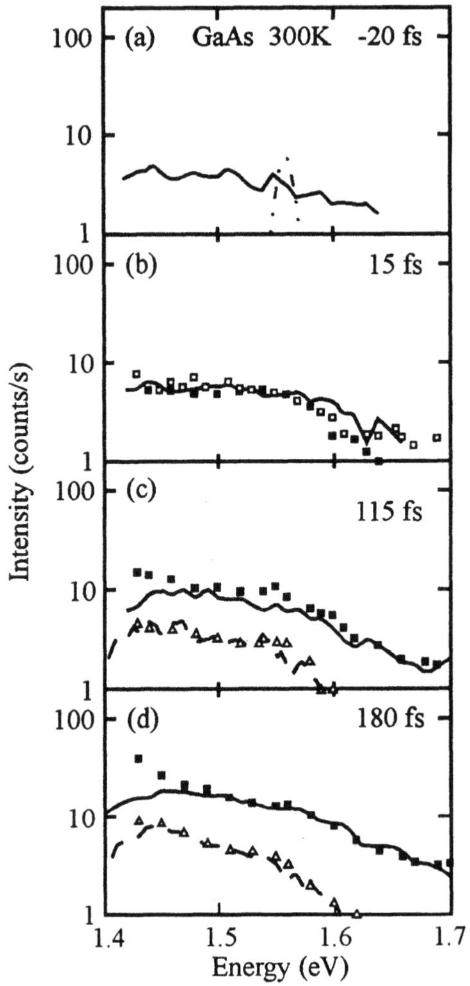

Fig. 12.8a–d. Transient photoluminescence spectra of GaAs at 300 K after excitation with 150 fs optical pulses at 1.93 eV. The excitation density was $7 \times 10^{17}\,\mathrm{cm}^{-3}$ (*solid line*) and $1.7 \times 10^{17}\,\mathrm{cm}^{-3}$ (*dashed line*). The *solid squares* and *triangles* correspond to results obtained from Monte Carlo simulations of the time evolution of the one-particle distribution. Taken from [346]

sities. When comparing different experiments one has to keep in mind that the dephasing times depend not only on the excited carrier density but also, independently, on other parameters such as the spectral width and central frequency of the excitation pulse. A characteristic dependence of the dephasing time on carrier density has been observed in four-wave-mixing experiments using optical pulses about 10 fs in duration. Results for bulk GaAs and GaAs quantum wells are depicted in Fig. 12.9 [348]. Here, the decay time τ of the photon echo signal measured as a function of the time delay is plotted versus the optically excited carrier density. For both bulk material and quantum wells the experimental dependence on the density can be fitted by means of $\tau^{-1} = \tau_0^{-1} + c n_{\mathrm{eh}}^{1/3}$. As much as quantum wells are concerned, this is different from conclusions based on previous data [68, 347]. One has to be

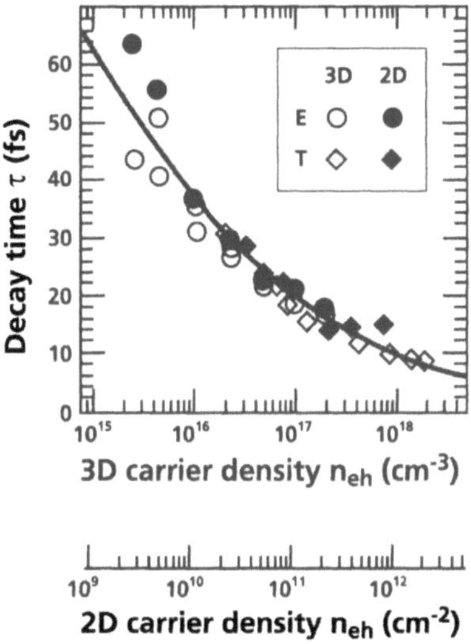

Fig. 12.9. Decay time constant of the transient four-wave-mixing signal as a function of carrier density for bulk GaAs and GaAs quantum wells held at room temperature. Assuming that T_2 is a meaningful description of the underlying physics and assuming an inhomogeneous ensemble of independent two-level systems, we have the relation $T_2 = 4\,\tau$. The *solid line* corresponds to a dependence $\tau^{-1} = \tau_0^{-1} + c\, n_{\mathrm{eh}}^{1/3}$. *Circles* correspond to experimental data, *triangles* to a microscopic quantum kinetic theory of electron–phonon and electron–electron scattering. Taken from [348]

cautious, however, with the interpretation of the measured quantity τ, especially when deducing a T_2 time from it. As we have discussed in Sect. 4.3, we have $T_2 = 2\,\tau$ for a homogeneous line and $T_2 = 4\,\tau$ for an inhomogeneously broadened ensemble of independent two-level atoms. Also, we have seen that one should obtain an ideal photon echo in the latter case. This aspect can be tested in real-time-resolved experiments under these conditions [349]. From the data, one can deduce the position of the echo maximum, $t = t_{\mathrm{max}}$, as a function of the time delay t_{21}. For an ideal photon echo we expect the relation $t = t_{\mathrm{max}} = t_{21}$ (here we have set the arrival time of the second pulse to $t = t_2 = 0$), for a homogeneous line, t_{max} should be constant, i.e. it should not depend on t_{21}. Figure 12.10 shows the experimental result. For intermediate carrier densities n_{eh}, one indeed finds an ideal photon echo behavior. However, for larger n_{eh}, significant deviations occur, which cannot be explained at the level of a T_2 time. This is already a hint that non-Markovian scattering processes play a significant role here.

What would be "smoking gun" evidence for such quantum kinetic effects in carrier–carrier scattering? In the phonon case (see Sect. 12.1.2), we have seen an oscillation which corresponds to the transfer of energy or excitation back and forth between the electrons – "the system" – and the phonons – "the bath". The frequency of this transfer was given by the phonon frequency

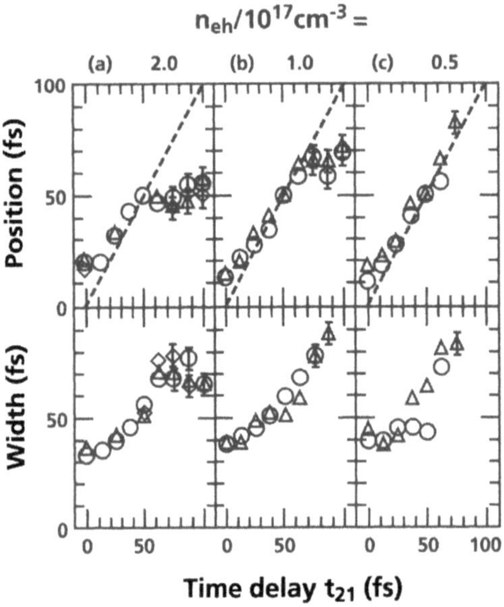

Fig. 12.10a–c. The position and width (full width at half maximum) of the photon echo signal versus time delay t_{21} as extracted from real-time resolved data. The *dashed straight lines* correspond to the expected temporal position t of the maximum of the ideal photon echo, i.e. to $t = t_{\max} = t_{21}$. The width of the echo would be constant for an ideal echo. Bulk GaAs held at room temperature. Taken from [349]

– *the* characteristic frequency of the problem. Two aspects seem to be different for carrier–carrier scattering. First, where is the bath? Second, at first glance, there is no single characteristic frequency. It is clear, however, that eventually the energy or excitation of an individual electron is transferred to all of the electrons, and that the collective excitation of the electrons is a plasmon. It turns out that electron–plasmon scattering is indeed a much more intuitive and simple way to think about this problem. While, in a Markovian picture, an electron is either in the state *electron has emitted a plasmon* or in the state *electron has not emitted a plasmon*, quantum mechanics also allows the superposition of both of these states. As the plasmon has a characteristic frequency, one again obtains an oscillation – a quantum beat – in perfect analogy to our discussion for phonons in Sect. 12.1.2. As the plasmon frequency is carrier-density dependent and as plasmons interact with LO phonons, one generally has a mixed plasmon–phonon mode (see Sect.11.5). While the quantum beating corresponding to this mode is barely visible in simple two-pulse photon echo experiments [349], it can be enhanced by preparation of specific initial-state wavefunctions. Experimentally, this can be achieved by means of coherent control [350]. Figure 12.11 shows the measured oscillation period versus carrier density. This behavior was actually *predicted* by a quantum kinetic theory [350].

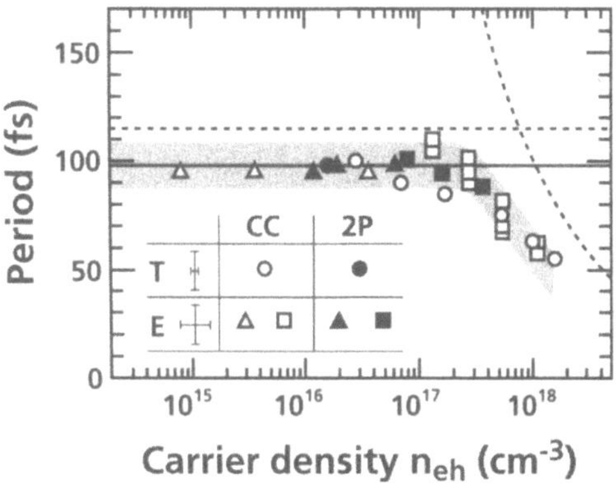

Fig. 12.11. Results of theory (T) and experiment (E) for bulk GaAs: oscillation period versus carrier density n_{eh}. The *filled symbols* are two-pulse (2P) photon echo results; the *open symbols* show results obtained by coherent control (CC) (see legend). Error bars are indicated. *Triangles* correspond to experiments at 77 K sample temperature, *squares* to experiments at 300 K. The *shaded area* is a guide to the eye. The two *dashed curves* correspond to the periods of the bare LO phonon and the plasmon oscillation, $2\pi/\omega_{\mathrm{LO}}$ and $2\pi/\omega_{\mathrm{pl}}(0)$, respectively. Taken from [350]

We have seen that non-Markovian quantum kinetics gives a consistent description of carrier–carrier scattering, i.e. of decay time constants (Fig. 12.9), of the real-time behavior of photon echo experiments (Fig. 12.10) and of electron–plasmon oscillations (Fig. 12.11). One aspect of this, the scaling of the decay time versus carrier density (Fig. 12.9), can be reproduced by a Markovian theory. Let us focus on this aspect as an instructive example.

Example (Dephasing after Excitation with Ultrashort Pulses). We focus our interest on the density range up to a few times $10^{17}\mathrm{cm}^{-3}$ in GaAs. We introduce excitonic units and measure the density according to[6] $\bar{n} = n\, r_{\mathrm{B}}^3$. Up to the range of $\bar{n} \approx 0.3$, saturation effects are negligible as, under the present excitation conditions, the distribution functions fulfill $f_{e,h}(\boldsymbol{k}) \ll 1$. These distributions can be obtained from numerical solutions of (10.153) or of a non-Markovian version based on (10.149). Their behavior is the key to understanding the decay of the polarization or of the four-wave-mixing signal. In the following, we consider the example of the dephasing rate due to the RPA electron–electron contribution in (10.150). This contribution to the dephasing rate is denoted by $S_{\mathrm{ee}}^{\mathrm{RPA}}$, and the result linear in the one-particle distribution function is[7]

---

[6] As an alternative, the mean particle spacing $r_{\mathrm{s}} = [3/(4\pi n)]^{1/3}/r_{\mathrm{B}}$ may be used.

[7] A quantitative analysis also requires, in principle, the spectral shift resulting from the real part of (10.150). In the present context we neglect this effect, as the shift is small compared with the spectral width of the pulse.

$$S_{\rm ee}^{\rm RPA}(\boldsymbol{k}) = 2e^4 \sum_{\boldsymbol{k}'\boldsymbol{q},} [w(\boldsymbol{q})]^2 \delta[E_{\rm e}(\boldsymbol{k}) - E_{\rm e}(\boldsymbol{k}+\boldsymbol{q}) + E_{\rm e}(\boldsymbol{k}'+\boldsymbol{q}) - E_{\rm e}(\boldsymbol{k}')] \\ \times \left[p_{\rm vc}(\boldsymbol{k})\, f_{\rm e}(\boldsymbol{k}'+\boldsymbol{q}) - p_{\rm vc}(\boldsymbol{k}+\boldsymbol{q})\, f_{\rm e}(\boldsymbol{k}')\right] . \qquad (12.17)$$

From the very beginning of the excitation process, the one-particle distributions differ not too much from high temperature thermal distribution functions, as consequence of the spectrally broad excitation of the band edge. The final temperature depends on the excess energy and the spectral width of the pulse. As long as saturation effects are negligible, the absolute magnitude of the final distribution scales with the excitation intensity and thus the density. The timescale on which a stationary distribution is reached, however, depends on the density. In order to obtain comparable one-particle distributions, we introduce the scaling $\boldsymbol{k} = \bar{\boldsymbol{k}} n^{1/3}$, which yields

$$\int \mathrm{d}\boldsymbol{k}^3\, f_{\rm e}(\boldsymbol{k}) = n \int \mathrm{d}\bar{\boldsymbol{k}}^3\, f_{\rm e}(\bar{\boldsymbol{k}} n^{1/3}) = n \int \mathrm{d}\bar{\boldsymbol{k}}^3\, \bar{f}_{\rm e}(\bar{\boldsymbol{k}}) = n . \qquad (12.18)$$

With this normalization, the properties of the distributions for different excitation intensities are very similar.

Let us turn now to the evaluation of (12.17). We express the δ function according to

$$\delta[E_{\rm e}(\boldsymbol{k}) - E_{\rm e}(\boldsymbol{k}+\boldsymbol{q}) + E_{\rm e}(\boldsymbol{k}'+\boldsymbol{q}) - E_{\rm e}(\boldsymbol{k}')] \\ = \int \mathrm{d}E\,\mathrm{d}E'\, \delta[E - E_{\rm e}(\boldsymbol{k}+\boldsymbol{q})]\delta[E' - E_{\rm e}(\boldsymbol{k}'+\boldsymbol{q})] \\ \times \delta[E_{\rm e}(\boldsymbol{k}) - E_{\rm e}(\boldsymbol{k}') - E + E'] \qquad (12.19)$$

and perform the integrations with respect to the angles enclosed by $\boldsymbol{k}, \boldsymbol{q}$ and by $\boldsymbol{k}', \boldsymbol{q}$. For the statically screened interaction (10.131) the q integration can also be performed analytically. Transforming back to the wave vector representation yields

$$S_{\rm ee}^{\rm RPA}(\boldsymbol{k}) = \frac{m_{\rm e} e^4}{8\hbar^2 \varepsilon_0^2 \pi^4} \int_0^\infty \mathrm{d}k_1\, \mathrm{d}k_2 [p_{\rm vc}(k)\, f_{\rm e}(k_2) - p_{\rm vc}(k_1)\, f_{\rm e}(k_3)] \\ \times \frac{k_1 k_2}{k\,\kappa^2} \left\{ \frac{k_{\rm u}}{k_{\rm u}^2 + \kappa^2} - \frac{k_{\rm l}}{k_{\rm l}^2 + \kappa^2} + \frac{1}{\kappa}\left[\mathrm{arctg}\left(\frac{k_{\rm u}}{\kappa}\right) - \mathrm{arctg}\left(\frac{k_{\rm l}}{\kappa}\right)\right]\right\} , \qquad (12.20)$$

where we have defined $k_{\rm u} = \min(k+k_1, k_2+k_3)$, $k_{\rm l} = \max(|k-k_1|, |k_2-k_3|)$, and $k_3 = (k^2 + k_2^2 - k_1^2)^{1/2}$. According to (10.130), the screening momentum is obtained as

$$\kappa^2 = \frac{e^2 m_{\rm e}}{2\pi^2 \varepsilon_0 \hbar^2} \int \mathrm{d}k\, f_{\rm e}(\boldsymbol{k}) = n^{1/3} \frac{e^2 m_{\rm e}}{2\pi^2 \varepsilon_0 \hbar^2} \int \mathrm{d}\bar{k}\, f_{\rm e}(\bar{\boldsymbol{k}}) \equiv n^{1/3}\, \bar{\kappa}^2 , \qquad (12.21)$$

where we have again performed the scaling of the distribution function. Now we make use of the same scaling in (12.20) and anticipate the $\boldsymbol{k}$ integration.

The result is given by

$$S_{\text{ee}}^{\text{RPA}}(\bar{\boldsymbol{k}}n^{1/3}) = n^{1/3}\frac{m_{\text{e}}e^4}{8\hbar^2\varepsilon_0^2\pi^4}\int_0^\infty \mathrm{d}\bar{k}_1\,\mathrm{d}\bar{k}_2[\bar{p}_{\text{vc}}(\bar{k})\,\bar{f}_{\text{e}}(\bar{k}_2) - \bar{p}_{\text{vc}}(\bar{k}_1)\,\bar{f}_{\text{e}}(\bar{k}_3)]$$

$$\times\frac{\bar{k}_1\bar{k}_2}{\bar{k}\,\bar{\kappa}^2}\left\{\frac{\bar{k}_u}{\bar{k}_u^2+\widetilde{\kappa}^2} - \frac{\bar{k}_l}{\bar{k}_l^2+\widetilde{\kappa}^2} + \frac{1}{\widetilde{\kappa}}\left[\operatorname{arctg}\left(\frac{\bar{k}_u}{\widetilde{\kappa}}\right) - \operatorname{arctg}\left(\frac{\bar{k}_l}{\widetilde{\kappa}}\right)\right]\right\}, \quad (12.22)$$

where we have defined $\widetilde{\kappa} = \bar{\kappa}/n^{1/6}$ and $\bar{p}_{\text{vc}}(\bar{k}) = p_{\text{vc}}(kn^{1/3})$. In addition to the prefactor $n^{1/3}$ the integrand exhibits two further density dependences. The first one is due to the change of the lineshape of $\bar{f}_{\text{e}}(\bar{k})$ with the density. For the present excitation conditions this variation is of minor influence on the final results. It may lead, however, to a slightly different behavior of the four-wave-mixing signal in comparison with the decay of the polarization. The second dependence results from $\widetilde{\kappa}$. In the density range of interest, however, this is also extremely weak owing to the exponent 1/6 and could come into play for small wave vectors $\bar{k}_u$ and $\bar{k}_l$ only. The strong compensation of self-energy and vertex corrections in this range leads to the suppression of the corresponding contributions to (12.22). For a dynamically screened interaction, this cancellation is even more effective.

The interaction with and among holes can be included along the same lines, together with a corresponding consideration of holes in the screening momentum. The exchange contributions in (10.150) exhibit the same behavior. Thus we find an asymptotic decay rate of the polarization proportional to $e^2n^{1/3}$, which is just a mean potential energy per particle. This result is in accordance with numerical solutions of (10.150) and (10.153) and is found also from a non-Markovian treatment restricted to the RPA contributions [349].

As we have seen, the origin of this dependence can be traced back to the scaling of the wave vector dependence of the one-particle distribution functions and the Coulomb interaction. Ultimately, it is the wave vector dependence of the carrier–carrier scattering rate, which translates into an energy dependence in optical spectra, and determines also the density dependence. In a strictly two-dimensional system, a corresponding analysis yields a decay rate $\sim n^{1/2}$. Numerical results for single quantum wells yield no unambigous power-law behavior owing to the more complicated structure of the Coulomb interaction. It is worth emphasizing that other excitation conditions which generate one-particle distributions far from equilibrium such as in the case of spectral hole burning, yield substantially different density dependences of the dephasing rates.

The wave vector dependence of carrier–carrier scattering, has been investigated also in modulation-doped GaAs quantum wells [351] and in optically pumped inverted semiconductors [352]. Both studies find long relaxation times at the Fermi energy, owing to the lack of final states to scatter into, and a rapid increase of scattering towards lower and higher energies.

12.2.1 The Limit of Quasi-Equilibrium

We have seen that scattering within the particle system and scattering from LO phonons and acoustic phonons eventually leads to carrier distributions which correspond to a carrier temperature equal to the lattice temperature. At this point, all coherent contributions have long ago died out. This does not mean, however, that the carrier system is already in thermal equilibrium with the lattice at this point; this only happens after recombination of electrons and holes. Thus, on a certain intermediate timescale, we are left with an excited but quasi-stationary system and we can study the linear optical properties of the inverted material. The description of these properties can again start with (10.150) and (10.153) or a non-Markovian version of these equations. As the one-particle distribution is time independent, the structure of the theory becomes much simpler than in the coherent range and we have a convenient opportunity to discuss the relationship between different formulations of the problem.

In fact, the stationary solution of (10.153) is easily shown to be the Fermi distribution. But the task of showing this is a very special proof, and this is a very special case only. The most general condition of *detailed balance* is obtained from the quantum kinetic equation (10.75). As the off-diagonal elements of the Green's function have died out and the diagonal elements depend on difference times only, we are left with the condition

$$\Sigma_{nn}^{+-}(\boldsymbol{k},\omega)\, G_{nn}^{-+}(\boldsymbol{k},\omega) = \Sigma_{nn}^{-+}(\boldsymbol{k},\omega)\, G_{nn}^{+-}(\boldsymbol{k},\omega)\,. \tag{12.23}$$

Together with the dependences (10.61) and (10.62) this implies

$$\frac{G_{nn}^{+-}(\boldsymbol{q},\omega)}{\mathrm{Im}[G_{nn}^{\mathrm{r}}(\boldsymbol{q},\omega)]} = \frac{\Sigma_{nn}^{+-}(\boldsymbol{q},\omega)}{\mathrm{Im}[\Sigma_{nn}^{\mathrm{r}}(\boldsymbol{q},\omega)]}\,. \tag{12.24}$$

Alternatively we can write this relation as (see also the derivation of (10.110))

$$G_{nn}^{+-}(\boldsymbol{q},\omega) = G_{nn}^{\mathrm{r}}(\boldsymbol{q},\omega)\, \Sigma_{nn}^{+-}(\boldsymbol{q},\omega)\, G_{nn}^{\mathrm{a}}(\boldsymbol{q},\omega)\,, \tag{12.25}$$

which can be regarded as a generalized *dissipation–fluctuation theorem* [355, 356, 357]. In its general form, we can represent this theorem as

$$A^{+-,-+}(12) = A_0^{+-,-+}(12) \int \mathrm{d}34\, A^{\mathrm{r}}(13) B^{+-,-+}(34) A^{\mathrm{a}}(42)\,, \tag{12.26}$$

where the quantities A and B correspond to response functions and fluctuating sources, respectively, and A_0 is related to initial correlations. We may identify A with G, w, or $\mathcal{D}$ and identify B with Σ, P^{l}, or P^{t}, as driving sources. Physically, these sources describe the scattering of particles (Σ) and the absorption or emission of longitudinal (P^{l}) and transverse (P^{t}) excitations of the system.

The ratios in (12.24) determine an occupation function $f(\omega)$, which is obviously the same for self-energies and Green's functions. As the self-energy is a functional of the Green's function, we can now make use of (12.23) to show that in an arbitrary approximation for the self-energy,

$$f(\omega) = \frac{1}{1 + e^{\beta(\hbar\omega - \mu)}} . \tag{12.27}$$

Provided the spectral function $\mathrm{Im}[G_r(\boldsymbol{q},\omega)]$ consists of a series of sharp and separated peaks, as in our common quasiparticle approximation, we can reduce (12.27) to the Fermi distribution $f(\boldsymbol{k})$.

In our case, the temperature is determined by the excitation conditions and the chemical potential by the total carrier density. When we consider the long-time behavior, we identify the carrier temperature with that of the lattice. In this way, we determine the distribution function and have only to deal with

$$\begin{aligned}
&\left(\mathrm{i}\hbar\frac{\partial}{\partial t} - E_{\mathrm{c}}(\boldsymbol{k}) + E_{\mathrm{v}}(\boldsymbol{k}) + e^2 \sum_{\boldsymbol{k}'} v(\boldsymbol{k}-\boldsymbol{k}')[f_{\mathrm{v}}(\boldsymbol{k}') - f_{\mathrm{c}}(\boldsymbol{k}') - 1] \right) p_{\mathrm{vc}}(\boldsymbol{k}) \\
&= -[f_{\mathrm{v}}(\boldsymbol{k}) - f_{\mathrm{c}}(\boldsymbol{k})] \left(\boldsymbol{d}_{\mathrm{cv}}\boldsymbol{E}(t) + e^2 \sum_{\boldsymbol{k}'} v(\boldsymbol{k}-\boldsymbol{k}') p_{\mathrm{vc}}(\boldsymbol{k}') \right) \\
&+ \mathrm{i}\hbar \frac{\partial}{\partial t} p_{\mathrm{vc}}(\boldsymbol{k}) \Big|_{\mathrm{corr}} .
\end{aligned} \tag{12.28}$$

The correlational contribution is given by (10.150), but now, as $\boldsymbol{E}(t)$ is a weak probe field, all third-order contributions of $p_{\mathrm{vc}}(\boldsymbol{k})$ can be omitted. In this linear range, we can again eliminate the field and, correspondingly, the transition amplitude in favor of the pair Green's function (compare Sect. 6.2). The result is a Bethe–Salpeter equation, which determines the linear response of the excited system. A non-Markovian version within the generalized Kadanoff–Baym ansatz can easily be derived also, by starting again from (10.149). The result can immediately be obtained from (10.150) by replacements of the form

$$\begin{aligned}
&g[E_a(\boldsymbol{k}) - E_a(\boldsymbol{k}+\boldsymbol{q}) + E_b(\boldsymbol{k}'+\boldsymbol{q}) - E_b(\boldsymbol{k}')] p_{\mathrm{vc}}(\boldsymbol{k},t) \\
&\longrightarrow \mathrm{i} \int_{-\infty}^{t} \mathrm{d}t' \, \mathrm{e}^{\mathrm{i}[E_a(\boldsymbol{k}) - E_a(\boldsymbol{k}+\boldsymbol{q}) + E_b(\boldsymbol{k}'+\boldsymbol{q}) - E_b(\boldsymbol{k}')](t-t')/\hbar} p_{\mathrm{vc}}(\boldsymbol{k},t') .
\end{aligned} \tag{12.29}$$

In both the Markovian and the non-Markovian case, we can solve the corresponding equation in either the time or the frequency domain.

Equilibrium allows a further simplification of the RPA contribution, which again is obtained ultimately again from a detailed-balance condition. From the dependences (10.61) and (10.62) together with (10.59), we have

$$w^{\mathrm{r,a}}(\boldsymbol{q},\omega) = \frac{v(\boldsymbol{q})}{1 - \varepsilon_0 V v(\boldsymbol{q}) \, P^{\mathrm{r,a}}(\boldsymbol{q},\omega)} , \tag{12.30}$$

which yields

$$w^{\mathrm{r}}(\boldsymbol{q},\omega) - w^{\mathrm{a}}(\boldsymbol{q},\omega) = \varepsilon_0 V \, |w^{\mathrm{r}}(\boldsymbol{q},\omega)|^2 \, [P^{\mathrm{r}}(\boldsymbol{q},\omega) - P^{\mathrm{a}}(\boldsymbol{q},\omega)] \, . \quad (12.31)$$

Making use of this relation in (10.110), we find

$$\frac{w^{+-}(\boldsymbol{q},\omega)}{\mathrm{Im}[w^{\mathrm{r}}(\boldsymbol{q},\omega)]} = \frac{P^{+-}(\boldsymbol{q},\omega)}{\mathrm{Im}[P^{\mathrm{r}}(\boldsymbol{q},\omega)]} \, , \quad (12.32)$$

which clearly corresponds to (12.24) and is equivalent to the detailed-balance condition

$$w^{+-}(\boldsymbol{q},\omega) \, P^{-+}(\boldsymbol{q},\omega) = w^{-+}(\boldsymbol{q},\omega) \, P^{+-}(\boldsymbol{q},\omega) \, . \quad (12.33)$$

In equilibrium, the ratios in (12.32) are determined by quantum statistics. By means of the identity

$$[1 - f(\omega+\omega')] f(\omega') = \mathrm{e}^{\beta\omega} f(\omega+\omega')[1 - f(\omega')] \, , \quad (12.34)$$

it is straightforward to show for an arbitrary approximation to the polarization propagator, that

$$w^{+-}(\boldsymbol{q},\omega) = 2\,\mathrm{i}N(\omega)\mathrm{Im}[w^{\mathrm{r}}(\boldsymbol{q},\omega)] \quad (12.35)$$

and, correspondingly,

$$w^{-+}(\boldsymbol{q},\omega) = 2\,\mathrm{i}\,(1 + N(\omega))\mathrm{Im}[w^{\mathrm{r}}(\boldsymbol{q},\omega)] \, , \quad (12.36)$$

where $N(\omega)$ is a Bose function. These relations reflect directly the detailed balance between emission and absorption of the quanta of longitudinal excitations, which are responsible for the screening, by the two–particle states represented by the polarization propagator. Applying (12.31) and (12.32) instead of (10.110) to evaluate the RPA contribution in (10.146) yields a completely equivalent formulation of the screened Hartree–Fock theory, which is first order in the screened Coulomb interaction [291].

Example (Quasi-Equilibrium Gain Spectra). Various formulations have been applied numerous times to study the gain spectra of semiconductors. In order to understand the basic physics, let us consider first he simplest limit of (12.28), in which we neglect the Coulomb interaction and introduce a constant dephasing rate. After Fourier transformation, we have the following for the polarization, from (12.28) in this limit:

$$\begin{aligned} \boldsymbol{P}(\omega) &= \frac{2\,\hbar\gamma_2\,|\boldsymbol{d}_{\mathrm{cv}}|^2 \boldsymbol{E}(\omega)}{V} \sum_{\mathrm{vc},\boldsymbol{k}} \frac{f_{\mathrm{v}}(\boldsymbol{k}) - f_{\mathrm{c}}(\boldsymbol{k})}{[E_{\mathrm{c}}(\boldsymbol{k}) - E_{\mathrm{v}}(\boldsymbol{k}) - \hbar\omega]^2 + \hbar^2\gamma_2^2} \\ &= 4\,|\boldsymbol{d}_{\mathrm{cv}}|^2 \boldsymbol{E}(\omega) \sum_{\mathrm{eh}} \int \mathrm{d}E \, \sinh[\beta(E - E_{\mathrm{g}} - \mu_{\mathrm{e}} - \mu_{\mathrm{h}})] \\ &\quad \times f_{\mathrm{e}}\left(\sqrt{\alpha}\,E\right) f_{\mathrm{h}}\left(\sqrt{\alpha}\,E\right) \frac{\hbar\gamma_2}{(\hbar\omega - E)^2 + \hbar^2\gamma_2^2} D_{\mathrm{eh}}(E) \, , \end{aligned} \quad (12.37)$$

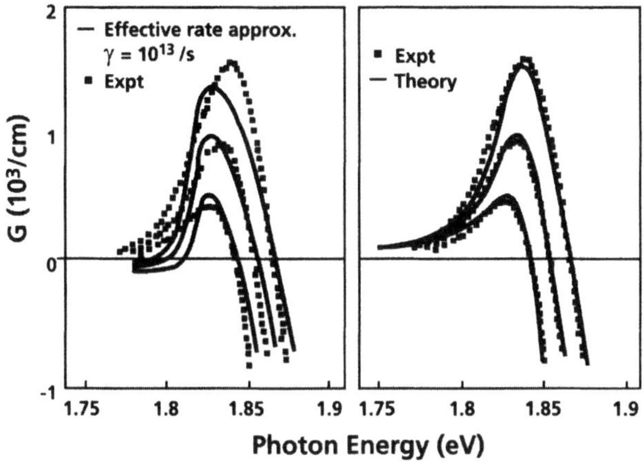

Fig. 12.12. Gain spectra for a 6.8 nm $Ga_{0.41}In_{0.59}P/(Al_{0.5}Ga_{0.5})_{0.51}In_{0.49}P$ quantum well. The points are experimental data at injection currents $I = 100$, 140, and 180 mA. The theoretical curves on the *left-hand side* were calculated with a constant dephasing rate at carrier densities $n = 2.6, 3.0$, and $3.4 \times 10^{12}\,\mathrm{cm}^{-2}$. The curves on the *right-hand side* were calculated from solutions of (12.28) with densities $n = 2.2, 2.7$, and $3.2 \times 10^{12}\,\mathrm{cm}^{-2}$. Taken from [359]

where $\alpha = 2m_r/\hbar^2$ and $D_{eh}(E)$ is the combined density of states. In the limit $\gamma_2 \to 0$, $\boldsymbol{P}(\omega)$ has a zero at $\hbar\omega = \mu_e - \mu_h \equiv \mu$ at sufficiently high densities. Thus, for energies below the chemical potential, the absorption becomes negative, i.e. we have *optical gain*. In fact, this is only a small portion of the truth and a quantitative understanding of gain spectra requires a much more elaborate theory. An unambiguous improvement is the inclusion of the screened Coulomb interaction in (12.28). This can be performed either numerically or by assuming model potentials, which allow an analytical construction of a Coulomb enhancement factor in (12.37) [358]. A corresponding calculation of gain-spectra of an InGaP quantum well is shown in the left part of Fig. 12.12 in comparison with experimental data [359]. The agreement is not too bad, but various experimental details are reproduced only rather roughly. The right-hand side shows, for comparison, the results of numerical solutions of (12.28), which lead to nearly perfect agreement with experimental data.[8] In particular, the blueshift of the gain peak and the absolute oscillator strength are well reproduced.

[8] Note that in the actual calculations the valence bands were treated within the framework of the Kohn–Luttinger theory (compare Sect. 3.5).

12.3 Scattering in the Presence of Bound States

If bound states come into play, a variety of scattering processes have to be distinguished. The most important channels are:

- Exciton–phonon scattering.
- Biexciton–phonon scattering.
- Exciton–electron scattering.
- Biexciton–electron scattering.
- Exciton–exciton scattering.
- Biexciton–exciton scattering.

In addition, we have to distinguish between coherent and incoherent occupations of excitons or biexcitons. The various scattering channels and their combinations give rise to a rich variety of dynamical properties. In the following, we shall only discuss a few selected features.

12.3.1 Exciton–Phonon Scattering

Scattering of excitons from phonons is different from scattering within bands in that any scattering event leading to energy relaxation needs to provide the exciton binding energy or more (unless it changes the center-of-mass motion of the exciton). As acoustic phonons cannot efficiently provide typical binding energies (a few meV to several tens of meV), scattering from optical phonons usually governs the exciton linewidth at room temperature. The scattering rate is simply proportional to the occupation of LO phonons, and gives rise to the absorption tail discussed in Sect. 11.4. In II–VI semiconductor quantum wells, however, the exciton binding energy for small well widths can become larger than the LO phonon energy (while the strength of the interaction is kept fixed), such that optical-phonon scattering is no longer efficient [353]. Exciton–LO-phonon scattering can be investigated more directly in the time domain; this was pioneered in [354]. Again, pump–probe experiments which resonantly excite the exciton and probe the time-delayed change in absorption are employed for this purpose. It turns out that the saturation of the absorption of the exciton is efficient as long as the excitons occupy the bound state, and decreases when excitons become ionized and thermally occupy the bands [354, 360]. Figure 12.13 compares linear absorption data and some results of such time-resolved experiments. From this comparison, we can conclude that a good fraction of the exciton linewidth is not associated with scattering at all but, rather, is governed by static imperfections of the system, in other words by a locally varying exciton energy, usually referred to as inhomogeneous broadening. For a careful investigation of the influence of exciton–acoustic-phonon scattering on the exciton linewidth at low temperatures, see, for example, [361].

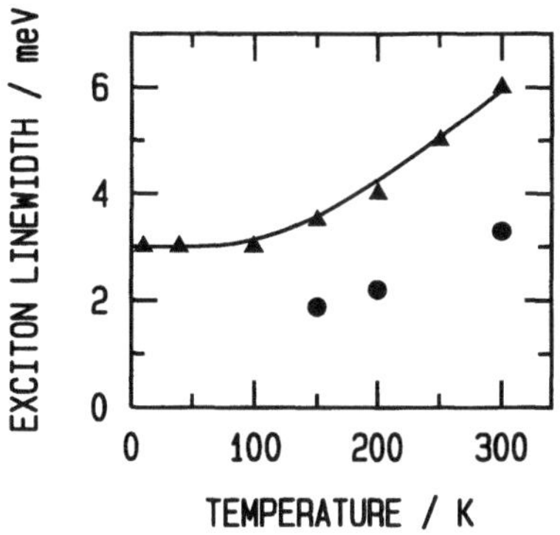

Fig. 12.13. *Triangles*: exciton linewidth (half width at half maximum), determined from linear absorption spectra, as a function of lattice temperature for an InGaAs/InAlAs multiple quantum well sample ($L_z = 10\,\mathrm{nm}$). The *filled circles* ($\hbar T_{\mathrm{decay}}^{-1}$) show the contribution to the linewidth due to exciton ionization as determined from a direct time-resolved pump–probe experiment ($L_z = 20\,\mathrm{nm}$). The *solid line* represents the Bose factor of LO phonons plus an offset. Data taken from [360]

12.3.2 Exciton–Exciton versus Exciton-Electron Scattering

Exciton–exciton and exciton–electron scattering becomes an important mechanism of phase relaxation of the 1s exciton at elevated carrier densities. In transient four-wave-mixing and pump–probe experiments, one finds at low excitation intensities a dependence of the phase relaxation rate $\gamma_2 = T_2^{-1}$ of the form $\gamma_2 = \gamma_2^0 + \gamma_2' n$, where n is either the exciton density or the electron density [362, 363, 364, 365]. Such a dependence is sometimes also referred to as collisional broadening or excitation-induced dephasing. Note that this dependence is quite different from that discussed for carrier–carrier scattering in Sect. 12.1.2. The scattering cross section γ_2' is about an order of magnitude larger for electron–exciton scattering than for exciton–exciton scattering [363], a fact which is intuitively clear, as continuum states have a much larger spatial extent. Even for exciton–exciton scattering, the scattering cross section is about an order of magnitude larger than πa_{B}^2 [363] (the geometrical cross section), demonstrating that excitons are an extremely sensitive species with respect to phase relaxation. The damping γ_2^0 at very low temperatures is usually governed by extrinsic effects, for example, by exciton–impurity scattering. Typical scattering times have been given in Table 6.3. In an ideal sample at zero temperature, γ_2^0 is determined by the rate of radiative exciton recombination only. For a discussion and overview of radiative recombination in semiconductors, which exhibits characteristic dependences on temperature, carrier density, and quantum well width, for example, see [366].

The features discussed above, represent two limiting cases. In the first case, an exciton is surrounded by an electron–hole plasma, in the second case, by other excitons. The problem to describe a bound electron–hole pair under the influence of a plasma of free electron–hole pairs is closely related to the problem of *metal insulator transitions* often referred to as *Mott transition* [367]. Owing to the screening of the electron–hole interaction, the binding energy decreases with increasing plasma density. According to a rule of thumb, the excitons become unstable, when the plasma screening length is comparable to the excitonic Bohr radius. The transition can be adequately described by means of a Bethe–Salpeter equation of the form (12.28) [368, 369].

The task to describe an exciton under the influence of other bound electron–hole pairs, is much more involved, and there exist a number of different theoretical approaches. Following the course of the optical excitation, we start with the coherent regime, where the theory described in Chap. 9 can be applied. Owing to the decay of the coherent transition amplitude, an incoherent distribution of electron–hole pairs is generated. We can treat this process again by the density matrix methods, but now, without assuming the coherent limit. From the factorization of the six-point functions (9.8), which enters (9.5), we obtain in this case coherent and incoherent contributions. The latter are proportional to products of the density–density correlation function (see Problem 9.1) and the transition amplitude. A quasi-equilibrium state is reached after all coherences have decayed, and we are left only with incoherent correlations described by the density–density correlation function. Beyond this restriction to four-particle correlations, we can consider the equation of motion for the six-point function (9.8). This approach has been used to study the influence of thermal electron–hole pair distributions on the lineshape of the optical absorption [221]. Equivalent formulations for the case of quasi-equilibrium can be obtained, if we start from an effective exciton Hamiltonian [370, 371] or apply equilibrium Green's function methods [372].

The correlation-induced signatures in the lineshape of optical spectra depend sensitively on the pair distribution, which enters the calculation of the density–density correlation function. In quasi-equilibrium this distribution function is determined by quantum statistics [373] and depends only on the density of pairs and the temperature. On the time scale of typical experiments, however, the assumption of quasi-equilibrium may have limited validity. In order to account for the dependence on the excitation conditions, we have to describe the transition from the coherently excited initial state to an incoherent quasi-equilibrium state. This requires, at least, the solution of the two-particle pendant to (10.153), or of a non-Markovian version, and is still a challenge. Needless to say, that the problem becomes more involved, if coherent or incoherent biexcitons contribute.

12.4 Problems

12.1. Apply the methods described in Sect. 10.3 to derive the equations of motion which determine the one-particle Green's function if the light field is treated quantum mechanically. The electron–photon interaction and the coupling to an external current are given by (2.105) and (2.120), respectively. Hint: Note that correlation functions of the type $\langle \mathbf{T}_c \boldsymbol{A}(\bar{1})\psi(\bar{1})\psi^\dagger(\bar{2})\rangle$ and $\langle \mathbf{T}_c \boldsymbol{A}^2(\bar{1})\psi(\bar{1})\psi^\dagger(\bar{2})\rangle$ can be expressed as functional derivatives of the one-particle Green's function with respect to an external current. Derive the generalization of (10.44) for the self-energy and of (10.60) for the transverse polarization propagator $\underline{P}^{\mathrm{t}}(\bar{1},\bar{2}) = \mu_0 V\, \delta\langle \hat{\boldsymbol{j}}^{\mathrm{t}}(\bar{1})\rangle / \delta\langle \hat{\boldsymbol{A}}(\bar{2})\rangle$.

12.2. Use the results of Problem 12.1 to show that, for a spatially weakly inhomogeneous system with slow variations on the macroscopic timescale, the self-energies Σ^{+-} and Σ^{-+} are given by the following, to first order in the photon Green's function:

$$\begin{aligned}\Sigma_{\mathrm{aa}}^{+-,-+}(\boldsymbol{k},\boldsymbol{r},\omega,t) &= \mathrm{i}\frac{\hbar\mu_0}{2\pi V} \\ \times \sum_{b,\boldsymbol{q}} j_{\mathrm{ab}} j_{\mathrm{ba}} \int \mathrm{d}\omega' G_{\mathrm{bb}}^{+-,-+}&(\boldsymbol{k}+\boldsymbol{q},\boldsymbol{r},\omega+\omega',t)\, \mathcal{D}^{-+,+-}(\boldsymbol{q},\boldsymbol{r},\omega',t)\,,\end{aligned} \tag{12.38}$$

where $j_{\mathrm{ab}} = e\, p_{\mathrm{ab}}/m_0$ is the interband current matrix element, and the Green's functions are Fourier transformed with respect to difference variables.

12.3. Derive the Dyson equation for the photon Green's function (12.5). Note that this function is proportional to the functional derivative of the expectation value of the vector potential with respect to an external current, i.e. $\delta\langle \hat{\boldsymbol{A}}(\bar{1})\rangle/\delta \boldsymbol{j}^{\mathrm{t}}_{\mathrm{ext}}(\bar{2})$ (compare Sect. 2.5). Hint: Make use of the chain rule and the definition of the current operator (2.123) to express the total current by means of the one-particle Green's function.

12.4. Consider (12.6) in the stationary limit, to derive a relation between the luminescence and the nonlinear absorption spectra for free-particle transitions.

12.5. Prove that a drifted Fermi distribution

$$f(\boldsymbol{k}) = \left\{1 + \exp\left[\beta\left(E(\boldsymbol{k}) - \boldsymbol{k}\boldsymbol{v} + \frac{m_{\mathrm{e}}}{2}\boldsymbol{v}^2 - \mu\right)\right]\right\}^{-1} \tag{12.39}$$

is the detailed-balance solution of (10.153). Discuss the physical meaning of the drift velocity $\boldsymbol{v}$ and possible applications of this distribution.

12.6. Derive (12.31) and (12.32) for the RPA polarization propagator.

12.7. Apply (12.31) and (12.32) to derive the Bethe–Salpeter equation of an inverted semiconductor within the dynamically screened Hartree–Fock approximation.

12.8. As discussed in Sect. 12.3, the exciton damping γ_2 often depends on exciton density $n = |p_1|^2$ (see Sect. 6.3.1) according to $\gamma_2 = \gamma_2^0 + \gamma_2' n$. Discuss the nonlinear optical response qualitatively. What does the response of a transient (two-beam) four-wave-mixing experiment look like? Compare this with the behavior of the local-field model. The result is discussed in [364].

13. The Semiconductor Laser

13.1 Introduction

Apart from light-emitting diodes, semiconductor lasers are the most important application of optoelectronic semiconductor devices at present. Compared with other laser media such as gases, molecules in solution or other solids, they offer by far the highest conversion efficiency from electric energy to emitted light energy (up to 50%). In a semiconductor laser the injected electric current is directly transformed into a population inversion without intermediate steps such as the optical pumping via a discharge used in gas lasers. Moreover, semiconductor lasers can be extremely compact, which is a crucial feature for applications in compact disc players, laser printers, and in telecommunication systems (where the most important emission wavelengths are around 1.3 μm and 1.55 μm). We can distinguish between two main (technologically relevant) geometric configurations: *edge emitters* and the even more compact *vertical-cavity surface-emitting lasers* (VCSELs), both of which are shown in Fig. 13.1.

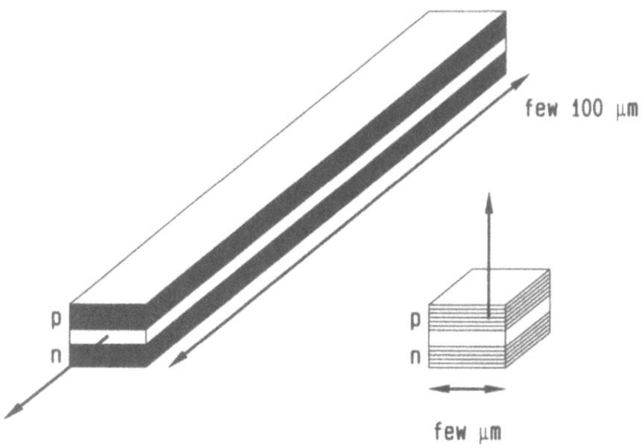

Fig. 13.1. Schematic illustration of a semiconductor edge-emitting laser (*left*) and a vertical-cavity surface-emitting laser (*right*). The current is injected through the n- and p-doped contacts, which, in case of the VCSEL, can also form the mirrors

In the case of edge emitters, the light field is guided by means of a planar waveguide, typically a few hundreds of microns in length. The cleaved (or etched) end facets serve as the two cavity mirrors. For GaAs parameters, the resulting (intensity) reflection coefficient R of the mirrors is rather low, approximately 30%. This is compensated by the large optical gain delivered by the long active region, which, in practice, can contain materials of different dimensionality. Most of today's CD players contain multiple quantum wells. The situation is quite different in vertical-cavity surface-emitting lasers. Here, the active medium has a very small physical length. Frequently, only a single quantum well is used, equivalent to merely $\approx 10\,\text{nm}$ of active material (about 20 lattice constants). The resulting optical gain is only 1% per pass or less, which necessitates mirrors with a reflectivity in excess of 99%. This can only be achieved by using dielectric mirrors, i.e. stacks of $\lambda/4$ layers with alternating low and high refractive index. VCSELs have a number of advantages for specific applications, as discussed in [374], for example.

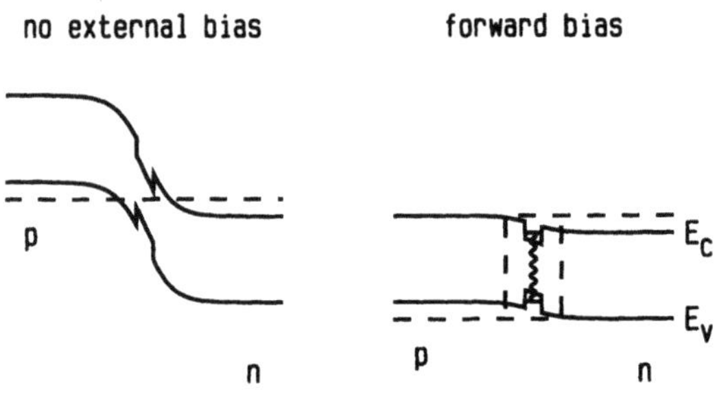

Fig. 13.2. Schematic band diagram of a semiconductor laser based on interband transitions. The *dashed lines* show the electrochemical potentials of electrons and holes (see Chap. 14). Under forward bias conditions of the p–n junction, both electrons and holes are present in the center region. In reality, the holes are usually not in the degenerate limit, i.e. their electrochemical potential is above the valence band edge

In the following sections we focus on the most common type of semiconductor laser, which is based on interband transitions in bipolar devices as sketched in Fig. 13.2. In contrast to this type, the recently developed *quantum cascade laser* [375, 376] relies on transitions within the conduction band of the semiconductor (more precisely, on intersubband transitions in a heterostructure). This unipolar edge-emitting device has its advantages and potential in the mid-infrared (several microns wavelength). One example of an InGaAlAs/InP heterostructure is depicted in Fig. 13.3. Population inversion between levels 3 and 2, which provide the laser transition, relies on two aspects. First, electrons injected from the left-hand side n-doped contact into level 3 cannot escape into the continuum owing to the electron Bragg reflector on the right, which creates a minigap. Secondly, fast scattering of electrons from level 2 to 1 is achieved via resonant LO phonon emission. From there, electrons enter the miniband on the right and escape into the contact. The quantum mechanical transmission spectrum of electrons (Fig. 13.3 b) illustra-

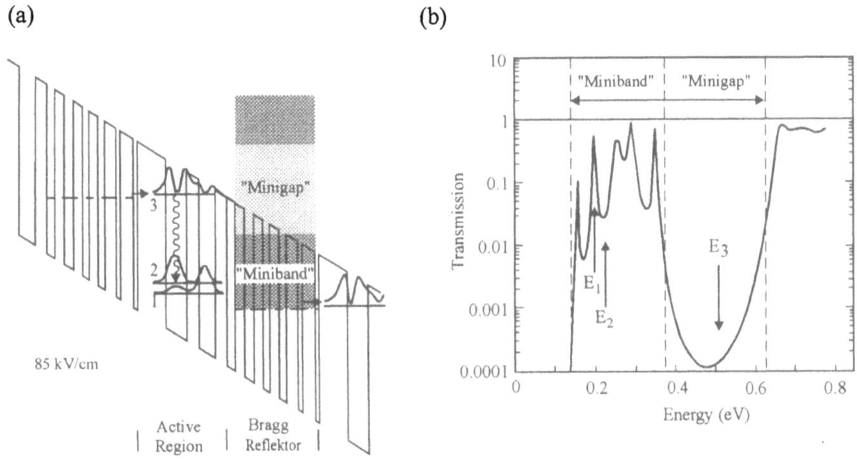

Fig. 13.3. The quantum cascade laser. (**a**) Layer sequence, energy levels, and important wavefunctions; (**b**) electron transmission spectrum. Taken from [376]

tes the design of the miniband and the minigap. The transfer matrix method discussed in Chap. 15 will enable us to compute such transmission spectra. If the difference between the effective in-plane masses of subbands 2 and 3 is neglected, the 2–3 transition becomes an atomic-like sharp line and the usual textbook description of three-level lasers is applicable.

13.2 Semiclassical Approach

In Chaps. 4, 6, and 10 we have introduced a suitable background for treating the dynamics of electrons within the semiconductor. In addition, we have to consider the dynamics of the light field, i.e. of the photons in the cavity. This allows two different levels of description: the semiclassical theory, in which the light field is treated via Maxwell's equations, and the quantum theory, where photons are treated fully quantum optically.

13.2.1 The Semiconductor Bloch Equations in a Cavity

Let us start by considering a one-dimensional cavity in the z direction. It is natural to expand the electric field $\boldsymbol{E}(z,t)$ and the polarization $\boldsymbol{P}(z,t)$ in terms of the cavity modes (standing waves) as follows:

$$\boldsymbol{E}(z,t) = \frac{1}{2}\sum_m \mathcal{E}_m(t)\mathrm{e}^{-\mathrm{i}[\omega_m t+\phi_m(t)]}\mathcal{Z}_m(z) + \text{c.c.}\,,$$
$$\boldsymbol{P}(z,t) = \frac{1}{2}\sum_m \mathcal{P}_m(t)\mathrm{e}^{-\mathrm{i}[\omega_m t+\phi_m(t)]}\mathcal{Z}_m(z) + \text{c.c.}\,, \tag{13.1}$$

where we have introduced the frequency ω_m, the time-dependent phase $\phi_m(t)$, and the standing-wave factor $\mathcal{Z}_m(z) = \sin(K_m z)$ of the mth cavity mode. The wavenumbers K_m follow directly from the boundary conditions on the cavity, with physical length L: $K_m = m\pi/L$. The coefficients $\mathcal{E}_m$ can be chosen to be real. From Maxwell's equations (2.124), we obtain the wave equation (see (2.212))

$$\frac{\partial^2}{\partial z^2}\boldsymbol{E}(z,t) - \frac{1}{c^2}\frac{\partial^2}{\partial t^2}\boldsymbol{E}(z,t) = \mu_0 \frac{\partial^2}{\partial t^2}\boldsymbol{P}(z,t) \tag{13.2}$$

where the velocity of light in the medium is c. Inserting (13.1) into the wave equation (13.2) and using the slowly-varying-amplitude approximation ($\ddot{\mathcal{E}}_m \approx 0$, $\ddot{\mathcal{P}}_m \approx 0$, $\dot{\mathcal{P}}_m \approx 0$, $\ddot{\phi} \approx 0$, $\dot{\mathcal{E}}_m\dot{\phi}_m \approx 0$) leads to

$$\left[\Omega_m^2 - \left(\omega_m + \dot{\phi}_m\right)^2\right]\mathcal{E}_m - 2\mathrm{i}\omega_m \frac{\partial}{\partial t}\mathcal{E}_m = \frac{\omega_m^2}{\epsilon_0\epsilon_\mathrm{b}}\mathcal{P}_m \,, \tag{13.3}$$

where we have introduced the eigenfrequency of the unexcited semiconductor $\Omega_m = cK_m$ and have expressed the velocity of light in the medium as $c = 1/\sqrt{\mu_0\epsilon_0\epsilon_\mathrm{b}}$, where the background dielectric constant is ϵ_b. Making the further approximation $\Omega_m + \omega_m \approx 2\omega_m$, i.e. considering only small shifts of the eigenfrequency, and separating (13.3) into real and imaginary parts leads to (for $\dot{\phi}_m^2 \approx 0$)

$$\begin{aligned}
\omega_m + \frac{\partial\phi_m}{\partial t} &= \Omega_m - \frac{\omega_m}{2\epsilon_0\epsilon_\mathrm{b}}\frac{\mathrm{Re}(\mathcal{P}_m)}{\mathcal{E}_m}\,, \\
\frac{\partial}{\partial t}\mathcal{E}_m &= -\frac{1}{2\epsilon_0\epsilon_\mathrm{b}}\omega_m\,\mathrm{Im}(\mathcal{P}_m)\,,
\end{aligned} \tag{13.4}$$

a set of equations referred to as the *self-consistency equations* [377]. These $2M$ differential equations for M modes (the first equation is for $(\omega_m + \dot{\phi}_m)$, the second for the mode amplitude $\dot{\mathcal{E}}_m$) look uncoupled at first sight. They are, however, coupled via the polarizations $\mathcal{P}_m$ and the electronic dynamics. Inspection of (13.4) shows that any decay of the intracavity field describing mirror losses via propagating field components has disappeared because of the approximations made above. Consideration of cavity losses, on the other hand, will prove to be a key physical ingredient. Hence we introduce them by a trick. We assume perfect mirrors at this point[1] and introduce the cavity losses later via an intracavity absorption coefficient α_R, which is chosen such that the loss per pass equals $\exp(-\alpha_\mathrm{R} L) = R$, the (intensity) reflection coefficient of the mirrors. At the moment, α_R is hidden in $\mathrm{Im}(\mathcal{P}_m)$. Physically, this replaces the actual losses taking place at the two cavity mirrors by a distributed loss.

We now restrict ourselves to the simple, educative case of only one mode of the light field, in which we can drop the index m. This is reasonable in

[1] Strictly speaking, lasers within this limit emit no light. We shall assume that the actual emission is proportional to the density of photons within the cavity.

VCSELs in particular, where the mode spacing is so large that only one longitudinal mode is supported. It can also be appropriate in edge emitters under suitable conditions. We consider only the evolution of the amplitude and are not interested in changes of the instantaneous laser frequency $(\omega_m + \dot{\phi}_m)$ due to changes of the refractive index described by the real part of the polarization in (13.4).

Let us first replace the field amplitude by the more intuitive photon density. We start from the electromagnetic energy density $u(z,t) = 1/2\epsilon_0\epsilon_b \boldsymbol{E}^2(z,t) + 1/2\epsilon_0 c_0^2 \boldsymbol{B}^2(z,t) = \epsilon_0\epsilon_b \boldsymbol{E}^2(z,t)$, which becomes

$$\langle u \rangle = \epsilon_0\epsilon_b \langle \boldsymbol{E}^2(z,t) \rangle = \epsilon_0\epsilon_b \frac{\mathcal{E}^2}{4} \,. \tag{13.5}$$

The factor 1/4 originates from the spatial and temporal average (with respect to the rapidly oscillating light frequency) $\langle u \rangle$ of $u(z,t) = \epsilon_0\epsilon_b\, \mathcal{E}^2 \cos^2(\omega_m t + \phi_m) \sin^2(K_m z)$. We now introduce the "classical" photon density $n_p(t)$,

$$n_p = \frac{\langle u \rangle}{\hbar\omega} = \frac{\epsilon_0\epsilon_b\, \mathcal{E}^2}{4\,\hbar\omega} \,, \tag{13.6}$$

with the corresponding equation of motion

$$\dot{n}_p = -\frac{1}{\epsilon_0\epsilon_b}\omega \frac{\mathrm{Im}(\mathcal{P})}{\mathcal{E}} n_p \,. \tag{13.7}$$

By using $\mathcal{P} = \epsilon_0 \chi^{(1)}(\omega)\mathcal{E}$, the imaginary part of the susceptibility can be expressed by means of the total absorption coefficient α. At this point we introduce the distributed loss α_R and make the replacement $\alpha(\omega) \to \alpha_R - g(\omega)$, where $g(\omega)$ is the gain coefficient (see e.g. (4.31)). In the limit of large background dielectric constant ϵ_b (see (2.223)), we have

$$\alpha(\omega) = \frac{\omega}{c_0\sqrt{\epsilon_b}} \mathrm{Im}[\chi^{(1)}(\omega)] = \frac{\omega}{c\,\epsilon_b} \mathrm{Im}[\chi^{(1)}(\omega)] \,, \tag{13.8}$$

which leads to

$$\dot{n}_p = -\frac{n_p}{\tau_p} + c g n_p \,, \tag{13.9}$$

where we have introduced the abbreviation $c\alpha_R = 1/\tau_p$, with the *photon lifetime* $\tau_p = -L/[c\ln(R)]$. For a typical VCSEL with $L = 1\,\mu$m, $c = c_0/n_b$, $n_b = 3.3$ and $R = 0.99$, we obtain $\tau_p = 1.1$ ps, for a typical edge emitter with $L = 200\,\mu$m and $R = 0.27$ (cleaved facets), we have $\tau_p = 1.7$ ps.

The first term in (13.9) describes an exponential decay of the photon density, while the second term is nothing but the exponential growth due to stimulated emission. This term can also be understood very intuitively. Beers law corresponds to an exponential growth of n_p in real space of the form $\partial n_p \partial z = g n_p$. The velocity of light relates the time and spatial coordinates via $z = ct$, and hence the photon density develops according to $\partial n_p/\partial t = c g n_p$. We could guess the corresponding equation of motion for the total electron–hole pair density n_{eh} using the well-known fact that stimulated emission conserves the combined number of photons and electron–hole pairs ($n_\mathrm{p} + n_\mathrm{eh} = \text{const.}$), and hence $\dot{n}_\mathrm{eh} = -cgn_\mathrm{p}$. It is, however, instructive to derive this result from the semiconductor Bloch equations.

The electron–hole pair density is given by[2]

$$\frac{\partial}{\partial t} n_\mathrm{eh} = \frac{\partial}{\partial t} n_\mathrm{e} = \frac{\partial}{\partial t} n_\mathrm{h} = \frac{1}{V} \sum_{\boldsymbol{k}} \frac{\partial}{\partial t} f_\mathrm{e}(\boldsymbol{k}) \,. \tag{13.10}$$

Employing the equation of motion (6.4) for $f_\mathrm{e}(\boldsymbol{k})$ with the scattering contribution defined in (10.153) leads to

$$\begin{aligned}\frac{\partial}{\partial t} n_\mathrm{eh} = & -\frac{2}{\hbar V} \,\mathrm{Im}\Big(\sum_{\boldsymbol{k}} \boldsymbol{d}_\mathrm{cv} \boldsymbol{E}(z,t) p^*_\mathrm{vc}(\boldsymbol{k}) \Big) + \Gamma_\mathrm{pump} + \frac{1}{V} \sum_{\boldsymbol{k}} \frac{\partial}{\partial t} f_\mathrm{e} \Big|_\mathrm{scatt} \\ & + \frac{\mathrm{i}\, e^2}{\hbar V} \sum_{\boldsymbol{k},\boldsymbol{k}'} v(\boldsymbol{k}-\boldsymbol{k}') [p^*_\mathrm{vc}(\boldsymbol{k}) p_\mathrm{vc}(\boldsymbol{k}') - p_\mathrm{vc}(\boldsymbol{k}) p^*_\mathrm{vc}(\boldsymbol{k}')] \,,\end{aligned} \tag{13.11}$$

where the last two contributions, i.e. the sum over the scattering contributions and the Hartree–Fock contributions, vanish, because both conserve the carrier density. In addition, we have included a generation rate of electron–hole pairs Γ_pump, describing the injection of carriers by either electric carrier injection or optical pumping. Inserting the mode expansion (13.1) for $\boldsymbol{E}(z,t)$ yields the following within the rotating-wave approximation:

$$\begin{aligned}\frac{\partial}{\partial t} n_\mathrm{eh} &= -\frac{2}{\hbar V} \mathrm{Im}\Big(\sum_{\boldsymbol{k}} \boldsymbol{d}_\mathrm{cv} \frac{\mathcal{E}}{2} \, \mathcal{Z}(z) \, \tilde{p}^*_\mathrm{vc}(\boldsymbol{k}) \Big) + \Gamma_\mathrm{pump} \\ &= \frac{1}{\hbar} \mathcal{E}^2 \, \mathcal{Z}^2(z) \, \frac{\mathrm{Im}[V^{-1} \sum_{\boldsymbol{k}} \boldsymbol{d}^*_\mathrm{cv} \tilde{p}_\mathrm{vc}(\boldsymbol{k})]}{\mathcal{E} \, \mathcal{Z}(z)} + \Gamma_\mathrm{pump} \\ &= \frac{1}{2\hbar} \mathcal{E}^2 \, \mathcal{Z}^2(z) \, \mathrm{Im}\left(\frac{\mathcal{P}}{\mathcal{E}} \right) + \Gamma_\mathrm{pump} = \frac{\mathcal{E}^2}{4\hbar} \, \epsilon_0 \, \mathrm{Im}(\chi^{(1)}) + \Gamma_\mathrm{pump} \,.\end{aligned} \tag{13.12}$$

In going from the second to the third line we have identified $V^{-1} \sum_{\boldsymbol{k}} \boldsymbol{d}^*_\mathrm{cv} \tilde{p}_\mathrm{vc}(\boldsymbol{k})$ with $1/2 \mathcal{P} \mathcal{Z}(z)$ using the two expressions for $\boldsymbol{P}(z,t)$ in (13.1) and (4.9), together with (4.46), i.e. $p_\mathrm{vc}(\boldsymbol{k}) = \tilde{p}_\mathrm{vc}(\boldsymbol{k}) \exp(-\mathrm{i}\omega t)$. In the last step we have averaged n_eh over the cavity length (consistently with the procedure for the photon density), leading to an additional factor of $1/2$ and have employed $\mathcal{P} = \epsilon_0 \chi^{(1)}(\omega) \mathcal{E}$. Inserting the gain (13.8) (clearly, α_R does not enter at this point) and the photon density (13.6) leads to the expected result

$$\dot{n}_\mathrm{eh} = -cgn_\mathrm{p} + \Gamma_\mathrm{pump} \,. \tag{13.13}$$

Note, that in our above model the cavity is completely filled with the active material. As in typical edge emitters, however, only a fraction of the

[2] We suppress the spin index s and the corresponding summation, consistently with the above suppression of the light polarization index on the photon density n_p.

optical waveguide contains the gain medium, the actual gain coefficient in the cavity is lower than g; this is usually expressed by a mode confinement factor $\Gamma \leq 1$, with the replacement $g \longrightarrow \Gamma g$. A related factor occurs for VCSELs containing only a few quantum wells in the cavity.

13.2.2 The Standard Rate Equations

Equations (13.9) and (13.13) describe the coupled evolution of the photon density and the electron–hole pair density. Even though they are both still exact with respect to the carrier kinetics, this set of equations cannot be closed in general, since the gain coefficient depends on the actual, unknown nonequilibrium distribution functions rather than being a function of the carrier density. *Hence we assume that the gain coefficient is a unique function of* n_{eh}. This is reasonable in the case of a slowly varying n_{eh} with respect to any type of scattering process of the carriers. Note that this does not necessitate Fermi distribution functions, which, however, are sometimes employed in order to obtain explicit expressions for $g(\omega, n_{\mathrm{eh}})$.

The classical approach contains *no spontaneous emission*, which, however, is a crucial ingredient of any laser. This is obvious from (13.9), and $n_{\mathrm{p}}(t) \equiv 0$ follows from $n_{\mathrm{p}}(t = -\infty) = 0$. On this level of description, spontaneous emission has to be introduced phenomenologically. While any spontaneous emission event reduces the carrier density by one electron–hole pair, not all of the corresponding photons couple into the laser mode, because of two reasons. For cavities long compared with the wavelength, most spontaneously emitted photons leave the cavity. Secondly, only a fraction of the spectrally broad spontaneous emission is resonant with the laser mode. Both processes are subsumed in an effective *coupling factor* $\beta \leq 1$ (and usually $\ll 1$) and we obtain the coupled, nonlinear rate equations in their final form as follows:

$$\dot{n}_{\mathrm{p}} = -\frac{n_{\mathrm{p}}}{\tau_{\mathrm{p}}} + cgn_{\mathrm{p}} + \beta\Gamma_{\mathrm{spont}} , \tag{13.14}$$

$$\dot{n}_{\mathrm{eh}} = -cgn_{\mathrm{p}} - \Gamma_{\mathrm{spont}} + \Gamma_{\mathrm{pump}} . \tag{13.15}$$

In order to further analyze the rate equations analytically, we expand the gain coefficient into a Taylor series around a carrier density n_{eh}^0 and truncate after first order:

$$g(n_{\mathrm{eh}}) = g_0 + g_0' \left(n_{\mathrm{eh}} - n_{\mathrm{eh}}^0\right) + \dots . \tag{13.16}$$

Often n_{eh}^0 is chosen as the transparency density, $g(n_{\mathrm{eh}}^0) = 0$, and hence $g_0 = 0$. For the sake of simplicity, spontaneous emission is assumed to be linear in the carrier density:[3] $\Gamma_{\mathrm{spont}} = -n_{\mathrm{eh}}/\tau_{\mathrm{eh}}$, where the carrier lifetime is τ_{eh}.

[3] In reality, the recombination rate is not linear in n_{eh}; we could, however, again employ a Taylor expansion around n_{eh}^0.

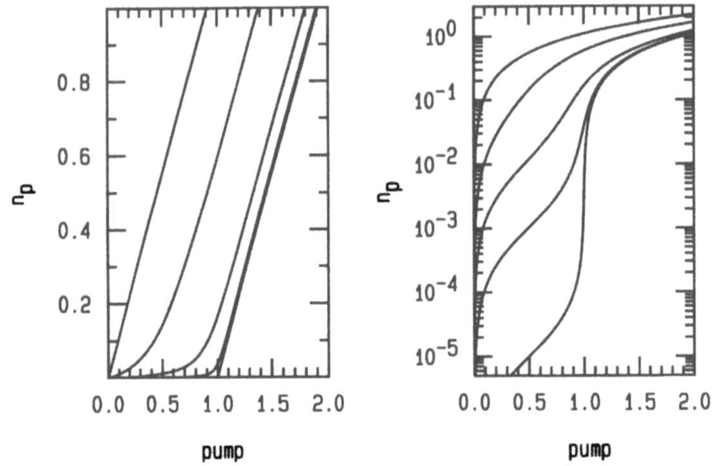

Fig. 13.4. Schematic illustration of photon density versus pump rate for various values of β: from *top* to *bottom*, $\beta = 1$, 0.5, 10^{-1}, 10^{-2}, and 10^{-4}. *Left*, linear scale; *right*, logarithmic scale

Stationary Case. For $\dot{n}_\text{p} = \dot{n}_\text{eh} = 0$, (13.14) and (13.15) are easily solved algebraically, delivering

$$n_\text{p}^2 + n_\text{p}\left(\frac{1}{\tau_\text{eh} g_0' c} - \frac{\tilde{g}_0 \tau_\text{p}}{g_0' \tau_\text{eh}} - \tau_\text{p} \Gamma_\text{pump} + \frac{\beta \tau_\text{p} \tilde{g}_0}{\tau_\text{eh} g_0'}\right) - \beta \frac{\tau_\text{p}}{\tau_\text{eh} g_0' c} = 0\ , \quad (13.17)$$

where we have used the abbreviation $\tilde{g}_0 = g_0 - g_0' n_\text{eh}^0$. Some solutions are visualized in Fig. 13.4 for various values of β. Edge emitters typically have values of β in the range of 10^{-5} and exhibit a pronounced threshold behavior. Empirically, the threshold current usually increases exponentially with device temperature $\sim \exp(T/T_0)$. Clearly, large values of the T_0 parameter ensure temperature-insensitive operation; $T_0 \approx 100$–200 K is typical for GaAs heterostructures. For large β the threshold gradually disappears, until one obtains *zero-threshold lasers* at $\beta = 1$. Such lasers with very low thresholds could prove useful in specific applications where large output powers are not required. Large coupling factors can be realized in small cavities which are comparable in volume to the cube of the wavelength of light, see Fig. 13.5. The largest coupling efficiencies reported to date have been achieved in *microdisk lasers* [378].

Above threshold, the optical gain is fixed at its threshold value via (13.14) if we neglect the small contribution from spontaneous emission. As the gain is a unique function of the carrier density within the standard rate equations, the carrier density also is fixed; this is often referred to as the *pinning of the carrier density.*

Dynamic Behavior. Let us consider small modulations of the variables related to the laser around their stationary values. All quantities are expres-

Fig. 13.5. Schematic illustration of a semiconductor microdisk laser. The light is reflected around the inside of a flat semiconductor disk via total internal reflection, leading to a *whispering-gallery mode* similar to the related phenomenon in acoustics. Devices as small as only a few microns in radius have been realized. An ideal laser of this type emits no light; in actual structures light is emitted owing to imperfections at the edges

sed in the following way: $\Gamma_{\mathrm{pump}} = \Gamma^0_{\mathrm{pump}} + \delta\Gamma_{\mathrm{pump}}$, $n_{\mathrm{eh}} = n^0_{\mathrm{eh}} + \delta n_{\mathrm{eh}}$ and $n_{\mathrm{p}} = n^0_{\mathrm{p}} + \delta n_{\mathrm{p}}$. The changes $\sim \delta$ are small compared with the respective stationary values. Introducing this ansatz into the rate equations, neglecting terms $\sim \mathcal{O}(\delta^2)$ and considering harmonic modulation with frequency ω, i.e. $\delta\Gamma_{\mathrm{pump}} = \delta\tilde{\Gamma}_{\mathrm{pump}} \exp(-\mathrm{i}\omega t)$ (with the ansatz $\delta n_{\mathrm{p}} = \delta\tilde{n}_{\mathrm{p}} \exp(-\mathrm{i}\omega t)$) leads to the following (for $\beta \ll 1$):

$$\frac{\delta\tilde{n}_{\mathrm{p}}}{n^0_{\mathrm{p}}} = \delta\tilde{\Gamma}_{\mathrm{pump}} \frac{g'_0 c}{\omega_{\mathrm{r}}^2 - \omega^2 - \mathrm{i}\gamma\omega} , \tag{13.18}$$

where we have used the abbreviations

$$\omega_{\mathrm{r}}^2 = \frac{g'_0 c}{\tau_{\mathrm{p}}} n^0_{\mathrm{p}} , \tag{13.19}$$

$$\gamma = \frac{1}{\tau_{\mathrm{eh}}} + \tau_{\mathrm{p}} \omega_{\mathrm{r}}^2 . \tag{13.20}$$

Obviously, the modulation amplitude of the photon density $\delta\tilde{n}_{\mathrm{p}}$ exhibits a resonance behavior around the *relaxation frequency* ω_{r}. During a relaxation oscillation the energy is transferred back and forth between the electronic and the photonic system. Starting with the gain, for example, stimulated emission reduces n_{eh} (and increases n_{p}) until the system becomes absorptive, which in turn increases n_{eh} (and decreases n_{p}) again, and so forth; ω_{r} is proportional to $\sqrt{n^0_{\mathrm{p}}}$, and hence proportional to the square root of the stationary value of the laser output. Such behavior is observed in all semiconductor lasers. It is, however, frequently masked by device parasitics (various RC time constants) to some extent for modulation of the electric injection current. Optical modulation allows a parasitic-free analysis, shown in Fig. 13.6 for a typical commercial semiconductor laser. The experimental data are in excellent agreement with the above rate-equation analysis. From this we can conclude that the shape of the carrier distribution functions follows the carrier density adiabatically on this timescale.

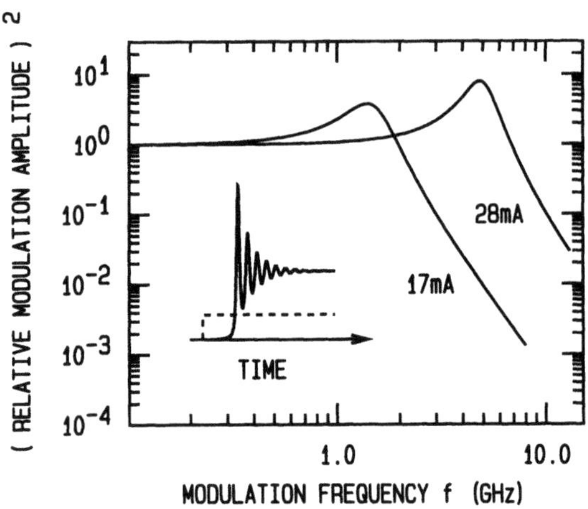

Fig. 13.6. Measured modulation response of a semiconductor edge-emitting laser ($\omega = 2\pi f$) at room temperature. The pump rate, controlled via the injection current, is the parameter. The *inset* shows schematically the emission in real time after switch-on of the pump rate (*dashed*). After [379]

13.2.3 Extended Rate Equations

As we have seen in the preceding section, the basic assumption of the standard rate equations for semiconductor lasers is that the gain coefficient is a unique function of the carrier density. Even up to relaxation frequencies of several tens of GHz, this basic assumption has proven to deliver physically meaningful results. On a timescale of the characteristic electron–electron or electron–phonon scattering processes, however, the shape of the carrier distribution functions is no longer a unique function of the carrier density, and hence physically different behavior may arise. We could go back to the beginning of this chapter and solve the (semiconductor) Bloch equations together with the electron–electron and electron–phonon scattering rates, as discussed in Chaps. 10 and 11, coupled to the equation of motion of the light field in the cavity. Parts of this procedure will in fact be performed in Sect. 13.3. First, however, we focus on a simple extension of the standard rate equations, which already covers most of the relevant semiconductor laser physics.

We consider the process of emission of a photon. Under operating conditions, the carrier distributions are characterized by some average kinetic energy, hence by a characteristic carrier temperature. If, for example, a photon is emitted in the center of the Brillouin zone, no kinetic energy is taken from the electronic system, but the number of electron–hole pairs is reduced by one. Consequently, the average kinetic energy per pair increases, and hence the carrier temperature rises. Such a process is clearly not accounted for in the standard rate equations. Similarly, the pump process, which inherently injects electron–hole pairs with above average energy, dynamically heats the carrier distribution functions. As electron–electron scattering conserves the total energy of the carrier system, the carrier temperature can only be reduced by emission of phonons; on a timescale of a few picoseconds, LO phonons are the dominant channel of energy loss. Let us assume at this point that carrier–carrier scattering is very fast, such that Fermi distribution functions $f_{\mathrm{e}}^{(0)}$ and $f_{\mathrm{h}}^{(0)}$ are obtained for the electrons and holes at any instant in time. In addition, we assume that identical temperatures T_{eh} for electrons and holes arise as a result of strong energy exchange between electrons and holes. If spontaneous emission and losses from the cavity are neglected, the total energy E_{tot} of the combined system of electron–hole pairs, photons in the cavity, and electron–hole pairs in the reservoir (from which pairs are injected in the pump process) is a conserved quantity:

$$\frac{E_{\mathrm{tot}}}{V} = \frac{1}{V}\sum_{\boldsymbol{k},s} E_{\mathrm{e}}(\boldsymbol{k}) f_{\mathrm{e}}^{(0)}(\boldsymbol{k}) + \frac{1}{V}\sum_{\boldsymbol{k},s} E_{\mathrm{h}}(\boldsymbol{k}) f_{\mathrm{h}}^{(0)}(\boldsymbol{k}) + \hbar\omega\, n_{\mathrm{p}} + E_{\mathrm{res}}\, n_{\mathrm{eh}}^{\mathrm{res}} , \tag{13.21}$$

where the photon energy of the laser mode is $\hbar\omega$ and the energy of an electron–hole pair injected from the reservoir is E_{res}. The total energy can be interpreted as a function of n_{eh}, T_{eh}, n_{p}, and the number of electron–hole pairs in the reservoir $n_{\mathrm{eh}}^{\mathrm{res}}$. Considering the total derivative

$$\mathrm{d}E_{\mathrm{tot}} = \frac{\partial E_{\mathrm{tot}}}{\partial n_{\mathrm{eh}}}\,\mathrm{d}n_{\mathrm{eh}} + \frac{\partial E_{\mathrm{tot}}}{\partial T_{\mathrm{eh}}}\,\mathrm{d}T_{\mathrm{eh}} + \frac{\partial E_{\mathrm{tot}}}{\partial n_{\mathrm{p}}}\,\mathrm{d}n_{\mathrm{p}} + \frac{\partial E_{\mathrm{tot}}}{\partial n_{\mathrm{eh}}^{\mathrm{res}}}\,\mathrm{d}n_{\mathrm{eh}}^{\mathrm{res}} , \tag{13.22}$$

where $\mathrm{d}E_{\mathrm{tot}} = 0$, immediately leads us to

$$\frac{\mathrm{d}T_{\mathrm{eh}}}{\mathrm{d}t} = \left(\frac{\partial E_{\mathrm{tot}}}{\partial n_{\mathrm{eh}}}\left(cgn_{\mathrm{p}} - \Gamma_{\mathrm{pump}}\right) - \hbar\omega V\left(cgn_{\mathrm{p}}\right) + E_{\mathrm{res}} V \Gamma_{\mathrm{pump}}\right)\left(\frac{\partial E_{\mathrm{tot}}}{\partial T_{\mathrm{eh}}}\right)^{-1} , \tag{13.23}$$

where we have replaced $\dot{n}_{\mathrm{eh}}$, $\dot{n}_{\mathrm{p}}$, and $\dot{n}_{\mathrm{eh}}^{\mathrm{res}}$ using the above approximations (no spontaneous emission, $\tau_{\mathrm{p}} = \infty$). The prefactors generally depend on the carrier density, the carrier temperature, and the dimensionality of the active region. Analytical evaluation of these prefactors is simple if the Boltzmann approximation is employed for both electrons and holes, in which case the terms (for a three-dimensional semiconductor) $V^{-1}\sum_{\boldsymbol{k},s}$ in (13.21) each become equal to $3/2\, k_{\mathrm{B}} T_{\mathrm{eh}} n_{\mathrm{eh}} + E_{\mathrm{c,v}}\, n_{\mathrm{eh}}$, the well-known result from the thermodynamics of ideal gases. The resulting expression exhibits all the important features qualitatively, but somewhat underestimates the heating:

$$\frac{\mathrm{d}T_{\mathrm{eh}}}{\mathrm{d}t} = \left(\frac{T_{\mathrm{eh}}}{n_{\mathrm{eh}}} - \frac{1}{3}\,\frac{\hbar\omega - E_{\mathrm{g}}}{k_{\mathrm{B}} n_{\mathrm{eh}}}\right)\left(cgn_{\mathrm{p}}\right) - \left(\frac{T_{\mathrm{eh}}}{n_{\mathrm{eh}}} - \frac{1}{3}\,\frac{E_{\mathrm{res}} - E_{\mathrm{g}}}{k_{\mathrm{B}} n_{\mathrm{eh}}}\right)\Gamma_{\mathrm{pump}} + \Gamma_{\mathrm{cool}} . \tag{13.24}$$

If both electrons and holes are treated within the degenerate limit, the heating is infinite as E_{tot} does not depend on the temperature. For GaAs-based lasers, for example, it is usually adequate to treat the holes in the

Boltzmann approximation and the electrons in the degenerate limit, leading to more lengthy expressions [380]. With the last term in (13.24), we have introduced a phenomenological cooling rate Γ_{cool} which describes the emission of LO phonons. In Sects. 11.4.2 and 12.1 we have seen that, on average, a typical electron–LO-phonon scattering event takes about 200 fs. As cooling requires a few scattering events, the cooling time constant $\tau_{(\mathrm{eh-LO})}$ is in the range of 0.5 to 1 ps (see Fig. 11.4) [381]. As each LO phonon emission event removes one LO phonon energy $\hbar\omega_{\mathrm{LO}}$ from the electronic system, we expect the following cooling rate:

$$\Gamma_{\mathrm{cool}} = -\frac{\hbar\omega_{\mathrm{LO}}}{k_{\mathrm{B}}} \frac{N_{\mathrm{LO}}(T_{\mathrm{eh}}) - N_{\mathrm{LO}}(T_{\mathrm{LO}})}{\tau_{(\mathrm{eh-LO})}}, \tag{13.25}$$

where the phonon temperature is T_{LO} and the Bose factor of LO phonons is

$$N_{\mathrm{LO}}(T) = \frac{1}{\exp\left(\hbar\omega_{\mathrm{LO}}/k_{\mathrm{B}}T\right) - 1}. \tag{13.26}$$

The energy taken from the electronic system heats up the LO phonons (hot phonons, see Sect. 12.1); this in turn slows down the electronic cooling [382, 383]. This increase in LO phonon temperature can be computed from the combined energy of electron–hole pairs and LO phonons (dispersion neglected), which is conserved in the scattering process:

$$\begin{aligned} &E_{\mathrm{eh,LO}} \\ &= \sum_{\boldsymbol{k},s} E_{\mathrm{e}}(\boldsymbol{k}) f_{\mathrm{e}}^{(0)}(\boldsymbol{k}) + \sum_{\boldsymbol{k},s} E_{\mathrm{h}}(\boldsymbol{k}) f_{\mathrm{h}}^{(0)}(\boldsymbol{k}) + \sum_{|\boldsymbol{q}|\le 2\,k_{\mathrm{F}}} \hbar\omega_{\mathrm{LO}} N_{\mathrm{LO}}(T_{\mathrm{LO}}) \,. \end{aligned} \tag{13.27}$$

Assuming degenerate carriers and the limit of large carrier density ($E_{\mathrm{F}} \gg \hbar\omega_{\mathrm{LO}}$), LO phonons can, at most, be emitted from the Fermi wave vector $+\boldsymbol{k}_{\mathrm{F}}$ to $-\boldsymbol{k}_{\mathrm{F}}$, leading to a maximum phonon wave vector $q_{\mathrm{max}} = 2\,k_{\mathrm{F}}$, which has already been introduced as a restriction in (13.27). It is now straightforward to evaluate the expressions in (13.27) using the degenerate limit for electrons and the Boltzmann approximation for holes, which leads us to

$$\frac{\mathrm{d}N_{\mathrm{LO}}(T_{\mathrm{LO}})}{\mathrm{d}t} = \frac{3}{8} \frac{N_{\mathrm{LO}}(T_{\mathrm{eh}}) - N_{\mathrm{LO}}(T_{\mathrm{LO}})}{\tau_{(\mathrm{eh-LO})}} - \frac{N_{\mathrm{LO}}(T_{\mathrm{LO}}) - N_{\mathrm{LO}}(T_0)}{\tau_{\mathrm{LO}}} \tag{13.28}$$

in three dimensions. In the second term on the right-hand side, we have introduced the decay of LO phonons brought about by anharmonicities of the phonons [382], leading to the equilibrium occupation of LO phonons $N_{\mathrm{LO}}(T_0)$ at the temperature of the surrounding heat bath T_0. For bulk GaAs, for example, at room temperature, the LO phonon lifetime is given by $\tau_{\mathrm{LO}} =$

2.5 ps. Obviously, the cooling rate reduces to the even simpler form $\Gamma_{\text{cool}} = -(T_{\text{eh}} - T_0)/\tau_{(\text{eh-LO})}$ in the limit $\tau_{\text{LO}} \to 0$ and $k_B T_{\text{eh}} \gg \hbar\omega_{\text{LO}}$.

In order to finally close the coupled set of equations (13.14), (13.15), (13.24), and (13.28), we can express the gain coefficient within the one-particle picture in the limit $\gamma_2 = 0$, see (4.31). For a three-dimensional semiconductor, we have

$$g(\omega, n_{\text{eh}}, T_{\text{eh}}) = g_0\, \Theta(\hbar\omega - E_g)\, \sqrt{\hbar\omega - E_g}\, \left(f_e^{(0)} + f_h^{(0)} - 1\right), \quad (13.29)$$

which is a function of n_{eh} and T_{eh} via the Fermi functions $f_e^{(0)}(\omega, n_{\text{eh}}, T_{\text{eh}})$ and $f_h^{(0)}(\omega, n_{\text{eh}}, T_{\text{eh}})$.[4] Corresponding results can be obtained similarly for two-dimensional semiconductor quantum wells [384].

Stationary Case. In Fig. 13.7, numerical solutions of the extended rate equations [383] are depicted for the stationary case, that is, $\dot{T}_{\text{LO}} = \dot{T}_{\text{eh}} = \dot{n}_{\text{eh}} = \dot{n}_p = 0$. It is obvious that the electron and hole temperature, as

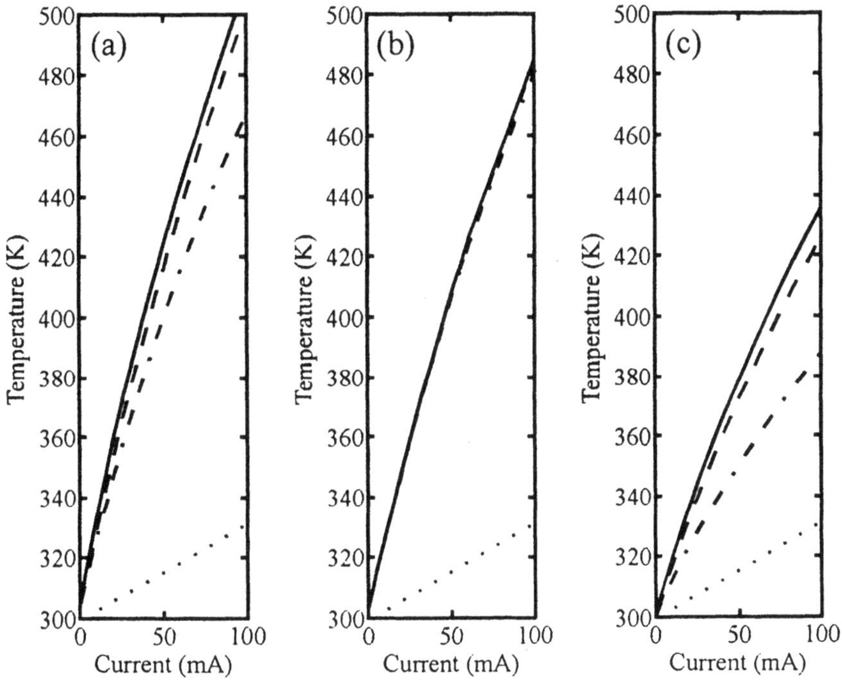

Fig. 13.7a–c. Temperature of electrons (*full lines*), holes, and LO phonons as a function of the pump rate, that is, the injection current. Notice the large rise in temperature, which can become larger than 100 K above room temperature under these steady-state conditions. Taken from [383]

[4] Usually, the relation between the chemical potential and n_{eh} cannot be evaluated analytically without further approximations, see Chap. 3. For details see [380].

well as the LO phonon temperature, experience a substantial increase with respect to room temperature as the electric injection current is increased. Consequently, the carrier density must rise even above threshold – in contrast to the standard rate equations – in order to keep the gain constant (see (13.14) and also Fig. 4.5). Such heating is clearly counterproductive to good laser operation, but it is intrinsic to semiconductor lasers as it even occurs for perfect heat sinking of the device. Imperfect heat sinking enhances the carrier heating.

Dynamic Behavior. An initial example is depicted in Fig. 13.8. This illustrates that not only quantitative but also severe qualitative differences with respect to the standard rate equations can arise because of dynamic changes of the carrier temperature. Here, a vertical-cavity surface-emitting laser containing four quantum wells as the active medium was studied [384]. The semiconductor laser, operating under steady-state conditions, was perturbed by a short optical pump pulse at $t = 0$. If the excess energy of this pump pulse is above a certain threshold value (see (13.24)), the rise in carrier temperature

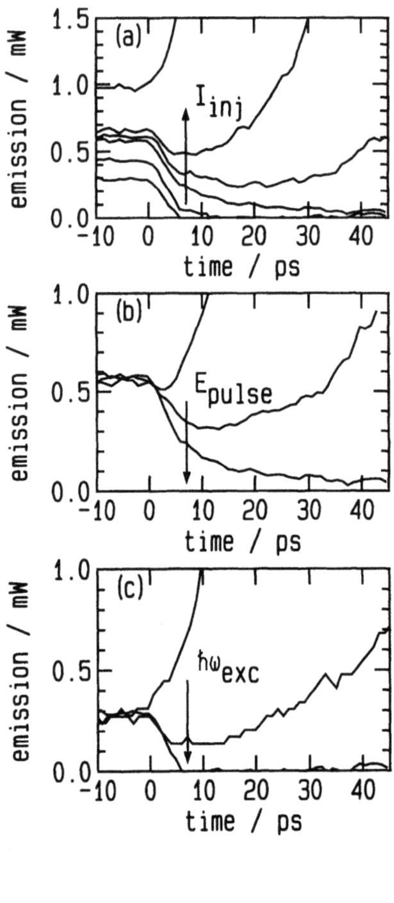

Fig. 13.8. Experimental results: transient response of a typical commercially available GaAs/AlGaAs vertical-cavity surface-emitting laser, electrically pumped under steady-state conditions at room temperature. The device was perturbed with a short optical pulse at t = 0. (**a**) Emission for a fixed pulse energy of 0.9 nJ and fixed excitation at 1.73 eV photon energy (corresponding to an excess energy of 0.275 eV), the stationary electrical injection current I_{inj} is a parameter (increasing from *bottom* to *top*, 3.50, 3.75, 4.00, 4.12, 4.25, 5.00 mA). (**b**) Emission for a fixed injection current of 4.0 mA, excitation at 1.73 eV photon energy; the pulse energy is a parameter (increasing from *top* to *bottom*, 0.3, 0.6, 0.9 nJ). (**c**) Emission for 3.5 mA injection current and a pulse energy of about 0.9 nJ; the excess photon energy is a parameter (increasing from *top* to *bottom*, 195, 230, 275 meV). Taken from [384]

becomes so large, that the corresponding reduction in gain overcompensates the increase in gain due to the increased carrier density (see Fig. 4.5). The gain is reduced with respect to its stationary value; for large pump pulse energies, even negative gain (absorption) can occur. Consequently, the laser switches off rather rapidly [384], a phenomenon which was predicted in [385]. Note that, in striking contrast to this result, the standard rate equations predict an initial rise. After the initial switch-off, the relaxation oscillation scenario – discussed in the previous section – follows. The rapid switch-off is specific to semiconductor lasers as no analogue of carrier heating exists in lasers employing gaseous or molecular gain media. The same phenomenon has also been observed in a bulk InGaAs VCSEL. Here (Fig. 13.9), a first pump pulse excites the device at t = 0 and a second pulse perturbs the laser after it has switched on (t = 13.3 ps). The switch-on is much faster than in Fig. 13.8 because of the larger effective optical gain in the cavity. Again, the extended rate equations depicted in Fig. 13.10 qualitatively reproduce the experimental findings [386].

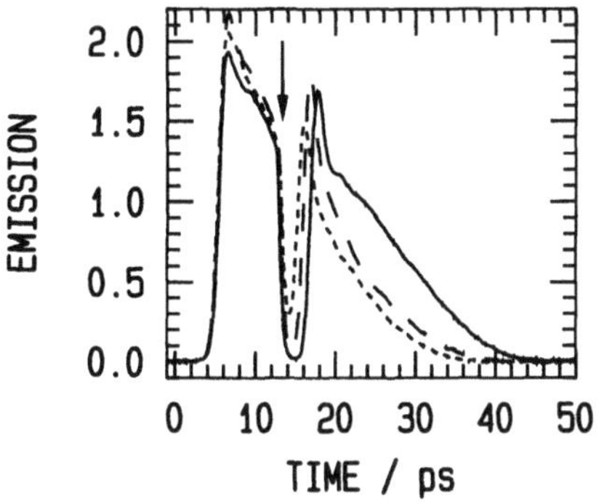

Fig. 13.9. Experimental results: emission of a bulk InGaAs VCSEL at 300 K versus time after excitation with a femtosecond pulse at $t = 0$ and a second pulse at $t = 13.3$ ps (see *arrow*). Emission at 0.84 eV, excitation at 1.56 eV photon energy. The energy of the second pulse is parameter. The time constants for switch-off and switch-on are 440 fs and 400 fs, respectively. After [386]

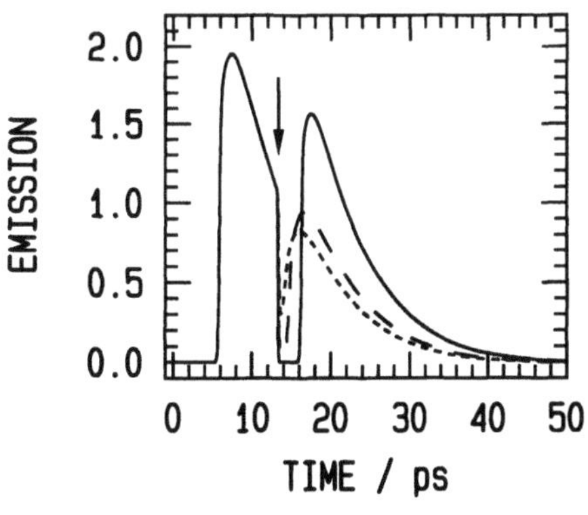

Fig. 13.10. Predictions of extended rate equations: laser emission after excitation with a first pump pulse at $t = 0$ and a second pulse at $t = 13.3$ ps (see *arrow*). The energy of the second pulse is a parameter. After [386]

13.2.4 Spectral Hole-Burning

It is natural to ask whether one of the basic assumptions of the extended rate equations, namely fast electron–electron scattering, is actually valid under typical semiconductor laser operation conditions. Intuitively, one expects deviations of the actual distribution functions $f_{\mathrm{e,h}}(\boldsymbol{k})$ from Fermi functions $f^{(0)}_{\mathrm{e,h}}(\boldsymbol{k})$, particularly at the laser photon energy $\hbar\omega$ as stimulated emission constantly removes energy from this spectral region. Neglecting Coulomb correlations at this point, the deviations from Fermi distributions can be estimated from the optical Bloch equations (4.20), (4.21).

Let us discuss here only a simple case which leads to analytical expressions. On a timescale long compared with γ_2^{-1} and γ_1^{-1} (typically $\approx 100\,\mathrm{fs}$), the adiabatic limit ($\partial/\partial t\, f_{e,h}(\boldsymbol{k}) = 0$) is well justified. We insert $p_{\mathrm{vc}}(\boldsymbol{k})$ from (4.20) into (4.21) and use the rotating-wave approximation. Under the very crude but simplifying assumption that the electron and hole distribution functions are equal, i.e. $f_{\mathrm{e}}(\boldsymbol{k}) = f_{\mathrm{h}}(\boldsymbol{k})$ and $f^{(0)}_{\mathrm{e}}(\boldsymbol{k}) = f^{(0)}_{\mathrm{h}}(\boldsymbol{k})$, we arrive at

$$\begin{aligned} &\hbar\gamma_1 \left[f_{\mathrm{e}}(\boldsymbol{k}) - f^{(0)}_{\mathrm{e}}(\boldsymbol{k})\right] \\ &= \frac{|\boldsymbol{d}_{\mathrm{cv}}|^2\mathcal{E}^2}{4}\left[1 - 2f_{\mathrm{e}}(\boldsymbol{k})\right] \frac{\hbar\gamma_2}{[E_{\mathrm{e}}(\boldsymbol{k}) + E_{\mathrm{h}}(\boldsymbol{k}) - \hbar\omega\,]^2 + \hbar^2\gamma_2^2} \,. \end{aligned} \tag{13.30}$$

Here we have made the replacement $|\tilde{\boldsymbol{E}}(\boldsymbol{r},t)|^2 \longrightarrow \mathcal{E}^2/2$. The factor of $1/2$ stems from the spatial average over the cavity length (see (13.1), (4.45), again for a single mode m and $\phi(t) = 0$). So far, we have computed the gain from the inversion factor $f^{(0)}_{\mathrm{e}} + f^{(0)}_{\mathrm{h}} - 1 = 2f^{(0)}_{\mathrm{e}} - 1$, (see (13.29)). A more realistic gain is obtained from the nonequilibrium distribution function f_{e} (at the laser mode) with inversion factor $2f_{\mathrm{e}} - 1$, which can be obtained directly from (13.30) as

$$2f_{\mathrm{e}} - 1 = (2f^{(0)}_{\mathrm{e}} - 1)\,\frac{1}{1 + n_{\mathrm{p}}/n_{\mathrm{p}}^{\mathrm{sat}}}\,, \tag{13.31}$$

where we have introduced the photon density according to (13.6) and have lumped the various prefactors into the saturation photon density[5]

$$n_{\mathrm{p}}^{\mathrm{sat}} = \frac{\epsilon_0\epsilon_{\mathrm{b}}\hbar}{2|\boldsymbol{d}_{\mathrm{cv}}|^2\omega}\,\gamma_1\,\gamma_2\,. \tag{13.32}$$

Obviously, the gain is obtained in the simple form

$$\begin{aligned} g(\omega, n_{\mathrm{eh}}, T_{\mathrm{eh}}) &\longrightarrow g(\omega, n_{\mathrm{eh}}, T_{\mathrm{eh}}, n_{\mathrm{p}}) \\ &= g(\omega, n_{\mathrm{eh}}, T_{\mathrm{eh}})\,\frac{1}{1 + n_{\mathrm{p}}/n_{\mathrm{p}}^{\mathrm{sat}}}\,. \end{aligned} \tag{13.33}$$

---

[5] Frequently, the completely equivalent spectral-hole burning parameter, or nonlinear gain saturation parameter, $\epsilon_{\mathrm{shb}} = \epsilon_{\mathrm{ngs}} = (n_{\mathrm{p}}^{\mathrm{sat}})^{-1}$ is used.

At large photon densities, and hence large rates of stimulated emission, a spectral hole occurs in the distribution functions, which reduces the gain with respect to the quasi-equilibrium case; at $n_p = n_p^{sat}$ the gain has decreased by a factor of two. As is obvious from (13.30), the width of the spectral hole in the distribution functions is directly related to the dephasing rate γ_2; its depth increases inversely with $\gamma_1\gamma_2$. In Fig. 13.11, the spectral hole is large compared with a typical case. Under stationary conditions, the nonequilibrium distribution functions usually merely deviate by a few % from Fermi functions.

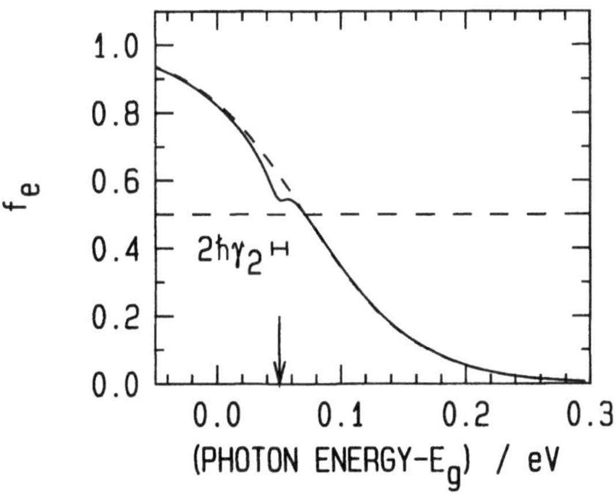

Fig. 13.11. Nonequilibrium carrier distribution function f_e according to (13.30). The *dashed line* corresponds to $f_e^{(0)}$, the laser photon energy is indicated by the *arrow*. GaAs quantum well parameters, $n_{eh} = 1.7 \times 10^{12}\,cm^{-2}$, $T_{eh} = 400\,K$, $n_p/n_p^{sat} = 1.75$, $\hbar\omega - E_g = 50\,meV$, and $\gamma_2^{-1} = 100\,fs$

More realistic dynamical calculations require numerical solution of the (semiconductor) Bloch equations (4.20), (4.21) coupled to the equation of motion of the light field (13.4). This has been discussed, neglecting carrier heating [387] and accounting for carrier heating [388]. For a truly detailed quantitative description of the semiconductor laser, a number of additional, more device-specific effects have to be accounted for also, such as the transverse mode profile [389, 390, 391], longitudinal inhomogeneities, the radial distribution of the pump rate, the lattice temperature distribution, nonradiative recombination at the edges of the device, carrier transport in the p–n junction and valence-band-mixing effects.

While for III–V-based semiconductor lasers the above picture of gain from an electron–hole plasma is well established, the situation is somewhat different for II–VI semiconductors, because the exciton binding energy E_R is much larger in these materials, and the exciton Bohr radius r_B^3 is much smaller. Hence the transition to an electron–hole plasma takes place at higher absolute densities, but at comparable values of $n_{eh}\, r_B^3$. The precise mechanism of gain is still the subject of some controversy but it may depend on the material and temperature. Inhomogeneous broadening [392], exciton–LO-phonon scattering [393], exciton–exciton scattering [394], biexcitonic gain [395], and electron–hole plasma gain [394, 396, 397] have been discussed. It seems, however, that under actual laser conditions – that, is room-temperature operation and large carrier densities – II–VI semiconductor lasers behave much the same as their III–V counterparts.

13.3 Quantum Theory

In general, laser emission differs from radiation from a light bulb in that it is coherent. The degree of coherence is limited by spontaneous emission and is reflected in a finite spectral linewidth or, equivalently, in a finite spatial coherence length. In Sect. 4.4.1 we have already shown that the *coherence length*, which can be measured in a Michelson interferometer, is directly related to the field–field correlation function, a quantity which will be evaluated in this section. A natural question that arises is: how spectrally narrow is the emission from a semiconductor laser? The semiclassical approach of the preceding section predicts the solution

$$\boldsymbol{E}(z,t) = \frac{\mathcal{E}}{2}\,\mathrm{e}^{-\mathrm{i}(\omega t+\phi)}\,\mathcal{Z}_m(z) + \text{c.c.} \tag{13.34}$$

from (13.1) for a single mode under steady-state conditions. The resulting spectral linewidth is zero, an obviously unphysical result. As will be seen, this shortcoming is one reason, but not the only one, for employing a quantum theoretical approach. The fundamentals of such a theory have been introduced in previous chapters and we shall discuss the basic ingredients in the following.

13.3.1 The Photon Kinetics

All information about the laser emission is contained in the photon Green's function (12.2) of the system, which is determined by the Dyson equation (12.3). Under strictly stationary and homogeneous conditions only field correlations contribute, as no coherent field amplitude occurs. This is obtained immediately from the wave equation for the expectation value of the vector potential $\langle \boldsymbol{A} \rangle$.[6] The transverse current is determined by the off-diagonal elements of the one-particle Green's function and is itself driven by $\langle \boldsymbol{A} \rangle$. For $\boldsymbol{j}_{\text{ext}} = 0$, we are left with a homogeneous integro-differential equation. Thus, unless $\langle \boldsymbol{A} \rangle \neq 0$ is introduced by initial conditions or boundary conditions, the coherent amplitude remains zero. This argument may be substantiated by considering the conditions which would be implied by the existence of a coherent amplitude [253].

By applying the methods introduced in Sect. 10.5, we can derive from (12.3) a kinetic equation for the photon propagator. In the same way as in

[6] Note that this equation is the same as in the case of a classical field, considered in Sect. 2.6.5.

the derivation of (10.45) and (10.49), we obtain two equations, which, after decomposition with respect to branch indices, take the form

$$\mathcal{D}_0^{-1}(1)\,\mathcal{D}^{+-}(12) = \int \mathrm{d}3\,[P^{\mathrm{t,r}}(13)\,D^{+-}(32) + P^{\mathrm{t},+-}(13)\,D^{\mathrm{a}}(32)] \tag{13.35}$$

and

$$\mathcal{D}_0^{-1}(2)\,\mathcal{D}^{+-}(12) = \int \mathrm{d}3\,[\mathcal{D}^{\mathrm{r}}(13)\,P^{\mathrm{t},+-}(32) + \mathcal{D}^{+-}(13)\,P^{\mathrm{t,a}}(32)]\,, \tag{13.36}$$

where $\mathcal{D}_0^{-1}(1) = \Delta_{\boldsymbol{r}_1} - c^{-2}\partial^2/\partial t_1^2$ and $c^2 = c_0^2/\varepsilon_{\mathrm{b}}$ is renormalized with the background dielectric function of the active material. Introducing again the macroscopic and microscopic timescales (see Sect. 10.4), we obtain a consistent quantum kinetic equation by expansion of the transverse polarization and photon Green's functions up to first order around the local macroscopic time[7] and Fourier transforming with respect to difference times. Since the dependence of the polarization propagator on the wave vector of the light is negligible, we have $P^{\mathrm{t,r}}(\boldsymbol{r},\boldsymbol{r}',\omega,t) = P^{\mathrm{t,r}}(\boldsymbol{r},\boldsymbol{r},\omega,t)\delta(\boldsymbol{r}-\boldsymbol{r}')$ in real space. Choosing $\boldsymbol{r}$ and $\boldsymbol{r}'$ inside the active region, we assume a homogeneous carrier distribution, and with the replacement $P^{\mathrm{t,r}}(\boldsymbol{r},\boldsymbol{r},\omega,t) \to P^{\mathrm{t,r}}(\omega,t)$, the difference between (13.35) and (13.36) can be cast into the form

$$\begin{aligned}
&\left(\frac{2\omega}{c^2} - \frac{\partial}{\partial\omega}\mathrm{Re}[P^{\mathrm{t,r}}(\omega,t)]\right)\frac{\partial}{\partial t}\mathcal{D}^{+-}(\boldsymbol{r},\boldsymbol{r}',\omega,t)\\
&= \{\mathrm{i}\,(\Delta_{\boldsymbol{r}} - \Delta_{\boldsymbol{r}'}) + 2\,\mathrm{Im}[P^{\mathrm{t,r}}(\omega,t)]\}\mathcal{D}^{+-}(\boldsymbol{r},\boldsymbol{r}',\omega,t)\\
&-2\,\mathrm{Im}[\mathcal{D}^{\mathrm{r}}(\boldsymbol{r},\boldsymbol{r}',\omega,t)]\,P^{\mathrm{t},+-}(\omega,t)\\
&-\frac{\partial}{\partial t}\mathrm{Re}[P^{\mathrm{t,r}}(\omega,t)]\frac{\partial}{\partial\omega}\mathcal{D}^{+-}(\boldsymbol{r},\boldsymbol{r}',\omega,t)\ \ldots\ .
\end{aligned} \tag{13.37}$$

Correspondingly, the sum of (13.35) and (13.36) yields, with neglect of the second derivative with respect to the macroscopic time,

$$\begin{aligned}
&\frac{\partial}{\partial\omega}\mathrm{Im}[P^{\mathrm{t,r}}(\omega,t)]\,\frac{\partial}{\partial t}\mathcal{D}^{+-}(\boldsymbol{r},\boldsymbol{r}',\omega,t)\\
&= -\left(\Delta_{\boldsymbol{r}} + \Delta_{\boldsymbol{r}'} + \frac{2\omega^2}{c^2} - 2\,\mathrm{Re}[P^{\mathrm{t,r}}(\omega,t)]\right)\mathcal{D}^{+-}(\boldsymbol{r},\boldsymbol{r}',\omega,t)\\
&+2\,\mathrm{Re}[\mathcal{D}^{\mathrm{r}}(\boldsymbol{r},\boldsymbol{r}',\omega,t)]\,P^{\mathrm{t},+-}(\omega,t)\\
&+\frac{\partial}{\partial t}\mathrm{Im}[P^{\mathrm{t,r}}(\omega,t)]\frac{\partial}{\partial\omega}\mathcal{D}^{+-}(\boldsymbol{r},\boldsymbol{r}',\omega,t)\ \ldots\ ,
\end{aligned} \tag{13.38}$$

[7] The same treatment was applied with respect to the nonlocal spatial coupling in the derivation of (10.69).

where the dots indicate non-Markovian corrections proportional to the derivatives $\partial P^{t,+-}/\partial t$ and $\partial P^{t,+-}/\partial\omega$. The contributions $\propto \mathcal{D}^{+-}$ describe *stimulated emission*, resulting from *optical gain* determined by $\mathrm{Im}[P^{t,r}(\omega,t)]$ and *cavity losses* due to light propagation. *Spontaneous emission* is proportional to the luminescence rate $P^{t,+-}$ and is treated strictly within the Markovian limit, as corrections become small well above threshold. The derivatives of $\mathrm{Re}(P^{t,r})$ and $\mathrm{Im}(P^{t,r})$ account for non-Markovian effects in the stimulated emission rates. The representation (13.37) and (13.38) is valid as long as the variations on the macroscopic timescale are slow in comparison with the lasing frequency. In actual numerical calculations it is often more meaningful to take the non-Markovian coupling of the polarization and photon Green's functions into account directly. If, in a transient stage of the system, the gain exceeds the cavity losses, this is in fact unavoidable, in order to guarantee causality.

If we relate the retarded transverse polarization function by means of (2.199) to the longitudinal response, we have approximately $\partial P^{t,r}/\partial\omega \simeq 2\omega\chi^{r}/c_0^2$. This reduces the prefactor on the left-hand side of (13.37) to $2\omega\,\mathrm{Re}[n^2(\omega,t)]/c_0^2$, where the renormalized complex refractive index inside the active material is defined by $n^2(\omega,t) = \varepsilon_b + \chi^{r}(\omega,t)$. In the same way, we can simplify the corresponding contributions in (13.38). The second line in (13.37) and (13.38) can be expressed by means of the inverse of the retarded and advanced photon Green's functions, respectively. All properties of the mode structure of the laser resulting from the features of the resonator enter by means of these functions ([399] and references therein), which are solutions of

$$\left(\Delta_{\boldsymbol{r}} + \frac{\omega^2}{c_0^2}\, n^2(\omega,t)\right)\mathcal{D}^{r}(\boldsymbol{r},\boldsymbol{r}',\omega,t) = \delta(\boldsymbol{r}-\boldsymbol{r}')\,. \tag{13.39}$$

Even in the range of ultrafast transient phenomena it is a good approximation to assume that $\mathcal{D}^{+-}$ changes slowly on the timescale on which a stationary mode structure in the resonator is established. This allows an adiabatic decoupling of (13.37) and (13.39). Thus only a knowledge of the stationary solution for $\mathcal{D}^{r}$ with a parametric dependence on $n(\boldsymbol{r},\omega,t)$ is required, which in turn can be used to approximate the spatial dependence of $\mathcal{D}^{+-}$. For Fabry–Pérot resonators, an eigenmode expansion allows one to map (13.37) onto a one-dimensional wave equation, which can be solved analytically [400] and allows further simplifications [399]. A knowledge of $\mathcal{D}^{+-}$ inside the cavity allows of course the calculation of the emission outside the resonator also, as can be shown by means of (13.35) and (13.36) again.

The kinetic equations for the photon Green's functions represent, in various respects, a generalization of the rate equation (13.14) for the photon density, which is obtained from the frequency integration of $\mathcal{D}^{+-}(\omega,t)$. In particular, the frequency dependence of the photon Green's functions, in contrast to the rate equation approach, allows one to take the spectral interplay between the properties of the light field and the active material into account.

This becomes obvious if we consider the intimate coupling to the material system in some detail.

13.3.2 The Carrier Kinetics

The laser emission obtained from (13.37) is governed by the gain spectrum and the luminescence spectrum of the active material. The corresponding polarization functions $P^{\mathrm{t,r}}(\boldsymbol{r},\omega,t)$ and $P^{\mathrm{t},+-}(\boldsymbol{r},\omega,t)$ have to be calculated from a Bethe–Salpeter equation or at least a suitable approximation which guarantees a quantitative description of the gain spectra (compare Sect. 12.2). The properties of the two functions are determined by the carrier dynamics, i.e. the one-paricle distributions, which in general are not simply determined by quantum statistics, but have to be determined from kinetic equations. In a shorthand notation, these equations have the structure

$$\frac{\partial}{\partial t}f_{\mathrm{n}}(\boldsymbol{k}) = \left.\frac{\partial}{\partial t}f_{\mathrm{n}}(\boldsymbol{k})\right|_{\mathrm{scatt}} + \left.\frac{\partial}{\partial t}f_{\mathrm{n}}(\boldsymbol{k})\right|_{\mathrm{phon}} + \left.\frac{\partial}{\partial t}f_{\mathrm{n}}(\boldsymbol{k})\right|_{\mathrm{pump}} + \left.\frac{\partial}{\partial t}f_{\mathrm{n}}(\boldsymbol{k})\right|_{\mathrm{stim}} + \left.\frac{\partial}{\partial t}f_{\mathrm{n}}(\boldsymbol{k})\right|_{\mathrm{spon}}, \tag{13.40}$$

where the distribution functions also may depend parametrically on the macroscopic variable $\boldsymbol{r}$. The first two contributions describe scattering within the particle system and with phonons, and are obtained from (10.153) and (11.147), respectively. The pump contribution which models either optical pumping or injection currents, is proportional to the Pauli blocking $1 - f_{\mathrm{e}}(\boldsymbol{k}) - f_{\mathrm{h}}(\boldsymbol{k})$. As new ingredients, we have the rates of *stimulated* and *spontaneous emission* owing to the transverse interactions. Within the simplest approximation, the corresponding self-energies are given by (12.4). In the various combinations of self-energies and one-particle Green's functions in the kinetic equations, the frequency integration is performed by means of a damped quasiparticle approximation for the Green's function, applying a Markovian approximation. Neglecting the photon momentum in comparison with the carrier momentum and summing all contributions according to (10.75), we find that the total scattering rate due to transverse interactions is a sum of a *stimulated-emission* contribution

$$\left.\frac{\partial}{\partial t}f_{\mathrm{e}}(\boldsymbol{k})\right|_{\mathrm{stim}} = \mathrm{i}\hbar\mu_0|j_{\mathrm{cv}}|^2\int\frac{\mathrm{d}\omega}{2\pi}G_{\mathrm{eh}}(\boldsymbol{k},\omega,t)[1-f_{\mathrm{e}}(\boldsymbol{k})-f_{\mathrm{h}}(\boldsymbol{k})]\mathcal{D}^{+-}(\boldsymbol{r},\boldsymbol{r},\omega,t) \tag{13.41}$$

and the *spontaneous emission*

$$\left.\frac{\partial}{\partial t} f_{\rm e}(\boldsymbol{k})\right|_{\rm spon} \tag{13.42}$$
$$= 2\hbar\mu_0 |j_{\rm cv}|^2 \int \frac{{\rm d}\omega}{2\pi} G_{\rm eh}(\boldsymbol{k},\omega,t) f_{\rm e}(\boldsymbol{k}) f_{\rm h}(\boldsymbol{k}) \,{\rm Im}[\mathcal{D}(\boldsymbol{r},\boldsymbol{r},\omega,t)]\,.$$

The electron–hole spectral function is defined as

$$G_{\rm eh}(\boldsymbol{k},\omega,t) = \frac{2\,[\Gamma_{\rm e}(\boldsymbol{k}) + \Gamma_{\rm h}(\boldsymbol{k})]}{[E_{\rm e}(\boldsymbol{k}) + E_{\rm h}(\boldsymbol{k}) - \hbar\omega]^2 + [\Gamma_{\rm e}(\boldsymbol{k}) + \Gamma_{\rm h}(\boldsymbol{k})]^2}\,, \tag{13.43}$$

where $\Gamma_{\rm e}(\boldsymbol{k})$ and $\Gamma_{\rm h}(\boldsymbol{k})$ define the inverse lifetimes of the carriers. The rates for stimulated and spontaneous emission may be further improved by taking vertex corrections to the self-energy into account, which result in a Coulomb enhancement factor. In (13.41) and (13.42), we have performed the $\boldsymbol{q}$ summation, which yields

$$\mathcal{D}^{+-}(\boldsymbol{r},\boldsymbol{r},\omega,t) = \frac{1}{V}\sum_{\boldsymbol{q}} \mathcal{D}^{+-}(\boldsymbol{q},\boldsymbol{r},\omega,t)\,, \tag{13.44}$$

and $\mathcal{D}^{+-}(\boldsymbol{q},\boldsymbol{r},\omega,t)$ is the Fourier transformation of $\mathcal{D}^{+-}(\boldsymbol{r},\boldsymbol{r}',\omega,t)$ with respect to $\boldsymbol{r}-\boldsymbol{r}'$. Since we have already assumed homogeneous carrier distributions, this propagator may be replaced by its average over the active region.

In spite of the variety of phenomena which can be understood in principle on the basis of rate equations, a reduction of the kinetic equations to equations of the type (13.14) and (13.15) or to their extension in Sect. 13.2.3 is not without pitfalls. A number of numerical studies of the microscopic theory show that the actual numbers are distinctly different in comparison with a rate equation approach.[8] The microscopic quantum optical approach, however, exhibits also characteristic qualitative differences, which can be traced back to the fact that the one-particle distributions and thus the gain are *not* uniquely determined by the particle density. As is obvious from (13.40) together with the various definitions of the scattering contributions, all relaxation processes in the particle system depend sensitively on their distribution in $\boldsymbol{k}$ space. In particular, on an ultrashort timescale, the dynamics are governed by the temporal evolution of the carrier distributions, which can no longer be assumed to adiabatically follow the carrier density. Together with the balance of stimulated emission, cavity losses, and spontaneous emission, this has the important consequence that a pinning of the carrier density (compare Sect. 13.2.2) does not occur. There are a good few features which can be understood in principle by means of extended rate equations, but a quantitative analysis, in particular one without adjustable parameters, requires a microscopic theory. This concerns, for instance, the threshold of lasing, the

[8] This is not in contradiction to the agreement between results based on rate equations and experimental data, owing to the variety of adjustable parameters.

relaxation oscillations, the buildup of side modes, the linewidth of the laser emission, and last but not least, the influence of the laser design. The system of equations introduced in this section represents a rather general starting point for studying such properties. Various applications to microcavity lasers show that the theory is well suited to yielding a quantitative understanding of experimental observations [378, 398]. As an instructive application of the theory, we consider in the following the spectral properties of the emission.

13.3.3 The Semiconductor Laser Linewidth

In order to turn to measurable quantities, we have to calculate the photon Green's function $\mathcal{D}(\boldsymbol{r},\boldsymbol{r},\omega)$ for $\boldsymbol{r}$ outside the active region [401] and apply a detector theory as discussed in Sect. 4.4. In the stationary limit, however, the problem of measuring the field–field correlation function becomes considerably simplified, as $\mathcal{D}^{+-}(t_1,t_2)$ is a function of the difference time Δt only. According to the Wiener–Kintchine theorem (4.88), the emission spectrum is obtained from the Fourier transform with respect to Δt, i.e. $\mathcal{I}(\omega) = \mathrm{i}\,\mu_0 |j_{\mathrm{cv}}|^2\,\mathcal{D}^{+-}(\omega)$. Also, if we have slow variations on the macroscopic timescale, it is often a reasonable approximation to define the emitted intensity in the same way. In this case the intensity is proportional to $\mathcal{D}^{+-}(\omega,t)$.

Stationary Case. Let us consider first the laser linewidth in the strictly stationary limit. For the intensity outside the cavity, we obtain the following from the integral representation obtained from (13.37):

$$\mathcal{D}^{+-}(\boldsymbol{r},\boldsymbol{r},\omega) = \frac{\omega^2}{c_0^2}\int \mathrm{d}^3\boldsymbol{r}'\,\mathcal{D}^{\mathrm{r}}(\boldsymbol{r},\boldsymbol{r}',\omega)\chi^{+-}(\omega,\boldsymbol{r}')\mathcal{D}^{\mathrm{a}}(\boldsymbol{r},\boldsymbol{r}',\omega)\,, \quad (13.45)$$

where $P^{+-}(\omega,\boldsymbol{r}') = P^{+-}(\omega)$ inside the cavity and zero otherwise. The relation (13.44) is again a variant of the *dissipation–fluctuation theorem* (compare Sect. 12.2.1) and can serve as a general starting point for calculating the lineshape of the emitted light in the strictly stationary case. We apply this result to a single dielectric layer of length L, for which the retarded photon Green's function can be represented as ([399] and references therein)

$$\mathcal{D}^{\mathrm{r}}(\boldsymbol{r},\boldsymbol{r}',\omega) \simeq \frac{F(\boldsymbol{r},\boldsymbol{r}',\omega)\,c_0}{4\,L\omega n'}\sum_m\left(\frac{\omega}{c_0}n' - \frac{\pi}{L}m + \mathrm{i}(\kappa + \frac{\omega}{c_0}n'')\right)^{-1}. \quad (13.46)$$

The round-trip losses in the Fabry–Pérot resonator are denoted by κ, $n = n' + \mathrm{i}n''$ is the refractive index of the active material, and $F(\boldsymbol{r},\boldsymbol{r}',\omega)$ is a form factor, which depends on the complex reflectivities and transmittances of the laser structure. This factor results from light propagation and determines the spatial intensity distribution both inside and outside the cavity. We restrict ourselves in the following to a single cavity mode with $m = k$ corresponding to a resonance at $\omega = \omega_{k0} = c_0 k\pi/[n'(\omega_{k0})\, L]$, which defines the laser frequency. Well above threshold, all quantities in (13.44) vary slowly in the range of the resonance frequency and can be approximated by their value at ω_{k0}. Using (13.45), we obtain

$$\mathcal{D}^{+-}(\boldsymbol{r},\boldsymbol{r},\omega) \propto \frac{\chi^{+-}(\omega_{k0})}{(\omega-\omega_{k0})^2+\Gamma_0^2(\omega_{k0})} \int \mathrm{d}^3\boldsymbol{r}' \, |F(\boldsymbol{r},\boldsymbol{r}',\omega_{k0})|^2 \,, \quad (13.47)$$

where $\Gamma_0(\omega_{k0}) = \kappa\, c_0/n'(\omega_{k0}) + \omega_{k0} n''(\omega_{k0})/n'(\omega_{k0})$. Calculating the stationary total intensity $\mathcal{I}$ from the frequency integration of (13.47), we find that

$$\mathcal{I} = \frac{\hbar\mu_0 |j_{\mathrm{cv}}|^2 c_0}{32\, L^2\, n'^3} \int \mathrm{d}^3\boldsymbol{r}' \, |F(\boldsymbol{r},\boldsymbol{r}',\omega_{k0})|^2 \, \frac{\chi^{+-}}{\Gamma_0} \,, \quad (13.48)$$

which shows that the intensity and linewidth are inversely proportional to one another. This narrowing of the linewidth is the essence of the *Schawlow–Townes formula* [402] and an immediate consequence of (13.45). As expected, the quantum limit of the laser linewidth is determined by the source of quantum fluctuations, i.e. the spontaneous emission. In the steady state the difference between the cavity losses κ and the optical gain $-\,\omega_{k0}n''(\omega_{k0})/c_0$ is just balanced by spontaneous-emission processes. Well above threshold, stimulated emission overwhelms spontaneous emission, and thus the relative weight of the noise decreases and the linewidth shrinks.

This finite linewidth in the stationary limit of the quantum theory is in contrast to the semiclassical approach, where a stable coherent amplitude is only possible if $\Gamma_0 = 0$ in the steady state. A finite laser linewidth is obtained only as a consequence of fluctuations of the amplitude and phase of the field. In stochastic laser theory the sources of these fluctuations are modeled by Langevin forces, which simulate spontaneous-emission processes [403, 404]. The coupling of phase and amplitude fluctuations gives rise to an additional phase shift and a substantial line broadening [405, 406].

Dynamic Behavior. Any modulation of the output intensity results primarily in a modulation of the density, according to (13.41) and (13.42), and this in turn results in changes of the complex refractive index. As we have seen in Sect. 13.2.2, the feedback to the intensity results in oscillations of the particle and photon densities around their steady-state values. These modulations also have a strong impact on the laser spectrum.

We start again from (13.37) and eliminate the frequency derivative of $\mathcal{D}^{+-}$ by means of (13.38). In order to simplify the notation, we do not take the spatial structure explicitly into account, but arrange, in accordance with the pole expansion (13.46), that the cavity losses are included in the imaginary part of the polarization. Introducing $\chi'' = \mathrm{Im}(\chi^{\mathrm{r}}) + 2\,\kappa c_0 n'/\omega_{k0}$ and the abbreviations $\chi' = \mathrm{Re}(\chi^{\mathrm{r}})$, $\varepsilon' = \mathrm{Re}(\varepsilon^{\mathrm{r}})$, the photon kinetics are determined by

$$\frac{\partial}{\partial t}\mathcal{D}^{+-}(\omega,t) = -\left(\frac{\omega\,\chi''}{\varepsilon'} - 2\alpha\,(\omega_k-\omega)\right)\mathcal{D}^{+-}(\omega,t)$$
$$-\frac{\omega}{\varepsilon'}\Big(\mathrm{Im}[\mathcal{D}^{\mathrm{r}}(\omega,t)] - \alpha\,\mathrm{Re}[\mathcal{D}^{\mathrm{r}}(\omega,t)]\Big)\,\chi^{+-}(\omega,t)\,. \tag{13.49}$$

In deriving (13.49), we have used an expansion in powers of $\alpha\chi''/\varepsilon'$ and have neglected products of small quantities such as χ''/ε'. The factor α is defined by the ratio

$$\alpha = \frac{\partial\chi'(\omega,t)/\partial t}{\partial\chi''(\omega,t)/\partial t} \simeq \frac{\partial\chi'(\omega,t)/\partial n_{\mathrm{eh}}}{\partial\chi''(\omega,t)/\partial n_{\mathrm{eh}}} \simeq \frac{\partial\chi'(\omega,t)/\partial\mathcal{I}}{\partial\chi''(\omega,t)/\partial\mathcal{I}}\,, \tag{13.50}$$

where the last two identifications are only approximately valid close to equilibrium. Otherwise, χ has to be considered as a functional of the distribution functions or the photon Green's function rather than as a function of the density or intensity.

Let us adopt for the moment the assumptions of stochastic laser theory. We consider modulations $\delta\chi'$ and $\delta\chi''$ of the steady-state values and assume $\chi''_0 = 0$ or $\Gamma_0 = 0$. The change $\delta\chi'$ gives rise to a shifted resonance frequency $\omega_k = \omega_{k0} - \omega_{k0}\,\delta\chi'/(2\,\varepsilon')$. We consider the vicinity of the resonance ω_{k0}, transform back from the (ω,t) to the $(t,\Delta t)$ representation, and shift t by $\Delta t/2$, in order to apply the Wiener–Kintchine theorem. As a result, we obtain the following from ((13.49):

$$\frac{\partial}{\partial t}\mathcal{D}^{+-}(t+\Delta t,t) = -\frac{\omega_{k0}}{\varepsilon'_0}(1+\alpha^2)\,\delta\chi''\left(t+\frac{\Delta t}{2}\right)\mathcal{D}^{+-}(t+\Delta t,t)$$
$$+\frac{\omega_{k0}}{2\,\varepsilon'_0}(1+\alpha^2)\,[\mathcal{D}^{\mathrm{r}}(t+\Delta t,t) - \mathcal{D}^{\mathrm{a}}(t+\Delta t,t)]\,\chi^{+-}\left(\omega_{k0},t+\frac{\Delta t}{2}\right). \tag{13.51}$$

A further analytical evaluation can be performed at different levels of sophistication. In the simplest approximation, the index change $\delta\chi''$ can be related to the intensity fluctuations according to $\omega_{k0}\,\delta\chi''(t)/\varepsilon'_0 \equiv \Gamma_{\mathrm{ST}}\,\delta\mathcal{I}/\mathcal{I}_0$, where $\delta\mathcal{I}$ is obtained from the solution of (13.51) for $t \geq 0$ with $\Delta t = 0$. Setting $\mathcal{I} = \mathcal{I}_0 + \delta\mathcal{I}$ and linearizing with respect to the fluctuations, we obtain

$$\delta\mathcal{I}(t) = \left(1 - \mathrm{e}^{-\Gamma_{\mathrm{ST}}(1+\alpha^2)t}\right)\frac{\omega_{k0}}{2\,\varepsilon'_0\Gamma_{\mathrm{ST}}}\,\chi_0^{+-}(\omega_{k0})\,. \tag{13.52}$$

We make use of this result in the formal solution of (13.51) (compare Problem 13.6) and linearize with respect to the exponential in (13.52), i.e. the expansion $\exp[\exp(-x)] \simeq 1+\exp(-x)$ is applied. The time-dependent spectrum can be calculated by means of (4.87), whereas the behavior for long times, is determined by the Wiener–Kintchine theorem (4.88). In the limit of large macroscopic times, almost all exponentials die down. What is left leads to a linewidth of

$$\Delta\omega = \Gamma_{\mathrm{ST}}\,(1+\alpha^2)\,, \tag{13.53}$$

This enhancement of the Schawlow–Townes linewidth, described by α, is well known from stochastic laser theory [404, 405, 406]. It may be improved by relating the index fluctuation $\delta\chi''$ to the electron–hole pair density n_{eh} rather than to the intensity. Taking the feedback of the particle system to the light field described by (13.41) into account, we obtain, as already discussed in Sect. 13.2.2, the result that the density executes relaxation oscillations with frequency ω_{r}. In the spectrum, these oscillations give rise to the formation of sidebands, which are shifted by $\pm n\omega_{\mathrm{r}}$ (where n is an integer) with respect to the laser frequency. The broadening of the laser line remains the same as in our simplified version.

Unfortunately, the assumptions of our analysis are not valid, as χ_0 does not vanish in the steady state. However, χ_0'' can in fact become small in comparison with $\delta\chi''$ and then (13.53) is indeed more than a rule of thumb for estimating the fundamental laser linewidth. While the linewidth enhancement factor α is typically small in gas lasers, it becomes important in semiconductors, where it exceeds unity [407] and can even take values up to the order of ten [405, 406]. In view of the not really controlled approximations which were necessary to derive (13.53), it is obvious that a numerical evaluation is preferable and an analytical treatment is not entirely satisfactory. Let us consider, therefore, another example, which leads by another line of reasoning to pronounced deviations of the linewidth from the Schawlow–Townes behavior.

As we have discussed in Sect. 13.2.2, spontaneous emission can contribute both to lasing and to nonlasing modes, depending on the coupling parameter β. The drastic increase of the *spontaneous emission coupling* from values of the order 10^{-5} in typical edge emitters to values not much smaller than unity in VCSELs results not only in a gradual disappearence of the laser threshold, but also in a considerable increase of the laser linewidth and a pronounced deviation from the Schawlow–Townes behavior. A microscopic treatment of the spontaneous-emission coupling requires the calculation of the full three-dimensional retarded photon Green's function of the actual laser structure [408]. To avoid this complicated problem, it is a reasonable procedure to take β as a parameter and to split the total photon density of states in the spontaneous emission (13.42) into a nonlasing and a lasing contribution with weights $1-\beta$ and β, respectively [409]. Numerical results for low ($\beta = 0.01$) and high ($\beta = 0.4$) spontaneous-emission coupling are shown in Fig. 13.12 for various pump rates. For low coupling, one obtains a high level of gain–loss compensation. Correspondingly, well above threshold the linewidth of the laser line is orders of magnitude smaller than the width of the cavity spectrum (dotted line), which is determined solely by the cavity losses. This behavior changes completely for high couplings, which result in a broad laser emission, owing to the drastic lowering of the gain–loss compensation. A comparison of the laser intensity and the linewidth shows that increasing β also results in pronounced deviations from the Schawlow–Townes line narrowing.

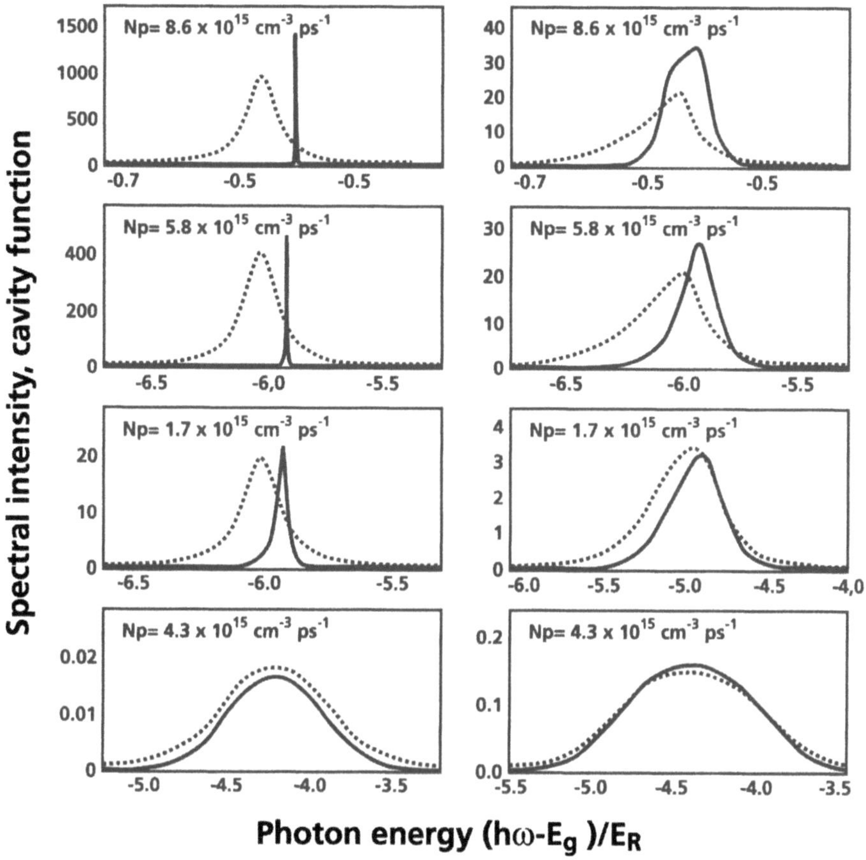

Fig. 13.12. Calculated laser emission spectrum (*full lines*) for various pump rates and low and high spontaneous-emission couplings, $\beta = 0.01$ (*left-hand side*) and $\beta = 0.4$ (*right-hand side*). The *dotted curves* show the cavity spectrum, which is normalized with an arbitrary factor. Taken from [409]

The few examples discussed in this chapter show that although semiconductor lasers work reliably in any CD player, the understanding of their fundamental quantum optical properties is still a field of current research and any new technological development is expected to give rise to new questions.

13.4 Problems

13.1. Discuss the temperature dependence of the (stationary) threshold pump rate of a (bipolar) semiconductor laser operating close to the band edge within the formalism of the rate equations. Employ the Boltzmann approximation for the distribution functions and use the gain coefficient given by (4.31). This result ($\sim T^{3/2}$ for bulk material and $\sim T$ for quantum wells) does not agree with the empirical finding $\sim \exp(T/T_0)$. What other mechanisms could be responsible?

13.2. Which performance parameters change in a II–VI semiconductor laser (larger $\boldsymbol{d}_{cv}$, larger effective masses, larger band gap) compared with a III–V laser? Assume that the lasing mechanisms are identical.

13.3. Consider the optical Bloch equations coupled to the dynamics of the light field (13.3) in the limit $\gamma_1 \to \infty$, $\gamma_2 \to \infty$. Some authors claim that the rate equations can be obtained in this strict limit of fast carrier scattering. Is this true? Hint: What happens to the gain spectrum in the limit $\gamma_2 \to \infty$?

13.4. Starting from (13.37) and (13.38), derive the rate equation (13.14) for the photon density. Discuss the approximations which are necessary for this reduction.

13.5. Make use of (13.40) to derive the rate equation (13.15) for the carrier density. Show that the density can be described by a damped harmonic oscillator which is driven by spontaneous-emission processes. Determine the relaxation frequency and the damping in terms of microscopic quantities.

13.6. Show that the solution of an equation of the type

$$\frac{\partial}{\partial t} f(t) = -\phi(t)\, f(t) + a(t) \tag{13.54}$$

can be written in the form

$$f(t) = f(t_0) + \int_{t_0}^{t} \mathrm{d}t' \exp\left(- \int_{t'}^{t} \mathrm{d}t'' \phi(t'') \right) a(t') \,. \tag{13.55}$$

14. Classical Transport

In this and the following chapters we consider transport of particles as a result of perturbations such as external voltages, magnetic fields, temperature gradients, and combinations of these. Particle transport results in electric currents and transport of heat via the charge and energy of the particles. The two corresponding physical quantities of interest are the *electric current density* $\boldsymbol{j}$ and the *heat current density* $\boldsymbol{w}$. In general, both quantities contain contributions from both electrons and holes:

$$\boldsymbol{j} = \boldsymbol{j}_{\mathrm{e}} + \boldsymbol{j}_{\mathrm{h}} , \tag{14.1}$$

$$\boldsymbol{w} = \boldsymbol{w}_{\mathrm{e}} + \boldsymbol{w}_{\mathrm{h}} . \tag{14.2}$$

Both the electromagnetic charge and the total energy are conserved quantities. Hence the following *continuity equations* hold for the charge density $\rho(\boldsymbol{r})$ and the energy density $u(\boldsymbol{r})$:

$$\frac{\partial \rho}{\partial t} + \nabla \cdot \boldsymbol{j} = 0 , \tag{14.3}$$

$$\frac{\partial u}{\partial t} + \nabla \cdot \boldsymbol{w} = \boldsymbol{j} \cdot \boldsymbol{E} . \tag{14.4}$$

If there are N_{e} electrons in a volume V with identical velocity $\boldsymbol{v}_{\mathrm{e}}$, and thus identical kinetic energy E_{e}, charge $-e$, and density $n_{\mathrm{e}} = N_{\mathrm{e}}/V$ (and hence charge density $\rho_{\mathrm{e}} = -en_{\mathrm{e}}$), these two quantities become

$$\boldsymbol{j}_{\mathrm{e}} = \rho_{\mathrm{e}} \boldsymbol{v}_{\mathrm{e}} = -\frac{e}{V} \boldsymbol{v}_{\mathrm{e}} N_{\mathrm{e}} , \tag{14.5}$$

$$\boldsymbol{w}_{\mathrm{e}} = \frac{1}{V} E_{\mathrm{e}} \boldsymbol{v}_{\mathrm{e}} N_{\mathrm{e}} . \tag{14.6}$$

In semiconductors, however, all electrons and all holes have different velocities and energies, corresponding to their wave vectors. Hence we generalize these equations to[1]

[1] Similar expressions hold for the holes, but with positive charge. Hence, from now on, we consider only electrons explicitly.

$$\boldsymbol{j}_\mathrm{e} = -\frac{e}{V}\sum_{\boldsymbol{k},s} \boldsymbol{v}_\mathrm{e} f_\mathrm{e}(\boldsymbol{k}) \,, \tag{14.7}$$

$$\boldsymbol{w}_\mathrm{e} = \frac{1}{V}\sum_{\boldsymbol{k},s} E_\mathrm{e}(\boldsymbol{k}) \boldsymbol{v}_\mathrm{e} f_\mathrm{e}(\boldsymbol{k}) \,, \tag{14.8}$$

which are summations over all states weighted with the occupation function $f_\mathrm{e}(\boldsymbol{k})$. Equation (14.7) is nothing but (2.128) in a basis diagonal in $\boldsymbol{p}$ with the terms containing $\boldsymbol{A}$, describing the photogalvanic effect, neglected. Without an external magnetic field, the occupation functions and energies usually do not depend on the spin index s; hence we have suppressed the index s above and in the following. Obviously, the entire transport problem is reduced to finding the actual occupation function $f_\mathrm{e}(\boldsymbol{k})$ and evaluating the above sums. If $f_\mathrm{e}(\boldsymbol{k})$ is given by the thermal equilibrium distribution $f_\mathrm{e}^0(\boldsymbol{k})$, both $\boldsymbol{j}$ and $\boldsymbol{w}$ are zero because the sums contain an asymmetric argument. Physically, for any electron moving with velocity $\boldsymbol{v}_\mathrm{e}$ there is another one moving with $-\boldsymbol{v}_\mathrm{e}$, resulting in zero net current.

In Chap. 10 (see (10.69) and (10.153)), we have already discussed the equation of motion of the occupation function under the influence of external perturbations (without magnetic fields), i.e. the *Boltzmann equation*[2]

$$\left(\frac{\partial}{\partial t} + \frac{\hbar\boldsymbol{k}}{m_\mathrm{e}}\nabla_{\boldsymbol{r}} - [\nabla_{\boldsymbol{r}} V(\boldsymbol{r})]\nabla_{\hbar\boldsymbol{k}} - [e\nabla_{\hbar\boldsymbol{k}} E_\mathrm{e}(\boldsymbol{k}) \times \boldsymbol{B}]\nabla_{\hbar\boldsymbol{k}}\right) f_\mathrm{e}(\boldsymbol{r},\boldsymbol{k},t)$$

$$= \left.\frac{\partial f_\mathrm{e}(\boldsymbol{r},\boldsymbol{k},t)}{\partial t}\right|_\mathrm{scatt} , \tag{14.9}$$

where we have employed the effective-mass approximation. From now on we restrict ourselves to stationary problems, i.e. $\partial f_\mathrm{e}/\partial t = 0$, and we drop t as an argument.

The main problem left in solving the Boltzmann equation (14.9) arises from the scattering term

$$\left.\frac{\partial f_\mathrm{e}(\boldsymbol{r},\boldsymbol{k})}{\partial t}\right|_\mathrm{scatt} . \tag{14.10}$$

[2] A shortcut derivation of the Boltzmann equation in the presence of a magnetic field can be given as follows: Let us consider a small phase space volume around coordinate $\boldsymbol{r}$ and wave vector $\boldsymbol{k}$. Owing to the external perturbations, the shape and position of this element in phase space may change, however, neglecting scattering, the total number of electrons in this volume is conserved. Hence, the total temporal derivative vanishes, i.e. $\mathrm{d}f_\mathrm{e}(\boldsymbol{r},\boldsymbol{k},t)/\mathrm{d}t = 0$, leading to $\left(\partial/\partial t + \dot{\boldsymbol{r}}\,\nabla_{\boldsymbol{r}} + \dot{\boldsymbol{k}}\,\nabla_{\boldsymbol{k}}\right) f_\mathrm{e}(\boldsymbol{r},\boldsymbol{k},t) = 0$. Using the effective-mass approximation, Newton's law, and the Lorentz force, we have $\hbar\dot{\boldsymbol{k}} = -\nabla_{\boldsymbol{r}} V(\boldsymbol{r}) - e\dot{\boldsymbol{r}} \times \boldsymbol{B}$; with this relation, the preceding equation can easily be arranged in the form (14.9).

In Chap. 10 we have shown that the scattering term can be simplified substantially if the *Markov approximation* is justified. Using (10.153) and assuming small deviations of $f_e(\boldsymbol{r},\boldsymbol{k})$ from its thermal equilibrium value $f_e^0(\boldsymbol{r},\boldsymbol{k})$, i.e.

$$f_e(\boldsymbol{r},\boldsymbol{k}) = f_e^0(\boldsymbol{r},\boldsymbol{k}) + \delta f_e(\boldsymbol{r},\boldsymbol{k}) \,, \tag{14.11}$$

where $\delta f_e \ll f_e^0$, the scattering rate becomes

$$\left.\frac{\partial f_e(\boldsymbol{r},\boldsymbol{k})}{\partial t}\right|_{\text{scatt}} = -\frac{\delta f_e(\boldsymbol{r},\boldsymbol{k})}{\tau(\boldsymbol{k})} = -\frac{f_e(\boldsymbol{r},\boldsymbol{k}) - f_e^0(\boldsymbol{r},\boldsymbol{k})}{\tau(\boldsymbol{k})} \,, \tag{14.12}$$

which describes a simple exponential relaxation of $f_e(\boldsymbol{r},\boldsymbol{k})$ towards thermal equilibrium (see also Chaps. 6 and 11). The $\boldsymbol{r}$ dependence of f_e^0 is usually referred to as the assumption of *local equilibrium*. The k dependence of τ is often neglected for simplicity. Note that τ is temperature dependent.

14.1 Transport Coefficients (Without Magnetic Field)

It is now straightforward to compute the electric current density and the heat current density by inserting (14.12) into the Boltzmann equation (14.9) and inserting the resulting occupation functions $f_e(\boldsymbol{r},\boldsymbol{k})$ into (14.7) and (14.8). The first step leads to

$$\frac{\hbar\boldsymbol{k}}{m_e}\nabla_{\boldsymbol{r}} f_e^0(\boldsymbol{r},\boldsymbol{k}) + e[\nabla_{\boldsymbol{r}}\phi(\boldsymbol{r})]\nabla_{\hbar\boldsymbol{k}} f_e^0(\boldsymbol{r},\boldsymbol{k}) = -\frac{\delta f_e(\boldsymbol{r},\boldsymbol{k})}{\tau} \,, \tag{14.13}$$

where we have expressed $\nabla_{\boldsymbol{r}} V(\boldsymbol{r}) = -e\nabla_{\boldsymbol{r}}\phi(\boldsymbol{r})$ via the electrostatic potential $\phi(\boldsymbol{r})$ and have neglected contributions from $\delta f_e \ll f_e^0$ on the left side of (14.9). The gradients of f_e^0 can be rewritten with the chain rule and the effective-mass approximation as follows:

$$\nabla_{\boldsymbol{r}} f_e^0 = \frac{\partial f_e^0}{\partial[E_e(\boldsymbol{k}) - \mu_e]}\left(\nabla_{\boldsymbol{r}}[E_e(\boldsymbol{k}) - \mu_e] - \frac{E_e(\boldsymbol{k}) - \mu_e}{k_B T} k_B \nabla_{\boldsymbol{r}} T\right) \,, \tag{14.14}$$

$$\nabla_{\hbar\boldsymbol{k}} f_e^0 = \frac{\partial f_e^0}{\partial E_e}\nabla_{\hbar\boldsymbol{k}} E_e(\boldsymbol{k}) = \frac{\partial f_e^0}{\partial E_e}\frac{\hbar\boldsymbol{k}}{m_e} \,. \tag{14.15}$$

Assuming a spatially constant energy dispersion[3] $E_e(\boldsymbol{k})$. we find for δf_e

$$\delta f_e = \tau\left(\frac{\hbar\boldsymbol{k}}{m_e}(E_e - \mu_e)\frac{\partial f_e^0}{\partial E_e}\frac{\nabla_{\boldsymbol{r}} T}{T} + \frac{\hbar\boldsymbol{k}}{m_e}\frac{\partial f_e^0}{\partial E_e}\nabla_{\boldsymbol{r}}\eta_e\right) \,, \tag{14.16}$$

where we have introduced the *electrochemical potential*[4] η_e,

$$\eta_e = \mu_e - e\phi \,, \tag{14.17}$$

[3] This is not appropriate for a semiconductor heterostructure where the band edge may vary locally.

[4] For a spatially varying $E_e(\boldsymbol{k})$, η_e has to be replaced by $\mu_e - e\phi - E_e(\boldsymbol{k})$.

as the sum of the chemical potential μ_e and the electrostatic energy $-e\phi$. Substituting δf_e in (14.7) and (14.8) by means of (14.16) leads to

$$\boldsymbol{j}_\mathrm{e} = \underbrace{\left(- \sum_{\boldsymbol{k},s} \tau \frac{e^2}{V} \left(\frac{\hbar \boldsymbol{k}}{m_\mathrm{e}}\right)^2 \frac{\partial f_\mathrm{e}^0}{\partial E_\mathrm{e}} \right)}_{a_{11}} \nabla_{\boldsymbol{r}} \left(\frac{\eta_\mathrm{e}}{e}\right)$$

$$+ \underbrace{\left(\sum_{\boldsymbol{k},s} \tau \frac{e}{V} \left(\frac{\hbar \boldsymbol{k}}{m_\mathrm{e}}\right)^2 [E_\mathrm{e}(\boldsymbol{k}) - \mu_\mathrm{e}] \frac{\partial f_\mathrm{e}^0}{\partial E_\mathrm{e}} \right)}_{a_{12}} \left(- \frac{\nabla_{\boldsymbol{r}} T}{T} \right) \tag{14.18}$$

and

$$\boldsymbol{w}_\mathrm{e} = \underbrace{\left(\sum_{\boldsymbol{k},s} \tau \frac{e}{V} \left(\frac{\hbar \boldsymbol{k}}{m_\mathrm{e}}\right)^2 E_\mathrm{e}(\boldsymbol{k}) \frac{\partial f_\mathrm{e}^0}{\partial E_\mathrm{e}} \right)}_{a_{21}} \nabla_{\boldsymbol{r}} \left(\frac{\eta_\mathrm{e}}{e}\right)$$

$$+ \underbrace{\left(\sum_{\boldsymbol{k},s} \frac{\tau}{V} \left(\frac{\hbar \boldsymbol{k}}{m_\mathrm{e}}\right)^2 E_\mathrm{e}(\boldsymbol{k}) \left(E_\mathrm{e}(\boldsymbol{k}) - \mu_\mathrm{e}\right) \frac{\partial f_\mathrm{e}^0}{\partial E_\mathrm{e}} \right)}_{a_{22}} \left(- \frac{\nabla_{\boldsymbol{r}} T}{T} \right) . \tag{14.19}$$

This final result, with coefficients a_{11}, a_{12}, a_{21}, and a_{22} can be rearranged into the more common form

$$\nabla_{\boldsymbol{r}} \left(\frac{\eta_\mathrm{e}}{e}\right) = \frac{\boldsymbol{j}_\mathrm{e}}{\sigma_\mathrm{e}} + \varepsilon_\mathrm{e} \nabla_{\boldsymbol{r}} T \,, \tag{14.20}$$

$$\boldsymbol{w}_\mathrm{e} = \Pi_\mathrm{e} \boldsymbol{j}_\mathrm{e} - \kappa_\mathrm{e} \nabla_{\boldsymbol{r}} T \,, \tag{14.21}$$

where the four *transport coefficients* are defined as follows:

electrical conductivity	σ_e	$= a_{11}$
thermopower	ε_e	$= a_{12}/(T a_{11})$
Peltier coefficient	Π_e	$= a_{21}/a_{11}$
thermal conductivity	κ_e	$= T^{-1}(a_{22}a_{11} - a_{12}a_{21})/a_{11}$.

Let us illustrate the physics of these coefficients with a few examples.

14.1.1 Electrical Conductivity

The electrical conductivity $\sigma_e = a_{11}$ can be evaluated via partial integration as follows:

$$\sigma_e = -\sum_{\boldsymbol{k},s} \tau \frac{e^2}{V} \left(\frac{\hbar \boldsymbol{k}}{m_e}\right)^2 \frac{\partial f_e^0}{\partial E_e} = -\sum_{\boldsymbol{k},s} \tau \frac{e^2}{V} \frac{\hbar \boldsymbol{k}}{m_e} \nabla_{\hbar \boldsymbol{k}} f_e^0$$

$$= -\tau \frac{e^2}{V} \left(\frac{\hbar \boldsymbol{k}}{m_e} f_e^0(\boldsymbol{k})\right)\Bigg|_{-\infty}^{+\infty} + \tau \frac{e^2}{m_e} \frac{1}{V} \sum_{\boldsymbol{k},s} f_e^0(\boldsymbol{k}) . \tag{14.22}$$

As the first contribution on the right-hand side vanishes, we obtain

$$\sigma_e = \frac{\tau e^2 n_e}{m_e} . \tag{14.23}$$

If the chemical potential is spatially constant, $\nabla_{\boldsymbol{r}}(\eta_e/e) = \boldsymbol{E}$ applies, and we obtain *Ohm's law*,

$$\boldsymbol{j}_e = \sigma_e \boldsymbol{E} . \tag{14.24}$$

Frequently, the *mobility* $\tilde{\mu}_e$ is defined via

$$\sigma_e = n_e e \tilde{\mu}_e . \tag{14.25}$$

Notice that the charge enters twice in the electrical conductivity. Consequently, the electron and hole contributions have the same sign.

14.1.2 Peltier Coefficient

Let us consider the arrangement sketched in Fig. 14.1 where three wires with cross section F made of two semiconductors A and B with different Peltier coefficients Π_e^A and Π_e^B are connected in a loop. The setup is held at constant temperature before a current j_e is sent through the arrangement. The electron contributions to the heat current density are given simply by $\boldsymbol{w}_e = \Pi_e^A \boldsymbol{j}_e$ and

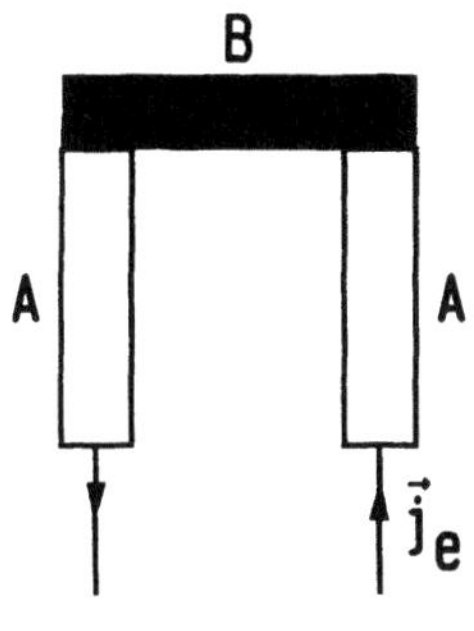

Fig. 14.1. Schematic illustration of a Peltier element

$\boldsymbol{w}_e = \Pi_e^B \boldsymbol{j}_e$ in materials A and B, respectively. Let us consider the energy $\mathcal{U}_{L,R}$ in a volume containing the left or right contact, respectively. We obtain

$$\frac{\partial \mathcal{U}_{L,R}}{\partial t} = -\int \boldsymbol{w}\, d\boldsymbol{F} = \pm I_e \left(\Pi_e^A - \Pi_e^B\right) , \tag{14.26}$$

where the total current I_e is given by $I_e = |\boldsymbol{j}_e| F$ if the contribution from holes is negligible. Depending on which Peltier coefficient is larger, the left contact is either heated or cooled, and the right contact is cooled or heated, respectively.

14.1.3 Thermal Conductivity

Via $\boldsymbol{w}_e = -\kappa_e \nabla_{\boldsymbol{r}} T$, a temperature gradient results in heat a current. It is straightforward to show that the electrical and thermal conductivities are closely related via the *Wiedemann–Franz law*

$$\frac{\kappa_e}{\sigma_e} = \frac{\pi^2}{3}\left(\frac{k_B}{e}\right)^2 T = LT , \tag{14.27}$$

where L is the *Lorentz number.* Intuitively, this important finding is clearly understandable because the particle transport of electrons and holes leads to both electric and heat currents; hence, the two currents have to be closely related.

14.2 Transport Coefficients (with Magnetic Field)

Let us now switch on a static magnetic field. The general procedure for computing the electric current density and the heat current density remains the same as in the case of a vanishing magnetic field. Now, however, we have to consider the complete, stationary Boltzmann equation[5] (14.9):

$$\left(\frac{\hbar \boldsymbol{k}}{m_e}\nabla_{\boldsymbol{r}} - [\nabla_{\boldsymbol{r}} V(\boldsymbol{r})]\nabla_{\hbar \boldsymbol{k}} - e[\nabla_{\hbar \boldsymbol{k}} E_e(\boldsymbol{k}) \times \boldsymbol{B}]\nabla_{\hbar \boldsymbol{k}}\right) f_e(\boldsymbol{r}, \boldsymbol{k}, t)$$
$$= -\frac{\delta f_e(\boldsymbol{r}, \boldsymbol{k})}{\tau} . \tag{14.28}$$

While we could safely replace $f_e = f_e^0 + \delta f_e$ by f_e^0 on the left-hand side in the case of $\boldsymbol{B} = 0$ (as $\delta f_e \ll f_e^0$), care has to taken for $\boldsymbol{B} \neq 0$ because of the last term inside the large parenthesis on the left-hand side, as the contribution

$$e[\nabla_{\hbar \boldsymbol{k}} E_e(\boldsymbol{k}) \times \boldsymbol{B}]\nabla_{\hbar \boldsymbol{k}}\, f_e^0(\boldsymbol{r}, \boldsymbol{k})$$

[5] In the presence of a magnetic field, suppression of the spin index in the distribution functions and energies is an approximation, because the Zeeman energy generally lifts the degeneracy of the spin states. This effect is rarely of importance in classical transport. We shall, however, find spin effects in the quantum Hall effect, discussed in Chap. 16.

$$= e[\nabla_{\hbar \boldsymbol{k}} E_{\mathrm{e}}(\boldsymbol{k}) \times \boldsymbol{B}] \frac{\partial f_{\mathrm{e}}^0}{\partial E_{\mathrm{e}}(\boldsymbol{k})} \nabla_{\hbar \boldsymbol{k}} E_{\mathrm{e}}(\boldsymbol{k}) = 0 \tag{14.29}$$

vanishes because $(\boldsymbol{a} \times \boldsymbol{b}) \cdot \boldsymbol{a} = 0$. This leads us to

$$\frac{\hbar \boldsymbol{k}}{m_{\mathrm{e}}} \nabla_{\boldsymbol{r}} f_{\mathrm{e}}^0(\boldsymbol{r}, \boldsymbol{k}) + e[\nabla_{\boldsymbol{r}} \phi(\boldsymbol{r})] \nabla_{\hbar \boldsymbol{k}} f_{\mathrm{e}}^0(\boldsymbol{r}, \boldsymbol{k})$$

$$-e[\nabla_{\hbar \boldsymbol{k}} E_{\mathrm{e}}(\boldsymbol{k}) \times \boldsymbol{B}] \nabla_{\hbar \boldsymbol{k}} \, \delta f_{\mathrm{e}}(\boldsymbol{r}, \boldsymbol{k}) = - \frac{\delta f_{\mathrm{e}}(\boldsymbol{r}, \boldsymbol{k})}{\tau} . \tag{14.30}$$

This differential equation for δf_{e} can be solved by means of an iterative scheme in orders of the magnetic field, where

$$\boldsymbol{B} \to \lambda \boldsymbol{B} \tag{14.31}$$

and

$$\delta f_{\mathrm{e}} = \lambda^0 \, \delta f_{\mathrm{e}}^{(0)} + \lambda^1 \, \delta f_{\mathrm{e}}^{(1)} + \ldots \tag{14.32}$$

with ordering parameter λ. The zero-order solution is obviously identical to that in the zero-field case, i.e. (14.16):

$$\delta f_{\mathrm{e}}^{(0)} = \tau \left(\frac{\hbar \boldsymbol{k}}{m_{\mathrm{e}}} [E_{\mathrm{e}}(\boldsymbol{k}) - \mu_{\mathrm{e}}] \frac{\partial f_{\mathrm{e}}^0}{\partial E_{\mathrm{e}}} \frac{\nabla_{\boldsymbol{r}} T}{T} + \frac{\hbar \boldsymbol{k}}{m_{\mathrm{e}}} \frac{\partial f_{\mathrm{e}}^0}{\partial E_{\mathrm{e}}} \nabla_{\boldsymbol{r}} \eta_{\mathrm{e}} \right) . \tag{14.33}$$

The first-order correction, which includes all contributions linear in the magnetic field, is given by

$$-\frac{\delta f_{\mathrm{e}}^{(1)}}{\tau} = -e[\nabla_{\hbar \boldsymbol{k}} E_{\mathrm{e}}(\boldsymbol{k}) \times \boldsymbol{B}] \nabla_{\hbar \boldsymbol{k}} \, \delta f_{\mathrm{e}}^{(0)} , \tag{14.34}$$

leading to[6]

$$\delta f_{\mathrm{e}}^{(1)} = \frac{e\tau^2}{m_{\mathrm{e}}} [\nabla_{\hbar \boldsymbol{k}} E_{\mathrm{e}}(\boldsymbol{k}) \times \boldsymbol{B}] \left([E_{\mathrm{e}}(\boldsymbol{k}) - \mu_{\mathrm{e}}] \frac{\partial f_{\mathrm{e}}^0}{\partial E_{\mathrm{e}}} \frac{\nabla_{\boldsymbol{r}} T}{T} + \frac{\partial f_{\mathrm{e}}^0}{\partial E_{\mathrm{e}}} \nabla_{\boldsymbol{r}} \eta_{\mathrm{e}} \right) . \tag{14.35}$$

This finally enables us to compute the electric current density (14.7):

$$\boldsymbol{j}_{\mathrm{e}} = -\frac{e}{V} \sum_{\boldsymbol{k},s} \boldsymbol{v}_{\mathrm{e}} f_{\mathrm{e}}(\boldsymbol{k}) = -\frac{e}{V} \sum_{\boldsymbol{k},s} \boldsymbol{v}_{\mathrm{e}} \left(\delta f_{\mathrm{e}}^{(0)}(\boldsymbol{k}) + \delta f_{\mathrm{e}}^{(1)}(\boldsymbol{k}) \right)$$

$$= a_{11} \nabla_{\boldsymbol{r}} \left(\frac{\eta_{\mathrm{e}}}{e} \right) + a_{12} \left(- \frac{\nabla_{\boldsymbol{r}} T}{T} \right)$$

$$+ \left[\boldsymbol{B} \times \nabla_{\boldsymbol{r}} \left(\frac{\eta_{\mathrm{e}}}{e} \right) \right] \underbrace{\left(- \sum_{\boldsymbol{k},s} \frac{e}{V} \frac{e^2 \tau^2}{m_{\mathrm{e}}} \left(\frac{\hbar \boldsymbol{k}}{m_{\mathrm{e}}} \right)^2 \frac{\partial f_{\mathrm{e}}^0}{\partial E_{\mathrm{e}}} \right)}_{= a_{13}}$$

[6] Again, some terms vanish because $(\boldsymbol{a} \times \boldsymbol{b}) \cdot \boldsymbol{a} = 0$.

$$+\left[\boldsymbol{B}\times\left(\frac{\nabla_{\boldsymbol{r}}T}{T}\right)\right]\underbrace{\left(-\sum_{\boldsymbol{k},s}\frac{e}{V}\frac{e\tau^2}{m_{\mathrm{e}}}\left(\frac{\hbar\boldsymbol{k}}{m_{\mathrm{e}}}\right)^2[E_{\mathrm{e}}(\boldsymbol{k})-\mu_{\mathrm{e}}]\frac{\partial f_{\mathrm{e}}^0}{\partial E_{\mathrm{e}}}\right)}_{=a_{14}}, \tag{14.36}$$

where a_{11} and a_{12} are as derived in the previous section. Interestingly, the electric current density has not only contributions in the direction of the gradient of the electrochemical potential or the temperature gradient, but also perpendicular contributions.

Let us keep in mind that similar contributions to $\boldsymbol{j}$ are obtained for the holes, with positive charge. Importantly, the charge enters three times in the contribution with coefficient a_{13}, and hence this term has opposite signs for holes and electrons.

14.2.1 Hall Effect and Hall Resistance

Let us choose the magnetic field to be in the z direction and consider a homogeneous semiconductor held at constant temperature by some sort of heat bath (see Fig. 14.2). In this case the electric field is given simply by

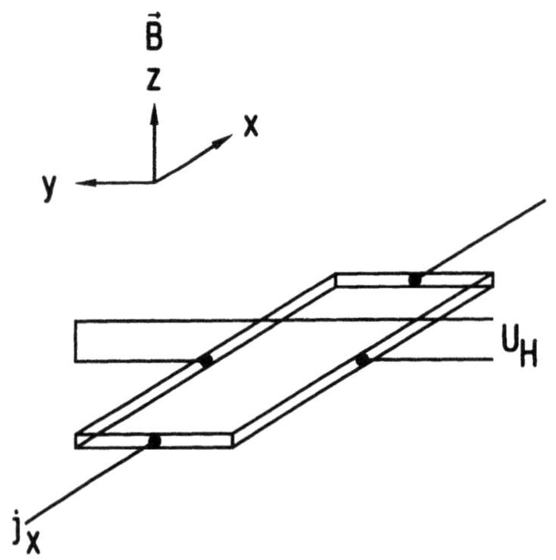

Fig. 14.2. Schematic illustration of a Hall element consisting of a finite slab of semiconductor. The spatial extent of the platelet in the x, y, and z direction is L_x, L_y and L_z, respectively

$\boldsymbol{E}=\nabla_{\boldsymbol{r}}(\eta_{\mathrm{e}}/e)$. For dominantly electronic transport ($\boldsymbol{j}\approx\boldsymbol{j}_{\mathrm{e}}$), we have, in components,

$$\begin{aligned} j_x &= a_{11}E_x - a_{13}B_zE_y\ , \\ j_y &= a_{11}E_y + a_{13}B_zE_x\ , \end{aligned} \tag{14.37}$$

or, equivalently,

$$\begin{pmatrix} j_x \\ j_y \end{pmatrix} = \begin{pmatrix} \sigma_{xx} & \sigma_{xy} \\ \sigma_{yx} & \sigma_{yy} \end{pmatrix} \begin{pmatrix} E_x \\ E_y \end{pmatrix} , \tag{14.38}$$

where we have introduced the *conductivity tensor* $\underline{\sigma}$. This is given by

$$\begin{pmatrix} \sigma_{xx} & \sigma_{xy} \\ \sigma_{yx} & \sigma_{yy} \end{pmatrix} = \begin{pmatrix} a_{11} & -B_z a_{13} \\ +B_z a_{13} & a_{11} \end{pmatrix} , \tag{14.39}$$

and is the inverse of the *tensor of specific resistivity* $\underline{\rho}$, where

$$\begin{pmatrix} \rho_{xx} & \rho_{xy} \\ \rho_{yx} & \rho_{yy} \end{pmatrix} = \begin{pmatrix} \sigma_{xx} & \sigma_{xy} \\ \sigma_{yx} & \sigma_{yy} \end{pmatrix}^{-1}$$
$$= \frac{1}{\sigma_{xx}\sigma_{yy} - \sigma_{xy}\sigma_{yx}} \cdot \begin{pmatrix} +\sigma_{yy} & -\sigma_{xy} \\ -\sigma_{yx} & +\sigma_{xx} \end{pmatrix} . \tag{14.40}$$

Let the electric field point in the x direction. If $\boldsymbol{B} = 0$, $\sigma_{xy} = \sigma_{yx} = 0$ and we have $\rho_{xx} = \sigma_{xx}^{-1}$. Consequently, for an perfect conductor, the conductivity σ_{xx} becomes infinite and ρ_{xx} goes to zero. If, on the other hand, $\boldsymbol{B}$ is finite, $\sigma_{xx} = 0$ is equivalent to $\rho_{xx} = 0$ as both $\sigma_{xy} \neq 0$ and $\sigma_{yx} \neq 0$. The xx component is often referred to as the longitudinal component and we summarize this somewhat unintuitive result as follows:

- For $\boldsymbol{B} \neq 0$, a vanishing longitudinal conductivity is equivalent to a vanishing longitudinal resistivity.
- For $\boldsymbol{B} = 0$, a vanishing (longitudinal) conductivity is equivalent to a diverging (longitudinal) resistivity.

Let us fix the current (not the field) in the x direction.[7] The resulting x component of the carrier velocity together with the magnetic field in the z direction, leads to a Lorentz force in the y direction. If the spatial extent in the y direction is finite, the carriers accumulate at the side contacts and give rise to an electric field in the y direction.[8] This field compensates the Lorentz field under steady-state conditions. This scenario is the *Hall effect*. Using $j_y = 0$ in the second line of (14.37), inserting the resulting $E_x = -(a_{11}E_y)/(a_{13}B_z)$ into the first line of (14.37), and consistently neglecting terms beyond linear order in $\boldsymbol{B}$ leads us to

$$E_y = -j_x B_z \frac{a_{13}}{a_{11}^2} = -\frac{j_x B_z}{e n_\mathrm{e}} , \tag{14.41}$$

where we have used $a_{13} = (e\tau/m_\mathrm{e})a_{11}$ and $a_{11} = e^2\tau n_\mathrm{e}/m_\mathrm{e}$ (see (14.23)). Introducing the Hall voltage $U_\mathrm{H} = -E_y L_y$ and expressing the current density by means of the absolute current, $j_x = I_x/(L_y L_z)$, we obtain the *Hall voltage*

$$U_\mathrm{H} = R_\mathrm{H} I_x , \tag{14.42}$$

[7] Usually, a constant-current source is used in experiments.

[8] In the limit of an infinite sample no electric field builds up, i.e. $E_y = 0$; however, we have a finite current, $j_y \neq 0$. This limit can be realized to a reasonable degree by a *Corbino disk*.

where the *Hall resistance* R_H is given by

$$R_\mathrm{H} = \frac{B_z}{en_\mathrm{e}L_z} =: \frac{B_z}{en_\mathrm{e}^{2\mathrm{d}}} \,. \tag{14.43}$$

Here we have replaced the three-dimensional electron density n_e by the sheet density $n_\mathrm{e}^{2\mathrm{d}} = n_\mathrm{e}L_z$ (carriers per unit area). The latter will prove useful for the discusion of Hall devices employing two-dimensional electron gases.

The Hall voltage (and consequently also the Hall resistance) has opposite signs for electrons and holes. Hence, one can derive the dominant type of conduction directly from a Hall experiment. In specific cases, the two contributions can add up to zero.

Semiconductor Hall devices are quite important for measuring magnetic fields, as the Hall voltage is simply proportional to the z component of the magnetic field for a given longitudinal current. The prefactor can easily be found in a calibration experiment; commercial Hall elements are usually already calibrated. In practice one is interested in large, and hence easy to detect, Hall voltages. This leads to two design criteria. As the Hall voltage is inversely proportional to the spatial extent of the platelet in the z direction, thin samples are usually preferred. As U_H is inversely proportional to the carrier concentration, semiconductors, with their lower values of n_e are preferred to metals.

If we do not fix generally the temperature of the Hall element, a corresponding temperature gradient builds up. In particular, a temperature gradient is obtained in the y direction, which can be derived from (14.36) and the corresponding expression for $\boldsymbol{w}_\mathrm{e}$. This effect, which we shall not discuss in detail, is called the *Ettinghausen effect.* It is usually of minor importance, as heat conduction is quite effective in semiconductors.

As can be seen from (14.36), a temperature gradient in the x direction, i.e. $\boldsymbol{B}\times(\nabla_{\boldsymbol{r}}T/T) \neq 0$ (while $\boldsymbol{B}\times\nabla_{\boldsymbol{r}}(\eta_\mathrm{e}/e) = 0$), together with a z component of the magnetic field, leads to an electric field in the y direction (for a finite sample), similarly to the Hall effect. This is the *Nernst effect.* Note that the charge enters twice in the corresponding coefficient a_{14} in (14.36), and hence the contributions of electrons and holes have the same sign – in contrast to the Hall effect. A secondary effect, similar to the Ettinghausen effect, is the *Righi–Leduc effect*, which is a minor correction to the Nernst effect.

14.3 Towards Ballistic Electrons: The Hot-Electron Transistor

A nice example of transport, in which the deviations from thermal equilibrium are large, but where the coherence of electrons is unimportant, is the *hot-*

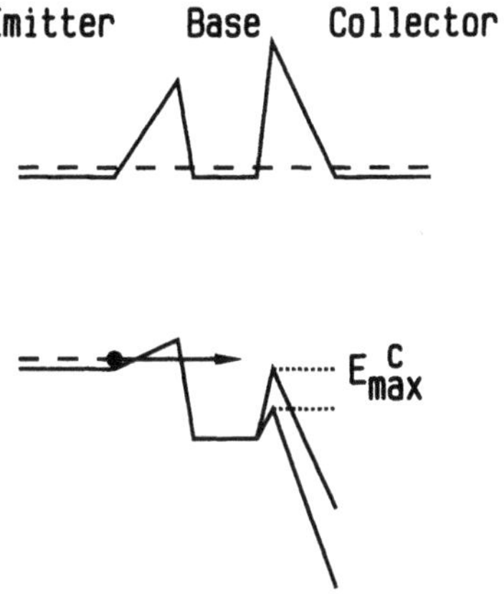

Fig. 14.3. Schematic illustration of the conduction band edge energy of a hot-electron transistor under two different bias conditions. *Top*, without applied bias voltage; *bottom*, with emitter–base bias voltage and two different base–collector bias voltages

electron transistor.[9] Some corresponding energy band diagrams are shown in Fig. 14.3. Such structures are realized by continuously grading the aluminum content of an $Al_xGa_{1-x}As$ alloy, grown by molecular-beam epitaxy. Electrons are injected from the left n-doped contact, the *emitter*, into the n-doped region in the middle, the *base* (typically ≈ 100 nm in thickness), and arrive at the right-hand-side n-doped contact, the *collector*. Bias voltages can be applied between the base and emitter (U_{be}) and the base and collector (U_{bc}). Electrons are injected via the left-hand-side potential barrier. For sufficiently thick right-hand-side potential barriers, only electrons with an energy larger than the right-hand-side potential maximum E^c_{max} can leave the base and enter the collector. In this limit the collector current density j^c_e is given by

$$\begin{aligned} j^c_e &= -\frac{e}{V} \sum_{\boldsymbol{k},s,E_e(\boldsymbol{k})>E^c_{max}} v_e f_e(\boldsymbol{k}) \\ &= -\frac{e}{V} \int_{E^c_{max}}^{\infty} v_e(E) D_e(E)\, f_e(E)\, dE\, , \end{aligned} \tag{14.44}$$

where $f_e(E)$ is the electron distribution at the right-hand end of the base region. Since the energy maximum E^c_{max} and the base–collector bias voltage are related linearly, we have

$$\frac{dj^c_e}{dE^c_{max}}(E) \sim -\frac{dj^c_e}{dU_{bc}}(E) \sim v_e(E)\, D_e(E)\, f_e(E)\, , \tag{14.45}$$

which predominantly reflects the carrier distribution function, as $v_e(E)$ and $D_e(E)$ are rather structureless functions above the conduction band edge (of the base). Obviously, this allows spectroscopy of the base electron distribution. The injection conditions can be varied by means of the emitter–base bias voltage. Some corresponding experimental results are exhibited in Fig. 14.4.

[9] Here, the term *hot electrons* refers to nonequilibrium carriers rather than hot thermal distributions.

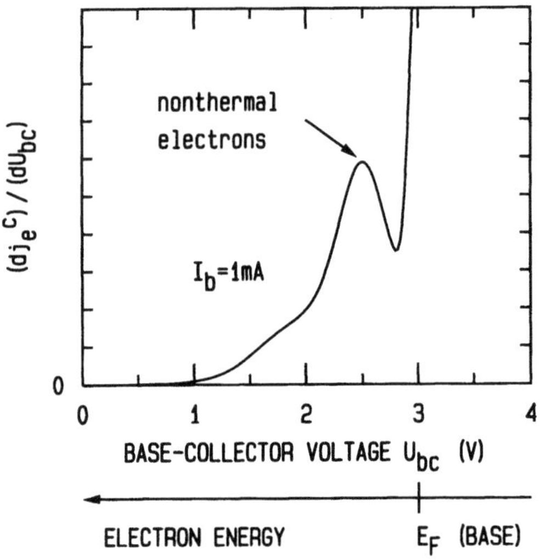

Fig. 14.4. Measured derivatives of collector current density j_e^c with respect to the base–emitter bias voltage U_{bc} of a hot-electron transistor. After [410]

Large base–collector bias voltages correpond to small electron energies and vice versa. Interestingly, a well-defined peak is observed for intermediate bias voltages, i.e. particle energies above the electrochemical potential of equilibrium base electrons. The corresponding electrons have traversed the base region with some scattering, but they are far from thermal equilibrium. Consequently, this peak gradually disappears as thicker base regions are used, where electrons have more time to thermalize. Such experimental results are still well described by the classical ansatz for the current density (14.7); however, one has to solve the full Boltzmann equation numerically (without assuming small deviations from thermal equilibrium and without using the relaxation time approximation). Both electron–electron and electron–phonon scattering have to be accounted for.

14.4 Problems

14.1. Consider a thermoelement made of two different semiconductors A and B (with different thermopowers ε_A and ε_B) in an arbitrary temperature field with temperatures T_1 and T_2 at the two contacts and equal temperatures at the two other ends of semiconductor A (Fig. 14.5). What voltage is obtained?

14.2. Rewrite the transport equations in the form of the diffusion equation

$$\Delta n_e - \frac{1}{D}\frac{\partial n_e}{\partial t} = 0$$

for the electron density n_e. What assumptions are necessary? Show that $D = (\sigma/e^2)(\partial\mu_e/\partial n_e)$, where D is the diffusion coefficient.

14.3. Derive the corresponding equation for the temperature,

$$\Delta T - \frac{1}{K}\frac{\partial T}{\partial t} = 0 \, .$$

What assumptions are necessary? (Introduce the specific heat c via $u = cT$.)

14.4. Derive the expression corresponding to (14.36) for the heat current density $\boldsymbol{w}_\mathrm{e}$.

14.5. Consider the heterostructure depicted in Fig. 14.5 and obtain an analytic expression for the shape of the conduction band edge by using the *abrupt-depletion-layer approximation*, i.e. assuming either fully depleted or undepleted regions in Poisson's equation.

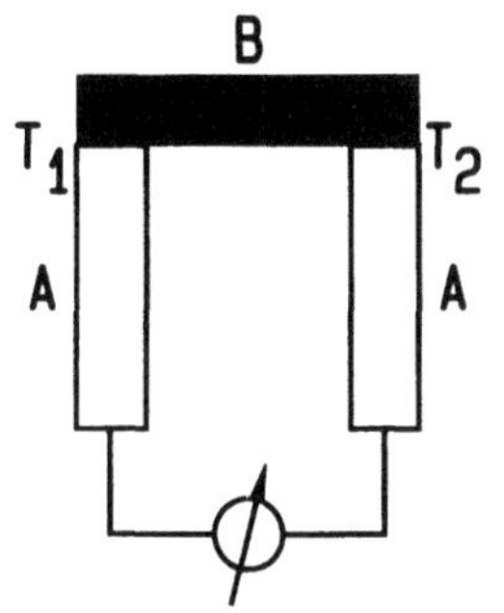

Fig. 14.5. Schematic illustration of a thermoelement

15. Electric Fields in Mesoscopic Systems

In the previous chapter we discussed systems in which the electron mean free path was short compared with typical device dimensions. The physics of transport becomes quite different in the opposite limit, i.e. when the system dimensions are small compared with the mean free path, requiring both high-mobility electrons and small systems. The high mobility of electrons can be achieved, for example, by modulation-doped structures at low temperatures, as discussed in Chap. 1. The typical scales of such structures are on the order of 1μm or below. Such small systems, which are macroscopic compared with one lattice constant, but still microscopic in actual size, are referred to as *mesoscopic systems.*

15.1 Elementary Approach

Let us neglect scattering altogether at this point. In this limit the dynamics of a single electron are purely coherent and are determined by Schrödinger's equation. Such a perfectly coherent situation is in conflict with the process of measuring an electric current (and also of a heat current), which inherently breaks the quantum mechanical phase.[1] Hence, we divide the entire system into three regions: one region in which electrons behave coherently and two regions called *contacts* (see Fig. 15.1). These contacts are treated as classical conductors. It remains to be seen whether this ad hoc separation of the problem is meaningful in specific situations.[2] In a realistic experimental situation we do not have a perfectly sharp borderline but a somewhat smeared-out transition region. Let us consider, for example, the transition from an undoped to a heavily n-doped region. In the n-doped region the electron mean free path is of the order of the average distance between doping

[1] This is nothing but the old problem of quantum mechanical measurement. Any measurement process represents an interaction between a microscopic subsystem and a macroscopic apparatus and leads to a reduction of the wave packet.

[2] This separation is rather natural for the devices used in the good old days of vacuum tubes. There, electrons were injected from one contact, the cathode, into an evacuated region (in which electrons move coherently) and were swallowed by the other contact, the anode.

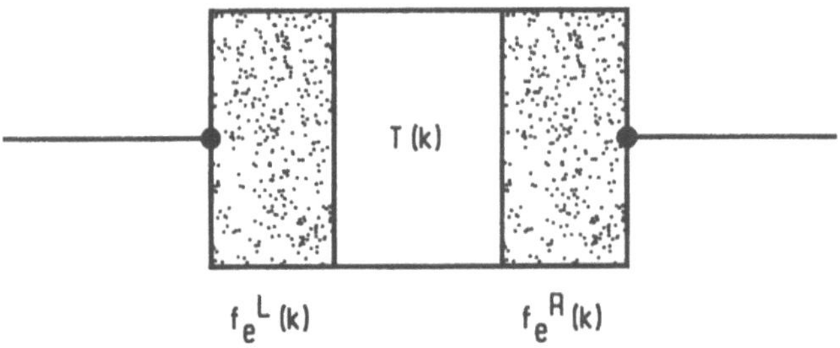

Fig. 15.1. The simplest case of transport in a mesoscopic systems. The system (*middle*) is connected to the ouside world by two contacts (*shaded*). These contacts are treated as classical conductors with corresponding equilibrium (Fermi) distribution functions

atoms. For a doping concentration of $10^{18}\,\mathrm{cm}^{-3}$ this corresponds to 10 nm, a width which can be neglected in many situations.

Electrons can now be injected from the left contact (see Fig. 15.1), say in the z direction with $k_z \geq 0$, and travel coherently through the middle section; quantum mechanical reflections may occur within this region. If the electron is transmitted, it reaches the right-hand contact; quantum mechanical reflections from this interface are assumed to be negligible at this point. Similarly, an electron can be transmitted from the right to the left. Clearly, such a process is proportional to the *quantum mechanical transmission probability* $T(\boldsymbol{k})$. Furthermore, the presence of transmission from left to right requires an occupied state at wave vector $\boldsymbol{k}$ in the left contact, and thus it is proportional to $f_\mathrm{e}^\mathrm{L}(\boldsymbol{k})\,T(\boldsymbol{k})$. Here $f_\mathrm{e}^\mathrm{L,R}(\boldsymbol{k})$ denote the *equilibrium* (Fermi) distributions of electrons in the left (L) and right (R) contacts with chemical potentials $\mu_\mathrm{e}^\mathrm{L}$ and $\mu_\mathrm{e}^\mathrm{R}$, respectively. Again, the opposite process is also possible, and is proportional to $f_\mathrm{e}^\mathrm{R}(\boldsymbol{k})\,T(\boldsymbol{k})$. Hence the net electric current is given by

$$\boldsymbol{j}_\mathrm{e} = -\frac{e}{V}\left(\sum_{\boldsymbol{k},k_z\geq 0} \boldsymbol{v}_\mathrm{e}(k_z)\, f_\mathrm{e}^\mathrm{L}(\boldsymbol{k})\, T(\boldsymbol{k}) + \sum_{\boldsymbol{k},k_z\leq 0} \boldsymbol{v}_\mathrm{e}(k_z)\, f_\mathrm{e}^\mathrm{R}(\boldsymbol{k})\, T(\boldsymbol{k})\right). \tag{15.1}$$

A similar expression holds for the holes. We generally have $T(k_z) = T(-k_z)$, i.e., for the transmission in such a two-terminal device, it does not matter whether we transfer an electron from the left to the right or from the right to the left. This statement remains true for a two-terminal device in the presence of a magnetic field (see Chap. 16); however, it is wrong for multiterminal devices in the presence of a magnetic field, such as those employed for the Hall effect. If $T(k_z) = T(-k_z)$, (15.1) simplifies to

$$\boldsymbol{j}_\mathrm{e} = -\frac{e}{V}\sum_{\boldsymbol{k},k_z\geq 0} \boldsymbol{v}_\mathrm{e}\,[f_\mathrm{e}^\mathrm{L}(\boldsymbol{k}) - f_\mathrm{e}^\mathrm{R}(\boldsymbol{k})]\, T(\boldsymbol{k})\,. \tag{15.2}$$

This heuristic derivation of the current density will be justified later within a microscopic theory. In particular, it will become evident that no additional

Pauli blocking factors ($(1-f_{\mathrm{e}}^{\mathrm{R}})$ and $(1-f_{\mathrm{e}}^{\mathrm{L}})$) occur. Interestingly, they would vanish exactly anyway for a two-terminal device.

It is instructive to compare (15.2) with the classical analogue (14.7)

$$j_{\mathrm{e}} = -\frac{e}{V}\sum_{\boldsymbol{k}} v_{\mathrm{e}} f_{\mathrm{e}}(\boldsymbol{k}) , \tag{15.3}$$

where the occupation functions f_{e} are evaluated in the limit of strong scattering. The occupation function in incoherent transport has to be replaced by the quantum mechanical transmission function for transport in the coherent limit.

The evaluation of (15.2) is rather straightforward. It requires knowledge of the quantum mechanical transmission probability of a single electron, $T(\boldsymbol{k})$. Let us consider two examples: resonant-tunneling diodes and conduction in a thin wire resulting in quantized conductance.

15.1.1 Resonant Tunneling I

Tunneling is one of the most striking phenomena arising from quantum mechanics. The amazing progress in epitaxial growth of semiconductor heterostructures enables us to realize almost any desired potential $V(z)$ as a function of the coordinate in the growth direction z. One example, a double-barrier structure, is depicted in Fig. 15.2.

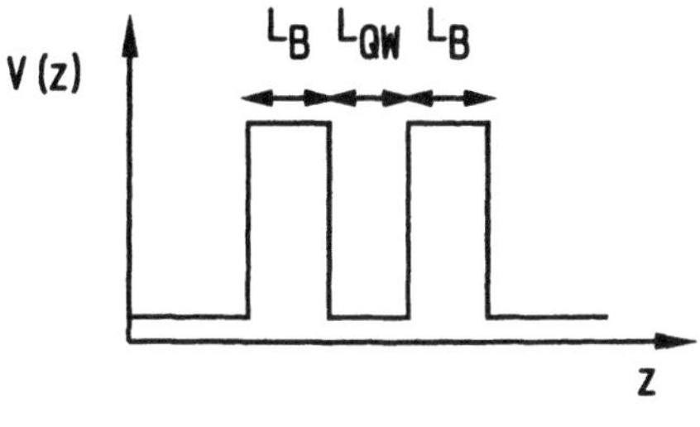

Fig. 15.2. Schematic illustration of a double-barrier heterostructure

Evaluating the quantum mechanical transmission probability T of such one-dimensional potentials is always possible with help of the transfer matrix method.

Transfer Matrix Method. Let us consider the one-dimensional Schrödinger equation for a piecewise constant potential $V(z)$, with potential values V_n, wave vectors k_z^n, and effective masses m_{e}^n in the interval $z \in [z_{n-1}, z_n]$,

$$\left(-\frac{\hbar^2}{2m_{\mathrm{e}}^n}\frac{\partial^2}{\partial z^2} + V_n\right) \Psi_n(z) = E_z \Psi_n(z) . \tag{15.4}$$

As is well known, the wavefunction $\Psi_n(z)$ within each section has the form

$$\Psi_n(z) = A_n \,\mathrm{e}^{\mathrm{i}k_z^n z} + B_n \,\mathrm{e}^{-\mathrm{i}k_z^n z} , \tag{15.5}$$

where

$$\frac{\hbar^2 (k_z^n)^2}{2m_{\rm e}^n} = E_z - V_n \,. \tag{15.6}$$

If $E_z - V_n > 0$, k_z^n is real and the wavefunction in this region is a plane wave; if $E_z - V_n < 0$, k_z^n is imaginary and the wavefunction consists of exponentially growing and decaying contributions. The boundary conditions between two sections are $\Psi_n(z_n) = \Psi_{n+1}(z_n)$ and $(m_{\rm e}^n)^{-1}\partial/\partial z \Psi_n(z_n) = (m_{\rm e}^{n+1})^{-1}\partial/\partial z \Psi_{n+1}(z_n)$ $(n = 1, ..., N-1)$. Hence we can relate the coefficients A and B in section n and section $n+1$ by a 2×2 matrix $\mathcal{M}_n$, *the transfer matrix*,[3] such that

$$\begin{pmatrix} A_n \\ B_n \end{pmatrix} = \mathcal{M}_n \begin{pmatrix} A_{n+1} \\ B_{n+1} \end{pmatrix} , \tag{15.7}$$

where the elements of $\mathcal{M}_n$ are given by

$$\begin{aligned}
\mathcal{M}_n\Big|_{11} &= \left(\frac{1}{2} + \frac{k_z^{n+1} m_{\rm e}^n}{2k_z^n m_{\rm e}^{n+1}}\right) \exp[{\rm i}(k_z^{n+1} - k_z^n) z_n] \,, \\
\mathcal{M}_n\Big|_{12} &= \left(\frac{1}{2} - \frac{k_z^{n+1} m_{\rm e}^n}{2k_z^n m_{\rm e}^{n+1}}\right) \exp[-\,{\rm i}(k_z^{n+1} + k_z^n) z_n] \,, \\
\mathcal{M}_n\Big|_{21} &= \left(\frac{1}{2} - \frac{k_z^{n+1} m_{\rm e}^n}{2k_z^n m_{\rm e}^{n+1}}\right) \exp[{\rm i}(k_z^{n+1} + k_z^n) z_n] \,, \\
\mathcal{M}_n\Big|_{22} &= \left(\frac{1}{2} + \frac{k_z^{n+1} m_{\rm e}^n}{2k_z^n m_{\rm e}^{n+1}}\right) \exp[-\,{\rm i}(k_z^{n+1} - k_z^n) z_n] \,.
\end{aligned} \tag{15.8}$$

Thus the coefficients A and B of the first and Nth sections are related via

$$\begin{pmatrix} A_1 \\ B_1 \end{pmatrix} = (\mathcal{M}_1 \cdot \mathcal{M}_2 \cdot ..... \cdot \mathcal{M}_{N-1}) \begin{pmatrix} A_N \\ B_N \end{pmatrix} . \tag{15.9}$$

For an electron incident from the left (section 1), only a transmitted wave will appear on the right (section N), i.e. $B_N = 0$. Using the quantum mechanical particle current density $\boldsymbol{j}$ given by

$$\boldsymbol{j} = \frac{{\rm i}\hbar}{2m_{\rm e}} \left(\Psi \nabla \Psi^* - \Psi^* \nabla \Psi\right), \tag{15.10}$$

the transmission probability

$$T = \frac{j_z^N}{j_z^{1,{\rm R}}} \tag{15.11}$$

(where $j_z^{1,{\rm R}}$ corresponds to the incident current, i.e. the right-going part of the current in section 1) finally becomes

[3] This transfer matrix approach is very closely related to the transfer matrices commonly used in optics to compute the transmission of a stack of dielectrics.

$$T = \frac{k_z^N m_e^1}{k_z^1 m_e^N} \frac{|A_N|^2}{|A_1|^2} . \tag{15.12}$$

For very simple potentials such as the one shown in Fig. 15.2, the resulting transmission probability can be evaluated analytically. In general, the above method provides a simple and effective way of solving the one-dimensional Schrödinger equation numerically. It merely requires multiplication of N 2×2 matrices with complex coefficients and one inversion of the resulting product for each injection energy. Usually, only a very few sections have to be considered in order to achieve convergence. A typical result for the transmission coefficient as a function of energy is shown in Fig. 15.3. Without an electric field, $T(E_z)$ exhibits peaks at energies corresponding to true eigenstates of the system; in the presence of electric fields, however, no truly bound states exist. The maxima in $T(E_z)$ correspond in this case to quasi-bound states (compare Sect. 8.3). The finite lifetime of these states leads to a *lifetime broadening* of the transmission resonance curve, which can be seen in Fig. 15.3 (compare the width of the second resonance with the first one). The lifetime can in fact be estimated via the full width at half maximum of the transmission resonance and the time–energy uncertainty principle.

Given the transmission probability, we can now evaluate the current density for, for example, a double-barrier resonant-tunneling diode, depicted in Fig. 15.4 for three different bias conditions. The outside sections are two heavily doped regions. Because of the strong scattering by impurities, these two

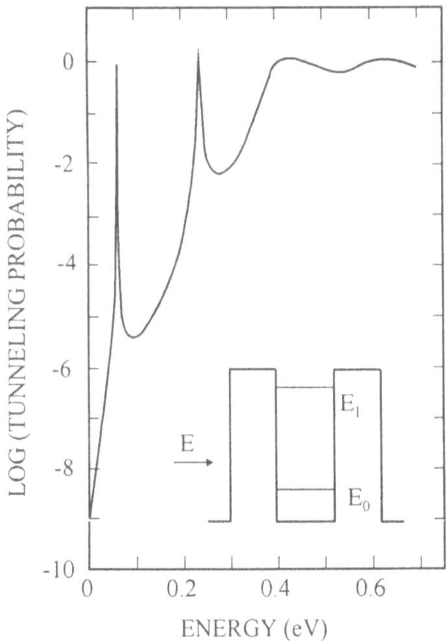

Fig. 15.3. Quantum mechanical transmission coefficient versus incident energy for the structure depicted in the *inset*. The parameters are: barrier height 0.3 eV, $L_B = 5$ nm, $L_{QW} = 6$ nm, quantum well mass $m_e^{QW} = 0.067\, m_0$ and barrier mass $m_e^B = 0.1\, m_0$. Taken from [412]

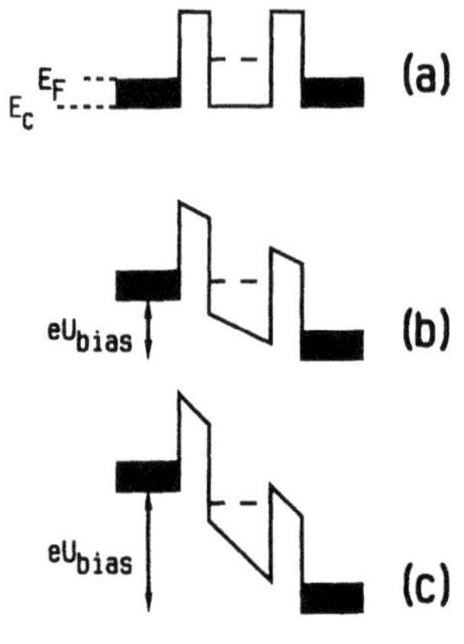

Fig. 15.4. Energy versus z coordinate for a resonant tunneling device under three different bias conditions (shown schematically). (**a**) Zero bias voltage U_{bias}. (**b**) At the peak of the current, the quantum well energy (*dashed line*) lies within the interval $[E_c, E_F]$ between the conduction band edge and the Fermi-energy. (**c**) Bias voltage above resonance

regions serve as the contacts. The undoped region in between, containing the double-barrier structure, is sufficiently short that our basic assumption of negligible scattering is a reasonable first approximation. Equation (15.2) can be rewritten as an integral, employing the definition of the velocity according to (3.87), $\hbar(\boldsymbol{v}_e)_z = \partial E_z/\partial k_z$:

$$(\boldsymbol{j}_{\text{e}})_z = -\frac{\text{e}}{4\pi^3\hbar}\int_{-\infty}^{+\infty} \mathrm{d}\boldsymbol{k}^3 \frac{\partial E_z}{\partial k_z}\Big(f_{\text{e}}^{\text{L}}(\boldsymbol{k}) - f_{\text{e}}^{\text{R}}(\boldsymbol{k})\Big)\, T(k_z)\,. \tag{15.13}$$

For low temperatures ($k_{\text{B}}T \ll E_{\text{F}}$) the Fermi functions can be replaced by step functions. Using $\mathrm{d}k_x\,\mathrm{d}k_y = (2\pi m_{\text{e}}/\hbar^2)\,\mathrm{d}E_{xy}$, where $E = E_{xy} + E_z$, choosing $E_{\text{c}} = 0$ for the left-hand side conduction band edge, and introducing the Fermi energy $E_{\text{F}} = \mu_{\text{e}}^{\text{L}}$ and the bias voltage U_{bias} via $eU_{\text{bias}} = \mu_{\text{e}}^{\text{L}} - \mu_{\text{e}}^{\text{R}}$, we arrive at

$$(\boldsymbol{j}_{\text{e}})_z = -\frac{em_{\text{e}}}{2\pi^2\hbar^3}\int_0^{E_{\text{F}}} \mathrm{d}E_z\,(E_{\text{F}} - E_z)\,T(E_z) \quad \text{for} \quad eU_{\text{bias}} \ge E_{\text{F}}\,, \tag{15.14}$$

$$(\boldsymbol{j}_{\text{e}})_z = -\frac{em_{\text{e}}}{2\pi^2\hbar^3}\Bigg(eU_{\text{bias}}\int_0^{E_{\text{F}}-eU_{\text{bias}}} \mathrm{d}E_z\,T(E_z) + \int_{E_{\text{F}}-eU_{\text{bias}}}^{E_{\text{F}}} \mathrm{d}E_z\,(E_{\text{F}} - E_z)T(E_z)\Bigg) \quad \text{for} \quad eU_{\text{bias}} \le E_{\text{F}}.\,. \tag{15.15}$$

For a numerically determined transmission probability $T(E_z)$ (which is a function of bias voltage), the tunneling current can immediately be evaluated via (15.15). The general behavior, however, can be seen without numerics. Electrons are injected only from the energetic interval between the conduction band edge and the Fermi energy of the left, n-doped region (the contact); see Fig. 15.4. If a quasi-bound state lies in this energy range, $T(E_z)$ exhibits a resonance in this interval and the current becomes large. For large bias voltages the resonant state drops below the left-hand conduction band edge

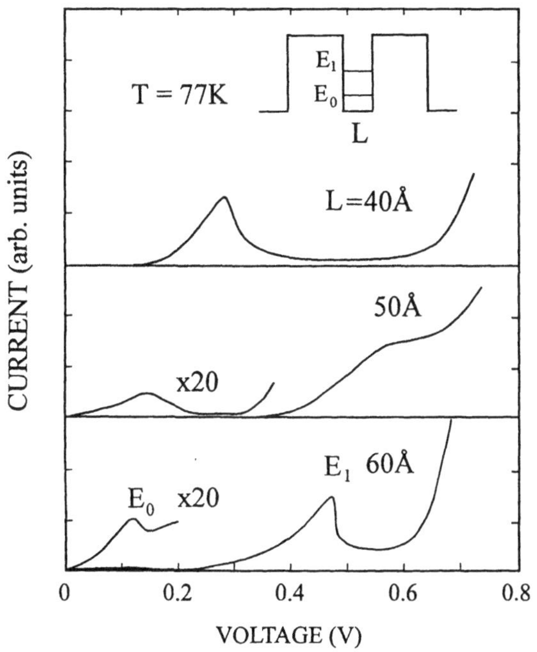

Fig. 15.5. Measured current–voltage characteristics of three resonant-tunneling devices with different quantum well widths as indicated. Taken from [412]

and tunneling is no longer effective. This leads to a current–voltage characteristic with negative slope, i.e. *negative differential conductivity*, which has been observed experimentally (see Fig. 15.5), years after the original proposal [411]. For N quasi-bound states, there are N regions of negative differential conductivity.

Before we come back to a more refined discussion of resonant-tunneling devices, adressing both some details and conceptual questions, we consider another example of transport described by (15.2).

15.1.2 Quantized Conductance

Let us consider a thin, homogeneous semiconductor wire (sketched in Fig. 15.6), again connected to two contacts. Note that this problem is closely related to optical waveguides.

The corresponding Hamiltonian is

$$H = \frac{\boldsymbol{p}^2}{2m_\mathrm{e}} + V_x(x) + V_y(y) \,. \tag{15.16}$$

If the spatial extent of the wire in the y direction is sufficiently small, only one value of k_y will contribute to the current. The spatial extent in the x direction, on the other hand, is considered as a parameter, and hence a small number of values of k_x (transverse modes) will contribute to the current.

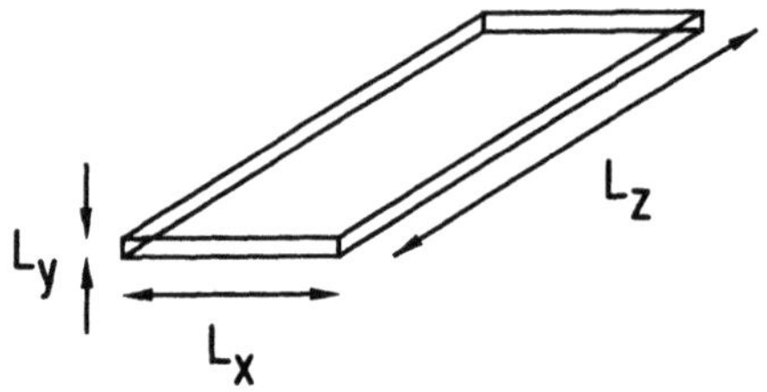

Fig. 15.6. Schematic illustration of a thin semiconductor wire

Under these conditions the electrons can propagate freely in the z direction with an associated one-dimensional density of states $D_{\mathrm{e}}^{1\mathrm{d}}(E_z)$, and (15.2) becomes

$$\begin{aligned} \boldsymbol{j}_{\mathrm{e}} &= -\frac{e}{V}\sum_{k_x}\sum_{k_z\geq 0}\boldsymbol{v}_{\mathrm{e}}[f_{\mathrm{e}}^{\mathrm{L}}(\boldsymbol{k}) - f_{\mathrm{e}}^{\mathrm{R}}(\boldsymbol{k})]\,T(\boldsymbol{k}) \\ &= -\frac{e}{2V}\sum_{k_x}\int_0^{\infty}\mathrm{d}E_z\,D_{\mathrm{e}}^{1\mathrm{d}}(E_z)\boldsymbol{v}_{\mathrm{e}}[f_{\mathrm{e}}^{\mathrm{L}}(E) - f_{\mathrm{e}}^{\mathrm{R}}(E)]\,T(E_z,k_x)\,, \end{aligned} \tag{15.17}$$

where $E = E_x + E_y + E_z$. The factor of 1/2 in the second line stems from the restriction $k_z \geq 0$; the spin summation is contained in the density of states. Within the effective-mass approximation, the one-dimensional density of states (3.140) (with $E_{\mathrm{c}} = 0$) is given by

$$D_{\mathrm{e}}^{1\mathrm{d}}(E_z) = \frac{L_z\,2\sqrt{2m_{\mathrm{e}}}}{h\sqrt{E_z}}\,, \tag{15.18}$$

and the product of electron velocity and one-dimensional density of states becomes

$$D_{\mathrm{e}}^{1\mathrm{d}}(E_z)\,(\boldsymbol{v}_{\mathrm{e}})_z = \frac{4\,L_z}{h}, \tag{15.19}$$

a result which, interestingly, *does not depend on any material parameter* such as the effective mass. This important finding is specific for physics in one dimension. At low temperatures the Fermi functions can be replaced by step functions, which leads to

$$(\boldsymbol{j}_{\mathrm{e}})_z = -\frac{2\,el_z}{hV}\sum_{k_x}\int_{\mu_{\mathrm{e}}^{\mathrm{R}}}^{\mu_{\mathrm{e}}^{\mathrm{L}}}\mathrm{d}E_z\,T(E_z,k_x)\,. \tag{15.20}$$

Here we have assumed $\mu_{\mathrm{e}}^{\mathrm{L}} > \mu_{\mathrm{e}}^{\mathrm{R}}$. Again, scattering is completely neglected. *Hence, the transmission coefficient is identical to unity, i.e.* $T(E,k_x) = 1$ for a homogeneous wire, independent of E. For small bias voltages, the remaining sum over k_x now reduces to a multiplication with an integer N, the number of transverse modes contributing to conduction, which is illustrated in Fig. 15.7.

In this example (Fig. 15.7), $N = 2$ subbands contribute to transport. Introducing the applied voltage U through $-e\,U = (\mu_{\mathrm{e}}^{\mathrm{L}} - \mu_{\mathrm{e}}^{\mathrm{R}})$ $((\mu_{\mathrm{e}}^{\mathrm{L}} - \mu_{\mathrm{e}}^{\mathrm{R}})$

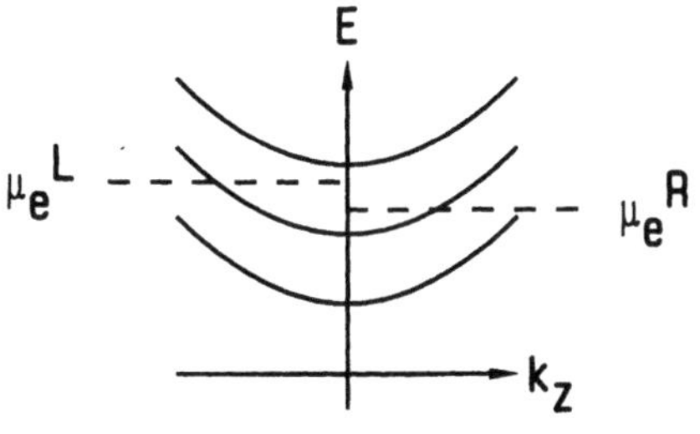

Fig. 15.7. Schematic illustration of one-dimensional subbands (corresponding to different k_x) contributing to transport at low temperatures. Here, the highest subband does not contribute to transport

being small compared with the subband spacing), we obtain the following for the electric current $I = (\boldsymbol{j}_\mathrm{e})_z\, L_x L_y$, where the volume is $V = L_x L_y L_z$:

$$I = GU\,. \tag{15.21}$$

The *conductance* G is given by

$$G = 2\frac{e^2}{h}N\,. \tag{15.22}$$

Equivalently, we could write, for the resistance R,

$$R = G^{-1} = \frac{h}{e^2}\frac{1}{2N}\,. \tag{15.23}$$

Interestingly, the conductance in a semiconductor wire is an integer multiple of a combination of fundamental constants, $2e^2/h$. The factor of two stems from the assumption of spin degeneracy.

Each subband contributes e^2/h to the conductance.

The value of e^2/h is $25.812807\,(k\mathrm{Ohm})^{-1}$. *Quantized conductance* is in striking contrast to classical transport, where the conductance can exhibit any value. The wider the wire in the x direction, the more subbands contribute. For very wide wires the steps in the conductance become relatively small, which eventually allow us to recover the classical limit. This finding can be tested experimentally using a GaAs/AlGaAs heterostructure with high mobility at very low temperatures [413, 414, 415]. The variation in wire width or, in other words, in the number of transverse modes can be realized experimentally by top gate contacts (see Figs. 15.8 and 15.9) that deplete certain regions depending on the applied gate voltage (which changes $V_x(x)$ and thus N). Such an arrangement is called a *point contact.*[4] At elevated temperatures (see Fig. 15.10), scattering by phonons can no longer be neglected, and

[4] Strict conductance quantization requires that the wire is smooth, i.e. that its width varies adiabatically on the scale of the Fermi wavelength. This condition can be fulfilled in semiconductor structures. In metal structures, on the other hand, the Fermi wavelength is on the order of the lattice constant and the wire is inherently rough. This mixes the subbands contributing to conduction, and thus integer quantization in metals is usually not observed.

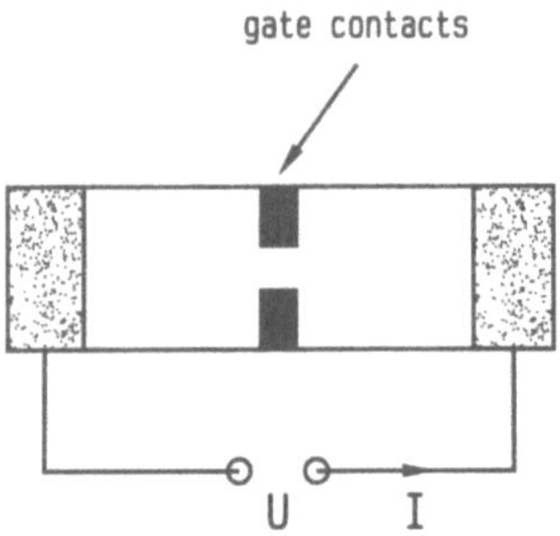

Fig. 15.8. Schematic illustration of a semiconductor heterostructure suitable for studying quantized conductance. In the normal direction the structure contains a two-dimensional electron gas and is identical to Fig. 1.5. Applying a voltage to the *split gate* modifies the potential in the center and thus effectively controls the width of the constriction

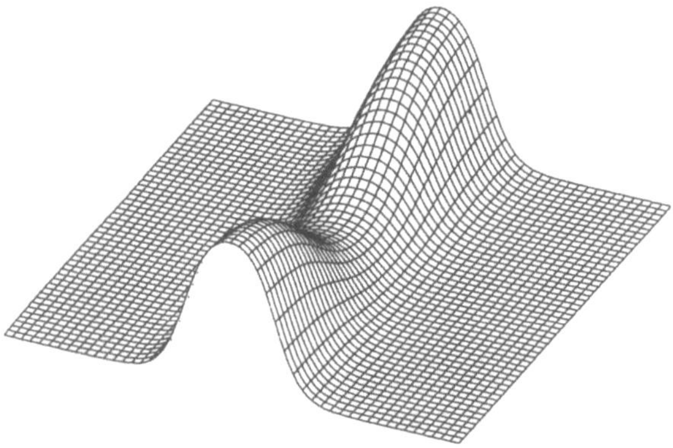

Fig. 15.9. Schematic illustration of the electrostatic energy seen by the two-dimensional electron gas underneath the negatively biased split gate, see Fig. 15.8

the steps are smeared out. Alternatively, the lateral potential can be controlled by an external magnetic field in the y direction. Without going into details at this point, it is clear that a magnetic field raises the energy levels because of the additional magnetic energy (cyclotron energy), which reduces the number of subbands contributing to the current, hence reducing N at a given bias voltage. In addition, the spin degeneracy is lifted by the Zeeman energy, which results in substructures at multiples of e^2/h rather than $2e^2/h$ (see [414]).

It is straightforward to recapitulate our derivation of the quantized conductance in the case that $T(k_z) = T(-k_z)$ does *not* hold, such for a multiterminal device in the presence of a magnetic field. In this case the electric current can, however, be expressed as

$$I = I_{\mathrm{L}} = -2\frac{e}{h}\left(N_{\mathrm{RL}}\mu_{\mathrm{e}}^{\mathrm{L}} - N_{\mathrm{LR}}\mu_{\mathrm{e}}^{\mathrm{R}}\right), \tag{15.24}$$

a form which will prove suitable for the extension to multicontact elements such as Hall devices. Clearly, if the number of modes N_{RL} leading from left to right equals the number for the inverse process, i.e. if $N_{\mathrm{RL}} = N_{\mathrm{LR}} = N$, the above result $I = -2e/h\, N\,(\mu_{\mathrm{e}}^{\mathrm{L}} - \mu_{\mathrm{e}}^{\mathrm{R}}) = 2e^2/h\, N\, U$ is recovered. For a two-terminal device, $N_{\mathrm{RL}} = N_{\mathrm{LR}}$ is already evident from the condition that

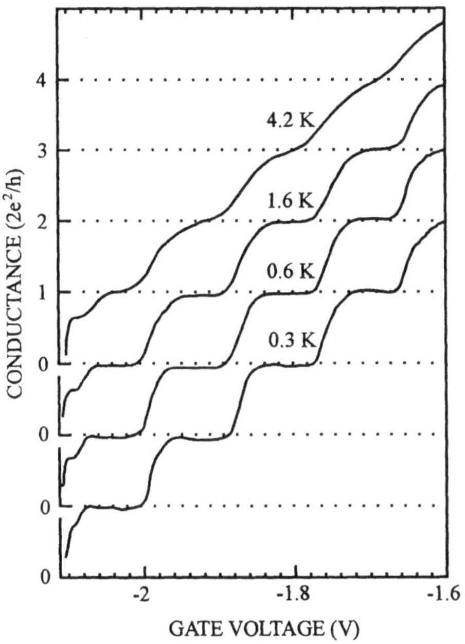

Fig. 15.10. Conductance of a structure similar to that shown in Fig. 15.8, as a function of gate voltage; the sample temperature is a parameter. Taken from [414]

the current vanishes, i.e. $I_{\mathrm{L}} = 0$, under equilibrium conditions ($\mu_{\mathrm{e}}^{\mathrm{L}} = \mu_{\mathrm{e}}^{\mathrm{R}}$), an argument which is also valid in the presence of a magnetic field.

15.1.3 Coulomb Blockade and the SET Transistor

So far, we have addressed the *wave nature of electrons* and the consequences for transport in the coherent regime. On the small physical scale of mesoscopic structures one might also expect effects arising from the *particle nature of electrons*, i.e. from the fact that the charge is quantized in multiples of the elementary charge e. The simplest effect of this kind is the *Coulomb blockade*, the basics of which was been discussed in the 1950s for granular metal films [416]. The underlying physics is very simple. Let us consider, for example, tunneling through an isolated small volume of undoped semiconductor packaged between two contacts that have a lower band gap (Fig. 15.11).

For large enough tunneling resistances, such a system can be considered as a leaky (tunneling) capacitor with capacitance C. Clearly, this capacitance decreases with some characteristic area of the contacts. Hence the charging energy $E(Q)$ of the capacitor for a total charge Q, which is simply given by

$$E(Q) = \frac{1}{2C}Q^2\,, \tag{15.25}$$

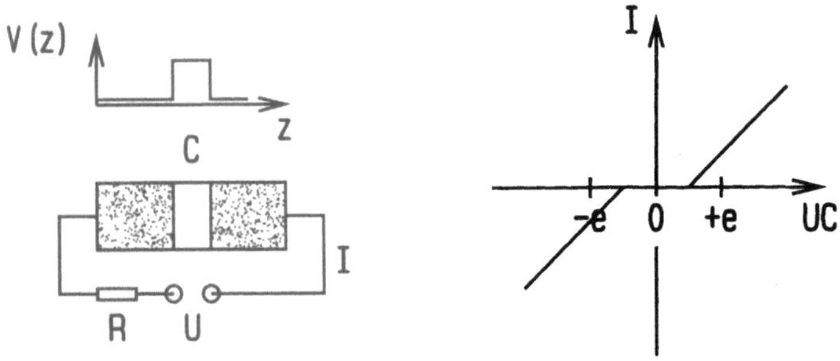

Fig. 15.11. Schematic illustration of a tunneling capacitance (*left*) exhibiting Coulomb blockade (*right*) at very low temperatures

increases with decreasing dimensions of the element. Adding one electron changes the charging energy to $E(Q-e)$, a process which only becomes favorable for negative energy differences:

$$E(Q-e)-E(Q)=\frac{1}{2C}[(Q-e)^2-Q^2]=\frac{1}{2C}\left(-2Qe+e^2\right)$$

$$=\frac{e^2}{2C}-Ue\leq 0\,. \tag{15.26}$$

That is, there is a voltage drop U across the capacitance, where $U>e/(2C)$. Hence the charging energy blocks current flow for $U<e/(2C)$ provided the thermal energy is (much) smaller than the charging energy, i.e.

$$k_{\mathrm{B}}T\ll E_{\mathrm{C}}=\frac{e^2}{2C}\,. \tag{15.27}$$

For example, $C=10^{-15}\,\mathrm{F}$ corresponds to $E_{\mathrm{C}}=80\,\mu\mathrm{eV}$ or $1\,\mathrm{K}\times k_{\mathrm{B}}$ and, equivalently, $C=10^{-18}\,\mathrm{F}$ corresponds to $E_{\mathrm{C}}=80\,\mathrm{meV}$ or $1000\,\mathrm{K}\times k_{\mathrm{B}}$. Experiments thus require small devices on the scale of few tens of nanometers and very low temperatures (typically several tens of millikelvin, obtained in dilution refrigurators) and have mostly been performed on metal junctions [417]. The occurrence of such current blocking requires a high-impedance voltage source [418] (or some series resistance R, see Fig. 15.11) connected to the element. For a low-impedance source, we simply have $E=CU^2/2$ and current can flow for arbitrarily small applied voltages U.

Instead of a series resistance, one can also consider the case of a series (tunnel) capacitance, depicted in Fig. 15.12. In addition, a true capacitance, the gate, is also introduced. The behavior of this device can easily be understood by considering the (free) energy of this arrangement, given by (for $U=0$)

$$E=\frac{1}{2C}\left(Q_{\mathrm{g}}-Ne\right)^2=E_{\mathrm{C}}\left(Q_{\mathrm{g}}/e-N\right)^2\,, \tag{15.28}$$

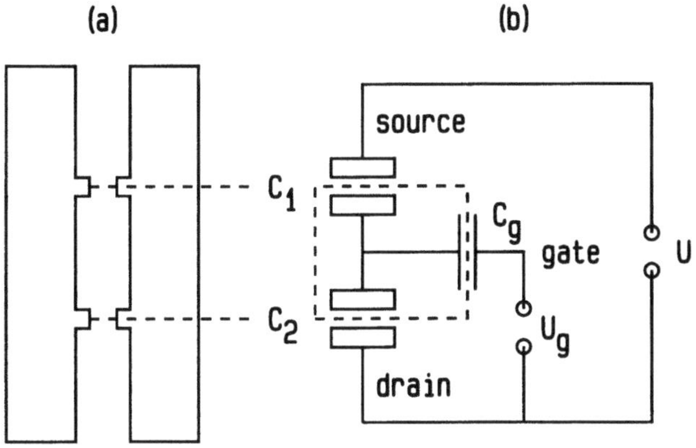

Fig. 15.12. Schematic illustration of a SET transistor; (**a**) top electrodes on the two-dimensional electron gas and (**b**) equivalent circuit with the two tunneling capacitances C_1 and C_2

where the total capacitance C seen from the "island" (Fig. 15.12) is given by $C = C_1 + C_2 + C_g$ and the charge $Q_g = C_g U_g + Q_0$ is given by the gate voltage U_g and some initial offset charge Q_0. Such an offset charge is unintended but is usually unavoidable. Unfortunately, it can also change when the temperature of the device is cycled. Let us assume here that electron quantization effects are not important in the central region; this is, however, an interesting subject on its own [419]. Applying a gate voltage U_g such that $Q_g = e/2$ makes the two configurations $N = 0$ and $N = 1$ energetically degenerate (see Fig. 15.13), which allows current to flow through the source and drain of this *SET transistor* (single-electron-tunneling transistor). If U_g is increased further, such degenerate configurations are obtained periodically and one obtains *conductance oscillations.*

Such double-junction elements have been realized in semiconductors in a controlled fashion [420, 421]. Again, a two-dimensional electron gas is created in a GaAs/AlGaAs heterojunction grown on a doped GaAs substrate (see Chap. 1). The top of the wafer is patterned with metal electrodes depicted schematically in Fig. 15.12. When a voltage is applied to these electrodes, the two-dimensional electron gas is laterally confined to a narrow channel interrupted by two constrictions, each about 100 nm in length. These constrictions induce saddle-shaped potential barriers (see Fig. 15.9) for electrons moving along the channel. In addition, the density of the electron gas can be controlled by applying a gate voltage, that is, a voltage applied to the conductive substrate. The two-dimensional electron gas on the sides forms the two contacts through which the current flows. The voltage $U \approx 0$ applied to these contacts is chosen to be small, such that it is within the Coulomb blockade regime. The resulting conductance through these two contacts as a function of the gate voltage U_g (minus some sample-dependent voltage $U_g^0 = C_g Q_0$) is

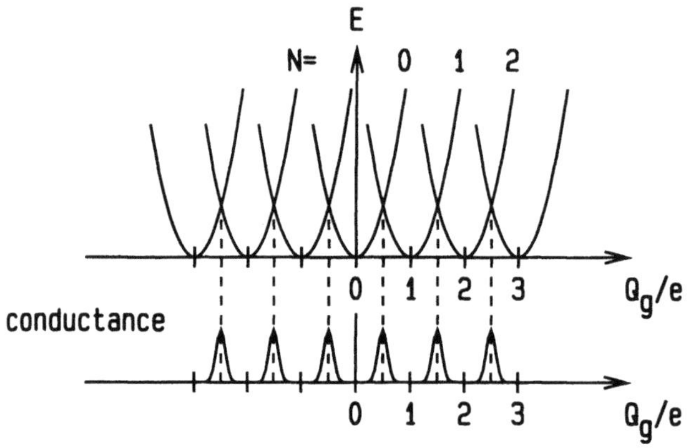

Fig. 15.13. Schematic illustration of the charging energy of an SET transistor versus $Q_{\rm g}/e$. The (source–drain) conductance exhibits an oscillatory behavior as a function of the gate voltage, and $Q_{\rm g}/e = C_{\rm g}(U_{\rm g} - U_{\rm g}^0)/e$

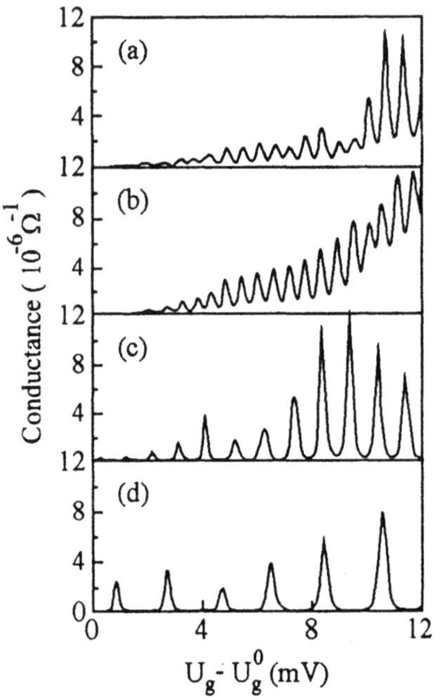

Fig. 15.14. Measured (source–drain) conductance oscillations as a function of the gate voltage. The peaks correspond to successive charging of the central region of the SET transistor by increments of one elementary charge e. Taken from [420]

depicted in Fig. 15.14. In particular, for small distances between the two constrictions (Fig. 15.14d), the conductance is only nonzero for particular gate voltages and one observes the incremental charging of the central region by $\Delta Q = \pm e$. This arrangement can also be looked at as a *nanoelectrometer* which exhibits an amazing sensitivity.

One can also vary the source–drain voltage, in which case steps in the conductance are obtained (if $C_1 \neq C_2$ in Fig. 15.12) owing to the incremental charging of the central region; this effect is called the *Coulomb staircase*. This effect has been observed in metallic nanostructures [422, 423] and discussed theoretically [418]. A nice overview of the theory concerning the physics of this subsection is given in [424].

15.2 Resonant Tunneling II

The simple approach of calculating currents via quantum mechanical transmission probabilities (15.2) has given us an initial idea about double-barrier resonant-tunneling devices and the physics of quantized conductance. In mesoscopic systems, the limit of negligible scattering can nowadays be realized to an amazing level of perfection. Consequently, electrons move coherently through some region of interest, which is connected to the incoherent outside world (and measurement process) by means of contacts. Owing to the strong coupling of the contacts to the coherent region, they become part of the problem. In the contacts, particles escape from the coherent region into the more highly conductive region of the external circuit and their distribution becomes incoherent owing to scattering. Thus, in general, we have to deal with an *open system* far from equilibrium, which is governed by irreversible processes. Irreversibility implies a finite lifetime of certain states in the system. Correspondingly, the particles cannot occupy eigenstates of a Hamiltonian, since in this case any time evolution of the system is reversible.

In the case of the tunneling structure, we have assumed in Sect. 15.1.1 that the electron states are stationary scattering solutions of a Schrödinger equation. Concerning the current density, this elementary approach seems to yield reasonable results. With respect to other physical observables, however, such as the charge density, which determines the self-consistent potential of a tunneling structure, this approach fails completely. This is because tunneling theory violates overall charge neutrality in the system. As a consequence the Poisson equation yields large fields at the boundaries, which make the violation of charge neutrality even worse. Thus, instead of the physically expected accumulation of electrons at the high-energy side of the structure, which is required to screen the external field, tunneling theory results in a depletion of the electron density [426]. In order to account for both the quantum interference in the coherent region and the irreversible scattering processes in the contacts, we have to turn to the quantum kinetic theory.

The basic formalism has been introduced in Sect. 10.5, where we have already derived the kinetic equations which govern transport theory. An appropriate starting point is (10.69). In the present problem, interband transitions vanish and only the particle propagators are required. Thus we have, for slowly varying external fields,[5]

$$\begin{aligned}
&\left(\mathrm{i}\hbar\frac{\partial}{\partial t} - \mathrm{i}\frac{\hbar^2}{m_\mathrm{e}}\boldsymbol{k}\boldsymbol{\nabla}_{\boldsymbol{r}}\right) G_{\mathrm{ee}}^{+-}(\boldsymbol{k},\boldsymbol{r},\omega,t) \\
&-e\sum_{\boldsymbol{q}} \mathrm{e}^{\mathrm{i}\boldsymbol{q}\boldsymbol{r}}\, V_{\mathrm{eff}}(\boldsymbol{q},t)\left[G_{\mathrm{ee}}^{+-}\left(\boldsymbol{k}-\frac{\boldsymbol{q}}{2},\boldsymbol{r},\omega,t\right) - G_{\mathrm{ee}}^{+-}\left(\boldsymbol{k}+\frac{\boldsymbol{q}}{2},\boldsymbol{r},\omega,t\right)\right] \\
&= G_{\mathrm{ee}}^{+-}(\boldsymbol{k},\boldsymbol{r},\omega,t)\,\Sigma_{\mathrm{ee}}^{-+}(\boldsymbol{k},\omega,t) - G_{\mathrm{ee}}^{-+}(\boldsymbol{k},\boldsymbol{r},\omega,t)\,\Sigma_{\mathrm{ee}}^{+-}(\boldsymbol{k},\omega,t)\,. \qquad (15.29)
\end{aligned}$$

The effective potential V_{eff} consists of the three contributions ϕ_{bar}, ϕ_{ext}, and ϕ_{ind}, which denote the potential of the barrier structure, the external potential, and the induced potential, respectively. The latter is determined by the Poisson equation, using the total charge densitiy of electrons and donors. Introducing the representation

$$\begin{aligned}
&U_{\mathrm{eff}}(\boldsymbol{k}-\boldsymbol{q},t) \\
&= \int \mathrm{d}^3\boldsymbol{r}'\,\sin[(\boldsymbol{k}-\boldsymbol{q})\boldsymbol{r}']\left[V_{\mathrm{eff}}\left(\boldsymbol{r}+\frac{\boldsymbol{r}'}{2},t\right) - V_{\mathrm{eff}}\left(\boldsymbol{r}-\frac{\boldsymbol{r}'}{2},t\right)\right] \qquad (15.30)
\end{aligned}$$

and performing the frequency integration, we find from (15.29) that the one-particle distribution is determined by

$$\begin{aligned}
&\frac{\partial}{\partial t} f_\mathrm{e}(\boldsymbol{k},\boldsymbol{r},t) + \frac{\hbar}{m_\mathrm{e}}\boldsymbol{k}\boldsymbol{\nabla}_{\boldsymbol{r}} f_\mathrm{e}(\boldsymbol{k},\boldsymbol{r},t) \\
&= \frac{e}{\hbar}\int\frac{\mathrm{d}^3\boldsymbol{q}}{(2\pi)^3}\,U_{\mathrm{eff}}(\boldsymbol{k}-\boldsymbol{q},t)\,f_\mathrm{e}(\boldsymbol{q},\boldsymbol{r},t) + \left.\frac{\partial}{\partial t} f_\mathrm{e}(\boldsymbol{k},\boldsymbol{r},t)\right|_{\mathrm{scat}}\,. \qquad (15.31)
\end{aligned}$$

The left-hand side corresponds to the undisturbed redistribution of carriers, driven by the inhomogeneity in the particle distribution. The first interaction contribution on the right-hand side describes potential scattering by the barrier structure and the self-consistent potential, whereas the second contribution accounts for scattering within the particle system, by phonons and impurities. When the $\boldsymbol{k}$-integration is performed, the right-hand side vanishes and we are left with the *continuity equation* (2.2), which guarantees local charge conservation. Assuming that the system is homogeneous in the plane perpendicular to the structure (chosen in the z direction), only

[5] Note that, in contrast to applications of the quantum kinetic theory in optics, the assumption of separability of the microscopic and macroscopic timescales is excellently justified and we can apply the Markovian approximation without any pangs of conscience. Further, we can neglect also the gradient corrections in the scattering contributions.

one-dimensional potentials occur. Nevertheless, because of the scattering contributions, the problem remains three-dimensional. There are, however, various reasonable approximation schemes which allow one, in the simplest case, to reduce the problem to the calculation of the distribution function $f_{\mathrm{e}\|}(k_z, z, t) = (1/A)\sum_{\boldsymbol{k}_\perp} f_{\mathrm{e}}(\boldsymbol{k}, z, t)$ (see below). These yield the current density and the particle density according to

$$j(z,t) = -e\int \frac{\mathrm{d}k_z}{2\pi}\frac{\hbar k_z}{m_{\mathrm{e}}} f_{\mathrm{e}\|}(k_z, z, t) \tag{15.32}$$

and

$$\varrho(z,t) = -e\int \frac{\mathrm{d}k_z}{2\pi} f_{\mathrm{e}\|}(k_z, z, t)\,. \tag{15.33}$$

Let us focus on this case, which is sufficiently complicated, in the following and consider next the boundary conditions which determine the solution of (15.31).

15.2.1 Boundary Conditions and Discretization

Sufficiently far from the barrier structure, i.e. in the reservoirs on both sides, the one-particle distribution has to approach a *Fermi distribution* for $\phi_{\mathrm{ext}} = 0$ or a *shifted Fermi distribution* for $\phi_{\mathrm{ext}} \neq 0$ (compare Problem 12.8). These distributions are averaged with respect to the wave vectors perpendicular to the structure and denoted by $f_{\mathrm{e}}^{\mathrm{L}}(k_z)$ and $f_{\mathrm{e}}^{\mathrm{R}}(k_z)$, respectively. Correspondingly, we have to distinguish between two types of carriers. The *upstream* carriers enter the barrier structure from the left and have a wave vector $k_z \geq 0$, whereas the *downstream* carriers, with $k_z \leq 0$, enter from the right. Only the distributions of the carriers which *leave the reservoirs* from either the left or the right, are fixed by the boundary conditions $f_{\mathrm{e}\|}(k_z \geq 0, z = -L/2) = f_{\mathrm{e}}^{\mathrm{L}}(k_z)$ and $f_{\mathrm{e}\|}(k_z \leq 0, z = L/2) = f_{\mathrm{e}}^{\mathrm{R}}(k_z)$, respectively. The distribution of the electrons which *enter the reservoirs* is then uniquely determined by the dynamics of the system. An immediate consequence of these boundary conditions is a coupling of the left- and right-hand sides of the system. This makes the solution of the problem, even without scattering contributions, intricate, and a few remarks concerning aspects of the methods are opportune.[6]

In order to solve (15.31), we have to introduce a discrete mesh with respect to the variables k_z and z. Let us assume a discretization of the form $k_{zm} = m\,\Delta k_z$, where $m_1 \leq m \leq m_1 - 1 + M$, and $z_n = n\,\Delta z$, where $n_1 \leq n \leq n_1 - 1 + N$. A key element is the discretization of the derivative of the distribution function. when introducing a finite-difference

[6] It should be noted that these considerations are not technical remarks of minor importance but fundamental conditions for obtaining numerical solutions of transport equations of the type (15.31), which guarantee particle number conservation.

approximation, we have, owing to the boundary conditions, to distinguish between the upstream and downstream directions. The simplest way is to introduce left-hand and right-hand difference approximations of the form $\pm [f_{\mathrm{e}}(k_{zm}, z_{n\pm 1}, t) - f_{\mathrm{e}}(k_{zm}, z_n, t)]/\Delta z$. Unfortunately, the accuracy of this simple difference quotient only becomes adequate if Δz is sufficiently small, and a better choice, accurate to order $\mathcal{O}(\Delta z^2)$, consists of the interpolation

$$\begin{aligned} &\frac{\partial}{\partial z} f_{\mathrm{e}\|}(k_z \geq 0, z, t) \\ &\rightarrow \frac{1}{2\Delta z}[f_{\mathrm{e}}(k_{zm}, z_{n+2}, t) - 4f_{\mathrm{e}}(k_{zm}, z_{n+1}, t) + 3f_{\mathrm{e}}(k_{zm}, z_n, t)] \end{aligned} \tag{15.34}$$

and

$$\begin{aligned} &\frac{\partial}{\partial z} f_{\mathrm{e}\|}(k_z \leq 0, z, t) \\ &\rightarrow -\frac{1}{2\Delta z}[f_{\mathrm{e}}(k_{zm}, z_{n-2}, t) - 4f_{\mathrm{e}}(k_{zm}, z_{n-1}, t) + 3f_{\mathrm{e}}(k_{zm}, z_n, t)]\,. \end{aligned} \tag{15.35}$$

An important further ingredient is local charge conservation, which requires that the k_z integral in the second line in (15.31) vanishes in the discretized version also. This requirement implies a condition on the discretization, which should satisfy $\Delta z\, \Delta k_z = 2\pi n/N$. Further, Δz and Δk_z need to be small enough to guarantee a sufficiently accurate reproduction of the potential and the distribution function. With a suitable choice of the mesh in the k_z–z plane, we can map (15.31) together with the boundary conditions discussed above onto a system of ordinary first-order differential equations. In the stationary limit, only a system of algebraic equations is left, which can in principle be solved by matrix inversion. Numerical studies, however, show [428] that the number of mesh points is crucial for the features of the current–voltage characteristics and convergent results *cannot* be attained even by means of optimized matrix inversion procedures. It is therefore advantageous, for several reasons, to switch on the potential for times $t \leq 0$ by means of a smooth function and to calculate the time evolution until the stationary state is reached. The initial state of this procedure is defined as a semiconductor layer with two heavily doped outside sections and an undoped region. In the course of the temporal evolution the self-consistency of the boundary conditions, i.e. the shift of the Fermi distributions, and the self-consistent potential can be taken into account directly. From a computational point of view, the storage required by this procedure increases only linearly with the number of mesh points and, in spite of the integration of the system of equations on the timescale one can easily handle matrices of the order $5 \times 10^4 \times 5 \times 10^4$. The results demonstrate that the peak-to-valley ratios in the current–voltage characteristics indeed depend sensitively on the discretization and the rate of convergence.[7]

[7] Needless to say that it would be completely meaningless to discuss the influence of scattering processes as long as the numerical uncertainties in calculations without scattering were much larger than the effects due to dissipative processes.

15.2.2 Scattering Contributions

A practical inclusion of the scattering contributions requires approximations, as any rigorous formulation increases the dimensions of the problem. Nevertheless, we cannot abandon scattering processes completely, as they not only determine the properties of the distribution functions at the boundaries but also influence the carrier dynamics within the system. The difficulty consists in the reduction of the scattering rates to one-dimensional quantities without having recourse to the three-dimensional distribution functions. An ostensible solution is the relaxation-time approximation, which requires, however, a knowledge of the stationary one-particle distribution throughout the entire structure, i.e. just the quantity to be determined. Plausible approximate choices are, for example, the thermal distributions of the bulk material [427] and the stationary distribution of the unbiased structure [425]. Moreover, the relaxation time remains an unknown parameter. A more systematic approximation scheme relies on the properties of the steady-state distribution at high temperatures, where Boltzmann statistics is a reasonable approximation. Within this limit, the one-particle distribution factorizes according to

$$f_{\mathrm{e}}(\boldsymbol{k}, z) = f_{\mathrm{e}\|}(k_z, z)\, f_{\mathrm{e}\perp}(\boldsymbol{k}_\perp, z) \tag{15.36}$$

and the evaluation of the scattering contributions from optical and acoustic phonons is straightforward [426]. This scheme can be improved for application at low temperatures by constructing an expansion of the form [428]

$$f_{\mathrm{e}}(\boldsymbol{k}, z) = \Phi_0(\boldsymbol{k}_\perp, k_z) \sum_{n=0} \Phi_n(\boldsymbol{k}_\perp, k_z)\, f_{n\|}(k_z, z)\,, \tag{15.37}$$

where the functions Φ_n are an orthogonal set of polynomials with respect to $k_\perp$, orthogonalized by means of the Gram–Schmidt method. Owing to our boundary conditions, we again have to distinguish between $k_z \geq 0$ and $k_z \leq 0$. Choosing the values of z corresponding to the left- or right-hand boundaries, we have the conditions $f_{0\|}(k_z \geq 0, z^{\mathrm{L}}) = f_{\mathrm{e}}^{\mathrm{L}}(k_z)$, $f_{0\|}(k_z \leq 0, z^{\mathrm{R}}) = f_{\mathrm{e}}^{\mathrm{R}}(k_z)$, and $f_{n\|} = 0$ for $n \neq 0$. This defines Φ_0 by

$$\Phi_0(k_\perp, k_z \geq 0) = \left(\frac{f_{\mathrm{e}}^{\mathrm{L}}(k_\perp, k_z)}{f_{\mathrm{e}}^{\mathrm{L}}(k_z)}\right)^{1/2} \tag{15.38}$$

and

$$\Phi_0(k_\perp, k_z \leq 0) = \left(\frac{f_{\mathrm{e}}^{\mathrm{R}}(k_\perp, k_z)}{f_{\mathrm{e}}^{\mathrm{R}}(k_z)}\right)^{1/2}, \tag{15.39}$$

where $f_{\mathrm{e}}^{\mathrm{L}}(k_\perp, k_z)$ and $f_{\mathrm{e}}^{\mathrm{R}}(k_\perp, k_z)$ are the shifted Fermi distributions at the left- and right-hand boundaries, respectively. The application of the expansion (15.37) allows again the reduction of (15.31) to a problem in the k_z–z plane, but now scattering contributions are also systematically included. Numerical studies show that the expansion converges rapidly and only a few functions Φ_n have to be taken into account.

15.2.3 Numerical Results

Let us consider a few numerical results for a double-barrier resonant tunneling-diode, characterized by the following parameters. The barriers have a width of 3.2 nm and a height of 225 meV, and the well width is 4.5 nm. This structure is embedded between heavily doped regions, which have a width of 17.5 nm and yield an electron density of $2 \times 10^{18} cm^{-3}$. The temperature dependence of the current–voltage characteristics is shown in Fig. 15.15 for three different temperatures. The striking feature is the rapid decrease of the resonance and, correspondingly, of the peak-to-valley ratio with increasing temperature. This behavior is expected, as the resonance structure in the one-particle distribution functions becomes strongly degraded because of the broadening of the distribution at the boundaries and the scattering within the structure. Also, the self-consistency of the potential has a strong influence on the resonance behavior. Compared with the case where only the external potential is taken into account, as in the elementary quantum mechanical treatment, the resonance occurs at much higher voltages (not shown). This effect is trivial, since the external potential is weakened by the induced potential. A nontrivial consequence is the sharpening of the resonance, which can be traced back to the charge accumulation on the upstream side. This accumulation layer results in an increase of the potential drop along the double barrier and thus in a more rapid detuning of the resonance condition with increasing voltage. The microscopic features of the tunneling device are best illustrated by means of the one-particle distributions. Two typical distributions obtained in the stationary limit at zero voltage and at resonance, are shown in Fig. 15.16.

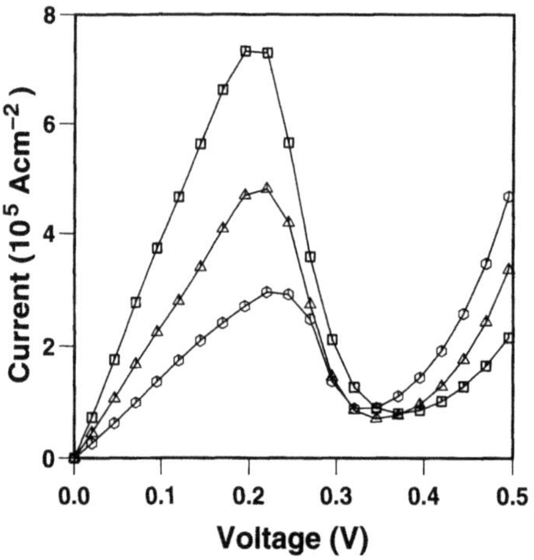

Fig. 15.15. Calculated current–voltage characteristics of a resonant tunneling diode (for details see text) at temperatures of 50 =K (*squares*), 100 K (*triangles*) and 300 K (*circles*)

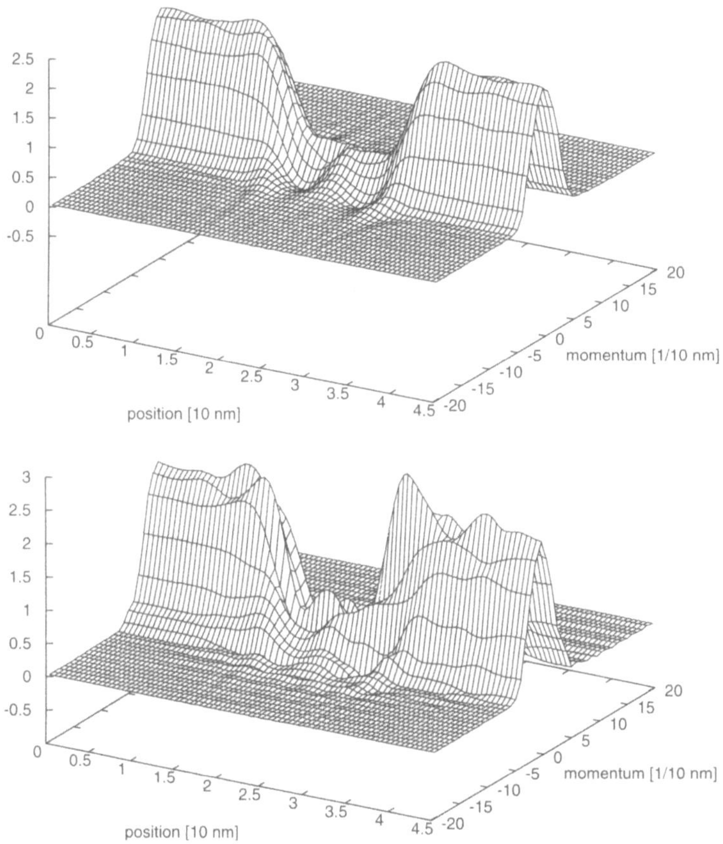

Fig. 15.16. Stationary one-particle distribution functions of a resonant tunneling diode at zero voltage (*top*) and at resonance (*bottom*). Taken from [428]

15.2.4 Time-Dependent Phenomena

The theory discussed in this section can be applied immediately to time-dependent phenomena such as resonant tunneling under the action of an additional time-varying modulation of the external voltage, and to more complicated structures. Actual investigations of time-varying phenomena, however, rely on simplifying assumptions such as the approximate application of the dissipation–fluctuation theorem (12.26), which allows one to describe the basic physics qualitatively [429]. Coherent photon-assisted tunneling has, for example, been studied in a double-quantum-dot system, which can be modelled by three barriers and two wells, with energy levels E_1 and E_2 [430]. This structure is again embedded between two heavily doped contact regions. If the frequency of the external field is resonant with the transition energy $\Delta E = [(E_1 - E_2)^2 + 4w^2]^{1/2}$, where w denotes the tunneling transition matrix element between the two wells, the system executes spatial Rabi oscillations.

The time-averaged currents for vanishing chemical potentials μ_e^L and μ_e^R in the doped regions are shown in Fig. 15.17 for several driving voltages. The current exhibits a series of resonances at energies $N\hbar\omega = \Delta E$, which are due to the following elementary processes. Without a resonant external field the tunneling probability of an electron from the first well into the neighboring contact region on the left vanishes. Owing to an N-photon-assisted tunneling process, the electron can perform Rabi oscillations between the wells, thus increasing the energy by $N\hbar\omega$. In this way, the energy in the second well becomes sufficiently high to enable tunneling into the other contact region on the right, and the system works as an electron pump. The pumping current reaches its maximum when the coupling to the reservoirs, which determines the lifetime of the one-particle states, equals the Rabi frequency.

Resonant tunneling phenomena and the quantization of the conductance in integer multiples of $2e^2/h$ are striking examples of coherent transport. Single-electron effects, such as the Coulomb blockade [431, 432, 433] or single-electron devices [434, 435, 436] are topics of current research.

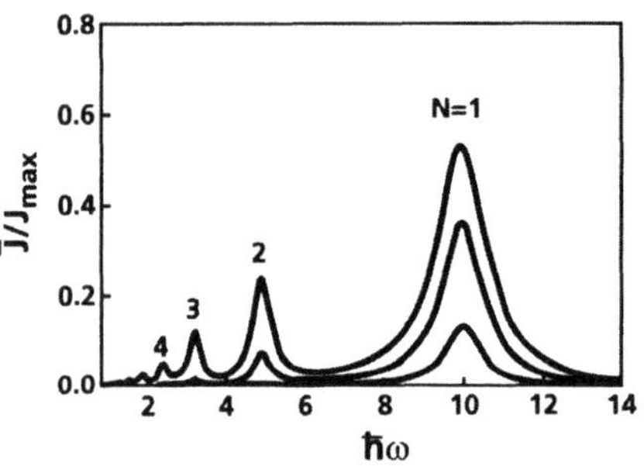

Fig. 15.17. Time-averaged currents J through a double quantum dot for $\mu_e^R = \mu_e^L = 0$ and several driving voltages. The current is measured in units of $J_{\max} = e\gamma/2\hbar$, where γ is the decay rate of one-particle states due to the coupling to the reservoirs. Energies are measured in units of the tunneling matrix element w and the amplitudes of the voltages are 2, 4, and 6 (with increasing J). Taken from [430]

15.3 Problems

15.1. Show that the quantization of the conductance according to (15.22) remains valid beyond the effective-mass approximation.

15.2. Estimate the physical size of a semiconductor tunneling junction that would show a Coulomb blockade at room temperature.

15.3. Consider coherent transport through two constrictions in series. The first transmits N_1 subbands, the second N_2. What would be the classical result?

15.4. Derive (15.24). Start from (15.1), not (15.2) and make no assumptions about the relation between the transmissions $T(k_z)$ and $T(-k_z)$ on the way.

15.5. Recapitulate the derivation of the quantized conductance (see Sect. 15.1.2) in the case that the symmetry relation $T(k_z) = T(-k_z)$ does not hold.

15.6. What happens to the current-voltage characteristic of a resonant tunneling diode if a magnetic field is applied in the direction of the current and if it is applied normal to the current? Discuss the behavior qualitatively.

15.7. Make use of the discretization discussed in Sect. 15.2.1 together with the upstream and downstream derivatives (15.34) and (15.35), to derive the discretized version of (15.31). Prove that the continuity equation is satisfied.

15.8. Make use of the expansion (15.37) to reduce the kinetic equation (15.31) to a representation in the k_z–z plane. Evaluate also the one-phonon self-energies for scattering from optical and acoustic phonons.

16. Electric and Magnetic Fields in Mesoscopic Systems

16.1 The Integer Quantum Hall Effect

The quantum Hall effect (QHE) has been one of the most prominent effects in semiconductor transport of recent years; von Klitzing received the Nobel prize in 1985 for its observation in a silicon MOSFET (metal–oxide–semiconductor field-effect transistor) [440]. We shall shortly be able to understand the basic quantization. The details of the QHE and of the fractional QHE, for which H. L. Stormer, D. S. Tsui, and R. B. Laughlin obtained the Nobel prize in 1998, however, still give rise to a number of open questions and are subjects of current research.

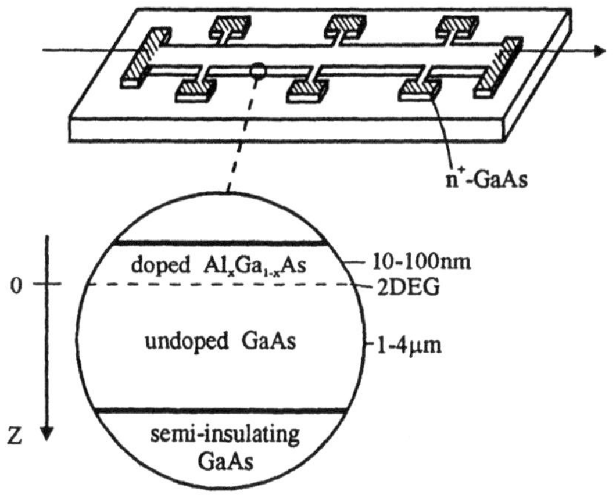

Fig. 16.1. Schematic illustration of a Hall bar, employing a modulation-doped GaAs heterostructure, subject to a perpendicular magnetic field. Such devices are used in experiments on the quantum Hall effect

While the original observation was made in a silicon MOSFET, even more striking results can be obtained in modulation-doped GaAs heterostructures [441]. A schematic illustration of such a structure, usually called a Hall bar, is shown in Fig. 16.1. This geometry is closely similar to that discussed for classical transport in Chap. 14. The mobility of the two-dimensional electron gas in the plane of the device can be as large as $10^6\,\mathrm{cm^2\,V^{-1}\,s^{-1}}$; temperatures down to several tens of mK and magnetic fields in excess of 20 T are accessible experimentally. The pronounced plateaus in the Hall resistance are accompanied by vanishing longitudinal resistance, as can be seen from Fig. 16.2. It is only in the transition regions between two adjacent plateaus that the longitudinal resistance becomes finite. Even more amazing than the quantization itself is the precision and stability with which different samples of somewhat different geometries, made of different materials (Si, GaAs, InGaAs) in various laboratories all over the world, deliver values of R_H given by

$$R_\mathrm{H} = \frac{h}{e^2}\frac{1}{N} \tag{16.1}$$

where N is an integer. The result obtained from the experimental data is

$$\frac{h}{e^2} = 25.812807\,\mathrm{k}\Omega\,. \tag{16.2}$$

This high precision has encouraged application of the integer QHE as a resistance standard [442]. Note that for the QHE, the quantization is far more pronounced than in the case of conductance quantization in thin wires (without magnetic field); see Chap. 15.

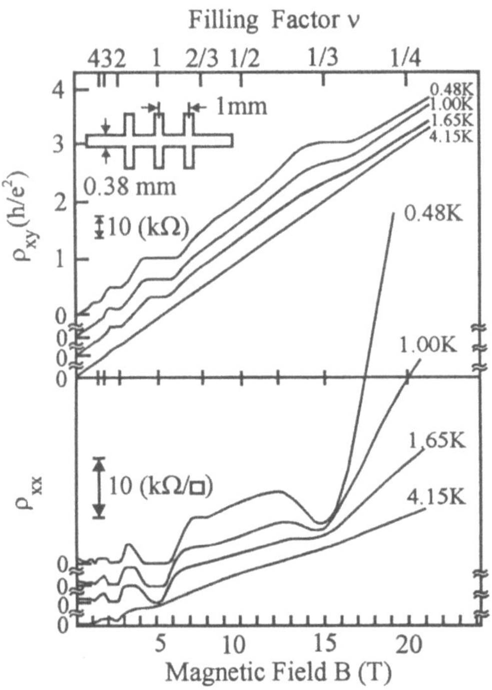

Fig. 16.2. Hall resistance $R_\mathrm{H} = \rho_{xy}$ and longitudinal resistance $R_\mathrm{L} = \rho_{xx}$ as a function of magnetic field. The structure of the Hall device is similar to that shown in Fig. 16.1, dimensions are depicted in the *inset*

16.2 Edge Channels and the Landauer–Büttiker Multiprobe Formula

In Chap. 15, we were first faced with the fact that the conductance of a semiconductor wire, sandwiched between two (ideal) contacts, is quantized in the limit of negligible scattering. Each state contributes e^2/h to the conductance. This important result of Chap. 15 is, however, only applicable directly to two-probe systems and not to the multiprobe problem related to Hall devices. Let us first generalize the result of Chap. 15 in a straightforward and intuitive fashion. This will lead us to the Landauer–Büttiker formula [437, 438, 439], which allows computation of multiprobe resistances at an amazingly simple level. A microscopic derivation within the framework of linear response theory will be given in Sect. 16.3

In a Hall device, a voltage is measured between the two probes m and n (hence $I_m = I_n = 0$), while a constant current is flowing through the two contacts k and l. As in Sect. 14.2, we define the corresponding resistance as

$$R_{kl,mn} = \frac{U_{mn}}{I}\,, \tag{16.3}$$

where U_{mn} is the voltage between contacts m and n or, in other words, the corresponding difference between the chemical potentials $-e\,U_{mn} = \mu_{\mathrm{e}}^{m} - \mu_{\mathrm{e}}^{n}$. The current I is given by

$$I = I_k = -\,I_l\,. \tag{16.4}$$

If, for example (see Fig. 16.3), we choose $m = 2$ and $n = 4$ as the voltage probes and let a constant current flow through the contacts $k = 1$ and $l = 3$, the Hall resistance becomes

$$R_{\mathrm{H}} = R_{13,24}\,. \tag{16.5}$$

The central idea is as follows. In Sect. 15.1.2, we have derived the result that the (two-probe) current between the left (L) and the right contact (R), with corresponding chemical potentials for electrons $\mu_{\mathrm{e}}^{\mathrm{L}}$ and $\mu_{\mathrm{e}}^{\mathrm{R}}$, is given by (15.24)

$$I = I_L = -\,2\frac{e}{h}\left(N_{\mathrm{RL}}\mu_{\mathrm{e}}^{\mathrm{L}} - N_{\mathrm{LR}}\mu_{\mathrm{e}}^{\mathrm{R}}\right).$$

Here N_{RL} is the number of modes leading particles from the left to the right. In the presence of a magnetic field, the spin degeneracy is generally lifted owing to the Zeeman energies, and the factor of 2 (which stems from the spin degeneracy) has to be replaced by 1. We can generalize this expression for the multiterminal problem to the current I_n through terminal n in a straightforward fashion:

$$I_n = -\,\frac{e}{h}\sum_{n'}\left(N_{n'n}\mu_{\mathrm{e}}^{n} - N_{nn'}\mu_{\mathrm{e}}^{n'}\right), \tag{16.6}$$

where, rather than having one (right) terminal, we have to sum over all terminals n'. The integers $N_{nn'}$ are the numbers of modes leading particles from terminal n' to terminal n. Under equilibrium conditions, all chemical potentials are equal and no current can flow through any of the terminals. From these conditions, we immediately obtain

$$\sum_{n'} N_{nn'} = \sum_{n'} N_{n'n} \,. \tag{16.7}$$

Introducing this result into (16.6), we can write, equivalently,

$$I_n = -\frac{e}{h} \sum_{n'} N_{nn'} \left(\mu_e^n - \mu_e^{n'} \right) \,. \tag{16.8}$$

Finally, using the above definition of the voltage $U_{nn'}$ between terminals n and n' ($-eU_{nn'} = \mu_e^n - \mu_e^{n'}$), we obtain the *Landauer–Büttiker formula* [438, 439]:

$$I_n = \frac{e^2}{h} \sum_{n'} N_{nn'} U_{nn'} \,. \tag{16.9}$$

The central question that remains is: how many states contribute to the conductance from probe n' to probe n, i.e. how can we obtain the integers $N_{nn'}$?

16.2.1 Edge Channels

Let us have a look at the classical trajectories of electrons in a Hall device in the limit of strong magnetic fields for an infinitesimaly small applied bias voltage. This situation is depicted in Fig. 16.3.

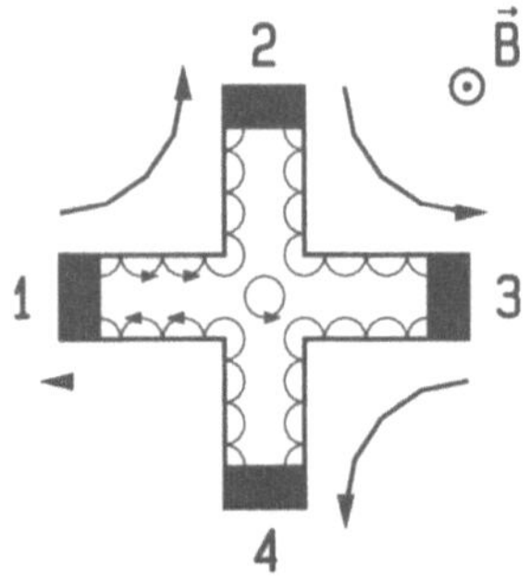

Fig. 16.3. Schematic illustration of a Hall element in a strong perpendicular magnetic field. The *black regions* are the contacts. The *arrows* indicate the direction of the electron motion, and hence the particle current; the *lines* within the device are classical trajectories

Trajectories separated from the side walls are clearly closed circles, while *trajectories at the edges* experience multiple reflections (skipping orbits) and thus contribute to transport between adjacent probe contacts. If, however, a finite electric field is applied, the trajectories in the center also contribute to conduction. This becomes different in the presence of disorder. Owing to the modulation doping, randomly distributed impurities are located at a distance of a few tens of nanometers from the two-dimensional electron gas. These impurities lead to a fixed random potential, which is characteristic of a given sample (as long as the impurities do not move). For typical electron sheet densities on the order of $10^{12}\,\mathrm{cm}^{-2}$, the average distance between impurities in the plane is merely 10 nm. In addition, imperfections such as surface roughness of the heterointerfaces on an atomic scale can contribute. Think of this random potential as a landscape of hills and valleys, as illustrated in Fig. 16.4.

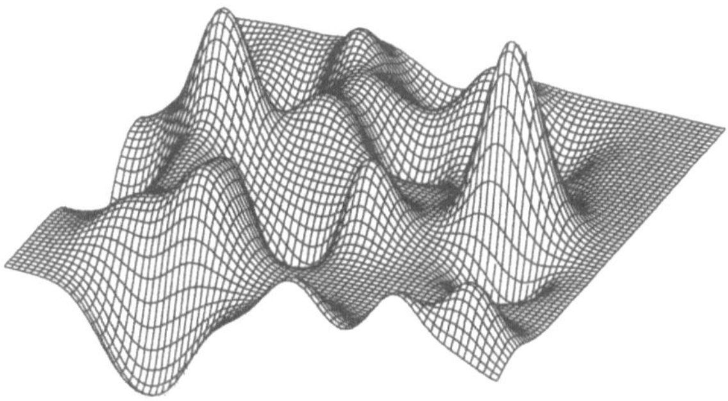

Fig. 16.4. Electrons in a device showing the quantum Hall effect move within a two-dimensional random potential landscape

Classical particles, in the presence of a strong perpendicular magnetic field, describe closed orbits in this landscape even in the presence of a finite electric field (bias voltage). Hence trajectories in the center region do not contribute to conduction. At the edges, however, skipping orbits can bounce against the walls many times, allowing conduction. Thus the trajectories at the edges are the only ones contributing to conduction.

One expects edge conduction in quantum mechanics also. Intuitively, the classical trajectories at the edges correspond to a finite number, say N, of *edge states* located more or less at the edges. These states are often referred to as *channels*. If we neglect the Zeeman contributions, the energy levels in the (homogeneous) central region of a Hall device are clearly given by equidistant Landau levels. Owing to the presence of some lateral confinement potential $V(x)$ (an etched edge or a negatively biased gate electrode), the energy levels bend up towards the edges of the device. The corresponding Hamiltonian,

$$H = \frac{1}{2m_\mathrm{e}} \left(\boldsymbol{p} + e\boldsymbol{A}\right)^2 + V(x) \tag{16.10}$$

becomes particularly simple for a parabolic confinement potential, which may be represented as

$$V(x) = \frac{1}{2} m_\mathrm{e} \omega_x^2 x^2 \,. \tag{16.11}$$

For a magnetic field in the z direction, we have (see Chap. 7) $\boldsymbol{B} = (0, 0, B_z) = \nabla \times \boldsymbol{A}$, where $\boldsymbol{A} = (0, B_z x, 0)$ (the Landau gauge). Substituting $x = q - ap_y$, and thus $p_x = p_q$, and then choosing $a = \omega_c/(\omega_{\text{eff}}^2 m_e)$, which lets the mixed terms $\sim p_y\, q$ vanish, we obtain

$$H = \frac{1}{2m_e} p_q^2 + \frac{1}{2} m_e \omega_{\text{eff}}^2 q^2 + \frac{1}{2m_e} p_y^2 \left(1 - \frac{\omega_c^2}{\omega_{\text{eff}}^2}\right) + \frac{1}{2m_e} p_z^2 \,. \tag{16.12}$$

This is nothing but a harmonic-oscillator Hamiltonian in the coordinate q with an effective frequency $\omega_{\text{eff}}^2 = \omega_c^2 + \omega_x^2$, where $\omega_c = eB_z/m_e$ is the cyclotron frequency. For a two-dimensional system, the kinetic energy from the motion in the z direction is a constant. As the lateral displacement x_0 of the harmonic-oscillator wavefunction is simply proportional to p_y (because $x = q - ap_y = q - x_0$), we have a quadratic dependence of the eigenenergy on x_0. Classically, x_0 is the center of the cyclotron orbit. In the case of an infinitely high potential barrier in $V(x)$, the Teller model, one has Weber functions as eigenfunctions [443]. Here, far away from the wall, the energy levels asymptotically become equal to the equidistantly spaced Landau ladder. Whatever the particular choice of other, more realistic confinement potentials may be, the general qualitative behavior of such a one-electron picture is as depicted in the left column of Fig. 16.5. Clearly, conduction can take place only at the crossings of the energy levels with the Fermi energy (see arrows in Fig. 16.5a), which determine the position of the edge channels [444, 445].[1]

As, however, the charge distribution arranges itself according to the confinement potential and in turn gives rise to an electrostatic potential (via Poisson's equation), which then modifies the confinement potential (and so forth), a quantitative description requires self-consistent solutions. This electrostatic problem can be solved analytically [446] if one assumes that the variations in the potential take place on a scale which is much larger than the cyclotron radius. Consequently, one has an equidistantly spaced Landau ladder at any point in space. The behavior is schematically depicted in the right-hand column of Fig. 16.5. Here regions with integer filling factors ν (see (7.18)) are separated from regions with noninteger filling factors (shaded), corresponding to *incompressible* and *compressible* strips, respectively. Only the compressible strips contribute to conduction (indicated by arrows). As the electrons within the compressible strips screen the electric field in the direction normal to the edge, the electrostatic potential is constant within these regions. Clearly, the scattering between channels is quite different in the self-consistent electrostatic picture from the situation in the one-electron picture [446].

Given the picture of edge channels we can now go ahead and compute the Hall resistance using the Landauer–Büttiker formula for two examples.

[1] Obviously, the Fermi energy lies between the Landau levels in the central part of the sample. It is often argued that such pinning of the Fermi energy is only possible if there are states between the Landau levels, which can be brought about by disorder.

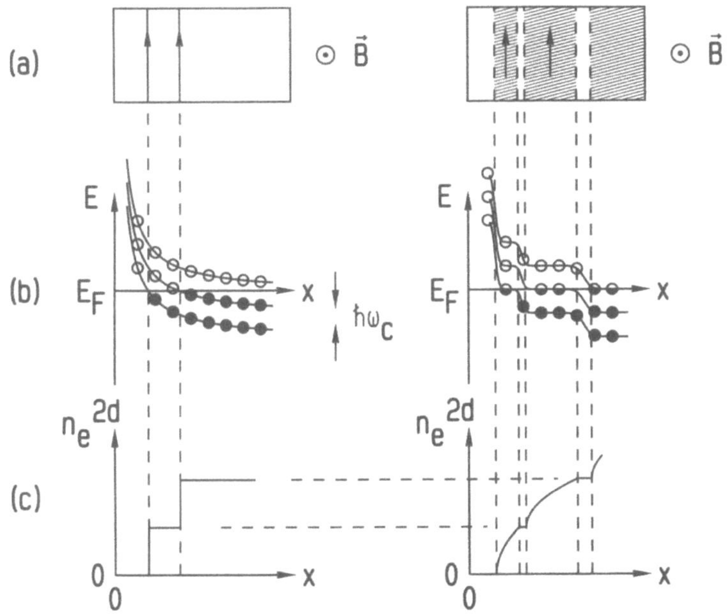

Fig. 16.5. Schematic illustration of the detailed behavior at the edge ($x = 0$) of a two-dimensional electron gas subject to a perpendicular magnetic field at low temperature. *Left column:* one-electron picture. (**a**) Top view of the two-dimensional electron gas, (**b**) bending of the energy levels at the edge, and (**c**) charge distribution. Notice that the horizontal energy dispersion for large x (**b**) and the finite electron density, and hence finite charge density, at the same x (**c**) are inconsistent with Poisson's equation, which would predict a *negative* curvature. The corresponding quantities for the self-consistent picture are depicted in the *right-hand* column. The *filled* and *open circles* represent occupied and unoccupied energy levels, respectively. Only the compressible strips with noninteger filling factors (*shaded*) contribute to conduction; see *arrows*

Example (1). We consider a Hall element of the usual type, as depicted in Figs. 16.3 and 16.6.

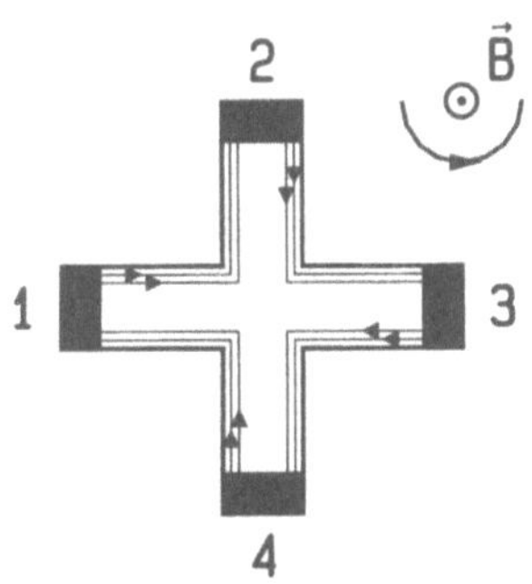

Fig. 16.6. Schematic illustration of a Hall element in a strong perpendicular magnetic field. For the edge channels depicted in this example, we have $N = 2$. Notice how far counterpropagating edge states are spatially separated, which particularly inhibits backscattering, e.g. that due to impurities. Compare with the classical trajectories shown in Fig. 16.3

Obviously, $N_{21} = N_{32} = N_{43} = N_{14} = N$ because of the fourfold rotational symmetry. All other off-diagonal $N_{nn'}$ are zero.[2] For the Hall resistance, we have $R_{13,24} = U_{24}/I_1$. Using $I_1 = e^2/hN_{14}U_{14} = e^2/hNU_{14}$ and $U_{21} = 0$, and hence $U_{14} = U_{24}$ (from the condition $I_2 = 0 = e^2/hN_{21}U_{21}$), we immediately obtain the quantization of the integer QHE resistance (16.1):

$$R_{\mathrm{H}} = R_{13,24} = \frac{U_{24}}{I_1} = \frac{h}{e^2}\frac{1}{N}\,. \tag{16.13}$$

Interestingly enough, one could in fact simply connect contact 1 with contact 2 and, equivalently 3 with 4, via external wires (thus $R_{13,13} = R_{13,24} = R_{\mathrm{H}}$). Even though the difference in chemical potential between contacts 1 and 2 is zero, there is a current flowing from 1 to 2 which compensates the current from 2 to 3 to produce the net result $I_2 = 0$. A similar argument holds for terminal 4. This finite current for vanishing difference in chemical potentials is closely related to the vanishing longitudinal resistance in the QHE (see discussion below).

On one hand, it is interesting that explicit consideration of the magnetic field is not necessary to understand the basic quantization. On the other hand, we get no idea about the relation between the integer N and the magnetic field from the Landauer–Büttiker formula alone. If we employ additionally the classical expression for the Hall resistance (14.36), $R_{\mathrm{H}} = B_z/(en_{\mathrm{e}}^{2\mathrm{d}})$, we obtain the following for the two-dimensional electron density:

$$n_{\mathrm{e}}^{2\mathrm{d}} = \frac{B_z e}{h} N\,. \tag{16.14}$$

The factor $B_z e/h$ is nothing but the number of states per Landau level, spin, and unit area (as shown in Chap. 7, (see 7.18)), which leads us to the result that *the plateaus in the Hall resistance correspond to completely filled Landau levels* or, in other words, to *integer filling factors* $\nu = N = n_{\mathrm{e}}^{2\mathrm{d}} h/(eB_z)$. This statement clearly refers to the bulk of the sample, but not necessarily to the edges.

The longitudinal resistance can also be obtained in a four-terminal measurement and is given by

$$R_{\mathrm{L}} = R_{14,23} = \frac{U_{23}}{I_1} = 0 \tag{16.15}$$

(or, equivalently, $R_{23,14}$ at this point). The longitudinal resistance is zero because $U_{23} = 0$ (from $I_3 = 0 = e^2/hN_{32}U_{32}$) while $I_1 \neq 0$, in accordance with experiment. Consequently, the longitudinal conductance also vanishes (see corresponding discussion in Sect. 14.2).

Example (2). We consider the modified Hall device shown in Fig. 16.7.

Here the fourfold rotational symmetry is broken and reduced to a twofold rotational symmetry because of the constriction in the middle. Considering

[2] This would not be the case for small magnetic fields.

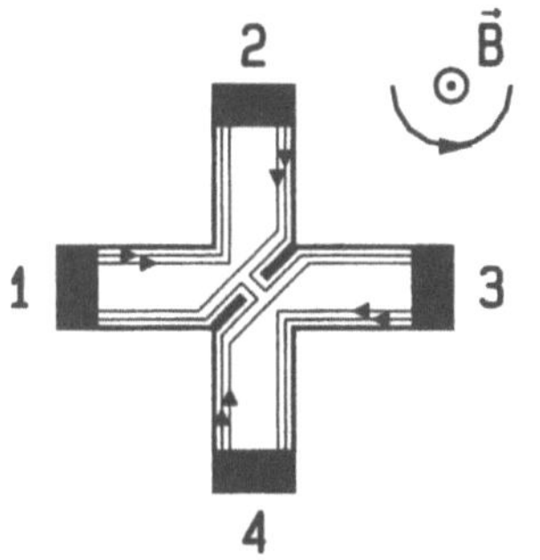

Fig. 16.7. Schematic illustration of a Hall element containing a constriction in the middle, in a strong perpendicular magnetic field. For the edge channels depicted in this example, we have $N = 2$ and $\tilde{N} = 1$

possible classical trajectories in Fig. 16.7, we see that N_{12} and N_{34} are no longer zero, because certain classical trajectories jump over the constriction if it is sufficiently narrow. This corresponds to an integer number of channels, say $\tilde{N}$, which are not transmitted by the constriction. We have $N_{14} = N_{32} = N - \tilde{N}$, $N_{12} = N_{34} = \tilde{N}$, and $N_{21} = N_{43} = N$. All other off-diagonal $N_{nn'}$ are zero. For the Hall resistance, we start from $I_1 = e^2/[h(N_{12}U_{12} + N_{14}U_{14})]$ and $U_{12} = 0$ (from $I_2 = 0$), and hence $U_{14} = U_{24}$, and immediately obtain

$$R_{13,24} = \frac{U_{24}}{I_1} = \frac{h}{e^2}\frac{1}{N - \tilde{N}} , \tag{16.16}$$

a result that looks very similar to the usual QHE. Here, however, the integer $N - \tilde{N} \geq 0$ no longer corresponds to an integer filling factor but, rather, is given by the width of the constriction [447, 448]. For a sufficiently wide constriction, $\tilde{N} = 0$ and the usual QHE is recovered. A corresponding experimental result is depicted in Fig. 16.8. Again (see Chap. 15), two negatively biased top gate contacts provide a constriction with variable width. For $B = 3.5\,\mathrm{T}$, for example, the resistance in Fig. 16.8 starts at $h/(e^2(5-0))$ for zero gate voltage and increases to $h/(e^2(5-1))$ around $-0.2\,\mathrm{V}$ gate voltage; for $B = 5.2\,\mathrm{T}$, the resistance is $h/(e^2(4-0))$ for zero gate voltage and $h/(e^2(4-2))$ around $-0.3\,\mathrm{V}$ gate voltage. The other plateaus are not very pronounced.

For the longitudinal resistance $R_{14,23} = U_{23}/I_1$, we can again start from $I_1 = e^2/[h\,(N_{12}U_{12} + N_{14}U_{14})]$ where $U_{12} = 0$ from $I_2 = 0$. Replacing $U_{14} = U_{24} = U_{23} + U_{34}$ with $U_{34} = U_{23}N_{32}/N_{34}$ (from the condition $I_3 = 0$), we obtain the longitudinal resistance

$$R_{14,23} = \frac{U_{23}}{I_1} = \frac{h}{e^2}\left(\frac{1}{N - \tilde{N}} - \frac{1}{N}\right) , \tag{16.17}$$

which is nonzero (for $\tilde{N} \neq 0$), in sharp contrast to the usual QHE. Note, however, that other nonequivalent longitudinal resistances exist, such as $R_{12,34}$ (which is zero even for $\tilde{N} \neq 0$ because $U_{34} = 0$, from the condition $I_4 = 0$), this could be suspected in any case from the lack of fourfould rotational symmetry.

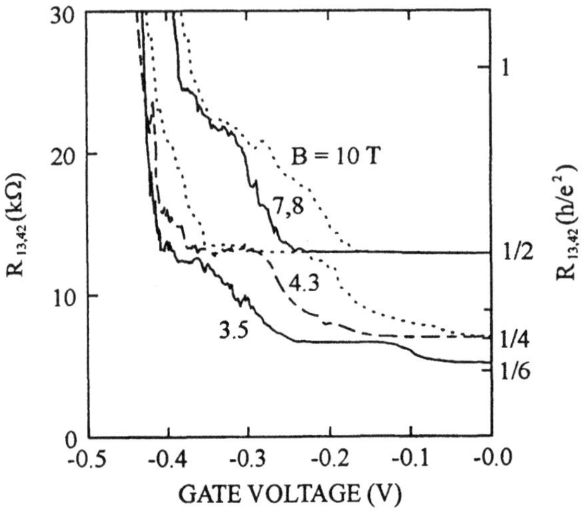

Fig. 16.8. $R_{13,42}$ versus gate voltage. ($R_{13,42}$ in the nomenclature used in this figure corresponds to $R_{13,24}$ in the text.) The constriction becomes narrower for increasing negative gate voltage. Hence $\tilde{N}$ increases from the *right* ($\tilde{N} = 0$) to the *left*. The device is similar to the one depicted schematically in Fig. 16.7. Taken from [447]

These two examples have shown that the concept of edge strips works very well for describing the physics of the QHE. Several experiments have directly measured the spatial position of these strips, e.g. [449, 450, 451]. Altogether, this strongly suggests that the compressible/incompressible strips at the sample edges are indeed real. For more details on the integer quantum Hall effect, we refer the reader to [452, 453].

16.3 Microscopic Derivation of the Landauer–Büttiker Formula

A microscopic derivation of the Landauer–Büttiker formula can be performed by means of various methods. Green's function techniques allow one easily to introduce a relation of the type (15.1) by applying the dissipation–fluctuation theorem (12.26) to the one-particle Green's function [454, 455]). The quantity obtained from the derivation through the dissipation–fluctuation theorem, however, is *not* immediately related to the transmission coefficient and if dissipative processes are involved, such a relation is completely absent. Thus we shall give a more elementary derivation from linear response theory, which shows clearly the unavoidable assumptions and limitations of the results.

16.3.1 Linear Response Theory

Let us consider, in addition to the magnetic field, an external perturbation of the form $\rho(\boldsymbol{r})\phi(\boldsymbol{r})$, corresponding to external voltages applied at the contacts. Anticipating a planar two-dimensional semiconductor heterostructure, we choose $\boldsymbol{r}$ to be two-dimensional. According to our general results in

Sect. 2.6.3, the quantity of interest is a current response function, which leads to the induced current density (compare (2.153))

$$\boldsymbol{j}_{\text{ind}} = -\frac{\mathrm{i}}{\hbar} \int_{-\infty}^{t} \mathrm{d}t' \int \mathrm{d}\boldsymbol{r}' \langle [\hat{\boldsymbol{j}}_0(\boldsymbol{r},t), \hat{\rho}(\boldsymbol{r}',t')] \rangle \phi(\boldsymbol{r}',t') \,. \tag{16.18}$$

We now introduce the current and density operators. Within the quasiparticle approximation, with eigenstates and eigenenergies φ_n and E_n, respectively, we find the following for the expectation value:

$$\langle [c_n^\dagger(t) c_m(t), c_{n'}^\dagger(t') c_{m'}(t')] \rangle = \delta_{nm'} \delta_{n'm} \, (f_n - f_m) \, \mathrm{e}^{-\mathrm{i}(E_n - E_m - \mathrm{i}\delta)(t-t')/\hbar} \,, \tag{16.19}$$

where the infinitesimal δ guarantees the existence of the integral. Within this approximation, we have, in the first step,

$$\begin{aligned} &\boldsymbol{j}_{\text{ind}}(\boldsymbol{r},t) \\ &= \frac{e^2}{m} \sum_{nm} \int_{-\infty}^{t} \mathrm{d}t' \, \boldsymbol{j}_{nm}(\boldsymbol{r}) (f_n - f_m) \mathrm{e}^{-\mathrm{i}(E_n - E_m - \mathrm{i}\delta)(t-t')/\hbar} \langle m | \phi(t') | n \rangle \,, \end{aligned} \tag{16.20}$$

where we have introduced

$$\boldsymbol{j}_{nm}(\boldsymbol{r}) = \varphi_n^*(\boldsymbol{r}) \nabla \varphi_m(\boldsymbol{r}) - (\nabla \varphi_n^*(\boldsymbol{r})) \varphi_m(\boldsymbol{r}) \,. \tag{16.21}$$

In the next step, we shall now show that matrix elements of the external potential can be reduced to line integrals within the leads only. This is accomplished by means of the identity

$$\langle m | \phi(t) | n \rangle = \frac{\langle m | [h, \phi(t)] | n \rangle}{E_m - E_n} = -\frac{\hbar^2}{2m} \frac{\langle m | [\Delta, \phi(\boldsymbol{r})] | n \rangle}{E_m - E_n} \tag{16.22}$$

and the commutator

$$[\Delta, \phi] = -\nabla \boldsymbol{E} - 2 \boldsymbol{E} \nabla \,. \tag{16.23}$$

Evaluating the matrix element of (16.23) allows us to split off the integral of a total divergence, which is transformed into a line integral along the boundary of the sample. As $\boldsymbol{E}$ vanishes within the leads and the one-particle wavefunction vanishes otherwise, this contribution vanishes completely and we obtain

$$\langle m | \nabla \boldsymbol{E} + 2 \boldsymbol{E} \nabla | n \rangle = \int \mathrm{d}^2 \boldsymbol{r}' \, \boldsymbol{j}_{mn}(\boldsymbol{r}') \boldsymbol{E}(\boldsymbol{r}',t') \,. \tag{16.24}$$

Now the induced current density takes the form

$$\begin{aligned} \boldsymbol{j}_{\text{ind}}(\boldsymbol{r},t) = \frac{e^2 \hbar^2}{4m^2} \sum_{nm} \int_{-\infty}^{t} \mathrm{d}t' \int \mathrm{d}^2 \boldsymbol{r}' \, \boldsymbol{j}_{nm}(\boldsymbol{r}) \frac{f_n - f_m}{E_n - E_m} \\ \times \mathrm{e}^{-\mathrm{i}(E_n - E_m - \mathrm{i}\delta)(t-t')/\hbar} \boldsymbol{j}_{mn}(\boldsymbol{r}') \boldsymbol{E}(\boldsymbol{r}',t') \,. \end{aligned} \tag{16.25}$$

In order to perform the integration with respect to t, we represent the electric field in the form

$$\boldsymbol{E}(\boldsymbol{r},t) = -\nabla\phi(\boldsymbol{r})\cos(\omega t)\,, \tag{16.26}$$

and make use of the integral

$$\begin{aligned}\int_{-\infty}^{t} & \mathrm{d}t'\,\mathrm{e}^{-\mathrm{i}(E_n-E_m-\mathrm{i}\delta)(t-t')/\hbar}\cos(\omega t') \\ &= -\frac{\mathrm{i}\hbar}{2}\cos(\omega t)\left(\frac{1}{E_n-E_m+\hbar\omega-\mathrm{i}\delta}+\frac{1}{E_n-E_m-\hbar\omega-\mathrm{i}\delta}\right) \\ &+\frac{\hbar}{2}\sin(\omega t)\left(\frac{1}{E_n-E_m+\hbar\omega-\mathrm{i}\delta}-\frac{1}{E_n-E_m-\hbar\omega-\mathrm{i}\delta}\right)\,.\end{aligned} \tag{16.27}$$

This results in a dissipative contribution to the induced current density which is in phase with the electric field, and to a reactive one which is out of phase. Our main interest here is in the stationary limit. When this limit is approached, the reactive contribution vanishes and we are left with the dissipative one. The latter is given by

$$\begin{aligned}\boldsymbol{j}_{\mathrm{ind}} &= -\frac{e^2\hbar^3\pi}{8m^2}\cos(\omega t)\sum_{nm}\int \mathrm{d}^2\boldsymbol{r}'\,\frac{f_n-f_m}{E_n-E_m}\boldsymbol{j}_{nm}(\boldsymbol{r}) \\ &\times\left[\delta(E_n-E_m+\hbar\omega)+\delta(E_n-E_m-\hbar\omega)\right]\boldsymbol{j}_{mn}(\boldsymbol{r}')\,\nabla\phi(\boldsymbol{r}')\,.\end{aligned} \tag{16.28}$$

Now we are ready to reduce the integration over the sample area to line integrals within the leads. By partial integration, we have

$$\begin{aligned}\int & \mathrm{d}^2\boldsymbol{r}'\,\boldsymbol{j}_{mn}(\boldsymbol{r}')\nabla\phi(\boldsymbol{r}') \\ &= \int \mathrm{d}^2\boldsymbol{r}'\,\nabla[\boldsymbol{j}_{mn}(\boldsymbol{r}')\phi(\boldsymbol{r}')] - \int \mathrm{d}^2\boldsymbol{r}'\,\phi(\boldsymbol{r}')\nabla\boldsymbol{j}_{mn}(\boldsymbol{r}')\,.\end{aligned} \tag{16.29}$$

The second term does not contribute, because

$$\nabla\boldsymbol{j}_{mn}(\boldsymbol{r}') = (E_n-E_m)\,\varphi_m^*(\boldsymbol{r}')\varphi_n(\boldsymbol{r}') \tag{16.30}$$

and because of the occurrence of a positive- and a negative-frequency part in (16.28). The total divergence in (16.29) is readily transformed to a line integral in the leads according to

$$\int \mathrm{d}^2\boldsymbol{r}'\,\nabla[\boldsymbol{j}_{mn}(\boldsymbol{r}')\phi(\boldsymbol{r}')] = \sum_k \phi_k J^k_{mn}\,, \tag{16.31}$$

where we have defined

$$J^k_{mn} = \int_{C_k}\mathrm{d}x_k\,\boldsymbol{e}_{y_k}\boldsymbol{j}_{mn}(x_k,y_k)\,. \tag{16.32}$$

Here C_k denotes the line of constant potential ϕ_k within the kth lead and represents the boundary between $\boldsymbol{E}=0$ and $\boldsymbol{E}\neq 0$. The unit vector $\boldsymbol{e}_{y_k}$

is parallel to the lead, and thus perpendicular to $\mathcal{C}_k$. All other contributions vanish, as the wavefunctions are zero at the boundaries outside the leads. We are left with having to evaluate the δ functions in (16.28) in the stationary limit. We consider the Fermi function $f_m = \Theta(E_F - E_m)$ at $E_m = E_n \pm \hbar\omega$ and expand for small ω. The result is

$$\begin{aligned}&\lim_{\omega\to 0} \frac{f_n - f_m}{E_n - E_m}[\delta(E_n - E_m + \hbar\omega) + \delta(E_n - E_m - \hbar\omega)] \\ &= 2\delta(E_n - E_m)\,\delta(E_F - E_n)\,. \end{aligned} \tag{16.33}$$

When we insert (16.31) and (16.32) in (16.28), the stationary dissipative current density takes the form

$$\boldsymbol{j}_{\mathrm{ind}}(\boldsymbol{r}) = \frac{\pi e^2\hbar^3}{4m^2}\sum_{nm}\sum_{l}\boldsymbol{j}_{nm}(\boldsymbol{r})\phi_l J^l_{mn}\delta(E_n - E_m)\,\delta(E_F - E_n)\,. \tag{16.34}$$

The total current I_l in the lth lead is found from (16.34) by integrating along a line $\mathcal{C}_l$ and results in

$$I_l = \sum_{l'} G_{ll'}\phi_{l'}\,, \tag{16.35}$$

where the conductance coefficients are

$$G_{ll'} = \frac{e^2\hbar^3\pi}{4m^2}\sum_{nm} J^l_{nm}J^{l'}_{mn}\delta(E_n - E_m)\,\delta(E_F - E_n)\,. \tag{16.36}$$

As we shall prove below, flux conservation implies

$$\sum_{l'} G_{ll'} = \sum_{l} G_{ll'} = 0\,, \tag{16.37}$$

which can be used to write (16.35) in the equivalent form

$$I_l = \sum_{l} G_{ll'}U_{ll'}\,. \tag{16.38}$$

With a much more lengthy derivation, this result can also be obtained from a current–current response function [456, 457, 458].[3] Equation (16.36) represents a rather general starting point for the discussion of conductivity in mesoscopic systems – at the level of the quasiparticle approximation. We would now like to connect (16.36) to the quantum mechanical transmission amplitudes which we have introduced in Sect. 15.1.

[3] Note that under the experimental conditions considered, i.e. external voltages applied to the leads, the current–current response is not the *natural* choice for calculating the induced current.

16.3.2 The Multiprobe Landauer–Büttiker Formula

It is obvious from (16.36) and (16.32) that calculation of the conductance coefficients requires knowledge of the asymptotic behavior of the wavefunction within the leads. The wavefunctions within any lead can be expanded in terms of the transverse solutions and the plane-wave solutions propagating along the lead. The propagating solutions consist of three partial waves in each lead: a part which is transmitted by the lead–sample boundary, a reflected part, and a third, incoming component from the sum of all the other leads. At any given energy, a finite number of such propagating solutions exists, which correspond to the channels, discussed in Sect. 16.2.1. Thus, the solutions in lead k depend on a two-dimensional wave vector $\boldsymbol{q}_{ki}$, with the component parallel to the lead denoted by q_{ki}, which determine the propagating solutions for each channel i. The corresponding wavefunctions can now be represented by

$$\Phi_{ki}(x_k, y_k) = \frac{g_{ki}(x_k)}{\sqrt{q_{ki}}}\mathrm{e}^{-\mathrm{i}q_{ki}y_k} + \sum_j R_{kk}^{ji}\frac{g_{kj}(x_k)}{\sqrt{q_{kj}}}\mathrm{e}^{\mathrm{i}q_{kj}y_k} . \tag{16.39}$$

The well functions $g_{ki}(x_k)$ account for the dependence on the transverse variable in the lead. After transmission through the sample, the plane-wave emitted from lead k is incident on the other leads $l \neq k$ in the form

$$\Phi_{ki}(x_l, y_l) = \sum_j T_{lk}^{ji}\frac{g_{lj}(x_l)}{\sqrt{q_{kj}}}\mathrm{e}^{\mathrm{i}q_{lj}y_l} . \tag{16.40}$$

The total wavefunction is then given by

$$\varphi_{ki}(\boldsymbol{r}) = \begin{cases} \Phi_{ki}(x_k, y_k) & \text{if } \boldsymbol{r} \text{ in lead } k \\ \Phi_{ki}(x_l, y_l) & \text{if } \boldsymbol{r} \text{ in lead } l \neq k . \end{cases} \tag{16.41}$$

All information on scattering by impurities in the sample, disorder, etc. is included in the reflection and transmission amplitudes R and T. Before we derive some basic relations between these quantities, we first consider the conductance coefficients (16.36) in the basis of scattering states. Owing to the existence of the continuum wave vectors, (16.36) becomes in detail

$$G_{ll'} = \frac{e^2\hbar^3\pi}{4m^2}\sum_{ijnm}\int\frac{\mathrm{d}^2\boldsymbol{k}_{ni}}{(2\pi)^2}\frac{\mathrm{d}^2\boldsymbol{k}_{mj}}{(2\pi)^2} J^l_{n\boldsymbol{k}_{ni},m\boldsymbol{k}_{mj}} J^{l'}_{m\boldsymbol{k}_{mj},n\boldsymbol{k}_{ni}} \times \delta[E(\boldsymbol{k}_{ni}) - E_\mathrm{F}]\,\delta[E(\boldsymbol{k}_{mj}) - E_\mathrm{F}] . \tag{16.42}$$

After the wave vector integration is performed, this yields

$$G_{ll'} = \frac{e^2}{16\pi\hbar}\sum_{ijnm} J^l_{ni,mj} J^{l'}_{mj,ni} , \tag{16.43}$$

where i and j label the channels at the Fermi energy in leads n and m, respectively. Obviously, there are a number of different combinations of indices in (16.43). These need to be distinguished. Making use of (16.39)–(16.41) and the orthogonality of the well functions g_{kj}, we find the following results. If all indices are different, i.e. $l \neq k \neq k'$, the matrix element is given by

$$J^l_{ki,k'j} = 2\mathrm{i} \sum_{j'} (T^{j'i}_{lk})^* T^{j'j}_{lk'} \,, \tag{16.44}$$

which is also valid for $k = k' \neq l$. If only the lead indices are different, we find, for $l \neq k$,

$$J^l_{ki,lj} = 2\mathrm{i} \sum_{j'} (T^{j'i}_{lk})^* R^{j'j}_{ll} \tag{16.45}$$

and

$$J^k_{ki,lj} = 2\mathrm{i} \sum_{j'} (R^{j'i}_{kk})^* T^{j'j}_{kl} \,. \tag{16.46}$$

Finally, if all indices are equal, we have

$$J^k_{ki,kj} = -2\mathrm{i} \left(\delta_{ij} - \sum_{j'} (R^{j'i}_{kk})^* R^{jj}_{kk} \right) . \tag{16.47}$$

The condition of flux conservation can be obtained from (16.30) for $E_n = E_m$ by considering an integral over the sample area, again transformed into a line integral:

$$\sum_l J^l_{ki,k'j} = 0 \,. \tag{16.48}$$

For $k = k'$, this result together with (16.44) and (16.47), implies

$$\sum_{j'} \left[(R^{j'i}_{kk})^* R^{j'j}_{kk} + \sum_{l \neq k} (T^{j'i}_{lk})^* T^{j'j}_{lk} \right] = \delta_{ij} \,. \tag{16.49}$$

Correspondingly, we find from (16.44)–(16.46) for $k \neq k'$

$$\sum_{j'} \left[(R^{j'i}_{kk})^* T^{j'j}_{kk'} + (T^{j'i}_{k'k})^* R^{j'j}_{k'k'} \right] = - \sum_{j', l \neq k, k'} (T^{j'i}_{lk})^* T^{j'j}_{lk'} \,. \tag{16.50}$$

As expected, flux conservation does not depend on wavenumber. This is guaranteed by the choice of the wavefunctions (16.41), which are normalized to unit longitudinal flux. Mathematically, (16.49) and (16.50) result from the unitarity of the scattering matrix defined by $S_{lk} = T_{lk}$ for $l \neq k$ and $S_{ll} = R_{ll}$ otherwise [457, 459].

Now we are ready to evaluate the conductance coefficients (16.43). Let us consider the case $l \neq l'$. We obtain

$$G_{ll'} = \frac{e^2}{16\hbar\pi} \sum_{ij} \Bigg(J^{l}_{li,lj} J^{l'}_{li,lj} + J^{l}_{l'i,l'j} J^{l'}_{l'i,l'j} + J^{l}_{li,l'j} J^{l'}_{l'i,lj} + J^{l}_{l'i,lj} J^{l'}_{li,l'j}$$
$$+ \sum_{k \neq l,l'} \left(J^{l}_{li,kj} J^{l'}_{ki,lj} + J^{l}_{ki,lj} J^{l'}_{li,kj} + J^{l}_{l'i,kj} J^{l'}_{ki,l'j} + J^{l}_{ki,lj} J^{l'}_{li,kj} \right)$$
$$+ \sum_{k \neq l,l' k' \neq l,l'} J^{l}_{ki,k'j} J^{l'}_{k'i,kj} \Bigg) . \tag{16.51}$$

Next, we insert (16.44)–(16.47) into the various contributions to (16.50). For the last five terms, we make use of (16.51). We find that all contributions involving the product of two reflection and two transmission coefficients vanish, and we are left with

$$G_{ll'} = \frac{e^2}{2h} \sum_{ij} \left(\left| T^{ij}_{ll'} \right|^2 + \left| T^{ij}_{l'l} \right|^2 \right) , \tag{16.52}$$

which is valid for $l \neq l'$. The case $l = l'$ can be obtained from flux conservation, which is obtained from (16.48) and (16.43) in the form (16.37). This yields, together with (16.49) and (16.50),

$$G_{ll} = \frac{e^2}{h} \sum_{ij} \left(\left| R^{ij}_{ll} \right|^2 - 1 \right) . \tag{16.53}$$

When we insert the relations (16.52) and (16.53) in (16.38), the derivation of the Landauer–Büttiker formula (16.9), which we introduced by phenomenological reasoning, is completed.

16.4 The Fractional Quantum Hall Effect

In the preceeding sections we were able to explain the plateaus in the transverse Hall resistance at integer filling factors $\nu\,(= N)$ in a picture of noninteracting electrons.[4] In the fractional QHE [441, 460, 461, 463, 464], one observes in addition plateaus in the transverse Hall resistance and minima in the longitudinal resistivity at noninteger filling factors ν, one example of which is depicted in Fig. 16.9.

A substantial part (not all) of the fractions that have been observed experimentally can be expressed by

$$\nu = \frac{N}{2MN \pm 1} , \tag{16.54}$$

where N and M are integers. There are, however, exceptions such as $\nu = 1/2$. Obviously, (16.54) gives only odd denominators. In close analogy to the

[4] Except for the compressible and incompressible strips.

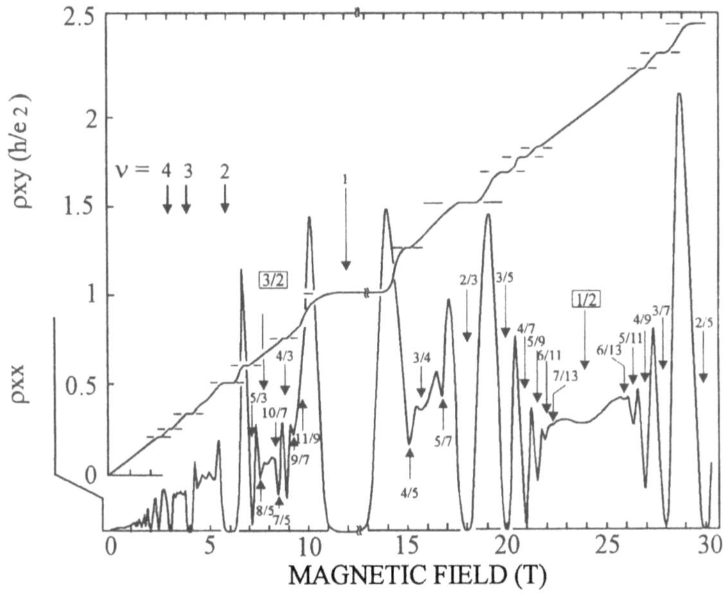

Fig. 16.9. Hall resistance ρ_{xy} and longitudinal resistivity ρ_{xx} in the fractional quantum Hall regime. The temperature is in the range of 100 mK. Notice that besides many odd-denominator fractions a few even denominator fractions appear. Taken from [462]

electron–hole picture for the conduction and valence bands of a semiconductor, a missing electron in a Landau level can also be interpreted as a hole. Because of electron–hole symmetry, this gives an additional series of fractions according to $\nu \longrightarrow 1 - \nu$. Given this, all odd-denominator fractions indicated in Fig. 16.9 for $\nu \leq 1$ ($\Leftrightarrow B \geq 12\,\mathrm{T}$ for a density of $n_{\mathrm{e}}^{2\mathrm{d}} = 3.0 \times 10^{11}\,\mathrm{cm}^{-2}$) can be reproduced directly by (16.54) (see Fig. 16.10).

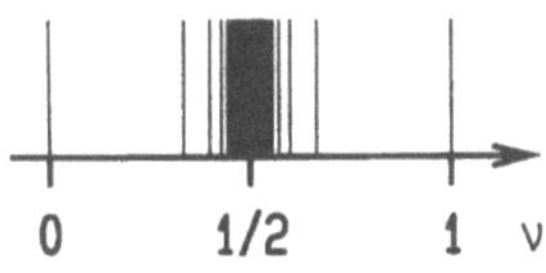

Fig. 16.10. The $M = 1$ series of (16.54)

Let us explain (16.54). By means of the filling factor of Landau levels ν (see (16.14)), where

$$n_{\mathrm{e}}^{2\mathrm{d}} = \frac{B_z e}{h}\,\nu\,, \tag{16.55}$$

the average magnetic flux per electron can be written as

$$\frac{B_z L_x L_y}{N_{\mathrm{e}}} = \frac{B_z}{n_{\mathrm{e}}^{2\mathrm{d}}} = \frac{h}{e}\,\frac{1}{\nu} = \frac{\Phi_0}{\nu}\,, \tag{16.56}$$

where we have introduced the *flux quantum* $\Phi_0 = h/e$. Hence, in the QHE (integer and fractional) each electron on the average carries a magnetic flux of Φ_0/ν. In [465] the following appealing picture is proposed, which is an extension of a trial many-body wavefunction discussed in [466, 467, 468, 469]. It was shown [465] that a system of two-dimensional electrons (fermions) interacting via the Coulomb interaction can – in a mean-field sense – be reduced to a noninteracting system of *composite fermions*. Each composite fermion consists of one electron plus an even number, say $2M$, of flux quanta. For $\nu = 1/3$, for example, we have an average of three flux quanta per electron. For $M = 1$, each composite fermion carries two flux quanta, and thus each composite fermion is subject to an effective magnetic field corresponding to only one flux quantum, i.e. $\nu_{\text{eff}} = 1$ (if the fluxes all point in the same direction). For other integers $\nu_{\text{eff}} = N$, each composite is on average subject to an external flux of $\pm\Phi_0/\nu_{\text{eff}}$ which translates into a flux of $(2M \pm 1/N)\Phi_0$ per electron; the + and − signs stand for magnetic fields parallel and antiparallel, respectively, to the flux quanta of the composite particles. From $\Phi_0/\nu = (2M \pm 1/N)\Phi_0 = \Phi_0(2MN \pm 1)/N$, the hierarchy of fractions (16.54) follows immediately. Thus *the fractional QHE of electrons is a manifestation of the integer QHE of composite fermions* [465]. The fractional charge of these composite fermions has indeed been observed experimentally [470, 471]. Somewhat similarly, *composite bosons* [472] have been introduced.

The concept of composite fermions is analogous to the concept of Cooper pairs in the BCS theory of superconductivity, where a system of strongly interacting electrons is reduced to a system of very nearly noninteracting Cooper pairs. It is also similar to the concept of noninteracting (to first order) excitons (see Chap. 6), which can replace a strongly interacting system of electrons. As many aspects of the fractional QHE and its highly structured phase diagram are, however, still subjects of current research, we shall not go into detail here but rather refer the reader to the original literature.

16.5 Magnetotransport Through Dot or Antidot-Lattices

In Chap. 7 we discussed the influence of a static magnetic field on the band structure of a semiconductor. For state-of-the-art static magnetic fields, the cyclotron radius is huge compared with one lattice constant. Consequently, the Hamiltonian is well described by a crystal electron with some effective mass immersed in the magnetic field. Some minor corrections to that picture are the field-dependent band-mixing effects (see Sect. 7.1.2). If the cyclotron orbit, however, were to decrease to the atomic scale, dramatic effects would be expected. Commensurability between the two scales would play an important role. While decreasing the cyclotron orbit that much is currently way out of reach, increasing the lattice constant by means of mesoscopic systems is possible. Let us consider a two-dimensional electron gas in the xy plane, for example in a GaAs heterostructure, subject to a perpendicular magnetic field. A square lattice of dots (or antidots) with lattice constant a can be realized by etching or by a patterned top gate electrode (especially well suited for antidot lattices). This leads to a periodic potential landscape (Fig. 16.11(a)) resulting in a two-dimensional miniband structure.

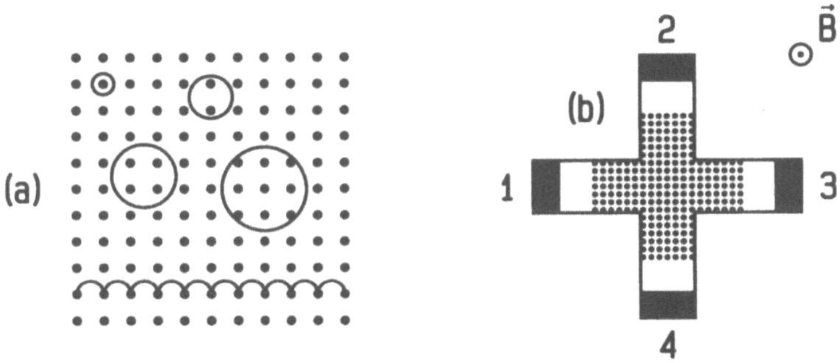

Fig. 16.11. (*a*) A two-dimensional electron gas is subject to a lattice of dots and is immersed in a perpendicular magnetic field. Selected commensurate classical cyclotron orbits and a so-called runaway trajectory are depicted. Generally, the classical trajectories are chaotic. (*b*) This artificial lattice can be introduced into a Hall bar (shown schematically) in order to study its properties

The detailed shape of the potential of such artifical lattices is generally not well known. Etched dots might lead to a periodic hard-sphere potential. To obtain an initial, crude idea of the general single-particle behavior in a magnetic field, we can assume a simple tight-binding form as in (3.78). For a single band (i.e. we drop the band index n), we have

$$E(\boldsymbol{k}) = E^1 [\cos(k_x a) + \cos(k_y a)] , \tag{16.57}$$

where the energy scale has been shifted ($E^0 = 0$) for our convenience. From here on we follow [473] closely. Substituting $\hbar \boldsymbol{k} \longrightarrow \boldsymbol{p} + e\boldsymbol{A}$ (the minimal substitution discussed in Chap. 2), we obtain an effective Hamiltonian for electrons. Again, the Landau gauge can be employed, i.e. $\boldsymbol{A} = (0, Bx, 0)$. Clearly we can decompose the cosines in (16.57) as $\cos(p_x a/\hbar) = 1/2\,[\exp(+\mathrm{i}p_x a/\hbar) + \exp(+\mathrm{i}p_x a/\hbar)]$. The action of the exponentials with operators p_x and p_y on the wavefunction $\varphi(x, y)$ results merely in a translation, e.g. $\exp(\pm\,\mathrm{i}p_x a/\hbar)\varphi(x, y) = \varphi(x \pm a, y)$. This immediately leads us to the eigenvalue equation

$$\begin{aligned} 2\frac{E}{E^1}\,\varphi(x, y) &= \varphi(x + a, y) + \varphi(x - a, y) \\ &+\varphi(x, y + a)\exp(-\mathrm{i}qBxa/\hbar) + \varphi(x, y - a)\exp(+\mathrm{i}qBxa/\hbar) . \end{aligned} \tag{16.58}$$

With the substitutions $\tilde{x} = x/a$, $\tilde{y} = y/a$, $2E/E^1 = \tilde{E}$, $\tilde{k}_y = k_y a$ and the ansatz

$$\varphi(x,y) = \mathrm{e}^{\mathrm{i}\tilde{k}_y\tilde{y}}\, g(\tilde{x}) \,, \tag{16.59}$$

we obtain the following difference equation, called Harper's equation [474]:

$$g(\tilde{x}+1) + g(\tilde{x}-1) + 2g(\tilde{x})\cos(2\pi\tilde{x}\alpha - \tilde{k}_y) = \tilde{E}g(\tilde{x}) \,. \tag{16.60}$$

Here we have introduced the dimensionless parameter

$$\alpha = \frac{qBa^2}{\hbar 2\pi} = \frac{1}{2}\frac{a^2}{\pi r_\mathrm{c}^2} = \frac{Ba^2}{\Phi_0} \,, \tag{16.61}$$

which is obviously proportional to the magnetic field strength. The quantity 2α can be interpreted as the ratio of the area of the two-dimensional unit cell a^2 to the area of the cyclotron orbit πr_c^2 (with cyclotron radius $r_c = \sqrt{\hbar/(qB)}$). Equivalently one can express α as the ratio of the flux through one unit cell Ba^2 to the elementary flux quantum $\Phi_0 = h/e$. The classical orbits depicted in Fig. 16.11 correspond to α =1, 1/4, 1/9, and 1/16. In a semiconductor crystal, e.g. GaAs, with a lattice constant of approximately $a = 0.5\,\mathrm{nm}$, $\alpha = 1$ corresponds to a huge magnetic field B of about 16000 T. In a mesoscopic superlattice with $a = 20\,\mathrm{nm}$, on the other hand, the corresponding magnetic field of about 10 T is accessible experimentally. For integers $m = \tilde{x}$, the difference equation (16.60) can easily be rewritten in matrix form as follows:

$$\begin{aligned}
\begin{pmatrix} g_{m+1} \\ g_m \end{pmatrix} &= \begin{pmatrix} [\tilde{E} - 2\cos(2\pi m\alpha - \tilde{k}_y)] & -1 \\ 1 & 0 \end{pmatrix} \begin{pmatrix} g_m \\ g_{m-1} \end{pmatrix} \\
&=: \underline{\mathcal{M}}_m \begin{pmatrix} g_m \\ g_{m-1} \end{pmatrix} \\
&= \underline{\mathcal{M}}_m \underline{\mathcal{M}}_{m-1} \underline{\mathcal{M}}_{m-2} \cdots \underline{\mathcal{M}}_1 \begin{pmatrix} g_1 \\ g_0 \end{pmatrix} \\
&= \underbrace{\underline{\mathcal{M}}_m \underline{\mathcal{M}}_{m-1} \cdots \underline{\mathcal{M}}_{m-M+1}}_{:=\, \underline{\mathcal{Q}}(\tilde{E},\tilde{k}_y)} \underbrace{\underline{\mathcal{M}}_m \underline{\mathcal{M}}_{m-1} \cdots \underline{\mathcal{M}}_{m-M+1}} \cdots
\end{aligned} \tag{16.62}$$

where we have introduced the 2×2 matrices $\underline{\mathcal{M}}_m$. Eigenfunctions of the Hamiltonian have to be bounded which immediately translates into the boundedness of the g_m for any $m = \tilde{x} = x/a$. One way in which this condition can be fulfilled is that the matrix product in the third line of (16.62) has a certain periodicity in m, say with period M. Looking at the argument of the cosine, periodicity necessitates

$$2\pi(m+M)\alpha - \tilde{k}_y = 2\pi m\alpha - \tilde{k}_y + 2\pi N \,, \tag{16.63}$$

where N is some integer. This is obviously equivalent to the condition that α is rational, i.e.

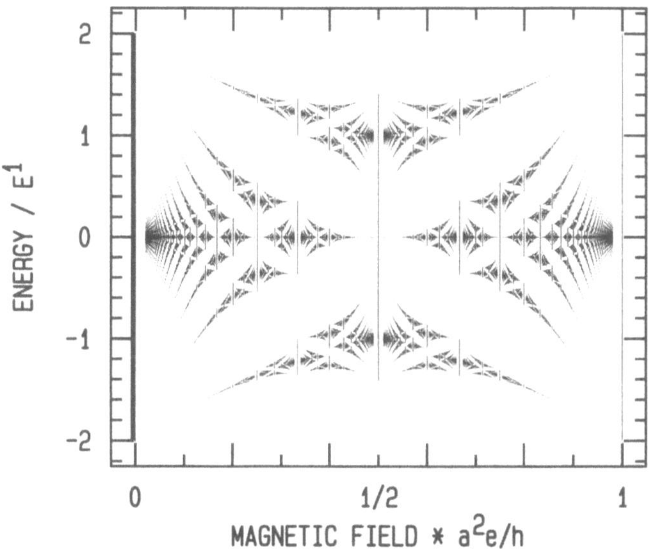

Fig. 16.12. The Hofstadter butterfly ($\alpha = N/M$, $M \le 50$, energy resolution $\approx 10^{-3}\,E^1$). The dispersion is periodic in $\alpha = Ba^2e/h$. The *thick bar* on the *left* indicates the width of the two-dimensional cosine-shaped miniband for zero field

$$\alpha = \frac{N}{M} \, . \tag{16.64}$$

Let the product of the M matrices be the 2×2 matrix $\underline{Q}(\tilde{E}, \tilde{k}_y)$. The g_m have to be bounded for any m. Thus $|\underline{Q}(\tilde{E}, \tilde{k}_y)\boldsymbol{g}| = |\boldsymbol{g}|$ has to hold for any vector $\boldsymbol{g}$. Hence each of the two eigenvalues of $\underline{Q}(\tilde{E}, \tilde{k}_y)$, say λ_1 and λ_2, has to have a magnitude of one, i.e. $|\lambda_1 = \exp(\mathrm{i}\varphi_1)| = |\lambda_2 = \exp(\mathrm{i}\varphi_2)| = 1$. The trace of $\underline{Q}(\tilde{E}, \tilde{k}_y)$ equals the sum of its two eigenvalues and, consequently, we have the equivalent condition

$$|\mathrm{Tr}[\underline{Q}(\tilde{E}, \tilde{k}_y)]| \le 2 \, , \tag{16.65}$$

which still depends on α. The set of all $\tilde{E}$ and $\tilde{k}_y$ which fulfill the trace relation (16.65) for any given $\alpha = N/M$ provides the unusual and beautiful energy spectrum of the problem, often called the Hofstadter butterfly. It is depicted in Fig. 16.12. Notice its fragile, self-similar structure. M simple bands are formed at magnetic field values $\alpha = 1/M$. A far more realistic numerical band structure calculation for an actual antidot lattice has been discussed [475]. Long-range disorder tends to smear out the fine structure of the Hofstadter-butterfly [476]. If electron interaction effects are accounted for at the Hartree level, the essential features of the Hofstadter butterfly remain; its symmetry, however, is somewhat reduced [477, 478].

Transport experiments have been performed using such dot or antidot lattices in two-dimensional electron gases. Most of these experiments, howe-

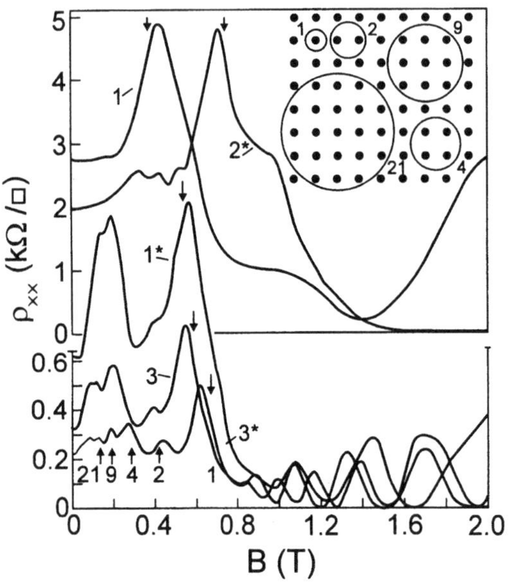

Fig. 16.13. Anomalies in the longitudinal Hall resistance ρ_{xx} observed in a Hall bar containing a mesoscopic superlattice. Taken from [479]

ver, have employed total sizes of the lattice which exceed the mean free path (several microns), and thus transport is not entirely coherent. Furthermore, the energy band width has mostly been quite small, owing to rather large lattice constants a. Nevertheless, unusual transport properties have been observed. Figure. 16.13 shows early results [479, 480] (here a=200–300 nm), which have been obtained by introducing mesoscopic lattices into Hall bars. For low fields, and thus large cyclotron orbits, anomalies in the longitudinal Hall resistance $\rho_{xx} = R_{13,24}$ are obtained. These anomalies are attributed to particular commensurate, periodic classical orbits as depicted in the inset. Certain magnetotransport peaks observed in antidot lattices have been ascribed to classically-chaotic orbits [481]. It has been argued that classical deterministic chaos might inhibit the observation of the Hofstadter butterfly [482]. On the other hand, transport calculations based on the Hofstadter energy spectrum have succeeded in reproducing the salient features of the above experiment [483]. More detailed work on magnetotransport in the QHE regime [484, 485, 486, 487, 488, 489], and in the fractional QHE regime [490], and on far-infrared properties [491, 492, 493, 494] of dot and antidot lattices has been done. Indeed, this is only a small and special selection of the comprehensive activities in this field and – this topic is not the only loose end of this book. *Things change faster in physics than in the book publishing business* [495].

16.6 Problems

16.1. Discuss the transverse and the longitudinal resistance of the Hall device depicted in Fig. 16.14 in the context of the Landauer–Büttiker formula. Show that such a device can exhibit nonvanishing longitudinal resistance. Discuss also the case of two point contacts in front of adjacent and nonadjacent terminals.

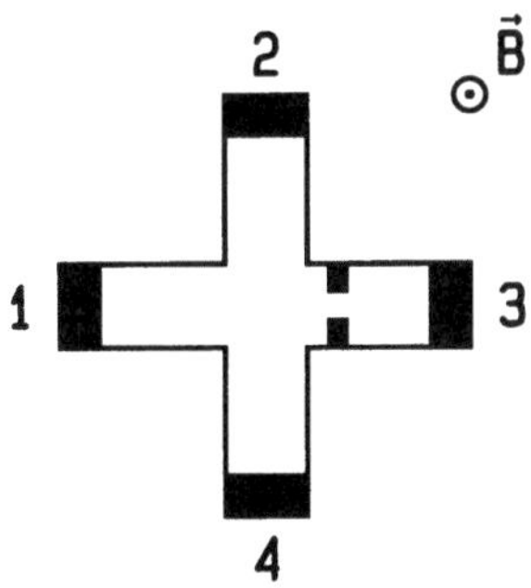

Fig. 16.14. Schematic illustration of a Hall element in a strong perpendicular magnetic field, containing an additional point contact in front of contact 3

16.2. Consider the normal Hall element depicted in Fig. 16.6. The measurement of the Hall voltage ϕ_{24} between terminals 2 and 4 is, however, performed with a *nonideal voltmeter* with an internal ohmic resistance R_i. Hence $I_2 = -I_4$ is no longer zero but given by $I_2 = -\phi_{24}/R_\text{i}$ (the minus sign indicates that ϕ_{24} is a voltage source to the outside world). Derive the following expression for the Hall resistance from the Landauer–Büttiker formula under these conditions:

$$R_\text{H} = R_{13,24} = \frac{h}{e^2} \frac{1}{N + h/(e^2 R_\text{i})} . \tag{16.66}$$

Clearly, this expression converges to the normal quantization (16.1) in the limit $R_\text{i} \to \infty$.

16.3. Derive (16.36) and (16.38) from the current–current response function of the two-dimensional semiconductor heterostructure discussed in Sect. 16.3.1.

16.4. What happens to the Hofstadter butterfly in the case of a rectangular lattice with lattice constants a_x and $a_y \neq a_x$? The zero-field dispersion is $E(\boldsymbol{k}) = E_x^1 \cos(k_x a_x) + E_y^1 \cos(k_y a_y)$.

References

1. J. R. Chelikowsky, M. L. Cohen, Phys. Rev. B **14**, 556 (1976)
2. Landolt-Börnstein, Vols. III/17a and III/17b, Springer, Berlin, Heidelberg
3. A. Ourmazd, D. W. Taylor, J. Cunningham, C. W. Tu, Phys. Rev. Lett. **62**, 933 (1989)
4. J. C. Maxwell, *Scientific papers*, Vol. 1, ed. W. D. Niven, pp. 526–597, Cambridge (1890)
5. E. Schrödinger, *Collected Papers on Wave Mechanics*, Blackie, London, Glasgow (1928)
6. P. A. M. Dirac, Proc. Roy. Soc. A **114**, 243 (1927)
7. W. Pauli, Phys. Rev. **58**, 716 (1940)
8. W. Pauli, Z. Phys. **31**, 765 (1925)
9. P. Jordan, E. P. Wigner, Z. Phys. **47**, 631 (1928)
10. A. Messiah, *Quantum Mechanics*, North-Holland, Amsterdam (1962)
11. J. C. Slater, Phys. Rev. **34**, 1293 (1929)
12. V. Fock, Z. Phys. **75**, 622 (1932)
13. J. von Neumann, *Collected Works*, Vol. 1, ed. A. H. Taub, Pergamon, New York (1961)
14. R. Feynman, *Statistical Mechanics*, W. A. Benjamin, Reading, MA (1972)
15. N. N. Bogolyubov, D. V. Shirkov, *Introduction to the Theory of Quantized Fields*, Wiley, New York, London (1959)
16. E. Schrödinger, Z. Phys. **14**, 664 (1926)
17. R. J. Glauber, Phys. Rev. Lett. **10**, 84 (1963)
18. E. C. Sudarshan, Phys. Rev. Lett. **10**, 227 (1963)
19. R. J. Glauber, in *Quantum Optics*, Scottish Summer School, eds. S. M. Kay, A. Maitland, Academic Press, London (1970)
20. M. S. Green, J. Chem. Phys. **22**, 398 (1954)
21. R. Kubo, J. Phys. Soc. Jpn. **12**, 570 (1957)
22. J. C. Slater, *Quantum Theory of Molecules and Solids*, Vol. 3, McGraw-Hill, New York (1967)
23. L. D. Landau, Sov. Phys. JETP **3**, 920 (1956)
24. J. M. Luttinger, P. Noziéres, Phys. Rev. **127**, 1423 (1962)
25. D. R. Hartree, Proc. Camb. Phil. Soc. **24**, 111 (1928)
26. V. Fock, Z. Phys. **61**, 126 (1930)
27. V. Fock, Z. Phys. **62**, 795 (1930)
28. T. Koopmans, Physica **1**, 104 (1934)
29. A. A. Abrikosov, L. P. Gorkov, I. E. Dzyaloshinski, *Methods of Quantum Field Theory in Statistical Physics*, Prentice Hall, Englewood Cliffs, NJ (1963)
30. J. W. Gibbs, *Collected Works*, Vol. 2, Longmans, Green, New York (1931)
31. L. H. Thomas, Proc. Camb. Phil. Soc. **23**, 542 (1927)
32. E. Fermi, Z. Phys. **48**, 73 (1928)
33. P. Hohenberg, W. Kohn, Phys. Rev. **136**, B864 (1964)

34. E. H. Lieb, Int. J. Quantum Chem. **24**, 243 (1983)
35. W. Kohn, L. J. Sham, Phys. Rev. **140**, A1133 (1965)
36. K. Schönhammer, O. Gunnarson, J. Phys C **20**, 3675 (1987)
37. L. J. Sham, M. Schlüter, Phys. Rev. B **32**, 3883 (1985)
38. J. C. Slater, Phys. Rev. **81**, 385 (1951)
39. D. Ceperley, B. Alder, Phys. Rev. Lett, **45** 566 (1980)
40. J. P. Perdew, A. Zunger, Phys. Rev. B **23**, 5048 (1981)
41. J. P. Perdew, Phys. Rev. Lett. **55**, 1665 (1985)
42. J. P. Perdew, Phys. Rev. B **33**, 8822 (1986)
43. E. Fermi, Nuovo Cimento **11**, 157 (1934)
44. H. J. Hellman, J. Chem. Phys. **3**, 61 (1935)
45. J. C. Phillips, Phys. Rev. **116**, 287 (1959)
46. G. B. Bachelet, D. R. Hamann, M. Schlüter, Phys. Rev. B **26**, 4199 (1982)
47. J. Ihm, A. Zunger, M. L. Cohen, J. Phys. C **12**, 4409 (1979)
48. U. Wenschuh, Ph. D. Thesis, University of Dortmund (1998)
49. M. L. Cohen, J. R. Chelikowsky, *Electronic Structure and Optical Properties of Semiconductors*, Springer Series in Solid-State Science, Vol. 75, Springer, Berlin, Heidelberg (1988)
50. G. B. Bachelet, H. S. Greenside, G. A. Baraff, M. Schlüter, Phys. Rev. B **24**, 4745 (1981)
51. *Handbook of Chemistry and Physics*, 62nd ed., ed. R. C. Weast, CRC Press, Boca Raton (1981)
52. J. C. Phillips, *Bonds and Bands in Semiconductors*, Academic Press, New York (1973)
53. U. Piesbergen, in *Semiconductors and Semimetals*, ed. R. K. Willardson, A. C. Beer, Academic Press, New York (1967), Vol. 2, p. 55
54. M. S. Hybertsen, S. G. Louie, Phys. Rev. B **34**, 5390 (1986)
55. R. W. Godby, M. Schlüter, L. J. Sham, Phys. Rev. Lett. **56**, 2415 (1986)
56. M. Rohlfing, P. Krüger, J. Pollmann, Phys. Rev. Lett. **75**, 3489 (1995)
57. W.-D. Schöne, A. G. Eguiluz, Phys. Rev. Lett. **81**, 1662 (1998)
58. W. G. Aulbur, L. Jönssen, J. W. Wilkins, Solid State Phys., **54**, 1 (1999)
59. F. Bloch, Z. Phys. **52**, 555 (1928)
60. G. Wannier, Phys. Rev. **52**, 191 (1937)
61. J. Des Cloizaux, Phys. Rev. **129**, 554 (1963)
62. E. O. Kane, in *Semiconductors and Semimetals*, ed. R. K. Willardson, A. C. Beer, Academic Press, New York (1967)
63. J. M. Luttinger, M. Kohn, Phys. Rev. **97**, 869 (1955)
64. J. M. Luttinger, Phys. Rev. **102**, 1030 (1956)
65. F. Bloch, Phys. Rev. **70**, 460 (1946)
66. R. P. Feynman, F. L. Vernon, R. W. Hellwarth, J. Appl. Phys. **28**, 49 (1957)
67. L. Allen, J. H. Eberly, *Optical Resonance and Two-Level Atoms*, Wiley, New York (1975)
68. P. C. Becker, H. L. Fragnito, C. H. Brito-Cruz, R. L. Fork, J. E. Cunningham, J. E. Henry, C. V. Shank, Phys. Rev. Lett. **61**, 1647 (1988)
69. P. C. Becker, H. L. Fragnito, C. H. Brito-Cruz, J. Shah, R. L. Fork, J. E. Cunningham, J. E. Henry, C. V. Shank, Appl. Phys. Lett. **53**, 2089 (1988)
70. W. H. Louisell, *Quantum Statistical Properties of Radiation*, Wiley, New York (1973)
71. J. H. Eberly, K. Wodkiewicz, J. Opt. Soc. Am. **67**, 1252 (1977)
72. S. L. Adler, Phys. Rev. **126** 413 (1962)
73. N. Wiser, Phys. Rev. **129**, 62 (1963)
74. W. Hanke, Adv. Phys. **27**, 287 (1978)
75. W. Schäfer, M. Schreiber, Solid State Commun., **38**, 1241 (1981)

76. W. Schäfer, J. Treusch, Solid State Commun., **43**, 949 (1982)
77. X. Zhu, S. G. Louie, Phys. Rev. B **43**, 14142 (1991)
78. H. Ehrenreich, M. H. Cohen, Phys. Rev. **115**, 786 (1959)
79. S. Albrecht, L. Reining, R. Del Sole, G. Onida, Phys. Rev. Lett. **80**, 4510 (1998)
80. M. Rohlfing, S. G. Louie, Phys. Rev. Lett. **81**, 2312 (1998)
81. L. X. Benedict, E. L. Shirley, R. B. Bohm, Phys. Rev. Lett., **80**, 4514 (1998)
82. G. A. Baker, J. L. Gammel, *Padè-Approximants in Theoretical Physics*, Academic Press, New York (1970)
83. J. L. Freeouf, Phys. Rev. B **7**, 3810 (1972)
84. M. Lindberg, S. W. Koch, Phys. Rev. B **38**, 3342 (1988)
85. W. Schäfer, in *Festkörperprobleme*, Advances in Solid State Physics, Vol. 28, ed. U. Rössler, Vieweg, Braunschweig (1988)
86. A. Stahl, I. Balslev, *Electrodynamics of the Semiconductor Band-Edge*, Springer Tracts in Modern Physics, Vol. 110, Springer, Berlin, Heidelberg (1986)
87. S. Schmitt-Rink, D. S. Chemla, H. Haug, Phys. Rev. B **37**, 941 (1988)
88. L. Hostler, R. H. Pratt, Phys. Rev. Lett. **10**, 469 (1963)
89. L. Hostler, Phys. Rev. **178**, 126 (1969)
90. J. Schwinger, J. Math. Phys. **5**, 1606 (1964)
91. H. Buchholz, *Die Konfluente Hypergeometrische Funktion*, Springer, Berlin, Göttingen, Heidelberg (1959)
92. E. T. Whittaker, G. N. Watson, *A Course in Modern Analysis*, 4th ed. p. 134, Cambridge University Press, Cambridge (1958)
93. R. Zimmermann, Phys. Status Solodi B **146**, 545 (1988)
94. S. M. Blinder, Int. J. Quantum Chem. S **18**, 293 (1984)
95. D. Dunn, J. Phys. C **10**, 2801 (1977)
96. R. G. Ulbrich, in *Materials Science and Technology*, ed. W. Schröter, Vol.4, VCH, Weinheim (1991)
97. R. Binder, S. W. Koch, M. Lindberg, W. Schäfer, Phys. Rev. B **43**, 6520 (1991)
98. M. Wegener, D. S. Chemla, S. Schmitt-Rink, W. Schäfer, Phys. Rev. A **42**, 5675 (1990)
99. J. J. Hopfield, Phys. Rev. **112**, 1555 (1958)
100. V. M. Agranovich, V. L. Ginzburg, *Crystal Optics with Spatial Dispersion and Excitons*, Springer, Berlin, Heidelberg (1984)
101. J. Tignon, T. Hasche, D. S. Chemla, H. C. Schneider, F. Jahnke, S. W. Koch, Phys. Rev. Lett. **84**, 3382 (2000)
102. D. Fröhlich, A. Kulik, B. Uebbing, A. Mysyrowicz, V. Langer, H. Stolz, W. von der Osten, Phys. Rev. Lett. **67**, 2343 (1991)
103. C. Weisbuch et al., Phys. Rev. Lett. **69**, 3314 (1992)
104. W. Schäfer, Phys. Scr. T **55**, 37 (1994)
105. W. Schäfer, F. Jahnke, S. Schmitt-Rink, Phys. Rev. B **47**, 1217 (1993)
106. K. Leo, M. Wegener, J. Shah, D. S. Chemla, E. O. Göbel, T. C. Damen, S. Schmitt-Rink, W. Schäfer, Phys. Rev. Lett. **65**, 1340 (1990)
107. E. Hecht, *Optics*, Addison-Wesley, Reading MA (1998)
108. V. F. Elesin, Sov. Phys. JETP **32**, 328 (1971)
109. C. Comte, G. Mahler, Phys. Rev. B **34**, 7164 (1986)
110. M. Hartmann, R. Zimmermann, H. Stolz, Phys. Status Solidi B **146**, 357 (1988)
111. K. Henneberger, K. H. Kühn, Physica A **150**, 439 (1988)
112. A. von Lehmen, D. S. Chemla, J. E. Zucker, J. P. Heritage, Opt. Lett. **11**, 609 (1986)
113. M. Joffre, D. Hulin, A. Antonetti, J. Phys. C **5**, 537 (1987).

114. N. Peyghambarian, H. M. Gibbs, J. L. Jewell, A. Antonetti, A. Migus, D. Hulin, A. Mysyrowicz, Phys. Rev. Lett. **53**, 2433 (19884)
115. J. Feldmann, T. Meier, G. von Plessen, M. Koch, E. O. Göbel, P. Thomas, G. Backer, C. Hartmann, H. Schweizer, W. Schäfer, H. Nickel, Phys. Rev. Lett. **70**, 3027 (1993)
116. F. Jahnke, M. Koch, T. Meier, J. Feldmann, W. Schäfer, P. Thomas, S. W. Koch, E. O. Göbel, H. Nickel, Phys. Rev. B **50**, 8114 (1994)
117. H. Stolz, V. Langer, E. Schreiber, S. Permogorov, W. von der Osten, Phys. Rev. Lett. **67**, 679 (1991)
118. M. Koch, J. Feldmann, G. von Plessen, E. O. Göbel, P. Thomas, K. Köhler, Phys. Rev. Lett. **69**, 3631 (1992)
119. R. Binder, S. W. Koch, M. Lindberg, N. Peygambarian, W. Schäfer, Phys. Rev. Lett. 65, 899 (1990)
120. S. W. Koch, N. Peyghambarian, M. Lindberg, J. Phys. C **21**, 5229 (1988)
121. L. Viña, G. E. W. Bauer, M. Potemski, J. C. Maan, E. E. Mendez, W. I. Wang, Phys. Rev. B **41**, 10767 (1990)
122. S. Schmitt-Rink, J. B. Stark, W. H. Knox, D. S. Chemla, W. Schäfer, J. Appl. Phys. A **53**, 491 (1991)
123. R. E. Peierls, Z. Phys. **80**, 763 (1933)
124. L. D. Landau, Z. Phys. **64**, 629 (1930)
125. C. R. Pidgeon, R. N. Brown, Phys. Rev. **146**, 575 (1955)
126. K. Suzuki, J. C. Hensel, Phys. Rev. B **9**, 4184 (1974)
127. J. C. Hensel, K. Suzuki, Phys. Rev. B **9**, 4219 (1974)
128. R. Ranvaud, H.-R. Trebin, U. Rössler, F. H. Pollack, Phys. Rev. B **20**, 701 (1979)
129. H.-R. Trebin, U. Rössler, R. Ranvaud, Phys. Rev. B **20**, 686 (1979)
130. E. Gornik, in *Landau Level Spectroscopy*, eds. G. Landwehr, E. I. Rashba, North-Holland, Amsterdam (1991)
131. M. Altarelli, N. O. Lipari, Phys. Rev. B **9**, 1733 (1974)
132. D. Bimberg, in *Festkörperprobleme*, Advances in Solid State Physics Vol. 17, ed. J. Treusch, Vieweg, Braunschweig (1977)
133. D. A. Broido, L. J. Sham, Phys. Rev. B **34**, 3817 (1986)
134. D. M. Wood, A. Zunger, Phys. Rev. B **53**, 7949 (1996)
135. F. Ancilotto, A. Fasolino, J. C. Maan, Phys. Rev. B **38**, 1788 (1988)
136. G. E. W. Bauer, T. Ando, Phys. Rev. B **38**, 6015 (1988)
137. A. Baldereschi, N. O. Lipari, Phys. Rev. B **3**, 439 (1970)
138. W. Ekardt, Solid State Commun. **16**, 233 (1975)
139. N. Lee, D. M. Larsen, B. Lax, J. Phys. Chem. Solids **34**, 1059 (1973)
140. I. S. Gradsteyn, I. M. Ryzhik, *Tables of Integrals, Series and Products*, Academic Press, London (1980)
141. C. Stafford, S. Schmitt-Rink, W. Schäfer, Phys. Rev. B **41**, 10000 (1990)
142. M. L. Glasser, N. J. Morgenstern Horing, Phys. Rev. B **31**, 4603 (1985)
143. O. Akimoto, H. Hasegawa, J. Phys. Soc. Jpn. **29**, 181 (1967)
144. M. Shinada, K. Tanaka, J. Phys. Soc. Jpn. **29**, 1258 (1970)
145. A. H. MacDonald, D. S. Ritchie, Phys. Rev. B **33**, 8336 (1986)
146. I. V. Lerner, Y. E. Lozovik, Sov. Phys. JETP **53**, 763 (1981)
147. A. B. Dzyubenko, Y. E. Lozovik, J. Phys. A **29**, 415 (1991)
148. D. Paquet, T. M. Rice, K. Ueda, Phys. Rev. B **32**, 5208 (1985)
149. N. N. Bogolyubov, Zh. Eksp. Teor. Fiz. **34**, 58 (1958)
150. C. Itzykson, Commun. Math. Phys. **4**, 92 (1967)
151. J. Bardeen, L. N. Cooper, J. Schrieffer, Phys. Rev. **108**, 1157 (1957)
152. L. V. Keldysh, A. N. Kozlov, Sov. Phys. JETP **27**, 521 (1968)
153. C. Comte, P. Nozieres, J. Physique **43**, 1069 (1982)

154. J. B. Stark, W. H. Knox, D. S. Chemla, W. Schäfer, S. Schmitt-Rink, C. Stafford, Phys. Rev. Lett. **65**, 3033 (1990)
155. T. Rappen, J. Schröder, A. Leiße, M. Wegener, W. Schäfer, Phys. Rev. B **44**, 13093 (1991)
156. W. Franz, Z. Naturforsch. A **13**, 484 (1958)
157. L. V. Keldysh, Sov. Phys. JETP **34**, 788 (1958)
158. L. V. Keldysh, Sov. Phys. JETP Lett. **29**, 658 (1979)
159. D. A. B. Miller, D. S. Chemla, T. C. Damen, A. C. Gossard, W. Wiegmann, T. H. Wood, C. A. Burrus, Phys. Rev. Lett. **53**, 2173 (1984)
160. G. Nenciu, Rev. Mod. Phys. **63**, 91 (1991)
161. G. H. Wannier, Phys. Rev. **52**, 191 (1937)
162. G. H. Wannier, Rev. Mod. Phys. **34**, 645 (1962)
163. G. H. Wannier, Phys. Rev. **181**, 1364 (1969)
164. J. Zak, Phys. Rev. Lett. **20**, 1477 (1968)
165. J. Callaway, Phys. Rev. **130**, 549 (1963)
166. E. E. Mendez, F. Agullo-Rueda, J. M. Hong, Phys. Rev. Lett. **60**, 2426 (1988)
167. P. Voisin, J. Bleuse, C. Bouche, S. Gaillard, C. Alibert, A. Regreny, Phys. Rev. Lett. **61**, 1639 (1988)
168. J. Feldmann, K. Leo, J. Shah, D. A. B. Miller, J. E. Cunningham, S. Schmitt-Rink, T. Meier, G. von Plessen, A. Schulze, P. Thomas, Phys. Rev. B **46**, 7252 (1992)
169. K. Leo, P. Haring Bolivar, F. Brüggemann, R. Schwedler, K. Köhler, Solid State Commun. **84**, 943 (1992)
170. C. Waschke, H. G. Roskos, R. Schwedler, K. Leo, H. Kurz, K. Köhler, Phys. Rev. Lett. **70**, 3319 (1993)
171. M. Abramowitz, I. A. Stegun, *Handbook of Mathematical Functions*, Dover, New York (1970)
172. L. D. Landau, E. M. Lifshitz, *Quantum Mechanics*, 3rd ed., Pergamon, Oxford (1974)
173. K. Tharmalingam, Phys. Rev. **130**, 2204 (1963)
174. J. D. Dow, D. Redfield, Phys. Rev. B **1**, 3358 (1970)
175. J. Fauchier, J. D. Dow, Phys. Rev. A **9**, 98 (1974)
176. F. L. Lederman, J. D. Dow, Phys. Rev. B **13**, 1633 (1976)
177. C. Lanczos, Z. Physik **68**, 204 (1931)
178. J. D. Dow, D. Redfield, Phys. Rev. B **5**, 594 (1972)
179. F. Urbach, Phys. Rev. **92**, 1324 (1953)
180. W. Martienssen, J. Phys. Chem. Solids **2**, 257 (1957)
181. D. A. B. Miller, D. S. Chemla, S. Schmitt-Rink, Phys. Rev. B **33**, 6976 (1985)
182. J. A. Brum, G. Bastard, Phys. Rev. B **31**, 3893 (1985)
183. S. Hong, M. Jaffe, J. Singh, IEEE J. Quant. Electron., **23**, 2182 (1987)
184. S. Nojima, Phys. Rev. B **37**, 9087 (1988)
185. D. A. B. Miller, D. S. Chemla, T. C. Damen, A. C. Gossard, W. Wiegmann, T. H. Wood, C. A. Burrus, Appl. Phys. Lett. **45**, 13 (1984)
186. D. A. B. Miller, D. S. Chemla, T. C. Damen, T. H. Wood, C. A. Burrus, A. C. Gossard, W. Wiegmann, IEEE J. Quantum Electron. **21**, 1462 (1985)
187. J. B. Krieger, G. J. Iafrate, Phys. Rev. B **33**, 5394 (1988)
188. J. Zak, Solid State Phys. **27**, 1 (1972)
189. W. V. Houston, Phys. Rev. **57**, 184 (1940)
190. C. Kittel, *Quantum Theory of Solids*, Wiley, New York (1963)
191. C. Zener, Proc. R. Soc. London, **145**, 523 (1934)
192. M. Glück, A. R. Kolovsky, H. J. Korsch, Phys. Rev. Lett. **83**, 891 (1999)
193. B. Rosam, D. Meinhold, F. Löser, V. G. Lyssenko, S. Glutsch, F. Bechstedt, F. Rossi, K. Köhler, K. Leo, Phys. Rev. Lett. **86**, 1307 (2001)

194. E. Gornik, R. Kersting, in *Ultrafast Physical Processes in Semiconductors* **67**, 389, Academic Press, San Diego (2001)
195. J. M. Lachaine, M. Hawton, J. E. Sipe, M. M. Dignam, Phys. Rev. B **62**, 4829 (2000)
196. P. Leisching, P. Haring Bolivar, W. Beck, Y. Dhaibi, F. Brüggemann, R. Schwedler, H. Kurz, Phys. Rev. B **50**, 14389 (1994)
197. P. Leisching, W. Beck, H. Kurz, W. Schäfer, K. Leo, K. Köhler, Phys. Rev. B **51**, 7962 (1995)
198. J. R. Haynes, Phys. Rev. Lett. **4**, 361 (1960)
199. A. Mysyrowicz, J. B. Grun, R. Levy, A. Bivas, S. Nikitine, Phys. Lett. A **26** 615 (1968)
200. B. Hönerlage, R. Levy, J. B. Grun, C. Klingshirn, K. Bohnert, Phys. Rep. **124**, 163 (1985)
201. S. Adachi, T. Miyashita, S. Takayama, Y. Takagi, A. Tackeuchi, M. Nakayama, Phys. Rev. B **55**, 1654 (1997)
202. A. Maruani, J. L. Oudar, E. Batifol, D. S. Chemla, Phys. Rev. Lett. **39**, 1372 (1985)
203. A. L. Ivanov, H. Haug, Phys. Rev. Lett. **74**, 438 (1995)
204. H. Wang, J. Shah, T. C. Damen, L. N. Pfeiffer, Phys. Rev. Lett. **74**, 3065 (1995)
205. E. Hanamura, Solid State Commun. **12**, 951 (1973)
206. G. M. Gale, A. Mysyrowicz, Phys. Rev. Lett. **32**, 727 (1974)
207. E. I. Rashba, Springer Tracts in Modern Physics, Vol. 73, p. 150, Springer, Berlin, Heidelberg (1975)
208. D. A. Kleinman, Phys. Rev. B **28**, 871 (1983)
209. J. Zhang, T. Pang, C. Chen, Phys. Lett. A **206**, 101 (1995)
210. D. Birkedal, J. Singh, V. G. Lyssenko, J. Erland, J. V. Hvam, Phys. Rev. Lett. **76**, 672 (1996)
211. I. Balslev, E. Hanamura, Solid State Commun. **72**, 843 (1989)
212. V. M. Axt, A. Stahl, Z. Phys. B **93**, 195 (1993)
213. V. M. Axt, A. Stahl, Z. Phys. B **93**, 205 (1993)
214. K. Victor, V. M. Axt, A. Stahl, Phys. Rev. B **51**, 14164 (1995)
215. M. Lindberg, R. Binder, Y. Z. Hu, S. W. Koch, Phys. Rev. B **50**, 18060 (1994)
216. W. Schäfer et al., Phys. Rev. B **53**, 16429 (1996)
217. V. M. Axt, G. Bartels, A. Stahl, Phys. Rev. Lett. **76**, 2543 (1996)
218. G. Bartels, G. C. Cho, T. Dekorsky, H. Kurz, A. Stahl, K. Köhler, Phys. Rev B **55**, 16404 (1997)
219. P. Kner, A. D. Bar, S. Marquezini, D. S. Chemla, W. Schäfer, Phys. Rev. Lett.**78**, 1319 (1997)
220. R. Lövenich, W. Schäfer, D. S. Chemla, Phys. Status Solidi A **164**, 347 (1997)
221. T. Meier, S. W. Koch, Phys. Rev. B **59**, 13202 (1999)
222. S. R. Bolton, U. Neukirch, L. J. Sham. D. S. Chemla, V. M. Axt, Phys. Rev. Lett. **85**, 2002 (2000)
223. J. Adamowski, S. Bednarek, M. Suffczinzki, Sol. State Commun. **9**, 2037 (1971)
224. O. Akimoto, E. Hanamura, J. Phys. Soc. Jpn. **33**, 1537 (1972)
225. W. J. Brinkmann, T. M. Rice, B. Bell, Phys. Rev. B **8**, 1570 (1973)
226. Y. Suzuki, K. Varga, A stochastic variational approach to quantum mechanical few-body problems, Springer, Berlin (1998)
227. J. L. Zhu, X. Chen, J. J. Xiong, J. Phys.: Condens. Matter **3**, 9559 (1991)
228. W. Heitler, F. London, Z. Physik **44**, 455 (1927)
229. Y. Z. Hu, R. Binder, S. W. Koch, S. T. Cundiff, H. Wang, D. G. Steel, Phys. Rev. B **49**, 14382 (1994)

230. T. Rappen, U. G. Peter, M. Wegener, W. Schäfer, Phys. Rev. B **49**, 7814 (1994)
231. W. Schäfer, R. Lövenich, N. A. Fromer, D. S. Chemla, Phys. Rev. Lett. **86**, 344 (2001)
232. C. Sieh, T. Meier, F. Jahnke, A. Knorr, S. W. Koch, P. Brick, M. Hübner, C. Ell, J. Prineas, G. Khitrova, H. M. Gibbs, Phys. Rev. Lett. **82**, 3112 (1999)
233. N. A. Fromer, P. Kner, D. S. Chemla, R. Lövenich, W. Schäfer, Phys. Rev. B **62**, 2516 (2000)
234. M. Lindberg, R. Binder, Y. Z. Hu, S. W. Koch, Phys. Rev. B **49**, 16942 (1994)
235. H. H. Yaffe, Y. Prior, J. P. Harbison, L. T. Florez, J. Opt. Soc. Am. B **10**, 578 (1993)
236. Y. Z. Hu. R. Binder, S. W. Koch, Phys. Rev. B **47**, 15679 (1993)
237. D. Hulin, M. Joffre, Phys. Rev. Lett. **65**, 3425 (1990)
238. M. Combescot, R. Combescot, Phys. Rev. B **40**, 3788 (1989)
239. E. J. Mayer, G. O. Smith, V. Heuckeroth, J. Kuhl, K. Bott, A. Schulze, T. Meier, D. Bennhardt, S. W. Koch, P. Thomas, Phys. Rev. B bf 50, 14730 (1994)
240. D. J. Lovering, R. T. Philips, G. J. Denton, G. W. Smith, Phys. Rev. Lett. **68**, 1880 (1992)
241. K. H. Pantke, D. Oberhauser, V. G. Lyssenko, J. M. Hvam, G. Weimann, Phys. Rev. B **47**, 2413 (1993)
242. B. F. Feuerbacher, J. Kuhl, K. Ploog, Phys. Rev. B **43**, 2439 (1991)
243. W. Schäfer, R. Lövenich, D. S. Chemla, Phys. Status Solidi B **221**, 195 (2000)
244. N. H. Kwong, R. Binder, Phys. Rev. B **61**, 8341 (2000)
245. S. Nakajima, Prog. Theoret. Phys. **20**, 948 (1958)
246. R. Zwanzig, J. Chem. Phys. **33**, 1338 (1960)
247. H. Grabert, *Projection Operator Techniques in Nonequilibrium Statistical Mechanics*, Springer, Berlin, Heidelberg (1982)
248. V. M. Galitskii, A. B. Migdal, Sov. Phys. JETP **7**, 96 (1958)
249. P. C. Martin, J. Schwinger, Phys. Rev. **115**, 1342 (1959)
250. J. Schwinger, J. Math. Phys. **2**, 407 (1961)
251. L. P. Kadanoff, G. Baym, *Quantum Statistical Mechanics*, W. A. Benjamin, New York (1962)
252. L. V. Keldysh, Sov. Phys. JETP **20**, 1018 (1965)
253. A. Korenman, Ann. Phys. **39**, 72 (1966)
254. D. F. DuBois, in *Lectures in Theoretical Physics*, Vol. 9, eds. W. E. Brittin, A. O. Barut, Gordon and Breach, New York (1967)
255. P. Danielewicz, Ann. Phys. **152**, 239 (1984)
256. S. Lundquist, S. Hedin, Solid State Phys. **23** (1969)
257. L. Hedin, Phys. Rev. **139**, 796 (1965)
258. W. Schäfer, J. Treusch, Z. Phys. B **63**, 407 (1986)
259. R. Binder, S. W. Koch, Prog. Quantum Electron. **19**, 307 (1995)
260. F. J. Dyson, Phys. Rev. **75**, 1736 (1949)
261. R. P. Feynman, Phys. Rev. **76**, 769 (1949)
262. E. E. Salpeter, H. A. Bethe, Phys. Rev. **84**, 1232 (1951)
263. D. Bohm, D. Pines, Phys. Rev. **92**, 609 (1953) Benjamin, New York (1966)
264. H. Weyl, Z. Phys. **46**, 1 (1927)
265. E. Wigner, Phys. Rev. **40**, 749 (1932)
266. K. Henneberger, Physica A **150**, 419 (1988)
267. P. Lipavsky, V. Spicka, B. Velicky, Phys. Rev. B **34**, 6933 (1986)
268. D. B. Tran Thoai, H. Haug, Z. Phys. B **91**, 199 (1993)
269. L. Banyai, D. B. Tran Thoai, C. Remling, H. Haug, Phys. Status Solidi B **173**, 149 (1992)

270. M Hartmann, W. Schäfer, Phys. Status Solidi B **173**, 165 (1992)
271. P. Gartner, L. Banyai, H. Haug, Phys. Rev. B **60**, 14234 (1999)
272. W. Schäfer, J. Opt. Soc. Am. **13**, 1291 (1996)
273. M. Bonitz, D. Kremp, D. C. Scott, R. Binder, W. D. Kraeft, H. S, Köhler, J. Phys. Condens. Matter **8**, 6057 (1996)
274. R. Binder, H. S. Köhler, M. Bonitz, N. H. Kwong, Phys. Rev. B **55**, 5110 (1997)
275. M. Bonitz, in *Physics of Strongly Coupled Coulomb Systems*, ed. G. Kalman, Plenum (1998)
276. N. H. Kwong, M. Bonitz, R. Binder, H. S. Köhler, Phys. Status Solidi B **206**, 197 (1998)
277. N. H. Kwong, M. Bonitz, Phys. Rev. Lett. **84**, 1768 (2000)
278. P. Danielewicz, Ann. Phys. **152**, 305 (1984)
279. H. S. Köhler, Phys. Rev. C **51**, 3232 (1995)
280. H. S. Köhler, Phys. Rev. C **53**, 3145 (1996)
281. L. Banyai, Q. T. Vu, B. Mieck, H. Haug, Phys. Rev. Lett. **81**, 882 (1998)
282. A. Schindlmayr, R. W. Godby, Phys. Rev. Lett. **80**, 1702 (1998)
283. G. Baym, Phys. Rev. **124**, 287 (1962)
284. J. M. Luttinger, J. C. Ward, Phys. Rev. **118**, 1417 (1960)
285. R. P. Feynman, Phys. Rev. **56**, 340 (1939)
286. G. Mahan, B. E. Sernelius, Phys. Rev. Lett. **62**, 2718 (1989)
287. L. Lindhard, K. Dan. Vidensk. Selsk., Mat. Fys. Medd. **28**, 8 (1954)
288. B. I. Lundquist, Phys. Kondens. Materie **6**, 206 (1967)
289. M. Rösler, R. Zimmermann, Phys. Status Solidi B **67**, 525 (1975)
290. H. Haug, D. B. Tran Thoai, Phys. Status Solidi B **85**, 561 (1978)
291. W. Schäfer, R. Binder, K. H. Schuldt, Z. Phys. **70**, 145 (1988)
292. K. El Sayed, S. Schuster, H. Haug, F. Herzel, K. Henneberger, Phys. Rev. B **49**, 7337 (1994)
293. M. U. Wehner, M. H. Ulm, D. S. Chemla, M. Wegener, Phys. Rev. Lett. **80**, 1992 (1998)
294. W. Schäfer, I. Brener, W. Knox, in *Coherent Optical Interactions in Semiconductors*, NATO Advanced Research Workshop, Plenum, New York (1993)
295. F. Jahnke, M. Kira, S. WS. Koch, Z. Physik **104**, 559 (1997)
296. K. Morawetz, G. Röpke, Phys. Rev. E **51**, 4246 (1995)
297. G. Baym, Ann. Phys. **14**, 1 (1961)
298. L. Migdal, Sov. Phys. JETP **7**, 996 (1958)
299. L. J. Sham, in *Dynamical Properties of Solids*, ed. G. K. Horton, A. A. Maradudin, North-Holland, Amsterdam (1974)
300. A. A. Maradudin, E. W. Montroll, G. H. Weiss, I. P. Ipatova, *Theory of Lattice Dynamics in the Harmonic Approximation*, Academic Press, New York (1971)
301. S. Baroni, P. Giannozzi, E. Molinari, Phys. Rev. B **41**, 3870 (1990)
302. P. E. van Camp, V. E. van Doren, J. T. Devreese, *Ab Initio Calculation of Phonon Spectra*, Plenum, New York (1983)
303. P. Giannozzi, S. de Gironcoli, P. Pavone, S. Baroni, Phys. Rev. B **43**, 7231 (1991)
304. L. J. Sham, Phys. Rev. **188**, 1431 (1969)
305. R. M. Pick, M. H. Cohen, R. M. Martin, Phys. Rev. B **1**, 910 (1970)
306. R. H. Lyddane, R. G. Sachs, E. Teller, Phys. Rev. **59**, 673 (1941)
307. H. Fröhlich, Adv. Phys. **3**, 325 (1954)
308. J. D. Zook, Phys. Rev. A **136**, 869 (1964)
309. V. F. Gantmakher, Y. B. Levinson, *Carrier Scattering in Metals and Semiconductors*, North-Holland, Amsterdam (1987)
310. J. Bardeen, W. Shockley, Phys. Rev. **80**, 72 (1950)

311. W. Shockley, *Electrons and Holes in Semiconductors*, Van Nostrand, Princeton, NJ (1950)
312. E. M. Conwell, *High Field Transport in Semiconductors*, Solid State Phys. Suppl. 9, Academic Press, New York (1967)
313. G. D. Mahan, J. Phys. Chem. Solids **26**, 751 (1965)
314. R. P. Feynman, Phys. Rev. **97**, 660 (1955)
315. W. Becker, B. Gerlach, H. Schliffke, Phys. Rev. B **28**, 5735 (1983)
316. B. A. Mason, S. Das Sarma, Phys. Rev. B **33**, 1412 (1986)
317. M. Ueta, H. Kanzaki, K. Kobayashi, Y. Toyozawa, E. Hanamura, *Excitonic Processes in Solids*, Springer, Berlin, Heidelberg (1986)
318. J. Fröhlich, Fortschr. Phys. **22**, 159 (1974)
319. H. Spohn, Ann. Phys. **175**, 278 (1987)
320. B. Gerlach, H. Löwen, Rev. Mod. Phys. **63**, 63 (1991)
321. J. A. Kash, J. C. Tsang, in *Light Scattering in Solids VI*, eds. M. Cardona, G. Günteroth, Springer, Berlin, Heidelberg (1991)
322. G. P. Srivastava, *The Physics of Phonons*, Adam Hilger, Bristol (1990)
323. D. von der Linde, J. Kuhl, H. Klingenberg, Phys. Rev. Lett. **44**, 1505 (1980)
324. U. Wenschuh, W. Schäfer, Ann. Phys. **4**, 289 (1995)
325. R. Zimmermann, Phys. Status Solidi B **159**, 317 (1989)
326. E. Mohler, B. Thomas, Phys. Rev. Lett. **11**, 543 (1980)
327. U. Haupt, Z. Phys. **159**, 232 (1959)
328. J. G. Liebler, H. Haug, Europhys. Lett. **14**, 71 (1991)
329. M. W. C. Dharma-Wardana, Phys. Rev. Lett. **66**, 197 (1991)
330. P. Debye, Ann. Phys. **39**, 189 (1912)
331. J. A. Kash, Phys. Rev. B **40**, 3455 (1989)
332. K. Hannewald, S. Glutsch, F. Bechstedt, Phys. Rev. Lett. **86**, 2451 (2001)
333. M. Kira, F. Jahnke, W. Hoyer, S. W. Koch, Prog. Quantum Electron. **23**, 189 (1999)
334. L. Banyai, D. B. Tran Thoai, E. Reitsamer, H. Haug, D. Steinbach, M. U. Wehner, M. Wegener, T. Marschner, W. Stolz, Phys. Rev. Lett. **75**, 2188 (1995)
335. M. Betz, G. Göger, A. Lauberau, P. Gartner, L. Banyai, H. Haug, K. Ortner, C. R. Becker, A. Leitensdorfer, Phys. Rev. Lett. **86**, 4684 (2001)
336. D. Steinbach, G. Kocherscheidt, M. U. Wehner, H. Kalt, M. Wegener, K. Ohakawa, D. Hommel, V. M. Axt, Phys. Rev. B **60** 12079 (1999)
337. K. Leo, W. W. Ruehle, H. J. Queisser, K. Ploog, Phys. Rev. B **37**, 7121 (1988)
338. H. Saito, E. O. Goebel, Phys. Rev. B **31**, 2360 (1985)
339. H. Roskos, B. Rieck, A. Seilmeier, W. Kaiser, Solid State Electron. **32**, 1437 (1989)
340. T. Rappen, U. Peter, M. Wegener, W. Schäfer, Phys. Rev. B **48**, 4879 (1993)
341. P. C. Becker, H. L. Fragnito, C. H. Brito Cruz, J. Shah, R. L. Fork, J. E. Cunningham, J. E. Henry, C. V. Shank, Appl. Phys. Lett. **53**, 2089 (1988)
342. J. Shah, B. Deveaud, T. C. Damen, W. T. Tsang, A. C. Gossard, P. Lugli, Phys. Rev. Lett. **59**, 2222 (1987)
343. J. Feldmann, R. Sattmann, E. O. Göbel, J.Kuhl, J. Hebling, K. Ploog, R. Muralidharan, P. Dawson, C. T. Foxon, Phys. Rev. Lett. **62**, 1892 (1989)
344. D. W. Snoke, W. W. Ruehle, Y.-C. Liu, E. Bauser, Phys. Rev. Lett. **68**, 990 (1992)
345. W. H. Knox, C. Hirliman, D. A. B. Miller, J. Shah, D. S. Chemla, C. V. Shank, Phys. Rev. Lett. **56**, 1191 (1986)
346. M. Elsässer, J. Shah, L. Rota, P. Lugli, Phys. Rev. Lett. **66**, 1757 (1991)
347. J. Y. Bigot, M. T. Portella, R. W. Schoenlein, J. E. Cunningham, C. V. Shank, Phys. Rev. Lett. **67**, 636 (1991)

348. B. Mieck, H. Haug, W. A. Hügel, M. F. Heinrich, M. Wegener, Phys. Rev. B **62**, 2686 (2000)
349. W. A. Hügel, M. F. Heinrich, M. Wegener, Q. T. Vu, L. Banyai, H. Haug, Phys. Rev. Lett. **83**, 3313 (1999)
350. Q. T. Vu, H. Haug, W. A. Hügel, S. Chatterjee, M. Wegener, Phys. Rev. Lett. **85**, 3508 (2000)
351. D. S. Kim, J. Shah, J. E. Cunningham, T. C. Damen, S. Schmitt-Rink, W. Schäfer, Phys. Rev. Lett. **68**, 2838 (1992)
352. K. Meissner, B. Fluegel, H. Giessen, G. Mohs, R. Binder, S. W. Koch, N. Peyghambarian, Phys. Rev. B **50**, 17647 (1994)
353. N. Pelekanos, J. Ding, M. Hagerott, A. V. Nurmikko, H. Luo, N. Samarth, J. K. Furdyra, Phys. Rev. B **45**, 6037 (1992)
354. W. H. Knox, R. L. Fork, M. C. Downer, D. A. B. Miller, D. S. Chemla, C. V. Shank, A. C. Gosssard, W. Wiegmann, Phys. Rev. Lett. **54**, 1306 (1986)
355. J. B. Johnson, Phys. Rev. **29**, 367 (1927)
356. H. Nyquist, Phys. Rev. **32**, 229 (1928)
357. H. B. Callen, R. F. Greene, Phys. Rev. **86**, 702 (1952)
358. L. Banyai, S. W. Koch, Z. Physik B **63**, 283 (1986)
359. W. W. Chow, P. M. Snowton, P. Blood, A. Girndt, F. Jahnke, S. W. Koch, Appl. Phys. Lett. **71**, 157 (1997)
360. M. Wegener, I. Bar-Joseph, G. Sucha, M. N. Islam, N. Sauer, T. Y. Chang, D. S. Chemla, Phys. Rev. B **39**, 12794 (1989)
361. J. Lee, E. S. Koteles, M. O. Vassell, Phys. Rev. B **33**, 5512 (1986)
362. G. W. Fehrenbach, W. Schäfer, J. Treusch, R. G. Ulbrich, Phys. Rev. Lett. **49**, 1281 (1982)
363. L. Schultheis, J. Kuhl, A. Honold, C. W. Tu, Phys. Rev. Lett. **57**, 1635 (1986)
364. H. Wang, K. Ferrio, D. G. Steel, Y. Z. Hu, R. Binder, S. W. Koch, Phys. Rev. Lett. **71**, 1261 (1993)
365. T. Rappen, U. Peter, M. Wegener, W. Schäfer, Phys. Rev. B **49**, 10774 (1994)
366. E. O. Göbel, *Festkörperpropleme*, Advances in Solid State Phys. **30**, 269 (1990)
367. N. F. Mott, *Metal Insulator Transitions*, Taylor & Francis, London (1974)
368. W. Schäfer, R. Binder, K. H. Schuldt, Z. Phys. B **70**, 145 (1988)
369. T. Schmielau, G. Manzke, D. Tamme, K. Henneberger, Phys. Status Solidi B **221**, 215 (2000)
370. E. Hanamura, J. Phys. Soc. Jpn. **37**, 1545 (1974)
371. G. Manzke, V. May, K. Henneberger, Phys. Status Solidi B **125**, 693 (1984)
372. G. Röpke, T. Seifert, H. Stolz, R. Zimmermann, Phys. Status Solidi B **100**, 215 (1980)
373. H. Stolz, R. Zimmermann, Phys. Status Solidi B **94**, 135 (1979)
374. J. L. Jewell, J. P. Harbison, A. Scherer, Y. H. Lee, L. T. Florez, IEEE J. Quantum Electron. **27**, 1332 (1991)
375. J. Faist, F. Capasso, D. S. Sivco, A. L. Hutchinson, C. Sirtori, S. N. G. Chu, A. Y. Cho, Appl. Phys. Lett. **65**, 2901 (1994)
376. J. Faist, F. Capasso, C. Sirtori, D. L. Sivco, A. L. Hutchinson, A. Y. Cho, Appl. Phys. Lett. **66**, 538 (1995)
377. P. Meystre, M. Sargeant III, *Elements of Quantum Optics*, Springer, Berlin. Heidelberg (1990)
378. U. Mohideen, R. E. Slusher, F. Jahnke, S. W. Koch, Phys. Rev. Lett. **73**, 1785 (1994)
379. M. A. Newkirk, K. J. Vahala, Appl. Phys. Lett. **52**, 770 (1988)
380. G. Pompe, T. Rappen, M. Wegener, Phys. Rev. B **51**, 7005 (1995)
381. G. Eisenstein, J. M. Wiesenfeld, M. Wegener, G. Sucha, D. S. Chemla, S. Weiss, G. Raybon, U. Koren, Appl. Phys. Lett. **58**, 158 (1991)

382. L. F. Lester, B. K. Ridley, J. Appl. Phys. **72**, 2579 (1992)
383. C. Y. Tsai, L. F. Eastman, Y. H. Lo, Appl. Phys. Lett. **63**, 3408 (1993)
384. S. Hense, M. Wegener, Phys. Rev. B **55**, 9255 (1997)
385. F. Jahnke, S. W. Koch, Appl. Phys. Lett. **67**, 2278 (1995)
386. M. Elsässer, S. G. Hense, M. Wegener, Appl. Phys. Lett. **70**, 853 (1997)
387. W. W. Chow, Opt. Lett. **20**, 2318 (1995)
388. S. G. Hense, M. Elsässer, M. Wegener, *Festkörperprobleme*, Adv. Solid State Phys. **37**, 207 (1998)
389. M. Sargeant III, Phys. Rev. A **48**, 717 (1993)
390. O. Hess, T. Kuhn, Phys. Rev. A **54**, 3347 (1996); Phys. Rev. A **54**, 3360 (1996)
391. O. D. Mücke, M. Wegener, Appl. Phys. Lett. **73**, 569 (1998)
392. J. Ding, M. Hagerott, T. Ishihara, H. Leon, A. V. Nurmikko, Phys. Rev. B **47**, 10528 (1993)
393. Y. Kawakami, I. Hauksson, H. Steward, J. Simpson, I. Galbraith, K. A. Prior, B. C. Cavenett, Phys. Rev. B **48**, 11994 (1993)
394. M. Umlauff, H. Kalt, C. Klingshirn, M. Scholl, J. Söllner, M. Heuken, Phys. Rev. B **52**, 5063 (1995)
395. F. Kreller, M. Lowisch, J. Puls, F. Henneberger, Phys. Rev. Lett. **75**, 2420 (1995)
396. R. Cingolani, R. Rinaldi, L. Calcagnile, P. Prete, P. Sciacovelli, L. Tapfer, L. Vanzetti, G. Mula, F. Bassani, L. Sorba, A. Franciosi, Phys. Rev. B **49**, 16769 (1994)
397. A. Diessel, W. Ebeling, J. Gutowski, B. Jobst, K. Schüll, D. Hommel, K. Henneberger, Phys. Rev. B **52**, 4736 (1995)
398. F. Jahnke, K. Henneberger, W. Schäfer, S. W. Koch, J. Opt. Soc. Am. B **10**, 2394 (1993)
399. F. Jahnke, S. W. Koch, Phys. Rev. A **52**, 1712 (1995)
400. H. W. Wyld, *Mathematical Methods for Physics*, Benjamin/Cummings, Reading, MA (1976)
401. K. Henneberger, S. W. Koch, Phys. Rev. Lett. **76**, 1820 (1996)
402. A. L. Schawlow, C. H. Townes, Phys. Rev. **112**, 1940 (1958)
403. M. Lax, Phys. Rev. **160**, 290 (1967)
404. H. Haug, H. Haken, Z. Phys. **204**, 262 (1967)
405. C. H. Henry, IEEE J. Quantum Electron. **18**, 259 (1982)
406. K. Vahala, A. Yariv, IEEE J. Quantum Electron. **19**, 1096 (1983)
407. R. Jin, D. Boggavarapu, G. Khitrova, H. M. Gibbs,Y. Z. Hu, S. W. Koch, N. Peyghambarian, Appl. Phys. Lett. **61**, 1883 (1992)
408. Q. deng, D. G. Deppe, Phys. Rev. A **53**, 1036 (1996)
409. F. Jahnke, S. W. Koch, in *Physics and Simulation of Optoelectronic Devices II*, SPIE International Symposium OE/LASE, Los Angeles (1994)
410. J R. Hayes, A. F. J. Levi, W. Wiegmann, Phys. Rev. Lett. **54**, 1570 (1985)
411. R. Tsu, L. Esaki, Appl. Phys. Lett. **22**, 562 (1973); L. L. Chang, L. Esaki, R. Tsu, Appl. Phys. Lett. **24**, 593 (1974)
412. E. E. Mendez, *Proceedings of a NATO Advanced Study Institute on Physics and Applications of Quantum Wells and Superlattices*, NATO ASI Series, Vol. 170, p. 159, eds. E. E. Mendez, K. von Klitzing (1987)
413. B. J. van Wees, H. van Houten, C. W. J. Beenakker, J. G. Wiliamson, L. P. Kouwenhoven, D. van der Marcel, C. T. Foxon, Phys. Rev. Lett. **60**, 848 (1988)
414. B. J. van Wees, L. P. Kouwenhoven, E. M. M. Willems, C. J. P. M. Harmans, J. E. Mooij, H. van Houten, J. G. Williamson, C. T. Foxon, Phys. Rev. B **43**, 12431 (1991)

415. D. Wharam, T. J. Thornton, R. Newbury, M. Pepper, H. Ahmet, J. E. F. Frost, D. G. Hasko, D. C. Peacock, D. A. Ritchie, G. A. C. Jones, J. Phys. C **21**, L209 (1988)
416. C. J. Gorter, Physica **17**, 777 (1951)
417. A. N. Cleland, M. A. Kastner, S. J. Wind, Phys. Rev. Lett. **64**, 1565 (1990)
418. H. Grabert, G.-L. Ingold, M. H. Devoret, D. Esteve, H. Pothier, C. Urbina, Z. Phys. B **84**, 143 (1991)
419. C. W. J. Beenakker, Phys. Rev. B **44**, 1646 (1991)
420. U. Meirav, M. A. Kastner, S. J. Wind, Phys. Rev. Lett. **65**, 771 (1990)
421. M. A. Kastner, Rev. Mod. Phys. **64**, 849 (1992)
422. L. S. Kuzmin, K. K. Likharev, Jpn. J. Appl. Phys. **26**, 1387 (1987)
423. T. A. Fulton, D. J. Dolan, Phys. Rev. Lett. **59**, 109 (1987)
424. G. Schön, A. D. Zaikin, Phys. Rep. **198**, 237 (1990)
425. F. A. Buot, K. L. Jensen, Phys. Rev B **42**, 9429 (1990)
426. W. R. Frensley, Rev. Mod. Phys. **62**, 745 (1990)
427. W. Frensley, Solid State Electron. **31**, 739 (1988); Phys. Rev. B **36**, 1570 (1987)
428. R. Lövenich, Ph.D. Thesis, University of Dortmund (1999); NIC Series, Vol. 5, Forschungszentrum Jülich (1999)
429. N. S. Wingreen, A. P. Jauho, Y. Meir, Phys. Rev. B **48**, 8487 (1993)
430. C. A. Stafford, N. S. Wingreen, Phys. Rev. Lett. **76**, 1916 (1995)
431. T. Brandes, B. Kramer, Phys. Rev. Lett. **83**, 3021 (1999)
432. L. I. Glazman, J. of Low Temp. Phys. **118**, 247 (2000)
433. S. de Franceschi, S. Sasaki, J. M. Elzerman, W. G. van der Wiel, S. Tarucha, L. P. Kouwenhoven, Phys. Rev. Lett. **86**, 878 (2001)
434. A. T. Tilke, F. C. Simmel, R. H. Blick, H. Lorenz, J. P. Kotthaus, Prog. Quantum Electron. **25**, 97 (2001)
435. D. E. Grupp, T. Zhang, G. J. Dolan, N. S. Wingreen, Phys. Rev. Lett. **87**, 6807 (2001)
436. I. G. Zacharia, D. Goldhaber–Gordon, G. Granger, M. A. Kastner, Y. B. Khavin, H. Shtriman, B. Mahala, U. Meirav, Phys. Rev. B **64**, 5311 (2001)
437. Philos. Mag. **21**, 863 (1970)
438. M. Büttiker, Phys. Rev. Lett. **57**, 1761 (1982)
439. M. Büttiker, Y. Imry, R. Landauer, S. Pinhas, Phys. Rev. B **31**, 6207 (1985)
440. K. von Klitzing, G. Dorda, M. Pepper, Phys. Rev. Lett. **45**, 494 (1980)
441. D. S. Tsui, H. L. Stormer, A. C. Gossard, Phys. Rev. Lett. **48**, 1559 (1982)
442. K. von Klitzing, G. Ebert, Metrologia **21**, 11 (1985)
443. M. Heuser, J. Hajdu, Z. Physik **270**, 289 (1974); E. Teller, Z. Phys. **64**, 311 (1930)
444. B. I. Halperin, Phys. Rev. B **25**, 2185 (1982)
445. A. H. MacDonald, P. Streda, Phys. Rev. B **29**, 1616 (1984)
446. D. B. Chlovskii, B. I. Shlovskii, L. I. Glazman, Phys. Rev. B **46**, 4026 (1992)
447. S. Washburn, A. B. Fowler, H. Schmid, D. Kern, Phys. Rev. Lett. **61**, 2801 (1988)
448. R. J. Haug, A. H. MacDonald, P. Streda, K. von Klitzing, Phys. Rev. Lett. **61**, 2797 (1988)
449. A. J. Kent, D. J. Mc Kitterick, L. J. Challis, P. Hawker, C. J. Mellor, M. Henini, Phys. Rev. Lett. **69**, 1684 (1992)
450. A. A. Shashkin, A. J. Kent, J. R. Owers-Bradley, A. J. Cross, P. Hawker, M. Henini, Phys. Rev. Lett. **79**, 5114 (1997)
451. Y. Y. Wei, J. Weis, K. von Klitzing, K. Eberl, Phys. Rev. Lett. **81**, 1674 (1998)
452. M. Janssen, O. Viehweger, U. Fastenrath, J. Hajdu, *Introduction to the Theory of the Integer Quantum Hall Effect*, VCH, Weinheim (1994)

453. R. E. Prange, S. M. Girvin (eds.), *The Quantum Hall Effect*, Springer, New York (1987)
454. R. Lake, S. Datta, Phys. Rev. B **45**, 6670 (1991)
455. Y. Meir, N. S. Wingreen, Phys. Rev. Lett. **68**, 2512 (1992)
456. C. L. Kane, P. A. Lee, D. P. Di Vincenzo, Phys. Rev. B **38**, 2995 (1988)
457. A. D. Stone, A. Szafer, IBM J. Res. Develop. **32**, 384 (1988)
458. H. U. Baranger, A. D. Stone, Phys. Rev. B **40**, 8169 (1989)
459. D. S. Fisher, P. A. Lee, Phys. Rev. B **23**, 6851 (1981)
460. H. L. Stormer, A. M. Chang, D. C. Tsui, J. C. M. Wang, A. C. Gossard, W. Weigmann, Phys. Rev. Lett. **50**, 1953 (1984)
461. A. M. Chang, P. Berglund, D. C. Tsui, H. L. Stormer, J. C. M. Hwang, Phys. Rev. Lett. **53**, 997 (1984)
462. R. Willett, J. P. Eisenstein, H. L. Stormer, D. C. Tsui, A. C. Gossard, J. H. English, Phys. Rev. Lett. **59**, 1776 (1987)
463. V. J. Goldman, M. Shayegan, T. C. Tsui, Phys. Rev. Lett. **61**, 881 (1988)
464. H. W. Jiang, H. L. Stormer, D. C. Tsui, L. N. Pfeiffer, K. W. West, Phys. Rev. B **40**, 12013 (1990)
465. J. K. Jain, Phys. Rev. Lett. **63**, 199 (1989)
466. R. B. Laughlin, Phys. Rev. Lett. **50**, 1395 (1983)
467. F. D. M. Haldane, Phys. Rev. Lett. **51**, 605 (1983)
468. B. I. Halperin, Phys. Rev. Lett. **52**, 1583 (1984)
469. R. B. Laughlin, Phys. Rev. Lett. **60**, 2677 (1988)
470. L. Saminadayar, D. C. Glattli, Y. Jin, B. Etienne, Phys. Rev. Lett. **79**, 2526 (1997)
471. R. de Picciotto, M. Reznikov, M. Heiblum, V. Umansky, C. Bunin, D. Mahalu, Nature **389**, 162 (1997)
472. S. Kivelson, D.-H. Lee, S.-C. Zhang, Phys. Rev. B **46**, 2223 (1992)
473. D. R. Hofstadter, Phys. Rev. B **14**, 2239 (1976)
474. P. G. Harper, Proc. Phys. Soc. Lond. A **68**, 874 (1955)
475. H. Silberbauer, J. Phys. C **4**, 7355 (1992)
476. U. Wulf, A. H. MacDonald, Phys. Rev. B **47**, 6566 (1993)
477. V. Gudmundsson, R. Gerhardts, Phys. Rev. B **52**, 16744 (1995)
478. V. Gudmundsson, R. Gerhardts, Phys. Rev. B **54**, R5223 (1996)
479. D. Weiss, M. L. Roukes, A. Menschig, P. Grambow, K. von Klitzing, G. Weimann, Phys. Rev. Lett. **66**, 2790 (1991)
480. D. Weiss, K. Richter, A. Menschig, R. Bergmann, H. Schweizer, K. von Klitzing, G. Weimann, Phys. Rev. Lett. **70**, 4118 (1993)
481. R. Fleischmann, T. Geisel, R. Ketzmerick, Phys. Rev. Lett. **68**, 1367 (1992)
482. T. Geisel, R. Ketzmerick, G. Petschel, Phys. Rev. Lett. **66**, 1651 (1991); **67**, 3536 (1991)
483. D. Pfannkuche, R. Gerhardts, Phys. Rev. B **46**, 12606 (1992)
484. A. Lorke, J. P. Kotthaus, K. Ploog, Phys. Rev. B **44**, 3447 (1991)
485. R. Schuster, G. Ernst, K. Ensslin, M. Entin, M. Holland, G. Böhm, W. Klein, Phys. Rev. B **50**, 8090 (1994)
486. F. Nihey, S. W. Hwang, K. Nakamura, Phys. Rev. B **51**, 4649 (1995)
487. T. Nakanishi, T. Ando, Phys. Rev. B **54**, 8021 (1996)
488. B. Mehlig, D. Boose, K. Müller, Phys. Rev. Lett. **75**, 57 (1995)
489. M. Koshino, H. Aoki, K. Kuroki, S. Kagoshima, T Osada, Phys. Rev. Lett. **86**, 1062 (2001)
490. J. H. Smet, K. von Klitzing, D. Weiss, W. Wegscheider, Phys. Rev. Lett. **80**, 4538 (1998)
491. A. Lorke, I. Jejina, J. P. Kotthaus, Phys. Rev. Lett. **64**, 2559 (1990)

492. K. Kern, D. Heitmann, P. Grambow, Y. H. Zhang, K. Ploog, Phys. Rev. Lett. **66**, 1618 (1991)
493. T. Darnhofer, M. Suhrke, U. Rössler, Phys. Rev. Lett. **77**, 2593 (1996)
494. E. Anisimovas, P. Jahanson, Phys. Rev. B **60**, 7744 (1999)
495. R. P. Feynman, QED, p. 152, Princeton University Press, NJ (1985)

Index

Location: http://www.springer.de/phys/

You are one **click** *away*
from a **world of physics** *information!*

Come and visit **Springer's**

Physics Online Library

Books

- Search the Springer website catalogue
- Subscribe to our free alerting service for new books
- Look through the book series profiles

You want to order? Email to: orders@springer.de

Journals

- Get abstracts, ToC's free of charge to everyone
- Use our powerful search engine LINK Search
- Subscribe to our free alerting service LINK *Alert*
- Read full-text articles (available only to subscribers of the paper version of a journal)

You want to subscribe? Email to: subscriptions@springer.de

Electronic Media

- Get more information on our software and CD-ROMs

You have a question on an electronic product? Email to: helpdesk-em@springer.de

Bookmark now:

http://www.springer.de/phys/

Springer · Customer Service
Haberstr. 7 · 69126 Heidelberg, Germany
Tel: +49 (0) 6221 - 345 - 217/8
Fax: +49 (0) 6221 - 345 - 229 · e-mail: orders@springer.de
d&p · 6437.MNT/SFb

Springer

MIX
Papier aus verantwortungsvollen Quellen
Paper from responsible sources
FSC® C105338

If you have any concerns about our products,
you can contact us on
ProductSafety@springernature.com

In case Publisher is established outside the EU,
the EU authorized representative is:
Springer Nature Customer Service Center GmbH
Europaplatz 3, 69115 Heidelberg, Germany

Printed by Libri Plureos GmbH
in Hamburg, Germany